THE METABOLISM OF CONTRACTION

Proceedings of the Seventh Annual Meeting of the
International Study Group for Research
on Cardiac Metabolism
18–21 June, 1974
Québec, CANADA

Jointly held by the
European and American Research Groups

Organized by the
Department of Pathology
and Medicine
of Laval University
and by the Institute of Cardiology
of Laval Hospital

Chairman of the Congress
ROY, Paul-E., Québec

Members of the Organization Committee:

Roy, Paul-E., Québec
Leblanc, Jacques, Québec
Dagenais, Gilles R., Québec
Rona, George, Montréal

Management:

Pageau, Laurette, Québec

Recent Advances in Studies on Cardiac Structure and Metabolism

Series Editor: G. Rona (*formerly with the late E. Bajusz*)

Volume 1: **MYOCARDIOLOGY** (edited by E. Bajusz and G. Rona)
Volume 2: **CARDIOMYOPATHIES** (edited by E. Bajusz and G. Rona, with A. J. Brink and A. Lochner)
Volume 3: **MYOCARDIAL METABOLISM** (edited by N. S. Dhalla)
Volume 4: **MYOCARDIAL BIOLOGY** (edited by N. S. Dhalla)
Volume 5: **MYOCARDIAL CELL DAMAGE** (edited by A. Fleckenstein and N. S. Dhalla)
Volume 6: **PATHOPHYSIOLOGY AND MORPHOLOGY OF MYOCARDIAL CELL ALTERATIONS** (edited by A. Fleckenstein and G. Rona)
Volume 7: **BIOCHEMISTRY AND PHARMACOLOGY OF MYOCARDIAL HYPERTROPHY, HYPOXIA, AND INFARCTION** (edited by P. Harris, R. I. Bing, and A. Fleckenstein)
Volume 8: **THE CARDIAC SARCOPLASM** (edited by P.-E. Roy and P. Harris)
Volume 9: **THE SARCOLEMMA** (edited by P.-E. Roy and N. S. Dhalla)
Volume 10: **THE METABOLISM OF CONTRACTION** (edited by P.-E. Roy and G. Rona)

Recent Advances in Studies on
Cardiac Structure and Metabolism, Volume 10

THE METABOLISM OF CONTRACTION

Edited by

Paul-Emile Roy, M.D.
Professor of Medicine
Laval University School of Medicine
Québec, Canada

and

George Rona, M.D., Ph.D., F.R.C.P.(C)
Professor of Pathology
McGill University
Montréal, Québec, Canada

UNIVERSITY PARK PRESS
Baltimore · London · Tokyo

UNIVERSITY PARK PRESS
International Publishers in Science and Medicine
Chamber of Commerce Building
Baltimore, Maryland 21202

Typeset by The Composing Room of Michigan, Inc.
Manufactured in the United States of America by
Universal Lithographers, Inc., and
The Maple Press Co.

Library of Congress Cataloging in Publication Data
Main entry under title:

The metabolism of contraction.

(Recent advances in studies on cardiac structure and metabolism; v. 10)
"Proceedings of the seventh annual meeting of the International Study Group for Research in Cardiac Metabolism, 18-21th June, 1974, Quebec, Canada, jointly held by the European and American research groups."
Includes index.
1. Heart—Muscle—Congresses. 2. Muscle contraction—Congresses. 3. Coronary heart disease—Congresses. 4. Heart—Muscle—Diseases—Congresses.
I. Roy, Paul-Emile, 1929- II. Rona, George, 1924- III. International Study Group for Research in Cardiac Metabolism. IV. Series.
[DNLM: 1. Myocardium—Metabolism—Congresses. 2. Myocardial contraction—Congresses. W3 RE354 v. 10 1974 / WG202 161m 1974]
RC667.R4 vol. 10 [RC682] 616.1'2'008s [612'.17]
ISBN 0-8391-0670-X 75-34253

Contents

Metabolic Manifestations of Ischemia in Experimental Animals

Exogenous and Endogenous Factors Influencing Ischemic Damage

Coronary Vessel Changes Related to Cardiac Ischemia

Investigative Methods in Cardiac Ischemia

The Cardiovascular System and Exercise

Vascular Biology–Recent Studies

Endocardial Reaction to Injury

Cardiomyopathies–Recent Studies

Contributing Authors

Abelmann, W. H., *Boston, Mass.* (USA)
Ahmad, M., *Aligarh, U.P.* (India)
Ahmed, S. S., *Newark, N.J.* (USA)
Angell, W. W., *San Jose, Calif.* (USA)
Apstein, C. S., *Boston, Mass.* (USA)
Badeer, H. S., *Omaha, Nebr.* (USA)
Badonnel, M.-C., *Geneva* (Switzerland)
Baker, L. D., *Berlin* (Germany)
Bär, H. P., *Edmonton, Alberta* (Canada)
Barletta, M., *Valhalla, N.Y.* (USA)
Barnett, J., *Houston, Texas* (USA)
van Bell, H., *Beerse* (Belgium)
Biamino, G., *Berlin* (Germany)
Bing, O. H. L., *Boston, Mass.* (USA)
Bishop, R. L., *Baltimore, Md.* (USA)
Bittar, N., *Madison, Wisc.* (USA)
Bodem, R., *Boston, Mass.* (USA)
Boutet, M., *Québec* (Canada)
Brawley, R. K., *Baltimore, Md.* (USA)
Bricknell, O., *Cape Town* (South Africa)
Brooks, W. W., *Boston, Mass.* (USA)
Calva, E., *Mexico City* (Mexico)
Cannon, P. T., *New York, N.Y.* (USA)
Casarella, W. J., *New York, N.Y.* (USA)
Challice, C. E., *Calgary, Alberta* (Canada)
Chau, R., *Valhalla, N.Y.* (USA)
Chiong, M. A., *Kingston, Ont.* (Canada)
Chohan, K. S., *Calgary, Alberta* (Canada)
Corday, E., *Los Angeles, Calif.* (USA)
Cordoba, M., *Los Angeles, Calif.* (USA)
Costantini, C., *Los Angeles, Calif.* (USA)
Crass, III, M. F., *Lubbock, Texas,* (USA)
Dagenais, G. R., *Québec* (Canada)
Degenring, F. H., *Charlottesville, Va.* (USA)
Delage, C., *Québec* (Canada)
Donahoo, J. S., *Baltimore, Md.* (USA)
Eich, R. H., *Syracuse, N.Y.* (USA)
Ellis, K., *New York, N.Y.* (USA)
Eu, H. Y., *Montréal, Qué.* (Canada)
Fenner, H., *Berlin* (Germany)
Flaherty, J. T., *Baltimore, Md.* (USA)
Flasterstein, A., *Houston, Texas* (USA)
Folkerth, T. L., *San Diego, Calif.* (USA)
Folts, J. D., *Madison, Wisc.* (USA)
Fosburg, R. G., *San Diego, Calif.* (USA)
Fowler, D. L., *New York, N.Y.* (USA)
Gabbiani, G., *Geneva* (Switzerland)
Gailis, L., *Québec* (Canada)
Ganote, C. E., *Chicago, Ill.* (USA)
van Gerven, W., *Beerse* (Belgium)
Gingell, R. L., *Baltimore, Md.* (USA)
Gold, H., *Los Angeles, Calif.* (USA)
Gott, V. L., *Baltimore, Md.* (USA)
Green, M., *Valhalla, N.Y.* (USA)
Greene, H. L., *Baltimore, Md.* (USA)
Greenfield, L. J., *New Orleans, La.* (USA)
Gross, U., *Berlin* (Germany)
Guideri, G., *Valhalla, N.Y.* (USA)
Haider, B., *Newark, N.J.* (USA)
Hawk, W. A., *Cleveland, Ohio* (USA)
Hearse, D. J., *London* (England)
Heinze, G., *Berlin* (Germany)
Hemker, H. C., *Leiden* (The Netherlands)
Hermens, W. T., *Leiden* (The Netherlands)
Hess, M. E., *Philadelphia, Pa.* (USA)
Hill, W. A., *Birmingham, Ala.* (USA)
Hinshaw, L. B., *New Orleans, La.* (USA)
Hirose, S., *Los Angeles, Calif.* (USA)
Hirota, Y., *Boston, Mass.* (USA)
Hjalmarson, Å. C., *Gotëborg* (Sweden)
Hollaar, L., *Leiden* (The Netherlands)
Hüttner, I., *Montréal, Qué.* (Canada)
Hwang, N. H. C., *Houston, Texas* (USA)
Jackson, D. H., *Birmingham, Ala.* (USA)
Jageneau, A. H. M., *Beerse* (Belgium)
James, T. N., *Birmingham, Ala.* (USA)
Jasmin, G., *Montréal, Qué.* (Canada)
Jennings, R. B., *Chicago, Ill.* (USA)
Juhász-Nagy, A., *Budapest* (Hungary)
Julien, P., *Québec* (Canada)
Kelliher, G. J., *Philadelphia, Pa.* (USA)
Kennedy, J. H., *Houston, Texas* (USA)
Khuri, S. F., *Baltimore, Md.* (USA)
Kloner, R. A., *Chicago, Ill.* (USA)
Koke, J. R., *Madison, Wisc.* (USA)
Koyama, T., *Sapporo* (Japan)
Kruger, R., *Beerse* (Belgium)
Kudász, J., *Budapest* (Hungary)
Kumazawa, T., *Rochester, N.Y.* (USA)
Lagace, R., *Québec* (Canada)
Lang, T.-W., *Los Angeles, Calif.* (USA)
Lehr, D., *Valhalla, N.Y.* (USA)
de Leiris, J., *Orsay* (France)
Levine, N. R., *Philadelphia, Pa.* (USA)
Limet, R., *Houston, Texas* (USA)
Lindström, L., *Gotëborg* (Sweden)
Liu, S.-K., *New York, N.Y.* (USA)
Lord, P. F., *Stockholm* (Sweden)
Lubbe, W. F., *Cape Town* (South Africa)
Marquis, Y., *Québec* (Canada)
Marsh, M., *Mexico City* (Mexico)
Maurer, B. J., *Birmingham, Ala.* (USA)
McDaniel, H. G., *Birmingham, Ala.* (USA)
Meerbaum, S., *Los Angeles, Calif.* (USA)
Merin, R. G., *Rochester, N.Y.* (USA)
Messer, J. V., *Chicago, Ill.* (USA)
Moisan, A., *Québec* (Canada)
Moorkherjee, S., *Syracuse, N.Y.* (USA)

Neffgen, J. F., *Toronto, Ont.* (Canada)
Neye, J., *Berlin* (Germany)
Nguyen, M.-H., *Québec* (Canada)
Norris, R. M., *Auckland* (New Zealand)
Nuñez, R., *Mexico City* (Mexico)
Obeid, A. I., *Syracuse, N.Y.* (USA)
Oehring, L., *Berlin* (Germany)
Oldewurtel, H. A., *Newark, N.J.* (USA)
Opie, L. H., *Cape Town* (South Africa)
O'Riordan, J., *Baltimore, Md.* (USA)
Osborne, D. Gary, *Cleveland, Ohio* (USA)
Oster, K. A., *Fairfield, Conn.* (USA)
Oury, J. H., *San Diego, Calif.* (USA)
Parker, J. O., *Kingston, Ont.* (Canada)
Pauly, T. J., *Madison, Wisc.* (USA)
Poupa, O., *Gotëborg,* (Sweden)
Pozefsky, T., *Baltimore, Md.* (USA)
Prasad, K., *Saskatoon, Sask.* (Canada)
Rakušan, K., *Ottawa, Ont.* (Canada)
Ramdohr, B., *Berlin* (Germany)
Reeves, T. J., *Birmingham, Ala.* (USA)
Regan, T. J., *Newark, N.J.* (USA)
Reimer, K. A., *Chicago, Ill.* (USA)
Reneman, R. S., *Beerse* (Belgium)
Roberts, J., *Philadelphia, Pa.* (USA)
Rona, G., *Montréal, Qué.* (Canada)
Ross, D. J., *Fairfield, Conn.* (USA)
Rowe, R. D., *Toronto, Ont.* (Canada)
Roy, P.-E., *Québec* (Canada)
Rubins, S., *Los Angeles, Calif.* (USA)
Salahuddin, *Aligarh, U.P.* (India)
Schenk, K. E., *Berlin* (Germany)
Scheuer, J., *Bronx, N.Y.* (USA)
Schmidt, D. H., *New York, N.Y.* (USA)
Schmidt, H., *Berlin* (Germany)
Schüren, K.-P., *Berlin* (Germany)
Schwabe, L., *Houston, Texas* (USA)
Shanfeld, J., *Philadelphia, Pa.* (USA)
Shirey, E. K., *Cleveland, Ohio* (USA)
Shug, A. L., *Madison, Wisc.* (USA)
Simonson, L. P., *Edmonton, Alberta* (Canada)
Singh, B. N., *Auckland* (New Zealand)
Smith, H. J., *Auckland* (New Zealand)
Solymoss, B., *Montréal, Qué.* (Canada)
Sonnenblick, E. H., *Boston, Mass.* (USA)
Soto, B., *Birmingham, Ala.* (USA)
Spear, R., *Syracuse, N.Y.* (USA)
Spitzer, J. J., *New Orleans, La.* (USA)
Sterrett, P. R., *Lubbock, Texas* (USA)
Stewart, D. A., *London* (England)
Sullivan, L., *Gotëborg* (Sweden)
Tajuddin, M., *Aligarh, U.P.* (India)
Tariq, M., *Aligarh, U.P.* (India)
Tashjian, R. J., *New York, N.Y.* (USA)
Thorp, K., *Boston, Mass.* (USA)
Tilley, L. P., *New York, N.Y.* (USA)
Vega, L. E., *New York, N.Y.* (USA)
Waldenström, A. P., *Gotëborg* (Sweden)
Weisfeldt, M. L., *Baltimore, Md.* (USA)
Weiss, M. B., *New York, N.Y.* (USA)
Wennmalm, Å., *Stockholm* (Sweden)
White, F., *Calgary, Alberta* (Canada)
Widmer, C., *Philadelphia, Pa.* (USA)
Wilkens, J. L., *Calgary, Alberta* (Canada)
Witteveen, S. A. G. J., *Leiden* (The Netherlands)

Editorial Foreword

The 7th Annual Meeting of the International Study Group for Research in Cardiac Metabolism was held in Québec City, Canada, in June, 1974. I consider it both an honor and a privilege to have been the chairman of such a remarkable scientific event.

The various presentations specifically dealing with cardiac metabolism have been assembled into this unprecedented volume. This unique collection of scientific observations should prove useful to scholars and research scientists wishing an up-to-date overview on the subject of cardiac metabolism.

We are particularly indebted to Professor Richard Bing, who chaired one of the sessions and played a determinant role in the general orientation of the scientific discussions.

We would also like to extend our gratitude to all the scientists who contributed their research achievements toward the success of this symposium.

The papers selected for this publication are representative of the state of knowledge in this field, and as such we think that the content is of exceptional scientific value.

Cardiac lipid metabolism, with and without coronary disease, ultrastructural morphology of the cardiovascular system, and cardiomyopathies complete this volume.

We do not attempt to present the different chapters with an exhaustive and comprehensive review of the subjects discussed. We would recommend therefore that the interested reader consult recent volumes on cardiac physiology and metabolism or the volumes that make up the series *Recent Advances in Studies on Cardiac Structure and Metabolism.*

Paul-E. Roy

Cardiac Lipid Metabolism

Recent Advances in Studies on
Cardiac Structure and Metabolism Volume 10
The Metabolism of Contraction
Edited by Paul-Emile Roy and George Rona

ASSESSMENT OF MYOCARDIAL FREE FATTY ACID METABOLISM IN HUMANS DURING HEPARIN INFUSION

GILLES R. DAGENAIS, YVES MARQUIS,
and LAIMONIS GAILIS

Institut de Cardiologie de Québec, Québec, Canada

SUMMARY

Free fatty acids (FFA) have been shown to be the major myocardial substrates during the postabsorptive state in humans. In order to document the utilization of these substrates by the heart while heparin was infused, the following study was designed. During heparin infusion, total and individual FFA were measured in arterial and coronary sinus plasma during the postabsorptive state in 13 patients with coronary artery disease and in 4 patients with no evidence of cardiac disease. Oleic-1-^{14}C acid bound to albumin was infused at a constant rate for at least 7 min into the left coronary artery. Left coronary blood flow was determined in all patients. It was assumed that oleic-1-^{14}C acid behaves like endogenous FFA. Arterial FFA concentration was high (mean and SEM: 1.52 ± 0.21 μmoles/ml) and varied by less than 10% during the course of the study in 14 of the 17 patients. Oleic acid constituted 44% of all arterial FFA. The relative concentrations of the various individual FFA were similar in arterial and coronary sinus plasma. The average FFA extraction was 19.9 ± 2.1%. In contrast, the oleic-1-^{14}C acid extraction was relatively more steady during the infusion, and the mean extraction of this tracer was 36.0 ± 2.1%. Myocardial uptake for the entire group was 52.3 ± 6.3 μmoles/min. Although a definite plateau for $^{14}CO_2$ release from oleic-1-^{14}C acid oxidation was not obtained, at least 31.8 ± 4.3% of the removed FFA was oxidized, representing 49.3 ± 6.4% of myocardial oxygen consumption. The failure of oleic-1-^{14}C acid oxidation to reach a steady state was not due to delay in the myocardial CO_2 pool, since $^{14}CO_2$ release from infused $NaH^{14}CO_3$ and lactate-1-^{14}C in 11 other subjects revealed that an equilibrium was obtained within 5 min. These findings indicate that the generation of $^{14}CO_2$ from oleic-1-^{14}C acid oxidation is relatively slow, and the uptake of labeled FFA is partly matched by release of unlabeled FFA, suggesting that the human heart has important lipid pools.

INTRODUCTION

Free fatty acids (FFA) constitute the major metabolic substrates of the human heart during the postabsorptive state. To date, no study has assessed how the human myocardium utilizes these substrates when their plasma concentrations are elevated. Increased arterial FFA are often encountered in clinical situations such as shock and heparin or catecholamine infusion and, according to some reports, may have deleterious effects on the ischemic myocardium (Oliver,

Kurien, and Greenwood, 1968; Mjøs, Kjekshus, and Lekven, 1974). Because of the clinical importance of these observations, we attempted to assess human myocardial FFA uptake, release, and oxidation during heparin administration using a constant left coronary infusion of oleic-1-^{14}C acid.

METHODS

Subjects

A total of 17 male patients aged 36–60, admitted for coronary arteriograms because of chest pain, agreed to participate in these evaluations. No patient had congenital or valvular heart disease, recent myocardial infarction, unstable angina, or diabetes mellitus. No patient was taking digitalis, and all antianginal drugs except nitroglycerin were discontinued at least 48 hours before the study. After a 12-hour fasting period, the patients received diazepam, 10 mg p.o., at least 1 hour before the procedure.

Procedure

An end-hole, bipolar pacing catheter was introduced into an antecubital vein and advanced up to the distal portion of the coronary sinus to obtain blood coming from the left coronary system as illustrated with the contrast medium injection from the tip of the catheter (Fig. 1). Following this procedure, a right coronary Judkins catheter was introduced into the right femoral artery and advanced into the aorta; the right coronary ostium was then engaged. Prior to each study, tests were made for contamination of the samples from the coronary sinus with blood coming from the right coronary artery. This was done both by angiographic studies in the right coronary artery and by indicator dilution technique with ascorbic acid injected into the right coronary artery and using the distal platinum tip of the bipolar coronary sinus catheter as detector for the indicator. The candidates for this study were only those who showed no evidence of right coronary blood at the sampling site as detected by angiograms and ascorbic acid (no detection or detection later than 5 sec after the bolus injection of ascorbic acid). A small catheter was then introduced into the left femoral artery and the right coronary catheter was replaced by a Judkins left coronary catheter. During these procedures, 1,000–2,000 units of heparin were injected.

The left coronary artery catheter was then engaged into the left coronary ostium and remained in the same position throughout the study. If there was no damping in coronary artery pressure in comparison with femoral artery pressure, the patient suffered no discomfort and the electrocardiogram remained unchanged, the injection was begun into the left coronary catheter at a constant rate of 3.84 ml/min for at least 7 min. The injection consisted of 35 ml of normal saline, 5 ml of Evans blue dye, 0.5 ml of oleic-1-^{14}C acid (2μc), and

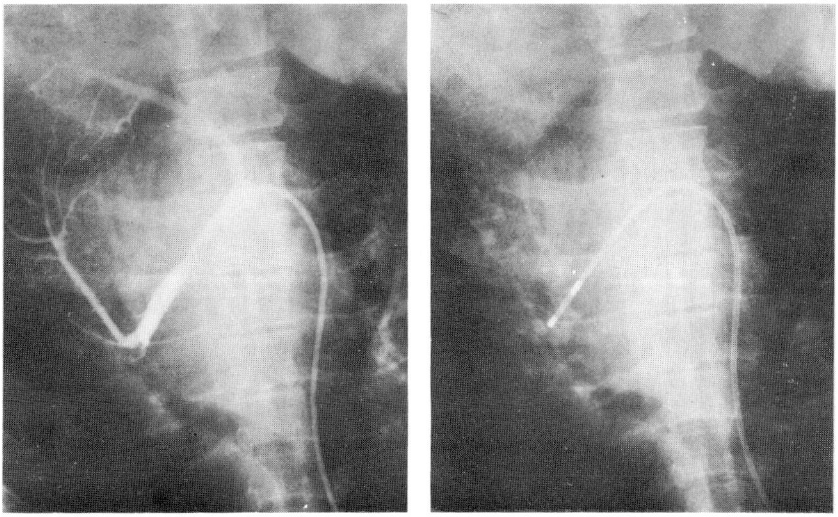

Figure 1. Left: Position of bipolar catheter in distal portion of coronary sinus. *Right:* Injection of renovist from catheter, to show venous drainage.

2,000 units of heparin. The oleic-1-^{14}C acid was bound to human albumin and had a specific activity of 40μc/μmole. No glucose solution was used during these studies. In order to estimate the steady state $^{14}CO_2$ release from the myocardium, $NaH^{14}CO_3$ and lactate-1-^{14}C instead of oleic-1-^{14}C acid were injected in seven and four other patients, respectively. In seven subjects, a second infusion of oleic-1-^{14}C acid was done 2–8 minutes after the first, in order to assess $^{14}CO_2$ release.

Arterial and coronary sinus blood samples were collected simultaneously in preheparinized syringes for measurements of hematocrit, Evans blue, oxygen, carbon dioxide, labeled carbon dioxide, FFA, and labeled FFA. There were two to four collection periods during each constant infusion for FFA and $^{14}CO_2$ and the same for O_2 and CO_2, while hemtocrit and Evans blue dye concentration were measured on each sample.

After this metabolic study, all patients were subjected to left ventriculograms and selective coronary arteriograms.

Biochemical analysis

The hematocrit was immediately determined and a correction factor of 6% was used for trapped plasma. The optical density of Evans blue dye in the plasma was then measured with a spectrophotometer.

Total oxygen and carbon dioxide were analyzed according to the method of Van Slyke and Neill (1924). Labeled carbon dioxide was measured as described by Saba and Di Luzio (1966), except that hyamine hydroxide was used instead of 10% potassium hydroxide as trapping agent on filter paper inserted into the scintillation vial. Recovery of $^{14}CO_2$ with this procedure assessed with $NaH^{14}CO_3$ in whole blood was 83 ± 3.3% (SEM 10 determinations). Each vial, containing filter paper and 20 ml of scintillation solution (5 g/liter 2,5-diphenyl-oxazole and 0.1 g dimethyl 2,2-phenylene bis-(5-phenyloxazole) in toluene), was counted for 100 min in a Liquimat liquid scintillation counter. The counts were corrected for quenching and the recovery rate and were expressed in dpm/ml of blood.

The plasma lipids were extracted by the method of Dole (1956) and separated by thin layer chromatography (TLC) as described by Brown and Johnston (1962) and Goldbrick and Hirsch (1963). The FFA spot was recovered and eluted from silica gel with repeat extraction by chloroform. The filtrate was collected in a glass counting vial and evaporated with nitrogen. FFA were then dissolved with the same scintillation solution used for labeled carbon dioxide and counted for 100 min. The concentration was expressed in dpm/ml of plasma after correction for quenching and for recovery rates from TLC and lipid extraction. The recovery rates from TLC in this study range from 93 to 98%. Radioactivity in the triglyceride fractions from TLC were analyzed similarly.

Individual FFA were measured using a gas chromatographic method with heptadecanoic acid as an internal standard added to the sample during the Dole extraction process. FFA obtained from TLC were eluted with chloroform which was subsequently evaporated and the residue was taken up in 5 ml of methanol. FFA were then esterified after addition of 0.2 ml of dimethyl acetal and 0.25 ml of 6.26 M H_2SO_4 by refluxing for 2 hours at 70°C. Distilled water was added to break the phases. The methylesters were then extracted in hexane, which was subsequently evaporated under nitrogen. An aliquot of the residue dissolved in heptane was injected into a Packard gas chromatograph with a column maintained at 170°C. The output of the chromatograph was displayed on a Honeywell recorder. Areas corrected for baseline were determined and concentrations and percentage distributions were calculated. Each determination for endogenous and labeled FFA was done in duplicate.

Calculations

Total left coronary plasma (PF) and blood (BF) flows were measured according to the description of Zierler (1962), modified for the left coronary system by Klassen *et al.* (1970).

1. Myocardial endogenous FFA uptake (μmoles/min) was calculated as follows: $(FFA\hat{a} - FFA\widehat{cs}) \times PF$, where $\hat{a}$ and $\widehat{cs}$ designate the FFA concentrations in arterial and coronary sinus plasma, respectively.

2. Myocardial FFA uptake based on oleic-1-^{14}C acid ($^{14}C18{:}1$) (μmoles/min) was measured as: $PF \times FFA\hat{a} \times E^*$ where E^* is the extraction of the $^{14}C18{:}1$.

3. Myocardial FFA oxidation based on $^{14}CO_2$ release (μmoles/min) was calculated as:

$$\frac{(^{14}CO_2\,cs - {}^{14}CO_2\,a) \times BF}{\text{FFA arterial specific activity}}$$

4. Myocardial FFA release (μmoles/min) was evaluated as:

$$[(\text{FFA}\hat{a} \times E^*) - (\text{FFA}\hat{a} - \text{FFA}\hat{cs})] \times \text{PF}$$

RESULTS

During these metabolic studies, there were no complications. No patient complained of any discomfort. There was no significant change in ECG, heart rate, or femoral arterial pressure.

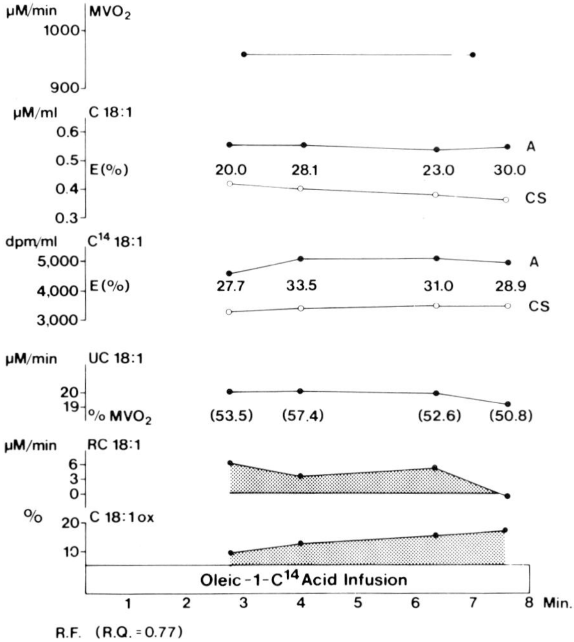

Figure 2. Data from individual experiment during continuous left coronary infusion of oleic-1-^{14}C acid. MVO_2 represents oxygen consumption of myocardium perfused by left coronary artery. C 18:1 and ^{14}C18:1 show endogenous and labeled oleic acid plasma concentration in artery (A) and coronary sinus (CS). *E* (%) is extraction ratio. UC 18:1 shows uptake of oleic acid. RC 18:1 and C 18:1 ox indicate release and percent oxidation of this substrate.

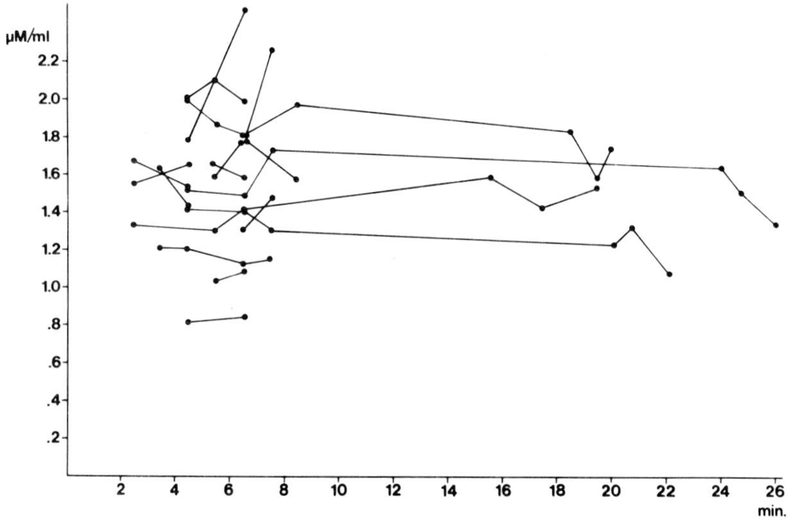

Figure 3. Arterial FFA concentration during course of study in each subject.

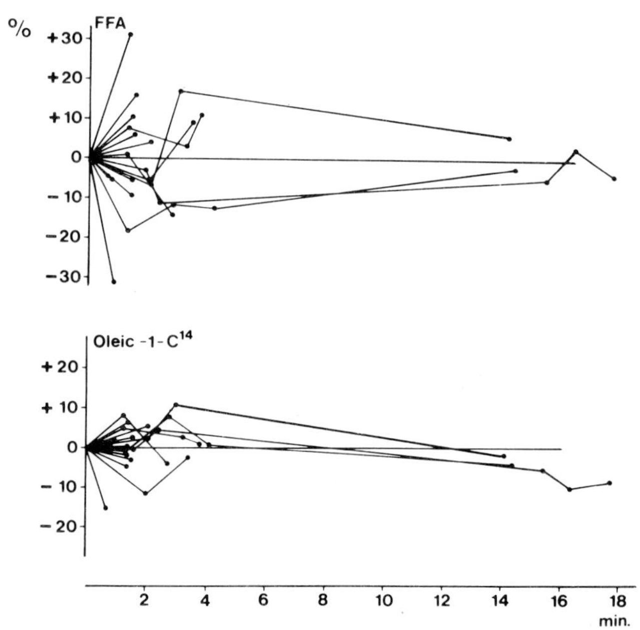

Figure 4. Changes (in percentages) from first determination of each subject in myocardial endogenous FFA and labeled oleic acid extraction.

Coronary artery disease was present in 13 patients, while 4 had normal arteriograms. Figure 2 illustrates an example of the metabolic study. The myocardial respiratory quotient was 0.76 ± 0.11 for the entire group, indicating that lipid was the major myocardial substrate. In a pilot study and in 4 of the subjects of the present study, the arterial FFA concentration before heparin infusion was 0.83 ± 0.21 μmole/ml; 5 min after heparin infusion (4,000 units), it was 1.73 ± 0.30 μmole/ml ($p < 0.01$).

Although the mean arterial FFA concentration during heparin infusion (Fig. 3) was relatively stable, this was not the case for all patients. In 3 of the 17 patients, the arterial FFA concentrations throughout the study varied by more than 10%. However, the myocardial endogenous FFA extractions showed more variation throughout the study (Fig. 4). When compared to the first determination of each patient, 9 of the 17 had an endogenous FFA extraction that varied by more than 10%. By contrast, similar variations for the myocardial extraction of oleic-1-^{14}C acid were observed in only 3 patients. These findings indicate that there was more variation in the coronary sinus concentration of endogenous FFA than of oleic-1-^{14}C acid. The mean value of endogenous FFA extraction for each patient is compared with their respective oleic-1-^{14}C extraction in Figure 5. Except in 1 subject, the labeled FFA extraction was higher. The mean extraction for endogenous FFA was 19.9 ± 2.1%, and significantly lower than the oleic-1-^{14}C extraction, which was 36.0 ± 2.1%.

The relative concentrations of individual plasma FFA expressed in percent of

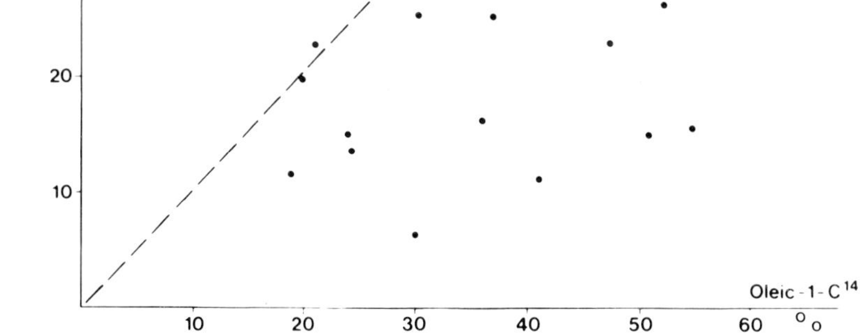

Figure 5. Comparison between fractional extraction of FFA and fractional extraction of labeled oleic acid (%). *Broken line* indicates line of identity.

Table 1. Relative concentrations of individual FFA (%)*

	Arterial	Coronary Sinus
Myristic acid	2.5 ± 0.1	2.6 ± 0.1
Palmitic acid	28.9 ± 0.5	30.0 ± 0.5
Palmitoleic acid	6.4 ± 0.2	6.3 ± 0.2
Stearic acid	7.9 ± 0.4	8.1 ± 0.4
Oleic acid	44.4 ± 0.5	43.1 ± 0.6
Linoleic acid	9.8 ± 0.5	9.9 ± 0.5

*Values expressed as mean ± SEM.

the sum of the 6 FFA analyzed is shown in Table 1. There was no difference between the arterial and venous relative concentrations of individual FFA. Oleic acid constituted 44% of all FFA.

Myocardial $^{14}CO_2$ production from oxidation of oleic-1-^{14}C acid is illustrated in Figure 6. At 7 min after the beginning of the continuous infusion, 31.8 ± 4.3% of the oleic-1-^{14}C acid taken up by the myocardium was oxidized. Considering the $^{14}CO_2$ production during the first 8 min or later on when the infusion was restarted, it is obvious that a steady state $^{14}CO_2$ plateau was not obtained. In order to find out if the failure to reach a plateau for $^{14}CO_2$ release

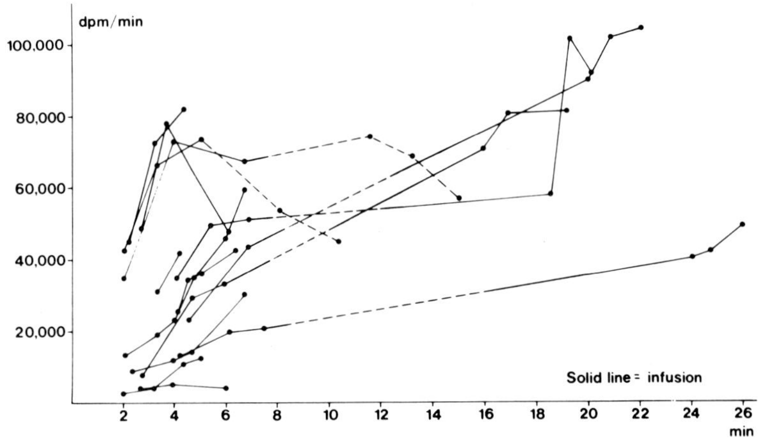

Figure 6. $^{14}CO_2$ myocardial production during (*solid line*) and after (*broken line*) infusion of oleic-1-^{14}C acid.

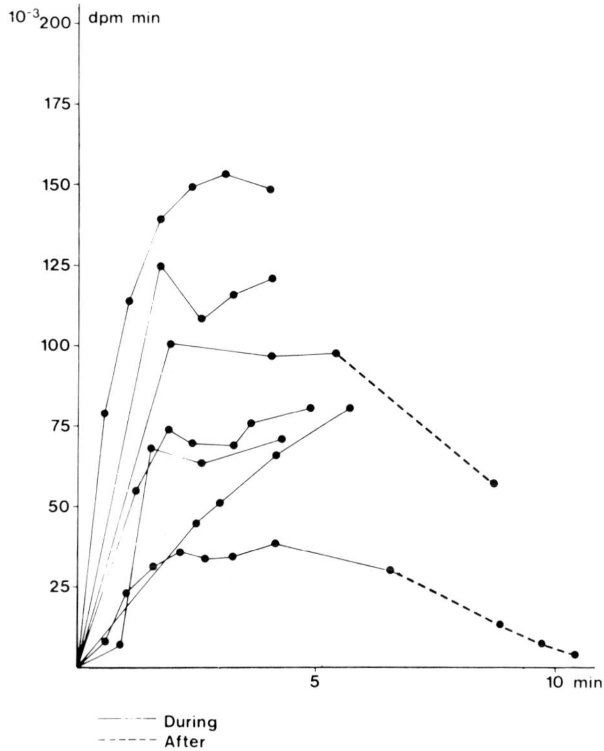

Figure 7. $^{14}CO_2$ myocardial release during (*solid line*) and after (*broken line*) $NaH^{14}CO_3$ infusion.

from oleic-1-^{14}C acid was not due to myocardial CO_2 pool, seven other patients received a constant infusion of $NaH^{14}CO_3$, while four others received a constant infusion of lactic-1-^{14}C acid. During $NaH^{14}CO_3$ infusion, a plateau of $^{14}CO_2$ production was reached within 4 min in six of the patients (Fig. 7). Similarly, $^{14}CO_2$ production during constant lactate-1-^{14}C acid infusion reached a relatively steady state within 5 min (Fig. 8).

In these studies, it is assumed that there is no preferential uptake of the tracer and that all FFA behave like the tracer. Based on the tracer extraction, the myocardium is taking up 52.3 ± 6.3 μmoles/min, which would require 180% of the oxygen consumed to be totally and immediately oxidized. Even though the calculated oxidation at 7 min represents an underestimation, 31.8 ± 4.3% of the FFA taken up by the myocardium have been oxidized. Such oxidation requires 49.3 ± 6.4% of myocardial oxygen consumption. However, an amount equal to 50% of the FFA taken up by the myocardium is released between the artery and

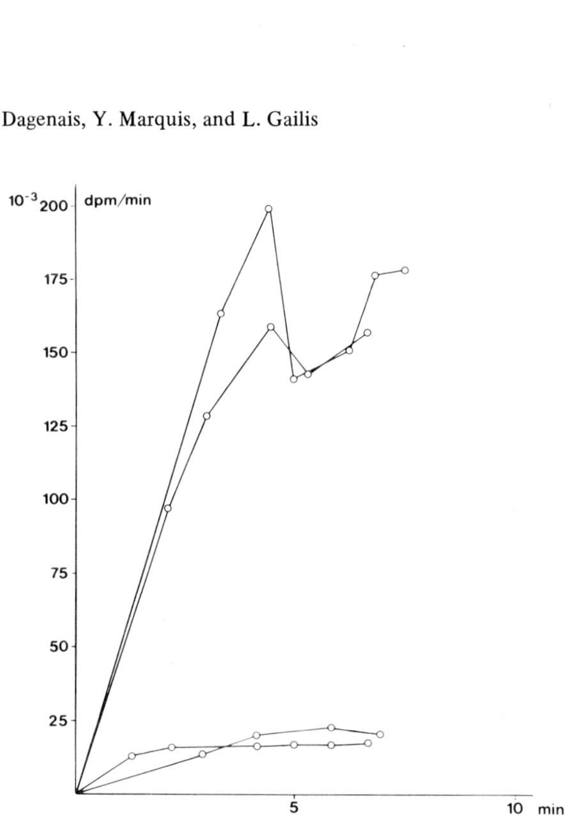

Figure 8. $^{14}CO_2$ myocardial production during lactic-1-^{14}C acid infusion.

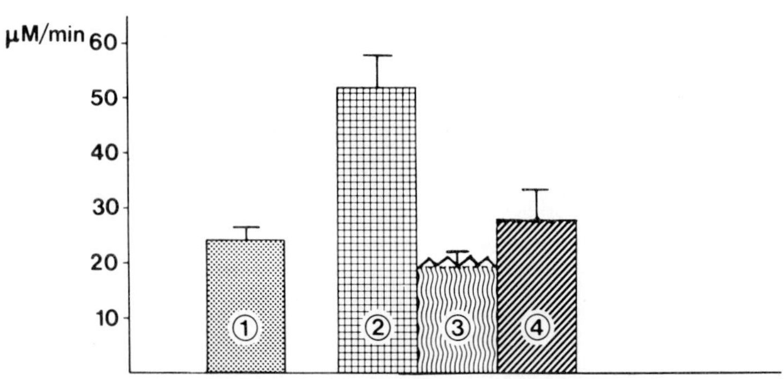

Figure 9. Myocardial uptake of endogenous FFA (*1*) and of total FFA based on labeled extraction (*2*) in 17 patients. At 7 min of oleic-1-^{14}C infusion, oxidation of FFA which represents an underestimation (*3*). "Release" of FFA in coronary sinus (*4*).

the coronary sampling site (Fig. 9). No significant labeled triglycerides were found during the course of this study.

DISCUSSION

It has been shown that the human heart may hydrolyze circulating triglycerides (Muir, 1968). Furthermore, it is probable that the human heart, like skeletal muscle, releases FFA in the coronary sinus (Most *et al.*, 1969; Lassers, Kaijser, and Carlson, 1972). Because of the contribution of triglyceride hydrolysis and possibly myocardial FFA release to the coronary sinus FFA concentration, the myocardial uptake assessed by endogenous FFA arteriocoronary sinus difference will represent an underestimation. To obviate this difficulty, labeled FFA tracer can be utilized. Most *et al.*, (1969) and Lassers *et al.* (1972) have used a constant intravenous infusion of labeled palmitate. The technique with labeled oleic acid used in the present study is derived from forearm studies and is identical to the procedure reported by Klassen *et al.* (1973). Although continuous intracoronary infusion is a more invasive approach, it offers several advantages over continuous intravenous infusion. With intracoronary infusion: 1) a smaller radioactive dose of the tracer can be used (2 μc instead of 30 μc); 2) a steady state extraction of the labeled tracer is obtained within 4 min instead of the 15–20 minutes needed during the intravenous infusion approach; 3) a shorter infusion time will prevent incorporation of the tracer into triglycerides, which occurred after 20 min of intravenous infusion (Most *et al.*, 1969); 4) the intracoronary infusion approach, although limited only to individuals who do not have evidence of right coronary blood draining at the coronary sinus sampling site, permits simultaneous measurement of coronary plasma flow with indicator dilution technique, which is necessary to quantitate myocardial FFA uptake and release.

All these isotopic techniques imply two assumptions. The oleic acid tracer is assumed to behave like endogenous oleic acid, so there is no preferential uptake of the tracer. To assure this condition, the tracer was prepared as oleic-1-^{14}C acid bound to human albumin. Despite these precautions, however, a possible preferential uptake of the tracer cannot be absolutely excluded. Secondly, it is assumed that all FFA analyzed behave like oleic acid. This will have to be confirmed by different labeled FFA studies. Nevertheless, such observations were documented in human skeletal muscle and the error inherent in this assumption is small (Havel *et al.*, 1964).

Heparin injected in a bolus has been shown to produce an elevation of FFA and a decrease of serum triglycerides (LaRosa *et al.*, 1971; Kongtahworm *et al.*, 1973). During constant heparin infusion in four subjects, we found that FFA had increased two-fold while triglycerides had decreased in the same proportion. The triglyceride myocardial balance after 5 min of heparin infusion was stable and a net uptake which was relatively low was observed in only two subjects.

The observed increase in plasma FFA is of the same order as those reported to produce detrimental effects, such as dysrythmia and excessive myocardial oxygen consumption, in animals and man. Since heparin is often used in clinical situations and deleterious cardiac effects are rarely encountered, the purpose of this study was to assess how the human myocardium handles FFA which are elevated.

In the present study, the variation of endogenous FFA, particularly in the coronary sinus, might be due to heparin. Although the arteriocoronary sinus difference in triglycerides was stable and slightly positive only in two of the four subjects, it is possible that some hydrolysis of circulating triglycerides or diglycerides had occurred, increasing the FFA coronary sinus concentration. However, when our results are compared with those of others (Carlson, Ekelund, and Orö, 1970; Lassers *et al.*, 1972), similar variations in FFA were observed, although heparin was not used in these studies. The changes in FFA concentration cannot be attributed to heparin alone.

For the entire group, the mean FFA myocardial uptake based on the labeled extraction was 52.3 μmoles/min for a mean arterial FFA of 1.52 μmoles/ml. These values are twice as high as those of Most *et al.* (1969), who did not utilize heparin in their study. To be totally and immediately oxidized, the amount of FFA taken up by the myocardium in the present study would necessitate 180% of the cardiac oxygen consumption. These findings indicate that not all FFA taken up by the myocaridum are oxidized. It is obvious from $^{14}CO_2$ release that a steady state was not reached after 7 min of continuous infusion of labeled oleic acid. This failure to reach a plateau was not due to myocardial CO_2 pool, since a steady release of $^{14}CO_2$ was obtained within 5 min during a constant infusion of $NaH^{14}CO_3$ or lactate-1-^{14}C. The generation of $^{14}CO_2$ from oleic-1-^{14}C is relatively slow and may explain why a plateau was not reached. Because of these findings, the 32% of FFA taken up by the myocardium which are oxided to $^{14}CO_2$ represent an underestimation of FFA oxidation. In these studies, it is assumed that FFA oxidation is total. However, there is some evidence that, in human exercising skeletal muscle, incomplete FFA oxidation occurred (Hagenfeldt and Wahren, 1968). Although Most *et al.* (1969) did not find it in the human heart, incomplete oxidation of FFA cannot be totally excluded. Most *et al.* (1969) found that 53.5% of FFA taken up by the myocardium were oxidized. However, because of their technical approach using continuous intravenous infusion of labeled palmitate, a myocardial uptake of labeled triglycerides was present. These investigators pointed out that such an uptake may contribute to $^{14}CO_2$ production. In the present study, no significant labeled triglycerides could be detected, indicating that $^{14}CO_2$ was generated from circulating labeled FFA alone.

It is possible that the nonoxidized fraction of FFA taken up by the myocardium may be incorporated into the cardiac lipid pool. According to studies done in animals, these lipid pools may be located in the myocardial cells (Olson and Hoeschen, 1967) or the cardiac interstitial space (Stokke *et al.*, 1974). However, it would seem unlikely that the normal human heart can continue to accumulate FFA at this rate. The lower specific activity of the coronary sinus FFA in all except 1 subject and, similarly, the higher extraction ratio of labeled oleic acid compared to endogenous FFA imply that some of the FFA in the coronary sinus are derived from the hydrolysis of circulating triglycerides and/or cardiac release. However, the triglycerides during heparin infusion are low and the arteriocoronary sinus difference is near zero. The heart may release FFA, which may originate from interstitial fluid, the myocardial cells, or even subepicardial fat tissue. In the present study, an amount equal to 50% of FFA uptake was released in the coronary sinus. In the absence of heparin, similar findings were reported in the human heart (Most *et al.*, 1969; Lassers *et al.*, 1972). Another possible way for the FFA to leave the myocardium would be through the lymphatics (Stokke *et al.*, 1974).

Although the human heart takes up large amounts of FFA during constant heparin infusion, the present findings suggest that the heart has important lipid pools which may play a role in FFA release.

Acknowledgments This work was supported by a grant-in-aid from the Medical Research Council of Canada. The technical assistance of Mr. Denis Simard and Miss Danielle Ouellet was much appreciated. Dr. Dagenais is a scholar of the Canadian Heart Foundation.

REFERENCES

ANDRES, R., K. L. ZIERLER, H. M. ANDERSON, W. N. STAINSBY, G. CADER, A. S. GHRAYYIB, and J. L. LILIENTHAL, Jr. 1954. Measurement of blood and volume in the forearm of man: With notes on the theory of indicator-dilution and on production of turbulence, hemolysis and vasodilatation by intra-vascular injection. J. Clin. Invest. 33: 482.

BROWN, J. L., and J. M. JOHNSTON. 1962. Radioassay of lipid components separated by thin layer chromatography. J. Lipid Res. 3: 480.

CARLSON, L. A., L. G. EKELUND, and L. ORÖ. 1970. Clinical, metabolic and cardiovascular effects of different prostaglandins in man. Acta Med. Scand. 188: 553.

DOLE, V. P. 1956. A relationship between non esterified fatty acids in plasma and the metabolism of glucose. J. Clin. Invest. 35: 150.

GOLBRICK, B., and J. HIRSCH. 1963. A technique for quantitative recovery of lipids from chromatoplates. J. Lipid Res. 4: 482.

HAGENFELDT, L., and J. WAHREN. 1968. Human forearm muscle metabolism during exercise. II Uptake, release and oxidation of individual FFA and glycerol. Scand. J. Clin. Lab. Invest. 21: 263.

HAVEL, R. J., L. A. CARLSON, L. G. EKELUND, and A. HOLMGREN. 1964. Turnover rate and oxidation of different free fatty acids in man during exercise. J. Appl. Physiol. 19: 613.

KLASSEN, G. A., J. B. AGARWAL, P. H. TANSER, S. P. WOODHOUSE, and D. MARPOLE. 1970. Blood flow and tissue space of the left coronary artery in man. Circulat. Res. 27: 185.

KLASSEN, G. A., P. H. TANSER, D. MARPOLE, and J. B. AGARWAL. 1973. Adaptation of the coronary circulation to primary myocardial disease. *In* N. S. Dhalla (ed.), Recent Advances in Studies on Cardiac Structure and Metabolism, Vol. 3: Myocardial Metabolism. University Park Press, Baltimore.

KONGTAHWORN, C., E. FOSTER, E. E. MASON, and K. J. PRINTEN. 1973. Heparin-induced elevation of free fatty acids in diabetics patients. Surgery 74: 30.

LAROSA, J. C., R. I. LEVY, W. G. BROWN, and D. S. FREDRICKSON. 1971. Changes in high-density lipoprotein composition after heparin-induced lipolysis. Am. J. Physiol. 220: 785.

LASSERS, B. W., L. KAIJSER, and L. A. CARLSON. 1971. Myocardial lipid and carbohydrate metabolism in healthy, fasting men at rest: Studies during continuous infusion of ^{3}H-palmitate. Europ. J. Clin. Invest. 2: 348.

MJØS, O. L., J. K. KJEKSHUS, and J. LEKVEN. 1974. Importance of free fatty acids as a determinant of myocardial oxygen consumption and myocardial injury during norepinephrine infusion in dogs. J. Clin. Invest. 53: 1290.

MOST, A. S., N. BRACHFELD, R. GORLIN, and J. WAHREN. 1969. Free fatty acid metabolism of the human heart at rest. J. Clin. Invest. 48: 1177.

MUIR, J. R. 1968. The regional production of lipoprotein lipase in man. Clin. Sci. 34: 261.

OLIVER, M. F., V. A. KURIEN, and T. W. GREENWOOD. 1968. Relation between serum-free-fatty-acids and arrhythmias and death after acute myocardial infarction. Lancet *i*: 710.

OLSON, R. E., and R. J. HOESCHEN. 1967. Utilization of endogenous lipid by the isolated perfused rat heart. Biochem. J. 103: 796.

SABA, T. M., and N. R. DI LUZIO. 1966. Method for collection and determination of $^{14}CO_2$ for in vitro metabolic studies. J. Lipid Res. 7: 566.

STOKKE, K. T., N. B. FJELD, T. H. KLUGE, and S. SKREDE. 1974. Lipid composition and cholesterol esterification in lymph. Scand. J. Clin. Lab. Invest. 33: 199.

VAN SLYKE, D. D., and J. M. NEILL. 1924. Determination of gases in blood and other solutions by vacuum extraction and manometric measurement. J. Biol. Chem. 61: 523.

ZIERLER, K. L. 1962. Circulation times and the theory of indicator-dilution methods for determining blood flow and volume. *In* W. F. Hamilton and P. Dow (eds.) Handbook of Physiology Vol. 1, Sec. 2. American Physiological Society, Washington, D.C.

Mailing address:
Doctor Gilles R. Dagenais,
Institut de Cardiologie de Québec,
2725, Chemin Ste-Foy,
Ste-Foy, Québec 10e,
Québec G1V 4G5 (Canada).

Recent Advances in Studies on
Cardiac Structure and Metabolism Volume 10
The Metabolism of Contraction
Edited by Paul-Emile Roy and George Rona

LIPID DROPLETS IN THE HEART INTERSTITIUM: CONCENTRATION AND DISTRIBUTION

PAUL-EMILE ROY

Institut de cardiologie de Québec, and Department of Medicine,
Laval University, Québec, Québec, Canada

SUMMARY

A morphological study of the sudanophilic lipids in dog and human hearts was carried out. The hearts of 5 mongrel dogs and 20 auricular appendices of human patients were utilized. The dog hearts were fixed *in situ* by perfusion with 2% glutaraldehyde, and the human tissue was fixed by immersion. For the ultrastructural study, some blocks of every portion of the specimen were postfixed in osmium and embedded in Epon. The topographical and quantitative study was done on frozen 20-μ sections of tissue stained with Sudan IV in propylene glycol and counterstained with hematein (Geigy). The results indicate that a significant quantity of visible triglycerides is extracellular and located in the interstitial spaces of the heart. The dimensions of these lipid droplets vary from 0.5 μ to more than 20 μ. The highest concentration was found in the epicardial region of the left ventricle, but a significant amount was found in the endocardium and in the middle portion of the myocardium. These observations indicate that a sizable portion of the heart interstitium is occupied by free droplets of triglycerides. This substrate pool is probably mobile and readily available as a source of energy.

INTRODUCTION

Lipids are the main substrates of cardiac metabolism (Bing, 1965; Opie, 1969; Evans, 1964a; Carlsten *et al.,* 1961). Under normal conditions in both humans and animals, the arteriovenous difference shows an important uptake of free and esterified fatty acids (Gordon and Cherkes, 1956; Dagenais, Marquis, and Gailis, 1974). This uptake may be appreciably larger than the quantity oxidized by the heart. The heart must, therefore, have a means of eliminating or putting into storage a part of this uptake.

The lipids are transported in plasma; since they are hydrophobic, they are found bound to lipoproteins or in the form of finely emulsified droplets with a diameter of 0.5–5 μ. Whatever the form in which the lipid molecules are present, they can diffuse across membranes very readily; even the chylomicrons can cross both the intestinal epithelium (Cardell, Badehausen, and Porter, 1967) and the peritoneum (Morris and Harding, 1955).

The uptake of lipids by the heart is a complex process. Nevertheless, in all the studies done until now (Dagenais *et al.*, 1974; Evans, 1964b), it seems that a portion of the free fatty acids is not immediately utilized by the heart and can be found in a homogenate of the cardiac tissue. Some of these lipids are eventually utilized for the production of energy (Shipp, Opie, and Challoner, 1961). However, the quantity of triglycerides eventually utilized and the form in which they are stored are not definitely known. It seems that the amounts stored vary continually with changes in the uptake and oxidation. Thus, in ischemia, the oxidation of lipids is necessarily decreased, but the tissue content of triglycerides is increased (Scheuer and Brachfeld, 1966). Lipid uptake is also increased after intestinal absorption of lipids. Under such circumstances, the uptake of lipids may be as much as five times that of carbohydrates (Bing *et al.*, 1954).

The present work attempts to define the morphological location and the structure of the lipid deposit. Normally, the myocardial cells as examined by histology (Bullard, 1916) or by electron microscopy (Fawcett and McNutt, 1969) contain practically no visible lipids. However, under abnormal conditions, when oxidation is defective, as during ischemia (Kent and Diseker, 1955; Wartman *et al.*, 1956) or when lipid uptake is increased (Beckett, 1963), the presence of lipid droplets can be observed in the cardiac muscle fibers. This may also occur under other conditions in which metabolism is probably disturbed (Maunsbach and Wirsen, 1966; Wittels and Bressler, 1964). From these observations, we cannot conclude that the lipids present in the muscular fibers can constitute a reserve. Lipid reserves must probably be sought outside the muscular fibers.

MATERIALS AND METHODS

The observations which follow are based on a study of the hearts of 5 mongrel dogs and of 20 right auricular appendices obtained from humans during operations for the correction of congenital malformations. The patients were fasting when the samples were obtained. The dogs had been fasting for an undetermined period.

Preparation of tissue

The dog hearts were fixed by perfusion. The cavities of the heart were drained and a cold solution (4°C) of Perfudex (solution for organ preservation, Pharmacia) was continuously injected through the aorta into the coronary arteries. This solution had been oxygenated by bubbling oxygen through it for 30 min. After 5 min of perfusion, the heart had stopped beating and the outflow from the coronary sinus was completely colorless. The perfusion was then continued for 10 min with a 2% solution of glutaraldehyde in phosphate buffer. Following the perfusion, the heart was opened and tissue samples were taken.

To simplify the interpretation of the results, all the samples were obtained from the same regions of the heart, either ventricular or auricular. In the same locations, we also made transmural sections that were stained specifically for lipids.

The preparation of the human atrial appendices, including fixation by immersion and a postfixation, has been described previously (Roy and Morin, 1971). All the samples, except those stained specifically for lipids, were postfixed with osmium and embedded in Epon 812. From all the blocks, sections 1 and 0.05 μ in thickness were made, and the tissues were examined by light and electron microscopy. The 1-μ sections were stained with toluidine blue and the 0.05-μ ones with lead citrate and uranyl acetate, according to the methods of Reynolds (1963) and Watson (1958).

These techniques enable us to see the triglycerides as well as the other constituents of the tissue. These lipids are clearly evident as easily identifiable rounded spheres stained in black. The method can usually show whether the triglycerides are intra- or extracellular. The tissues used for the lipid-specific stains were kept in neutral 10% formaldehyde until they were used (Thompson, 1966).

Samples were obtained from a complete transverse section of the middle portion of the two ventricles and the interventricular septum. The sections were numbered as shown in Figure 1. The blocks measuring approximately 1 cm^2 were frozen at -20°C and cut with a microtome into slices 22 μ thick. The slices

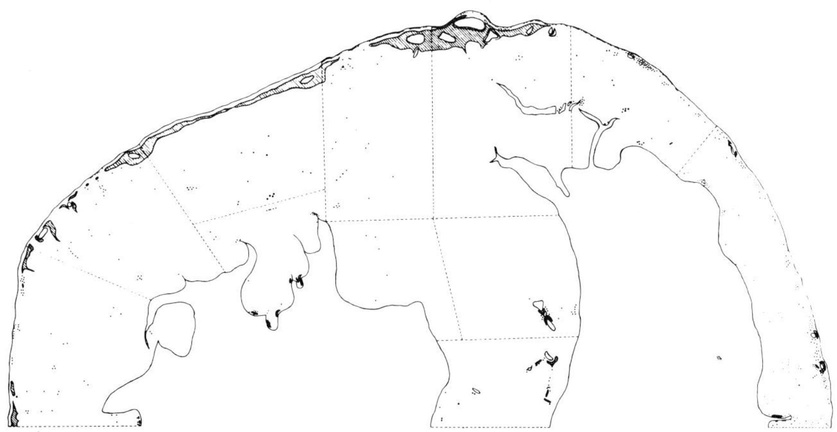

Figure 1. Partial reconstruction of the ventricular wall. The *hatched areas* represent the epicardial adipose tissue. The *dots* represent individual lipid droplets as they were observed under the microscope; the localization is approximate. The droplets are concentrated near the epicardium and are least frequent in the interventricular septum. Samples were obtained from a complete transverse section of the middle portion of the two ventricles and the interventricular septum. The blocks, measuring approximately 1 cm^2, were frozen at -20°C and stained with Sudan IV.

were picked up with a fine glass rod, washed in distilled water, passed through 70% ethanol, and stained with Sudan IV (20 min). The slices were again passed through 70% ethanol and distilled water, then counterstained with hemalun (5 min). After a 5-min wash with distilled water, the slices were placed on a glass slide and mounted with glycerogel for examination.

It should be noted that these techniques will only show triglycerides if they are present in the form of sufficiently large aggregates, such as chylomicrons (0.2–5μ) or larger droplets. The individual lipid molecules cannot be observed, even under the electron microscope. The total lipids of different portions of the heart have been studied previously (Denton and Randle, 1967) as has the distribution of lipids in the organelles and the subcellular fractions (Stein and Stein, 1968). The lipid deposits that can be observed are more likely to be utilized by the heart as energy reserves. An understanding of these substrate deposits may be indispensable for the evaluation of the cardiac metabolism of lipids.

RESULTS

The triglycerides present in the hearts are located in two sites. The first consists of adipose tissue, where the triglycerides are located inside the adipocytes. The second location is extracellular.

Localization of adipocytes

We will not describe the locations of the adipocytes in detail; their distribution in human and dog hearts is approximately the same. A similar distribution is also found in beef, sheep, and pig hearts. In smaller animals, such as cats, rabbits, and rats, the lipids are not very abundant and are thus harder to locate with precision. Precise descriptions of this distribution can be found in anatomy textbooks.

However, it may be important to note that the most significant lipid deposits are almost invariably found in locations where a space is available. These deposits thus tend to round out the heart and to eliminate surface irregularities. Hence, they are found alongside the coronary arteries and veins and in the auriculoventricular furrows. They can also be found on structures forming crests, such as the auricular appendages or the ventricular angles of fishes, reptiles, or birds.

This preferential location can also be produced artificially. Figure 2 shows a dog heart to which a Silastic disc was added to form a rigid projection. After 6 months, the adjoining hard surface is covered with adipose tissue so as to fill the space surrounding the disc. This adipose deposit will probably disappear if the object responsible for its formation is removed. It is thus conceivable that the changes in the form, the dimension or the orientation of coronary arteries, or

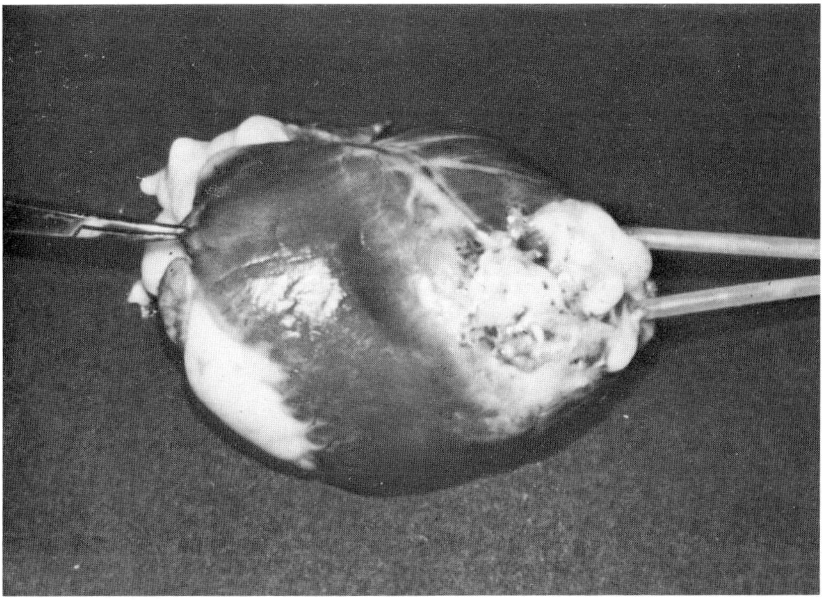

Figure 2. The adipose tissue on the surface of the heart tends to accumulate where depressions are created by the anatomical structure. As seen in the figure, this tendency can be reproduced artificially. An implanted pacemaker was kept on the surface of this dog heart for 6 months. The adipose tissue proliferated on every side of this hard, bulging obstacle to make the surface smooth.

any other anatomical changes affecting the surface of the heart or its anatomical relation with external surfaces can modify adipose deposits, resulting in their increase, decrease, or rearrangement.

Extracellular triglycerides

We observed numerous free lipid particles in the interstitial spaces, both in the human auricular appendices and in all parts of the dog hearts. It may sometimes be difficult to differentiate extracellular and intracellular lipids by light microscopy, especially if the droplets are large. Under these conditions, a fine layer of cytoplasm around the droplets may be overlooked and a lipid globule erroneously identified as being extracellular. For this reason, it is safer to consider only those particles that, because of dimension and shape, are obviously extracellular. Figure 3 shows dog cardiac tissue containing clearly interstitial triglycerides; the particles are small and are found between the muscle fibers. Figure 4 shows the same phenomenon in the human heart auricular appendix. Figures 5 and 6 show the same structure obtained with an entirely different technique, that of frozen sectioning.

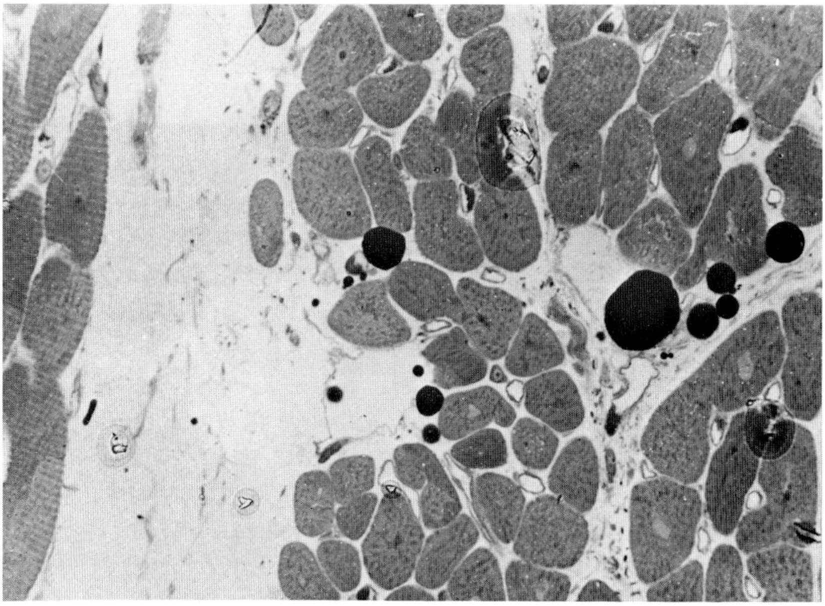

Figure 3. The *black spots* are free lipid droplets dispersed in a small area of the left ventricle. This normal dog heart was fixed by perfusion with 2% glutaraldehyde, postfixed in osmium, and embedded in Epon 812. The triglycerides are strongly osmiophilic. × 700.

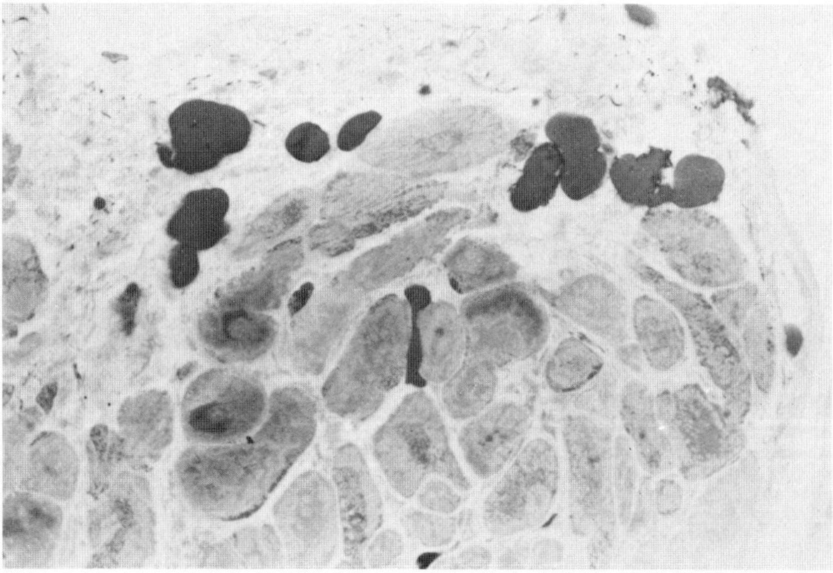

Figure 4. The *dark spots* are free lipid droplets in human myocardium. These droplets are concentrated in the endocardium, but they can also be present in the bundles of muscle fibers. × 700.

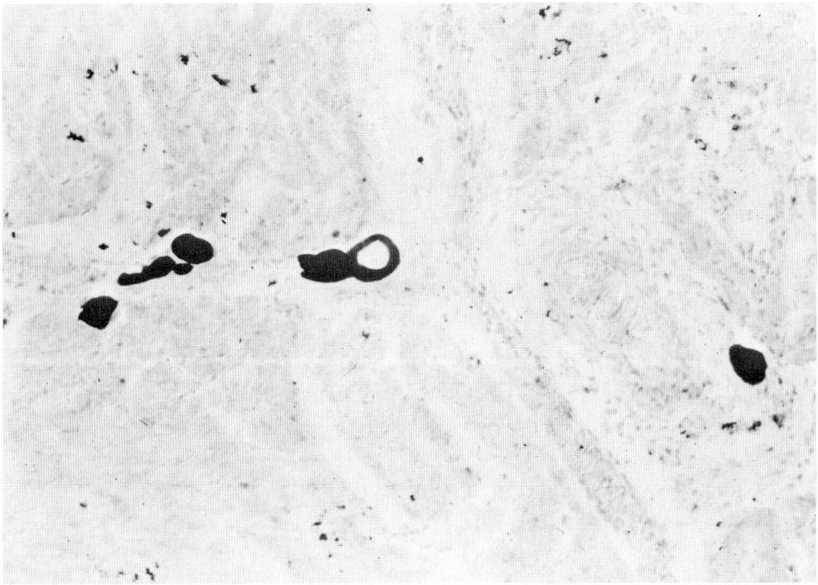

Figure 5. Fat droplets in the left ventricle of a normal mongrel dog. This tissue was fixed by perfusion with glutaraldehyde. Frozen sections were obtained and stained with Sudan IV. The droplets are dispersed between the muscle fibers; they can be differentiated from fat cells by their small size and their shape. × 500.

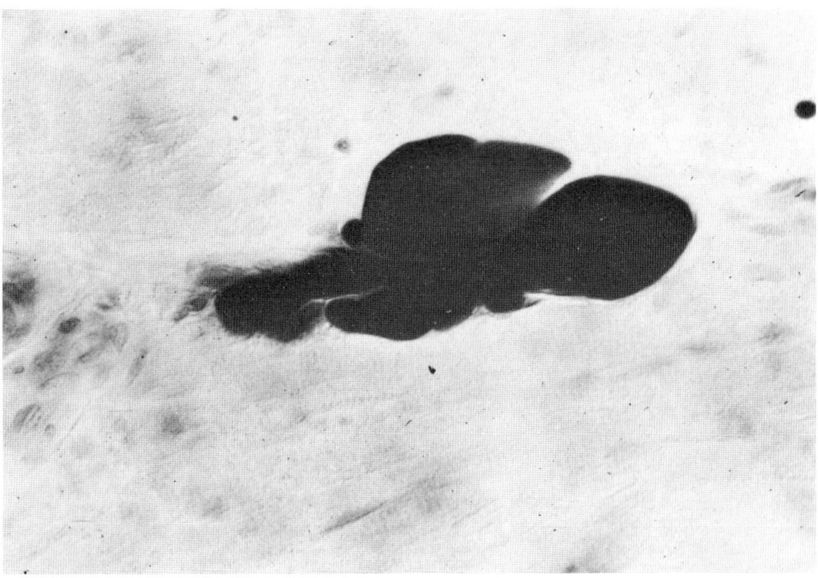

Figure 6. Group of lipid droplets in a left ventricle of a dog. This tissue received the same treatment as in Figure 5. The droplets are often found in small clusters, but they are mainly solitary. × 800.

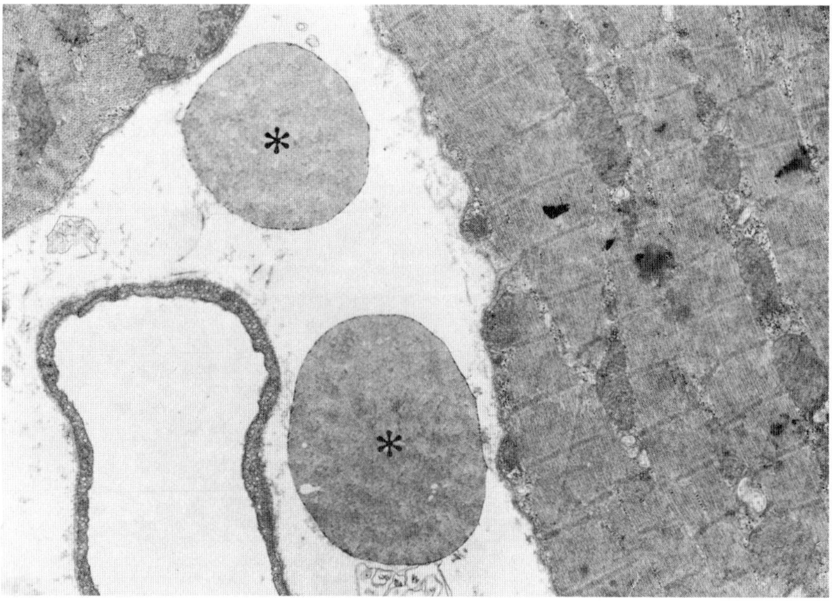

Figure 7. Two free lipid droplets are present in this electron micrograph (*) of myocardium from a normal mongrel dog. Portions of two muscle fibers and a capillary can be seen. The fat droplets measure 4 and 5 μ, respectively. They are usually surrounded by a thin rim of black-stained material, which may be a lipoprotein. × 10,000.

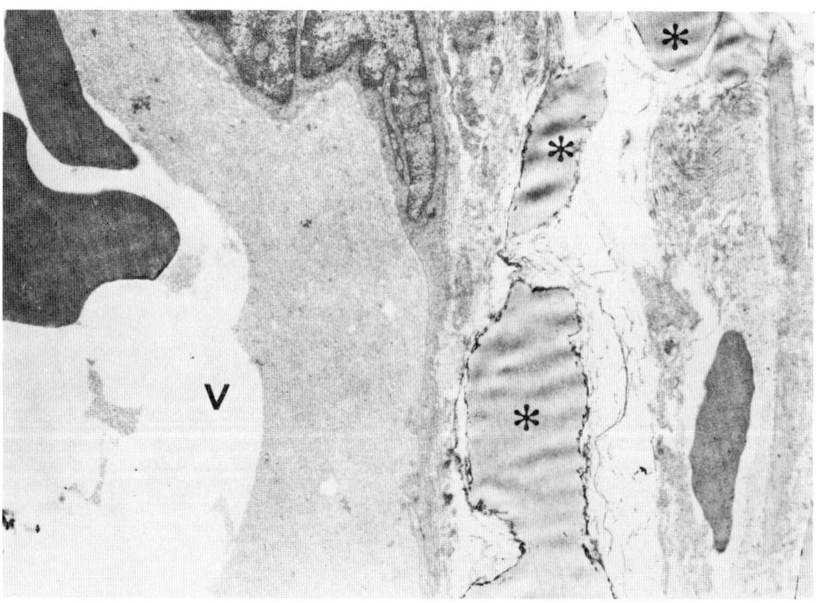

Figure 8. In the human myocardium, the same free lipid droplets are also seen. This electron micrograph shows three droplets (*) in the connective tissue surrounding a small vein (V). Note the thin rim of black-stained material around the fat droplets. × 7,500.

It is surprising to observe that many droplets can be seen by electron microscopy, droplets which we had not observed during histological examination, either because they were too small or because they had the size and the shape of an adipocyte and were thus erroneously identified as being intracellular. Triglycerides are easily identified by electron microscopy, since they form an osmiophilic mass with a diffuse outline (Figs. 7 and 8). Heterogeneous granules can often be found on the surface, but no identifiable cytoplasmic element can be seen on the periphery of the free droplets.

In the dog hearts, where a more systematic study was conducted, a quantitative estimation of the number of free lipid droplets was obtained. Figure 1 illustrates the results. It can be seen that every 1-cm section of the left and right ventricles contains some free lipid droplets in the interstitia. The concentration is highest near the epicardium, and lowest in the interventricular septum.

DISCUSSION

The results of this study indicate that, in dog and human hearts, a considerable amount of fat in the form of free triglyceride droplets is present in the interstitium.

From these observations, we cannot identify the mechanism of formation of these particles. The fact that they are present in greatest concentration near the epicardium may be an indication that they are formed there from adipose tissue. This would occur by the release of nonesterified fatty acids from the adipocytes after the hydrolysis of triglycerides; this mechanism is well known (White and Engel, 1958) and seems to be the predominant one in the mobilization of adipose tissue lipid reserves. If this mechanism is producing the effect, extracellular resynthesis of triglycerides must necessarily occur, which is not impossible. The other possibility is that the triglycerides can be excreted directly by the adipocytes. This mechanism is not presently known to occur, but it is plausible since there are examples of the passage of whole triglyceride particles across membranes (Courtice and Morris, 1955).

The important aspect of these observations is that the lipid particles that we have observed can readily be hydrolyzed by the lipoprotein lipases that are abundantly present in the myocardium (Robinson and Wing, 1970). Thus, they provide a source of readily utilizable substrate. Since this source consists of free particles, it must necessarily be mobile. Although we do not know what quantity of lipids may be transported in this fashion, we may suppose that they represent an important aspect of the cardiac metabolism of lipids.

REFERENCES

BECKETT, E. B. 1963. Histochemistry of cardiac muscle. Biochem. Clin. 1: 51–60.
BING, R. J. 1965. Cardiac metabolism. Physiol. Rev. 45: 171.

BING, R. J., A. SIEGEL, I. UNGER, and M. GILBERT. 1954. Metabolism of the human heart. II. Metabolism of fats, proteins and ketones. Am. J. Med. 16: 504.

BULLARD, H. H. 1916. On the occurrence and physiological significance of fat in the muscle fibers of the normal myocardium and atrioventricular system: Interstitial granules (mitochondria) and phospholipines in cardiac muscle. Am. J. Anat. 19: 1.

CARDELL, R. R. Jr., S. BADENHAUSEN, and K. R. PORTER. 1967. Intestinal triglyceride absorption in the rat. J. Cell Biol. 34: 123–155.

CARLSTEN, A., B. HALLGREN, R. JAGENBURG, A. SVANBORG, and L. WERKO. 1971. Myocardial metabolism of glucose, lactic acid, amino acids and fatty acids in healthy human individuals at rest and at different work loads. Scand. J. Clin. Lab. Invest. 13: 418.

COURTICE, F. C., and B. MORRIS. 1955. The exchange of lipids between plasma and lymph of animals. Quart. J. Exp. Physiol. 40: 138.

DAGENAIS, G. R., Y. MARQUIS, and L. GAILIS. 1974. Assessment of myocardial FFA metabolism in humans during heparin infusion. Abstract, 7th Annual Meeting of the International Study Group for Research in Cardiac Metabolism, Quebec, June 18th–21st.

DENTON, R. M., and P. J. RANDLE. 1967. Concentrations of glycerides and phospholipids in rat heart and gastrocnemius muscles. Effects of alloxan diabetes and perfusion. Biochem. J. 104: 416.

EVANS, J. R. 1964a. Cellular transport of fatty acids. Canad. J. Biochem. 42: 955.

EVANS, J. R. 1964b. Importance of fatty acids in myocardial metabolism. Circulat. Res. XIV-XV (Suppl.): 96.

FAWCETT, D. W., and N. S. McNUTT. 1969. The ultrastructure of the cat myocardium. I. Ventricular papillary muscle. J. Cell Biol. 42: 1–46.

GORDON, R. S., Jr., and A. CHERKES. 1956. Unesterified fatty acids in human blood plasma. J. Clin. Invest. 35: 206.

KENT, S. P., and M. DISEKER. 1955. Early myocardial ischemia. Study of histochemical changes in dogs. Lab. Invest. 4: 398.

MAUNSBACH, A. B., and C. WIRSEN. 1966. Ultrastructural changes in kidney, myocardium and skeletal muscle of the dog during excessive mobilisation of free fatty acids. J. Ultrastruct. Res. 16: 35–54.

MORRIS, B., and J. HARDING. 1955. Unpublished data.

OPIE, L. H. 1969. Metabolism of the heart in health and disease. Am. Heart J. 77: 100.

REYNOLDS, E. S. 1963. The use of lead citrate at high pH as an electron-opaque stain in electron microscopy. J. Cell Biol. 17: 208–212.

ROBINSON, D. S., and D. R. WING. 1970. Regulation of adipose tissue clearing factor lipase activity. *In* R. Levine and E. F. Pfeiffer (eds.), Adipose tissue, p. 41. Academic Press, New York.

ROY, P. E., and P. J. MORIN. 1971. Variations of the Z-band in human auricular appendages. Lab. Invest. 25: 422.

SCHEUER, J., and N. BRACHFELD. 1966. Myocardial uptake and fractional distribution of palmitate 1-C^{14} by the ischemic dog heart. Metabolism 15: 945.

SHIPP, J. C., L. H. OPIE, and D. CHALLONER. 1961. Fatty acid and glucose metabolism in the perfused heart. Nature (London) 189: 1018.

STEIN, O., and Y. STEIN. 1968. Lipid synthesis, intracellular transport, and storage. III. Electron microscopic radioautographic study of the rat heart perfused with tritiated oleic acid. J. Cell Biol. 36: 63.

THOMPSON, S. W. 1966. Sudan colorants in propylene glycol. *In* Selected Histochemical and Histopathological Methods, p. 30. Charles C Thomas Publisher, Springfield, Ill.

WARTMAN, W. B., R. B. JENNINGS, H. O. YOKOHAMA, and G. F. CLABAUGH. 1956. Fatty change of the myocardium in early experimental infarction. Arch. Path. 62: 318.

WATSON, M. L. 1958. Staining of tissue sections for electron microscopy with heavy metals. II. Application of solutions containing lead and barium. J. Biophys. Biochem. Cytol. 4: 727.

WHITE, J. E., and F. L. ENGEL. 1958. A lipolytic action of epinephrine and norepinephrine on rat adipose tissue *in vitro*. Proc. Soc. Exp. Biol. Med. 99: 375.

WITTELS, B., and R. BRESSLER. 1964. Biochemical lesion of diphtheria toxin in the heart. J. Clin. Invest. 43: 630.

Mailing address:
Paul-E. Roy, M.D.,
Institut de cardiologie de Québec,
2725, chemin Sainte-Foy,
Québec, Québec G1V 4G5 (Canada).

Recent Advances in Studies on
Cardiac Structure and Metabolism Volume 10
The Metabolism of Contraction
Edited by Paul-Emile Roy and George Rona

EFFECT OF COMBINATION OF STRESSES ON MYOCARDIAL METABOLISM

M. TARIQ, M. AHMAD,
SALAHUDDIN, and M. TAJUDDIN

Pharmacology Laboratory, A. K. Tibbiya College; and Experimental
and Metabolic Laboratory, Department of Medicine,
Aligarh Muslim University, Aligarh, India

SUMMARY

The effect of hypoxia, exercise, and thermal stress on myocardial metabolism have been widely investigated, but little attention has been paid to studying the effects of a combination of these stresses. The influence of hypoxia as modified by physical exertion (swimming) and cold stress was therefore studied. The parameters investigated included myocardial glycogen and noradrenaline, serum free fatty acid, adrenal ascorbic acid, and adrenal weight. It was observed that maximal stress was produced when hypoxia was combined with physical exertion. No suppression of cold-induced lipolysis by hypoxia was observed, in contrast to previously reported observations. Maximal depletion of cardiac glycogen and cardiac noradrenaline was noted in hypoxic exercise. Adrenal overactivity was not found to be related to any particular stress but was seen to be proportional to the severity of the stress applied.

INTRODUCTION

The peculiar oxygen supply and the high work load to which the heart is constantly exposed makes it extremely vulnerable to anoxic conditions. Great attention has therefore been paid to studying the effects of hypoxia on myocardial metabolism. However, the effects of hypoxia on myocardial metabolism as modified by hypothermia and physical exertion have not received due attention to date. Only a few reports dealing with the combination of stresses are available in the literature (Hackel, Goodale, and Kleinerman, 1954; Baum and Porte, 1969; Baum, Anthony, and Seattle, 1971; Jones *et al.*, 1972). Such studies are necessary to evaluate the metabolic changes occurring at high altitude, where all the factors operate in combination. An experimental study was therefore planned to evaluate myocardial metabolism in albino rats under hypoxia, as modified by hypothermia and physical exertion.

MATERIAL AND METHODS

The animals selected for the experiments were healthy male albino rats weighing 100–150 g. A total of 192 animals were used in the study; they were fed a standard cube diet, and tap water was available *ad libitum.* In all the experiments, the animals were fasted overnight before being subjected to various stresses. Hypoxia was achieved by lowering the pressure of a hypoxia chamber equivalent to an altitude of 20,000 feet. Exercise consisted in swimming in a metallic tank 18 inches in diameter and 24 inches in height, and maintained at 32 ± 1°C; 6 rats were made to swim at a time. This arrangement ensured enough interaction between the rats to prevent them from floating passively (Scheuer *et al.,* 1970). Cold stress was produced in an incubator maintained at 4°C. The combined effect of hypoxia and exercise (hypoxic exercise) was studied by reducing the pressure of the exercise chamber to simulate an altitude of 20,000 feet. The combined effect of hypoxia and cold stress was studied by placing the hypoxic chamber inside the incubator. The duration of various stresses was 1 hour. Biochemical parameters included cardiac glycogen (Montgomery, 1957), cardiac noradrenaline (Crout, 1961), serum triglycerides (TGS) by the method of Van Handel and Zilversmith (1957), and adrenal ascorbic acid (Roe and Keuther, 1943).

RESULTS

The results of the biochemical investigations are shown in Table 1. Normal cardiac glycogen was 7.47 ± 0.5mg/g of tissue. The levels decreased to 5.17 ± 0.46 and 5.37 ± 0.52 mg/g after hypoxia and cold stress, respectively; the value was 4.75 ± 0.29 mg/g after hypoxic cold stress. Exercise caused a further reduction of cardiac glycogen (3.75 ± 0.51 mg/g of tissue), while maximal depletion of cardiac glycogen occurred with hypoxic exercise (3.15 ± 0.62 mg/g). The findings are statistically significant. Cardiac noradrenaline was also reduced after the animals were exposed to different stresses. Normal cardiac glycogen was 1.01 ± 0.069 μg/g. The value decreased to 0.77 ± 0.038 μg/g with hypoxia and to 0.80 ± 0.05 μg/g with cold stress. Combination of stresses caused a greater reduction of cardiac noradrenaline, the findings in hypoxic cold stress and hypoxic exercise being 0.67 ± 0.05 and 0.58 ± 0.03 μg/g of cardiac tissue, respectively. Maximal depletion was noted in hypoxic exercise. There was a rise in serum FFA after the animals were exposed to different stresses, increasing from the normal level of 99.3 ± 3.26 μEq/100 ml to 109.1 ± 4.0 μEq/100 ml after hypoxia, 121.6 ± 4.77 μEq/100 ml after cold stress, 133.8 ± 4.70 μEq/100 ml after exercise, 141.1 ± 8.95 μEq/100 ml after cold stress, and 148.0 ± 7.58 μEq/100 ml after hypoxic exercise. The findings relating to exercise, hypoxic exercise, and hypoxic cold stress were highly significant ($p < 0.09$). The

Table 1. Metabolic changes in stress conditions*

Group	Cardiac Glycogen (mg/g)	Cardiac Noradrenaline (μg/g)	Serum FFA (μEq/100 ml)	Serum Triglycerides (mg/100 ml)	Adrenal Ascorbic Acid (μg/mg)
Controls	7.47 ± 0.51	1.01 ± 0.069	99.3 ± 3.26	68.7 ± 4.21	5.75 ± 0.46
Hypoxia	5.17 ± 0.46†	0.77 ± 0.038‡	109.1 ± 4.90	81.6 ± 6.26	3.86 ± 0.27†
Cold	5.37 ± 0.52‡	0.80 ± 0.05	121.6 ± 4.77†	91.8 ± 7.45†	3.25 ± 0.21†
Exercise	3.75 ± 0.51§	0.73 ± 0.08‡	133.8 ± 4.70§	79.1 ± 7.29	3.17 ± 0.15§
Hypoxic cold	4.75 ± 0.29§	0.67 ± 0.05†	141.1 ± 8.95§	104.0 ± 7.72§	2.60 ± 0.20†
Hypoxic exercise	3.15 ± 0.62§	0.58 ± 0.03†	148.0 ± 7.58§	82.2 ± 10.1	2.33 ± 0.35§

*Results are mean ± SE. Significance was calculated by Student's *t*-test.
†$p < 0.01$.
‡$p < 0.05$.
§$p < 0.001$.

maximal rise occurred in hypoxic exercise. No significant increase in serum TGS was noted with hypoxia, exercise, or hypoxic exercise; however, a significant increase in TGS was observed with cold stress, the levels being 68.7 ± 4.21 mg/100 ml in the control group and 91.8 ± 7.45 mg/100 ml in animals subjected to cold stress. Hypoxic cold stress caused a highly significant increase of TGS (104 ± 7.72 mg/100 ml; $p < 0.001$). Adrenal ascorbic acid was significantly depleted after the animals were exposed to various stresses, the findings with exercise and hypoxic exercise being highly significant ($p < 0.001$), with maximal depletion occurring after hypoxic exercise. An increase in adrenal weight resulted from all the stresses studied; maximal increase occurred with hypoxic exercise.

DISCUSSION

The metabolic changes occurring as a result of various stresses profoundly affect the energetics and function of the myocaridum. Since oxygen is mandatory for adequate energy liberation, hypoxia will result in deficient energy production. Very low temperatures have been shown to adversely affect the myocardium (Tajuddin, Ahmad, and Tariq, 1973). Acute exercise also disturbs the balance of oxygen demand and supply through the liberation of catecholamines (Taggart, Parkinson, and Carruthers, 1972). Combination of stresses tends to produce a much more severe affect as compared to any single stress. Furthermore, combination of stresses may alter the response produced by a single stress (Baum *et al.*, 1971). Studies conducted in this respect consist mostly of experiments on isolated hearts which are first subjected to one stress and then to another (Scheuer, 1970).

In the present study, a fall in cardiac glycogen was observed after hypoxia; this is in agreement with the observations of other workers (Merrick and Meyer, 1954; Ballinger and Vollenweider, 1962; Wollenberger and Krause, 1968; Scheuer, 1972). Cardiac glycogen was also depleted by exercise, a not unexpected finding since, in the event of vigorous exercise, glycogenolysis becomes the major energy-yielding process and results in glycogen depletion (Blount and Meyer, 1959; Scheuer, 1972). The myocardial glycogen was also decreased by cold stress. Opie (1965) suggested that cardiac glycogen should increase after cold stress, because hypothermia has been found to be associated with increased glucose uptake (Edward *et al.*, 1954; Jude, Haroutunian, and Folse, 1957) which would tend to increase cardiac glycogen since the greater percentage of glucose taken up by the heart is converted into glycogen. However, other investigators have reported a decrease in the cardiac glycogen level (Itzhaki and Wertheimer, 1957; Sharma and Parmar, 1967). The findings of the present study are, therefore, in agreement with the latter workers. This suggests that the utilization of glycogen is much in excess of the increased formation of glycogen from enhanced glucose uptake under cold stress. Maximal depletion of cardiac glycogen was observed in hypoxic exercise. This may signify a greater load on the heart when hypoxia is combined with exercise in comparison to cold. Greater glucose uptake during cold stress may also be considered to account for a lesser fall of cardiac glycogen in hypoxic cold stress in comparison to hypoxic exercise.

The fall in cardiac noradrenaline noted with all the stresses may be due either to the inability of the myocardium to take up noradrenaline or to its inability to retain noradrenaline. Bhagat and Dhalla (1973) observed that the uptake of noradrenaline by the myocardium after isoproterenol-induced cardiac injury is not hampered, but the anoxic myocardium is unable to retain noradrenaline. Utilization of catecholamines for glycogenolysis may also account for the fall of cardiac noradrenaline resulting from hypoxia and cold stress. Maximal depletion of cardiac noradrenaline was observed with hypoxic exercise. The inability of hypopoxic myocardium to retain noradrenaline, together with the markedly increased demand for glycogenolysis, will explain the marked depletion of cardiac noradrenaline with hypoxic exercise. Myocardial noradrenaline was also depleted after cold stress. There are conflicting reports of noradrenaline changes in the myocardium after cold exposure. Many investigators have reported a fall in the noradrenaline content (Moore, Calvert, and Brody, 1961; Khanna and Sharma, 1971); however, Gordon *et al.* (1966) observed no decrease in cardiac noradrenaline. These authors suggested that the increased synthesis of noradrenaline, which is associated with such stressful conditions, is responsible for the maintenance of the noradrenaline content of the heart. It seems that, even if there is an increase in the synthesis of catecholamines, it does not appear to be sufficient to affect the greatly increased utilization of catecholamines during stressful conditions. The greater fall of myocardial noradrenaline in hypoxic cold stress was to be expected, since hypoxia and cold stress both tend to lower myocardial noradrenaline and would exert a greater effect while acting in combination. Reduction of cardiac noradrenaline was highly significant in the animals subjected to hypoxic exercise. No report of changes in cardiac noradrenaline resulting from hypoxic exercise is available in the literature. With the combined stress of hypoxia and exercise, there is a further increase in anaerobic glycogenolysis in an attempt to provide sufficient energy and this may account for the low levels of noradrenaline observed in this group. A rise in serum FFA was observed after the animals were exposed to hypoxia, but these findings were not stastically significant. The reports concerning FFA changes after hypoxia are controversial. Baum *et al.* (1971) observed impairment of lipolysis, but the mechanism of this impairment is not clearly understood. Enhanced anaerobic glycolysis, high lactate levels, diminished store of adenylcyclase and phosphodiesterase activity have been considered as possible factors (Baum *et al.,* 1971). On the other hand, many investigators have observed an increase in serum FFA during hypoxic exposure (Bing, 1965; Jones *et al.,* 1972). This has been considered to be due to the stimulating effect of a high level of circulating catecholamines on lipoprotein lipase, resulting in lipolysis in adipose tissue. Although the results of the present study reveal a rise in serum FFA induced by hypoxia, the findings are not statistically significant. The increased FFA levels observed after exposure to cold are attributable to the cold-stimulated increased activity of tissue lipase reported by Maickel, Westerman, and Brodie (1961). The increase in FFA was much higher with hypoxic cold stress. Obviously, the increased mobilisation of fat from adipose tissue coupled with the effect of hypoxia is responsible for the higher FFA level observed under combined stress. There was no significant elevation in serum triglyceride (TGS) with hypoxia, exercise, or hypoxic exercise. The failure of this parameter, to rise may be due to the short duration of exposure, because it takes some time to enhance the production of triglycerides in the liver (Brachfeld, 1969). However, there was a significant rise in serum TGS after the animals were exposed to cold stress and hypoxic cold stress. This rise may be due to increased mobilization of lipid from adipose tissue for thermogenesis.

Increase in adrenal weight was observed after the animals were subjected to various stresses. The dynamic adrenal hypertrophy indicates intense activation of the adrenal cortex (Wexler and Kittinger, 1963). Several factors have been considered, including hypoxia per se, changes in circulatory and respiratory functions (Fishman, Fritts, and Cournard, 1960), physical exertion (Mackinon, Monkjones, and Fotherby, 1963), and adverse emotional or psychic influences from changes in the environment (Timaras, Pav, and Hwang, 1957; Moncloa *et al.*, 1965), to explain the cause of adrenal overactivity observed at high altitude. Since adrenal overactivity was observed under all types of stresses in the present study, it can be concluded that it is not related to any particular stress. It was found, rather, to be related to the severity of stress as measured in terms of depletion of adrenal ascorbic acid (Selye, 1952). Maximal depletion of adrenal ascorbic acid occurred in hypoxic exercise, thus signifying maximal stress under these conditions. The increase in adrenal weight was also found to be maximal under the same conditions.

REFERENCES

BALLINGER, W. F., and H. VOLLENWEIDER. 1962. Anaerobic metabolism of heart. Circulat. Res. 11: 681.

BAUM, D., C. L. ANTHONY, JR., and C. S. SEATTLE. 1971. Impairment of cold lipolysis by acute hypoxia. Am. J. Dis. Child. 121: 115.

BAUM, D., and D. PORTE, Jr. 1969. Effect of acute hypoxia on circulating insulin level. J. Clin. Endocrin. 29: 991.

BHAGAT, B., and N. S. DHALLA. 1973. Alterations in norepinephrine (NE) pattern in damaged myocardium of the rat. Abstract VI Annual Meeting of the International Study Group for Research In Cardiac Metabolism, Freiburg, Sept. 25–28.

BING, R. J. 1965. Cardiac metabolism. Physiol. Rev. 45: 2: 171.

BLOUNT, D. H., and D. K. MEYER. 1959. Effects of cardiac work on glycogen fraction of the heart. Am. J. Physiol. 23: 1013.

CROUT, R. J. 1961. Catecholamines in urine. *In* D. Seligson (ed.), Standard Methods of Clinical Chemistry, Vol. 3, pp. 62–80. Academic Press, New York.

FISHMAN, A. P., H. W. FRITTS, Jr., and A. COURNARD. 1960. Circulation. 22: 204.

GORDON, S. S., A. S. SPECTOR, JOERDAMA, and S. UDENFRIEND. 1966. Increased of norepinephrine in the intact rat during exercise and exposure to cold. J. Pharm. Exp. Therap. 153: 440.

HACKEL, D. B., W. T. GOODALE, and J. KLEINERMAN. 1954. Effect of hypoxia on myocardial metabolism of intact dog. Circulat. Res. 2: 169.

ITZHAKI, S., and E. WERTHEIMER. 1957. Changes in carbohydrate metabolism of cardiac and striated muscle in hypothermic rats. Circulat. Res. 5: 451.

JONES, L. N., D. G. ROBERTSON, J. W. KANE, and R. N. HART. 1972. Effect of hypoxia on free fatty acid metabolism during exercise. J. Appl. Physiol. 33: 733.

JUDE, J. R., L. M. HAROUTUNIAN, and R. FOLSE. 1957. Hypothermic myocardial oxygenation. Am. J. Physiol. 19: 57.

KHANNA, N. K., and V. N. SHARMA. 1971. Effect of cold stress on myocardial and plasma catecholamine level of rat and its modification by reserpine, Mecamylamine and Eserine. Indian J. Med. Res. 59: 627.

MERRICK, A. W., and D. K. MEYER. 1954. Glycogen function of cardiac muscle in the normal and anoxic heart. Am. J. Physiol. 177: 441.

MONCLOA, F. J., L. A. DONAYRE, SOBREVILLA, and GNORRA-GRACIA. 1965. J. Clin. Endocrin. 25: 1640.

MONTGOMERY, R. 1957. Determination of glycogen. Arch. Biochem. Biophys. 67: 376.

MOORE, K. E., D. N. CALVERT, and T. M. BRODY. 1961. Tissue catecholamine content of cold acclimated rats. Proc. Soc. Exp. Biol. Med. 106: 816.

OPIE, L. H. 1965. Cardiac metabolism: The effect of sone physiological, pharmacological and pathological influences. Am. Heart. J., 69: 401.

ROE, J. H., and C. A. KUETHER. 1943.: The determination of ascorbic acid in the whole blood and urine through the 2,4-dinitrophenyl hydrazine derivative of dehydroascorbic acid. J. Biol. Chem. 147: 399.

SCHEUER, J. 1970. Metabolism of heart in cardiac failure. Prog. Cardiovasc. Dis. XIII: I.

SCHEUER, J. 1972. Pathogenic significance of disproportion between myocardial oxygen supply and consumption. *In* E. Bajusz and G. Rona (eds.), Recent Studies on Advances in Cardiac Structure and Metabolism, Vol. 1: Myocardiology, pp. 721–728. University Park Press, Baltimore.

SCHEUER, J., L. KAPNER, C. A. STRINGFELLOW, C. A. ARMSTRONG, and S. PENPARGKUL. 1970. Glycogen, lipid and high energy phosphate stores in hearts from conditioned rats. J. Lab. Clin. Med. 75: 6: 924.

SELYE, H. 1952. The Story of the Adaptation Syndrome, p. 44. Medical Publishers, Montreal.

SHARMA, V. N., and N. S. PARMAR. 1967. Effect of chlorpromazine, mecamylamine and guanithidine on the cold stress induced changes in myocardial glycogen and acetyl choline content in rabbits. Indian J. Med. Res. 55: 60.

TAGGART, P., J. D. H. SLATER, M. E. CARRUTHERS, and I. K. McMILLAN. 1971. Effect of transient occlusion of coronary circulation on catecholamine and electrolyte gradient in human atrium. Proc. British Cardiac Society. Brit. Heart J. 33: 609.

TAJUDDIN, M., M. AHMAD, and M. TARIQ. 1973. Myocardial metabolism in myocardial infarction at different environmental temperatures. Abstract VI. Annual Meeting of the International Study Group For Research in Cardiac Metabolism, Freiburg., September, 25–28.

TIMARAS, P. S. N., N. S. PAV, and C. A. HWANG. 1957. Plasma and urine 17-hydroxy corticosteroid and urine 17-ketosteroid levels in man during acclimatization to high altitude. Fed. Proc. 16: 340.

TROUT, P. L., E. H. ESTERS, and S. J. FRIEDBERG. 1960. Titration of free fatty acids of plasma. A study of current method and a new modification. J. Lipid. Res. 1: 199.

VAN HANDEL, E., and D. B. ZILVERSMITH. 1957. Micromethod for the direct determination of serum triglycerides. J. Lab. Clin. Med. 50: 152.
WEXLER, B. C., and G. W. KITTINGER. 1963. Myocardial necrosis in rats: Serum enzymes, adrenal steroids and histopathological alterations. Circulat. Res. 13: 159.
WOLLENBERGER, A., and E. G. KRAUSE. 1968. Metabolic control of characteristics of the acutely ischaemic myocardium. Am. J. Cardiol. 22: 349.

Mailing Address:
Col. M. Tajuddin,
Professor and Head, Department of Medicine,
J. N. Medical College, Aligarh Muslim University,
Aligarh, U.P. (India).

Recent Advances in Studies on
Cardiac Structure and Metabolism Volume 10
The Metabolism of Contraction
Edited by Paul-Emile Roy and George Rona
Copyright 1975 University Park Press Baltimore

LIPID COMPOSITION OF CARDIAC LYMPH

PIERRE JULIEN and PAUL-EMILE ROY

Institut de Cardiologie de Québec, Québec, Québec, Canada

SUMMARY

Cardiac lymph was obtained from fasting dogs by cannulating the efferent collecting trunks and the interventricular (subepicardial) trunk. The total lipid content of the lymph was 450 mg/100 ml and consisted of 16–20% cholesterol, 30–33% triglycerides, 41% phospholipids, 2% free fatty acids, and 7–8% fatty acids esterified with cholesterol.

The lipid composition of the lymph obtained from the two collecting sites was identical. We can demonstrate that the loss of total fatty acids by the myocardium due to the drainage of the interstitial fluid into the lymph represents only 2–8% of the fatty acids used.

With the exception of triglycerides, the concentration of all the lipids was higher in the arterial plasma than in the lymph. This concentration of triglycerides in the lymph suggests that triglycerides are contributed from endogenous sources.

INTRODUCTION

From the physiological point of view, the lymphatic system is an important homeostatic mechanism. In the fasting organism, energy production by the heart depends mostly on lipids, especially fatty acids (Bing *et al.*, 1954). The study of the lipid composition of the cardiac lymph is thus of considerable interest, since it provides a better approximation of the interstitial environment than any other readily available sample.

Ungar *et al.* (1955) and Ballard *et al.* (1960) determined the myocardial utilization of fatty acids, using the arteriovenous differences. Our main interest was to determine the lipid composition of the cardiac lymph relative to the arterial plasma composition. From these observations, we can estimate the importance of lymphatic drainage on the uptake and utilization of lipids by the heart.

MATERIALS AND METHODS

Twelve healthy dogs weighing between 15 and 22 kg were used for this study. The fasting animals were anesthetized by intravenous injection of 30 mg/kg sodium pentobarbital (Nembutal, Abbott) and were artificially respired. The surgical approach was by median sternotomy; bleeding was controlled by cautery. The pericardium was cut anteriorly to free the apex and the interventricular lymphatic trunk. The cardiac lymphatic ganglion was then located by lifting the brachocephalic artery.

The efferent cardiac lymphatic trunks were cannulated by the method described by Kluge and Ullal (1971), placing the cannulas as close as possible to the heart. The thoracic lymphatic canal was ligated both anterior and posterior of the operating field. The interventricular anterior trunk (subepicardial) was carefully cannulated (Julien, Blouin, and Roy, 1974) after it was located by an injection of Evans blue into the apex. The catheters used were Silastic tubing of the appropriate diameter attached to No. 25 or No. 27 hypodermic needles; this tubing has the advantage of being extremely flexible and, thus, being able to follow the respiratory and cardiac movements.

The lymph was obtained from the catheters, while the arterial blood was obtained at the beginning and at the end of each experiment from a femoral artery. The tubing was heparinized. The lipids were analyzed with the Technicon AutoAnalyzer (Technicon Instruments Corporation, Tarrytown, N.W.). Cholesterol and triglycerides were measured by the method of Rush, Leon, and Turrell (1970) and the cholesterol esters by the technique of Sullivan (1969). Free fatty acids were measured by the method of Itaya and Ui (1965) and the phospholipids by that of Zilversmit and Davis (1950).

RESULTS

Lymph flow

We obtained lymph stained with Evans blue during the 20–30 min following insertion of the catheters into the lymphatic trunks. The flow of the cardiac lymph was 6–10 ml/hour for the efferent trunks and 0.5–1 ml/hour for the subepicardial trunk. The lymphatic flow remains constant for at least 2 hours.

Lipid composition of the cardiac lymph and arterial plasma

The lipid content of the arterial plasma was practically constant during the sampling period. The mean lipid concentration in the plasma was 725 mg/100 ml, while in the cardiac lymph it was 450 mg/100 ml.

The mean concentrations of the individual lymphatic and plasmatic lipids obtained are shown in Table 1. The concentration of lymphatic triglycerides was 141 mg/100 ml, while the plasma concentration was 155 mg/100 ml. For the free fatty acids, we only found a difference of 1 mg/100 ml between the plasmatic and lymphatic concentrations. These differences between the plasma and lymph concentrations of the triglycerides and the free fatty acids were not statistically significant.

On the other hand, we obtained significant differences ($p<0.01$) in the

Table 1. Mean concentrations of lymph and plasma lipids*

	Lymph (mg/100 ml)		Plasma (mg/100 ml)		Statistical Significance of Differences	
	Mean	99% Confidence Limits	Mean	99% Confidence Limits	*p*	Conclusion
Triglycerides	141	88–194	155	116–194	>0.4	NS
Cholesterol (total)	81	47–115	154	118–190	<0.01	S
Phospholipids	175	101–249	326	263–389	<0.01	S
Fatty acids esterified with cholesterol	39	21–57	79	63–95	<0.01	S
Free fatty acids	10	4–16	9	8.4–7.6	>0.5	NS

**p* values were calculated by Student's *t*-test. S = significant; NS = not significant.

Table 2. Comparison of lipid composition of cardiac lymph obtained from efferent trunks with that obtained from subepicardial trunk

Lipids	Efferent Main Trunks (%)	Subepicardial Trunks (%)	Statistical Significance of Differences	
			p	Conclusion
Triglycerides	33	30	>0.5	NS
Cholesterol (total)	16	20	>0.3	NS
Phospholipids	41	41	–	NS
Fatty acids esterified with cholesterol	8	7	>0.5	NS
Free fatty acids	2	2	–	NS

concentrations of cholesterol, fatty acids esterified with cholesterol, and phospholipids. The plasma concentrations of these compounds are approximately twice the lymph concentration.

Table 2 shows the comparison of the composition of the cardiac lymph obtained from the efferent trunks with that obtained from the subepicardial trunk. The distribution of lipid in the lymph from the two sampling sites was almost identical. The differences in the distribution of cholesterol, triglycerides, and fatty acids esterified with cholesterol are not statistically significant. In all the samples of cardiac lymph, the hematocrit was very low, less than 1%, and the hemoglobin content was less than 4 mg/ml.

Correlation between the concentration of lymphatic and plasmatic lipids

As is seen from Figure 1, significant correlation exists between the lymphatic and plasmatic concentrations of triglycerides in the lymphatic concentration range of 80–250 mg/100 ml. Each point on the graph represents the value from 1 dog. The graph indicates that the lymphatic concentrations are essentially proportional to the plasmatic concentrations.

A significant correlation can also be obtained for phospholipids, cholesterol, and the fatty acids esterified with cholesterol (Fig. 1). These correlations hold for lymphatic concentrations between 70 and 353 mg/100 ml for the phospholipids, 29 and 168 mg/100 ml for cholesterol, and 11 and 83 mg/100 ml for fatty acids esterified with cholesterol. The graphs show that the plasma concentration of these three compounds is higher than the lymph concentration.

On the other hand, we could not find any significant correlations between the lymphatic and plasmatic free fatty acids concentrations.

Lipid elimination by cardiac lymph

Ungar *et al.* (1955) and Ballard *et al.* (1960) measured the arterial concentration as well as the arteriovenous difference for the total fatty acids (Table 3). They also measured the myocardial utilization of these acids, which varied between 11 and 49 μEq/100 g/min. At an arterial concentration similar to that found by the above authors, we estimate that the loss of total fatty acids by the myocardium into the lymph amounts to 0.9 μEq/100 g/min.

DISCUSSION

From our anatomical observations, we can identify the cardiac ganglion and the efferent trunks. We have observed anatomical differences from animal to animal in the localization and the number of efferent trunks, which can vary from 2 to 4. Certain efferent trunks leave the heart and proceed directly to the cardiac ganglion (Miller, Ellis, and Katz, 1964) without passing by the pretracheal ganglion; others originating in the pretracheal ganglion proceed towards a cardiac ganglion, where they are found parallel to the cardiac ganglion. These variations of the anatomical descriptions of Patek (1939) and Drinker *et al.* (1940) are important if we wish to estimate the total lymphatic flow of the myocardium. However, since we cannot cannulate all the efferent lymphatic vessels of the

Table 3. Comparison of arterial concentration, myocardial utilization, and lymphatic drainage of total fatty acids

	Total Fatty Acids			
Reference	Arterial Level (μEq/100 ml)	Arteriovenous Difference (μEq/100 ml)	Myocardial Utilization (μEq/100 g/min)	Loss in Cardiac Lymph (μEq/100 g/min)
Ungar *et al.* (1955)	868	18.7	11.0	
Ballard *et al.* (1960)	959 928	42.0 52.7	38.4 49.3	
Julien *et al.* (1974)	848			0.9

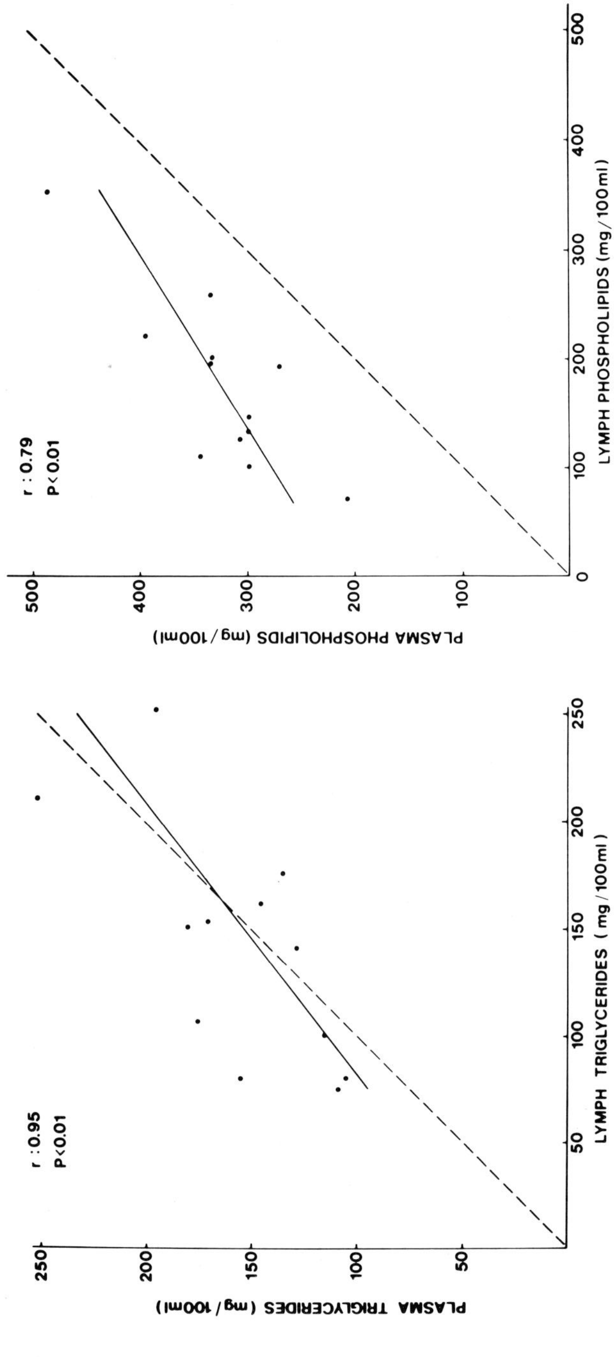
r : 0.79
P < 0.01
PLASMA PHOSPHOLIPIDS (mg / 100ml)
LYMPH PHOSPHOLIPIDS (mg/100ml)
r : 0.95
P<0.01
PLASMA TRIGLYCERIDES (mg/ 100ml)
LYMPH TRIGLYCERIDES (mg / 100ml)

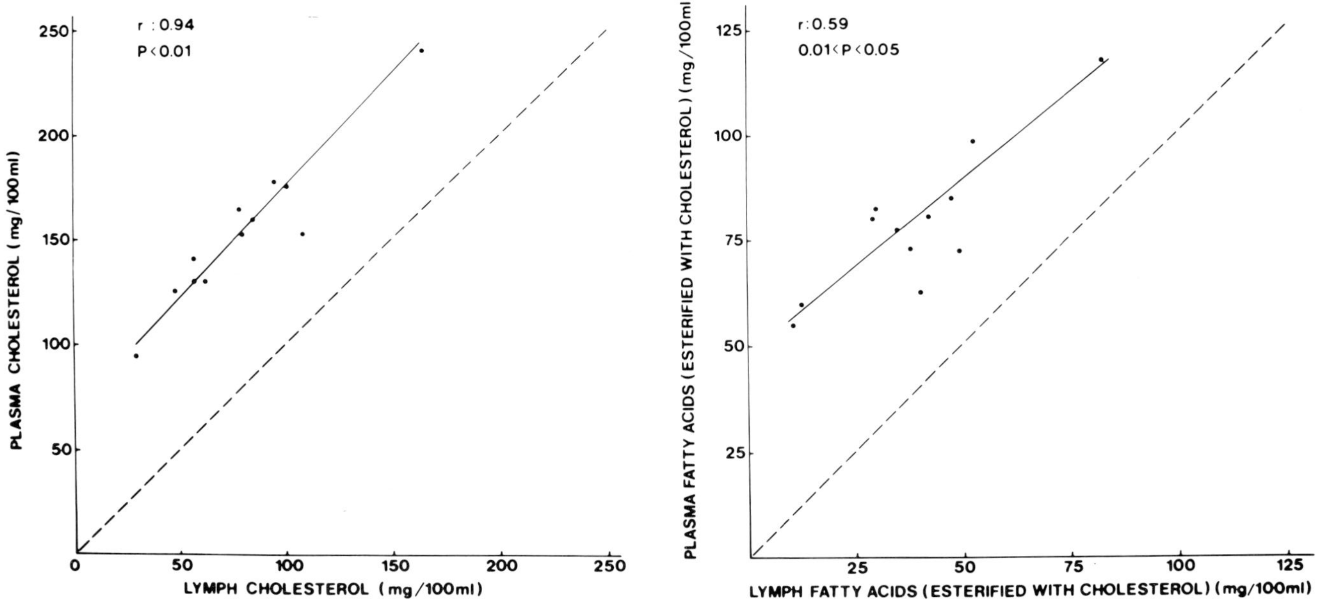

Figure 1. Relationship between plasma and lymph concentrations of triglycerides, phospholipids, cholesterol, and fatty acids esterified with cholesterol. Coefficients of correlation (r) were calculated from *points* shown. *Interrupted diagonal line* shows plot that would be obtained if plasma and lymph concentrations were equal.

myocardium, we ligated the secondary lymphatic vessels visualized with Evans blue injected into the myocardium and cannulated the efferent trunks. With this technique, we obtained a lymphatic flow of 6–10 ml/hour. This flow is higher than those usually reported, which vary between 0.8 and 6.8 ml/hour (Miller *et al.*, 1964; Leeds and Uhley, 1971; Miller, Pick, and Johnson, 1972).

On the basis of the anatomical studies of Ullal *et al.* (1972), we can postulate that the lymph obtained from the efferent trunks originates in the myocardium. To eliminate any possible reflux of lymph from other ganglions, we ligated the thoracic lymphatic canal, the secondary lymphatic vessels, and the efferent vessels of the cardiac ganglion. The lipid composition of the cardiac lymph obtained from the efferent trunks and the subepicardial trunk is identical (Table 2).

Lipid composition of the cardiac lymph

In every case, the concentration of all the lipids except triglycerides was higher in the plasma than in the lymph. For the triglycerides, we found that the plasmatic and lymphatic concentrations were identical and that a correlation exists between these two concentrations (Fig. 1). Since triglycerides are utilized by the heart, as reported by Opie (1969), two hypotheses are possible: 1) the triglycerides passed freely through the capillary walls, and the concentrations in either side are comparable; 2) they are filtered like the phospholipids and cholesterol, but their concentration is increased by a contribution from endogenous triglycerides.

The latter hypothesis seems the most probable, since the triglycerides have a high molecular weight and they can be hydrolyzed and taken up by the myocardium as fatty acids (Carlson *et al.*, 1973). The lymphatic triglycerides can then originate: 1) partially from triglycerides having penetrated the capillary walls; 2) from cardiac adipose tissue and lipid droplets present in the interstitial spaces (Roy, 1974); 3) from a synthesis of triglycerides from fatty acids and other substrates.

The absence of a correlation between the arterial and lymphatic concentrations of the free fatty acids can be explained by the fact that the lymphatic fatty acids can originate from the free fatty acids which readily cross the capillary walls, from the hydrolysis of triglycerides, or from fat reserves.

Importance of the cardiac lymphatic drainage

The study of cardiac metabolism by measuring the arteriovenous differences can give useful results; however, several prerequisites must be met. Among these is the absence of lymphatic drainage.

When we compare the quantity of total fatty acids drained by the lymph to the amount used by the myocardium, as estimated by Ungar *et al.* (1955) and

Ballard *et al.* (1960), we find that the amount lost in the lymph represents only 2–8% of the fatty acids utilized (Table 3). Thus, the drainage of the interstitial liquid by the lymph cannot markedly change the estimate of the myocardial utilization obtained from arteriovenous differences.

Acknowledgment We would like to thank Doctor Laimonis Gailis, Institut de Cardiologie de Québec, for his continued advice.

REFERENCES

BALLARD, F. B., W. H. DANFORTH, S. NAEGLE, and R. J. BING. 1960. Myocardial metabolism of fatty acids. J. Clin. Invest. 39: 717–723.

BING, R. J., A. SIEGEL, I. UNGAR, and M. GILBERT. 1954. Metabolism of the human heart. II. Studies on fat, ketone and amino acid metabolism. Am. J. Med. 16: 504–515.

CARLSON, L. A., L. KAIJSER, S. ROSSNER, and M. L. WAHLQVIST. 1973. Myocardial metabolism of exogenous plasma triglycerides in resting man. Acta Med. Scan. 193: 233–245.

DRINKER, C. K., M. F. WARREN, F. W. MAURER, and J. D. McCARRELL. 1940. The flow, pressure and composition of cardiac lymph. Am. J. Physiol. 130: 43–55.

ITAYA, K., and M. UI. 1965. Colorimetric determination of free fatty acids in biological fluids. J. Lipid Res. 6: 16–20.

JULIEN, P., A. BLOUIN, and P. E. ROY. 1974. Technique de canulation des vaisseaux lymphatiques cardiaques. Union Méd. Canad. 103: 1941–1943.

KLUGE, T., and S. R. ULLAL. 1971. A standardized technique for drainage and collection of cardiac lymph. Acta Physiol. Scand. 83: 433–436.

LEEDS, S. E., and H. N. UHLEY. 1971. Measurement of lymph flow of the heart. Lymphology 4: 31–34.

MILLER, A. J., A. ELLIS, and L. N. KATZ. 1964. Cardiac lymph: flow rates and composition in dogs. Am. J. Physiol. 206: 63–66.

MILLER, A. J., R. PICK, and P. J. JOHNSON. 1972. The rates of formation of cardiac lymph and pericardial fluid after the production of myocardial venous congestion in dogs. Lymphology 5: 156–160.

OPIE, L. H. 1969. Metabolism of the heart in health and disease. Part II. Am. Heart J. 77: 100–122.

PATEK, P. R. 1939. The morphology of the lymphatics of the mammalian heart. Am. J. Anat. 64: 203–249.

ROY, P.-E. 1975. Lipid droplets in the heart interstitium: concentration and distribution. *In* P.-E. Roy and G. Rona (eds.), Recent Advances in Studies on Cardiac Structure and Metabolism, Vol. 10: The Metabolism of Contraction, pp. 17–27. University Park Press, Baltimore.

RUSH, R. L., L. LEON, and J. TURRELL. 1970. Automated simultaneous cholesterol and triglyceride determination on Auto-Analyzer II Instrument. *In* Advances in automated analysis. Technicon International Congress 1: 503–507.

SULLIVAN, D. A. 1969. A semi-automated method for cholesterol esters. *In* Advances in automated analysis. Technicon International Congress: 127.

ULLAL, S. R., T. H. KLUGE, W. J. KERTH, and F. GERBODE. 1972. Anatomical studies on lymph drainage of the heart in dogs. Ann. Surg. 175: 305–310.

UNGAR, I., M. GILBERT, A. SIEGEL, J. M. BLAIN, and R. J. BING. 1955. Studies on myocardial metabolism. IV. Myocardial metabolism in diabetes. Am. J. Med. 18: 385–396.

ZILVERSMIT, D. G., and A. K. DAVIS. 1950. Microdetermination of plasma phospholipids by trichloroacetic acid precipitation. J. Lab. Clin. Med. 35: 155–160.

Mailing address:
Pierre Julien,
Institut de Cardiologie de Québec,
2725, Chemin Sainte-Foy,
Québec, G1V 4G5 (Canada).

Recent Advances in Studies on
Cardiac Structure and Metabolism Volume 10
The Metabolism of Contraction
Edited by Paul-Emile Roy and George Rona

REGIONAL DIFFERENCES IN FATTY ACID OXIDATION BY THE ADULT RABBIT MYOCARDIUM

ROBERT L. GINGELL,[1] THOMAS POZEFSKY,
and RICHARD D. ROWE[2]

Departments of Pediatrics and Medicine, Johns Hopkins Hospital,
Baltimore, Maryland, USA

SUMMARY

Differences in coronary perfusion and tissue oxygenation between the epicardium and endocardium have been shown. Thus, intrinsic capacity for substrate oxidation may also exhibit regional variation. Accordingly, homogenates of left ventricular papillary muscle ("endocardium") and left ventricular free wall trimmed of endocardium ("epicardium") were compared for their capacity to oxidize palmitic acid-1-^{14}C (PA) and octanoic acid-1-^{14}C (OcA). Homogenates from one to three rabbits were combined for each study; 18 studies were performed using PA, and 10 using OcA. The protein content of endo- and epicardial homogenates was the same (Biuret method). At ambient PA concentrations of 0.2 mM, oxidation of PA by "endocardium" was 64 ± 6.3% that of the "epicardium" ($p <$ 0.001). At ambient concentrations of 1.0 mM, oxidation of OcA by "endocardium" was not different from that of "epicardium" (118 ± 10.5%; $p > 0.1$). Thus, the oxidative capacity for long chain FFA by the endo- and epicardial homogenates are at variance. Differences in carnitine-mediated transmitochondrial transport of long chain fatty acids may be the explanation. Such regional metabolic differences could play a role in regional alterations in myocardial function under adverse hemodynamic and metabolic circumstances.

INTRODUCTION

Long chain fatty acids are the preferential metabolic fuel for the myocardium (Bing, 1965). However, other substrates, principally glucose, may also provide caloric energy. Glycolysis becomes an important energy source when oxidative metabolism is restricted, as in ischemia. Recent studies indicate that the sub-endocardial blood supply may be compromised under certain adverse hemodynamic conditions (Buckberg *et al.*, 1972). It was felt that such alterations in epicardial/endocardial perfusion ratio might be reflected by different regional

[1] Present address: The E. C. Lambert Department of Cardiology, Buffalo Children's Hospital, 219 Bryant Street, Buffalo, New York 14222.

[2] Present address: Division of Pediatric Cardiology, Hospital for Sick Children, Toronto, Ontario, Canada.

patterns of substrate utilization. The present study was undertaken to delineate possible variations in regional myocardial utilization of fatty acids.

METHODS

New Zealand white rabbits, male or female, fed *ad lib* on a standard laboratory diet were sacrificed. The heart was removed and opened; the left ventricular papillary muscles was excised at the base and fibrous tissue was removed. A section of left ventricular free wall was obtained and trimmed free of the trabeculae, leaving approximately the outer half (epicardium). The samples were maintained in iced saline until homogenized.

Palmitate -1- ^{14}C and octanoate -1- ^{14}C were chosen as representative of the long and short chain fatty acids, respectively. The palmitate was bound to bovine serum albumin (essentially fatty acid free) at a 3:1 fatty acid to albumin ratio. The octanoate was carried in a dilute ethanol solution (2% of the final reaction volume).

To measure the capacity of the representative myocardial samples to oxidize the selected substrates, the tissues from one to three animals were homogenized in a calcium-free Krebs-Ringer phosphate buffer, pH 7.3, at approximately 2–4°C in a conical ground glass homogenizer. These homogenates were then filtered through surgical gauze and used immediately. The reaction was carried out in a 25-ml Erlenmeyer flask fitted with a plastic center well, inserted through a collared rubber serum stopper. Specified amounts of the carbon-14 labeled substrates were added to each flask, the flask was sealed, and the reaction was begun by adding 1 ml of homogenate. To collect the CO_2 generated, hyamine was added to the center well. The reaction was allowed to proceed for 30 min at 30°C with agitation. Subsequently, the reaction was stopped by acidification with sulfuric acid, and the CO_2 generated was collected for 60 min at 30°C with agitation. The center well containing the hyamine was then added to scintillation fluid and counted in a spectro-photometer. Aliquots of the homogenates were analyzed for protein content by the Biuret method (Layne, 1957). The results were expressed as μmoles of substrate utilized per mg protein. Each papillary muscle homogenate was analyzed relative to its matched epicardial sample (oxidative capacity ratio).

RESULTS

The results obtained are summarized in Table 1. The protein content of the homogenates had no statistically significant difference. At ambient concentrations of 0.2 mM, palmitate-1- ^{14}C was oxidized by 18 homogenates of the left ventricular papillary muscles at significantly lower rates than the homogenates of matched left ventricular free wall epicardium. Basic differences in mitochondrial

Table 1. Comparison of oxidative capacity ratios (PAP/LV) for palmitate-1-^{14}C and octanoate-1-^{14}C

Substrate	Substrate Concentration (mM)	PAP/LV (%)
Palmitate-1-^{14}C	0.2	64 ± 6.34*
Octanoate-1-^{14}C	1.0	118 ± 10.45†

*Papillary muscle versus LV epicardium, $p < 0.001$.
†Oxidative capacity ratio, palmitate versus octanoate, $p < 0.005$.

mass or β-oxidative capacity of the papillary muscle and epicardium were tested by assaying their ability to oxidize the short chain fatty acid octanoate-1-^{14}C. Using this substrate at ambient concentrations of 1.0 mM, homogenates of papillary muscle and matched left ventricular epicardium showed no difference in utilization.

In addition, comparison of the papillary muscle/epicardial oxidative capacity ratios for the two substrates was also found to be significantly different.

DISCUSSION

Regional aspects of myocardial biochemistry and coronary perfusion have been investigated. Such studies indicate alterations in the redox potential, high energy phosphate depletion, and decrease in perfusion of the endocardium under adverse hemodynamic conditions (Buckberg *et al.,* 1972; Griggs, Chen, and Tchokoev, 1973; Fleckenstein *et al.,* 1974). The results obtained in the present study suggest a transmural gradient in long chain fatty acid oxidation may exist even under control circumstances. Fatty acid utilization by myocardium appears to be controlled to a large extent by substrate availability and work load by the heart (Neely, Rovetto, and Oram, 1972). Under basal conditions, the perfusion of the endocardium is not altered. Thus, the differences noted appear not to be explained on the basis of substrate level alone.

The long chain fatty acids are dependent on a carnitine-mediated transmitochondrial transport system for β-oxidation (Fritz and Marquis, 1965). However, the short chain fatty acids are not dependent on this mechanism. These data suggest that the carnitine carrier mechanism may be responsible for the variation observed. This may be due to lower tissue levels of carnitine and/or the enzyme acyl carnitine-CoA transferase itself. It has been shown that stimuli such as ischemia and hypoxia as well as toxic substances divert fatty acids from β-oxidation to deposition as intracellular lipid (Wittels and Bressler, 1964; Scheuer and Brachfeld, 1966). Reduced utilization under control conditions may be further compromised by hypoxia/ischemia leading to the accumulation of fatty acids within the myocardial cell. Such accumulations may exert a deleterious effect through the inhibition of ATP formation by increased levels of fatty acid CoA esters (Shug and Shrago, 1973). Subsequent high energy phosphate depletion could contribute to myocardial cell dysfunction.

REFERENCES

BING, R. J. 1965. Cardiac metabolism. Physiol. Rev. 45: 171–213.

BUCKBERG, G. D., D. E. FIXLER, J. P. ARCHIE, and J. I. E. HOFFMAN. 1972. Experimental subendocardial ischemia in dogs with normal coronary arteries. Circulat. Res. 30: 67–80.

FLECKENSTEIN, A., J. JANKE, H. J. DORING, and O. LEDER. 1974. Myocardial fiber necrosis due to intracellular Ca overlead–a new principle in cardiac pathophysiology. *In* N. S. Dhalla and G. Rona (eds.), Recent Advances in Studies on Cardiac Metabolism, Vol. 4: Myocardial Biology, pp. 563–580. University Park Press, Baltimore.

FRITZ, I. B., and N. R. MARQUIS. 1965. The role of acyl carnitine esters and carnitine palmityltransferase in the transport of fatty acyl groups across mitochondrial membranes. Proc. Nat. Acad. Sci. 54: 1226–1233.

GRIGGS, D. M., C. C. CHEN, and V. V. TCHOKOEV. 1973. Subendocardial anerobic metabolism in experimental aortic stenosis. Am. J. Physiol. 224: 607–612.

LAYNE, E. 1957. Biuret method. *In* S. P. Colowick and W. O. Kaplan (eds.), Methods in Enzymology, Vol. III, pp. 450–457. Academic Press, New York.

NEELY, J. R., M. J. ROVETTO, and J. F. ORAM. 1972. Myocardial utilization of carbohydrates and lipids. Progr. Cardiovasc. Dis. 25: 288–329.

SCHEUER, J., and N. BRACHFELD. 1966. Myocardial uptake and fractional distribution of palmitate -1-^{14}C by the ischemic dog heart. Metabolism 15: 945–954.

SHUG, A. L., and E. SHRAGO. 1973. A proposed mechanism for fatty acid effects on energy metabolism of the heart. J. Lab. Clin. Med. 81: 214–220.

WITTELS, B., and R. BRESSLER. 1964. Biochemical lesion of diphtheria toxin in the heart. J. Clin. Invest. 43: 630–637.

Mailing address:
Robert L. Gingell, M.D.,
The E. C. Lambert Department of Cardiology,
Buffalo Children's Hospital,
219 Bryant Street,
Buffalo, New York 44222 (USA).

Pharmacological Action of Commonly Used Drugs on the Heart

Recent Advances in Studies on
Cardiac Structure and Metabolism Volume 10
The Metabolism of Contraction
Edited by Paul-Emile Roy and George Rona

SIMILARITY OF ACETALDEHYDE AND SULFHYDRYL REAGENT ACTIONS ON THE PERFUSED GUINEA PIG HEART

MINH-HAU NGUYEN[1] and LAIMONIS GAILIS

Institut de cardiologie de Québec and Department of Medicine,
Laval University, Québec, Québec, Canada

SUMMARY

The effects of acetaldehyde (1 mM) and two known sulfhydryl reagents, diamide (0.25 mM) and cystamine (0.5 mM), on the perfused guinea pig heart were compared and were found to be markedly similar. All compounds increased heart rate and coronary flow. Dichloroisoproterenol blocked the increase in heart rate but not the increase in coronary flow. It is concluded that acetaldehyde may produce its effects by reaction with cellular thiols.

INTRODUCTION

Acetaldehyde is pharmacologically highly active, especially on the cardiovascular system (Handovsky, 1934; Moller, 1971a; Gailis and Verdy, 1971). Since several other aldehydes can produce similar results (Wingard, Hitchcock, and Teague, 1955), the activity probably resides in the aldehyde group, which is highly reactive chemically. It has not been determined at which of several potentially active cellular sites the aldehydes act.

A common functional group that reacts with aldehyde is the sulfhydryl (SH) group (Schubert, 1936). It seems to react more readily than the amino group (Ratner and Clarke, 1937), which is also attacked by aldehydes. It has been suggested that aldehydes may exert their biological effect by reaction with SH groups (Wald, Greenblatt, and Brown, 1953).

We compared the effects of acetaldehyde on the isolated guinea pig heart with those of two known sulfhydryl reagents–diamide and cystamine–and found that they produced markedly similar results.

---

[1] Present address: Département d'Endocrinologie-Nutrition Centre Hospitalier de l'Université Laval, Québec, Québec, Canada.

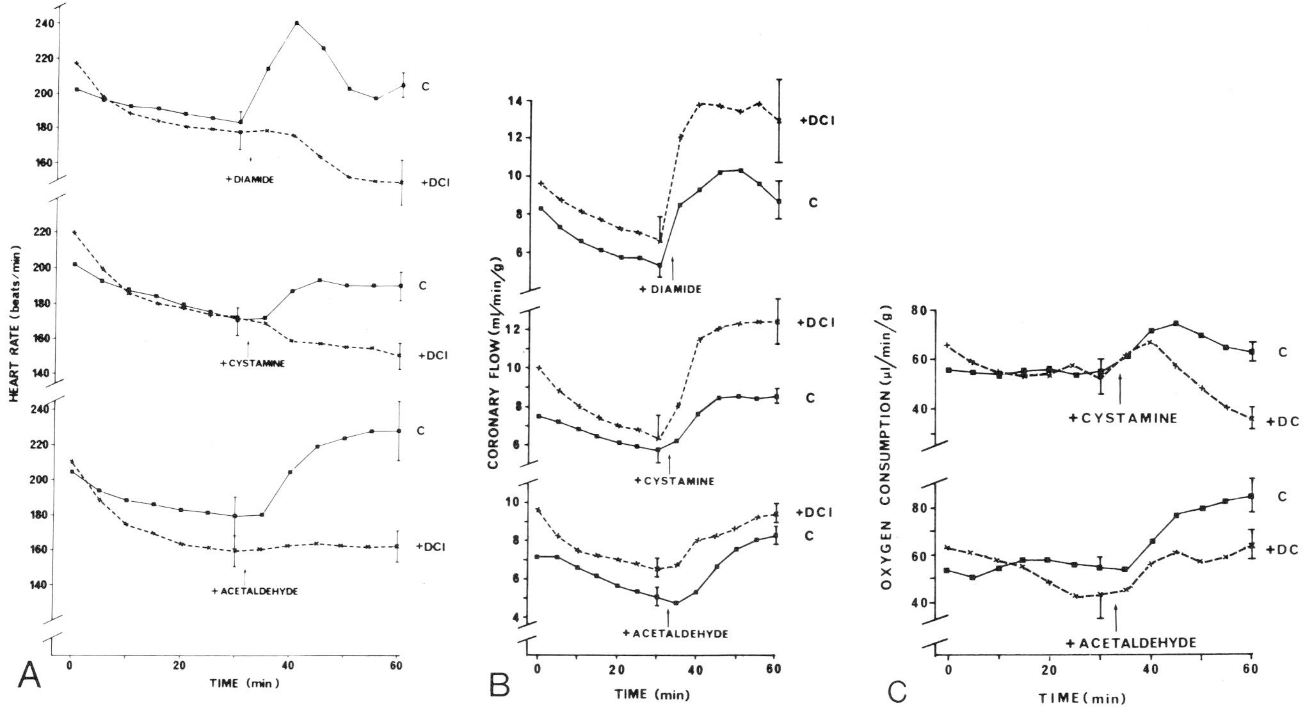

Figure 1. A, effect of diamide (0.25 mM), cystamine (0.5 mM), and acetaldehyde (1 mM) on the heart rate of the isolated guinea pig heart, perfused in the presence and absence of dichloroisoproterenol (DCI). Dichloroisoproterenol was added at the beginning of the experiment (0 min). The *vertical bars* at 30 and 60 min indicate 1 standard error. Each *point* is the mean of 4 to 7 experiments. Figure 1*B*, effect of diamide, cystamine, and acetaldehyde on the coronary flow of the perfused guinea pig heart. Conditions as in *part A*. Figure 1*C*, effect of cystamine and acetaldehyde on the oxygen consumption of the perfused guinea pig heart. Conditions as in *part A*.

METHODS

The perfusion method has been described previously (Nguyen, Mohiuddin, and Gailis, 1971). Hearts from fed male guinea pigs were perfused at a constant pressure of 50 mm Hg with Krebs-Henseleit bicarbonate buffer (Ca^{++} and Mg^{++} reduced to half) containing 5 mM glucose and gassed with 95% O_2/5% CO_2. Heart rate, coronary flow, and oxygen consumption were recorded at 5-min intervals. Thiols were measured using Ellman's reagent (Beutler, Duron and Kelly, 1963), and amino groups by the ninhydrin method of Soep and Janssen (Lee and Takahashi, 1966).

RESULTS AND DISCUSSION

As reported previously (Gailis and Verdy, 1971; Nguyen and Gailis, 1974), acetaldehyde produced an increase in heart rate and coronary flow (Fig. 1, *A* and *B*). It also increased oxygen consumption (Fig. 1*C*); however, the increase was probably secondary to the other effects (Nguyen and Gailis, 1974). Dichloroisoproterenol, a catecholamine β-blocker, completely blocked the increase in heart rate (Fig. 1*A*), but not coronary flow (Fig. 1*B*).

Both cystamine, a disulfide, and diamide, an SH oxidizing agent (Kosower *et al.*, 1969), produced effects very similar to those of acetaldehyde (Fig. 1, *A*–*C*). They increased heart rate and coronary flow; however, diamide tended to have a slight depressive effect after the initial stimulation. A further similarity with acetaldehyde was that dichloroisoproterenol blocked their effect on heart rate (Fig. 1*A*), but not on coronary flow (Fig. 1*B*). The only difference between the actions of acetaldehyde and the oxidizing agents was that, after β-blockade by dichloroisoproterenol, cystamine and diamide produced a decrease in heart rate. This is probably a function of the concentrations used and not a qualitative difference. Kumar and Sheth (1962), using higher concentrations of acetaldehyde, found that, in the presence of dichloroisoproterenol, it decreased the beat rate of isolated atria.

The best way of demonstrating that thiols are involved in the action of acetaldehyde would be to show a decrease in thiols. Unfortunately, this is only possible with thiols that have a free adjacent amino group (e.g., cysteine or cysteamine) which form a thiazolidine ring with aldehydes. Other thiols only form addition compounds (thiohemiacetals) which give a color reaction similar to that of the free thiols (Schubert, 1936; Ratner and Clarke, 1937). We measured the effect of acetaldehyde on thiols using Ellman's reagent. As expected, it showed a decrease in thiols when acetaldehyde was added to a solution of cystamine and cysteine, but not when glutathione was used. Since glutathione constitutes most of the low molecular weight thiols in the cell, we would not expect acetaldehyde to produce a large measurable decrease in heart acid-soluble SH. Indeed, perfusion with 1 mM acetaldehyde for 30 min did not produce any change in SH or amino groups (Table 1). After perfusion with 5 mM acetaldehyde, however, there was a slight but significant decrease in SH. The possibility that the decrease was caused by the loss of peptide from the heart cannot be excluded.

We have also tried a number of different thiol reagents, and they all produce effects on the isolated heart similar to those of acetaldehyde (Gailis and Nguyen, 1972). Other similarities between the actions of acetaldehyde and thiol reagents have been reported in the literature. One is their action on neurotransmitter release. Acetaldehyde produces the release of catecholamines in a number of tissues (Perman, 1958; Moller, 1971b). It is probably this release that is responsible for its positive chronotropic action (Kumar and Sheth, 1962). Similarly, catecholamines can be released from adrenal medulla granules by mercurials and other SH reagents (D'Iorio, 1957), and from isolated atria by *N*-ethylmaleimide and ethacrynic acid (Yamamoto *et al.*, 1973; Pousti *et al.*, 1973). Diamide has also been reported to promote the release of acetylcholine from nerve terminals (Werman *et al.*, 1971).

Obviously, the fact that known SH reagents produce the same effects as acetaldehyde is not proof that acetaldehyde acts on intracellular thiols. Since acetaldehyde can also react with amino groups, we could not completely eliminate the possibility that it may exert its action in this way; we could not find any amino group reagent that did not also react with SH groups. However, for a number of reasons, we tend toward the theory that acetaldehyde effects are mediated by reaction with SH groups. First, as we have shown, compounds that do not react with amino groups but do react with SH groups exactly mimic the action of acetaldehyde. Secondly, acetaldehyde seems to react faster with SH than with amino groups (Ratner and Clarke, 1937). In addition, we have

Table 1. Effect of acetaldehyde on acid-soluble SH and NH_2 groups in the beating and K^+-arrested guinea pig heart

	-SH		$-NH_2$	
Conditions	μmoles/g	% Change	μmoles/g	% Change
6 mM K^+				
Control (6)*	2.16 ± 0.08†			
Acetaldehyde (1 mM) (2)	2.10	− 3		
Acetaldehyde (5 mM) (5)	1.82 ± 0.06	−16‡		
20 mM K^+ (5)				
Control (5)	2.09 ± 0.06		23.2 ± 1.1	
Acetaldehyde (1 mM) (5)	2.26 ± 0.14	+ 8	25.1 ± 0.7	+8

*Number of experiments.
†Standard error of the mean.
‡$p<0.05$.

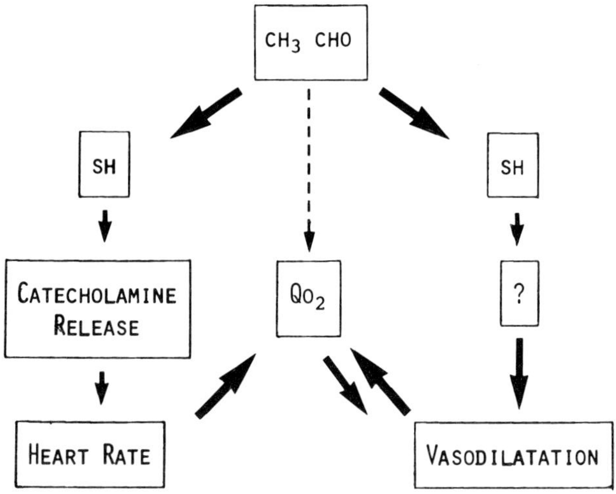

Figure 2. Postulated mechanism of action of acetaldehyde on the perfused heart. See text for details.

found that acetaldehyde is bound to reduced glutathione, but not to the oxidized form, although its amino groups are intact. Finally, thiols can protect against the effect of acetaldehyde both *in vivo* (Sprince *et al.*, 1974) and *in vitro* (Gailis and Beaulieu, 1974).

Our present conception of the action of acetaldehyde on the isolated heart is shown in Figure 2. Acetaldehyde reacts with intracellular thiols. This leads to the release of catecholamines, which increase the heart rate. The reaction with the same or different thiols produces a vasodilation, either directly or through the release of a transmitter. Both the increase in heart rate and the increase in flow can increase the oxygen consumption; however, higher concentrations of acetaldehyde can have a direct depressive effect on the mitochondrial oxidation.

Acknowledgments We would like to thank Mr. G. Huard for his capable assistance and Miss T. Turcotte for the typing of the manuscript. The illustration was produced by the Department of Photography of Laval Hospital under the direction of Mr. M Giguère. A Fellowship from the Ministry of Education of Quebec is gratefully acknowledged. This work was supported by grants from the Joseph C. Edwards Foundation and the Medical Research Council of Canada (MA-4597).

REFERENCES

BEUTLER, E., O. DURON, and B. M. KELLY. 1963. Improved method for the determination of blood glutathione. J. Lab. Clin. Med. 61: 882.

D'IORIO, A. 1957. The release of catecholamines from the isolated chromaffine granules of the adrenal medulla using sulfhydryl inhibitors. Canad. J. Biochem. Physiol. 35: 395.

GAILIS, L., and L. BEAULIEU. 1974. Relaxing effect of acetaldehyde on the norepinephrine-contracted guinea-pig aorta. Proc. Canad. Fed. Biol. Soc. 7th Annual Meeting. 17: 137.

GAILIS, L., and M. H. NGUYEN. 1972. Vasodilators as SH reagents: a possible mechanism of action. Abstracts of Volunteer Papers, Fifth International Congress on Pharmacology, p. 75.

GAILIS, L., and M. VERDY. 1971. The effect of ethanol and acetaldehyde on the metabolism and vascular resistance of the perfused heart. Canad. J. Biochem. 49: 227.

HANDOVSKY, H. 1934. Au sujet des propriétés biologiques et pharmacologiques de l'acétaldéhyde. C. R. Soc. Biol. 117: 238.

KOSOWER, N. S., E. M. KOSOWER, B. WERTHEIM, and W. CORREA. 1969. Diamide, a new reagent for the intracellular oxidation of glutathione to the disulfide. Biochem. Biophys. Res. Commun. 37: 593.

KUMAR, M. A., and U. K. SHETH. 1962. The sympathomimetic action of acetaldehyde on isolated atria. Arch. Int. Pharmacodyn. 137: 188.

LEE, Y. P., and T. TAKAHASHI. 1966. An improved colorimetric determination of amino acids with the use of ninhydrin. Anal. Biochem. 14: 71.

MOLLER, S. E. 1971a. The effects of acetaldehyde on the mechanical and electrical properties of the isolated heart from normal and disulfiram-treated guinea-pigs. Acta Pharmacol. Toxicol. 30: 429.

MOLLER, S. E. 1971b. The effect of acetaldehyde on the noradrenaline content in the isolated heart from normal and disulfiram-treated guinea-pigs. Acta Pharmacol. Toxicol. 30: 440.

NGUYEN, M.-H., and L. GAILIS. 1974. Effect of acetaldehyde on the isolated, non-working guinea-pig heart: Independence of the coronary flow increase from changes in heart rate and oxygen consumption. Canad. J. Physiol. Pharmacol. 52:65.

NGUYEN, M.-H., S. M. MOYHUDDIN, and L. GAILIS. 1971. Effet de l'acétaldéhyde sur le coeur de cobaye perfusé sans recirculation du perfusat. Union Méd. Can. 100: 745.

PERMAN, E. S. 1958. The effect of acetaldehyde on the secretion of adrenaline and noradrenaline from the suprarenal gland of the cat. Acta Physiol. Scand. 43: 71.

POUSTI, A., M. ZARRINDAST, D. SADEGHI, and M. A. KHOYI. 1973. Effect of ethacrynic acid on spontaneous contractions of isolated guinea-pig atria. Europ. J. Pharmacol. 22: 351.

RATNER, S., and H. T. CLARKE. 1937. The action of formaldehyde upon cysteine. J. Am. Chem. Soc. 59: 200.

SCHUBERT, M. P. 1936. Compounds of thiol acids with aldehyde. J. Biol. Chem 114: 341.

SPRINCE, H., C. M. PARKER, G. G. SMITH, and L. GONZALES. 1974. Protection against acetaldehyde toxicity in the rat by L-cysteine, thiamin, and L-2-methylthiazolidine-4-carboxylic acid. Agents Actions 4: 125.

WALD, G., C. GREENBLATT, and P. K. BROWN. 1953. Aldehydes as sulfhydryl reagents. Fed. Proc. 12: 285.

WERMAN, R., P. L. CARLEN, M. KUSHNIR, and E. M. KOSOWER. 1971. Effect on the thiol-oxidizing agent, diamide, on acetylcholine release at the frog endplate. Nat. New Biol. 233: 120.

WINGARD, C., P. HITCHCOCK, and R. S. TEAGUE. 1955. A survey of aldehydes with respect to their action on the blood pressure. Arch. Int. Pharmacodyn. 102: 65.

YAMAMOTO, H., T. KITANO, H. NISHINO, and T. MURANO. 1973. Studies on pharmacodynamic action of N-ethylmaleimide (NEM). Its report: Influence of NEM on synaptic transmission of sympathetic nerves in guinea-pigs. Jap. J. Pharmacol. 23: 151.

Mailing address:
L. Gailis,
Institut de cardiologie de Québec, 2725, chemin Ste-Foy,
Québec, Québec G1V 4G5 (Canada).

Recent Advances in Studies on
Cardiac Structure and Metabolism Volume 10
The Metabolism of Contraction
Edited by Paul-Emile Roy and George Rona

COMPARATIVE *IN VITRO* AND *IN VIVO* STUDIES ON THE EFFECTS OF TRICYCLIC ANTIDEPRESSANTS ON MYOCARDIAL CONTRACTILITY

GIANCARLO BIAMINO, HELMUT FENNER,
JÜRGEN NEYE, KLAUS-P. SCHÜREN,
and BERND RAMDOHR

Department of Cardiology, Klinikum Steglitz, and Department of Pharmacy,
Freie Universität Berlin, Berlin, Germany

SUMMARY

The effects of the similarly structured antidepressant drugs imipramine and dimetacrine and their demethylated derivates on myocardial performance were studied *in vitro* on 42 right ventricular papillary muscles of cats. The hemodynamic effects of intravenous administration of 22 mg of free base imipramine and dimetacrine were investigated in 12 patients with coronary disease.

The results of the *in vitro* studies show that imipramine > dimetacrine >> desmethyl compounds have a marked dose-dependent negative inotropic effect on all parameters of myocardial contractility, i.e., afterloaded isotonic contraction (ΔL), dL/dt_{max}, dL/dt_{min}, dT/dt. The hemodynamic investigations indicate that imipramine >> dimetacrine induces a significant increase in end-diastolic left ventricular pressure. Both drugs increase mean arterial blood pressure, correlated to a peripheral vasoconstriction and corresponding to an increase in the plasma level of catecholamines. It is concluded that both effects of these antidepressant drugs (reduction of myocardial performance and an increase in the plasma level of catecholamines) must be considred a risk in patients predisposed to myocardial failure.

INTRODUCTION

Tricyclic antidepressants of the imipramine and amytriptiline type are being used more and more in the clinics and by practitioners. Despite their well established antidepressant activity, few well controlled studies are available in the literature regarding their cardiovascular side effects (Sigg, Osborne, and Korol, 1963; Greeff and Wagner, 1969; Langslet *et al.*, 1971). In addition to their anticholinergic activity, the influence exerted on the cardiovascular system by this group of drugs is of striking importance.

Imipramine and dimetacrine are widely used in Europe when a stimulant effect is indicated in the treatment of depressive states. In spite of the structural similarity of the two drugs (Fig. 1), some significant differences have been

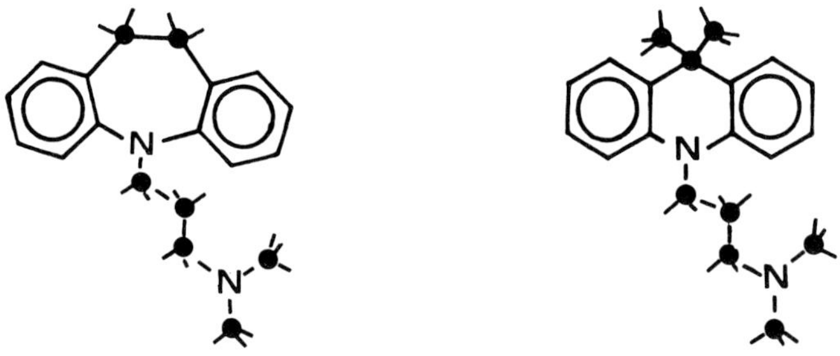

Figure 1. Structural similarity of imipramine and dimetacrine. Both antidepressant drugs contain a diphenylamine moiety and the same side chain. They differ in bridge between the aromatic nuclei.

established on the basis of animal experiments and clinical data (Abuzzahab, 1973). In depressive states, a similar central nervous system (CNS) activity has been demonstrated; however, the severity and the frequency of side effects are different. Relationships between the structure and the CNS activity have been studied by many authors (Fenner, 1974). Nevertheless, although cardiotoxic effects of these drugs have been reported repeatedly as regards electrocardiographic (ECG) changes (Sacks *et al.*, 1968; Alexander and Niño, 1969; Freeman *et al.*, 1969; Raisfeld, 1972), hemodynamic parameters have not been investigated in sufficient detail in man (Freyschuss *et al.*, 1970; Carlsson *et al.*, 1971; Moccetti *et al.*, 1971).

The aim of the present study was to determine the difference in the negative inotropic effects of imipramine and dimetacrine *in vitro* as well as *in vivo*. Despite the importance regarding the CNS activity of the main metabolites of these drugs, i.e., their demethylated derivatives, nothing is known about the effect of these compounds on the cardiovascular system. We therefore compared, *in vitro,* the effect on myocardial contractility exerted by imipramine and dimetacrine with that of their metabolites.

MATERIALS AND METHODS

In vitro studies

For these experiments, 42 right ventricular papillary muscles of cats were used. After dissection, the preparation was mounted in a mantled organ bath (37°C) containing 25 ml of a modified Krebs solution, pH 7.4, continuously bubbled with a gas mixture containing 95% O_2 and 5% CO_2. The preparation was driven at a constant frequency of 30/min, using supramaximal square stimuli of 5-msec duration (Tektronix). Preload (200 mg) and afterload (300 mg) remained con-

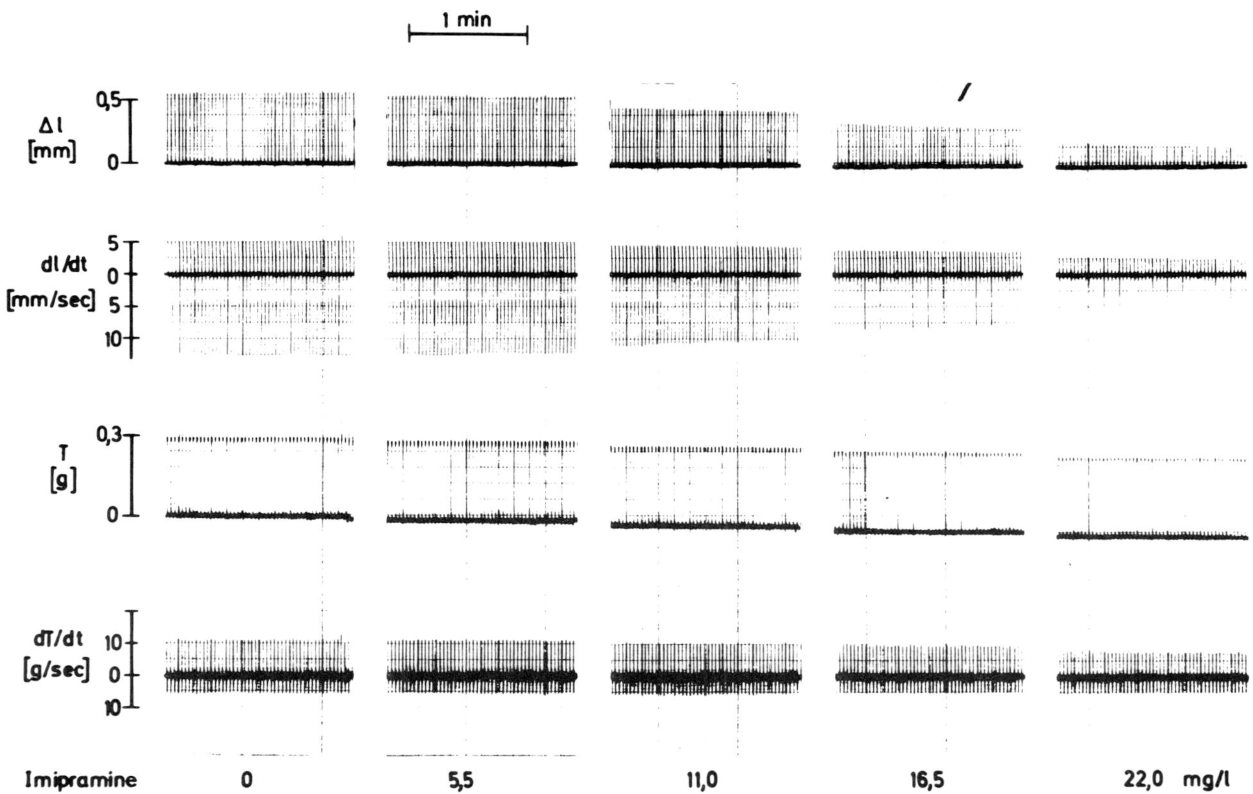

Figure 2. Original tracings of the dose dependency of imipramine-induced effects on the isolated cat papillary muscle.

stant during all experiments. Parameters simultaneously recorded were: changes in the amplitude of afterloaded isotonic contractions (ΔL), contraction velocity (dl/dt), tension (T), and the rate of force development (dT/dt). The recorder used was a Schwarzer Physioscript.

The following drugs were studied: imipramine, dimetacrine, desmethylimipramine, and desmethyldimetacrine. The dose-response correlation was calculated in mg/liter of the free base of the drugs.

Hemodynamic studies

These investigations were conducted on 12 patients (aged 38–65 years) with a history of angina pectoris and/or preceding myocardial infarction. Left ventricular and arterial as well as pulmonary pressures were recorded simultaneously. Cardiac output was calculated, using the dye dilution method. Following a control measurement, either 25 mg of imipramine-HCl (Tofranil), corresponding to 22.1 mg of free base, or 33.5 mg of dimetacrine-hydrogentartrate (Istonil), corresponding to the same dose of free base of imipramine, were injected intravenously. Hemodynamic parameters were recorded 15, 30, 45, or 60 min after drug administration.

RESULTS

Figure 2 illustrates a typical experiment on pre- and afterloaded cat papillary muscle. The stepwise increase of imipramine concentration caused a dose-dependent decrease of the amplitude of isotonic contractions (ΔL). A corresponding impairment of isotonic contraction velocity and of isotonic relaxation velocity was observed. Furthermore, imipramine decreased the rate of force development (dT/dt). These data, recorded in the same manner in all preparations, indicate a striking negative inotropic effect of imipramine, one of the most widely used tricyclic antidepressant drugs.

In Figures 3 and 4 are summarized the following data: percentage changes of the initial values of afterloaded (300 mg) isotonic contractions (ΔL); isotonic contraction velocity (dl/dt_{max}); isotonic relaxation velocity (dl/dt_{min}); and rate of force development (dT/dt) at constant preload (200 mg) induced by increasing concentrations (mg/liter free base) of imipramine and dimetacrine and their main metabolites, desmethylimipramine and desmethyldimetacrine.

As shown in Figure 3*a*, the decrease in the amplitude of afterloaded isotonic contractions induced by the desmethyl compounds is significantly ($p < 0.001$) different from the imipramine- and dimetacrine-induced effect on the isotonic contraction. Between imipramine and dimetacrine, a significant difference in the decrease of the amplitude can be demonstrated at a dose of 20 mg/liter or higher. The difference in the decrease of isotonic contraction produced by imipramine as compared with the other drugs and its more potent activity on the

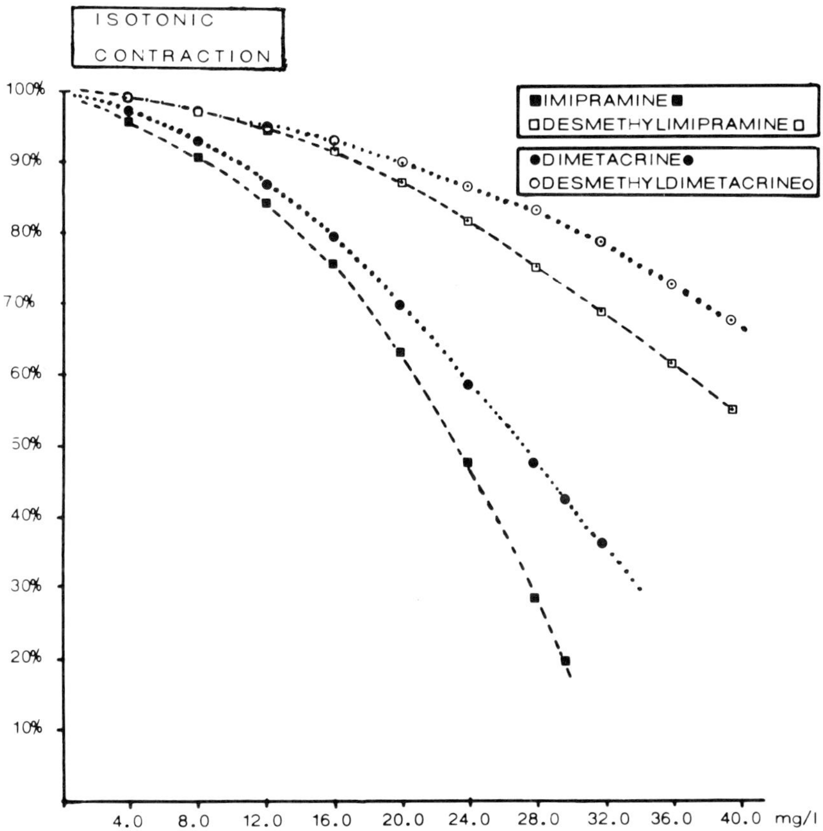

Figure 3. Effect (per cent changes) of imipramine, dimetacrine, desmethylimipramine, and desmethyldimetacrine an *a:* the afterloaded isotonic contraction (ΔL); *b* (p. 64): the maximal isotonic contraction (dL/dt_{max}); and *c* (p. 65): relaxation velocity (dt/dt_{min}).

myocardial contractility are evident from the other parameters, also. The isotonic contraction velocity and the isotonic relaxation velocity curves demonstrate the same effect, as is seen in Figure 3, *b* and *c*.

The decrease in the myocardial parameters induced by dimetacrine is significantly less than that of imipramine. This different potency of the two drugs is particularly notable in the decrease of dl/dt_{min}. No differences were observed among the metabolites. In contrast to these effects, imipramine, dimetacrine and their desmethyl compounds do not induce a corresponding percentage decrease in the rate of force development, using a constant preload of 200 mg (Fig. 4).

Our aim was to investigate the influence of the antidepressant drugs on the myocardial contractility *in vitro* and *in vivo* and to correlate the results obtained

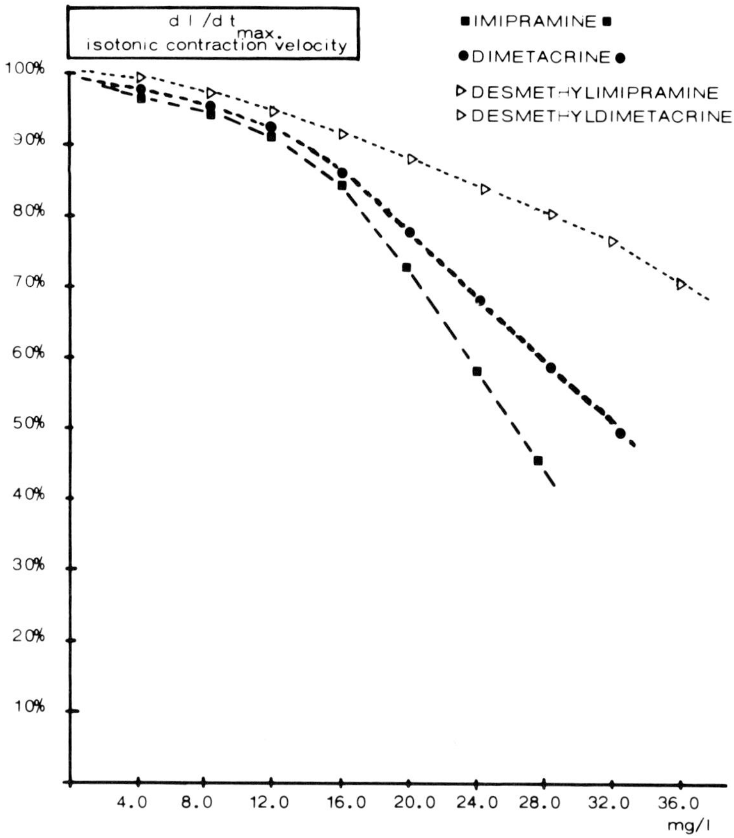

Figure 3b.

from animal data to the corresponding effects on hemodynamic parameters in man. From our experiments with cat papillary muscles, it was evident that the desmethyl compounds have a significantly smaller effect on myocardial contractility; therefore, we limited our studies in man to imipramine and dimetacrine.

The results obtained from our measurements of hemodynamic parameters are shown in Figure 5. Shortly after intravenous administration of both drugs, drowsiness occurred regularly, but no heart rhythm abnormalities or ECG changes were observed in any instance. No subjective anticholinergic effects or increase of heart rate were encountered.

The left ventricular end-diastolic pressure (LVEDP) increased significantly 30 min after injection of imipramine. At this time interval, dimetacrine did not give rise to this effect, although at 60 min, a similar increase of LVEDP (5 mm Hg) could be seen. A slight but not significant increase in cardiac (CI) and stroke

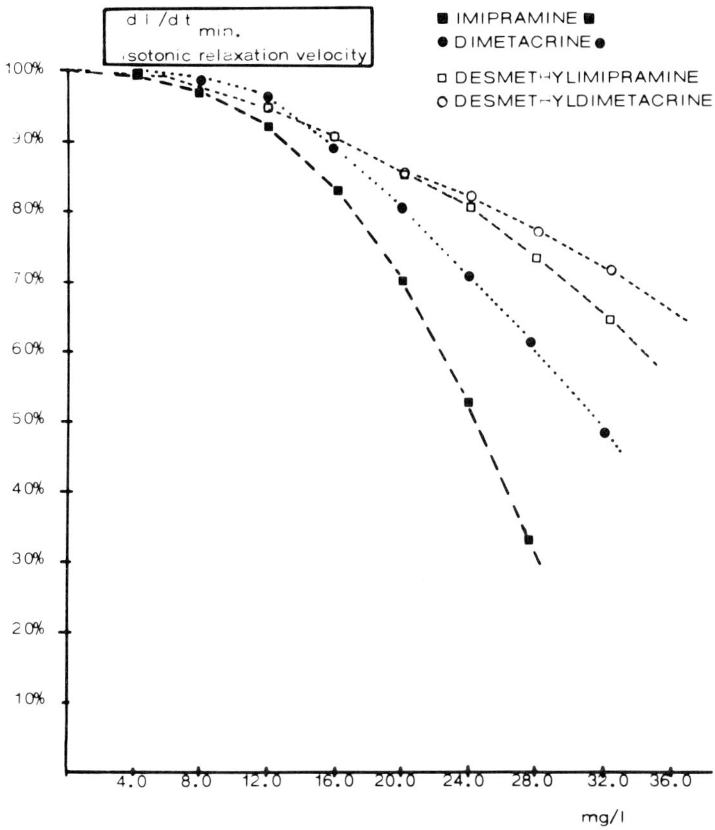

Figure 3c.

volume index (SVI) was caused by both imipramine and dimetacrine. In addition, both drugs caused a significant increase in the mean arterial blood pressure, corresponding to an adequate increase of peripheral resistance. Consequently, a slight but not significant augmentation of stroke work index (SWI) was observed.

DISCUSSION

The results of our *in vitro* experiments demonstrate an influence of imipramine on all parameters of myocardial contractility. The effects of dimetacrine are less than those of imipramine; however, they are more pronounced than those of the desmethyl compounds. The marked difference observed with imipramine and dimetacrine can be correlated with their general difference regarding data of

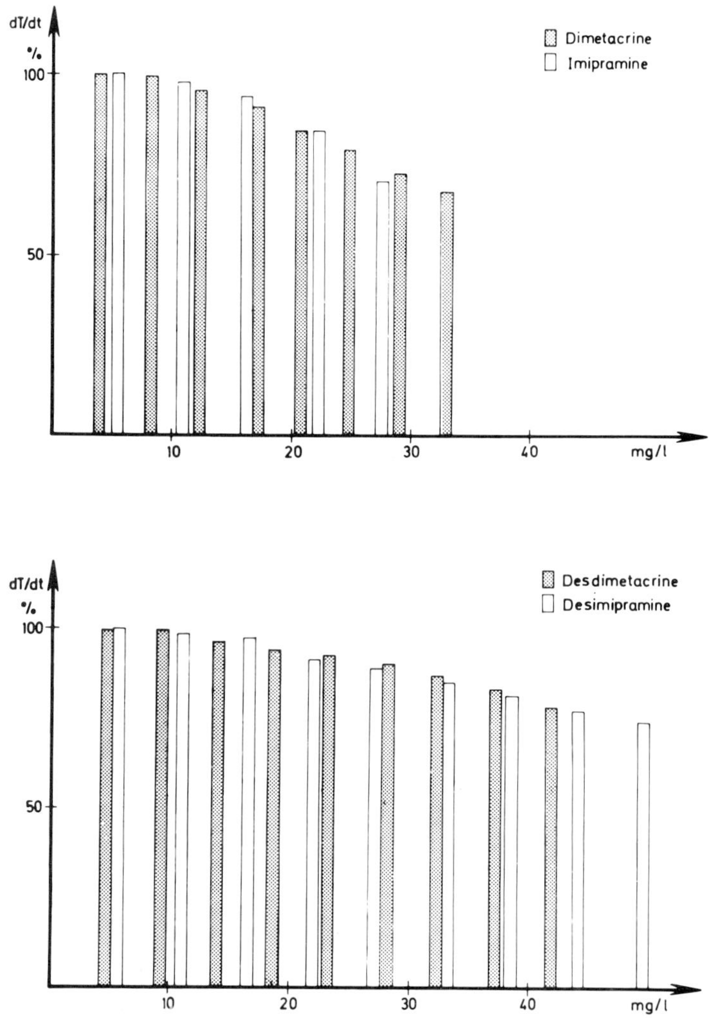

Figure 4. Changes (in percent) of the maximal rate of force development (dT/dt).

biopharmaceutical importance. Their difference in the pK value of about 1 unit has striking consequences in physiological pH ranges: at a pH of 7.4, the ratio of free base to the protonated species is 75%:25% in the case of dimetacrine and 25%:75% for imipramine. These differences are important as regards the activity of both drugs at tissue-binding sites.

The *in vivo* results are not representative with regard to the normal dosage format of these drugs; the oral dose of 200–400 mg of both drugs normally

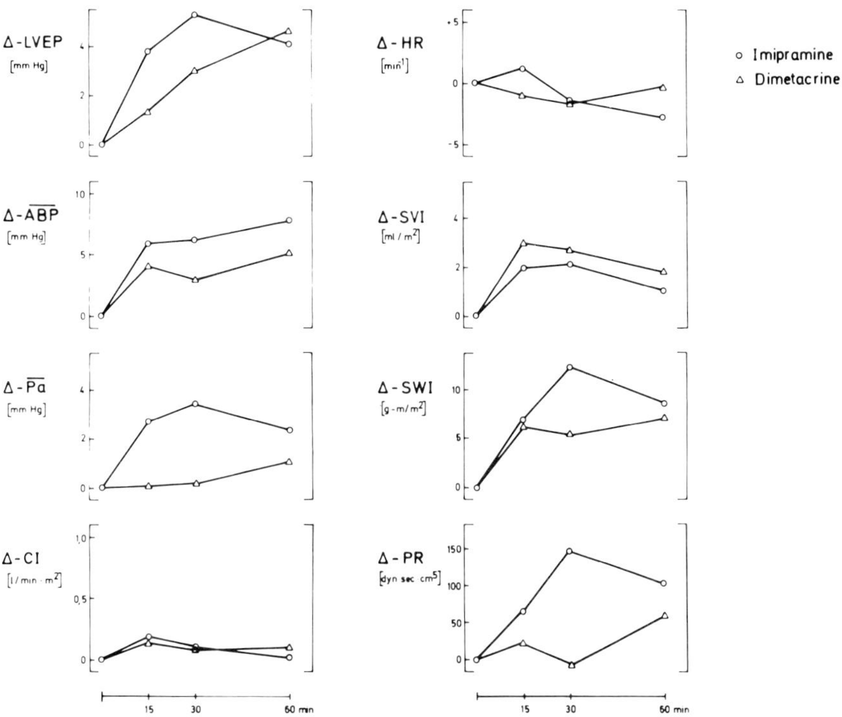

Figure 5. Changes in hemodynamic parameters following intravenous administration of equivalent doses of imipramine and dimetacrine.

administered in the treatment of depressions should be correlated with the plasma levels obtained by intravenous injection to the dose range used in our studies. On the other hand, in severe depressive states, infusion of antidepressant drugs is a useful method of drug administration.

However, our studies on the influence of these drugs on hemodynamic parameters indicate a negative inotropic effect in patients without manifest myocardial failure; in fact, the increase of LVEDP reflects a decrease of myocardial performance. In the case of imipramine, this effect is more pronounced as compared with that of dimetacrine. The difficulties of conducting these studies in man and the relatively small number of patients who are available for such investigations must be pointed out with regard to the significance of the difference between the two drugs.

Imipramine and dimetacrine increase the mean arterial blood pressure as a consequence of peripheral vasoconstriction. These results suggest that the antidepressant drugs induce a marked increase of norepinephrine plasma levels, as demonstrated in studies on dogs and cats (Sigg *et al.,* 1963; Eble, 1964;

Thoenen, Huerlimann, and Haefely, 1964). Preliminary investigations in man have confirmed these results. The mechanism of this effect consists in a decreased reuptake of liberated catecholamines in tissues (Thoenen *et al.*, 1964; Holtz and Palm, 1966; Axelrod and Weinshilboum, 1973; Davis and McNeill, 1973). The increase of catecholamine levels induced by these drugs may partially mask their negative inotropic effect in man. In any event, the combined effect, i.e., a negative inotropic action of these drugs and their influence on the catecholamine plasma level, must be due to an increase of myocardial oxygen consumption. Therefore, uncontrolled use of antidepressant drugs must be considered a risk in patients predisposed to myocardial failure.

Further studies on the influence of these drugs on hemodynamic parameters in man are in progress. These investigations should contribute to the elucidation of the negative inotropic effect of tricyclic antidepressants in man during exercise, as should comparison of these data with our results obtained from patients at rest.

REFERENCES

ABUZZAHAB, F. S. 1973. A double blind investigation of dimetacrine versus imipramine in hospitalized depressive states. Int. J. Clin. Pharmacol. B: 244.

ALEXANDER, C. S., and A. NINO. 1969. Cardiovascular complications in young patients taking psychotropic drugs. Am. Heart J. 78: 757–769.

AXELROD, J., and R. WEINSHILBOUM. 1972. Catecholamines. New Engl. J. Med. 287: 237–242.

CARLSSON, C., S. J. DENCKER, G. GRIMBY, J. HÄGGENDAL, and G. JOHNSSON. 1971. Effects on hemodynamics and plasma noradrenaline levels in man of long term treatment with imipramine, haloperidol and chlorpromazine. Europ. J. Clin. Pharmacol. 3: 163–171.

DAVIS, R. S., and J. H. McNEILL. 1973. The cardiac effects of cocaine and certain antihistamines and antidepressants. Arch. Int. Pharmacodyn. 201: 262–279.

EBLE, N. J. 1964. A study of the mechanisms of the modifying actions of cocaine, ephedrine and imipramine on the cardiovascular response to norepinephrine and epinephrine. J. Pharmacol. Exp. Ther. 144: 76–82.

FENNER, H. 1974. EPR studies on the mechanism of biotransformation of tricyclic neuroleptics and antidepressants. *In* I. S. Forrest, C. J. Carr, and E. Usdin (eds.), The Phenothiazines and Structurally Related Drugs. Raven Press, New York.

FREEMAN, J. W., G. R. MUNDY, R. R. BEATTIE, and C. RYAN. 1969. Cardiac abnormalities in poisoning with tricyclic antidepressants. Brit. Med. J. 2: 610–611.

FREYSCHUSS, U., F. SJOQVIST, D. TUCK, and M. ASBERG. 1970. Circulatory effects in man of nortriptyline, a tricyclic antidepressant drug. Pharmacol. Clin. 2: 68.

GREEFF, K., und J. WAGNER. 1969. Cardiodepressive und lokalanaesthetische Wirkungen der Thymoleptica. Arzneimittelforsch. 19: 1662–1664.

HOLTZ, P., und D. PALM. 1966. Brenzkatechinamine und andere sympathicomimetische Amine, Biosynthese und Inaktivierung, Freisetzung und Wirkung. Ergebn. Physiol. 58: 210.

LANGSLET, A., W. GRINI JOHANSEN, M. RYG, and I. ØYE. 1971. Effects of dibenzepine and imipramine on the isolated rat heart. Europ. J. Pharmacol. 14: 333–339.

MOCCETTI, T., P. LICHTLEN, H. ALBERT, E. MEIER, and P. IMBACH. 1971. Kardiotoxizität der trizyklischen Antidepressiva (Phenothiazine und Imipraminderivate). Schweiz. Med. Wschr. 101: 1.

RAISFELD, I. H. 1972. Cardiovascular complications of antidepressant therapy. Am. Heart J. 83: 129–133.

SACKS, M. H., R. J. BONFORTE, R. P. LASSER, and I. DIMICH. 1968. Cardiovascular complications of imipramine intoxication. J. Am. Med. Ass. 205: 588–590.

SIGG, E. B., M. OSBORNE, and B. KOROL. 1963. Cardiovascular effects of imipramine. J. Pharmacol. Exp. Ther. 141: 237–243.

THOENEN, H., A. HUERLIMANN, and W. HAEFELY. 1964. Mode of action of imipramine and 5- (3-methylaminopropyliden)-dibenzo [a,e] cyclohepta [1,3,5] trien hydrochloride (RO 4-6011), a new antidepressant drug, on peripheral adrenergic mechanisms. J. Pharmacol. Exp. Ther. 144: 405–414.

Mailing address:
Dr. G. Biamino,
D 1000 Berlin 45, Hindenburgdamm 30,
Klinikum Steglitz, Cardiol. Abt. der Med. Klinik (Germany).

Recent Advances in Studies on
Cardiac Structure and Metabolism Volume 10
The Metabolism of Contraction
Edited by Paul-Emile Roy and George Rona

EFFECTS OF INHALATION ANESTHETICS ON CARDIAC FUNCTION AND METABOLISM IN THE INTACT DOG

TERUO KUMAZAWA and ROBERT G. MERIN

Department of Anesthesiology, University of Rochester School of Medicine, Rochester, New York, USA

SUMMARY

In healthy, closed-chest dogs, dose-dependent depression of ventricular function was produced by the anesthetics halothane, methoxyflurane, and fluroxene, as evidenced by decreases in left ventricular stroke volume, stroke work, dP/dt, and an increased end-diastolic pressure. Myocardial blood flow and oxygen consumption decreased concomitantly and were correlated with aortic blood pressure decreases. There was no change in myocardial lactate extraction with halothane and methoxyflurane, suggesting that myocardial oxygenation was adequate in spite of the decrease in blood flow. However, even with marked increases in arterial lactate concentration during fluroxene anesthesia, extraction did not change and, in fact, tended to decrease. The hemodynamic effects of halothane and methoxyflurane are similar to those previously reported in man, but those of fluroxene are different. Consequently, clinical speculation from these results is not justified at this time.

INTRODUCTION

Inhalation anesthetics depress cardiac muscle in a dose-dependent fashion *in vitro* (Brown and Crout, 1971). In intact animals and man, however, several of these drugs, including diethyl ether, cyclopropane, and fluroxene, maintain ventricular function at deep levels of anesthesia through sympathetic stimulation, while the halogenated anesthetics halothane and methoxyflurane depress ventricular function and produce hypotension in intact animals just as in muscle preparations (Price, 1960; Eger *et al.,* 1971). In order to assess the effect of the cardiac depression and arterial hypotension arising from these latter drugs on myocardial oxygenation, we have measured myocardial blood flows, oxygen, and substrate uptake in an intact closed-chest dog. Two levels of anesthesia were studied: "light," which was just enough to keep the animals from moving or responding to stimulation, and "deep," which was as high a concentration as would maintain a systolic aortic pressure of 75 mm of mercury. In addition, the halogenated anesthetic, fluroxene, which preserves ventricular function in man at deep levels of anesthesia (Cullen *et al.,* 1970) was also studied.

METHODS

Our methods have been published previously in detail (Merin, 1969, 1972). Briefly, healthy mongrel dogs were fasted overnight. Without premedication, anesthesia was induced with 10 mg/kg thiopental and 0.5 mg/kg succinyl choline, intravenously. The tracheas were intubated with a cuffed tube and ventilation was controlled using 100% oxygen through a nonrebreathing valve with a Bird ventilator. Esophageal temperature was monitored and controlled by external heating. With the aid of fluoroscopy, catheters were placed in the coronary sinus, left ventricle, right atrium, and descending thoracic aorta. Anesthetic concentration and blood gases were stabilized for at least 30 min before any measurements were made. Vascular and cardiac pressures were measured through the cardiac catheters, using Statham P23 transducers. Left ventricular dP/dt was differentiated, using an R-C circuit. Cardiac output was determined by the dye dilution technique in duplicate. Myocardial blood flows were measured by ^{85}Kr or ^{133}Xe wash-out methods. Simultaneous arterial and coronary venous samples were taken for gas and substrate determinations.

RESULTS

Arterial CO_2 and body temperature remained constant during both levels of all three anesthetics (Table 1). Hematocrit was lower during deep anesthesia with all anesthetics. Arterial pH and pO_2 were not significantly different during halothane and methoxyflurane anesthesia, but there was metabolic acidosis at deep levels of fluroxene anesthesia.

During light anesthesia, cardiovascular hemodynamics were well sustained with all anesthetics, although heart rate, cardiac output, and cardiac work were appreciably higher with fluroxene (Table 2).

Left ventricular function was markedly depressed by deep anesthesia with all anesthetics as shown by the decreases in stroke volume, stroke work, and dP/dt, and the increases in diastolic pressure, although the latter were not significant for fluroxene (Fig. 1).

In like manner, aortic blood pressure and cardiac output were decreased significantly (Fig. 2). Heart rate changed only with fluroxene. Systemic vascular resistance tended to be lower, but not significantly so.

Myocardial blood flow decreased, consistent with the decreases in cardiac output and aortic pressure (Fig. 3). There was no change in the ratio of myocardial blood flow to cardiac output in deep levels of anesthesia compared to light. The previously shown decreases in arterial hematocrit in deep anesthesia are reflected in the small decreases in arterial oxygen content. Oxygen consumption by the heart decreased to the same extent as myocardial blood flow.

Table 1. Controlled parameters

	Halothane		Methoxyflurane		Fluroxene	
	Light	Deep	Light	Deep	Light	Deep
Alveolar anesthetic concentration (%)	0.63 ± 0.04	1.64 ± 0.07	0.20 ± 0.01	0.44 ± 0.01	6.5 ± 0.17	13.6 ± 0.58
Arterial pH	7.41 ± 0.02	7.39 ± 0.07	7.37 ± 0.02	7.36 ± 0.02	7.30 ± 0.01	7.20 ± 0.01*
Arterial oxygen (torr)	487 ± 9.5	485 ± 8.6	466 ± 20	448 ± 21	352 ± 35	304 ± 28
Arterial carbon-dioxide (torr)	33 ± 1.3	34 ± 2.1	31.3 ± 0.9	32.4 ± 1.4	36.3 ± 1.1	37.8 ± 0.32
Arterial hematocrit	42.6 ± 1.58	39.2 ± 1.75*	44.0 ± 1.00	41.7 ± 0.80	49.9 ± 1.11	46.3 ± 1.62*
Esophageal temperature (°C)	35.9 ± 0.18	36.0 ± 0.24	37.2 ± 0.2	37.4 ± 0.3	38.0 ± 0.22	37.8 ± 0.32

$*p < 0.05$.

Table 2. Hemodynamics during light anesthesia

	Halothane	Methoxyflurane	Fluroxene
LVSV (ml)	23.2 ± 2.05	20.7 ± 1.9	29.12 ± 2.3
LVSW (GmM)	37.3 ± 2.05	31.01 ± 3.0	45.46 ± 5.0
LV dP/dt	16 ± 1.5	18.4 ± 1.1	22.1 ± 2.1
LVEDP (torr)	4.1 ± 1.1	6.1 ± 0.9	4.3 ± 1.4
Heart rate	120 ± 6.8	140 ± 6.2	161 ± 5.7
MAP (torr)	119.6 ± 5.6	116.1 ± 3.3	121.3 ± 8.9
CO (cc/min)	2630 ± 170	2870 ± 280	4530 ± 300
LVW (GmM/min)	4177 ± 316	4251 ± 379	7079 ± 745
MBF (cc/100 g/min)	96.3 ± 7.9	101.3 ± 7.6	95.3 ± 24.7

*LVSV, left ventricular stroke volume; LVSW, left ventricular stroke work; LV dP/dt, left ventricular dP/dt; LVEDP, left ventricular end-diastolic pressure; MAP, mean aortic pressure; CO, cardiac output; LVW, left ventricular work; MBF, myocardial blood flow.

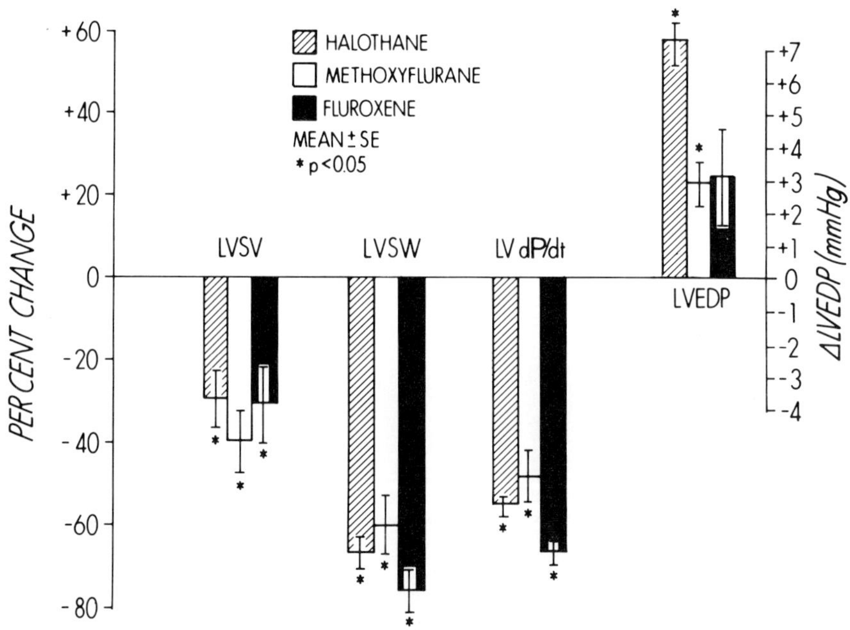

Figure 1. Change in left ventricular dynamics during deep anesthesia. LVSV, left ventricular stroke volume; LVSW, left ventricular stroke work; LV dP/dt, left ventricular dP/dt; LVEDP, left ventricular end-diastolic pressure.

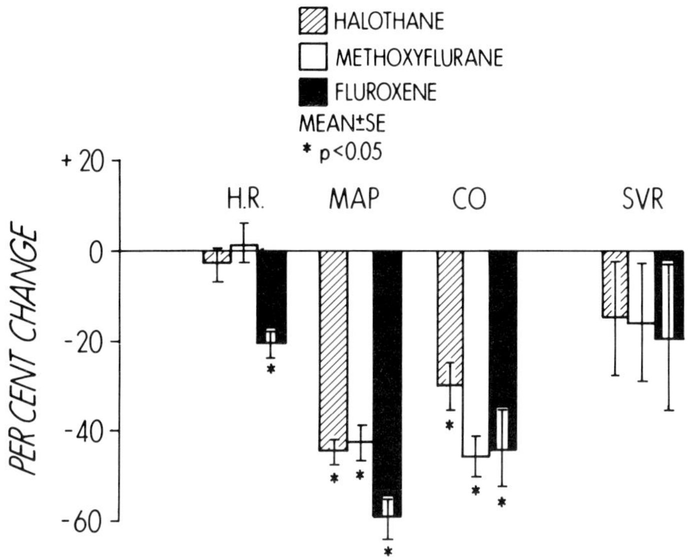

Figure 2. Change in hemodynamics during deep anesthesia. H.R., heart rate; MAP, mean aortic pressure; CO, cardiac output; SVR, systemic vascular resistance.

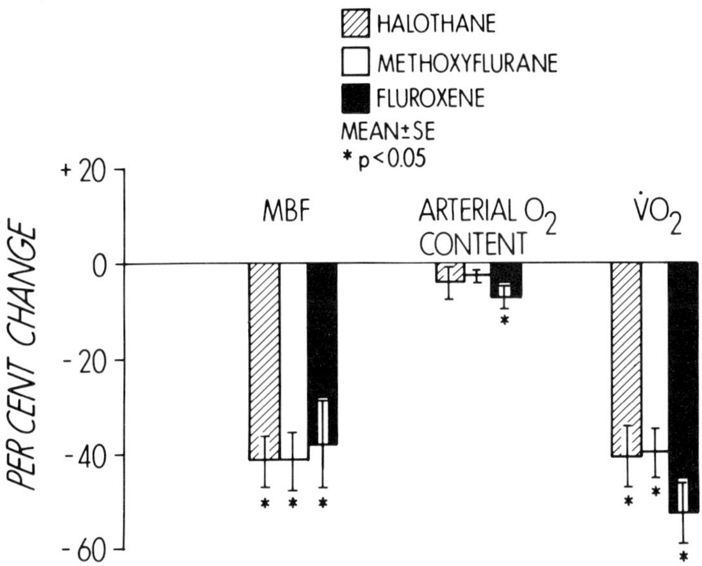

Figure 3. Change in myocardial blood flow and oxygen kinetics during deep anesthesia. MBF, myocardial blood flow; VO_2, myocardial oxygen uptake.

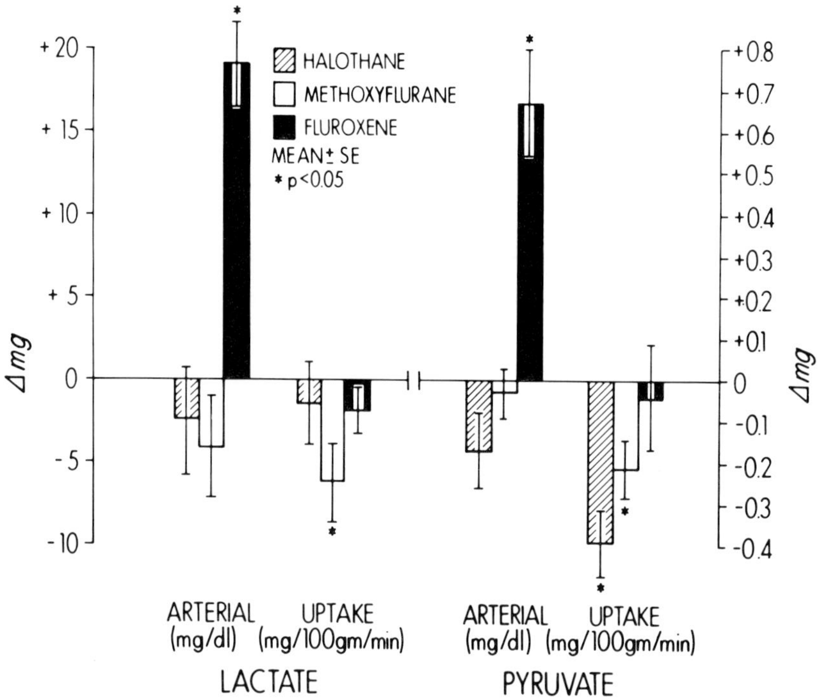

Figure 4. Change in myocardial lactate and pyruvate kinetics during deep anesthesia.

Arterial lactate and pyruvate levels were not different between light and deep anesthesia during halothane and methoxyflurane anesthesia (Fig. 4). However, deep fluroxene anesthesia produced marked increases in both arterial lactate and pyruvate concentrations. Although lactate uptake was decreased during deep methoxyflurane anesthesia and pyruvate uptake was decreased in both deep halothane and methoxyflurane anesthesia, these changes were the result of the decreases in myocardial blood flow. There were no changes in arteriovenous differences of these two substrates. In spite of the marked increases in arterial levels of lactate and pyruvate during deep fluroxene anesthesia, the uptake of these substrates did not change.

There was a significant correlation between aortic blood pressure and myocardial blood flow during halothane (Fig. 5), methoxyflurane (Fig. 6), and fluroxene anesthesia (Fig. 7). Consequently, coronary vascular resistance was not changed at the different levels of anesthesia.

DISCUSSION

The cardiac depression produced by deep levels of halothane and methoxyflurane anesthesia was accompanied by marked decreases in ventricular work,

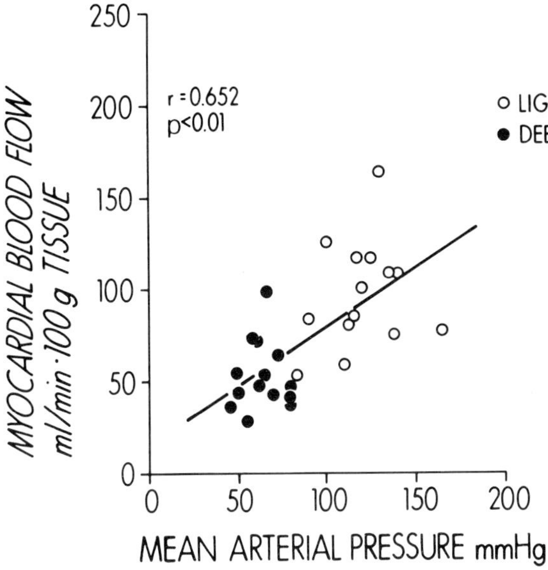

Figure 5. Relationship between arterial pressure and myocardial blood flow during halothane anesthesia.

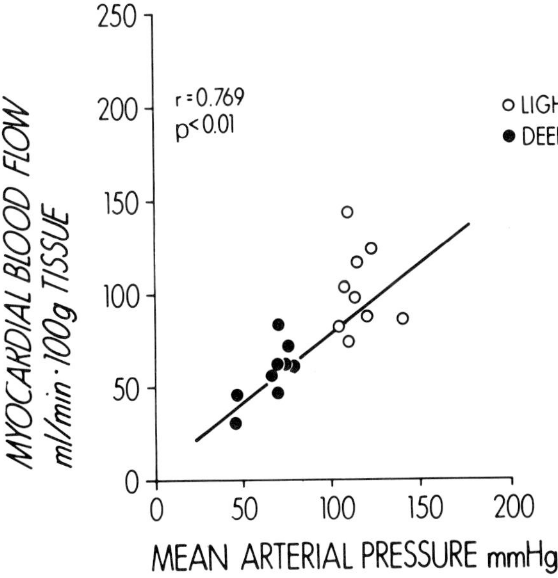

Figure 6. Relationship between arterial pressure and myocardial blood flow during methoxyflurane anesthesia.

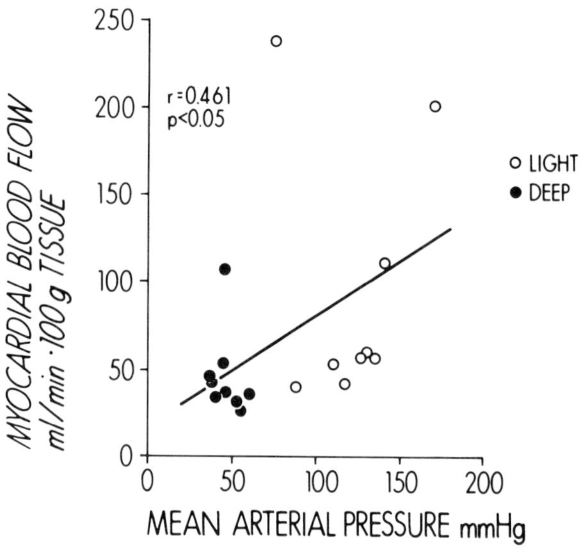

Figure 7. Relationship between arterial pressure and myocardial blood flow during fluroxene anesthesia.

myocardial blood flow, and oxygen consumption. The fact that the myocardial arteriovenous oxygen extraction did not increase and that significant lactate and pyruvate extraction continued suggests that myocardial oxygenation was adequate (Scheuer, 1967). From similar data, Hackel, Sancetta, and Kleinerman (1956) came to the same conclusion 20 years ago, concerning the hypotension produced by spinal anesthesia in man.

The effects of fluroxene are puzzling. Unlike the hemodynamic effects reported in man (Cullen *et al.*, 1970), we saw significant cardiac depression along with metabolic acidosis with deep fluroxene anesthesia. Preliminary studies have indicated that correction of the acidosis with sodium bicarbonate does not improve cardiac function. In spite of the marked increases in arterial lactate and pyruvate levels, there was no increase in myocardial extraction of these substrates. Indeed, in several animals, there was "production" of both substrates. This would suggest that these hearts were hypoxic. However, myocardial arteriovenous oxygen extraction decreased significantly in this series of animals. The combination of hypoxia and decreased oxygen extraction is inconceivable. Another possibility is that endogenous catecholamines stimulated glycolysis and glycogenolysis, markedly increasing the amounts of lactate and pyruvate present in blood and heart muscle (Merin, 1973).

Although the hemodynamic effects of halothane and methoxyflurane in the dog and man are very similar, markedly different effects of fluroxene give pause to making clinical speculation from our data. It is tempting to suggest that halothane and methoxyflurane might be beneficial for the patient with ischemic heart disease because of the decreased myocardial oxygen demand seen with these anesthetics. On the other hand, mean aortic pressure (the coronary perfusion pressure) is also decreased. The balance between these two effects is crucial. Definitive statements concerning the anesthetic management of patients with ischemic heart disease will have to await studies in both healthy and diseased humans.

Acknowledgments The conscientious and able technical assistance of Ms. Sally Basch made this work possible. Supported by United States Public Health Service Grant HL13257 and Research Career Development Award HL31752 (Robert G. Merin).

REFERENCES

BROWN, B. R., and J. R. CROUT. 1971. A comparative study of the effects of five general anesthetics on myocardial contractility: I. Isometric conditions. Anesthesiology 34: 236–245.

CULLEN, D. J., E. I. EGER, N. T. SMITH, D. C. SAWYER, G. A. GREGORY, and T. C. JOAS. 1970. Cardiovascular effects of fluroxene in man. Anesthesiology 32: 218–226.

EGER, E. I., N. T. SMITH, D. J. CULLEN, B. F. CULLEN, and G. A. GREGORY. 1971. Comparison of the cardiovascular effects of halothane, fluroxene, ether and cyclopropane in man. A resume. Anesthesiology 34: 25–41.

HACKEL, D. B., S. M. SANCETTA, and J. KLEINERMAN. 1956. Effect of hypotension due to spinal anesthesia on coronary blood flow and myocardial metabolism in man. Circulation 13: 92–97.

MERIN, R. G. 1969. Myocardial metabolism in the halothane-depressed canine heart. Anesthesiology 31: 20–26.

MERIN, R. G. 1972. Myocardial metabolism and dynamics after practolol during light halothane anaesthesia. Brit. J. Anaesth. 44: 437–445.

MERIN, R. G. 1973. Inhalation anesthetics and myocardial metabolism: possible mechanisms for functional effects. Anesthesiology 39: 216–255.

PRICE, H. L. 1960. General anesthesia and circulatory homeostasis. Physiol. Rev. 40: 187–218.

SCHEUER, J. 1967. Myocardial metabolism in cardiac hypoxia. Am. J. Cardiol. 19: 385–392.

Mailing address:
Dr. Robert Merin,
University of Rochester Medical Center,
Rochester, New York 14642 (USA).

Recent Advances in Studies on
Cardiac Structure and Metabolism Volume 10
The Metabolism of Contraction
Edited by Paul-Emile Roy and George Rona
Copyright 1975 University Park Press Baltimore

METABOLIC AND INOTROPIC EFFECTS OF VERAPAMIL IN PERFUSED RAT HEART[1]

MARILYN E. HESS, JOSEPH SHANFELD,[2]
and NEIL R. LEVINE

Department of Pharmacolgy,
University of Pennsylvania Medical School, Philadelphia, Pennsylvania, USA

SUMMARY

The effect of verapamil on myocardial contractility, heart adenosine 3′,5′-monophosphate (cyclic AMP), and cardiac phosphorylase *a* activity was studied in isolated perfused rat hearts. In a concentration of 0.025 μg/ml, verapamil decreased force of contraction 50% and caused a significant fall in heart cyclic AMP within 4 to 5 min after perfusion with the drug was begun. When perfusion of the heart with medium containing verapamil was continued for 60 min, contractile force gradually returned to control. At the end of 60 min of perfusion with verapamil, the myocardial concentration of cyclic AMP was not different from that measured in hearts perfused with control medium for a similar time period. Isoproterenol, given at the point of maximal contractile depression induced by verapamil, restored normal force of contraction and raised cardiac cyclic AMP to the same level as that observed when the catecholamine was given to untreated hearts. When methoxamine was administered to hearts depressed by verapamil, contractility returned to normal, but cyclic AMP content remained below control values.

INTRODUCTION

It is generally accepted that the presence of extracellular calcium is a prerequisite for contraction of cardiac muscle to occur (Niedergerke, 1956; Winegrad, 1961; Nayler, 1963). Less well established is the role of adenosine 3′,5′-monophosphate (cyclic AMP) in mediating the inotropic effect of certain drugs, such as the catecholamines. Several investigators have proposed that cyclic AMP not only mediates the metabolic effects of sympathomimetic amines but is also the "messenger" responsible for the observed increase in contractile force of hearts

---

[1] Supported by a grant-in-aid from the American Heart Association and Grant 1C-73-16 from the National Science Foundation. Portions of the work described in this paper have been submitted for publication in the Journal of Pharmacology and Experimental Therapeutics.

[2] Present address: University of Pennsylvania, School of Dental Medicine, Philadelphia, Pennsylvania.

given epinephrine (Robison, Butcher, and Oye, 1965; Robison, Butcher, and Sutherland, 1967). The purpose of the work reported in this paper was to study myocardial contractility, heart cyclic AMP, and cardiac phosphorylase activity in relation to alterations in the transmembrane movement of calcium across heart cells. The drug verapamil, shown by Fleckenstein (1971) to prevent the influx of calcium ions into myocardial cells and thereby stop contractions, was used in our studies in amounts insufficient to completely inhibit contractility, but in concentrations adequate to produce significant diminution in force of contraction.

MATERIALS AND METHODS

All of the experiments were performed with isolated perfused rat heart preparations exactly as described in a previous publication (Shanfeld, Frazer, and Hess, 1969). The hearts were perfused either with control Krebs-Ringer bicarbonate solution (KRB) or with KRB containing verapamil (0.025 mg/liter = 5.73×10^{-8} M). The following drugs were infused through a catheter inserted into the cannula tied into the aorta: isoproterenol 0.075 μg/0.15 ml/min; methoxamine 7.5 μg/0.15 ml/min; mephentermine 7.5 μg/0.15 ml/min; sotalol 16.7 μg/0.34 ml/min. Force of contraction was measured as reported by Shanfeld *et al.* (1969).

Cyclic AMP was assayed according to the method of Gilman (1970, 1972), with minor modifications as suggested by Shanfeld, Hess, and Levine (1974). Cardiac phosphorylase activity was measured by the method of Cori and Illingworth (1956), with small changes described by Aronson and Hess (1967). Each heart was allowed to equilibrate for 6 min and then perfused for varying periods of time with control KRB solution or KRB containing verapamil. At the end of each perfusion period, the heart was frozen with Wollenberger tongs cooled previously in liquid nitrogen and the tissue used subsequently for determination of cyclic AMP and phosphorylase activity.

RESULTS

The results of these experiments are presented in Table 1. It can be seen that, within 2 to 3 min after beginning a perfusion with verapamil, the contractile force (ISF) was significantly decreased and reached a maximal depression within the 4- to 5-min period after the perfusion of verapamil was begun. When perfusion with verapamil was continued for extended periods (20–60 min), the force of contraction returned to normal levels in spite of the fact that the heart had been continuously exposed to the drug.

Cyclic AMP content in the heart followed changes in myocardial contractility with the exception of one time period (Table 1). Verapamil caused a depression

Table 1. Effect of continuous perfusion of verapamil on isometric contractile force, cardiac cyclic AMP, and heart phosphorylase *a* activity

	Isometric Contractile Force (mm)						Cardiac Cyclic AMP (pmoles/mg wet wt)			Phosphorylase *a* (% of total)		
	Control			Verapamil†								
Time (min)	Initial	After infusion of perfusion fluid	p*	Initial	After infusion verapamil	p*	Control	Verapamil†	p‡	Control	Verapamil†	p‡
2–3	11.6 ± 0.69 (5)§	11.7 ± 0.83 (5)	>0.05	11.7 ± 0.71 (5)	6.2 ± 1.58 (5)	<0.05	1.06 ± 0.09 (5)	0.94 ± 0.06 (5)	>0.05	12.4 ± 3.5 (5)	9.2 ± 1.3 (5)	>0.05
4–5	11.4 ±1.54 (9)	11.4 ± 1.20 (9)	>0.05	11.0 ± 0.88 (10)	5.0 ± 0.72 (10)	<0.05	0.97 ± 0.03 (9)	0.83 ± 0.05 (10)	<0.05	8.5 ± 0.7 (9)	6.4 ± 0.6 (10)	<0.05
8–12	10.5 ± 0.73 (8)	10.8 ± 0.61 (8)	>0.05	11.8 ± 0.35 (8)	8.2 ± 0.33 (8)	<0.05	1.05 ± 0.06 (8)	0.83 ± 0.05 (8)	<0.05	7.5 ± 0.7 (8)	8.5 ± 1.0 (8)	>0.05
20	8.9 ± 1.04 (4)	11.8 ± 1.45 (4)	>0.05	10.7 ± 1.69 (5)	10.9 ± 1.71 (5)	>0.05	0.94 ± 0.09 (4)	0.87 ± 0.07 (5)	>0.05	8.1 ± 0.3 (4)	12.8 ± 2.6 (5)	>0.05
60	12.1 ± 0.68 (7)	13.9 ± 0.68 (7)	>0.05	10.7 ± 0.85 (7)	10.4 ± 0.98 (7)	>0.05	0.85 ± 0.05 (7)	0.74 ± 0.05 (7)	>0.05	10.3 ± 0.7 (7)	10.7 ± 0.9 (7)	>0.05

*Compared to initial.
†0.025 μg/ml.
‡Compared to control.
§Number of experiments is shown in parentheses.

in contractile force before significant diminution in cyclic AMP content of the heart could be observed. We believe the difference in the time of onset of mechanical depression and the appearance of decreased cyclic nucleotide in the heart to be genuine. However, we are aware that the sensitivity of the method used to measure cyclic AMP may preclude obtaining precise enough determinations to pick up minute changes in concentration of the nucleotide.

Alterations in cardiac phosphorylase *a* activity induced by verapamil were minimal in that only during the period of maximal contractile depression (4–5 min) was there a significant decrease in the enzyme.

Catecholamines are among the agents known to antagonize the effects of verapamil on transmembrane movement of calcium (Fleckenstein, 1971). It seemed appropriate, therefore, to determine whether the metabolic effects of verapamil were also overcome by sympathomimetic drugs possessing a positive inotropic action; isoproterenol was the drug we used to study this aspect of the problem. Further, we wanted to investigate the effects of beta-adrenergic receptor blockade on the interaction between verapamil and isoproterenol.

When administered to the isolated perfused rat heart, isoproterenol caused the expected increase in force of contraction and rise in heart cyclic AMP and phosphorylase *a* activity. All of these responses to isoproterenol were prevented by the beta-adrenergic receptor blocking drug, sotalol.

The effects of these drugs given to rat hearts being perfused with verapamil are shown in Table 2. Isoproterenol, infused into the hearts at the point of maximal depression of contractility, caused a complete reversal of the effects of verapamil; i.e., force of contraction was restored to normal and cyclic AMP content in the heart rose to approximately the same level as that observed when the catecholamine was given to non-verapamil-treated hearts. Sotalol alone infused into hearts depressed by verapamil had no effect on myocardial contractility, cyclic AMP, or heart phosphorylase *a* activity and blocked the ability of isoproterenol to reverse the inhibitory effects of verapamil.

In order to determine whether sympathomimetic amines having little inotropic effect on the heart could counteract the depression of cardiac cyclic AMP produced by verapamil, we used methoxamine and mephentermine. Neither methoxamine nor mephentermine infused into control rat hearts caused any change in force of contraction, cyclic AMP, or phosphorylase *a* activity.

However, within 3–4 min, methoxamine produced a marked increase in contractility when given to hearts at the time of maximal depression of contractility obtained with the dose of verapamil we used. Unlike isoproterenol, methoxamine did not cause a rise in cyclic AMP nor was phosphorylase *a* activity elevated.

Mephentermine had no effect on the decrease in myocardial contractility or cyclic AMP produced by verapamil; phosphorylase *a* activity was similarly unaffected by mephentermine in verapamil-treated hearts.

Table 2. Effect of isoproterenol and sotalol on rat hearts treated with verapamil*

	Isometric Contractile Force (mm)				Cardiac Cyclic AMP (pmoles/mg wet wt.)		Phosphorylase *a* (% of total)	
Drugs	Initial	Verapamil (maximal depression)	Postinfusion	*p*		*p* (compared to verapamil)		*p* (compared to verapamil)
Verapamil	12.6 ± 0.64 (6)	5.3 ± 0.75 (6)		<0.05[†]	0.85 ± 0.08 (6)		8.0 ± 1.07 (6)	
Verapamil + isoproterenol	11.9 ± 0.79 (7)	5.4 ± 1.07 (7)	13.8 ± 0.90 (7)	<0.05[†] >0.05[‡]	1.38 ± 0.08 (7)	<0.05	23.5 ± 2.28 (7)	<0.05
Verapamil + sotalol	11.8 ± 0.79 (7)	6.2 ± 0.59 (7)	6.4 ± 0.66 (7)	<0.05[†,‡]	0.80 ± 0.06 (7)	>0.05	6.2 ± 1.22 (7)	>0.05
Verapamil ± sotalol + isoproterenol	13.2 ± 0.93 (6)	6.9 ± 0.63 (6)	8.1 ± 0.89 (6)	<0.05[†,‡]	0.93 ± 0.03 (6)	>0.05	6.1 ± 1.35 (6)	>0.05

*Concentrations used: verapamil, 0.025 μg/ml; isoproterenol, 0.075 μg/0.15 ml/min; sotalol, 16.7 μg/0.34 ml/min. Number of experiments in parentheses.

[†]Verapamil compared to initial.

[‡]Postinfusion compared to initial.

Table 3. Effect of methoxamine and mephentermine on rat hearts treated with verapamil*

	Isometric Contractile Force (mm)				Cardiac Cyclic AMP (pmoles/mg wet wt.)		Phosphorylase *a* (% of total)	
Drugs	Initial	Verapamil	Postinfusion	*p*		*p* (compared to verapamil)		*p* (compared to verapamil)
Verapamil	11.8 ± 0.82 (6)	4.5 ± 0.78 (6)		<0.05†	0.66 ± 0.03 (6)		5.8 ± 0.61 (6)	
Verapamil + methoxamine	12.2 ± 1.21 (7)	5.4 ± 0.48 (7)	11.4 ± 1.11 (7)	<0.05† >0.05‡	0.68 ± 0.02 (7)	>0.05	7.2 ± 0.57 (7)	>0.05
Verapamil + mephentermine	12.8 ± 0.84 (7)	6.1 ± 1.15 (7)	6.7 ± 1.08 (7)	<0.05†,‡	0.71 ± 0.01 (7)	>0.05	7.0 ± 1.00 (7)	>0.05

*Concentrations used: verapamil, 0.025 μg/ml; methoxamine, 7.5 μg/0.15 ml/min; mephentermine, 7.5 μg/0.15 ml/min. Number of experiments in parentheses.
†Verapamil compared to initial.
‡Postinfusion compared to initial.

DISCUSSION

Some of the results reported in this investigation warrant special comment. The finding that contractile changes preceded the diminution in the content of cyclic AMP caused by verapamil is unique. In studies with cardiac stimulants, such as the catecholamines, an increase in cyclic AMP occurred prior to the mechanical stimulation (Cheung and Williamson, 1965; Robison *et al.*, 1965).

In view of earlier reports that calcium inhibits adenyl cyclase (Rasmussen, 1970), our demonstration that verapamil, which blocks transmembrane movement of calcium, caused a decrease in cyclic AMP was somewhat unexpected. That is to say, inhibiting the movement of calcium across the cell membrane would supposedly allow greater adenyl cyclase activity and promote a rise in cyclic AMP, such as reported by Namm, Woods, and Zucker (1972) using calcium-free perfusion medium.

An intriguing question arising from observations made in this investigation is why does the heart recover normal contractile force even though verapamil was continuously perfused over a 1-hour period? The same inquiry could be made concerning the restoration of normal cardiac cyclic AMP levels with prolonged perfusion with verapamil.

Finally, the reason for the difference in the contractile response to methoxamine in hearts perfused with and without verapamil deserves further study. To paraphrase the words of an eminent pharmacologist, Carl F. Schmidt, "The most exciting research problems are those which pose more questions than they answer." Certainly, further study of the pharmacology of verapamil is such a problem.

REFERENCES

ARONSON, C. E., and M. E. HESS. 1967. Metabolic effects of bretylium and phentolamine in euthyroid and thyrotoxic rats. Europ. J. Pharmacol. 2: 113–118.

CHEUNG, W. Y., and J. R. WILLIAMSON. 1965. Kinetics of cyclic adenosine monophosphate changes in rat heart following epinephrine administration. Nature (London) 207: 979–981.

CORI, G. T., and B. ILLINGWORTH. 1956. The effect of epinephrine and other glycogenolytic agents on the phosphorylase *a* content of muscle. Biochim. Biophys. Acta 21: 105–110.

FLECKENSTEIN, A. 1971. Specific inhibitors and promoters of calcium action in the excitation-contraction coupling of heart muscle and their role in the prevention or production of myocardial lesions. *In* P. Harris and L. Opie (eds.), Calcium and the Heart, pp. 135–188. Academic Press, New York.

GILMAN, A. 1970. A protein binding assay for adenosine 3',5'-cyclic monophosphate. Proc. Nat. Acad. Sci. 67: 305–312.

GILMAN, A. 1972. Protein binding assays for cyclic nucleotides. *In* Advances in Cyclic Nucleotide Research, Vol. 2, pp. 9–24. Raven Press, New York.

NAMM, D. H., E. L. WOODS, and J. L. ZUCKER. 1972. Incorporation of the terminal phosphate of ATP into membranal protein of rabbit cardiac sarcoplasmic reticulum:

Correlation with active calcium transport and study of the effects of cyclic AMP. Circulat. Res. 31: 308–316.
NAYLER, W. G. 1963. The significance of calcium ions in cardiac excitation and contraction. Am. Heart J. 65: 404–411.
NIEDERGERKE, R. 1956. The 'staircase' phenomenon and the action of calcium on the heart. J. Physiol. 134: 569–583.
RASMUSSEN, H. 1970. Cell communication, calcium ion, and cyclic adenosine monophosphate. Science 170: 404–412.
ROBISON, G. A., R. W. BUTCHER, and I. OYE. 1965. The effect of epinephrine on adenosine 3′, 5′-phosphate levels in the isolated perfused rat heart. Mol. Pharmacol. 1: 168–177.
ROBISON, G. A., R. W. Butcher, and E. W. Sutherland. 1967. Adenyl cyclase as an adrenergic receptor. Ann. N. Y. Acad. Sci. 139: 703–723.
SHANFELD, J., A. FRAZER, and M. E. HESS. 1969. Dissociation of the increased formation of cardiac adenosine 3′, 5′-monophosphate from the positive inotropic effect of norepinephrine. J. Pharmacol. Exp. Ther. 169: 315–320.
SHANFELD, J., M. E. HESS, and N. R. LEVINE. 1974. Effects of verapamil on myocardial contractility, cardiac adenosine 3′, 5′-monophosphate and heart phosphorylase. J. Pharmacol. Exp. Ther. (Submitted for publication).
WINEGRAD, S. 1961. The possible role of calcium in excitation-contraction coupling of heart muscle. Circulation 24: 523–529.

Mailing address:
Dr. Marilyn E. Hess,
University of Pennsylvania Medical School, Department of Pharmacology,
Philadelphia, Pennsylvania 19174 (USA).

Mechanical Activity of Cardiac Muscle

Recent Advances in Studies on
Cardiac Structure and Metabolism Volume 10
The Metabolism of Contraction
Edited by Paul-Emile Roy and George Rona

TIME COURSE OF MECHANICAL ACTIVITY IN MAMMALIAN CARDIAC MUSCLE: DEPENDENCE ON SPECIES, LOADING, AND DISPLACEMENT[1]

RAINER BODEM[2] and EDMUND H. SONNENBLICK[3]

Cardiovascular Research Laboratories, Department of Medicine,
The Peter Bent Brigham Hospital, Harvard Medical School, Boston, Massachusetts, USA

SUMMARY

We have found that animal species is of great importance when mechanics of contraction are analyzed in mammalian cardiac muscle. Mechanics of contraction have been studied comparatively in the isolated papillary muscles of the rat, cat, and rabbit under control conditions and following alterations of loading and inotropic interventions. The time course of the shortening ability measured in terms of the velocity of shortening at constant contractile element length and load following quick releases to preload peaked at 20, 30, and 70% of time to peak force (TPF) in the rat, cat, and rabbit, respectively, and declined more slowly in the rabbit. The shortening ability peaked progressively later in contraction at increasing loads in the cat, but not in the rabbit. Shortening ability measured after releases to zero load or after imposing "load clamps" of variable magnitude and duraton reached the maximum much later in the rabbit than in the cat. The contractile response to changes of the contraction mode from isometric to isotonic was different in cat and rabbit. Upon increase in the stimulation rate from 6 to 60/min, force increased to a greater extent in the rabbit than in the cat, while TPF did not decrease in the rabbit as seen in the cat. Caffeine (10 mM) increased developed force more in the rabbit than in the cat, while even 1 mM caffeine decreased force in the rat. The onset of the shortening ability was greatly delayed by caffeine in the cat and rat but not in the rabbit. Afterloaded force-peak velocity relations were curvilinear in the cat and rat; these relations approached linearity in the rabbit in the control state, and in both cat and rabbit in the presence of caffeine. The differences in the mechanics of contraction, in the contractile response to alterations in loading, and to caffeine in the ventricular muscle of the rat, cat, and rabbit are viewed as indicators of differences in excitation-contraction coupling of ventricular cardiac muscle from these different mammalian species.

---

[1] Supported in part by Grants HL 11306 and HL 05890.

[2] Research fellow of the Deutsche Forschungsgemeinschaft. Present address: Medizinische Poliklinik der Universität D-6900 Heidelberg, Hospitalstr. 3, W. Germany.

[3] Present address: Albert Einstein College of Medicine, Department of Medicine, 1300 Morris Park Ave., Bronx, New York 10461.

INTRODUCTION

When mechanics of contraction are studied in isolated cardiac muscle, the loading conditions under which the muscle contracts require careful consideration (Parmley, Brutsaert, and Sonnenblick, 1969; Parmley and Sonnenblick, 1969; Jewell and Rovell, 1973). Thus, as in skeletal muscle (Jewell and Wilkie, 1960), the duration of a given contraction in cardiac muscle is prolonged by increasing afterload (Parmley and Sonnenblick, 1969). If the load is abruptly changed from one level to another after the first half of the rising phase of the twitch by means of a stretch or a release, the contractile state of the same contraction may also be changed during the course of contraction, apparently owing to inactivation (Brady, 1965, 1968; Edman and Nilsson, 1972; Bodem and Sonnenblick, 1974). Moreover, inactivation of contraction after quick releases is greater in the cat than in the rabbit papillary muscle (Bodem and Sonnenblick, 1975).

The loading mode of a given contraction may also alter the contractile state of the subsequent contractions. Thus, if a cat papillary muscle is contracting isotonically with a small load, the contractile state increases in the subsequent beats until a stable level is reached (Parmley *et al.*, 1969). This phenomenon has been related to changes in excitation-contraction coupling (Parmley *et al.*, 1969; Kaufmann *et al.*, 1971) but could not be demonstrated in the rabbit (Jewell and Rovell, 1973).

Mechanical activity of the rabbit heart, measured either as tension-bearing ability or shortening ability of the contractile element, is slow in onset, changes continuously with time during contraction, and peaks near the peak of the isometric twitch (Brady, 1965, 1968; Edam and Nilsson, 1968, 1969, 1972). In the cat, however, mechanical activity of the contractile element rises early to a peak at about 18% of time to peak force and was considered to be constant and independent of time during the major portion of the rising phase of the twitch solely determined by the relationship of force, velocity of shortening, and length (Brutsaert, Claes, and Sonnenblick, 1971a,b; Brutsaert and Sonnenblick, 1971; Brutsaert and Henderson, 1973; Henderson, Claes, and Brutsaert, 1973a; Henderson, van Ocken, and Brutsaert, 1973b).

The contractile response to caffeine, an agent which alters excitation-contraction coupling in cardiac muscle (Blinks *et al.*, 1972), is also different in cat, rat, and rabbit (Bodem and Sonnenblick, 1975).

Accordingly, we have studied comparatively the mechanics of contraction in the isolated papillary muscles of the cat, rat, and rabbit to define more precisely possible differences in the mechanics of contraction in these species. In addition, the contractile responses were studied to changes in loading, caffeine, and altered pacing. Further, we attempted to relate our results to data obtained with different techniques which suggest differences in excitation contraction coupling in different mammalian species (Fabiato and Fabiato, 1972, 1973a,b). Some of these results reported here have been published in part elsewhere (Bodem and Sonnenblick, 1975).

METHODS

New Zealand rabbits (2–3 kg) were either sacrificed by fracture of the cervical spine or anesthetized with intravenous sodium pentobarbital (10 mg/kg). Wistar rats (200–300 g) were either decapitated or given intraperitoneal sodium pentobarbital (20 mg/kg). The performance of the isolated papillary muscles *in vitro* was not altered by the method of sacrifice. Mongrel adult cats were anesthetized with sodium pentobarbital (25 mg/kg i.p.). The heart was rapidly removed and the papillary muscles of the right ventricle (left ventricle in the rat) dissected free. The papillary muscles were then mounted vertically in a bath containing a modified Krebs bicarbonate solution of the following composition (mM): Na+, 142; K^+, 3.5; CA^{2+}, 2.5; Mg^{2+}, 1.2; Cl^-, 125.3; HCO_3^-, 24; $H_2PO_4^-$, 1.2; $SO_4^=$, 1.2; glucose, 5.0. This solution was bubbled with a mixture of 95% O_2 and 5% CO_2 and maintained at a constant temperature of 30 ± 0.5°C at pH 7.4.

The muscles were stimulated with rectangular pulses 5 msec in duration and about 10% above threshold, provided through two platinum electrodes arranged longitudinally along opposite sides of the muscle. The basic rates of electrical stimulation were, unless otherwise stated: rats, 6/min; cats, 12/min; rabbits, 36/min. The rates of stimulation used in the muscles of the different species were selected to provide a stable mechanical performance of the muscles over several hours while producing as much active force as possible.

The lower nontendinous end of the muscle was held by a spring-loaded stainless steel clip which was directly attached by cement to a DSK-3BB4-103 (DSC Inc., Bellevue, Wa.) force transducer which was immersed in the bath solution. The transducer was connected via a plastic and a metal bar to a Palmer stand which allowed fine length adjustments of the transducer. The tendinous end of the muscle was tied by a knot of surgical 4-0 braided silk to a carefully straightened nonmagnetic stainless steel wire (mass 40 mg) extending upwards to a lever system which was mounted on a second Palmer stand. In all experiments, the preload on the muscle was adjusted so that the initial muscle length corresponded to L_{max}, the length where actively developed force was maximum. The muscles were allowed to stabilize while contracting isometrically for at least 2 hours.

Essentially the same mode of altering load as described from this laboratory (Parmley *et al.*, 1969) has been used to measure the effect of altered loading on subsequent contractions of the papillary muscles of the three mammalian species. In short, the muscle was allowed to stabilize contracting isometrically, then it was made to contract isotonically for a period of time until a stable performance was obtained under isotonic conditions, and then the muscle was again made to contract isometrically in subsequent beats. The changes which occurred during these maneuvers in the developed isometric force, in the duration of the isometric and isotonic contractions, in the rate of force development, in the extent of shortening, and in the velocity of shortening were measured.

The electromagnetic lever system, described in detail elsewhere (Brutsaert *et al.,* 1971a; Bodem and Sonnenblick, 1974), was designed to measure shortening and also to impose a force on the muscle. The lever, 30 mm long, was fashioned from magnesium and attached to a coil, which was suspended in the field of a permanent magnet. The total equivalent moving mass of the lever, coil, and connections was 225 mg. The current through the coil determined the load. Displacement of the lever was measured with a photoelectric transducer which was linear over a range of 2 mm. Velocity of displacement of the lever was determined with an active differentiator. The electrical stimulus, force, shortening, and velocity of shortening were simultaneously displayed relative to time on a storage oscilloscope (Tektronix 5103) and a multichannel oscillograph (Sanborn 7700). On the second oscilloscope (Tektronix 564B), velocity of shortening and force were simultaneously displayed versus shortening.

The muscle was made to contract isometrically when the lever was held against an upper stop by the current from one source or by an airjet (Sonnenblick, 1965, 1967) directed against the base of the lever. At a selected time, the lever could be released and the muscle permitted to shorten with a load that was set by a second current source. This release was triggered after various delays, either by switching off the first current source or by a solenoid-controlled valve which directed the airjet away from the lever. The timing of the events was controlled by the two units of a S88 Grass stimulator. The quick release was modified by damping in order to slow the rapid passive length change resulting from the recoil of the series elastic (SE) element, and thus to reduce the inertial oscillations of the lever, and deactivation of contraction. Damping was achieved electronically by feeding back the output voltage of the dl/dt signal of the length transducer to the coil of the lever. Mechanical damping was obtained by modifying the duration of time to direct the airjet away from the lever. The time required to deflect the air from the lever is dependent on the capacitance and elasticity of the tubing connecting the valve with the jet (Sonnenblick, 1965, 1967). In order to achieve damping of the release, this was adjusted to occur over about 10 msec.

For the purpose of calculation, a two-component muscle model was assumed, consisting of a contractile element (CE) and a series elastic element (Hill, 1938). Resting force at L_{max} did not exceed 10–20% of total force. Velocities were measured after quick releases at about 7–10% below L_{max}, where resting tension was negligible. The parallel elastic (PE) element was unloaded at the time of velocity measurements. Consequently, the CE was assumed to be shortening against total load and the influence of the PE was neglected. The rapid shortening after the quick release of the isometrically contracting muscle reflects primarily the amount of stretch of the SE at that time and, thus, the internal shortening of the CE during isometric contraction before the release. Subsequently, shortening reflects the CE alone. The instantaneous velocity of shortening, at constant CE length, after quick releases to a given load was used to define the intensity of the shortening ability as a function of time during contraction (Edam and Nilsson, 1968; Bodem and Sonnenblick, 1974). The constant length of the CE at which all velocity measurements were made corresponded to that to which the CE had shortened at 15–20 msec after a damped quick release made at the peak of the isometric twitch. The time course of the shortening ability in a particular twitch was then given by the velocities of shortening of the CE after damped quick releases to a constant load at multiple times after stimulation.

Unloaded velocity of shortening was determined as a function of time after stimulation after damped quick releases to zero load of the muscle which was contracting either isometrically (Bodem and Sonnenblick, 1975) or isotonically with the preload (Brutsaert *et al.*, 1971a). The time when the peak velocity of shortening at zero load was attained was measured relative to time to peak force.

At various delays after stimulation, "load clamps" of variable magnitude and duration were applied on the preloaded contracting muscle as described in detail by Brutsaert *et al.* (1971b). By displaying simultaneously shortening and force versus time, and velocity of shortening and force versus shortening in the phase plane it is possible to assess with the "load clamp" technique the relationship of force-length-velocity of shortening at any time during contraction. Since after a "load clamp" the muscle may reach the same length with the same load at different times during contraction, it is possible to measure the effect of the variable time on the force-velocity-length relationship during contraction (Brutsaert *et al.*, 1971b).

The force-frequency relationship in the range of 12–60/min was determined in four cat and eight rabbit papillary muscles. The muscles were mounted vertically between stainless steel clips, which were attached via stainless steel rods to a force transducer and a rigid metal bar, respectively. Resting force, active force, and its first derivative dF/dt were recorded.

RESULTS

Force-frequency-relation

It is well known that the force-frequency relation of rat ventricular muscle differs greatly from that of other mammalian heart muscles, in that force decreases upon increase of the driving rate, at least in a range of 6–60 beats/min. Figure 1 shows the differing force-frequency relation of the cat and rabbit

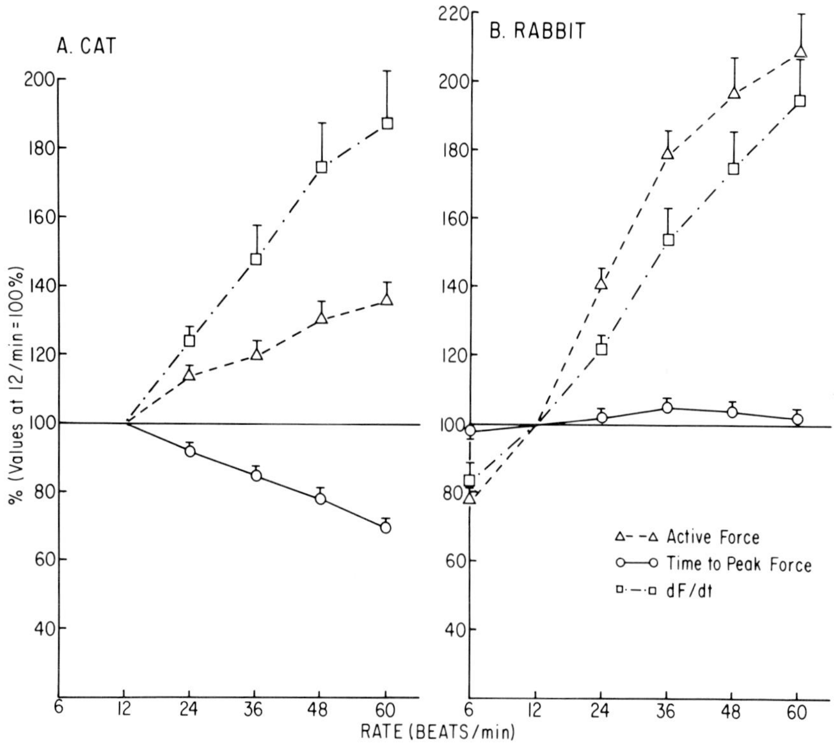

Figure 1. Force-frequency relations in cat (*N*=4) and rabbit (*N*=8) papillary muscle. The percentage change of isometric active force, rate of force development (dF/dt), and time to peak force are shown as functions of the frequency of contraction; values are means + 1 SEM (*vertical bars*). Values measured at 12/min = 100%. Temperature, 30°C. Figure 1 and parts of Figures 4,5, and 9 are taken from Bodem and Sonnenblick (1975), with permission of the publisher.

papillary muscle. In the cat (*A*), active force increased by 35%, time to peak force decreased by 30%, and the rate of force development, dF/dt, increased by 90% when the stimulation rate was increased from 12 to 60/min. However, in the rabbit (*B*), active force increased by 110%, while time to peak force actually increased slightly. The rate of force development increased to a similar level in the rabbit as in the cat when the pacing rate was changed from 12 to 60/min. Of note, the average developed force at a rate of 12/min was smaller in the rabbit (2.4 g/mm^2) than in the cat (4.6 g/mm^2). Nevertheless, not only the relative but also the absolute increase in developed force was greater in the rabbit than in the cat, presumably because the time to peak force did not decrease, as is usually seen in mammalian cardiac muscle (Koch-Weser and Blinks, 1963).

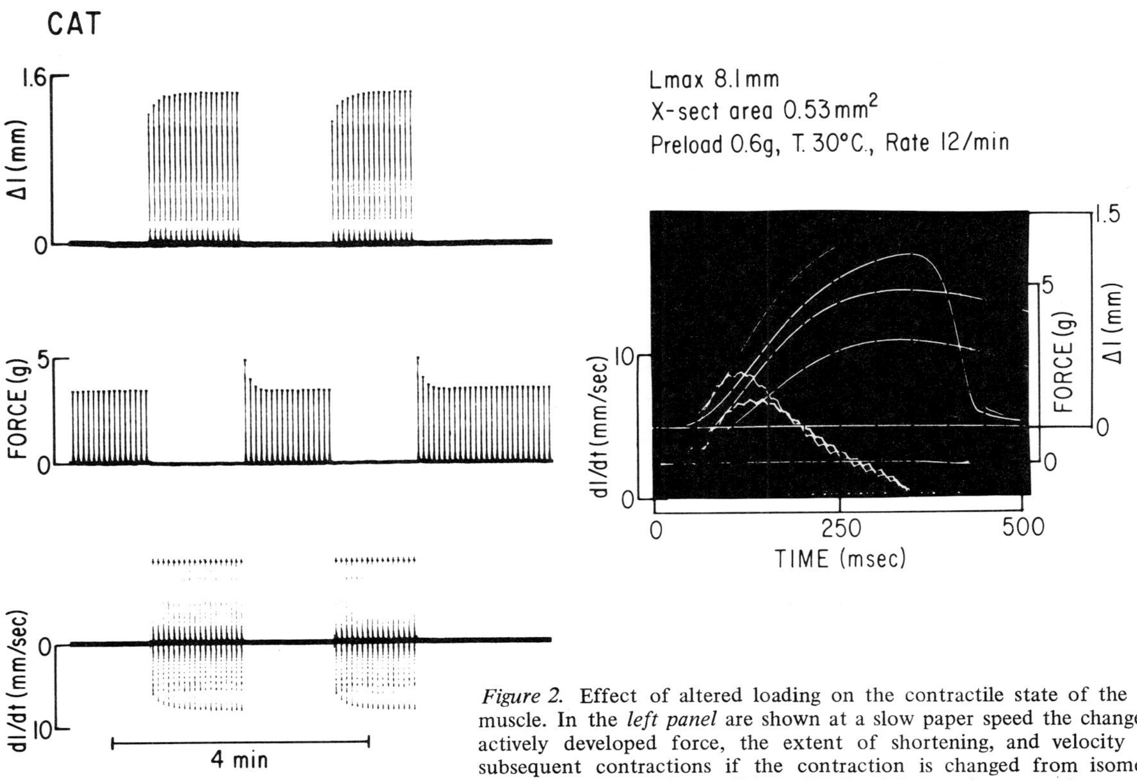

Figure 2. Effect of altered loading on the contractile state of the isolated cat papillary muscle. In the *left panel* are shown at a slow paper speed the changes which occur in the actively developed force, the extent of shortening, and velocity of shortening in the subsequent contractions if the contraction is changed from isometric to isotonic and returned to isometric conditions. In the *right panel* are superimposed the stable isometric and isotonic contractions and the first isometric and isotonic contractions immediately after the change in loading had been made.

Effects of alteration in loading on the contractile state of subsequent contractions

Figure 2 shows the response of a representative cat papillary muscle to changes in loading from isometric to isotonic and vice versa. At a low paper speed (left panel), the series started from stable isometric contractions. When the muscle was made to contract isotonically with only the preload, the extent of shortening (upper trace) and the velocity of shortening (lower trace) gradually increased until, after about eight contractions, a new stable level of contractile state was again reached. If the muscle was then made to contract isometrically again, it developed considerably more force than under the previous stable isometric conditions. This state of augmented contractility decayed in the subsequent isometric beats. At a faster time base (right panel) are shown superimposed the stable isometric beat, the first isotonic preloaded beat, the stable isotonic beat, and the first isometric beat after change to the isometric contraction mode. One notes that, in the potentiated state, the duration of the isotonic and isometric contraction and the rate of force development were increased. Thus, in the potentiated state, at any given time during the rising phase of contraction, the active force was greater than in the stable beat, while time to peak force was little changed (Parmley *et al.*, 1969).

In rabbit myocardium, the contractile response to altered loading was different from that in the cat (Fig. 3). When the isometrically contracting muscle was made to contract isotonically (left panel, slow paper speed), the extent of shortening increased only slightly, while the velocity of shortening did not change. At return to isometric conditions, the first isometric contraction developed more active force than in the previous stable condition. As seen in the cat, the state of increased force decayed in the subsequent beats. The middle panel shows that this phenomenon was reproducible at each change of the mode of contraction. At the faster time base in the right panel, it is apparent that the extent of preloaded shortening was little changed; however, the duration of contraction was relatively more prolonged than in the cat during the potentiated state. Of note, the potentiated isometric contraction was increased in duration owing to a prolongation of both time to peak force and the relaxation phase and not to an increase in the velocity of shortening. In the subsequent contractions, developed force transiently fell below the stable level. Thus, in the rabbit, time to peak force increased while the rate of force development and active force was not augmented at any time in the rising phase of contraction. This stands in contrast to the observations in the cat.

Load and species dependency of the time course of the shortening ability

In Figure 4 is shown the effect of loading on the time course of the shortening ability relative to the isometric force development in the cat (*A*) and in the rabbit (*B*). The shortening ability has been determined as the velocity of

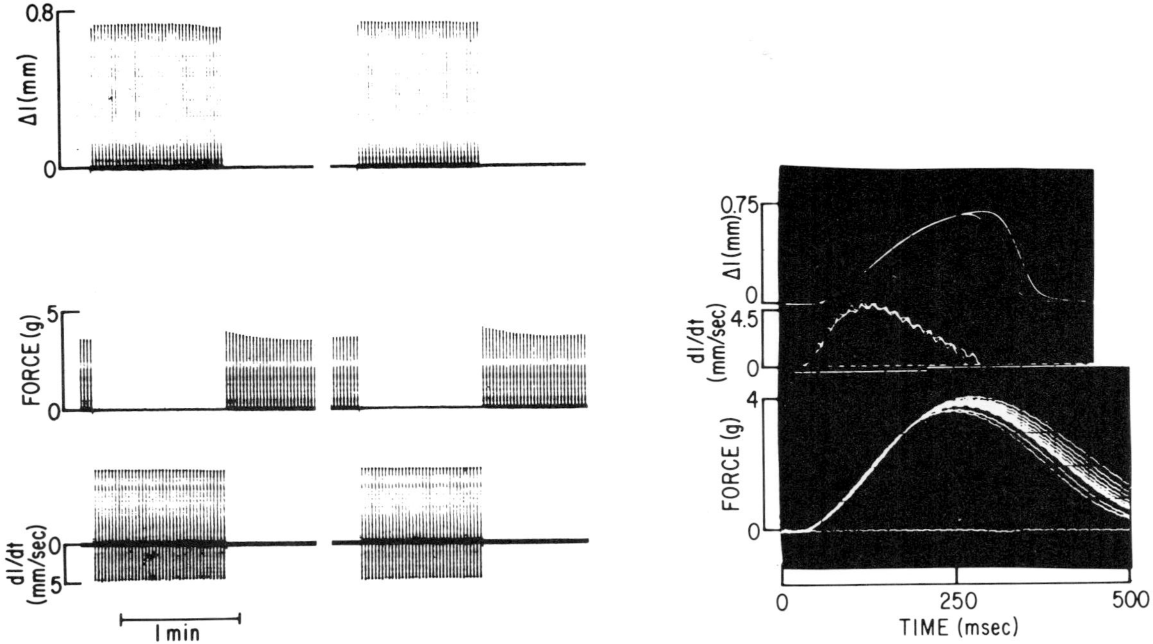

Figure 3. Effect of altered loading on the contractile state of the isolated papillary muscle of the rabbit; same presentation as in Figure 2. In the *right panel,* after return from isotonic to isometric conditions, the following isometric contractions are shown superimposed until a stable contractile state was reached again. L_{max} 5.0 mm; X-section area, 0.62 mm^2; preload 0.5 g, temperature, 30 C, rate, 36/min.

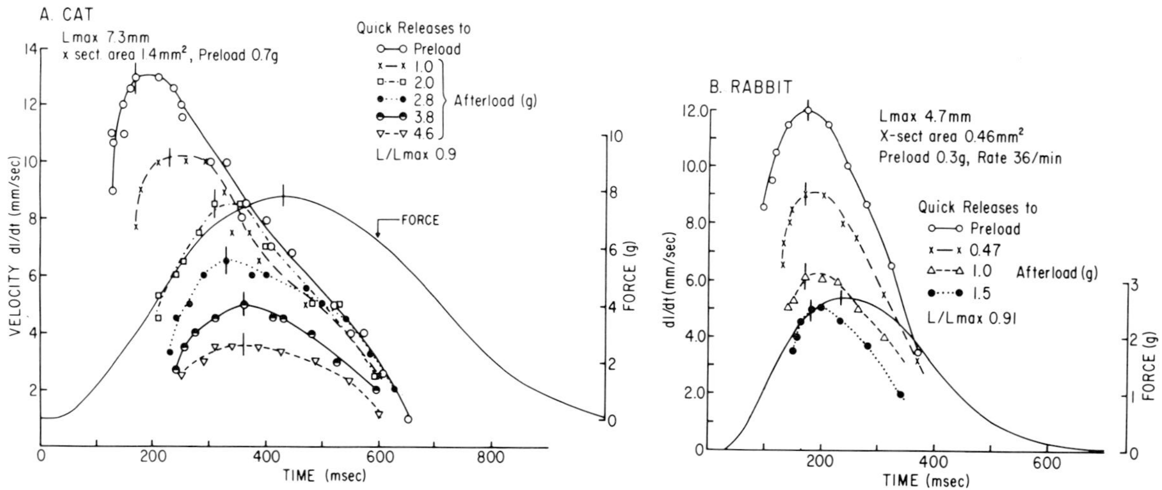

Figure 4. Effect of loading on the time course of the ability to shorten during contraction relative to isometric force development, in the cat and rabbit papillary muscle. Each *point* represents the velocity of shortening corrected for constant contractile element length, L/L_{max}, after damped quick releases to the preload and different afterloads. The velocities measured after multiple releases throughout the twitch define the time course of the shortening ability. The key of the symbols used is noted on the figure.

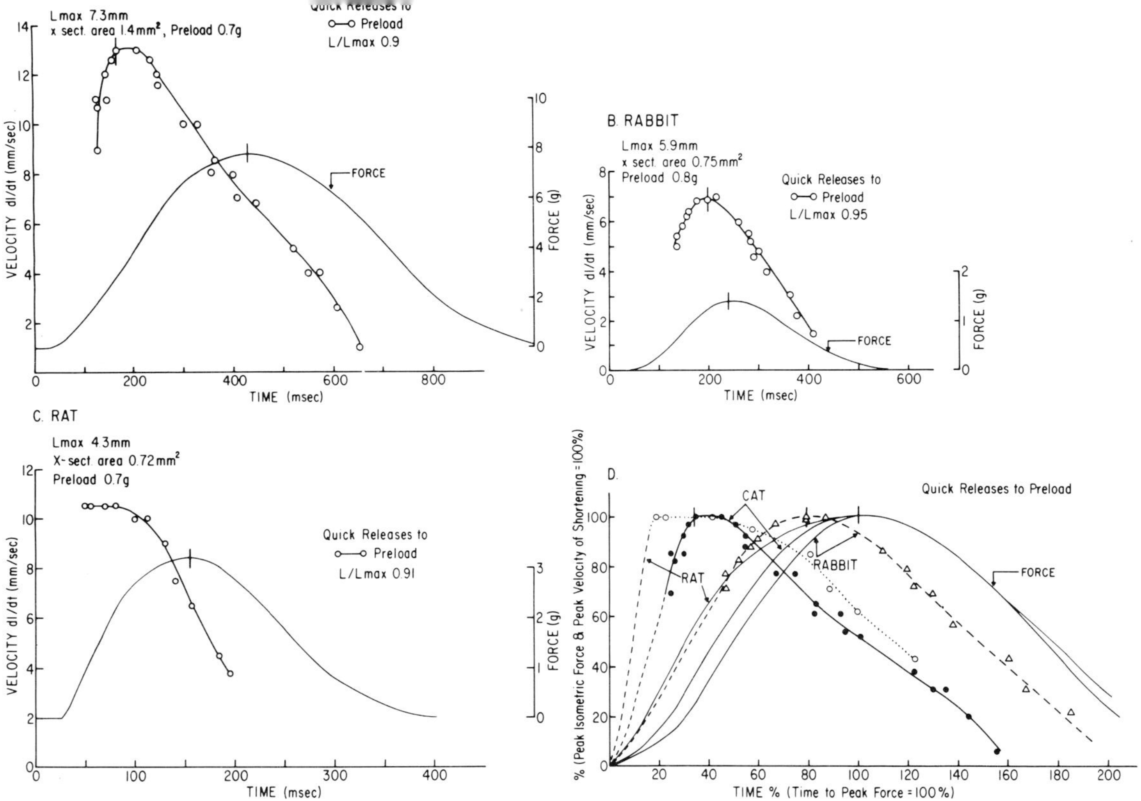

Figure 5. Time course of the ability to shorten with the preload relative to isometric active force development in the cat, rabbit, and rat papillary muscle. In *panels A, B,* and *C,* same presentation as in Figure 4. In *panel D* the active force and velocities of shortening of *panels A, B,* and *C* at any time during contraction have been expressed as percentages of peak force and peak velocity of shortening on the *ordinate.* On the *abscissa,* the time scale has been normalized as a percentage of the time to peak force of each contraction. Rate: 12/min. The *dashed lines* indicate the estimated onset of the shortening ability.

shortening after a quick release measured at a constant load corrected for a constant contractile element (CE) length. In the cat, the shortening ability with the preload alone rose to a peak early in contraction at about 30% of time to peak force and then declined rapidly. The shortening ability with increasing afterloads rose and declined more slowly, while the peak was attained progressively later in contraction, so that the curves actually converged in the relaxation phase. In the rabbit (*B*), the shortening ability measured with the preload peaked later in contraction, at about 70% of time to peak force; as the afterload was increased, the peak of shortening ability was not shifted to the right, while the rate of rise and decline was little altered. Thus, in the rabbit, the curves with different loads did not converge to the same extent in the relaxation phase of contraction as was noted with the cat.

The ability to shorten with the preload alone was studied as a function of time during contraction, using the velocity of shortening of the CE after quick release, and related to isometric force development (Fig. 5). In this experiment, each muscle was stimulated at the same rate of 12/min. In the cat (*A*), the peak ability to shorten was reached at about 30% of time to peak force. In the rabbit (*B*), the ability to shorten rose more slowly, reaching a peak at about 75% of the time to peak force, and declined more slowly as compared with the cat. However, in the rat, (*C*), shortening ability rose to a peak considerably faster than in the former species. These results have been superimposed in panel *D*, where velocity of shortening and isometric force at any time are expressed as percentages of peak velocity and peak developed force on the ordinate; the time on the abscissa is expressed as a percentage of time to peak force of each contraction. This figure illustrates the rapid attainment of the peak shortening ability in the rat, while in the cat shortening ability rose somewhat slower than in the rat, but considerably faster than in the rabbit. The rate of rise of the shortening ability correlated with the rate of force development in the three species; in the rabbit, normalized force showed an inflection at the beginning of contraction. Furthermore, the decline of the shortening ability was more delayed in the rabbit than in other species.

Unloaded velocity of shortening (Brutsaert *et al.*, 1971a) has been measured as a function of time during contraction to exclude possible effects of different preloads. In Figure 6, a representative cat papillary muscle, contracting isotonically with the preload (left panels) or isometrically (right panels), has been released to zero load at progressively later times in contraction. The unloaded velocity of shortening reached a peak at about 30 and 35% of time to peak force, after releases from the isotonically or isometrically contracting muscle, respectively. Peak unloaded velocity of shortening increased only by 11% above the level measured after the earliest release. If the releases were made from the preloaded shortening muscle (left panels), the velocities measured later in time could be determined only at short length where the preloaded shortening velocity had already declined.

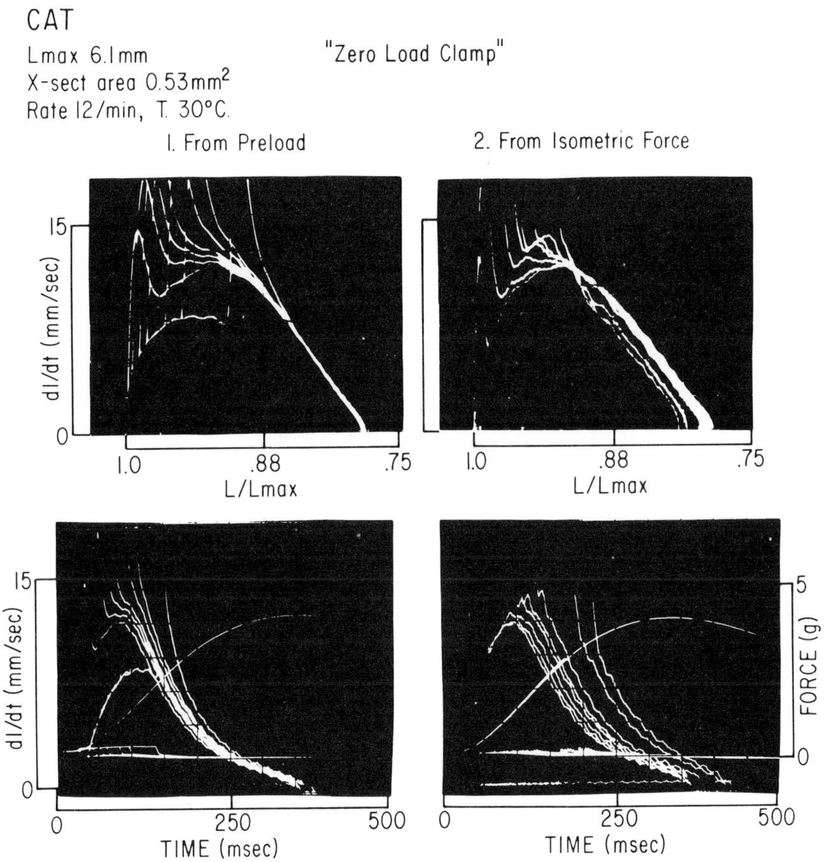

Figure 6. Velocity of shortening at zero load as a function of time after stimulation, in the cat papillary muscle. In *left panels,* releases to zero load have been performed from the preloaded shortening muscle at increasing times after stimulation. In the *right panels,* the releases were made from the isometrically contracting muscle. In the *upper panels,* velocity of shortening is displayed relative to shortening and, in the *lower panels,* relative to time together with the isometric twitch.

Unloaded velocity of shortening was measured in the rabbit (Fig. 7) after releases from the isometrically contracting muscle in order to prevent the muscle from shortening to short lengths where velocity of shortening rapidly declines. This would have occurred in the rabbit, since the shortening ability peaked later in contraction (Fig. 5). Figure 7 shows an experiment in a representative rabbit papillary muscle. Unloaded velocity of shortening reached the peak only at about 55% of time to peak force. Moreover, in the rabbit, peak unloaded velocity increased by about 45% above the velocity measured after the earliest release. Thus, in the rabbit, unloaded velocity of shortening peaked late and

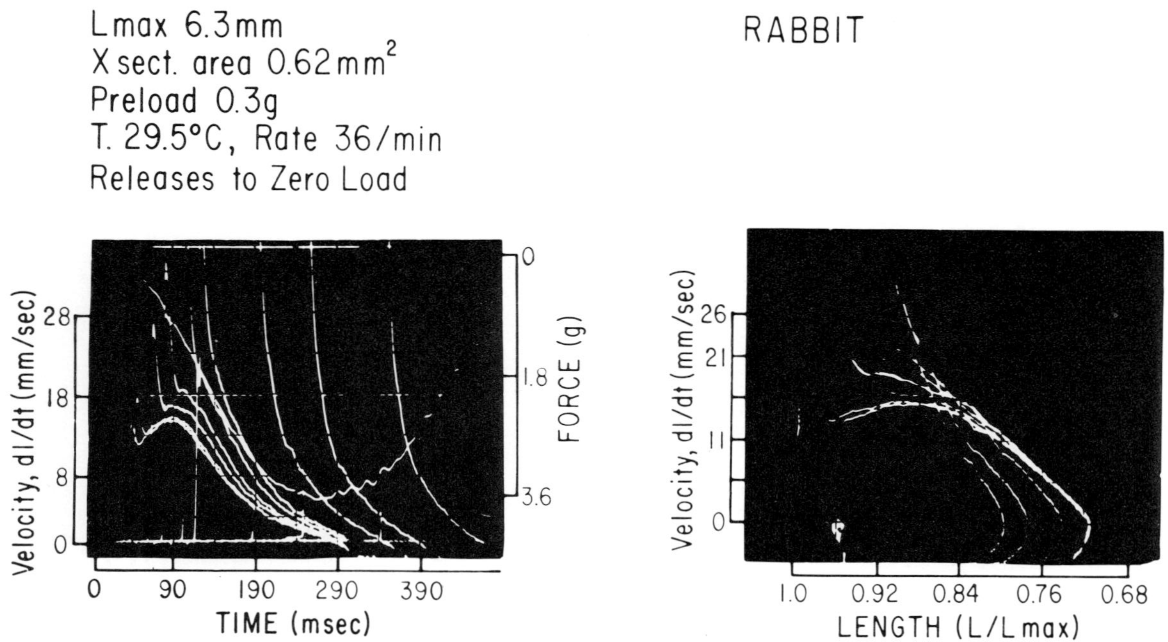

Figure 7. Velocity of shortening at zero load as a function of time after stimulation, in the rabbit papillary muscle; same presentation as in Figure 6. The unloaded velocities are displayed simultaneously versus time and versus shortening in the *left* and *right panels*, respectively. The velocities after multiple releases of the isometrically contracting muscle have been superimposed.

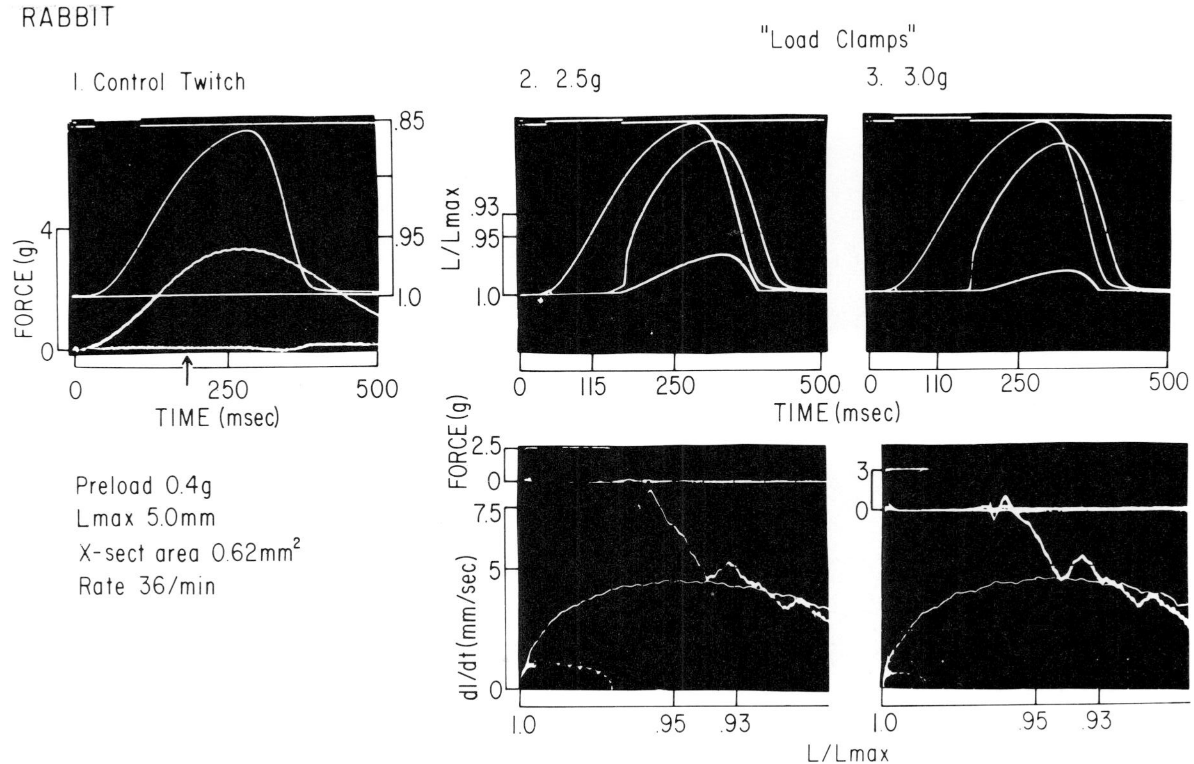

Figure 8. Effects of "load clamping" on the time course of the shortening ability, in the rabbit papillary muscle. *Panel 1:* An isometric and a preloaded isotonic contraction have been superimposed. *Panels 2* and *3:* A preloaded contraction, a contraction in which, 40 msec after stimulus for a duration of 130 msec, a load of 2.5 and 3 g, respectively, was imposed on the preloaded shortening muscle. In a third contraction, the muscles carries the load throughout the twitch. The three contractions have been displayed as shortening relative to time in the *upper panels,* and as velocity of shortening relative to shortening in the *lower panels.* The *arrow* in the *left panel* denotes the time where the peak shortening velocity has been measured after the "load clamps" in *panels 2* and *3.*

increased to a much greater extent with time than was observed in the cat. The results in the rat were similar to those obtained in the cat.

A relatively new technique, called "load clamping," developed by Brutsaert *et al.* (1971b), has been used to study the shortening ability as a function of time. "Load clamps" of variable magnitude and duration have been imposed on the preloaded shortening muscle of the cat, rat, and rabbit at variable delays after stimulation. Essentially the same results were obtained with this technique as with the quick release technique with releases to loads above preload and with "load clamps" to zero load. The shortening ability of the contractile element measured as velocity of shortening as a function of time at constant load and length rose to a peak early in rat, somewhat later in cat, but was markedly delayed in the rabbit. Figure 8 shows an experiment in a representative rabbit papillary muscle. In the left panel, the isometric contraction and isotonic shortening with the preload are superimposed. In the middle and right panels, load clamps of 2.5 and 3 g have been imposed on the preloaded shortening muscle at a delay of 40 msec after stimulus and then maintained for 130 msec, so that the muscle after "load clamp" reached the same length later in contraction. In the lower panels, in which velocity of shortening is displayed relative to shortening, it is apparent that velocity of shortening after the load clamp had increased by about 50% in the time which had elapsed between the control and the clamped contraction. The arrow in the left panel indicates the time where the velocity of shortening after the clamped contraction has been measured, which was about 65% of time to peak force.

Effect of caffeine on the time course of the ability to shorten and isometric active force development

In Figure 9 are shown the effects of caffeine on the time course of the isometric twitch and the ability to shorten with the preload in the papillary muscles of the cat, rat, and rabbit. The control contractions are shown in Figure 5. The response to caffeine of the cat, rabbit (10 mM caffeine), and the rat (1 mM caffeine) were different. In the cat, the rate of force development (dF/dt) rose slowly to a maximum later in contraction, the time to peak force was prolonged, while peak active force was increased by about 10%. Shortening ability reached a maximum later in contraction, occurring at about 60% of the time to peak force. The peak velocity of shortening was increased by about 15%. In the rabbit (Fig. 9*B*), 10 mM caffeine led to a 170% increase in developed force and 85% increase in shortening velocity, effects which were much greater than in the cat; moreover, time to peak force and to peak shortening ability were not changed by caffeine. In contrast, in the rat, 10 mM caffeine led to a drastic decrease in developed force and shortening ability. With only 1 mM caffeine, developed

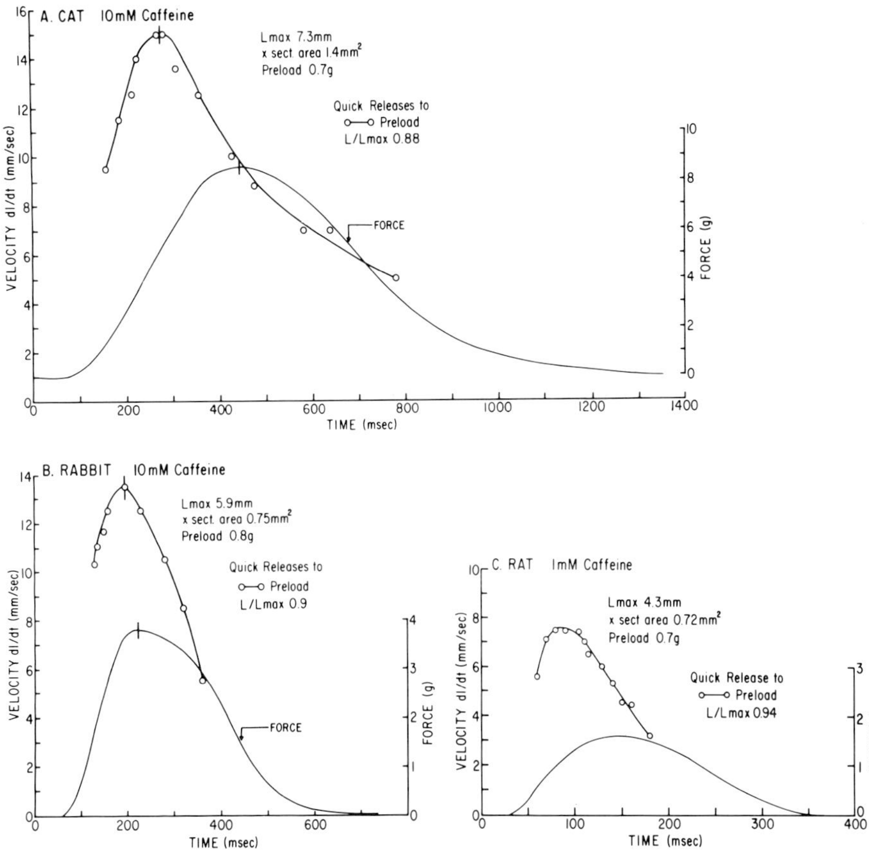

Figure 9. Effect of caffeine on the time course of the ability to shorten with the preload relative to isometric active force development, in the cat, rabbit, and rat papillary muscle; same presentation and same muscles as in Figure 5. Rate: 12/min.

force and shortening ability were decreased by 100 and 29%, respectively, while the time to peak force was slightly decreased and the time to peak shortening ability was increased.

In Figure 10, the effects of caffeine shown in Figure 9 have been normalized. Caffeine slowed the onset of the shortening ability in the cat and rat relative to force development, but not in the rabbit. In the presence of caffeine, the peak of the shortening ability in the cat, and less in the rat, is reached considerably later in contraction; in the rabbit, however, peak shortening ability reached the maximum at about the same time in the absence and presence of caffeine. Of note, the rate of force development was the same in the cat and rabbit in the

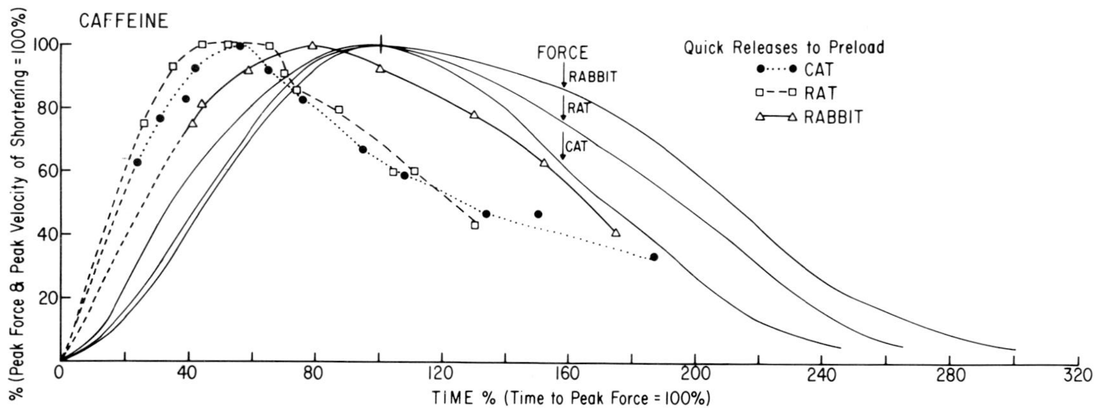

Figure 10. The data shown in Figure 9 have been normalized in same way as shown in Figure *5D*. Active force and velocities of shortening at any time during contraction in the presence of caffeine have been expressed on the *ordinate* as percentages of the peak force and velocity of each contraction. On the *abscissa,* the time scale has been normalized as a percentage of the time to peak force of each contraction.

presence of caffeine, while the relaxation time was relatively prolonged in the rabbit.

Table 1 summarizes the relations of time peak shortening ability with the preload to time to peak force in the cat, rat, and rabbit. In the presence of caffeine, the time to reach peak shortening was prolonged by about a factor of 2 in the cat and rat, but not in the rabbit, while time to peak force was slightly increased in the cat and decreased in the rat; as a result, the ratio of these two parameters increased in the rat and cat but remained unchanged in the rabbit. In the cat, this ratio increased to a similar value in the presence of caffeine as that seen in the rabbit in the control state.

The effect of caffeine on the isometric twitch and on the preloaded isotonically shortening muscle of the rabbit, cat, and rat are shown in Figure 11. In the rabbit (upper panel), caffeine (10 mM) increased developed force, extent of shortening, and the preloaded shortening velocity, which peaked even earlier in time than in the control contraction; in the cat, however, preloaded peak shortening velocity decreased in the presence of caffeine; this was probably due to the delayed attainment of peak velocity of shortening at a time when the muscle had already shortened to a length where velocity falls off rapidly; the extent of shortening was increased as seen in the rabbit. In contrast, developed force, extent of shortening, and velocity of shortening with the preload decreased drastically in the rat in the presence of only 1 mM caffeine; moreover, the peak shortening velocity was attained later in time, as seen in the cat.

The afterloaded force-peak velocity relation in the cat was curvilinear in the control state (Fig. 12), which would have allowed a fairly accurate measurement of the velocity of shortening at zero load (V_{max}) by extrapolation to the ordinate. The values of V_{max} shown in Figure 12 have been directly measured after releases of the muscle to zero load. In the presence of 10 mM caffeine, velocities of shortening at small loads were lower than in the control state. This could result from the induced delay in development of the peak of shortening ability, so that measurements were made after the muscle had shortened to a length where velocity of shortening decreases rapidly with shortening (see Fig. 11), and at times when the shortening ability is far below the maximum. With heavier loads, velocity of shortening is somewhat higher than in the control; thus, the afterloaded force-peak velocity relation of the cat approached linearity in the presence of caffeine and extrapolation to the ordinate would have suggested that caffeine decreased V_{max}. However, direct measurement of V_{max} later in contraction revealed that caffeine increased V_{max} in the cat. In the rabbit (Fig. 12), the afterloaded force-velocity relation was more linear in the control state and was shifted in parallel upward and to the right by caffeine; extrapolations to the ordinate would not have allowed a measurement of V_{max} because, as shown in the presence of caffeine in the cat, velocities at low loads were measured at a time when the shortening ability was below the maximum and at short lengths

Table 1. Effect of caffeine on differences in time to peak shortening ability at preload relative to time to peak force in three mammalian species*

Species	*N*	Latent Period (LP) (msec)	Time to Peak Force (TPF)-LP (msec)	Time to Peak Shortening Ability (TTPA)-LP (msec)	$\frac{\text{TTPA-LP}}{\text{TPF-LP}}$
			Control		
Cat	17	33.7 ± 1.5	340.3 ± 6.9	103.0 ± 5.1	0.30 ± 0.02
Rat	5	19.0 ± 2.6	135.8 ± 4.5	29.4 ± 4.1	0.22 ± 0.03
Rabbit	8	33.8 ± 2.3	220.6 ± 7.4	154.4 ± 7.0	0.70 ± 0.03
			Caffeine (10 mM) ×		
Cat	8	69.5 ± 5.2	376.4 ± 10.2	262.9 ± 11.7	0.70 ± 0.03
Rat	4	30.0	111.6 ± 6.0	48.3 ± 6.0	0.43 ± 0.04
Rabbit	6	49.2 ± 9.1	206.6 ± 10.6	148.6 ± 10.9	0.72 ± 0.04

*Values are means ± 1 SEM, × rat: 1 mM; Latent period = time from stimulation to first evidence of active force or shortening.

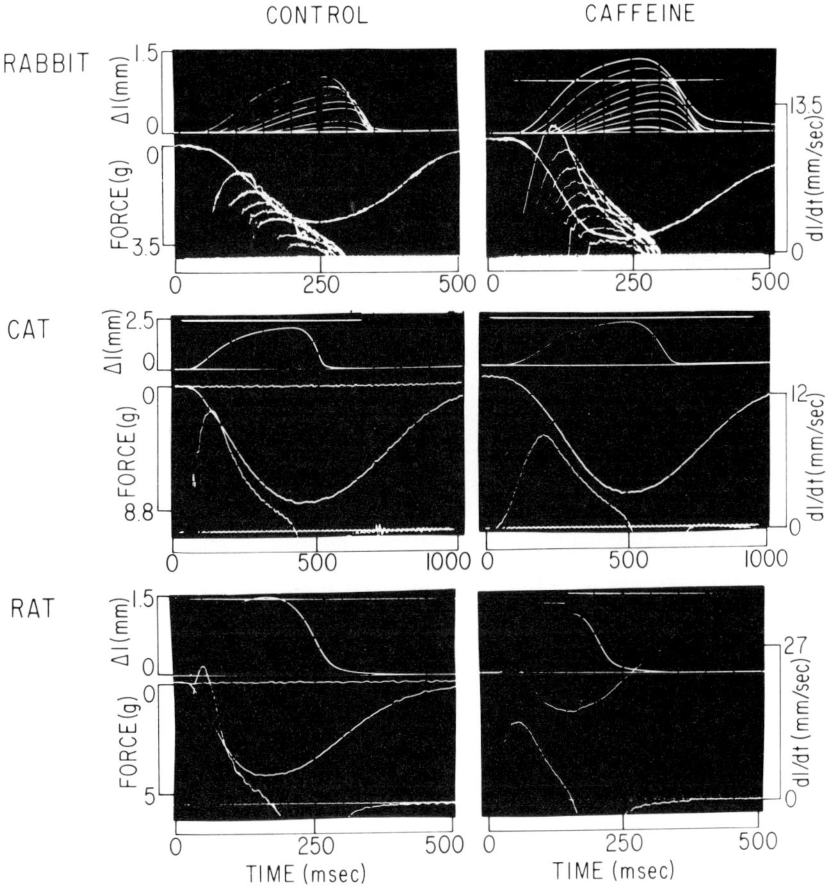

	Lmax (mm)	X-sect area (mm^2)	Rate (beats/min)	Preload (g)	T.
RABBIT	5.9	0.65	36	0.6	30°C.
CAT	7.3	1.4	12	0.7	" "
RAT	5.5	1.4	12	0.3	" "

Figure 11. Effect of caffeine on isometric and isotonic contractions of the rabbit, cat, and rat papillary muscles. In each panel, *upper trace,* shortening; *middle trace,* isometric force; *lower trace,* velocity of shortening. The isometric and isotonic contractions (preloaded in cat and rat, increasing afterloads in rabbit) have been superimposed. The baseline of the force trace of the rat was changed in the presence of caffeine (1 mM). Note that, in the rabbit, caffeine (10 mM) increased both preloaded velocity of shortening and isometric force; in the cat, isometric force increased and preloaded velocity decreased in the presence of 10 mM caffeine while, in the rat, both preloaded velocity of shortening and active force were reduced by 1 mM caffeine.

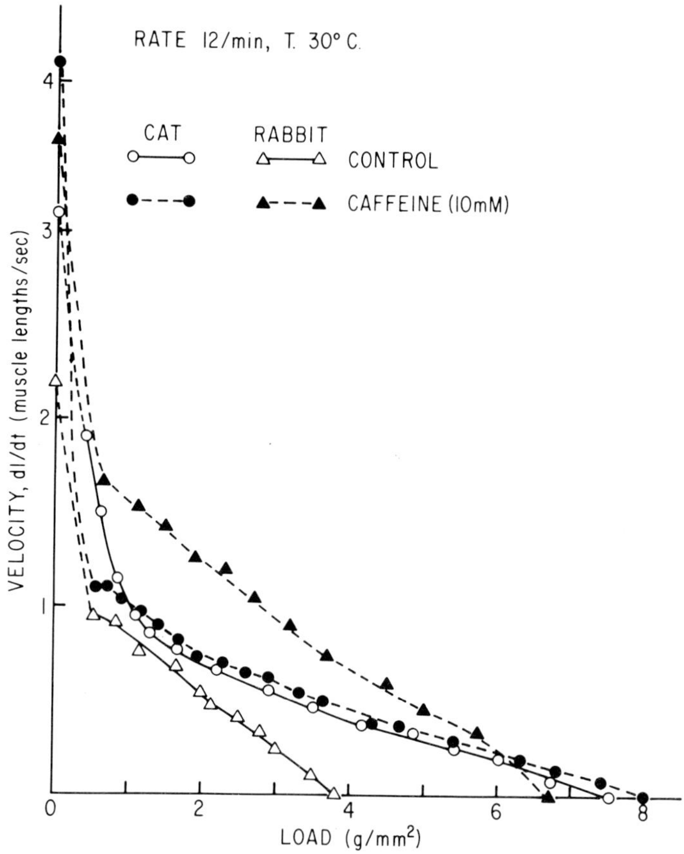

Figure 12. Effect of caffeine (10 mM) on the afterloaded force-peak velocity relation of the cat and rabbit. The velocities have been normalized as muscle lengths/sec and the force as g/mm^2. The velocity of shortening at zero load was directly measured after releasing the muscle to zero load. Curves have been fitted by eye.

because of the delayed peak of the shortening ability in the rabbit in the control state as well as in the presence of caffeine. In the rat, caffeine (1 mM) decreased V_{max} to a lesser extent than isometric force, so that the afterloaded force-velocity relation was shifted downward and to a greater degree to the left.

DISCUSSION

The present study has confirmed the fact that the mode of loading affects the mechanical activity of the same, as well as of the subsequent contractions in isolated cardiac muscle (Parmley and Sonnenblick, 1969; Parmley *et al.*, 1969).

Thus, the time course of the shortening ability in the cat measured as the velocity of shortening of the contractile element after quick releases at constant length peaked early in contraction and declined rapidly at low loads; with increasing loads, however, the shortening ability peaked later in time and declined more slowly (Sonnenblick, 1967; Bodem and Sonnenblick, 1974) so that the curves measured with different loads converged in the relaxation phase of contraction. This finding may result from abbreviation of the duration of contraction when shortening occurs with a small load (Parmley and Sonnenblick, 1969), and the greater deactivation when releasing to small loads due to a larger change in length and/or load with the release (Bodem and Sonnenblick, 1974). In the rabbit, shortening ability with small and heavier loads declined more slowly and almost in parallel, presumably owing to less deactivation after quick release (Bodem and Sonnenblick, 1975).

The effect of altered loading on the contractile performance of the cat papillary muscle in the subsequent beats was the same as that described in a previous study from this laboratory (Parmley *et al.*, 1969) and a more recent report from Jewell and Rovell (1973). The increased contractility after isotonic contractions at low loads in the cat has been explained by an increased Ca^{2+} availability for contraction, the mechanism by which this occurs remaining still obscure (Parmley *et al.*, 1969; Kaufmann *et al.*, 1971; Jewell and Rovell, 1973).

We could not, however, confirm the findings of Jewell and Rovell (1973) that altered loading does not affect the mechanical activity of subsequent contractions in the rabbit. Instead, we found that, in the rabbit, isotonic contraction with low loads led to an increased duration of subsequent isotonic and isometric contractions. Moreover, the active isometric force was increased after isotonic contractions solely because of the increase in time to peak force. Unlike in the cat, the rate of force development remained unchanged in the rabbit. This finding suggests that, in the rabbit, isotonic contractions prolong the duration of the Ca^{2+} availability while, at any time earlier than time to peak force, Ca^{2+} is not increased in its availability. This would be consistent with the assumption that, in the rabbit, an increased amount of Ca^{2+} is stored at sites, which, however, may not release this Ca^{2+} to the myoplasm early in contraction. The increase in time to peak force may then, however, be due to a prolonged supply of Ca^{2+} from these sites to the myoplasm and/or to a slower removal of Ca^{2+} from the myoplasm

In this context, the difference of the force frequency relationship in the rat, cat, and rabbit is of interest. When the pacing rate is increased in the rat, active force and time to peak force decrease (Hajdu, 1969); in the cat, active force increases and time to peak force decreases while, in the rabbit, time to peak force does not change and active force increases to a much greater extent than in the cat. The increase in active force has been explained ultimately by an increased Ca^{2+} availability resulting from an enhanced supply from the extracellular space entering the cell during the action potentials (Hajdu, 1969). The decrease in time to peak force has been attributed to a faster removal of Ca^{2+} from the myoplasm by the sarcoplasmic reticulum. Thus, the different force-frequency relation in the rabbit may be due to a greater Ca^{2+} influx during the action potential and/or a slower removal of Ca^{2+} from the myoplasm. Why the duration of contraction of the rabbit myocardium is not shortened when frequency is increased is unknown.

The ability to shorten at preload has been measured in terms of the shortening velocity after quick releases as a function of time. The results confirmed the data of previous studies that, in the rabbit, the ability to shorten of the contractile element corrected for constant length ("velocity active state") is slow in its onset and rises to a peak at about 70–80% of time to peak force (Edam and Nilsson, 1968, 1969, 1972; Bodem and Sonnenblick, 1975). However, the shortening ability measured after releases to the preload peaked much earlier in contraction at about 20% of time to peak force in the rat, and at about 30% of time to peak force in the cat.

In the cat, the shortening ability measured in terms of unloaded velocity of shortening or velocity of shortening after "load clamping" (Brutsaert *et al.*, 1971a,b) reached a peak rapidly at about 18% of time to peak force and then remained constant during the major portion of the rising phase of contraction, relatively independent of time and largely determined by the interrelation with force and length (Brutsaert and Sonnenblick, 1971; Brutsaert *et al.*, 1971b; Brutsaert and Henderson, 1973; Henderson *et al.*, 1973a, b). In addition, it was claimed that this was also true in the rabbit cardiac muscle (Brutsaert and Henderson, 1973). Using the same techniques, we found that peak shortening ability was attained at about 20% and 30% of time to peak force in the rat and cat, respectively. However, in the rabbit, the shortening ability was continually changing with time, reaching the peak at about 80% of time to peak force. In our study, the results obtained with "zero load clamping" and "load clamping" in the rat, cat, and rabbit were consistent with those obtained with the quick release technique; i.e., the shortening ability rose early to the peak in the rat, somewhat later in the cat, and markedly delayed near to time to peak force in the rabbit.

As a result of the slower onset of the shortening ability in rabbit papillary muscle as compared with rat and cat, peak isotonic velocities measured with small afterloads are low because they are determined early in contraction when mechanical activity is far below its maximum level. This finding explains why force-peak velocity curves from afterloaded isotonic contractions are not curvilinear or hyperbolic at low loads in the rabbit (Brady, 1965; Edman and Nilsson, 1972) whereas, in cat and rat papillary muscles, this has been shown to be the case (Sonnenblick, 1962; Henderson *et al.*, 1969).

In the presence of caffeine, the onset of the ability to shorten may be delayed in the cat and rat papillary muscle (Henderson *et al.*, 1973a; Bodem and Sonnenblick, 1975). The action of caffeine in cardiac muscle has been attributed to a partial inhibition of the sequestration of Ca^{2+} by the sarcoplasmic reticulum (SR) with a subsequent reduction of the Ca^{2+} which is released by the SR (Blinks *et al.*, 1972). Furthermore, caffeine is thought to increase the influx of Ca^{2+} across the sarcolemma (Blinks *et al.*, 1972). In the presence of caffeine, the peak of the ability to shorten is attained later in contraction in the cat papillary muscle, similar to the control state in the rabbit heart. As a result, in afterloaded contractions with small loads, the peak velocities of shortening are low since they are measured early in contraction when the shortening ability is far below the maximum. Extrapolations from afterloaded force-peak velocity curves to zero load would lead to an underestimation of the velocity of shortening at zero load (" V_{max} "). That " V_{max} " in the cat was increased in the presence of caffeine could be demonstrated by measuring " V_{max} " directly after releases to zero load at later times in contraction. In the rabbit, caffeine caused little alteration in the time when peak shortening ability occurred and when peak active force was obtained. This is consonant with the finding that caffeine did not increase time to peak force of the perfused rabbit interventricular septum (Shine and Langer, 1971). Thus, in the rabbit, the afterloaded force-peak velocity curve was not altered in its shape by caffeine and shifted almost in parallel upward and to the right.

In the rat, unlike the other species, 10 mM or even 1 mM caffeine reduced active force and velocity of shortening. The negative staircase and the high "rest state" contraction of rat ventricular muscle have suggested that the Ca^{2+} which initiates and maintains contraction comes mainly from internal stores such as the SR (Hajdu, 1969), while the Ca^{2+} coming from outside the cell or superficial sites would play a minor role in activation. Thus, an inhibition of the Ca^{2+} sequestration by the SR in the presence of caffeine will lead eventually to a reduction of the amount of Ca^{2+} the SR may release into the myoplasm upon membrane depolarization. This would explain the observed decrease in active force.

Thus, in cat and rat, caffeine would tend to increase the Ca^{2+} fraction coming from more superficial sites and to decrease the Ca^{2+} fraction deriving from the SR, so that in the presence of caffeine, the source of activating Ca^{2+} would become similar to that of the rabbit heart in the control state. We would hypothesize that the foregoing species differences in the time course of mechanical activity, in the contractile response to alterations in loading, in response to increasing frequency of contraction and to caffeine, are related to differences in excitation-contraction coupling in mammalian cardiac muscle. The results are consistent with the view that activating Ca^{2+} primarily derives from internal stores in the rat and from more superficial sites in the rabbit; in the cat, the fraction of activating Ca^{2+} from internal stores would be greater than the fraction coming from superficial sites.

Studies in the interventricular septum of the rabbit have suggested a superficial location of activating Ca^{2+} (Shine, Serena, and Langer, 1971; Langer, 1973). However, in cat ventricular muscle, electrophysiological studies using the voltage clamp technique supported the view that early after excitation activating Ca^{2+} was abruptly released from internal stores of the cell (most likely the SR) (New and Trautwein, 1972). In skinned ventricular fibers of rat and cat, contraction may be induced by a regenerative release of Ca^{2+} from the SR (Fabiato and Fabiato, 1972, 1973a, b). It was more difficult to obtain a regenerative release of Ca^{2+} in the rabbit, implying that contraction may be induced by Ca^{2+} from more superficial sites (Fabiato and Fabiato, 1972, 1973a, b). Differences in the delay of Ca^{2+} release to the myoplasm and in the reaction time of the troponin-tropomyosin complex may also be factors. Isolated SR of the rat and rabbit cardiac muscle differ considerably in the rate of uptake and release of Ca^{2+}; this difference could well explain the differences in the mechanics of contraction (W. G. Nayler, personal communication). Our hypothesis is an attempt to correlate our data with results obtained with techniques which may allow a more precise definition of the role of Ca^{2+} in excitation-contraction coupling of cardiac muscle. Clearly, mechanics of contraction can be only indirect indicators of Ca^{2+} pathways during contraction.

The qualitative differences in the mechanics of contraction in the ventricular myocardium of different species help to explain some of the controversies in the current literature concerning the afterloaded force-peak velocity curve, and the time course of mechanical activity in mammalian cardiac muscle. The results clearly indicate that any measurements of myocardial mechanics and contractility need to consider these differences among mammalian species (Bodem and Sonnenblick, 1975). The consideration of species-dependent differences in the mechanics of contraction is also important for the measurements of the effects of inotropic interventions in mammalian cardiac muscle. Evaluations of myocardial mechanics which do not consider species differences in mammalian cardiac muscle may be misleading (Jewell and Blinks, 1968; Blinks and Jewell, 1972).

REFERENCES

BLINKS, J. R., and B. R. JEWELL. 1972. The meaning and measurement of myocardial contractility. *In* D. H. Bergel (ed.), Cardiovascular Fluid Dynamics, pp. 225–260, Academic Press, New York.

BLINKS, J. R., C. B. OLSON, B. R. JEWELL, and P. BRAVENY. 1972. Influence of caffeine and other methylxanthines on mechanical properties of isolated mammalian heart muscle. Circulat. Res. 30: 367–392.

BODEM, R., and E. H. SONNENBLICK. 1974. Deactivation of contraction by quick releases in the isolated papillary muscle of the cat: Effects of lever damping, caffeine and tetanization. Circulat. Res. 33: 214–225.

BODEM, R., and E. H. SONNENBLICK. 1974b. Mechanical activity of mammalian heart muscle: variable onset, species differences and the effect of caffeine. Am. J. Physiol. 228: 250–261.

BRADY, A. J. 1965. Time and displacement dependence of cardiac contractility: Problems in defining the active state and force velocity relation. Fed. Proc. 24: 1410–1420.

BRADY, A. J. 1966. Onset of contractility in cardiac muscle. J. Physiol. (Lond.) 184: 560–580.

BRADY, A. J. 1968. Active state in cardiac muscle. Physiol. Rev. 48: 570–600.

BRUTSAERT, D. L., V. A. CLAES, and M. A. GOETHALS. 1973. Effect of calcium on force-velocity-length-relations of heart muscle of the cat. Circulat. Res. 32: 385–392.

BRUTSAERT, D. L., V. A. CLAES, and E. H. SONNENBLICK. 1971a. Velocity of shortening of unloaded heart muscle and the length-tension relation. Circulat. Res. 29: 63–75.

BRUTSAERT, D. L., V. A. CLAES, and E. H. SONNENBLICK. 1971b. Effect of abrupt load alterations on force-velocity-length and time relations during isotonic contractions of heart muscle: load-clamping. J. Physiol. (Lond.) 216: 319–330.

BRUTSAERT, D. L., and A. H. HENDERSON. 1973. Time course of mechanical activation in cardiac muscle. Europ. J. Cardiol. 1/2: 201–208.

BRUTSAERT, D. L., and E. H. SONNENBLICK. 1971. The early onset of maximum velocity of shortening in heart muscle of the cat. Pflügers Arch. 324: 91–99.

EDMAN, K. A. P., and E. NILSSON. 1968. The mechanical parameters of myocardial contraction studied at a constant length of the contractile element. Acta Physiol. Scand. 72: 205–219.

EDMAN, K. A. P., and E. NILSSON. 1969. The dynamics of the inotropic change produced by altered pacing of rabbit papillary muscle. Acta Physiol. Scand. 76: 236–247.

EDMAN, K. A. P., and E. NILSSON. 1972. Relationships between force and velocity of shortening in rabbit papillary muscle. Acta Physiol. Scand. 85: 488–500.

FABIATO, A., and F. FABIATO. 1972. Excitation-contraction coupling of isolated cardiac fibers with disrupted or closed sarcolemmas. Calcium-dependent cyclic and tonic contractions. Circulat. Res. 31: 293–307.

FABIATO, A., and F. FABIATO. 1973a. Regenerative release of calcium within skinned cardiac cells of different species (abstr.). Physiologist 16: 308.

FABIATO, A., and F. FABIATO. 1973b. Activation of skinned cardiac cells. Subcellular effects of cardioactive drugs. Europ. J. Cardiol. 1/2: 143–155.

HAJDU, S. 1969. Mechanism of the Woodworth phenomenon in heart and skeletal muscle. Am. J. Physiol. 216: 206–216.

HENDERSON, A. H., D. L. BRUTSAERT, W. W. PARMLEY, and E. H. SONNENBLICK. 1969. Myocardial mechanics in papillary muscles of the rat and cat. Am. J. Physiol. 217: 1273–1279.

HENDERSON, A. H., V. A. CLAES, and D. L. BRUTSAERT. 1973a. Onset of maximum unloaded velocity of shortening in mammalian heart muscle as influenced by caffeine and other inotropic interventions: time course of activation. Circulat. Res. 33: 291–302.

HENDERSON, A. H., E. van OCKEN, and D. L. BRUTSAERT. 1973b. A reappraisal of force-velocity measurements in isolated heart muscle preparations. Europ. J. Cardiol. 1/1: 105–118.

HILL, A. V. 1938. Heat of shortening and dynamic constants of muscle. Proc. Roy. Soc. Ser. B. 126: 136–195.

JEWELL, B. R., and J. R. Blinks. 1968. Drugs and the mechanical properties of heart muscle. Ann. Rev. Pharmacol. 8: 113–130.

JEWELL, B. R., and J. M. ROVELL. 1973. Influence of previous mechanical events on the contractility of isolated cat papillary muscle. J. Physiol. 235: 715–740.

JEWELL, B. R., and D. R. WILKIE. 1960. Mechanical properties of relaxing muscle. J. Physiol. (Lond.) 152: 30–47.

KAUFMANN, R. L., M. J. LAB, R. HENNEKES, and H. KRAUSE. 1971. Feedback interaction of mechanical and electrical events in the isolated mammalian ventricular myocardium (cat papillary muscle) Pflüger's Arch. 324: 100–123.

KOCH-WESER, J., and J. R. BLINKS. 1963. The influence of the interval between beats on myocardial contractility. Pharmacol. Rev. 15: 601–642.

LANGER, G. A. 1973. Heart: Excitation-contraction coupling. Ann. Rev. Physiol. 35: 55–86.

NEW, W., and W. TRAUTWEIN. 1972. The ionic nature of slow inward current and its relation to contraction. Pflügers Arch. 334: 24–39.

PARMLEY, W. W., D. L. BRUTSAERT, and E. H. SONNENBLICK. 1969. Effects of altered loading on contractile events in isolated cat papillary muscle. Circulat. Res. 24: 521–532.

PARMLEY, W. W., and E. H. SONNENBLICK. 1969. Relation between mechanics of contraction and relaxation in mammalian cardiac muscle. Am. J. Physiol. 216: 1084–1091.

SHINE, K. I., and G. A. LANGER. 1971. Caffeine effects upon contraction and calcium exchange in rabbit myocardium. J. Mol. Cell. Cardiol. 3: 255–270.

SHINE, K. I., S. D. SERENA, and G. A. LANGER. 1971. Kinetic localization of contractile calcium in rabbit myocardium. Am. J. Physiol. 221: 1408–1417.

SONNENBLICK, E. H. 1962. Force-velocity relations in mammalian heart muscle. Am. J. Physiol. 202: 931–939.

SONNENBLICK, E. H. 1965. Determinants of active state in heart muscle. Force, velocity, instantaneous muscle length, time. Fed. Proc. 24: 1396–1409.

SONNENBLICK, E. H. 1967. Active state in heart muscle: Its delayed onset and modification by inotropic agents. J. Gen. Physiol. 50: 661–676.

Mailing address:
Rainer Bodem, M.D.,
Medizinische Poliklinik der Universtät,
D-6900 Heidelberg, Hospitalstr. 3
(W. Germany).

Recent Advances in Studies on
Cardiac Structure and Metabolism Volume 10
The Metabolism of Contraction
Edited by Paul-Emile Roy and George Rona

CARDIAC METABOLISM AND ELECTROMECHANICS OF HUMAN HEART

K. PRASAD

Department of Physiology, College of Medicine, University of Saskatchewan, Saskatoon, Saskatchewan, Canada

SUMMARY

The effects of substrates on the metabolic inhibitor-induced changes in the action potential and contraction of papillary muscles obtained from patients undergoing corrective open-heart surgery were studied. Anoxia produced a marked shortening of the action potential duration and a decrease in the resting potential, rate of rise of action potential, effective refractory period, and contractility. In anoxic muscle, although glucose completely restored the action potential duration, effective refractory period, and resting potential to control levels, it was unable to completely restore the contractility to the control level. Substrate depletion and metabolic inhibitors (iodoacetate, dinitrophenol) produced effects similar to that of anoxia, but at a faster rate. Glucose restored the action potential and, to a lesser extent, contractility to the control level in dinitrophenol-treated muscle but was ineffective in so doing in the iodoacetate-treated muscle. Pyruvate, however, was effective in restoring the action potential and contractility in iodoacetate-treated muscle. Pretreatment of the muscle with glucose and, particularly, with glucose plus insulin prevented the combined effects of anoxia and lack of glucose on the action potential and contractility for a prolonged period. These results suggest that intravenous infusion of glucose and insulin before and during surgery might prevent or reduce the effect of anoxia on the electrical and mechanical activity of the heart during open-heart surgery.

INTRODUCTION

Myocardial ischemia has been reported to produce ventricular fibrillation in dogs (Harris *et al.*, 1954; Regan *et al.*, 1967). Ventricular ectopic beats are the most common arrhythmias in patients with acute myocardial infarction and these extrasystoles very often initiate fatal ventricular tachycardia and fibrillation (Kimball and Killip, 1968; Meltzer and Kitchell, 1966). Loss of myocardial potassium as a result of myocardial infarction has been observed in dogs (Cummings, 1964; Jennings *et al.*, 1964) and man (Iseri *et al.*, 1952). Efflux of potassium during ischemia has been implicated in the production of cardiac arrhythmias in dogs (Harris *et al.*, 1954; Regan *et al.*, 1967). Shortening of the action potential duration and a decrease in the force of contraction in cardiac muscle have been observed during anoxia (Prasad and Callaghan, 1969; Prasad and McLeod, 1969). Shortening of the action potential duration during anoxia has been suggested to be due to an increase in the efflux of potassium (Trautwein and Dudel, 1956; McLeod and Prasad, 1969; Prasad and McLeod, 1969). Shortening of the action potential duration and effective refractory period and a decrease in the resting potential during anoxia might lead to the production of cardiac arrhythmias.

Recently, ischemic contracture of the heart (stone-heart) during open-heart surgery, using elective anoxic cardiac arrest in combination with cardiopulmonary bypass, has been observed by Cooley, Reul, and Wukasch (1972). Adenosine triphosphate and creatine phosphate have been proposed as major indices and possible determinants of the ability of myocardial tissue to survive during and to recover from cellular anoxia (Hearse and Chain, 1972). Anoxic cardiac arrest, in combination with cardiopulmonary bypass, is extensively used during open-heart surgery. The success of open-heart surgery depends to a large extent upon adequate protection of the myocardium during the period of cardiopulmonary bypass. Failure to preserve the biochemical and morphological integrity of the cardiac cells during cardiopulmonary bypass can severely impair the functional recovery of the heart.

The present paper deals with the study of various substrates on the effects of metabolic inhibitors on the electromechanical activity of human papillary muscles.

METHODS

The methods used in this study were similar to those described earlier (Prasad, 1972). The experiments were carried out on papillary muscles obtained from patients undergoing corrective open-heart surgery. Immediately after excision, the muscles were transferred to a chilled Krebs-Ringer solution. The muscle strips (3–5 mm long, 1–2 mm wide, and 0.5–0.75 mm thick) were tied horizontally in a jacketed 100-ml constant temperature bath at 37°C containing Krebs-Ringer solution of the following composition in mEq/liter: Na, 138.5; K, 4.6; Ca, 4.9; Mg, 2.3; HCO_3, 21.91; PO_4, 3.48; Cl, 125; glucose, 20 mM or otherwise mentioned. The solution in the bath was constantly bubbled with a mixture of 95% oxygen and 5% carbon dioxide. This will be referred to in the text as oxygenated solution. One end of the muscle was attached to a fixed holder, while the other end was tied to a Grass FT-03 force displacement transducer through a movable holder in order to permit isometric contraction. The muscle was stimulated supramaximally with square wave stimuli of 5 msec duration at the rate of 60/min by a Grass stimulator through platinum electrodes placed at one end of the muscle.

Single cell electrical activity was recorded with flexibly mounted glass microelectrodes of tip diameter less than 0.5 μ filled with 3 M KCl, having a resistance of 5–10 megohms. A negative capacitance electrometer was used and action potentials were displayed on a Tektronix dual beam 564A storage oscilloscope.

Action potentials and contractions were simultaneously monitored on the storage oscilloscope and photographed on polaroid film.

The muscles were equilibrated in Krebs-Ringer solution for at least 2 hours before normal action potential and contraction were recorded. The muscles were then exposed to various interventions for the period of observation. Force of contraction and duration of action potential at 25% level of repolarization were measured from the photographic records.

The methods of determination of the rate of rise of 0 phase of action potential (depolarization phase) was the same as that described by Bigger, Bassett, and Hoffmann (1968) and Prasad and Callaghan (1971). The output of the differentiator was linear, with rates of potential change from 0 to 1000 v/sec. The effective refractory period was determined by the method described by Prasad and Callaghan (1971). Driving stimuli were delivered to the muscle by means of bipolar platinum electrodes placed at one end of the muscle. The stimuli were generated by a combination of two SD5 Grass stimulators arranged such that the basic drive stimulus (S_1) and the intermittent single or trains of test stimuli (S_2) could be delivered to the preparation through the same electrodes. The basic drive rate, the drive and test stimulus duration and amplitude, the recurrence of test stimuli, and the interval between the drive stimuli and test stimuli were independently variable. The duration of S_1 was usually 3 msec and its amplitude was twice threshold. The duration of S_2 was usually 5 msec and its amplitude was 3–4 times threshold. S_2 was used to determine the effective refractory period. The test stimuli were applied at selected intervals during every seventh drive cycle. Responses were recorded through microelectrodes placed in a cell within 1 mm of the stimulating electrodes. The test stimulus was applied during the phase of repolarization of action potential until propagated action potential was obtained, to determine the effective refractory period. Action potential and rate of rise of 0 phase of action potential were monitored on the storage oscilloscope and photographed on the polaroid film. The effective refractory period, duration of action potential at 25% level of repolarization, and the rate of rise of 0 phase of action potential were measured from the photographic records.

Anoxia, whenever mentioned, was induced by bubbling the Krebs-Ringer solution with a mixture of 95% nitrogen and 5% carbon dioxide. This solution is referred to in the text as nonoxygenated solution.

RESULTS

Effects of anoxia and glucose on duration of action potential and contraction

Twelve experiments were carried out in which the muscles, after equilibration in oxygenated Krebs-Ringer containing 20 mM glucose for 1 hour, were exposed in succession to the following solutions: a) Krebs-Ringer solution without glucose

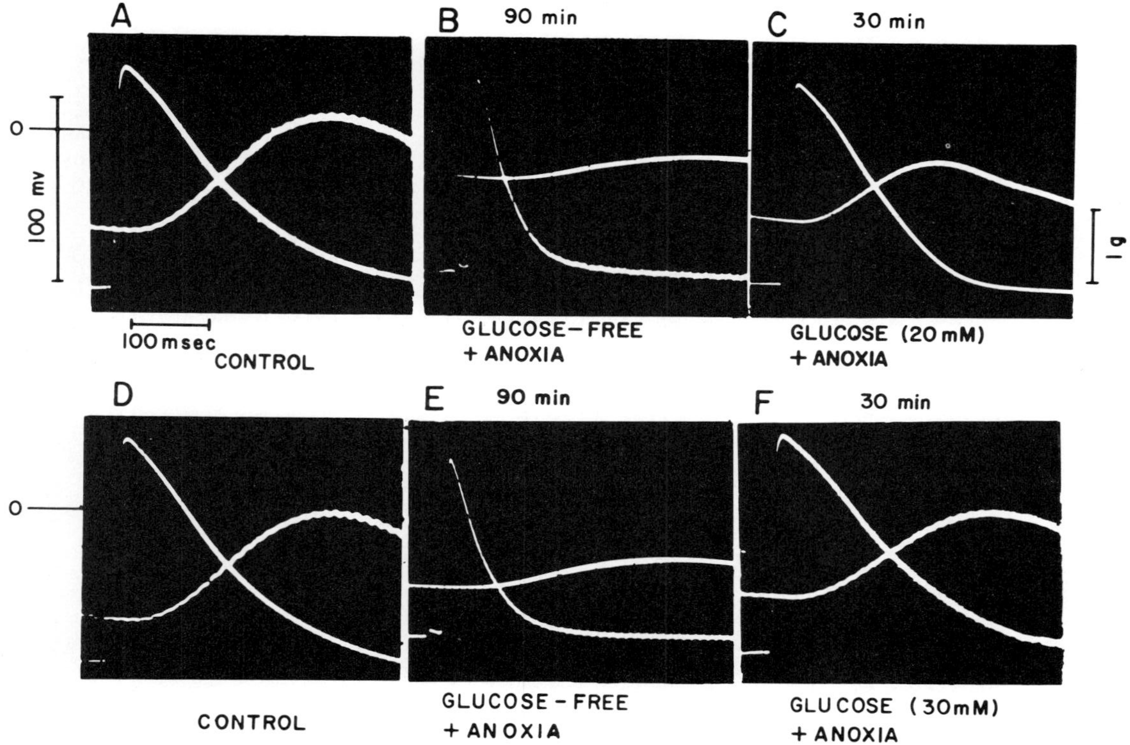

Figure 1. Effect of glucose on simultaneously recorded action potential and contraction of human papillary muscles during anoxia. *A* and *D*: control in oxygenated Krebs-Ringer solution containing 20 mM glucose; *B* and *E*: 90 min after exposure of muscles in nonoxygenated glucose-free Krebs-Ringer solution; *C* and *F*: after 30 min in 20 and 30 mM glucose, respectively, during anoxia. In each tracing in this and other figures, *upper recording* is force of contraction and *lower recording* is action potential. Calibration is indicated in figure.

equilibrated with a mixture of 95% nitrogen and 5% carbon dioxide for 90 min; and b) 6 muscles each in Krebs-Ringer solution containing either 20 or 30 mM glucose equilibrated with a mixture of 95% nitrogen and 5% carbon dioxide for another 30 min. A representative recording of the changes in the duration of action potential and contraction from a single muscle with such treatment is shown in Figure 1 and the results are summarized in Figure 2. Anoxia in the absence of glucose produced a marked shortening of the duration of action potential and a decrease in contractility. The effects of these interventions on the duration of action potential were less than their effects on the contractility. Glucose (20 and 30 mM), in the muscles preincubated in glucose-free Krebs-Ringer solution equilibrated with a mixture of 95% nitrogen and 5% carbon dioxide, produced a marked lengthening of the duration of action potential and an increase in the contractility. These concentrations of glucose, although they were able to bring back the reduced action potential duration to control levels, were unable to bring back the contractility to control levels. The increase in the contractility was more with 30 mM glucose than with 20 mM glucose.

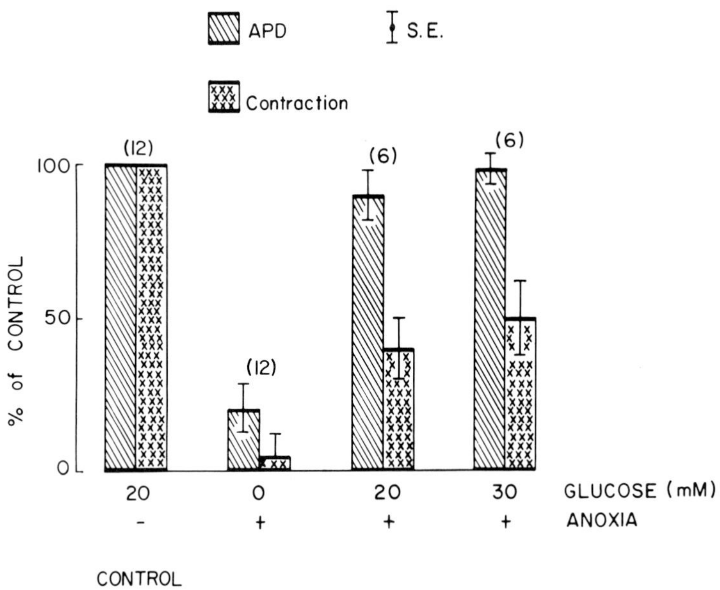

Figure 2. Histogram showing changes in action potential duration (APD) at 25% level of repolarization and contraction of anoxic papillary muscles in different concentrations of glucose shown *at bottom of each bar.* Results are expressed as percentage of control in oxygenated Krebs-Ringer solution containing 20 mM glucose taken as 100%. *Numerals over bars* show number of experiments; (+) and (–) respectively indicate presence and absence of anoxia. Note marked decrease in contractility and action potential duration (APD) in nonoxygenated glucose-free Krebs-Ringer solution. Also note reversal of effects of anoxia on APD and contraction with glucose.

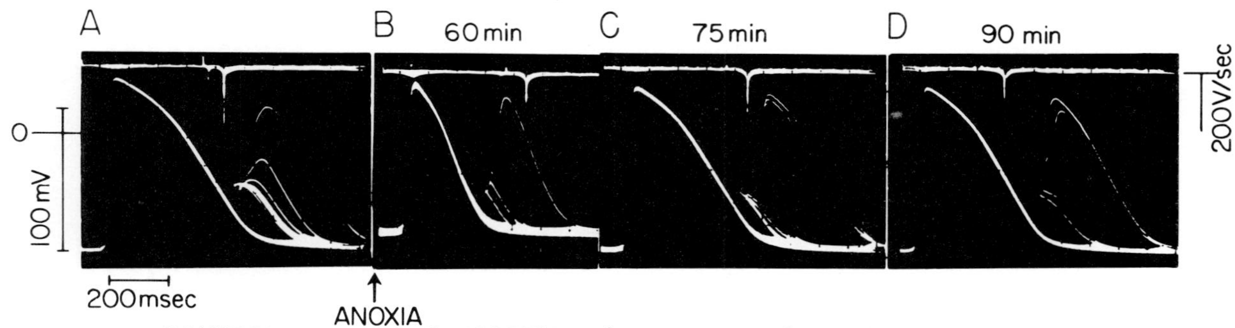

Figure 3. Effects of glucose on action potential, effective refractory period, and rate of rise of 0 phase of action potential during anoxia. *A:* control in oxygenated Krebs-Ringer solution containing 5 mM glucose; *B:* 60 min after exposure to nonoxygenated glucose-free solution; *C* and *D:* after 15 and 30 min, respectively, in nonoxygenated Krebs-Ringer solution containing 50 mM glucose. *Upper trace* in this figure is differentiated record (rate of rise) of 0 phase of action potential recorded at much faster speed (5 msec/min). First action potential is due to driving stimuli, and subsequent responses on repolarization phase of first action potential are due to test stimuli to find out effective refractory period. Effective refractory period ends where propagated action potential is obtained.

Effects of anoxia and glucose on effective refractory period, and rate of rise and duration of action potential

Strips of human papillary muscles after equilibration in oxygenated Krebs-Ringer solution containing 5 mM glucose were exposed in succession to: a) glucose-free Krebs-Ringer solution equilibrated with 95% nitrogen and 5% carbon dioxide for 60 min; and b) Krebs-Ringer solution containing 10, 20, or 50 mM glucose equilibrated with 95% nitrogen and 5% carbon dioxide for 30 min. A representative recording of the changes in the duration of action potential with such treatments is shown in Figure 3 and the results are summarized in Figure 4. Anoxia in the absence of glucose produced a marked shortening of the duration of action potential, and a decrease in the effective refractory period and rate of rise of 0 phase of action potential. Nonoxygenated glucose-free solution exerted a greater effect on the duration of action potential and effective refractory period than on the rate of rise of 0 phase of the action potential. Glucose in muscles so treated produced a lengthening of the duration of action potential, an increase in the effective refractory period, and rate of rise of 0 phase of the action potential. The duration of action potential and the effective refractory period during anoxia were dependent upon the extracellular concentrations of glucose. The rate of rise of 0 phase of action potential was affected to a very small degree.

Effect of iodoacetate on duration of action potential and contractility

Iodoacetate in low concentration is known to inhibit triosephosphate dehydrogenase, thereby blocking glycolysis and inhibiting the formation of pyruvate

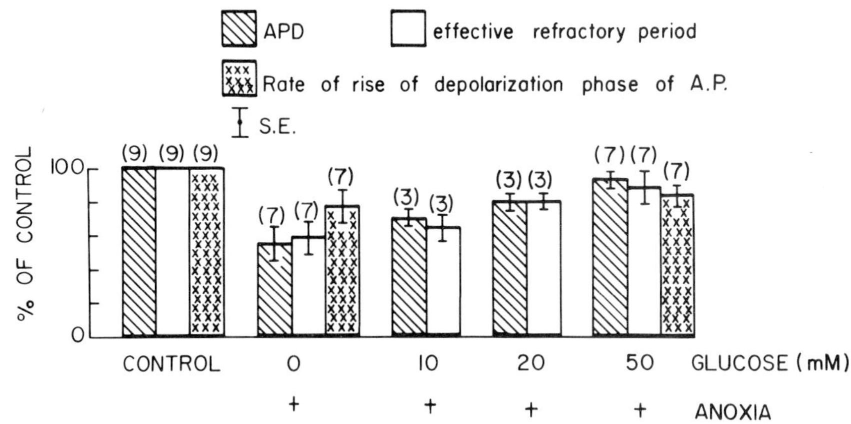

Figure 4. Histogram showing effects of various concentrations of glucose, shown *at the bottom of each bar,* on action potential duration (APD), effective refractory period, and rate of rise of depolarization phase of action potential (AP) of muscles during anoxia. Control, in oxygenated Krebs-Ringer solution containing 5 mM glucose. *Numerals over each bar* show number of experiments; (+) indicates presence of anoxia.

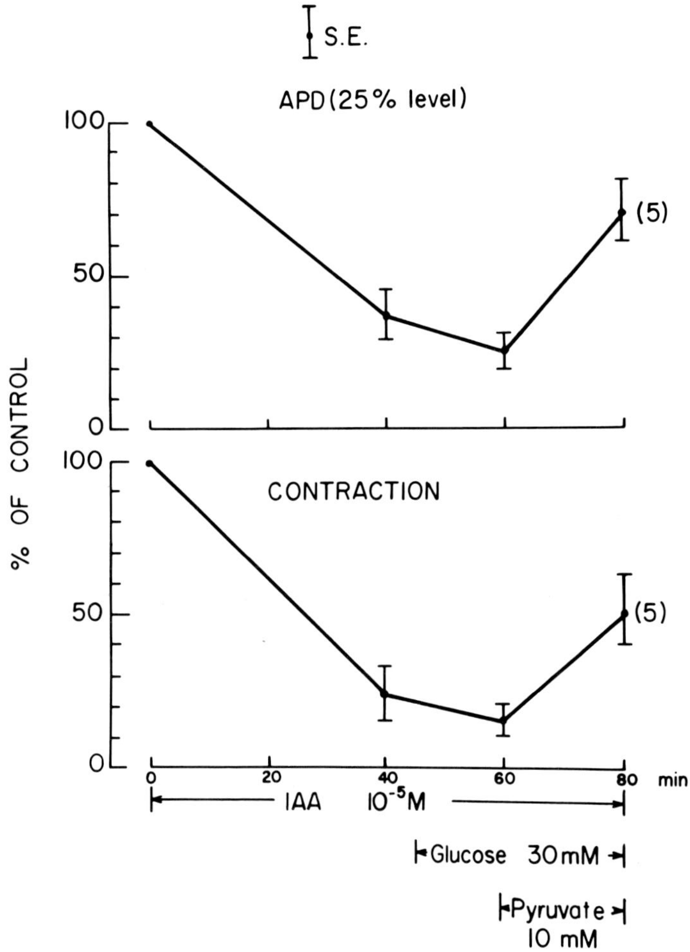

Figure 5. Effects of glucose and pyruvate on iodoacetate (IAA)-induced changes in action potential duration (APD) and contraction of human papillary muscles. *Figures in parentheses* indicate number of experiments. Note decrease in APD and contractility with iodoacetate, and ineffectiveness of glucose and effectiveness of pyruvate in reversing iodoacetate-induced shortening of APD and decrease in contractility.

from glycogen or glucose (Dixon, 1937). Therefore, in the presence of iodoacetate, glucose should be ineffective in producing any changes in the duration of action potential and contraction, while pyruvate should be able to affect the duration of action potential and contraction. With this view in mind, experiments were conducted in which the strips of human papillary muscles after equilibration in oxygenated Krebs-Ringer solution for 60 min were exposed in succession to: a) a solution containing iodoacetate (10^{-5} M) for 40 min; b) a solution containing iodoacetate (10^{-5} M) and glucose (30 mM) for 20 min; and c) solution as in b) plus sodium pyruvate (10 mM) for a further period of 20 min. The results are shown in Figure 5. Iodoacetate produced a marked shortening of the duration of action potential and a decrease in the contractility. There was a decrease in the resting and amplitude of action potential. Iodoacetate, within 40 min, reduced the duration of action potential and the contractility to 37 and 24% of control levels, respectively. Glucose was unable to reverse the effects of iodoacetate on the duration of action potential and contraction in the papillary muscles, and the shortening of the duration of action potential and decrease in the contractility continued. Pyruvate, however, was effective in reversing the effects of iodoacetate on the duration of action potential and contractility toward control levels, although not completely. Pyruvate, within 20 min, increased the contractility from 15 to 50% of control and the duration of action potential from 25 to 75% of control.

Effect of dinitrophenol on APD and contraction

2,4-Dinitrophenol (DNP), which is known to uncouple oxidative phosphorylation (Loomis and Lipmann, 1948), was used to investigate the role of glycolysis in the maintenance of duration of action potential and contraction. DNP was used at a concentration of 10^{-5} M in three experiments. Within 10 min, this concentration of DNP almost completely abolished the action potential and reduced the contraction drastically (Fig. 6); it was therefore decided to use DNP at a concentration of 10^{-6} M. Experiments were carried out in which the muscles, after equilibration in oxygenated Krebs-Ringer solution for 60 min, were exposed in succession to: a) a solution containing DNP (10^{-6} M) for 40 min; b) a solution containing DNP (10^{-6} M) and sodium pyruvate (10 mM) for 20 min; and c) a solution as in b) plus glucose (30 mM). The results of such interventions are summarized in Figure 7. At a concentration of 10^{-6} M, DNP produced a shortening of the duration of action potential to 50% and a decrease in the contractility to 30% of the control values. Pyruvate could not reverse the DNP-induced shortening of the duration of action potential or decrease in the contractility toward control values. However, glucose in the DNP-treated muscle produced a lengthening of the duration of action potential from 45 to 90% and an increase in the contractility from 19 to 40% of control values.

Effect of nonoxygenated glucose-free solution on muscles pretreated with glucose

Assuming that a high glucose level could increase the glycogen level of the muscle, the muscles would be able to withstand anoxia for a prolonged period. With this view in mind, the muscles after equilibration in oxygenated Krebs-Ringer solution containing various concentrations of glucose for 30 min were exposed to a nonoxygenated glucose-free Krebs-Ringer solution for 90 min.

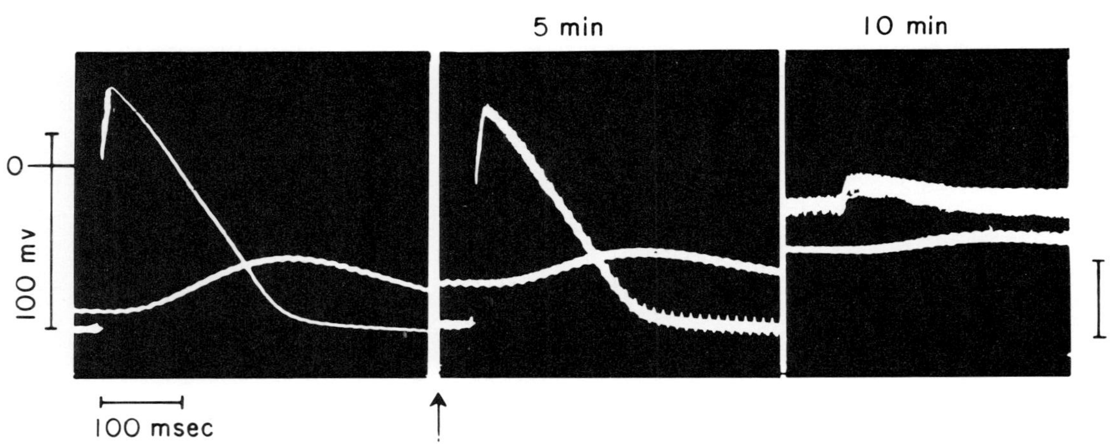

Figure 6. Effects of DNP (10^{-5} M) on simultaneously recorded action potential and contraction in single papillary muscle equilibrated in oxygenated Krebs-Ringer solution. Control, in oxygenated Krebs-Ringer solution. At *arrow*, muscle was exposed to a solution containing DNP (10^{-5} M); subsequent tracings are at intervals shown *at top of each tracing*. Note disappearance of action potential and contractility within 10 min of exposure of muscle to DNP (10^{-5} M).

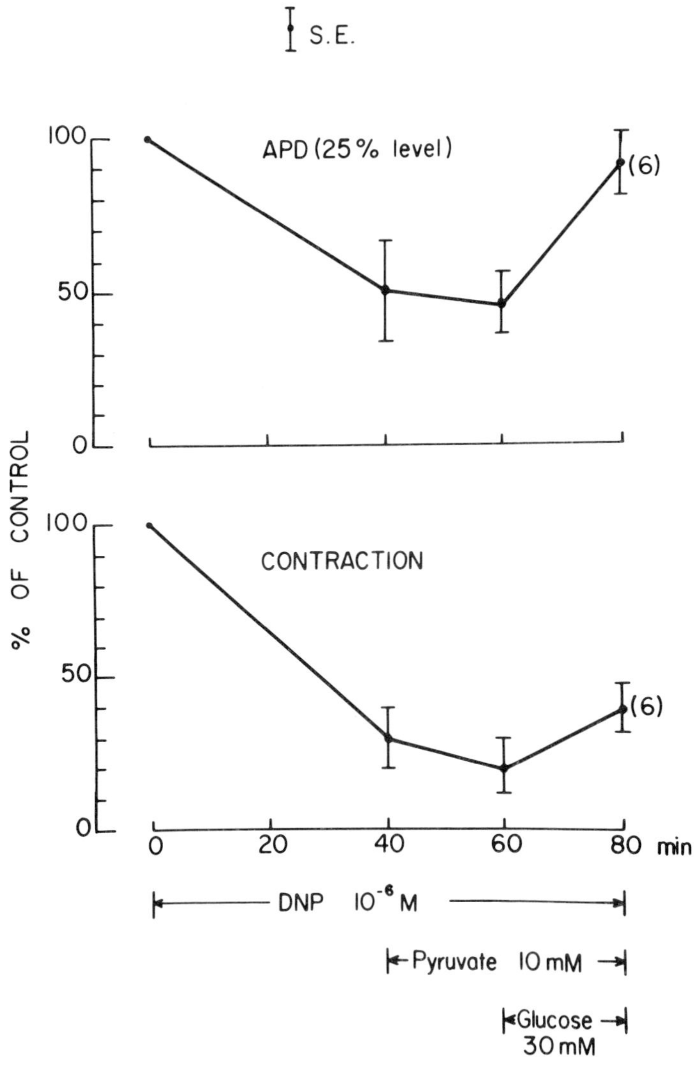

Figure 7. Effects of glucose and pyruvate on dinitrophenol (DNP)-induced changes in action potential duration (APD) at 25% level of repolarization and contraction of human papillary muscles. *Figures in parentheses* indicate number of experiments. Note decrease in APD and contractility with DNP, ineffectiveness of pyruvate, and effectiveness of glucose in reversing DNP-induced shortening of APD and decrease in contractility.

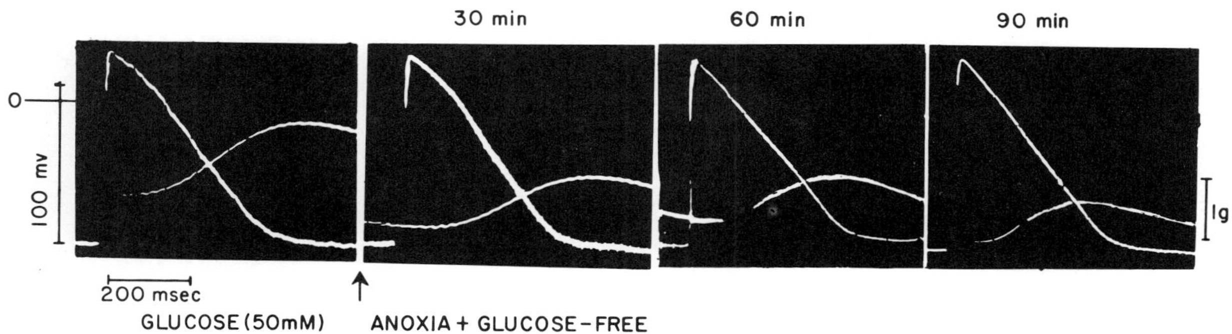

Figure 8. Effects of nonoxygenated glucose-free solution on simultaneously recorded action potential and contraction of human papillary muscles preincubated in oxygenated Krebs-Ringer solution containing 50 mM glucose for 30 min. First tracing is in oxygenated Krebs-Ringer solution containing 50 mM glucose. At *arrow,* muscle was exposed to a nonoxygenated glucose-free Krebs-Ringer solution; subsequent tracings are at intervals shown *at top of each tracing.*

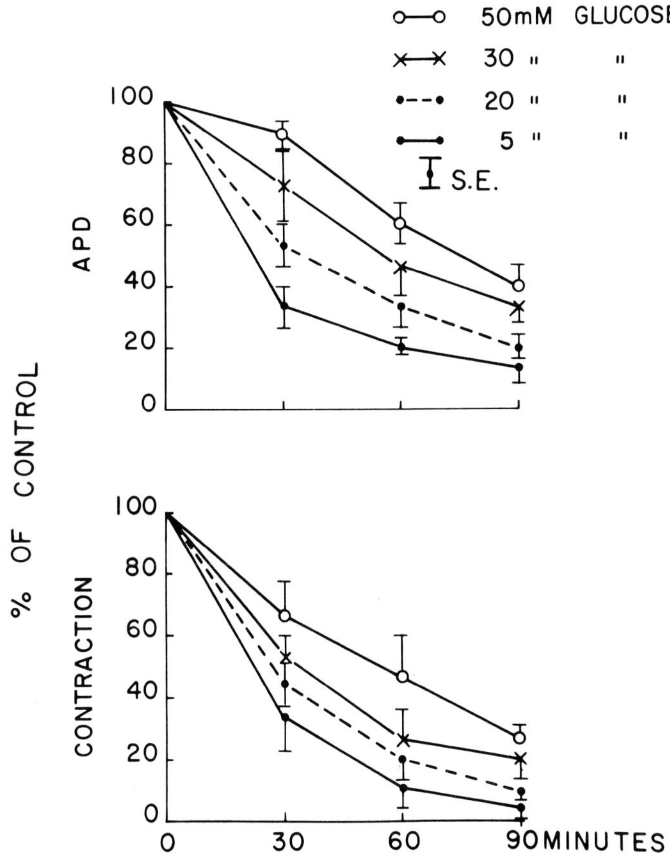

Figure 9. Effects of nonoxygenated glucose-free solution on action potential duration (APD) and contraction of human papillary muscles preincubated in oxygenated Krebs-Ringer solution containing various concentrations of glucose. Note glucose concentration-dependent protection of effects of nonoxygenated glucose-free solution on APD and contraction.

Four experiments were conducted with each concentration (5, 20, 30, or 50 mM) of glucose. A representative recording of the changes in the action potential and contraction with such treatment is shown in Figure 8, and the results are summarized in Figure 9. Nonoxygenated glucose-free Krebs-Ringer solution produced a gradual shortening of the duration of action potential and a decrease in the contractility, the latter effect being more marked than the former. The shortening of the duration of action potential and the decreases in the contractility were inversely proportional to the concentrations of glucose to which the muscles were pre-exposed.

Effect of nonoxygenated glucose-free solution on muscles pretreated with insulin and glucose

Insulin is known to increase the transport of glucose and the synthesis of glycogen (Bessman, 1966; Ellis, 1966). The muscles should be able to withstand the effects of nonoxygenated glucose-free solution for a prolonged period if pretreated with insulin and glucose in combination. After 30 min of equilibration in oxygenated Krebs-Ringer solution containing either 50 mM glucose or 50 mM glucose and 5 mU/ml insulin, five muscles each were exposed to a nonoxygenated glucose-free Krebs-Ringer solution for 90 min. A representative recording of the duration of action potential and contraction with such treat-

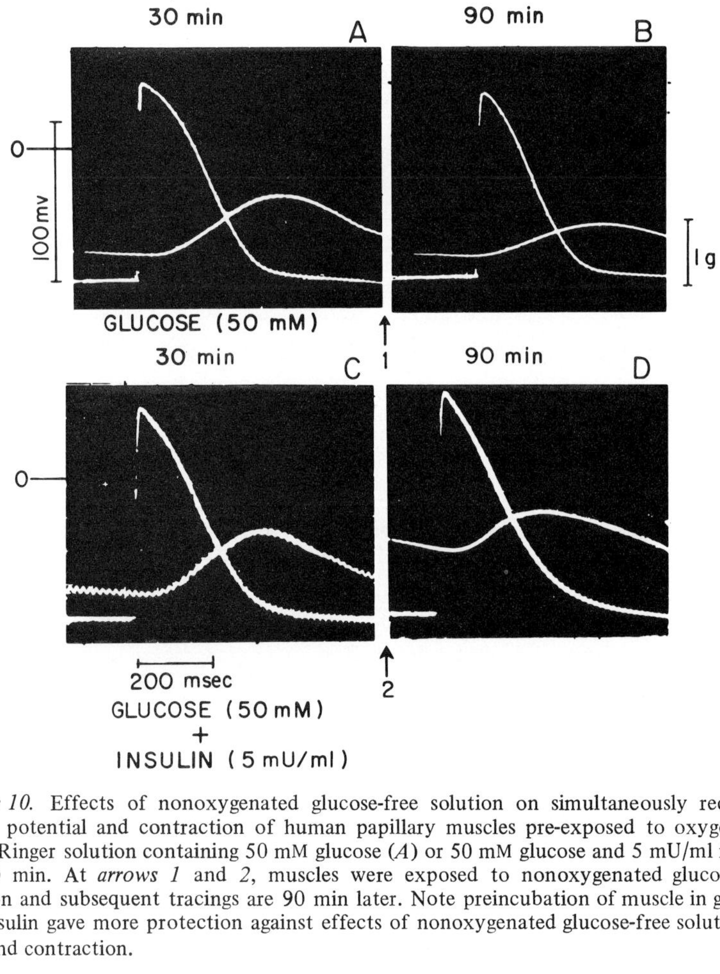

Figure 10. Effects of nonoxygenated glucose-free solution on simultaneously recorded action potential and contraction of human papillary muscles pre-exposed to oxygenated Krebs-Ringer solution containing 50 mM glucose (*A*) or 50 mM glucose and 5 mU/ml insulin for 30 min. At *arrows 1* and *2*, muscles were exposed to nonoxygenated glucose-free solution and subsequent tracings are 90 min later. Note preincubation of muscle in glucose and insulin gave more protection against effects of nonoxygenated glucose-free solution on APD and contraction.

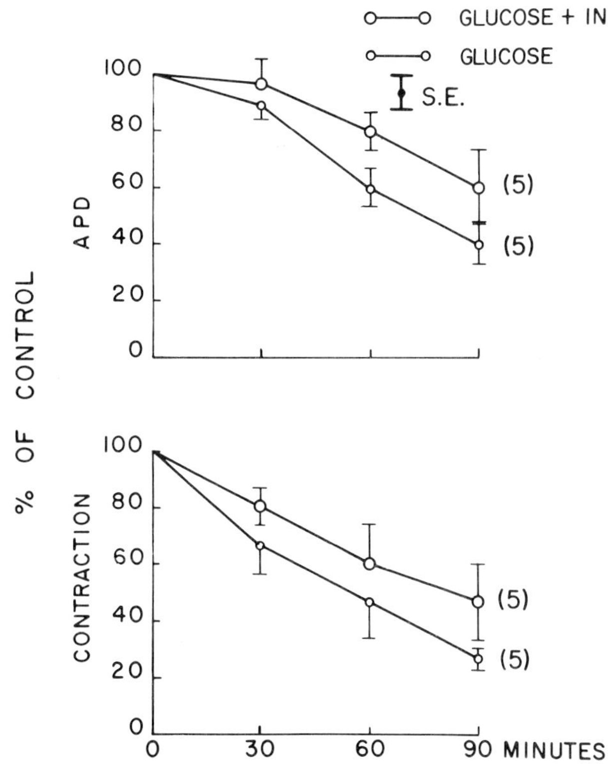

Figure 11. Effects of nonoxygenated glucose-free solution on action potential duration (APD) and contraction of human papillary muscles pretreated with oxygenated Krebs-Ringer solution containing glucose (50 mM) or glucose (50 mM) and insulin (5 mU/ml).

ment is shown in Figure 10 and the results are summarized in Figure 11. It is apparent from the results that the rate of shortening of the duration of action potential and the decrease in contractility in the muscles pretreated with glucose and insulin in combination were slower than those pretreated with glucose alone.

DISCUSSION

The results demonstrated that metabolic inhibitors (anoxia, iodoacetate, and DNP) produced a shortening of the action potential duration and a decrease in the contractility in the human papillary muscles. Anoxia also produced a decrease in the rate of rise of depolarization phase of action potential and in the effective refractory period.

Shortening of the action potential duration and a decrease in the contractility in the cardiac muscle during anoxia (Trautwein and Dudel, 1956; Prasad and

Callaghan, 1969, 1971; Prasad and McLeod, 1969) and in the presence of iodoacetate (Prasad and McLeod, 1969) or DNP (Prasad and Callaghan, 1970) have been reported earlier. The effects of anoxia on the electromechanical activity of the papillary muscle might be due to a reduced amount of adenosine triphosphate (ATP). During myocardial ischemia, there is a decrease in the glycogen, ATP, and creatine phosphate content of heart (Michal *et al.,* 1959; Hearse and Chain, 1971). During anaerobic glucose metabolism, the net yield of ATP from 1 molecule of glucose is only 2 molecules. Under anoxic condition, low glucose does not provide enough energy to maintain the action potential duration and contraction to control levels. In the present series of experiments, glucose not only abolished but also prevented the effects of anoxia on the various parameters of action potential and contraction. Since uptake of glucose in the heart is dependent upon extracellular glucose concentrations (Morgan *et al.,* 1965), an increased extracellular glucose would provide an increased amount of ATP through glycolysis. During anoxia, certain enzymes of the glycolytic pathway of glucose metabolism are also stimulated (Morgan, Randle, and Regan, 1959). Thus, during anoxia, glucose can provide enough energy to maintain action potential duration and, to a lesser extent, contraction at normal values.

Iodoacetate-induced shortening of the action potential duration and decrease in the contractility is probably due to an inhibition of the formation of pyruvate from glucose, attributable to an inhibition of the triose phosphate dehydrogenase (Dixon, 1937). DNP-induced changes in the action potential and contraction in the muscle is probably due to an interference with oxidative metabolism by DNP (Loomis and Lipmann, 1948).

Restoration of the action potential duration and contraction by glucose in anoxic or in DNP-treated muscle might be the result of an increase in the amount of ATP through glycolysis. Ineffectiveness of glucose in the iodoacetate-treated muscle is probably due to blockade of glycolysis by iodoacetate (Dixon, 1937). Effectiveness of pyruvate in restoring the action potential duration and contraction in the iodoacetate-treated muscle might be attributable to sparing of the pathway of the Krebs cycle by iodoacetate.

It is surprising that pretreatment of the muscle with glucose or glucose plus insulin delayed the effect of anoxia and glucose-free solution on the action potential and contraction. This fact has its advantage in open-heart surgery. If the heart is primed with high glucose and insulin before anoxic arrest during open-heart surgery, there is a better chance that the heart will withstand anoxic effects. The chances that the heart will go into contracture will also be reduced, because enough energy might be available for removing calcium from the active sites to the sarcoplasmic reticulum and, hence, relaxation. It has been reported that a decrease in the myocardial content of ATP causes a strong linkage to develop between actin and myosin filaments (Katz 1970, 1972). This phenomenon can produce rigor (Bate-Smith and Bendall, 1947). During anoxia, Ca^{++} influx (Lehr and Chau, 1974; Prasad, 1974; Prasad *et al.*, 1974) is assumed to increase and this increase in intracellular calcium would also increase the linkages between actin and myosin filament and, hence, lead to production of stone-heart. During anoxia, glucose might provide enough energy to reduce the efflux of potassium to maintain normal action potential duration. It has been suggested that shortening of the action potential duration in the presence of metabolic inhibitors is due to an increased efflux of potassium (Trautwein and Dudel, 1956; McLeod and Prasad, 1969; Prasad and McLeod, 1969).

A decreased efflux of potassium from the heart during hyperglycemia under anoxic conditions has been reported by Regan *et al.* (1967). The increase in the action potential duration and effective refractory period produced by glucose during anoxia would reduce the chances of cardiac arrhythmias.

Based on the results in this paper, it is suggested that sufficient glucose or glucose plus insulin should be added to the priming fluid used in cardiopulmonary bypass and coronary artery perfusion during open-heart surgery in order to prevent the untoward effects of anoxia on the electromechanical activity of the human heart.

Acknowledgments This work was supported partly by Medical Research Council Grant MA-5164 and by the Alberta Heart Foundation. The author is very much indebted to Dr. John C. Callaghan, Director of Thoracic and Cardiovascular Surgery, University Hospital, Alberta, for supplying muscles used in these studies. The author is also grateful to Dr. B. Bhardwaj, Department of Surgery, University Hospital, Saskatoon, for supplying muscles for investigation.

REFERENCES

BATE-SMITH, E. C., and J. R. BENDALL. 1947. Rigor mortis and adenosine triphosphate. J. Physiol. 106: 177–185.

BESSMAN, S. P. 1966. A molecular basis of the mechanism of insulin action. Am. J. Med. 40: 740–749.

BIGGER, J. T., Jr., A. L. BASSETT, and B. F. HOFFMAN. 1968. Electrophysiological effects of diphenylhydantoin on canine Purkinje fibres. Circulat. Res. 22: 221–236.

COOLEY, D. A., G. J. REUL, and D. C. WUKASCH. 1972. Ischemic contracture of the heart: "Stone Heart". Am. J. Cardiol. 29: 575–577.

CUMMINGS, J. R. 1960. Electrolyte changes in heart tissue and coronary arterial and venous plasma following coronary occlusion. Circulat. Res. 8: 865–870.

DIXON, M. 1937. Action of iodoacetate on dehydrogenases and alcoholic fermentation. Nature 140: 806–000.

ELLIS, S. 1966. The physiological interactions of epinephrine and insulin on metabolism. Israel J. Med. Sci. 2: 673–676.

HARRIS, A. S., A. BISTENI, R. A. RUSSEL, J. C., BRIGHAM, and J. E. FIRESTONE. 1954. Excitatory factors in ventricular tachycardia resulting from myocardial ischemia: Potassium a major excitant. Science 119: 200–203.

HEARSE, D. J., and E. B. CHAIN. 1972. The role of glucose in the survival and 'recovery' of the anoxic isolated perfused rat heart. Biochem. J. 128: 1125–1133.

ISERI, L. T., L. C. ALEXANDER, R. S. McGAUGHEY, A. J. BOYLE, and G. B. MYERS. 1952. Water and electrolyte content of cardiac and skeletal muscle in heart failure and myocardial infarction. Am. Heart J. 43: 215–227.

JENNINGS, R. B., H. M. SOMMERS, J. P. KALTENBACH, and J. J. WEST. 1964. Electrolyte alterations in acute myocardial ischemic injury. Circulat. Res. 14: 260–269.

KATZ, A. M. 1970. Contractile proteins of the heart. Physiol. Rev. 50: 63–158.

KATZ, A. M. 1972. Myocardial metabolism in ischemic heart disease. *In* W. Likoff, B. L. Segal, W. Insull, Jr., and J. H. Moyer (eds.), Atherosclerosis and Coronary Heart Disease, pp. 190–199, Grune & Stratton, New York.

KIMBALL, J. H., and T. KILLIP. 1968. Aggressive treatment of arrhythmias in acute myocardial infarction: Procedures and results. Progr. Cardiovasc. Dis. 10: 483–504.

LEHR, D., and R. CHAU. 1974. Changes of the cardiac electrolyte content during development and healing of experimental myocardial infarction. *In* N. S. Dhalla (ed.), Recent Advances in Studies on Cardiac Structure and Metabolism, Vol. 3: Myocardial Metabolism, pp. 721–751. University Park Press, Baltimore.

LOOMIS, W. F., and F. LIPMANN. 1948. Reversible inhibition of the coupling between phosphorylation and oxidation. J. Biol. Chem. 173: 807–808.

McLEOD, D. P., and K. PRASAD. 1969. Influence of glucose on the transmembrane action potential and contraction of papillary muscle. Effects of concentration, phlorizin, and insulin, non-metabolized sugars, and stimulation of glycolysis. J. Gen. Physiol. 53: 792–815.

MELTZER, L. E., and J. B. KITCHELL. 1966. The incidence of arrhythmias associated with acute myocardial infarction. Progr. Cardiovasc. Dis. 9: 50–63.

MICHAL, G., S. NAEGLE, W. H. DANFORTH, F. B. BALLARD, and R. J. BING. 1959. Metabolic changes in heart muscle during anoxia. Am. J. Physiol. 197: 1147–1151.

MORGAN, H. E., J. R. NEELY, R. E. WOOD, C. LIEBERG, H. LIEBERMEISTER, and C. R. PARK. 1965. Factors affecting glucose transport in heart muscle and erythrocytes. Fed. Proc. 24: 1040–1045.

MORGAN, H. E., P. J. RANDLE, and D. M. REGEN. 1959. Regulation of glucose uptake by muscle: III. Effects of insulin, anoxia, salicylate, 2,4-dinitrophenol on membrane transport and intracellular phosphorylation of glucose in the isolated rat heart. Biochem. J. 73: 573–579.

PRASAD, K. 1972. Effect of glucagon on the transmembrane potential, contraction and ATPase activity of the failing human heart. Cardiovasc. Res. 6: 684–695.

PRASAD, K. 1974. Membrane Na^+-K^+-ATPase and electromechanics of human heart. *In* N. S. Dhalla (ed.), Recent Advances in Studies on Cardiac Structure and Metabolism, Vol. 4: Myocardial Biology, pp. 91–105. University Park Press, Baltimore.

PRASAD, K., and J. C. CALLAGHAN. 1969. Effects of glucose metabolism on the transmembrane action potential and contraction of human papillary muscle during surgical anoxia. Ann. Thorac. Surg. 7: 571–581.

PRASAD, K., and J. C. CALLAGHAN. 1970. Influence of glucose metabolism on ouabain-induced changes in the transmembrane potential and contraction of human heart in vitro. Canad. J. Physiol. Pharmacol. 48: 801–812.

PRASAD, K., and J. C. CALLAGHAN. 1971. Electrophysiological basis of use of polarizing solution in the treatment of myocardial infarction. Clin. Pharmacol. Therap. 12: 666–675.

PRASAD, K., and D. P. McLEOD. 1969. Influence of glucose on the transmembrane action potential of guinea-pig papillary muscle. Metabolic inhibitors, ouabain, $CaCl_2$ and their interaction with glucose, sympathomimetic amines, and aminophylline. Circulat. Res. 24: 939–950.

PRASAD, K., S. S. NAKAI, J. C. RUSSELL, L. P. STERNS, and J. C. CALLAGHAN. 1974. Intracellular ionic and adenosine triphosphate changes in the myocardium during cardiopulmonary bypass and anoxia. Canad. J. Surgery 17: 80–83.

REGAN, T. J., M. A. HARMAN, P. H. LEHAN, W. M. BURKE, and H. A. OLDEWUTEL. 1967. Ventricular arrhythmias and K^+ transfer during myocardial ischemia and intervention with procaine amide, insulin or glucose solution. J. Clin. Invest. 46: 1657–1668.
TRAUTWEIN, W., und J. DUDEL. 1956. Aktion Potential und Kontraktion des Herzmuskels in Sauerstoffmangel. Pfluegers Arch. Ges. Physiol. 263: 23–32.

Mailing address:
K. Prasad, M.D., Ph.D., F.I.C.A.,
Dept. of Physiology, College of Medicine,
University of Saskatchewan,
Saskatoon, Saskatchewan S7N OWO (Canada).

Metabolic Manifestations of Cardiac Ischemia in Man

Recent Advances in Studies on
Cardiac Structure and Metabolism Volume 10
The Metabolism of Contraction
Edited by Paul-Emile Roy and George Rona

METABOLIC INDICATORS OF MYOCARDIAL ISCHEMIA IN MAN[1]

M. A. CHIONG and J. O. PARKER

Division of Cardiology, Department of Medicine, Queen's University
and Kingston General Hospital, Kingston, Ontario, Canada

SUMMARY

Myocardial metabolism has been studied in 54 patients with continuous sampling of arterial (A) and coronary sinus (CS) blood during 8- to 10-min periods of control in sinus rhythm, rapid atrial pacing, and recovery. The results showed that 17 subjects were normal or had insignificant coronary artery disease (CAD; nonischemic group = NI); 37 patients had significant CAD (ischemic group = I) and developed clinical, hemodynamic, and electrocardiographic evidence of myocardial ischemia during pacing, characterized by angina, elevated left ventricular end-diastolic pressure, and depressed ST segments. During pacing-induced ischemia the following metabolic abnormalities were detected: (1) myocardial anaerobiosis indicated by lactate % uptake ((A–CS)/AS × 100) of –17.2±5.0% (mean±SE); (2) myocardial loss of K^+ suggested by an A–CS difference of –0.25±0.08 mEq/liter (N=18); (3) small but significant loss of inorganic phosphorus (P_i) of –1.0±1.4% (N=18); and (4) elevation of CS blood creatine phosphokinase activity (N=5). These metabolic abnormalities were temporally related to the other manifestations of myocardial ischemia and were not seen in the NI. Lactate production and P_i loss occurred in 75 and 55% of the IG, respectively, suggesting that accelerated anaerobic glycolysis was the best indicator of myocardial ischemia in man. K^+ loss was an unreliable index in this experimental situation, since tachycardia alone caused significant K^+ egress from the heart. Lactate production and K^+ loss were reduced by nitroglycerin, which abolished angina and improved hemodynamics and electrocardiographic manifestations. That these metabolic abnormalities were not observed in all I patients may have been related to methodology, the random distribution of CAD, and the fact that the chemical composition of the CS blood reflects the metabolic balance of both well oxygenated and ischemic areas of the myocardium.

INTRODUCTION

The metabolic responses of the heart to ischemia have been extensively investigated in animal species, utilizing a variety of preparations including tissue slices and homogenates, heart-lung preparations, and open-chest animals (Olson, 1962). Twenty-five years ago, the technique of coronary sinus (CS) catheterization was deliberately used for the study of CS blood flow and myocardial

[1] Supported by Ontario Heart Foundation Grants 2-13 and 2-16 and Grant MT-3062 from the Medical Research Council of Canada.

metabolism in the closed-chest dog (Goodale *et al.,* 1948) and in man (Bing *et al.,* 1947). This approach greatly stimulated studies in this field (Bing, 1965; Opie, 1968). During the past few years, we have been engaged in delineating the metabolic responses of the human heart before, during, and immediately after reversible myocardial ischemia (Parker *et al.,* 1969a, b; Chiong, West, and Parker, 1972, 1974). These studies were carried out in patients with coronary artery disease (CAD) during atrial pacing-induced myocardial ischemia. A resumé of our results is reported below.

METHODS

The studies were carried out in 54 patients with known or suspected CAD, who were referred for diagnosis and evaluation of severity of the disease. None of these patients showed evidence of heart failure, cardiomegaly, or severe arterial hypertension, nor were they receiving digitalis, antihypertensive drugs, or beta-adrenergic blockade. The complete study lasted 24–30 min and included sequential 8- to 10-min periods of control, atrial pacing, and recovery. Under the stress of pacing, 37 patients developed chest pain (ischemic group = I). The remaining 17 asymptomatic subjects constitute the nonischemic group (NI), in turn composed of 8 subjects with normal coronary arteries and 9 with significant CAD (more than 75% stenosis). Clinical details of these groups and the results of selective cinecoronary angiography and left ventriculography appear in Table 1.

Right and left heart catheterization procedures were employed and are described in detail elsewhere (Parker *et al.,* 1969a, 1970). This approach permitted the recording of brachial and pulmonary artery pressures, left ventricular pressure and the electrocardiogram, and the measurement of cardiac output by the dye dilution technique. The following variables were calculated: cardiac and stroke index, left ventricular stroke work index, modified tension time index (Monroe, 1964), and the rate of rise of the first derivative of the left ventricular pressure. In addition, a bipolar No. 8 Gorlin catheter was placed in the midportion of the CS for the purpose of pacing and blood sampling.

Throughout the study, blood for metabolic studies was drawn at the rate of 5.0 or 2.5 ml/min simultaneously from the brachial artery and CS, using a continuous sampling technique (Nasser, Wachter, and Case, 1969), and collected in heparinized tubes placed in a fraction collector set to rotate every 1 or 2 min. Since the length of the arterial line compensated for the dead space of the CS catheter and the transit time of the blood from the collection sites to the receiving tubes was known, it was easy to correlate exactly the chemical changes in the blood with hemodynamic events. Initially, blood was drawn at a rate of 5.0 ml/min, and the total blood loss for chemical analyses was between 250 and 300 ml. To prevent hypovolemia, in approximately half of the patients, 300 ml of blood were drawn into a bag with acid-citrate-dextrose solution 2 days prior

Table 1. Clinical summary (N=54)*

	Nonischemic		Ischemic	
	N	%	N	%
Number of patients	17	100	37	100
Sex				
Male	15	88.2	34	91.9
Female	2		3	
Age (mean and range)	45.8 (36–57)		49.8 (32–68)	
Male (mean)	45.1		49.4	
Female (ages)	(47 and 55)		(49, 54, and 59)	
Coronary angiography				
Normal	8	47.1	0	
Unsatisfactory	0		2	
Nonsignificant CAD	0		3	
Significant CAD	9	52.9	32	86.5
1VD	6	35.3	7	18.9
2VD	3	17.6	18	48.6
3VD	0		7	18.9
Left ventriculogram (↓LV contractility)	0		20	55.1
ST_i depression (>0.1 mv)	4	23.5	28	75.7
$LVEDP_i$ (>15 mmHg)	5	29.4	33	89.2

*Abbreviations: N = number of patients; CAD = coronary artery disease; significant disease = more than 75% narrowing; VD = vessel disease; LV = left ventricle; ST_i = ST segment depression during interruption of pacing; mv = millivolt; LVEDPi = left ventricular end-diastolic pressure during interruption of pacing; ↓=decreased.

to the study and reinfused at the same rate during the procedure. No untoward reactions were observed. In the remainder of the patients, the rate of sampling was halved, and the latter part of the protocol was omitted. This continuous sampling technique made it possible to study the time course changes of myocardial metabolism before, during, and after stress. The effects of nitroglycerin (NG) were studied in a separate group of ischemic patients (N = 15) in whom only two paired blood samples were taken at the end of the control, pacing, and recovery periods which were then repeated after the administration of 0.5 mg of chewable NG (Chiong *et al.*, 1972).

Plasma was separated by centrifugation at 3°C within 30 min of collection and kept at −30°C until analyzed for lactate by automated colorimetric and fluorometric methods (Apstein, Puchner, and Brachfield, 1970; Hochella and Weinhouse, 1965), potassium (K^+) by standard automated flame photometry, inorganic phosphate (P_i) by a Technicon automated colorimetric method (N-4C I/II), creatine phosphokinase (CPK) by the Sigma kit 661 (normal values <12 units/ml), and glutamic oxaloacetic transaminase (GOT) using a Technicon automated fluorometric method (N-68 I/II; normal levels 10–40 Karman units). Lactate was measured in all patients ($N = 54$), but the other determinations were carried out in smaller groups of patients to prevent excessive blood loss (for K^+, $N = 30$; for P_i, $N = 29$; for CPK, $N = 15$; for GOT, $N = 16$). Arterial-coronary sinus (A-CS) differences and extraction ratios ((A–CS)/A × 100) were used to assess myocardial metabolism, since data on CS blood flow were not available and arterial blood levels of substrates remained reasonably constant during the study.

RESULTS AND DISCUSSION

In these studies, atrial pacing was used to stress the heart and increase its metabolic demands in the presence of a restricted coronary circulation. Other stresses, such as exercise and isoproterenol infusion (Baker *et al.*, 1969), were not considered suitable in this experimental situation. The remarkable effects of supine leg exercise on both arterial and CS blood lactate levels are shown in Figure 1 and contrasted with those of atrial pacing (Parker *et al.*, 1969a, b). It is obvious that the rapidly changing arterial blood levels preclude the calculation of myocardial extraction ratios. The use of isoproterenol infusion to induce myocardial ischemia was also avoided because of possible unwanted metabolic effects (Mueller and Horwitz, 1962).

In patients in the I group, atrial pacing caused chest pain with the characteristics of angina pectoris, ST segment depression during pacing interruption (ST_i >0.1 mv), abnormal elevation of the left ventricular end-diastolic pressure also when pacing was interrupted ($LVEDP_i$ >15 mm Hg), and metabolic changes (Table 2). These clinical, electrocardiographic, and hemodynamic abnormalities plus the metabolic changes to be described below will be considered part of an "ischemic syndrome" temporally associated with reversible pacing-induced myocardial stress. The significance and correlations of the electrical and hemodynamic components of this syndrome have been discussed by other investigators (Case, Roselle, and Crampton, 1966; Scheuer and Brachfield, 1966). The remainder of this discussion will be restricted to the metabolic responses.

Lactate metabolism

Arterial and CS blood lactate levels in the NI group and arterial levels in the I group remained constant during the study (Figs. 2 and 3). However, in the ischemic patients, the CS blood lactate levels rose during pacing-induced myocardial ischemia in close temporal relationship with the other components of the "ischemic syndrome" (Fig. 3) and reversed toward prepacing levels during

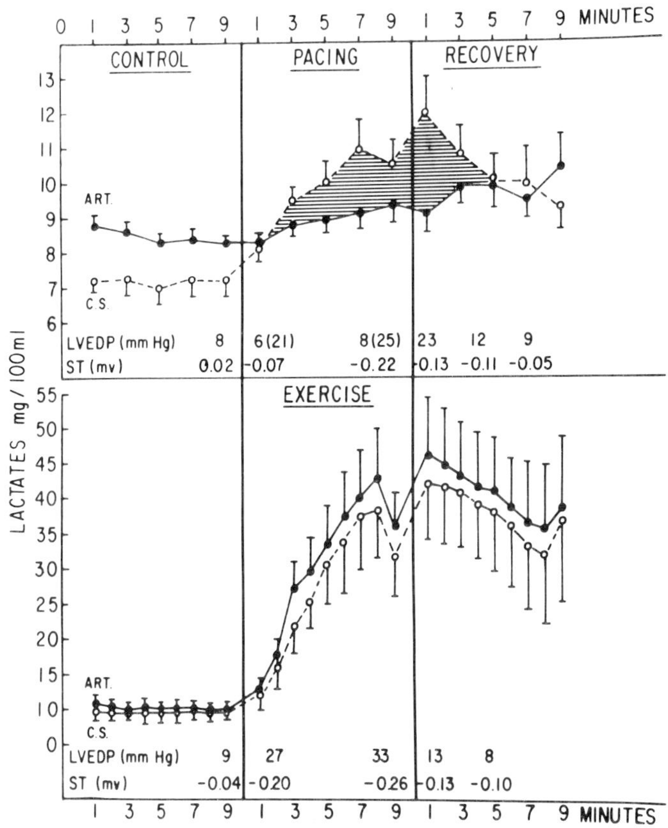

Figure 1. Effect of atrial pacing and supine leg exercise on blood lactate levels in patients with ischemic heart disease.

Table 2. Atrial pacing-induced "ischemic syndrome"*

1. Clinical:	Chest pain (angina)
2. Electrocardiographic:	ST_i depression (>0.1 mv)
3. Hemodynamic:	$LVEDP_i$ elevation (>15 mm Hg)
4. Metabolic:	Abnormal lactate metabolism Loss of K^+ Loss of phosphate Loss of CPK (?)

*For abbreviations, see Table 1.

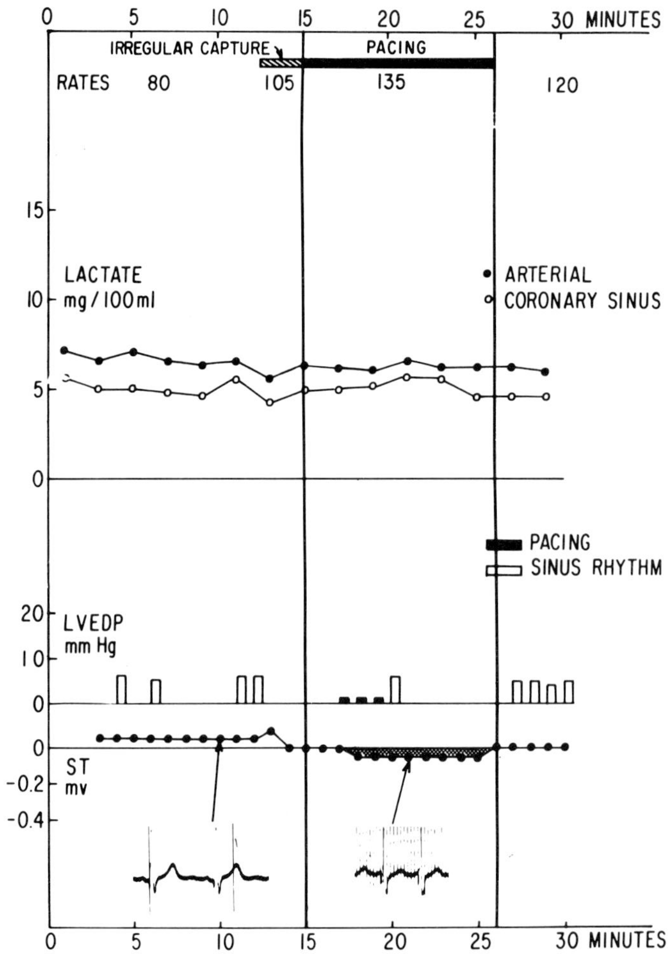

Figure 2. Effect of atrial pacing in a patient of the nonischemic group. For details see text. Reproduced by permission of Circulation.

recovery. Myocardial lactate uptake during control was significantly lower in the I group (+10.7±2.0, mean ± SE) than in the NI group (+20.6±1.9, $p<0.005$). At the peak of myocardial ischemia, lactate was produced (−17.2±5.0) in contrast to the normal uptake observed in the NI group (+21.7±2.1, $p<0.001$). During recovery, lactate production gradually disappeared; but uptake was not yet normal at the end of the study, despite the absence of chest pain and correction of the electrocardiographic and hemodynamic abnormalities (Fig. 4). Therefore, in this study, mean lactate uptake was useful in separating the NI and I groups. Of the 37 patients in the I group, only 4 had normal lactate metabo-

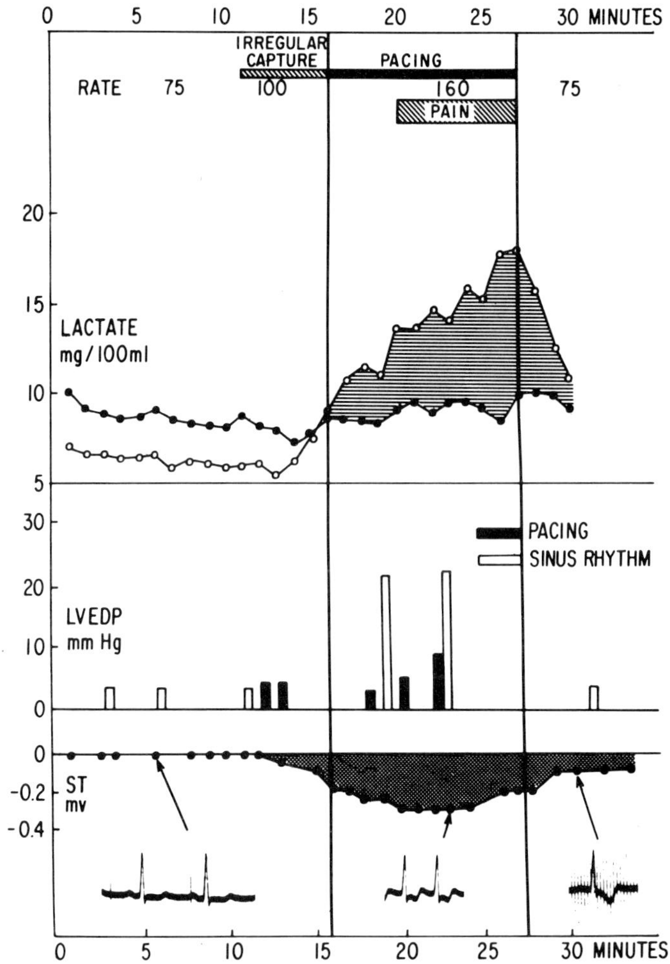

Figure 3. Effect of atrial pacing in a patient of the ischemic group. For details see text. Reproduced by permission of Circulation.

lism, whereas 5 (14.3%) had uptakes below 10%, and 28 (75.7%) showed frank lactate production. It is possible that, with more sensitive techniques, myocardial lactate production could be demonstrated in all patients with significant CAD. Studies on regional myocardial metabolism show that when an area of the dog heart is made ischemic, anaerobic glycolysis is always stimulated and lactate is demonstrated in the local venous effluent (Owen *et al.*, 1970). Since in our subjects selective venous sampling of the ischemic area was not feasible, it is possible that blood with high lactate concentration was diluted by the larger stream of low-lactate blood draining the well perfused areas of the myocardium.

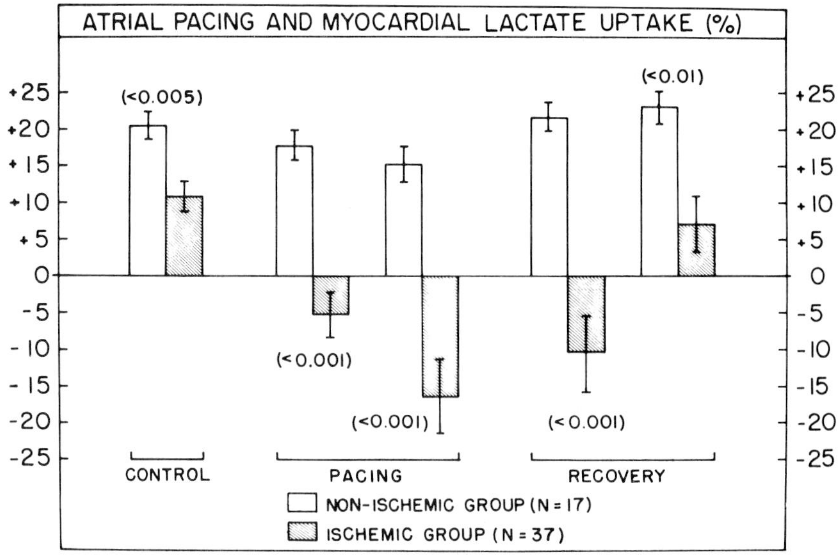

Figure 4. Summary of myocardial lactate uptakes in the nonischemic and ischemic groups.

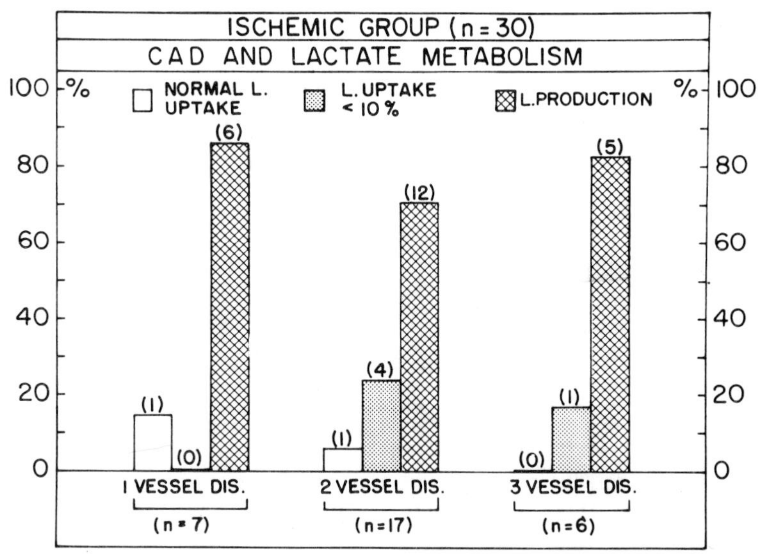

Figure 5. Correlation between lactate metabolism and number of coronary arteries with significant obstructive lesions.

Although the mean myocardial lactate uptake in the NI group was normal, four of the patients with CAD showed lactate production and no angina during pacing. Although it is known that lactate production can occur in the apparently well oxygenated heart (Opie *et al.*, 1973), the fact that two of these patients also had abnormal $LVEDP_i$ and that another showed changes in both ST_i and $LVEDP_i$ suggests that myocardial ischemia was indeed present but not so severe as to produce the full ischemic syndrome.

There was no correlation between the numbers of coronary arteries involved in the I group and lactate production as shown in Figure 5. Of seven patients with 1 vessel disease, six showed lactate production. Apparently, the number of vessels involved is not as important as an obstructive lesion strategically located to cause myocardial ischemia and release of lactate in the CS blood.

These results indicate that, in man, myocardial ischemia stimulates lactate production and that this metabolic abnormality is a good indicator of such a state. The biochemical changes stimulated by ischemia have been discussed by other investigators in animal experiments (Scheuer, 1967; Opie, 1971/72) and in man (Gorlin, 1972; Guabjarnason, 1972).

Myocardial K^+ balance

Arterial K^+ levels were not affected by atrial pacing in either group. During the control period, there was a small loss of this cation by the heart as indicated by a negative A–CS difference (Fig. 6). At the onset of atrial pacing, a sudden

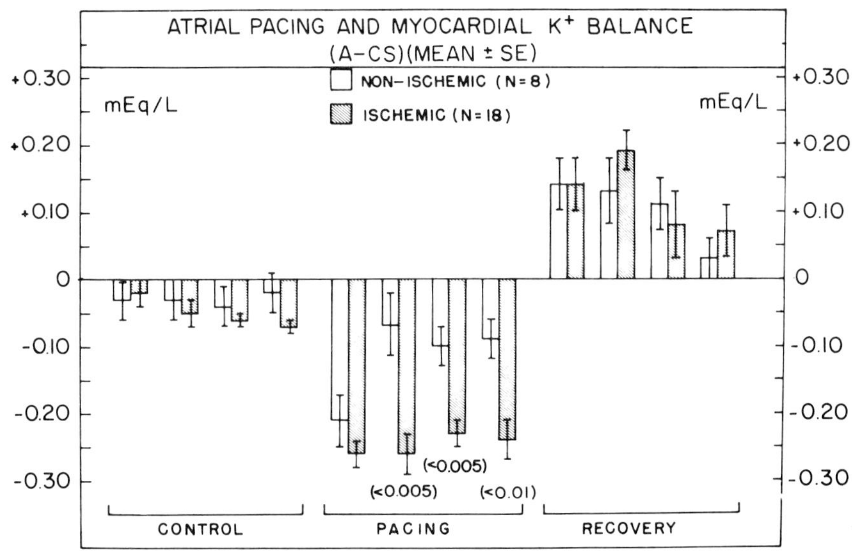

Figure 6. Time course of mean A–CS K^+ differences in nonischemic and ischemic groups.

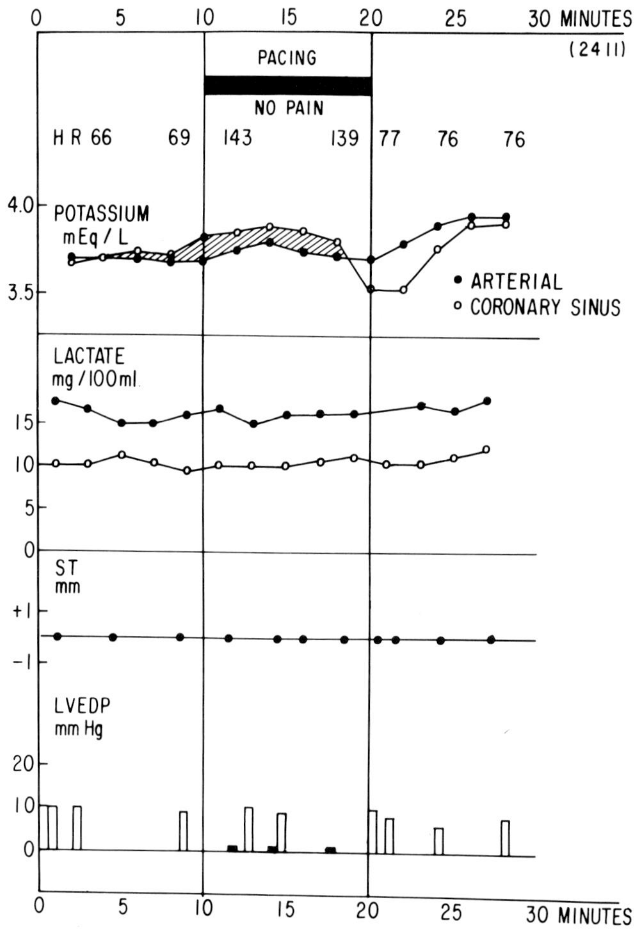

Figure 7. Effect of atrial pacing on myocardial K^+ balance and correlation with other components of the ischemic syndrome in a nonischemic patients. Reproduced by permission of Circulation.

increment in CS blood K^+ levels was observed in the NI group (mean A–CS difference of -0.21 ± 0.04 mEq/liter), which declined to -0.10 mEq/liter or less during the last 6 min of pacing. Similar A–CS K^+ differences immediately after the onset of pacing were observed in the I group (-0.26 ± 0.02). However, the CS blood levels remained elevated throughout the remainder of the ischemic period, and the difference between the two groups was highly significant (Fig. 6). During the immediate postpacing period, the CS K^+ levels suddenly fell below the arterial levels (A–CS difference of $+0.14 \pm 0.04$ mEq/liter) in both groups, indicating myocardial uptake of K^+. This change gradually declined during the

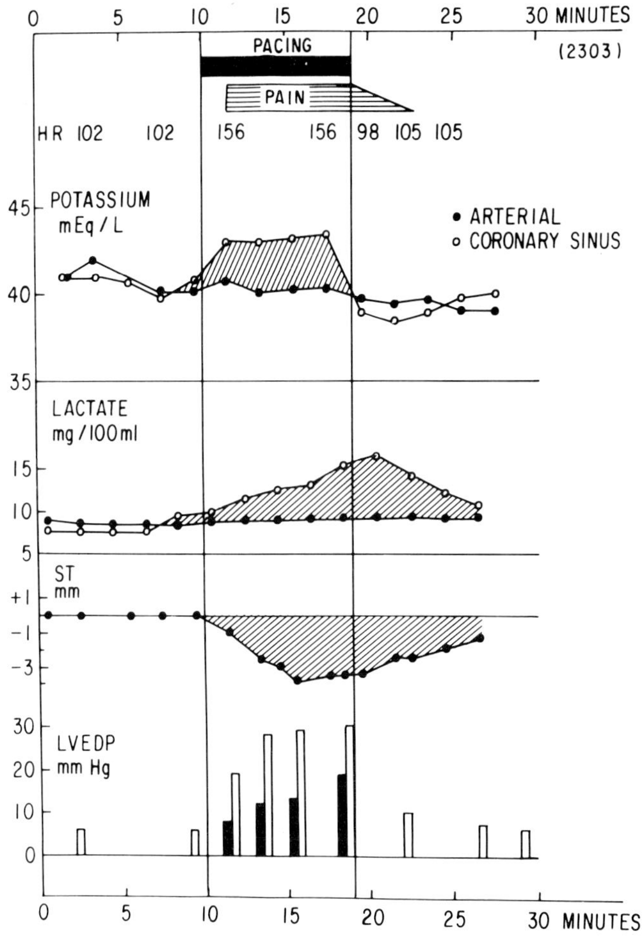

Figure 8. Same as in Figure 7, in a patient of the ischemic group. Reproduced by permission of Circulation.

recovery period. The time course of K^+ blood changes and their relationships with the other components of the "ischemic syndrome" are shown in Figures 7 and 8. These observations indicate that tachycardia per se causes K^+ egress from the human heart as it does in animals (Grupp, Acheson, and Charles, 1967; Gerlings and Gilmore, 1968; Gilmore and Gerlings, 1969) and that this effect is aggravated by ischemia (Gerlings, Miller, and Gilmore, 1969). It has been shown in dogs that reduction of coronary blood flow alone below a critical level causes reversible myocardial efflux of K^+ and lactate production (Case, 1971). Since the relative contribution of the rate and ischemic effects to the total K^+ loss is

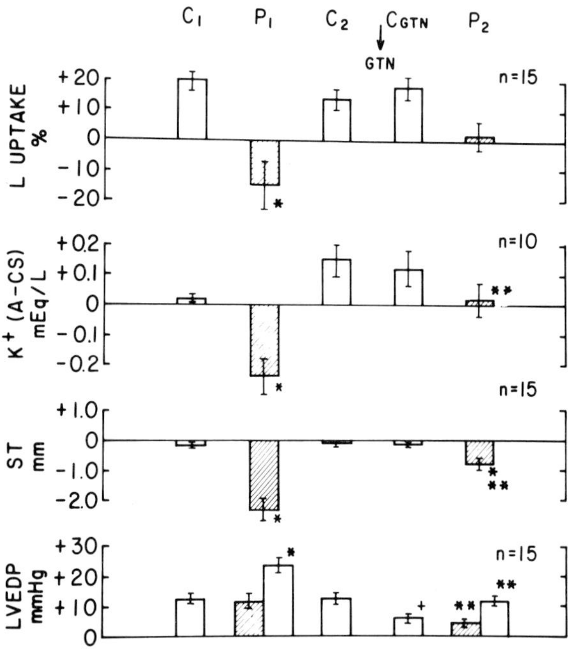

Figure 9. Summary of the effects of nitroglycerin (GTN) on the effects of artrial pacing in a group of ischemic patients. For details see text. Reproduced by permission of Circulation.

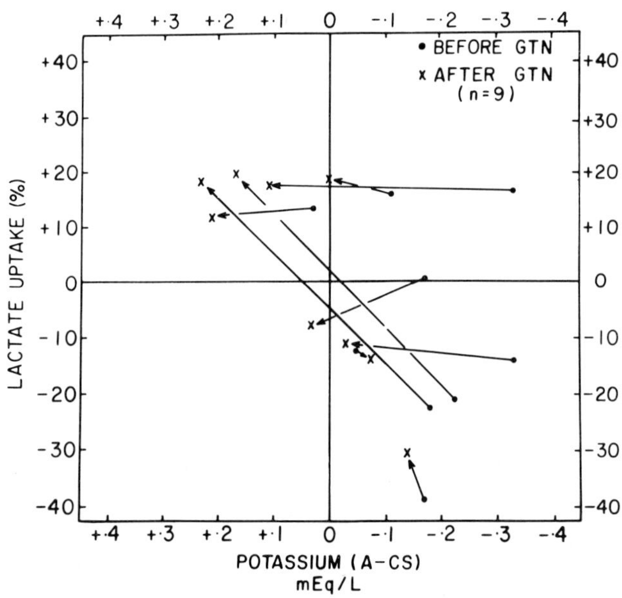

Figure 10. Effect of nitroglycerin (GTN) on lactate uptake and K^+ balance in nine ischemic patients during pacing. For details see text. Reproduced by permission of Circulation.

unknown, this metabolic abnormality is not a reliable index of myocardial ischemia in man during pacing-induced ischemia.

An attempt was made to modify the metabolic responses to atrial pacing in the I group by NG administration (Chiong *et al.*, 1972). It was found that this drug prevented the occurrence of chest pain; in addition, it not only diminished the severity of changes in ST_i and $LVEDP_i$, but also reduced the average myocardial lactate production and K^+ loss (Fig. 9). However, in a group of nine patients, normalization of lactate metabolism was observed in only two subjects, while myocardial K^+ loss was abolished or reversed in seven (Fig. 10). These NG-induced changes are suggestive of improved ischemia, probably secondary to changes in hemodynamics involving decreased venous return and venous pooling, smaller cardiac volumes, decline in cardiac wall tension and LVEDP, and a fall in myocardial oxygen requirements (Burton, 1957; Mason and Braunwald, 1965; Campion, Frye, and Zitnik, 1970; Lee, Sung, and Zaragoza, 1970; Braunwald, 1971).

Myocardial P_i balance

Blood levels of P_i were not influenced by atrial pacing in the NI patients (N = 11), and the uptake averaged 5% throughout the study (Fig. 11). However, in the I group (N = 18), myocardial uptake fell during ischemia owing to a release of this ion in the CS blood. Small but significant P_i losses were observed during the last 2 min of pacing and first 2 min of recovery (Fig. 11). This metabolic abnormality has been reported in the dog after coronary artery ligation (Owen *et al.*, 1970; Opie *et al.*, 1972) and reversible restriction of coronary blood flow (Case, 1971; Mathur and Case, 1973), and was closely related to the other components of the "ischemic syndrome." It most likely represents the breakdown of high energy phosphate compounds and intracellular accumulation and discharge of this ion in the CS blood (Opie *et al.*, 1972). Whereas lactate production was observed in 75% of the patients in group I, P_i loss was detected in only 55%, making this abnormality a less reliable index of myocardial ischemia in man. This discrepancy may be explained by the smaller changes in P_i uptake compared to those seen with lactate. This assumption is supported by the observation that, in the ischemic dog heart and on a molar basis, P_i loss is approximately only one-fifth of that of lactate (Case, 1971).

Myocardial enzymes

A preliminary study of the possible release of CPK and GOT from the heart during pacing-induced ischemia was carried out in 14 NI patients and 17 I subjects. Arterial and CS levels of GOT were practically identical and not affected by pacing or ischemia (Fig. 11). During the control and recovery periods, CPK levels in the CS blood were lower than in arterial blood, indicating a small uptake by the heart. At the onset of pacing, there was a small but abrupt

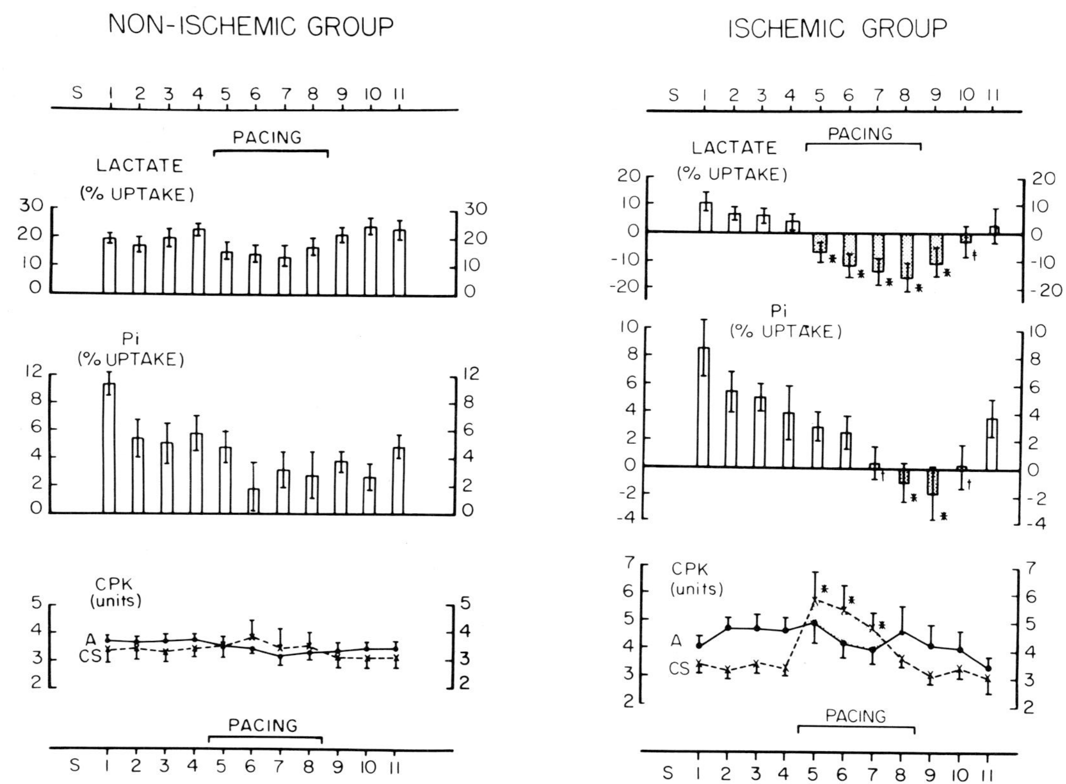

Figure 11. Effect of atrial pacing on lactate uptake, inorganic phosphate loss, and CPK blood levels in the nonischemic and ischemic groups. For details see text.

increment in CS CPK above arterial levels in 5 of 9 ischemic patients, suggesting myocardial release of this enzyme (Fig. 11). This change was not sustained, but declined while pacing was continued and, during the last 2 min of ischemia, CPK levels had fallen below those in the arterial blood. These findings suggest that this enzyme may escape from the paced ischemic human heart in the absence of demonstrable necrosis. The combination of tachycardia and ischemia may be another situation capable of facilitating the escape of intracellular enzymes, similar to those reported by other investigators during exercise (Halonen and Konttinen, 1962; Loegering and Critz, 1971). The observation that CS CPK levels declined even when ischemia was still present may be related to the preferential clearance of the enzyme through cardiac lymphatics as has recently been discussed (Malmberg, 1972). Amsterdam *et al.* (1971) failed to detect the release of myocardial enzymes during ischemia in man, and this discrepancy remains unexplained. Additional studies are required to clarify this point.

In conclusion, these studies show that the human heart responds to pacing-induced ischemia by either decreasing lactate extraction or producing lactate, by the loss of K^+ and P_i ions, and possibly also by the release of CPK. Myocardial lactate production appears to be the most reliable index of pacing-induced ischemia, followed by P_i loss, whereas K^+ efflux is of no value in situations associated with increased heart rate.

Acknowledgments These studies were performed in collaboration with Drs. R. O. West and R. B. Case.

REFERENCES

AMSTERDAM, E. A., R. ZELIS, J. A. BONANO, D. T. MASON, and R. GORLIN. 1971. Relation of cardiac ischemia and necrosis during angina pectoris: comparison of lactate metabolism and enzyme release by the human myocardium. Circulation 43–44 (Suppl. II): II–130.

APSTEIN, C. S., E. PUCHNER, and N. BRACHFELD. 1970. Improved automated lactate determination. Anal. Biochem. 38: 20–34.

BAKER, L. D., S. J. LESHIN, G. V. R. K. SHARMA, and J. V. MESSER. 1969. Relative effects of exercise and isoproterenol in eliciting regional metabolic abnormalities in cineangiographically quantitated human coronary artery disease. Am. J. Cardiol. 23: 104 (Abstr.).

BING, R. J. 1965. Cardiac metabolism. Physiol. Rev. 45: 171–213.

BING, R. J., L. D. VANDAM, F. GREGOIRE, J. C. HANDELSMAN, W. T. GOODALE, and J. E. ECKENHOFF. 1947. Catheterization of coronary sinus and middle cardiac vein in man. Proc. Soc. Exp. Biol. Med. 66: 239–240.

BRAUNWALD, E. 1971. Control of myocardial oxygen consumption: physiologic and clinical considerations. Am. J. Cardiol. 27: 416–432.

BURTON, A. C. 1957. The importance of the shape and size of the heart. Am. Heart J. 54: 801–810.

CAMPION, B., R. L. FRYE, and R. S. ZITNIK. 1970. Effects of nitroglycerin on capacitance vessels: a mechanism for reduction of left ventricular end-diastolic pressure. Mayo Clin. Proc. 45: 573–578.

CASE, R. B. 1971. Ions alterations during myocardial ischemia. Cardiology 56: 245–262.

CASE, R. B., H. A. ROSELLE, and R. S. CRAMPTON. 1966. Relation of ST-depression to metabolic and hemodynamic events. Cardiologia (Basel) 48: 2–41.

CHIONG, M. A., R. O. WEST, and J. O. PARKER. 1972. Influence of nitroglycerin on myocardial metabolism and hemodynamics during angina induced by atrial pacing. Circulation 45: 1044–1056.

CHIONG, M. A., R. O. WEST, and J. O. PARKER. 1974. Myocardial balance of inorganic phosphate and enzymes in man. Effects of tachycardia and ischemia. Circulation 49: 283–290.

GERLINGS, E. D., and J. P. GILMORE. 1968. Continuous potassium balance analysis in dog heart. J. Appl. Physiol. 25: 316–318.

GERLINGS, E. D., D. T. MILLER, and J. P. GILMORE. 1969. Oxygen availability: a major determinant of myocardial potassium balance. Am. J. Physiol. 216: 559–562.

GILMORE, J. P., and E. D. GERLINGS. 1969. Influence of interstimulus interval on myocardial potassium balance. Am. J. Physiol. 217: 136–141.

GOODALE, W. T., M. LUBIN, J. E. ECKENHOFF, J. H. HAFKENSCHIEL, and W. G. BANFIELD, Jr. 1948. Coronary sinus catheterization for studying coronary blood flow and myocardial metabolism. Am. J. Physiol. 152: 340–355.

GORLIN, R. 1972. Assessment of hypoxia in the human heart. Cardiology 57: 24–34.

GRUPP, G., G. H. ACHESON, and A. CHARLES. 1967. Heart rate and paired ventricular stimulation: effects on contractile force and K^+ exchange. Am. J. Physiol. 212: 607–611.

GUDBJARNASON, S. 1972. The use of glycolytic metabolism in the assessment of hypoxia in human hearts. Cardiology 57: 35–46.

HALONEN, P. I., and A. KONTTINEN. 1962. Effect of physical exercise on some enzymes in the serum. Nature, Lond. 193: 942–944.

HERMAN, M. V., W. C. ELLIOTT, and R. GORLIN. 1967. An electrocardiographic, anatomic and metabolic study of zonal myocardial ischemia in coronary heart disease. Circulation 35: 834–846.

HOCHELLA, N. J., and S. WEINHOUSE. 1965. Automated lactic acid determination in serum and tissue extracts. Anal. Biochem. 10: 304–317.

LANGER, G. A., and A. J. BRADY. 1966. Potassium in dog ventricular muscle: kinetic studies of distribution and effects of varying frequency of contraction and potassium concentration of perfusate. Circulat. Res. 18: 164–171.

LEE, S. J. K., Y. K. SUNG, and A. J. ZARAGOZA. 1970. Effects of nitroglycerin on left ventricular volumes and wall tension in patients with ischemic heart disease. Brit. Heart J. 32: 790–794.

LINHART, J. W., and F. J. HILDNER. 1973. Right atrial pacing in the evaluation of patients with atherosclerotic heart disease. *In* P. Samet (ed.), Cardiac Pacing, pp. 281–308. Grune and Stratton, New York.

LOEGERING, D. J., and J. B. CRITZ. 1971. Effects of hypoxia and muscular activity on plasma enzyme levels in dogs. Am. J. Physiol. 220: 100–104.

MALMBERG, P. 1972. Time course of enzyme escape via heart lymph following myocardial infarction in the dog. Scand. J. Clin. Lab. Invest. 30: 405–409.

MASON, D. T., and E. BRAUNWALD. 1965. The effects of nitroglycerin and amyl nitrite on arteriolar and venous tone in the human forearm. Circulation 32: 755–766.

MATHUR, P. P., and R. B. CASE. 1973. Phosphate loss during reversible myocardial ischemia. J. Mol. Cell. Cardiol. 5: 375–393.

MONROE, R. G. 1964. Myocardial oxygen consumption during ventricular contraction and relaxation. Circulat. Res. 14: 294–300.

MUELLER, P. S., and D. HORWITZ. 1962. Plasma free fatty acids and blood glucose responses to analogues of norepinephrine in man. J. Lipid Res. 3: 251–255.

NASSER, M. G., M. WACHTER, and R. B. CASE. 1969. A continuous automated sampling technique. J. Appl. Physiol. 26: 501–503.

OLSON, R. E. 1962. Physiology of cardiac muscle. *In* Handbook of Physiology. Section 2: Circulation 1: 199–235. American Physiological Society, Washington D.C.

OPIE, L. H. 1968. Metabolism of the heart in health and disease. Part I. Am. Heart J. 76: 685–698.
OPIE, L. H. 1971/72. Substrate utilization and glycolysis in the heart. Cardiology (Basel) 56: 2–21.
OPIE, L. H., P. OWEN, M. THOMAS, and R. SAMSON. 1973. Coronary sinus lactate measurements in assessment of myocardial ischemia. Comparison with changes in lactate/pyruvate and beta-hydroxybutyrate/aceto-acetate ratios and with release of hydrogen, phosphate and potassium ions from the heart. Am. J. Cardiol. 32: 295–305.
OPIE, L. H., M. THOMAS, P. OWEN, and G. SHULMAN. 1972. Increased coronary venous inorganic phosphate concentrations during experimental myocardial ischemia. Am. J. Cardiol. 30: 503–513.
OWEN, P., M. THOMAS, V. YOUNG, and L. H. OPIE. 1970. Comparison between metabolic changes in local venous and coronary sinus blood after experimental coronary artery occlusion. Am. J. Cardiol. 25: 562–570.
PARKER, J. O., M. A. CHIONG, R. O. WEST, and R. B. CASE. 1969a. Sequential alterations in myocardial lactate metabolism, ST segments and left ventricular function during angina induced by atrial pacing. Circulation 40: 113–131.
PARKER, J. O., M. A. CHIONG, R. O. WEST, and R. B. CASE. 1970. The effect of ischemia and alterations of heart rate on myocardial potassium balance in man. Circulation 42: 205–217.
PARKER, J. O., R. O. WEST, R. B. CASE, and M. A. CHIONG. 1969b. Temporal relationships of myocardial lactate metabolism, left ventricular function and ST segment depression during angina precipitated by exercise. Circulation 40: 97–112.
SCHEUER, J. 1967. Myocardial metabolism in cardiac hypoxia. Am. J. Cardiol. 19: 385–392.
SCHEUER, J., and N. BRACHFELD. 1966. Coronary insufficiency: relations between hemodynamic, electrical and biochemical parameters. Circulat. Res 18: 178–189.

Mailing address:
M. A. Chiong, M. D.,
Etherington Hall, Queen's University,
Kingston, Ontario K7L 2V7 (Canada).

Recent Advances in Studies on
Cardiac Structure and Metabolism Volume 10
The Metabolism of Contraction
Edited by Paul-Emile Roy and George Rona

HEMODYNAMIC AND METABOLIC RESPONSES TO PACING IN PATIENTS WITH CHEST PAIN AND NORMAL CORONARY ARTERIOGRAMS

W. A. HILL, D. H. JACKSON, B. J. MAURER,
H. G. McDANIEL, BENIGNO SOTO, T. J. REEVES,
and T. N. JAMES

University of Alabama Medical Center and Veterans Administration Hospital,
Department of Medicine, Division of Cardiology, Birmingham, Alabama, USA

SUMMARY

Hemodynamic and metabolic responses to pacing from either the coronary sinus or right atrium were evaluated in 41 patients with chest pain and normal coronary arteriograms. A group of patients (group II) with angina, lactate production, or significant ST segment depression had a significantly higher mean pulmonary capillary pressure on peak pacing or angina than did a group of patients considered to have a normal pacing response (group I). In 6 of 9 group II patients, the left ventricular end-diastolic pressure either rose abnormally with pacing or was greater than 14 mm Hg immediately after pacing and resembled that of a group of patients with coronary artery disease. Patients with a prolapsing mitral valve (group III) also had a significantly higher mean pulmonary capillary pressure on peak pacing as compared to those of group I, although abnormal left ventricular pressure responses occurred in only 2 of 9 of these patients. The stroke index was significantly lower in group III on peak pacing, while group II was no different from group I. Lactate production occurred in 6 of 9 group II patients. However, only 1 of 6 patients with a prolapsing mitral valve who were studied for lactate production was found to produce lactate, suggesting a different mechanism for their pain.

INTRODUCTION

Since the introduction of selective coronary arteriography, the dilemma of chest pain in patients with normal coronary arteriograms has become widely recognized (Kemp, Elliott, and Gorlin, 1967; Likoff, Segal, Kasparian, 1967). Various explanations have been advanced for this phenomenon, including difficulties in relating the arteriographic appearance to the true pathology of the coronary vessels and the possibility of small vessel disease (James, 1970; Hermon, 1973). Cardiac origin of the pain has also been questioned (Waxler, Kimbiris, and Dreifus, 1971).

Atrial pacing has been widely used in the evaluation of angina pectoris and hemodynamic alterations produced by it have been well documented in the presence of disease of the major coronary arteries (Sowton *et al.*, 1967; O'Brien *et al.*, 1969; Helfant *et al.*, 1970; Linhart, 1971, 1972; Parker, Khaja, and Case, 1971). Such information as exists with respect to pacing in patients with chest pain and normal coronary arteriograms suggests that they have a normal hemodynamic response; however, myocardial lactate production similar to that in patients with large vessel coronary artery disease has been found in a number of these patients (Kemp *et al.*, 1967; Arbogast and Bourassa, 1973). We have studied 41 patients using the technique of atrial or coronary sinus pacing. Our observations presented in this paper indicate that hemodynamic changes do occur which can be correlated with the occurrence of angina on pacing or with lactate production.

METHODS AND MATERIALS

Patient population

Forty-one patients who were found to have normal coronary arteriograms and a history of chest pain having many features of angina pectoris agreed to a second cardiac catheterization. Patients with rheumatic valvular heart disease, aortic valvular or subvalvular disease, cardiomegaly, congestive heart failure, severe hypertension by history, or a diastolic blood pressure of > 100 mm Hg were excluded. The research nature of the procedure was fully explained, and informed consent was obtained from each patient. All coronary arteriograms were reviewed by the arteriographers (WAH, DHJ, and BJM), a cardiovascular radiologist (BS), and an independent observer (TNJ).

Catheterization techniques

Cardiac catheterization was performed with patients in a fasting state without premedication. Previous medications were discontinued at least 3 days before the procedure. The pulmonary artery was catheterized using either a No. 7F Goodale-Lubin catheter of a No. 5F or 7F Swan-Ganz catheter. A No. 5F bipolar pacing catheter was positioned at the junction of the superior vena cava and the right atrium or a No. 7F or 8F Gorlin coronary sinus pacing catheter was passed to the coronary sinus. A No. 4F Teflon cannula, inserted percutaneously into the brachial artery, was passed to the aorta in 15 patients or to the left ventricle in 2 patients. In the remaining 24 patients, a No. 8F Cook pigtail catheter was passed to the left ventricle via the femoral artery and a No. 18 Cournand needle was inserted in the brachial artery. Pressures were recorded through the fluid filled catheters, which were connected with a Stratham P23Db transducer with a zero reference point 10 cm above the tabletop. Pressures were filtered at 12 Hz. Cardiac output was measured by the Fick method, with collection of expired gas over a 3-min period and pulmonary artery and arterial blood sampling during the 1st to 2nd minute. Pressures and electrocardiogram

(L2 or V5) were recorded on an Electronics for Medicine DR8 photographic recorder.

Experimental protocol

Patients were observed for a control period of 15 min after placement of the catheters. During the last 10 min of the control period, pulmonary artery, pulmonary capillary wedge, and aortic or brachial artery and left ventricular pressures were recorded at a paper speed of 25 and 100 mm/sec on a 10 or 20 mm Hg/cm scale before and after a cardiac output measurement. Pacing was begun at a rate of 100–110 beats/minute or 10 beats/minute above the control rate and increased in increments of 10–20 beats/minute until the development of chest pain thought to be angina, Wenckebach periodicity, or until a peak pacing rate which varied between 146 and 179 beats/minute had been reached. In 4 patients, 1.2 mg of atropine were administered intravenously in order that a pacing rate over 146 beats/minute could be achieved. Each level of pacing was maintained for at least 2 min, with pressure recordings between the 1st and 2nd minute. Mean pressures were determined electronically. Cardiac outputs were always measured at a heart rate of approximately 130 beats/min and at the maximal pacing rate achieved or during angina when this occurred. At these levels, pressures were measured before and after the output determination. On cessation of pacing, left ventricular or aortic, pulmonary capillary wedge, and pulmonary artery pressures were recorded continuously for at least 1 min or until they returned to control levels. Arterial and coronary sinus blood lactate concentrations were measured at rest, before and after the cardiac output determination at 130 beats/min, at the peak pacing rate, and 5 min after pacing stopped (revovery). In 14 patients, glucose and free fatty acids were also determined. Lactate and glucose were measured enzymatically on perchlorate extracts of whole blood (Hohorst, 1963). Free fatty acids were measured in plasma by the Duncombe copper method (Iraya and Ui, 1965).

The same protocol with the exception of coronary sinus catheterization was carried out in 11 patients with coronary artery disease before coronary arteriography.

Analysis of data

Pressures were measured and averaged over at least 10 beats. The left ventricular end-diastolic pressure was measured where the downslope of the left ventricular (A) coincided with the initial upstroke of left ventricular pressure. This usually occurred approximately 0.05 sec after the Q-wave and near the peak of the R-wave of the electrocardiogram. The time relationship between the Q-wave and the left ventricular end-diastolic pressure at the control rate was used as the reference point for measurement of the left ventricular end-diastolic pressure at rapid pacing rates. The stimulus to Q interval (S-Q interval) was measured from

Table 1. Criteria for classification of chest pain during pacing

Angina	Diffuse, steady with gradual↑ and↓ without abrupt fluctuations, deep dull quality, initiated by work of heart (including pacing), consistently and rapidly relieved by rest (discontinuing pacing), and above the diaphragm
Possible angina (atypical)	Variations of angina as defined above in one or more ways but not in terms of locations of chest pain
Not angina	Shooting or darting a few seconds or several hours long, aggravated by respiration, and precipitated by a single motion or by pressure over the chest

the pacing spike to the beginning of the Q-wave on the ECG. This measurement was taken during pacing rather than PR-interval since, frequently, it was not possible to clearly identify a P wave on the electrocardiogram (ECG) tracing. Significant ST segment depression was defined as depression of greater than 20% of the total QRS amplitude. This was done, since it was not possible to calibrate the ECG recording. It was required that the ST segment depression be present immediately after pacing to insure that it was not artifactual. Chest pain was classified during the pacing procedure as either angina, possible angina, or not angina according to criteria outlined by Harrison and Reeves (1968) (Table 1).

Table 2. Criteria for assignment of patients to the various groups

Group I
1) Not angina
2) Absence of lactate production
3) Slight or no ST↓
4) Possible angina in absence of lactate production or significant ST↓

Group II
1) Angina on pacing
2) Lactate production
3) Significant ST↓

Group III
1) A prolapsing mitral leaflet

Group IV
1) Patients with > 50% stenosis of at least 1 vessel

Group IV_a
1) Angina or significant ST↓

Patients were divided into groups according to their response to pacing. Group I (defined as having a normal pacing response) consisted of patients paced to at least 146 beats/min without either the development of chest pain classified as angina, production of lactate, or significant ST segment depression. Group II (defined as having an abnormal pacing response) consisted of patients who developed pain thought to be angina on pacing, whose myocardium produced lactate at some point during pacing, or who had significant ST segment depression. Group III consisted of patients with a prolapsing mitral leaflet documented by angiography. None of these patients had significant mitral regurgitaton; four patients who could not be paced above 135 beats/min because of Wenckebach periodicity were excluded from the groupings. Group IV consisted of patients with coronary artery disease of greater than 50% stenosis in one or more vessels; three patients had single vessel disease, and eight had two or three vessel disease (Table 2). Group IV_a consisted of only those patients with angina or significant ST segment depression on pacing (one with single vessel disease and seven with two- or three-vessel disease).

Statistical comparisons were made of the variables using the *t*-test.

Calculations

Derived variables were calculated using the following formulas:

Total pulmonary resistance (dynes sec cm^{-5}) = mean pulmonary artery pressure $\times$ 1,332/cardiac output (liter/min) $\times$ $10^3/60$

Pulmonary arteriolar resistance (dynes sec cm^{-5}) = (mean pulmonary artery pressure − mean pulmonary capillary pressure) $\times$ 1,332/cardiac output $\times$ $10^3/60$

Stroke index = cardiac index/heart rate (beats/min)

RESULTS

The mean age was similar for the first three groups: 42.5 years (range 32–54), group I; 44.7 years (range 25–60), group II; 48.1 years (range 38–61), group III. Group I consisted of 7 males and 7 females; group II, 8 males and 5 females; and group III, 3 males and 7 females. Group IV was older than group I ($p<0.05$; mean age 53.3 years, range 38–68) and consisted of 10 males and 1 female. The patients who were not grouped because they could not be paced at rates above 135 beats/min consisted of 1 male and 3 females (mean age 51.5 years, range 41–66).

The control heart rate in group I was higher than in the other groups ($p<0.05$); however, there was no difference among the groups in the peak heart rates reached. The systolic blood pressure $\times$ heart rate product was similar at peak pacing in all groups. All patients showed a slight rise in the mean aortic or

Table 3. Heart rate, blood pressure, and PR (or S-Q) interval during the control and during peak pacing*

Group	Control PR (msec)	Peak Pacing SQ	Heart Rate (beats/min) Control	Heart Rate (beats/min) Peak Pacing	SBP × HR Peak Pacing
I	150±21	220±52	87±12	159±8	20,000±4,500
II	160±20	241±44	77±14 $p < 0.05$	157±16	19,700±3,400
III	173±28 $p<0.05$	275±46 $p < 0.05$	77±9 $p < 0.05$	160±8	19,100±3,300
IV	146±19	242±46	67±9 $p < 0.05$	155±14	20,000±4,800

*SBP = Systolic blood pressure. HR = Heart rate. Values are expressed as means; statistical comparisons are described in the text.

brachial artery pressure with pacing, except for one group II patient and one group III patient, both of whom developed pulsus alternans. The PR interval during the control period and the S-Q interval at peak pacing were both significantly longer in Group III than in group I ($p<0.05$). In groups II and IV, these parameters were not significantly different from those of group I. (Table 3).

In group I, 10 patients had discomfort during the pacing study which was not felt to be angina, and 4 denied symptoms. Coronary sinus lactates were measured in 6 patients (nonproducers by definition). Slight ST segment depression was observed in 1 patient (Table 4).

In group II (abnormal response to pacing), 11 patients developed chest pain during pacing which was thought to be angina. Coronary sinus lactate measurements were obtained in 9 of these patients; 6 changed from myocardial lactate extraction to production and, in 1 patient, myocardial lactate extraction fell to zero. Of 7 patients in group II who exhibited ST segment depression with pacing, 3 were classified as significant. Only 1 patient in group II was included solely on the basis of ST segment depression, and this was a patient who developed chest pain during pacing which was not felt to resemble angina and in whom coronary sinus lactate blood was not obtained. Another patient developed chest pain which was labeled as possible angina, but produced lactate during pacing.

In group III, three patients developed chest pain during pacing which was felt to be angina, four developed possible angina, and two had discomfort which was not felt to be angina; one patient denied chest pain during the procedure but experienced pain in the postpacing period which was felt to represent angina. ST segment depression was observed in six of these patients, significant in two. Only one of six patients changed from myocardial lactate extraction to lactate production.

In group IV, eight patients developed angina or significant ST segment depression on pacing (Table 4).

Of the patients excluded from the groupings, two had prolapsing mitral valves but could not be paced above 110 beats/min because of Wenckebach periodicity; one of these patients extracted lactate during the control period. A third patient who could not be paced above 120 beats/min produced lactate during the control period when she experienced spontaneous pain. A fourth patient, a 40-year-old white male with left bundle branch block in the ECG and a history of classical angina, could not be paced above 130 beats/min and no coronary sinus blood was obtained.

Table 4. Frequency of the abnormal responses to pacing within each group

	Angina	Possible Angina	Not Angina
Group I			
Chest pain			14 (4 no chest pain)
Lactate			1 zero extraction 5 extraction 8 not obtained
ST			1 slight↓
Group II			
Chest pain	11	1	1
Lactate	3 not obtained 5 production 1 zero extraction 2 extraction	1 production	1 not obtained
ST	2 significant↓ 3 slight↓	1 slight↓	1 marked↓
Group III			
Chest pain	4 (1 patient postpacing)	4	2
Lactate	3 not obtained; 1 production	4 extraction	1 not obtained 1 extraction
ST	2 significant; 1 slight↓	3 slight↓	
Group IV			
Chest pain	7		4 no chest pain
Lactate	–	–	–
ST	5 marked↓		1 significant↓

Hemodynamic findings

The mean pulmonary capillary wedge pressure was no different in the four groups during the control period. During pacing, it rose in all groups but was significantly higher in groups II and III than group I (Table 5). The difference in the maximal pulmonary capillary wedge pressure between groups I and IV was not significant; but when only patients with angina or ST depression were included (IV_a), the difference was significant. In only one patient in group I was the mean pulmonary capillary wedge pressure above 14 mm Hg (16 mm Hg), while only three patients in group II had a maximal mean pulmonary capillary wedge pressure of less than 15 mm Hg (Fig. 1). Of these three subjects, one was a patient who underwent two separate pacing studies and the pulmonary pressures were measured at a time other than that when he had his angina and lactate production. The other two were patients whose pain on pacing was felt to be angina but in whom coronary sinus blood was not obtained. A maximal mean pulmonary capillary wedge pressure of less than 15 mm Hg was recorded in four patients of group III, and all of these were patients with a normal pacing response by the criteria outlined previously for group I. Of seven group II patients with lactate production on pacing in whom a simultaneous pulmonary capillary wedge pressure was recorded, none had a mean pulmonary capillary wedge pressure of less than 15 mm Hg. Of six patients in group I in whom lactate measurements were obtained, only one patient had a mean pulmonary

Table 5. Pulmonary capillary wedge pressure response to pacing*

Group	$\overline{PCP}$ (mm Hg) Control	Maximum	Change in $\overline{PCP}$ (mm Hg) (Maximum − Control)
I	4.8±2.3	10.3±4.3	5.4±2.8
II	6.4±3.8	18.5±6.2 $p < 0.005$	12.1±6.0 $p < 0.001$
III	5.0±2.5	16.7±6.0 $p < 0.01$	11.7±4.1 $p < 0.001$
IV	4.8±2.3	13.1±6.5	8.5±5.9
IV_a (angina or ST↓)	5.1±2.6	14.6±7.1 $p < 0.05$	9.5±6.9 $p < 0.05$

*Mean pulmonary capillary wedge pressure during control and the maximal pulmonary capillary wedge pressure observed with pacing. The change in the mean pulmonary capillary wedge pressure with pacing. $\overline{PCP}$ = mean pulmonary capillary pressure. Maximum − control = mean maximum pulmonary capillary pressure minus mean pulmonary capillary pressure during control.

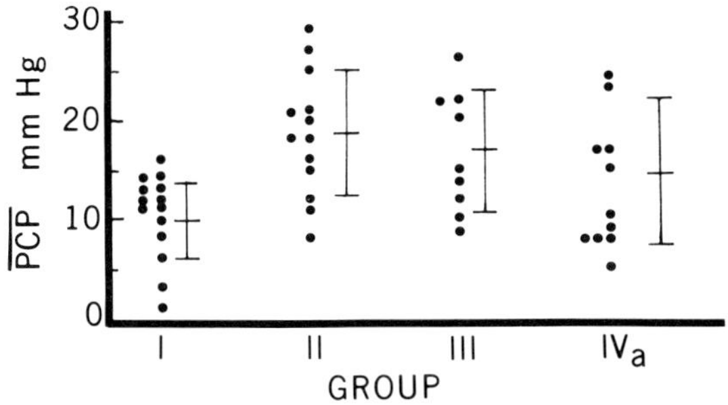

Figure 1. Maximal mean pulmonary capillary pressure for patients according to group. Mean and standard deviation for each group indicated by *bars.* $\overline{PCP}$ = mean pulmonary capillary pressure.

capillary pressure of greater than 14 mm Hg (16 mm Hg). No difference was noted in the pulmonary capillary pressure response within the groups between patients paced from the coronary sinus and those paced from the right atrium. Approximately equal numbers were paced from the coronary sinus in groups I, II, and III: Six group I, eight group II, and six group III patients.

The difference between the maximal and the control pulmonary capillary pressure was also significantly higher in groups II, III, and IV_a as compared to group I (Table 5). Although the maximal pulmonary artery pressure was also higher in groups II, III, and IV as compared to group I, it was significantly higher only in group II ($p<0.005$). The change in total pulmonary resistance (dynes sec cm^{-5}) on peak pacing or angina was significantly higher in groups II (149±152; $p<0.05$) and III (163±66; $p<0.005$) as compared to group I (58±61), while there was no difference between groups IV (56±29) and I. The pulmonary arteriolar resistance (dynes sec cm^{-5}) decreased slightly in all groups on peak pacing, and there was no difference between the groups (I, −11±26; II, −26±10; III, −20±42; IV, −29±49).

The left ventricular pressure was measured in 9 patients in group I, 9 patients in group II, 7 patients in group III, and in all 11 patients in group IV. There was no difference in the left ventricular end-diastolic pressure between the groups during the controlled period (Table 6). On peak pacing, the left ventricular end-diastolic pressure fell in groups I, II, and III, while it was unchanged from control in group IV. The left ventricular end-diastolic pressure at peak pacing was significantly higher in groups II and IV than in groups I or III. In all patients in group I, the left ventricular end-diastolic pressure was less than or equal to 4 mm Hg on peak pacing. No patient in this group had a left ventricular end-diastolic pressure immediately after pacing greater than 14 mm Hg. In group II, 4 patients had a left ventricular end-diastolic pressure of greater than 14 mm Hg immediately after pacing. In 1 patient, the left ventricular end-diastolic pressure

Table 6. Left ventricular end-diastolic pressure response to pacing*

	LVEDP (mm Hg)			Change in LVEDP (mm Hg)		
	Control	Peak Pacing	Postpacing	(Postpacing	–	Control)
I	7.8±2.8	2.2±1.0	10.5±3.0	2.7	±	2.0
II	9.4±4.1	6.8±5.3 $p < 0.05$	15.2±5.5 $p < 0.05$	5.6	±	5.4
III	6.6±3.1	2.6±2.8	12.3±7.7	5.7	±	5.0
IV	6.4±2.9	5.7±5.7 $p < 0.05$	14.2±8.9	7.8	±	7.5
IVa	6.6±3.1	5.8±9.5 $p < 0.05$	15.8±9.5 $p < 0.05$	9.1	± $p < 0.05$	8.2

*Mean left ventricular end-diastolic pressure at control, peak pacing, and immediately postpacing. Change in mean left ventricular end-diastolic pressure. Postpacing – control = left ventricular end-diastolic pressure immediately postpacing minus control left ventricular end-diastolic pressure.

rose to 25 mm Hg on pacing with the occurrence of angina; but when the angina decreased despite continued pacing, the left vetricular end-diastolic pressure fell to 12 mm Hg and was 13 mm Hg immediately after pacing. In another patient, the left ventricular end-diastolic pressure fell initially and rose to 7 mm Hg with the occurrence of angina on peak pacing, but postpacing left ventricular pres-

Table 7. Cardiac index and stroke volume index response to pacing

	Cardiac Index (liters/min/m^2)		Stroke Volume Index	
Group	Control	Peak	Control	Peak
I	3.5 ±.83	3.66±.70	39.7±9.7	22.9±4.1
II	3.04±.54	3.21±.94	39.4±9.3	20.5±6.1
III	2.99±.90	2.60±.87 $p < .005$	43.6±6.9	19.4±4.4 $p < 0.05$
IV	2.53±.43 $p < 0.005$	2.86±.66 $p < 0.05$	38.1±4.3	18.9±3.6 $p < 0.05$

sures were not obtained. Left ventricular end-diastolic pressures immediately after pacing greater than 14 mm Hg were recorded in 2 patients of group III and in 5 of group IV. Only in group IV was the change in the postpacing left ventricular end-diastolic pressure from control significantly different from group I.

The difference in the cardiac index between groups I, II, and III was not significantly different during the control period while, in group IV, it was lower than in group I. The cardiac index in groups I, II, and IV rose slightly on peak pacing. In group III, the cardiac index fell with peak pacing and, at this point, was significantly lower than in group I. The stroke volume index was no different between the groups during the control period. At peak pacing, it was significantly lower in groups III and IV than in group I (Table 7).

Left ventriculograms were done at peak pacing in three group I patients, two group II patients, and two group III patients. No mitral regurgitation was seen in group I or II patients. Slight mitral regurgitation was present in the two group III patients, but this was no different from previous angiograms done when the subjects were in normal sinus rhythm.

Other metabolic and hemodynamic findings

In 14 patients, arterial and coronary sinus glucose and free fatty acids were measured as well as lactate. The mean rise in arterial glucose on pacing is shown in Figure 2, and the mean increase in glucose extraction (arterial–coronary sinus) on pacing is demonstrated in Figure 3. The mean free fatty acid responses are illustrated in a similar fashion in Figure 4; the most prominent feature in this figure is the marked variability in the arterial free fatty acid level with pacing.

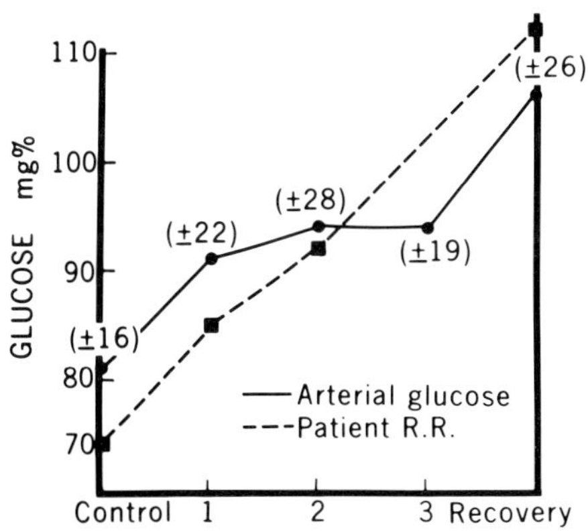

Figure 2. Mean arterial glucose with pacing. Standard deviation in parentheses. *Interrupted line* indicates response in patient R.R. *1* = pacing 130 beats/min before cardiac output measurement. *2* = pacing 130 beats/min after cardiac output measurement. *3* = peak pacing or angina.

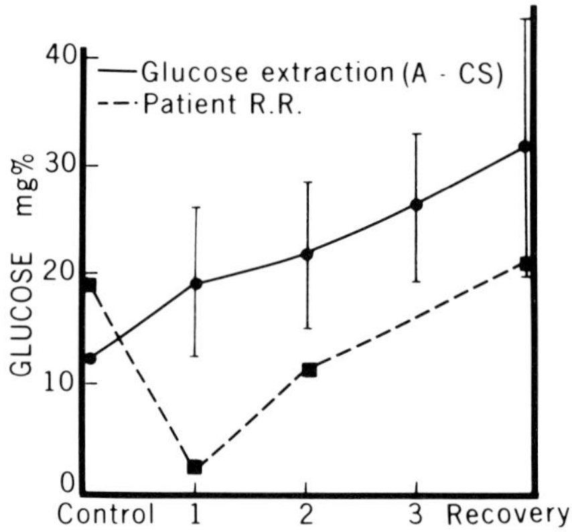

Figure 3. Mean glucose extraction. Extraction in patient R.R. represented by *interrupted line. A – CS = arterial minus coronary sinus. 1, 2,* and *3* = pacing rates as in Figure 2.

In two patients, lactate extraction fell to zero. A group I patient (pI in Fig. 5) had markedly elevated free fatty acids at the time. As the free fatty acid level fell on pacing, the percentage of lactate extraction increased. This is in contrast to a group II patient (pII, Fig. 5) who had a normal free fatty acid level when lactate extraction was zero.

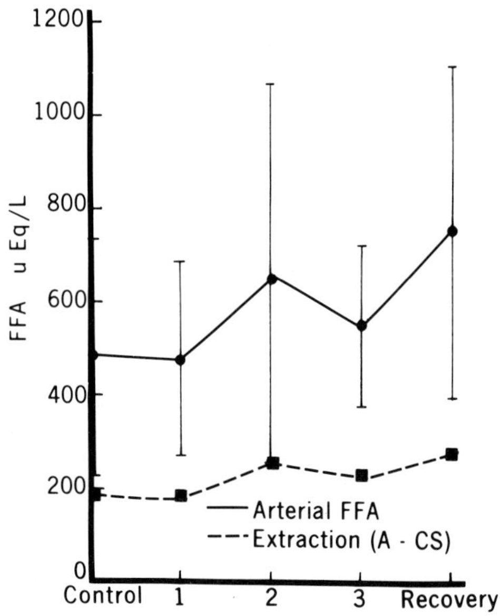

Figure 4. Mean arterial free fatty acid and mean free fatty acid extraction with pacing. *Bars* indicate mean and standard deviation. FFA = free fatty acid. A – *CS = arterial minus* coronary sinus. *1, 2,* and *3* = pacing rates as in Figure 2.

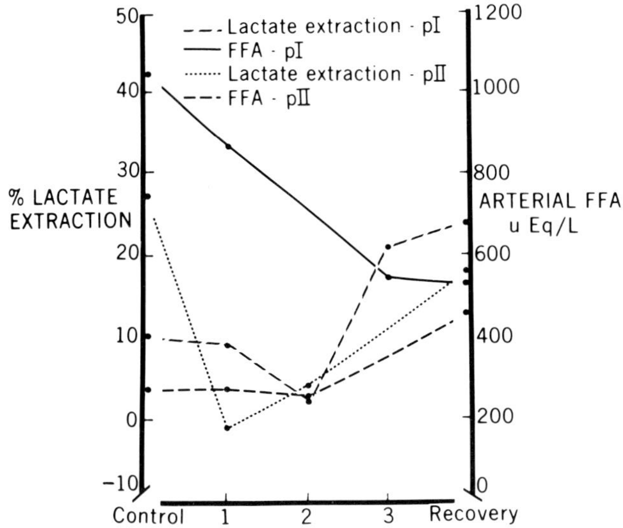

Figure 5. Lactate extraction and arterial free fatty acids for 2 patients whose lactate extraction fell to zero. pI = group I patient with a normal pacing response. pII = group II patient with an abnormal pacing response. % Lactate extraction = arterial lactate – coronary sinus lactate/ arterial lactate × 100. FFA = free fatty acids. *1, 2,* and *3* = pacing rates as in Figure 2.

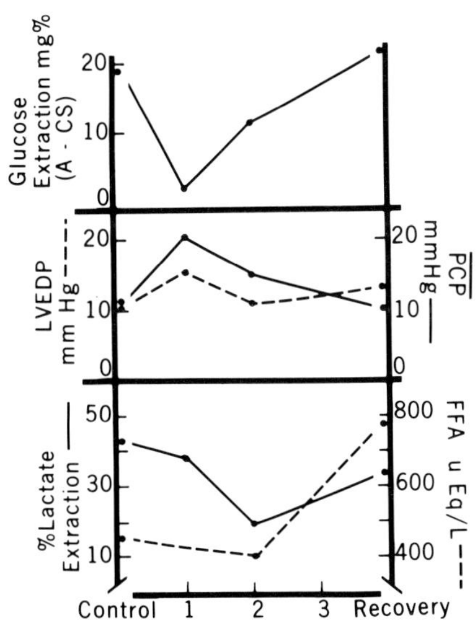

Figure 6. Hemodynamic and metabolic responses in patient R.R.; % lactate extraction as in Figure 5. FFA = free fatty acid. LVEDP = left ventricular end-diastolic pressure. $\overline{\text{PCP}}$ = mean pulmonary capillary pressure. A – CS = arterial minus coronary sinus.

Both group II patients who extracted lactate throughout pacing were untreated chemical diabetics. The hemodynamic and metabolic responses of one of these patients (R.R.) are represented in Figures 2,3, and 6. Glucose extraction by the heart fell to virtually zero at a point when his chest pain was most severe and the pulmonary capillary pressure was at its maximum. With continued pacing, the chest pain diminished and the pulmonary capillary pressure and the left ventricular end-diastolic pressure fell. This corresponded with the return to glucose extraction by the heart. In three other group II patients, the pulmonary capillary wedge pressure and left ventricular end-diastolic pressure decreased following the cardiac output measurement on peak pacing. In all of these patients, the chest pain had either diminished or disappeared completely and, in one the myocardial response reverted back to lactate extraction. These types of changes were not observed in patients from other groups.

In group II, five patients (38%) had abnormal glucose tolerance curves by criteria of the American Diabetes Association (1969). This compared to two patients in group I (14%) and one patient in group III (10%).

DISCUSSION

Patients were divided into groups for the purpose of analyzing the data. An abnormal response was defined as pain resembling angina on pacing, lactate production, or significant ST segment depression. Two types of hemodynamic abnormalities were observed. First, patients with an abnormal pacing response had a significantly higher mean pulmonary capillary pressure on peak pacing or angina as compared with patients considered to be normal. Secondly, the left ventricular end-diastolic pressure was higher in the abnormal group at peak pacing and in the immediate postpacing period and resembled the group with coronary artery disease. In 12 patients who had a prolapsing mitral valve, the mean pulmonary capillary pressure on peak pacing was also significantly higher than in the normal group, but an abnormal left ventricular end-diastolic pressure response was seen in only 2 of these patients. Thus, a significant segment of patients with chest pain and normal coronary arteriograms have hemodynamic and metabolic responses to pacing stress resembling those of patients with large vessel coronary artery disease.

The rise in the mean pulmonary capillary wedge pressure with rapid atrial pacing has been attributed in part to a relatively reduced time for egress of blood from the atrium in diastole (Mitchell, Gilmore, and Sarnoff, 1962). Since the mean peak heart rates were similar for all groups, this explanation should not account for the difference in the mean pulmonary capillary wedge pressure observed between the groups. Fusion of the A-wave on the preceding V-wave as A-V conduction time increases and, hence, atrial contraction against a partially closed mitral valve have also been used as an explanation for the rise in the pulmonary capillary pressure with pacing (Mitchell, Gilmore, and Sarnoff, 1962). Diastolic waves on the pulmonary capillary wedge tracing which have previously been attributed to fusion of the A and V waves were noted in patients of all groups (Leighton *et al.*, 1969). In four patients given atropine, the S-Q interval shortened and the height of the diastolic waves decreased as did the mean pulmonary capillary wedge pressure, thus supporting the second concept although an alteration of vagal tone is a possibility.

In group III patients, the greater prolongation of S-Q time with rapid pacing suggests that this mechanism may account, at least partially, for the higher mean pulmonary capillary wedge pressure in some of these patients. However, in groups II and IV, the S-Q intervals were similar to those of group I for the various pacing rates. Since the speed of electrical conduction from atrium to ventricle is determined mainly by the transmission rate through the A-V node, the S-Q interval should be a reasonable estimate of the time lag between atrial and ventricular conduction (Detweiler, 1973). Thus, the relationship between the atrial and ventricular contraction should not account for the differences in the mean pulmonary capillary wedge pressure between groups I and II.

Pacing left ventriculograms suggest that mitral regurgitation is not the cause of the higher pulmonary capillary wedge pressure in group II patients, although it could possibly contribute to the higher pulmonary capillary pressure and lower stroke volume index in patients with a prolapsing mitral valve.

Other studies have shown only minor hemodynamic differences between pacing from the coronary sinus and the right atrium (Befeler *et al.*, 1971). We did not observe any differences within the groups between patients paced from the coronary sinus and right atrium, making it unlikely that this had any effect on our hemodynamic observations.

The correct labeling of chest pain in this group of patients is obviously a difficult task, and observer bias in labeling pain as angina is a possibility. Nevertheless, the fact that our labeling was correct is supported by the finding that, in six of eight group II patients felt to have angina, the myocardium failed to extract lactate at some point during the pacing study while this was not seen in any of the group I patients who were said not to have angina, except for the patient referred to in Figure 5. The lactate results were not known at the time the pain was classified. In group II, five patients were classified as such solely on the basis of angina; one of these was the untreated chemical diabetic (R.R.) referred to in Figure 6. The left ventricular end-diastolic pressure of 22 mm Hg immediately postpacing in this particular patient is additional strong evidence of abnormality. Another was the patient (pII) seen in Figure 5, whose myocardial lactate extraction fell to zero. This patient had an abnormal rise in his left ventricular end-diastolic pressure to 7 mm Hg on peak pacing, but a postpacing left ventricular pressure was not obtained. In one patient, lactates were not obtained and the left ventricular pressure was not measured. The electrocardiographic (ECG) response could not be interpreted because of the presence of left bundle branch block. The other two patients did not have significant ECG changes with pacing, and lactates were not obtained. While it is possible that in these patients chest pain may have been mislabeled, their exclusion from group II would only have made the rise in the pulmonary capillary pressure more significant, since maximal wedge pressures in these patients were only 8 and 11 mm Hg.

Although the left ventricular end-diastolic pressure and S-Q intervals were similar for groups II and IV on peak pacing, the mean pulmonary capillary wedge pressure was not quite as high in group IV_a patients. Barry *et al.* (1974) have demonstrated an increase in left ventricular diastolic stiffness during angina pectoris in patients with coronary artery disease. It is possible to explain our observations by alteration of atrial as well as ventricular compliance. Assuming a generalized abnormality, involvement of both atrial and ventricular myocardium would be expected. This would account for the higher mean pulmonary capillary wedge pressure in group II patients while the left ventricular end-diastolic pressure changes were slightly higher in group IV_a. The fact that heart rates immediately postpacing were no different from control rates argues against invoking increased vagal tone with pacing to account for the higher mean pulmonary capillary wedge pressure in group II patients. In addition, alteration of vagal tone cannot account for the changes seen in the left ventricular end-diastolic pressure.

Endomyocardial biopsy specimens taken from patients with chest pain and normal coronary arteriograms have been reported as showing either endothelial, smooth muscle, or basement membrane thickening consistent with small vessel disease (Rider, Billingham, and Harrsion, 1974). The fact that a higher percentage of group II patients had evidence of abnormal glucose tolerance suggests that small vessel disease may be playing a part in their syndrome. The fact that in four patients the mean pulmonary capillary wedge pressure fell as the chest pain diminished and that in one patient the myocardium reverted back to lactate extraction indicates that, at least in some patients, the abnormality is reversible even with the continued stress of pacing. The return of the heart to glucose extraction in one of these four patients (R.R.) suggests that in this particular case a reversible metabolic defect is playing a major role.

A recent study using atrial pacing in patients with chest pain and normal coronary arteriograms suggests that these patients have normal myocardial function with the stress of pacing (Arbogast and Bourassa, 1973). Several factors can account for the differences in our observations. First, our patients were paced to a much higher heart rate. Although lactate production generally was observed at a pacing rate of 130 beats/min in our study, hemodynamic changes were not usually significant until higher pacing rates were reached. Secondly, owing to the reversibility of the abnormality in some patients with continued pacing, unless frequent measurements were made, abnormalities could be missed. Thirdly, only three patients in the previously published study produced lactate, and no effort was made to examine these patients separately. Fourthly, pacing at a given heart rate continuously for 8 min may not be suitable to stress the heart maximally. We chose to increase the pacing rate every 2–3 min, since it is known that maximal myocardial oxygen requirements are reached after exercising for 4 min at any particular level (Anderson, 1968).

The use of myocardial lactate production in rapid atrial pacing as an indicator of abnormality is supported by studies in both dogs and humans (Dear, Harman, and Reeves, 1971; Neill and Kremkau, 1974). Dear *et al.* (1971) demonstrated that lactate production in dogs does not occur even at heart rates greater than 300/min unless coronary perfusion pressure drops to unphysiological levels concomitant with maximal oxygen extraction. Neill and Kremkau (1974), using pacing rates similar to those used by us, demonstrated lactate extraction in 16 patients without evidence of heart disease.

However, it is evident from our observations that, in certain instances, other parameters in addition to lactate are needed in order to properly interpret the metabolic response of the heart to stress. The lactate produced by the heart during ischemia is derived either from glycogen stores in the heart or from glucose taken up by the heart (Weissler *et al.*, 1968). A defect in glycogen storage or in glucose uptake by the heart because of diabetes mellitus could result in lactate values during pacing that do not correlate with myocardial ischemia. This may be what is taking place in the group II patient with untreated chemical diabetes (R.R.) who extracts lactate at all times. Competition between substrates for utilization by the heart may also influence the percentage of arterial lactate that is extracted (Spitzer, 1974). Although the extraction of lactate and free fatty acids by the heart is a linear function of the arterial concentration, this relationship is altered by the competition between free fatty acids and lactate for utilization by the heart (Stanley *et al.*, 1974). An increase in arterial free fatty acid level would increase the myocardial extraction of free fatty acids and correspondingly depress the myocardial extraction of lactate even in a normal heart. This appears to be the case in the group I patient (pI) shown in Figure 5. Thus, in assessing the significance of lactate masurements across the heart, it is important to know free fatty acid and glucose values as well as lactate.

Whether the lactate production in patients with normal coronary arteriograms represents true ischemia or a metabolic abnormality cannot be answered from our data as yet. It was initially thought that the calculation of excess lactate from the lactate-to-pyruvate ratio would be a good judge of the degree of ischemia (Huckabee, 1958). This has not proven to be true. As shown by Opie, the lactate-to-pyruvate ratio simply reflects the redox state of cytoplasm rather than the degree of oxygenation of the mitochondria and may be influenced by factors other than the oxygen state of the cell (Opie *et al.*, 1973). Recent work has shown that adenosine is released into the coronary circulation during ischemia, and this, or its derivative inosine, may prove useful in evaluating this group of patients (Fox *et al.*, 1974).

It became evident early in the study that a significant number of patients with chest pain and normal arteriograms had a prolapsing mitral leaflet, and for that reason it was decided to include this group of patients in the study. Review of previous articles on chest pain with normal coronary arteriograms reveals a significant number of patients with ST-T wave changes in the inferior leads, a finding commonly associated with a prolapsing mitral valve (Kemp *et al.*, 1967; Engle, 1969). Thus, it is likely that these previous studies included a number of patients with this syndrome. Failure of lactate production in the majority of patients with a prolapsing mitral valve suggests a different mechanism for the chest pain in this syndrome, although hemodynamic changes were also found in this group of patients.

CONCLUSIONS

This study indicates that a significant number of patients with chest pain and normal coronary arteriograms develop chest pain resembling angina pectoris during pacing. That the pain is of cardiac origin is supported in the majority of these patients by myocardial lactate production. They can best be separated hemodynamically by a significantly higher elevation of the mean pulmonary capillary pressure at peak pacing or angina. Changes also occur in the left ventricular end-diastolic pressure but are not as useful in separating the groups. Alteration of atrial and ventricular compliance may account for these observations, and some of these patients may have small vessel disease.

A significant number of patients included in the study (12 of 41) had a prolapsing mitral valve. A greater rise in the mean pulmonary capillary pressure was also generally noted in these patients. However, only 1 of 6 patients in this group studied for lactate production was found to produce lactate, suggesting a different mechanism for their pain.

Acknowledgments This work was supported by the following grants: Social and Rehabilitation Service, Department of Health, Education, and Welfare, 13-P-55477/4-01. Cardiovascular Research and Training Center Program Project Grant HL 11, 310. Myocardial Infarction Research Unit Program, Contract PH 43-67-1441. Grant RR-32 from the General Clinical Research Centers Program of the Division of Research Resources, National Institutes of Health. VA Research Service, Birmingham, Alabama. Dr. W. A. Hill was sponsored by Training Grant HL 5737. Dr. B. J. Maurer was sponsored by American Heart Association through the British-American Research Fellowship. Dr. H. G. McDaniel is a V.A. Clinical Investigator.

REFERENCES

ANDERSEN, K. L. 1968. The cardiovascular system in exercise. *In* Exercise Physiology, pp. 79–128. Academic Press, New York and London.

ARBOGAST, R., and M. G. BOURASSA. 1973. Myocardial function during atrial pacing in patients with angina pectoris and normal coronary arteriograms: comparison with patients having significant coronary artery disease. Am. J. Cardiol. 32: 257–263.

BARRY, W. H., J. Z. BROOKER, E. L. ALDERMAN, and D. C. HARRISON. 1974. Changes in diastolic stiffness and tone of the left ventricle during angina pectoris. Circulation 49: 255–263.

BEFELER, B., F. J. HILDNER, R. P. JAVIER, L. S. COHEN, and P. SAMET. 1971. Cardiovascular dynamics during coronary sinus, right atrial, and right ventricular pacing. Am. Heart J. 81: 372–380.

Committee on Statistics of the American Diabetic Association. 1969. Standardization of the oral glucose tolerance test. Diabetes 18: 299–310.

DEAR, H. D., M. A. HARMAN, and T. J. REEVES. 1971. Pacing induced maximal stress in the normal canine circulation. Cardiovasc. Res. 5: 566–571.

DETWEILER, D. K. 1973. Electrophysiology of the heart. *In* Best and Taylor's Physiological Basis of Medical Practice, pp. 3–55. The Williams and Wilkins Company, Baltimore.

ENGLE, M. A. 1969. The syndrome of apical systolic click, late systolic murmur and abnormal T-waves. Circulation 39: 1–2.

FOX, A. C., G. E. REED, E. GLASSMAN, A. J. KALTMAN, and B. B. SILK. 1974. Release of adenosine from human hearts during angina induced by rapid atrial pacing. J.C.I. 53: 1447–1457.

HARRISON, T. R., and T. J. REEVES. 1968. Principles and Problems of Ischemic Heart Disease, p. 200. Year Book Medical Publishers, Chicago.

HELFANT, R. H., J. S. FORRESTER, J. R. HAMPTON, J. I, HAFT, H. G. KEMP, and R. GORLIN. 1970. Coronary heart disease: Differential hemodynamic, metabolic and electrocardiographic effects in subjects with and without angina pectoris during atrial pacing. Circulation 42: 601–610.

HERMON, M. V. 1973. Angina-like chest pain without identifiable cause. Ann. Int. Med. 79: 445–447.

HOHORST, H. J. 1963. L (+) lactate: determination with lactate dehydrogenase and DPN. *In* M. V. Bergmeyer (ed.), Method of Enzymatic Analysis, p. 266. Academic Press, New York.

HUCKABEE, W. E. 1958. Relationships of pyruvate and lactate during anaerobic metabolism. I. Effects of infusion of pyruvate or glucose and of hyperventilation. J.C.I. 37: 244–254.

ITAYA, K., and M. UI. 1965. Colorimetric determination of free fatty acids in biological fluids. J. Lipid Res. 6: 16–20.

JAMES, T. N. 1970. Angina without coronary disease. Circulation 42: 189–191.

KEMP, H. G., ELLIOTT, W. C., and GORLIN, R. 1967. The anginal syndrome with normal coronary arteriography. Trans. Ass. Am. Physicians 80: 59–70.

KEMP, H. G., P. S. VOKONAS, P. F. COHN, and R. GORLIN. 1973. The anginal syndrome associated with normal coronary arteriograms. Am. J. Med. 54: 735–742.

LEIGHTON, R. F., S. J. ZARON, J. L. ROBINSON, and A. M. WEISSLER. 1969. Effects of atrial pacing on left ventricular performance in patients with heart disease. Circulation 40: 615–622.

LIKOFF, W., B. L. SEGAL, and H. KASPARIAN. 1967. Paradox of normal selective coronary arteriograms in patients considered to have unmistakable coronary heart disease. New Engl. J. Med. 76: 1063–1066.

LINHART, J. W. 1971. Myocardial function in coronary artery disease determined by atrial pacing. Circulation 44: 203–212.

LINHART, J. W. 1972. Atrial pacing in coronary artery disease, including preinfarction angina and postoperative studies. Am. J. Cardiol. 30: 603–610.

MITCHELL, J. H., J. P. GILMORE, and S. J. SARNOFF. 1962. The transport function of the atrium: Factors influencing the relation between mean left atrial pressure and left ventricular end-diastolic pressure. Am. J. Cardiol. 9: 237–247.

NEILL, W. A., and E. L. KREMKAU. 1974. Criteria for detecting ischemic myocardial hypoxia from lactate and pyruvate data during atrial pacing in humans. J. Lab. Clin. Med. 83: 428–435.

O'BRIEN, K. P., L. M. HIGGS, D. L. GLANCY, and S. E EPSTEIN. 1969. Hemodynamic accompaniments of angina: A comparison during angina induced by exercise and by atrial pacing. Circulation 39: 735–743.

OPIE, L. H., P. OWEN, M. THOMAS, and R. SAMSON. 1973. Coronary sinus lactate measurements in assessment of myocardial ischemia. Am. J. Cardiol 32: 295–305.

PARKER, J. O., F. KHAJA, and R. B. CASE. 1971. Analysis of the left ventricular function by atrial pacing. Circulation 43: 241–252.

RIDER, A. K., M. E. BILLINGHAM, and D. C. HARRISON. 1974. Endomyocardial biopsy in the diagnosis of ischemic myocardial disease with normal coronary arteries. (Abstract). Clinical Res. 22: 299A.

SOWTON, G. E., R. BALCON, D. CROSS, and M. H. FRICK. 1967. Measurement of the angina threshold using atrial pacing: A new technique for the study of angina pectoris. Cardiovasc. Res. 1: 301–307.

SPITZER, J. 1974. Effect of lactate infusion on canine myocardial free fatty acid metabolism in vivo. Am. J. Physiol. 226: 213–217.

STANLEY, A., R. MORASKI, R. RUSSELL, C. RACKLEY, and H. G. McDANIEL. 1974. Alteration of myocardial fuel and oxygen extraction by glucose-insulin-potassium. (Abstract). Clinical Res. 22: 13A.

WAXLER, E. B., D. KIMBIRIS, and L. S. DREIFUS. 1971. The fate of women with normal coronary arteriograms and chest pain resembling angina pectoris. Am. J. Cardiol. 28: 25–32.

WEISSLER, A. M., F. A. KRUGER, N. BABA, D. G. SCARPELLI, R. F. LEIGHTON, and J. K. GALLIMORE. 1968. Role of anaerobic metabolism in the preservation of function capacity and structure of anoxic myocardium. J. Clin. Invest. 47: 403–415.

Mailing address:
W. A. Hill, M.D.,
University of Alabama, University Station, 1919 South 7th Avenue,
Birmingham, Alabama 35294 (USA).

Recent Advances in Studies on
Cardiac Structure and Metabolism Volume 10
The Metabolism of Contraction
Edited by Paul-Emile Roy and George Rona

COMPARISON OF CHANGES IN MYOCARDIAL BALANCES OF LACTATE, GLUCOSE, POTASSIUM, AND INORGANIC PHOSPHATE DURING PACING-INDUCED ANGINA

GILLES R. DAGENAIS, YVES MARQUIS,
and ANDRÉ MOISAN

Institut de Cardiologie de Québec, Québec, Canada

SUMMARY

During an atrial pacing test, a correlative study in myocardial lactate, glucose, potassium, and inorganic phosphate balances was done in 34 patients with clinical evidence of ischemic heart disease. Electrocardiogram was continuously monitored while left ventricular end-diastolic pressure (LVEDP) was measured before and immediately after pacing. Coronary angiograms performed after the pacing test revealed atherosclerotic narrowings in all patients. During pacing, 16 patients developed anginal pain, and their LVEDP increased significantly. The 18 other patients had no angina and no significant change in LVEDP. In these 18 patients, there were no significant changes throughout the pacing study in myocardial balances of lactate, glucose, potassium, and inorganic phosphate. In contrast, the 16 patients with induced angina during pacing showed a significant myocardial production of lactate and a loss of potassium. Myocardial inorganic phosphate loss was not statistically significant. There was no significant change in myocardial glucose extraction during angina, although a slight increase was observed during the 1st min after pacing. There was no correlation between the arterial concentration and the myocardial extraction of these substances. No stoichiometric relationship was found between glucose and lactate or between potassium and inorganic phosphate balance. Myocardial extraction and production of lactate correlated best with inorganic phosphate uptake and loss. In the present study, lactate was a more reliable metabolic indicator of myocardial ischemia than potassium and inorganic phosphate, although these last two substances may be helpful in achieving a greater accuracy for biochemical diagnosis of ischemia. Myocardial glucose balance was of no value as a metabolic indicator of ischemia in this pacing study.

INTRODUCTION

During pacing-induced angina, myocardial balances of potassium (Parker et al., 1970; Dagenais et al., 1973), glucose (Most, Gorlin, and Soeldner, 1972), and inorganic phosphate (Chiong, West, and Parker, 1974) have been evaluated as metabolic indicators of ischemia. However, the value of each of these substances has been assessed independently by comparison with the changes observed simultaneously in myocardial lactate metabolism. The present study was designed to determine the relationship between lactate, glucose, potassium, and inorganic phosphate myocardial balance and to assess which of these four substances was the most reliable metabolic indicator of ischemia in humans during pacing-induced angina.

METHODS

Subjects

A total of 33 males and 1 female aged 35–63 years (mean 50.2), with either angina pectoris and/or a previous history of a well documented myocardial infarction and admitted for coronary arteriographic studies, agreed to participate in these evaluations. No patient had congenital or valvular heart disease, unstable angina, or recent myocardial infarction. No patient had diabetes mellitus. All antianginal drugs except nitroglycerin were discontinued for at least 36 hours before the study. After an overnight fast period of 10 hours, the patient received 2 tablets of diazepam, 5 mg, as premedication at least 1 hour before the procedure.

Procedure

A bipolar Zucker catheter was introduced into the basilic vein and advanced to the midportion of the coronary sinus. The position of the catheter, which was verified with a manual injection of contrast medium, did not change during the study. The Zucker catheter was then connected to a pacing Medtronic 5800 generator. A second catheter was introduced percutaneously into the femoral artery and advanced to the left ventricle. Left ventricular pressures, measured with a Statham P23Db pressure transducer, were recorded on a photographic recorder (DR-8 recorder, Electronics for Medicine). Zero reference pressure was at midchest level. Left ventricular end-diastolic pressure (LVEDP) was determined at the point where the downslope of the "a" wave coincided with the upstroke of the left ventricular pressure. LVEDP, recorded at high sensitivity at a paper speed of 75 mm/sec, was measured during eight consecutive cardiac cycles before and immediately after pacing.

Leads II and V5 were continuously monitored throughout the study. A positive electrocardiogram is defined as a flat or downsloping ST segment depression of at least 1 mm for a period of 0.08 sec, in comparison to the control tracing.

Coronary sinus pacing was done by increasing the heart rate by 20 beats/min until anginal pain occurred or until the maximal heart rate of 150–180 beats/min was reached. Pacing was maintained during 4 min when heart rate reached 130 beats/min and also when maximal rate was attained. In 6 of these 34 patients, atropine sulfate (0.6–0.8 mg i.v.) was given to correct an atrioventricular block during pacing. However, in spite of this medication, it was impossible to reach a heart rate of 150 beats/min in 3 patients.

Blood was collected simultaneously from the coronary sinus and the left ventricle in heparinized syringes for lactate, glucose, and potassium and in nonheparinized syringes for inorganic phosphate. Blood samples were taken in triplicate over a 2-min period during the control period, after 2 min of pacing at 130 beats/min and 2 min at maximal heart rate, and 1 min after pacing.

In all patients, left ventriculograms and selective coronary arteriograms were recorded following the metabolic study.

Biochemical analysis

Blood samples were immediately put in ice water and centrifuged within 30 min. Lactate was measured by the enzymatic technique of Horn and Bruns (1956) and potassium by flame photometry (Atomsorb, Jarrell Ash). The standard deviation of repeated pooled plasma analysis for potassium was 0.03 mEq/liter and for lactate 0.02 mmole/liter. Glucose was determined according to the technique of Raabo and Terkildsen (1960), and inorganic phosphate was analyzed according to the technique of Dryer, Tammes, and Routh (1957). The standard deviation of repeated pooled plasma analysis for glucose was 0.20 mmole/liter and for inorganic phosphate, 0.03 mEq/liter.

The myocardial extraction ratio of each of the substances was calculated as the difference between the arterial and coronary sinus concentration divided by the arterial concentration. An extraction equal or inferior to zero was defined as myocardial production or loss.

RESULTS

All 34 patients with clinical evidence of ischemic heart disease had coronary atherosclerotic narrowing greater than 50%; 25 of these patients had at least two coronary vessel diseases. During pacing, 18 patients did not have any chest pain and constituted the asymptomatic group, while the 16 patients who had anginal pain were designated as the group with induced angina. A higher incidence of angina and more severe coronary artery disease was present in the patients with induced angina. Electrocardiographic and hemodynamic findings are shown in Figure 1. In the asymptomatic group, the heart rate attained was higher; 4 of these patients had a positive electrocardiogram, while there was no significant change between the mean pre- and postpacing LVEDP for the entire group. In contrast, 9 of the 16 patients with pacing-induced angina had a positive electrocardiogram and the postpacing LVEDP increased significantly.

Table 1 summarizes the metabolic data. No significant change in the arterial concentrations of each substance was observed throughout the study. There was no significant relationship between myocardial uptake and the arterial concentration of each of these substances.

In the asymptomatic group, no significant change in myocardial extraction of

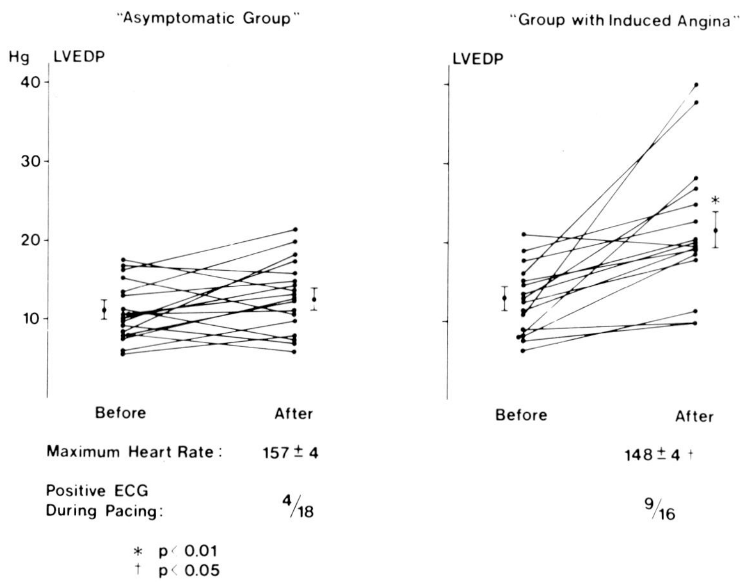

Figure 1. Maximal heart rate during pacing, number of positive electrocardiograms (ECG), and left ventricular end-diastolic pressure (LVEDP) before and immediately after pacing in each group (mean and SEM).

lactate and glucose was seen. During the last minute of pacing, there was a myocardial loss of potassium which was not statistically different from the control value. However, 1 min after pacing, the myocardium extracted this electrolyte. Although there was some variation in inorganic phosphate throughout the study, there was no significant change of the myocardial balance of this substance. Table 2 shows the changes in myocardial extraction during pacing at maximal heart rate from the control values in each patient of the asymptomatic group, together with their respective electrocardiographic, hemodynamic, and angiographic findings. Although 10 of the 18 patients showed decreased myocardial lactate extraction, only 4 had a myocardial lactate production. Among these 4 patients, an increased myocardial glucose extraction was observed in only 2, while myocardial potassium loss occurred in 2. In this asymptomatic group, only 4 had a myocardial potassium extraction during pacing, while a positive myocardial balance in inorganic phosphate was documented in 11 patients.

In the 16 patients with pacing-induced angina (Table 1), a significant myocardial lactate production was observed during the last minute of pacing and immediately after pacing. The myocardial extraction of glucose did not vary significantly, except during the postpacing period when the extraction increased significantly. A significant myocardial loss of potassium occurred only during

Table 1. Atrial (A) and arteriocoronary sinus difference (A-CS) concentrations of lactate, glucose, potassium, and inorganic phosphate (mean and SEM)

	Lactate (mmole/liter)		Glucose (mmole/liter)		Potassium (mEq/liter)		Inorganic phosphate (mEq/liter)	
	A	A-CS	A	A-CS	A	A-CS	A	A-CS
Asymptomatic group*								
1. Control	1.02 ± 0.08	0.06 ± 0.03	4.80 ± 0.08	0.22 ± 0.05	3.86 ± 0.13	−0.02 ± 0.06	2.01 ± 0.07	0.08 ± 0.03
2. Pacing, 130	0.90 ± 0.07	0.13 ± 0.03	4.84 ± 0.09	0.29 ± 0.07	4.13 ± 0.15	0.02 ± 0.08	1.99 ± 0.08	0.04 ± 0.03
3. Pacing, max.	0.98 ± 0.08	0.07 ± 0.03	4.79 ± 0.09	0.16 ± 0.06	3.90 ± 0.14	−0.09 ± 0.05	2.01 ± 0.07	0.04 ± 0.03
4. Postpacing	0.91 ± 0.08	0.07 ± 0.03	4.82 ± 0.10	0.23 ± 0.06	3.97 ± 0.16	0.08 ± 0.06	1.97 ± 0.07	0.03 ± 0.04
Group with induced angina*								
1. Control	0.95 ± 0.06	0.03 ± 0.04	4.83 ± 0.14	0.25 ± 0.06	4.05 ± 0.10	−0.04 ± 0.05	1.99 ± 0.09	0.02 ± 0.03
2. Pacing, 130	0.94 ± 0.08	0.01 ± 0.05	4.91 ± 0.17	0.23 ± 0.09	4.00 ± 0.12	−0.22 ± 0.06	2.02 ± 0.13	0.02 ± 0.03
3. Pacing, max.	0.93 ± 0.06	−0.17 ± 0.07	4.84 ± 0.13	0.26 ± 0.06	4.08 ± 0.10	−0.21 ± 0.04	2.01 ± 0.09	−0.07 ± 0.05
4. Postpacing	0.92 ± 0.06	−0.12 ± 0.09	4.98 ± 0.14	0.40 ± 0.07	4.12 ± 0.12	0.13 ± 0.08	2.00 ± 0.10	−0.06 ± 0.06
*p*1	NS	NS	NS	NS	NS	NS	NS	NS
*p*2	NS	NS	NS	NS	NS	0.01	NS	NS
*p*3	NS	0.01	NS	NS	NS	0.01	NS	0.05
*p*4	NS	0.01	NS	0.05	NS	NS	NS	NS

*The number indicates the metabolic values at: 1=rest; 2=paced heart rate of 130 beats/min; 3=maximal paced heart rate; and 4=1 min after pacing. *p* refers to the difference between the 2 groups. NS=no significant change.

Table 2. Changes in myocardial extraction (%) during pacing (maximal) from control values in asymptomatic group*

Patients	Lactate	Glucose	Potassium	Inorganic Phosphate	ECG	ΔLVEDP	CAD
C.O.	−7.5	−1.6	−6.8†	−10.0	−	+5	2
J.C.	+11.6	−5.2	+4.9	−1.2	+−	+4	1
D.G.	−0.7	+0.1	−7.2†	−1.4	−	+6	1
P.R.	−7.9	−2.2	−	−17.2†	−	−2	3
C.D.	−0.8†	−0.9	+12.8	+8.7	−	−4	2
J.I.	−2.8	−2.4	−2.2†	−6.4	−	−5	1
H.H.	+9.3	−3.6	+6.2†	+7.4	−	+2	2
M.B.	−4.7	−	−11.2†	−1.9	+	+6	1
M.G.	−21.9†	+6.6	−4.9†	+8.9	+	+1	2
P.B.	−3.4†	+3.3	−12.6	−	−	+10	2
L.P.	−9.8	+0.9	−2.6	+3.5	+	−1	2
A.F.	+10.9†	−3.9†	+0.5†	+3.7	−	+2	2
E.S.	+6.5	−0.2	−7.1†	−15.5†	+	−2	2
P.L.	−1.2	+2.1	−3.6†	−1.5†	−	+3	1
R.D.	+7.4	+6.3	−1.4†	+7.3	−	−3	1
R.T.	+5.7	−0.3	−7.5†	−18.5†	−	+5	2
C.P.	+13.4	−16.9†	+5.9†	+0.5†	−	+8	3
N.G.	+1.3	−1.9	+2.4†	+2.2†	−	+1	3

*ECG, electrocardiographic findings during pacing: −negative; +positive; +−doubtful. ΔLVEDP, absolute change between post- and prepacing left ventricular end-diastolic pressure. CAD, number of vessels involved with atherosclerotic narrowings > 50%.

†Indicates a negative arteriocoronary sinus difference.

Table 3. Changes in myocardial extraction (%) during pacing (maximal) from control values in patients with induced angina*

Patients	Lactate	Glucose	Potassium	Inorganic Phosphate	ECG	ΔLVEDP	CAD
J.D.	$-26.2^{\dagger}$	−0.3	$-10.1^{\dagger}$	$-13.4^{\dagger}$	−	+14	2
H.C.	+0.2	+4.1	$-1.9^{\dagger}$	−1.2	+	+22	3
G.N.	$+3.5^{\dagger}$	−1.8	$-7.1^{\dagger}$	$-9.2^{\dagger}$	−	+8	2
R.G.	$-80.1^{\dagger}$	−2.2	$-9.5^{\dagger}$	+3.0	+	+7	2
A.Mc.	$-4.6^{\dagger}$	—	$-16.1^{\dagger}$	$-1.8^{\dagger}$	−	+5	2
W.M.	$-20.6^{\dagger}$	−0.8	$+11.2^{\dagger}$	−0.6	−	+1	3
E.P.	$-84.9^{\dagger}$	+3.3	$-8.6^{\dagger}$	$-24.3^{\dagger}$	+	+29	3
A.L.	−28.7	—	$-3.9^{\dagger}$	$-15.6^{\dagger}$	+	+5	1
H.G.	−8.2	$-8.4^{\dagger}$	$+1.7^{\dagger}$	$-8.5^{\dagger}$	+ −	+7	2
E.R.	$-34.6^{\dagger}$	−9.0	$-8.5^{\dagger}$	+1.3	+	+9	1
L.B.	−7.7	−2.3	$-2.2^{\dagger}$	$+0.8^{\dagger}$	+	+2	2
E.L.	$-37.4^{\dagger}$	+5.9	$-2.7^{\dagger}$	$-1.2^{\dagger}$	+−	−1	2
S.D.	$-26.7^{\dagger}$	−0.1	$-5.5^{\dagger}$	$-5.6^{\dagger}$	+−	+6	3
D.R.	$-22.4^{\dagger}$	+4.2	$-14.0^{\dagger}$	$-5.0^{\dagger}$	+	+5	2
H.C.	+10.2	+5.4	+10.1	+21.9	+	+20	1
L.B.	$-29.4^{\dagger}$	+0.1	$-4.6^{\dagger}$	$-7.3^{\dagger}$	+	+3	2

*Symbols and abbreviations as in Table 2.

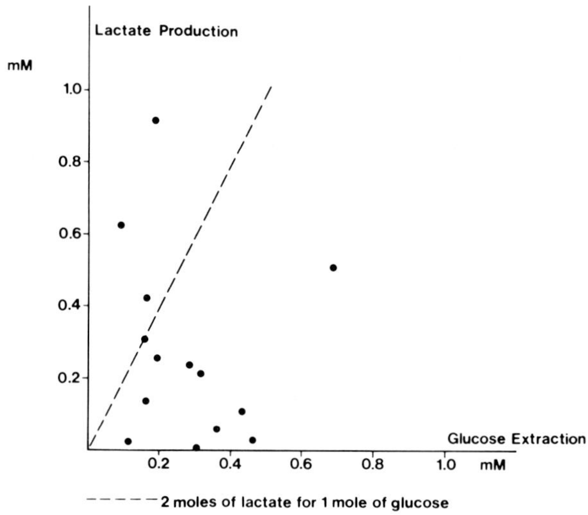

Figure 2. Comparison between myocardial lactate production (coronary sinus-arterial concentration) and myocardial glucose extraction (arteriocoronary sinus concentration) during maximal pacing heart rate. The correlation is not statistically significant.

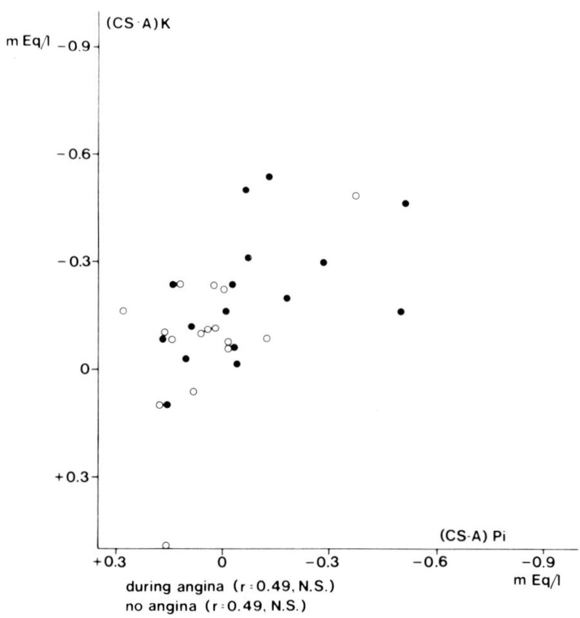

Figure 3. Comparison between myocardial balance in potassium (K) and inorganic phosphate (P_i) during maximal pacing heart rate: •, patient with angina; ○, asymptomatic patient.

pacing; but after pacing, in contrast to lactate, the myocardium extracted this electrolyte. On the other hand, myocardial inorganic phosphate loss increased during and immediately after pacing in a manner similar to that of lactate, but this loss was not statistically significant. Table 3 indicates the individual angiographic, hemodynamic, and electrocardiographic findings, and the changes in the extraction values during induced angina compared to the resting values. During pacing angina, lactate was the most reliable metabolic indicator: 11 of the 16 patients with pacing-induced angina had myocardial lactate production and 8 of them had a simultaneous myocardial loss of potassium and inorganic phosphate. Among these 11 patients with lactate production, only 4 increased their myocardial glucose extraction. Myocardial potassium loss was observed in all except 1 patient, while myocardial inorganic phosphate loss was documented in 11 patients.

The metabolic data were examined for stoichiometric relationships among the 4 substances. Figure 2 illustrates a comparison between the arteriocoronary sinus

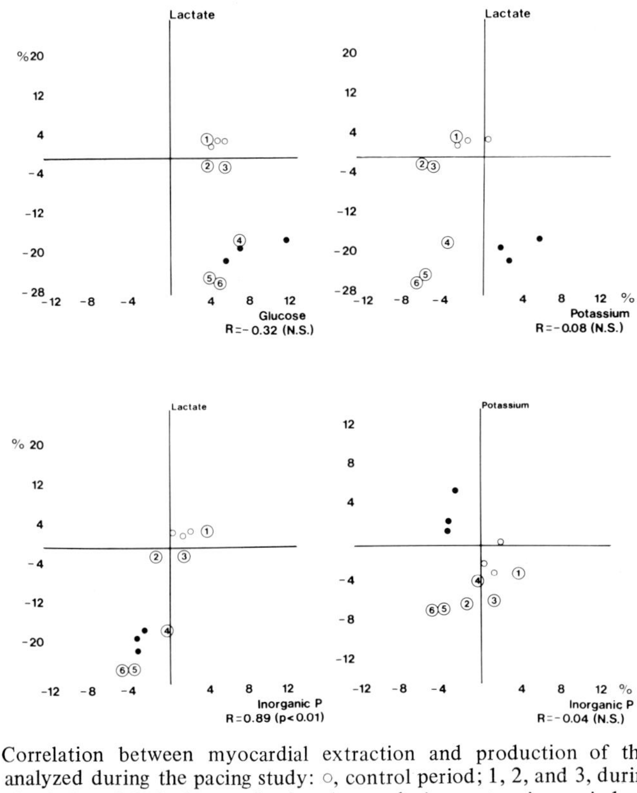

Figure 4. Correlation between myocardial extraction and production of the different substances analyzed during the pacing study: ○, control period; 1, 2, and 3, during pacing at 130 beats/min; 4, 5, and 6, during maximal pacing; •, during postpacing period.

difference in glucose and the coronary sinus-arterial difference in lactate during maximal pacing heart rate. No molar relationship could be found between these two substances. No significant correlation on equivalent basis could be detected between inorganic phosphate and potassium (Fig. 3). In the group with pacing-induced angina, a good correlation was observed between myocardial extraction and loss of lactate and inorganic phosphate, while a barely significant correlation was found between lactate and potassium only during maximal heart rate (R = 0.52, p<0.05) (Fig. 4).

DISCUSSION

The metabolic changes occurring in animals during myocardial ischemia induced either by acute coronary obstruction or by gradual decrease in total coronary blood flow have been well evaluated by analyses of blood obtained from venous drainage and/or of myocardial tissues of the ischemic region. It has been shown in animals that the ischemic myocardium takes up more glucose (Owen, Thomas, and Opie, 1969) and releases lactate (Owen *et al.*, 1969; Case, 1971; Opie *et al.*, 1973), potassium, and inorganic phosphate (Case, 1971; Opie *et al.*, 1973). Opie *et al.* (1973) have also compared the changes in all these metabolic indicators during induced myocardial ischemia in dogs. The present study was designed to assess these metabolic changes in man. However, there are several difficulties in assessing myocardial balance of any substance in patients with coronary artery disease. 1) Coronary atherosclerotic disease is often associated with focal myocardial fibrosis and heterogenous blood flow (Dwyer, 1972). Thus, fixed coronary sinus sampling may not reveal regional ischemic changes as well as selective sampling from a coronary vein draining the ischemic region, as was pointed out by Opie *et al.* (1973). 2) During these metabolic studies based on the arteriocoronary sinus difference, it is assumed that coronary blood flow during the sampling period is stable. If this is not the case, the value of metabolic evaluations under these conditions may be erroneous, as was pointed out by Zierler (1961). 3) The metabolic findings may vary according to the duration and extent of ischemia. 4) It is possible that other substrates, such as free fatty acids, may interfere with the myocardial metabolism of glucose and lactate (Lassers *et al.*, 1971; Brachfeld and Apstein, 1972).

These inherent difficulties in assessing metabolic changes in patients with coronary artery disease may limit the detection of all ischemic manifestations but would not affect the main purpose of this study, which is based on the correlative evaluation of the myocardial balances of glucose, lactate, potassium, and inorganic phosphate done simultaneously during 12 successive sampling periods under the same conditions.

In the present study, the myocardial extraction of lactate at rest was low. However, it should be recalled that all patients in this study were selected on the basis of clinical evidence of ischemic heart defined as angina pectoris and/or previous myocardial infarction. Of the 34 patients studied, 11 had myocardial lactate production at rest, severe angina, and at least two vessels involved with coronary atherosclerosis. During pacing, 18 patients did not have anginal pain but 4 had myocardial lactate production. On the other hand, among the 16 patients with induced angina and with associated hemodynamic and electrocardiographic changes, 11 showed or increased myocardial lactate production during pacing. These results are similar to previous observations (Parker *et al.*, 1969; Helfant *et al.*, 1970; Dagenais *et al.*, 1973).

In accordance with the findings of Most *et al.* (1972), the present study reveals no relationship between myocardial glucose uptake and arterial glucose concentration. It has been shown that ischemia in the canine heart significantly enhances glucose uptake (Owen *et al.*, 1969). Such enhancement was not seen in patients with pacing-induced angina by Most *et al.* (1972) or in the present study. In contrast to the findings of the latter investigators, no definite correlation was observed between the amount of lactate produced by the myocardium and myocardial glucose uptake during pacing angina. However, 1 min after interruption of pacing, a significant increase in glucose extraction was observed and correlated well, at that time, with myocardial lactate production. It seems likely that both myocardial glycogen and glucose uptake contribute to lactate production. Quantitative analysis of this anaerobic metabolism would be only a gross approximation because of the inherent difficulties in such a study and the marked variation of glucose uptake. In the present study, no stoichiometric relationship between glucose and lactate could be obtained, and glucose could not be used as a metabolic indicator during pacing-induced angina.

Myocardial potassium loss has been documented in the canine heart during ischemia and correlated well with myocardial lactate production and electrocardiographic findings compatible with ischemia (Case, 1971). Parker *et al.* (1970) observed that pacing-induced tachycardia was associated with myocardial potassium loss, a loss that was significantly increased during pacing angina; immediately after interruption of pacing, the human myocardium extracted this electrolyte. Similar observations were made during the present study. However, in contrast with Parker's findings, an excellent correlation between potassium loss and lactate production could not be demonstrated. Under the conditions of the present study, lactate was a better metabolic indicator of ischemia than potassium. This is mainly due to the small arteriocoronary sinus difference in potassium during pacing-induced tachycardia and pacing-induced angina.

In the canine myocardium during induced ischemia, Wollenberger and Krause (1968) documented an intracellular increase of adenosine diphosphate, adenosine monophosphate, and inorganic phosphate due to a decrease of adenosine triphosphate and creatine phosphate. Case (1971) and Opie *et al.* (1973) demonstrated inorganic phosphate release during induced myocardial ischemia in dogs.

During pacing-induced angina, Chiong *et al.* (1974) found that, of 18 patients, 10 showed myocardial inorganic phosphate release, while myocardial lactate production occurred in 13. Although in the present study 11 of the 16 patients with induced angina showed a myocardial inorganic phosphate loss, the anion loss was not statistically significant for the entire group. However, the inorganic phosphate loss correlated well with myocardial lactate production but not with myocardial potassium release.

In conclusion, this study reveals the absence of a constant molar relationship between the release of lactate, potassium, and inorganic phosphate during pacing-induced angina. As pointed out by Opie *et al.* (1973), the absence of a definite stoichiometric relationship may be due to unequal rates of production and/or diffusion of these substances from ischemic myocardium. During this pacing study, lactate was the most reliable indicator of myocardial ischemia, while myocardial glucose balance was of no value in detecting anaerobic metabolism. Although potassium and inorganic phosphate are less valuable as metabolic indicators, their association with lactate may be helpful in achieving a better biochemical diagnosis of ischemia.

Acknowledgments This work was supported by grants from the Medical Research Council of Canada. The technical skill of Mr. Denis Simard was appreciated. Doctor Dagenais is a scholar of the Canadian Heart Foundation.

REFERENCES

BRACHFELD, N., and C. S. APSTEIN. 1972. The influence of available substrate on contraction of the hypoxic myocardium. Cardiology 57: 47.

CASE, R. B. 1971. Ion alterations during myocardial ischemia. Cardiology 56: 245.

CHIONG, M. A., R. O. WEST, and J. O. PARKER. 1974. Myocardial balance of inorganic phosphate and enzymes in man. Effects of tachycardia and ischemia. Circulation 49: 283.

DAGENAIS, G. R., E. BENMOUYAL, J. ALSAC, L. GAILIS, Y. MARQUIS, and A. MOISAN. 1973. Effects of pacing-induced angina on myocardial α-glycerophosphate balance in humans. *In* N. S. Dhalla (ed.), Recent Advances in Studies on Cardiac Structure and Metabolism, Vol. 3, p. 805. University Park Press, Baltimore.

DRYER, R. L., A. R. TAMMES, and J. I, ROUTH. 1957. The determination of phosphorus and phosphatase with N-phenyl-p-phenylenediamine. J. Biol. Chem. 225: 177.

DWYER, E. M. 1972. Regional myocardial blood flow. Circulation 45: 4.

HELFANT, R. F., J. S. FORRESTER, J. R. HAMPTON, J. I. HAFT, H. G. KEMP, and R. GORLIN. 1970. Coronary heart disease: differential hemodynamic, metabolic and electrocardiographic effects in subjects with and without angina pectoris during atrial pacing. Circulation 42: 601.

HORN, H. D., und BRUNS, F. H. 1956. Quantitative Bestimmung L (+) milchsaure mit Milchsauredehydrogenase. Biochim. Biophys. Acta 21: 376.

LASSERS, B. W., M. L. WAHLQVIST, L. KAIPER, and L. A. CARLSON. 1971. Relationship in man between plasma free fatty acids and myocardial metabolism of carbohydrate substrate. Lancet *ii:* 448.

MOST, A. S., R. GORLIN, and J. S. SOELDNER. 1972. Glucose extraction by the human myocardium during pacing stress. Circulation 45: 92.

OPIE, L. H., P. OWEN, M. THOMAS, and R. SAMSON. 1973. Coronary sinus lactate measurements in assessment of myocardial ischemia. Comparison with changes in lactate/pyruvate and beta-hydroxybutyrate/acetoacetate ratios and with release of hydrogen, phosphate and potassium ions from the heart. Am. J. Cardiol. 32: 295.

OWEN, P., M. THOMAS, and L. OPIE. 1969. Relative changes in free fatty acid and glucose utilisation by ischemic myocardium after coronary arterial occlusion. Lancet *i:* 1187.

PARKER, J. O., M. A. CHIONG, R. O. WEST, and R. B. CASE. 1969. Sequential alterations in myocardial lactate metabolism, S-T segments and left ventricular function during angina induced by atrial pacing. Circulation 40: 113.

PARKER, J. O., M. A. CHIONG, R. O. WEST, and R. B. CASE. 1970. The effect of ischemia and alterations of heart rate on myocardial potassium balance in man. Circulation 42: 205.

RAABO, E., and T. C. TERKILDSEN. 1960. On the enzymatic determination of blood glucose. Scand. J. Clin. Lab. Invest. 12: 402.

WOLLENBERGER, A., and E. G. KRAUSE. 1968. Metabolic control characteristics of the acutely ischemic myocardium. Am. J. Cardiol. 22: 349.

ZIERLER, K. L. 1961. Theory of the use of arteriovenous concentration differences for measuring metabolism in steady and non-steady states. J. Clin. Invest 40: 2111.

Mailing address:
Doctor Gilles R. Dagenais,
Institut de Cardiologie de Québec, 2725, Chemin Ste-Foy,
Ste-Foy, Québec, Québec G1V 4G5 (Canada).

Metabolic Manifestations of Ischemia in Experimental Animals

Recent Advances in Studies on
Cardiac Structure and Metabolism Volume 10
The Metabolism of Contraction
Edited by Paul-Emile Roy and George Rona

SOME METABOLIC APPROACHES TO IMPROVING MYOCARDIAL PERFORMANCE DURING CARDIAC HYPOXIA

JAMES SCHEUER

Division of Cardiology, Montefiore Hospital and Medical Center,
Albert Einstein College of Medicine, Bronx, New York, USA

SUMMARY

Four possible metabolic approaches to improving cardiac function in the presence of myocardial hypoxia have been considered.

1. It appears that there is increasing evidence which suggests that free fatty acids are harmful to the ischemic heart.
2. Although it has been demonstrated that Krebs cycle intermediates can result in anaerobic energy formation by the mitochondria, and under certain extreme conditions can lead to improved performance of the heart, the potential for a physiologically important effect of this approach is probably limited.
3. The protection of the ischemic or hypoxic heart by alkalosis may be a feasible approach. The major beneficial effect appears to be exerted through more efficient conversion of energy that is already available to contractile performance rather than by increasing energy supply.
4. There appears to be some real potential for improving cardiac energy delivery via the glycolytic pathway. Calculations based on isolated rat heart studies indicate that, at 50% oxygenation, glycolytic ATP generation could totally correct for the deficit in mitochondrial ATP formation. Therefore, it is in the area of overcoming the inhibition of glycolytic ATP formation and tapping this potential metabolic pathway that energy delivery may be restored toward normal in the hypoxic and perhaps the borderline zone of underperfusion in the ischemic heart. The problem of the ischemic inhibition of glycolysis may partially be overcome by creating extracellular alkalosis, but this presumption will have to be tested.

In this communication, we will examine some methods that have been explored in recent years in an attempt to improve cardiac function and maintain cell viability during hypoxia. Stress will be placed upon studies in which myocardial function was measured simultaneously with the metabolic evaluation.

GLYCOLYSIS AND GLYCOGENOLYSIS

Studies as early as 1933 led to the suggestion that increases in glycolytic flux could enhance energy formation and contractile performance in the mammalian heart (Evans *et al.,* 1933). This topic has been well reviewed in a recent publication, and we will present only some new and perhaps unique insights (Opie, 1971/72).

The potential rates of adenosine triphosphate (ATP) generation through the glycolytic system must be considered. Calculations made from the enzymatic activities of heart would predict that the maximal glycolytic flux should be in the range of 200–300 μmoles/g, dry weight/min of hexose converted to pyruvate (Opie, 1971/72). The rates of glycolysis that we and others have observed in the perfused heart are considerably less than that: about 2.2–4.2 μmoles/g dry weight/min in aerobic perfused rat heart (Morgan *et al.,* 1961; Scheuer and Stezoski, 1970). In the oxygenated perfused heart, there is a doubling of glucose utilization in the presence of insulin and an additional increase with anoxia (Morgan *et al.,* 1961). This is further potentiated with anoxia plus insulin. The levels of anaerobic glycolysis achieved are about 3.5 times higher in the hearts that are paced (Scheuer and Stezoski, 1970) than when pacing is omitted (Morgan *et al.,* 1961). In the very early phase of anoxia, when glycogenolysis is maximal, the glycolytic flux is much greater, about twice the steady state flux achieved after several minutes, and reaches about 50 μmoles/g dry weight/min (Scheuer and Stezoski, 1970).

Figure 1 shows glycolytic ATP production as a percentage of the maximal achievable anaerobic production with varying levels of oxygenation calculated from our previously published reports (Scheuer and Stezoski, 1970, 1972a). The maximal achievable level is approximately 109 μmoles/g dry weight/min early in

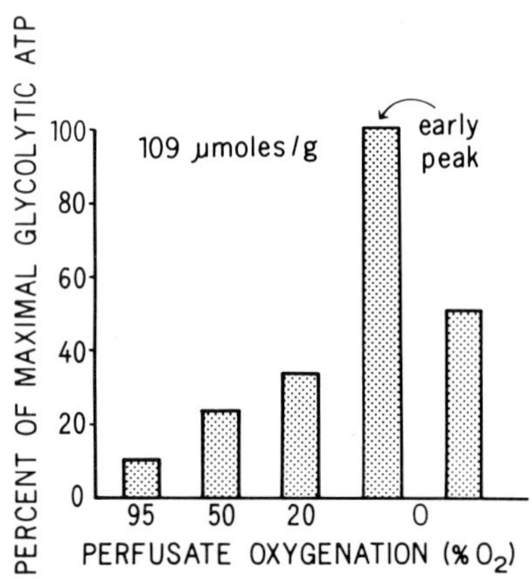

Figure 1. Effects of graded hypoxia upon glycolytic ATP formation in isolated perfused rat hearts. Hearts were perfused aerobically and then subjected to abrupt hypoxia. In all experiments in this and subsequent figures, hearts were paced at 330 beats/min, the medium was Krebs bicarbonate buffer containing 5 mM glucose, and the temperature was 37°C. The early peak with 0% oxygenation (anoxia) is taken as the maximal glycolytic potential (109 μmoles/g weight/min). The other values are taken after 5 min of anoxia or hypoxia. All metabolic values are based upon dry heart weights.

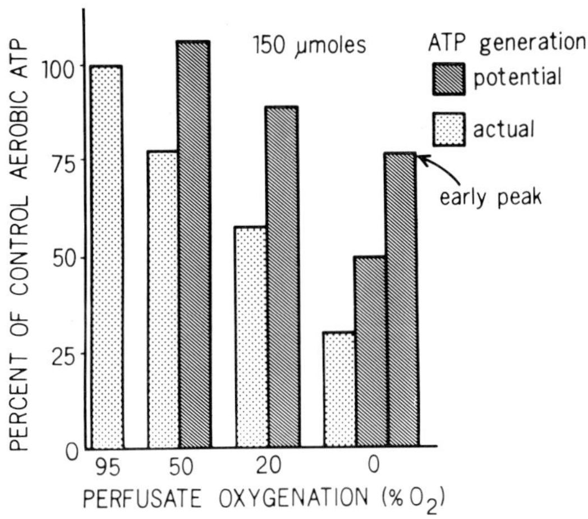

Figure 2. Effects of graded hypoxia upon total ATP generation in isolated perfused rat hearts. The conditions of perfusion were the same as in Figure 1. The stippled bars represent values calculated from observed glycolytic rates. Potential rates are those that would be calculated if the maximal potential glycolytic flux had always been achieved.

glycolysis. This degree of glycolytic ATP production is not reached at 50 or 20% oxygenation, which are levels of hypoxia that cause declines in mechanical performance and a rise in the lactate/pyruvate ratio; nor is the peak level of glycolytic ATP production sustained after the earlier burst in glycolytic flux. Even these rates of glycolytic ATP production would be markedly attenuated in ischemia as opposed to hypoxia (Neely, Whitmer, and Rovetto, 1974).

Figure 2 shows the percentage of total ATP generation during hypoxia, that is, glycolytic plus oxidative ATP in isolated heart experiments; 100% is taken as the rate of ATP production in aerobic hearts or approximately 150 μmoles/g dry weight/min. At 50% oxygenation, ATP production is reduced by 25%; yet if maximal glycolysis had occurred, ATP generation would have been sufficent to remain at or above the control level. At 20% oxygenation, the ATP generation rate is reduced over 40%, but with maximal glycolysis it would be reduced by less than 15%. With anoxia, ATP generation is about 30% of control, but the maximal potential could reach over 50% if all the ATP came from glucose or 75% if all came from glycogen.

These theoretical considerations can be tested in systems that permit evaluation of the mechanical effects of enhanced glycolysis.

STUDIES WITH HIGH GLYCOGEN STRESS

In order to create high glycogen stores, rats were reserpinized 24 hours prior to perfusing their hearts (Scheuer and Stezoski, 1970).

Figure 3 shows the mechanics of reserpinized hearts compared to controls in a

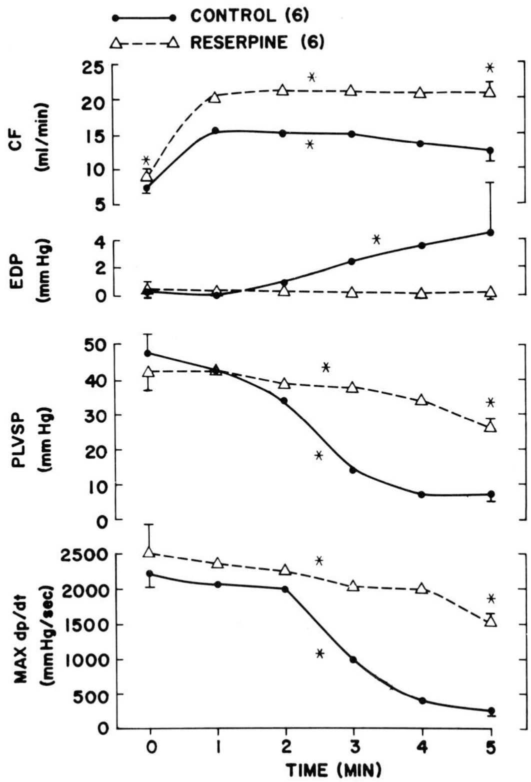

Figure 3. Effects of anoxia upon the mechanical performance of isolated perfused rat hearts from normal rats and rats reserpinized 24 hours prior to the study. Hearts were perfused under aerobic conditions for 15 min and were then subjected to abrupt hypoxia. Conditions of perfusion are the same as for Figure 1. CF = coronary flow; EDP = LV end-diastolic pressure; PLVSP = peak left ventricular systolic pressure. From Scheuer and Stezoski (1970), with permission of the publisher.

Langendorff system when both were subjected to 5 min of anoxia. In the control hearts but not in the reserpinized hearts, the end-diastolic pressures rose. Peak left ventricular systolic pressure declined in the control hearts to a greater degree than in reserpinized hearts, as did maximal dP/dt. The point at which performance began to separate occurred after 2 min of anoxia.

Figure 4 shows the glycogen levels and rates of lactate production in these two sets of hearts. After 15 min of aerobic perfusion, the glycogen levels were much higher in the reserpinized hearts. With anoxia, glycogenolysis was markedly increased in the reserpinized hearts. At the same time, during anoxia there was a marked rise in lactate production that was sustained for the 5-min anoxic period in the reserpinized hearts but fell off sharply in the control hearts. The

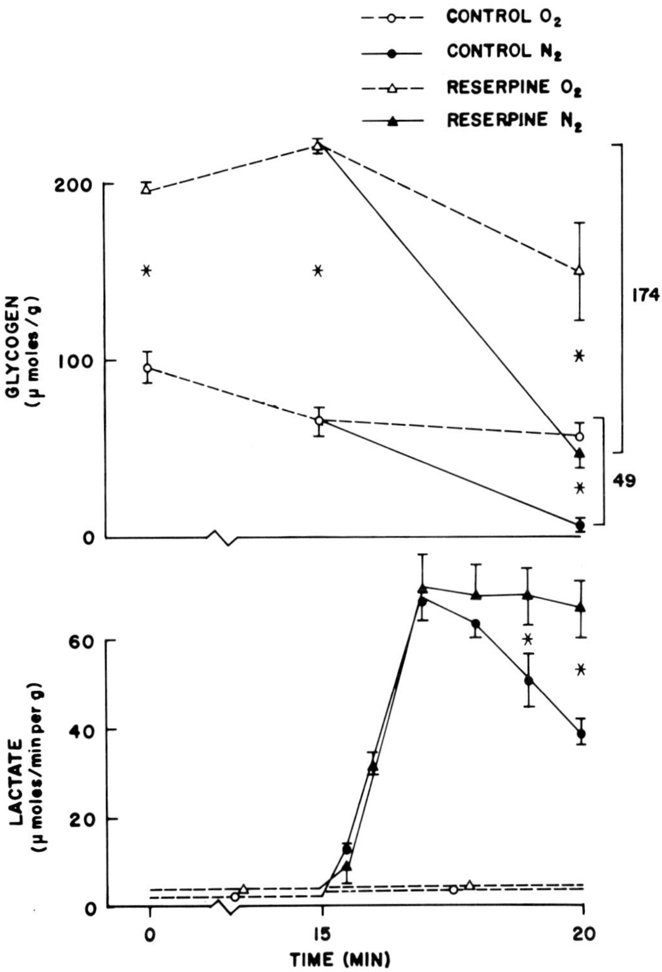

Figure 4. Effects of anoxia on cardiac glycogen stores and upon lactate production in the hearts of normal and reserpinized rats. Conditions are given in Figure 3. Abrupt anoxia was begun after 15 min of aerobic perfusion. From Scheuer and Stezoski (1970), with permission of the publisher.

point of separation was at 2 min, which was the time at which the mechanical difference began to develop between the two groups.

When the conversion of ^{14}C-glucose to ^{14}C-lactate was studied, even during the 1st min of anoxia controls had a higher rate of conversion of glucose to lactate than the reserpinized hearts, and this difference became magnified over the 5-min anoxic period. It was calculated that, during anoxia, approximately

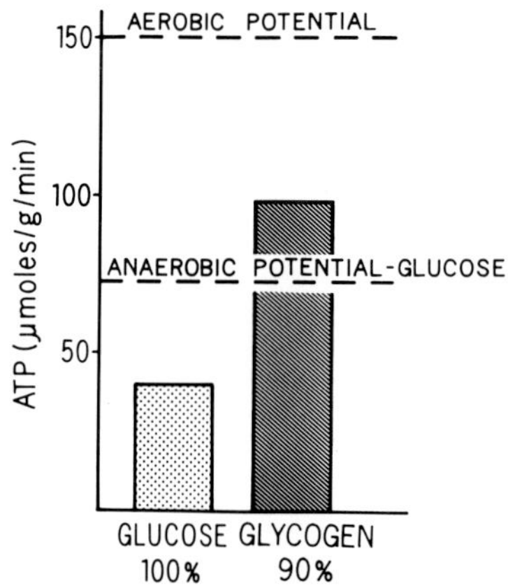

Figure 5. Calculated rates for ATP generation after 5 min of anoxia in hearts from normal rats (*stippled bars*) and rats with high cardiac glycogen stores (*shaded bars*). These calculations are based upon glycogenolytic rates and rates of glucose conversion to lactate in hearts from normal rats and hearts from rats treated 24 hours earlier with reserpine.

90% of the lactate came from glycogen in the reserpinized hearts whereas, in the control hearts, approximately 40% came from glycogen.

During anoxia, the creatine phosphate levels fell markedly and became almost zero in both groups, but ATP at the end of 5 min of anoxia was significantly higher in the reserpinized hearts than in the control hearts. Thus, since the reserpinized hearts were doing more mechanical work yet maintained higher high energy phosphate levels, they must have been producing more high energy phosphate compounds.

Figure 5 shows the calculated ATP generation rates in these two sets of hearts at the end of 5 min of anoxia, assuming that the control hearts were using a 100% glucose at that time and the reserpinized hearts continued to depend upon 90% glycogen. Hearts with high glycogen stores would have ATP production rates over twice those of hearts with normal stores. However, hearts could only reach approximately two-thirds of their aerobic potential with glycogen as substrate and less than one-third of the aerobic potential with glucose. Thus, in these experiments, the increased glycolytic potential was responsible for the higher levels of ATP production, and this could be translated into a sustained contractile performance in the hearts with high glycogen content.

We have shown with other methods of increasing glycogen that the same effects can be achieved (Scheuer and Stezoski, 1970), and others have reported that one possible mechanism by which propranolol protects against ischemia may be through enhancing glycogen stores and the glycogenolytic response (Smithen, Christodoulou, and Brachfeld, 1974).

STUDIES WITH ALKALOSIS

Another method to increase glycolytic flux is by raising the pH of the perfusion medium. We previously demonstrated that a rise in extracellular pH from 7.45 to 8.5 in the isolated perfused aerobic rat heart increased lactate production approximately three-fold (Scheuer and Berry, 1967). This is, in part, due to release of ATP inhibition of the phosphofructokinase step.

Figure 6 shows the glycolytic ATP that would be produced under these aerobic conditions. A marked increase in ATP would be generated by the glycolytic pathway in the alkalotic heart. This would also provide significantly more pyruvate for oxidative metabolism should the cardiac energy demands increase. If alkalosis has a similar effect in the hypoxic heart, we could expect a significant increase in ATP generation.

Figure 7 shows the effects of respiratory alkalosis in the isolated hypoxic working rat heart (Scheuer and Stezoski, 1972b). There were small but statistically significant differences in pressure and dP/dt early during the period of hypoxia, with a slight advantage in the alkalotic hearts. The major difference was in terms of cardiac output during hypoxia, where the alkalotic hearts had cardiac outputs 80–100% greater than hearts perfused at normal pH.

Figure 8 shows the metabolic values for these hearts. Myocardial oxygen consumption was the same during hypoxia in the presence of normal or alkalotic pH, and lactate production also did not differ. The lactate/pyruvate ratio in the effluent medium was higher with the normal pH, but this can be accounted for

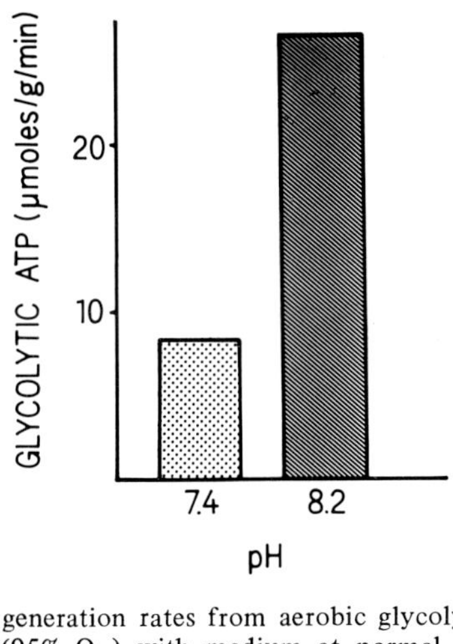

Figure 6. Calculated ATP generation rates from aerobic glycolysis in rat hearts perfused under aerobic conditions (95% O_2) with medium at normal and high pH (respiratory alkalosis).

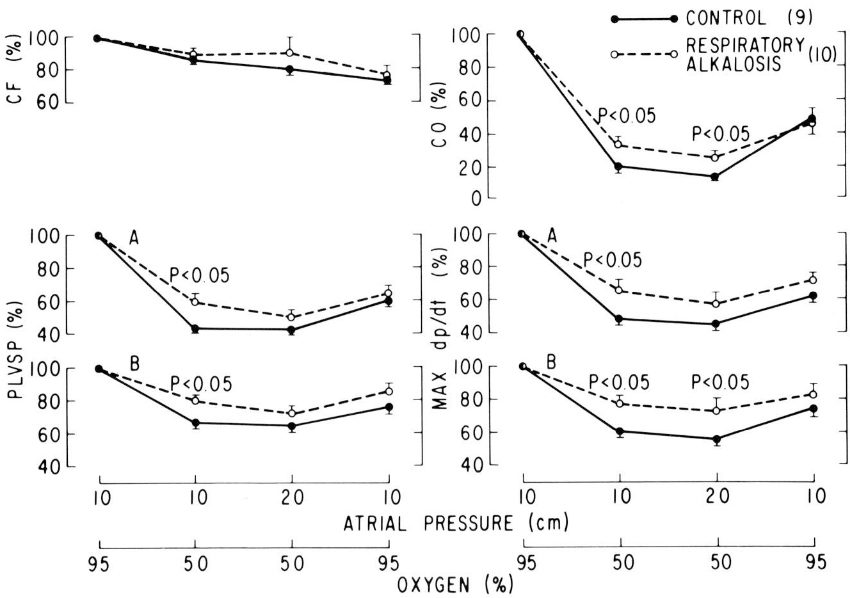

Figure 7. Effect of respiratory alkalosis (pH 7.8) upon the performance of isolated working rat hearts during hypoxia. Alkalosis was present only during the hypoxic (50% O_2) periods. Hearts were perfused for 5 min under each condition before mechanics were measured. CO = cardiac output; all other abbreviations are the same as for Figure 3. *A* indicates performance during isovolumic beats with the aorta clamped. *B* indicates performance during ejecting beats. From Scheuer and Stezoski (1972b), with permission of the publisher.

by the difference in hydrogen ion concentrations and does not imply a difference in the cytoplasmic nicotinamide adenine dinucleotide/reduced nicotinamide adenine dinucleotide (NAD/NADH) ratio. External work efficiency was significantly greater in the alkalotic hearts during hypoxia.

Therefore, there was no evidence here of an increase in glycolytic or oxidative metabolism in these hearts, and the increased efficiency and performance of the heart as a pump must have been due to improved mechanisms of energy utilization rather than of energy delivery. This is consistent with the hypothesis of other workers (Katz and Hecht, 1969; Nakamura and Schwartz, 1970) that there is a defect in excitation-contraction coupling in hypoxia or ischemia that is mediated by acidosis, and that this can, therefore, be reversed to a certain degree by creating alkalosis. Metabolic alkalosis was also found to afford some protection against the mechanical effect of hypoxia. The failure of alkalosis to increase glycolytic rate in these hypoxic hearts with normal coronary flow does not insure the same findings in ishcemic hearts where acid products would be retained.

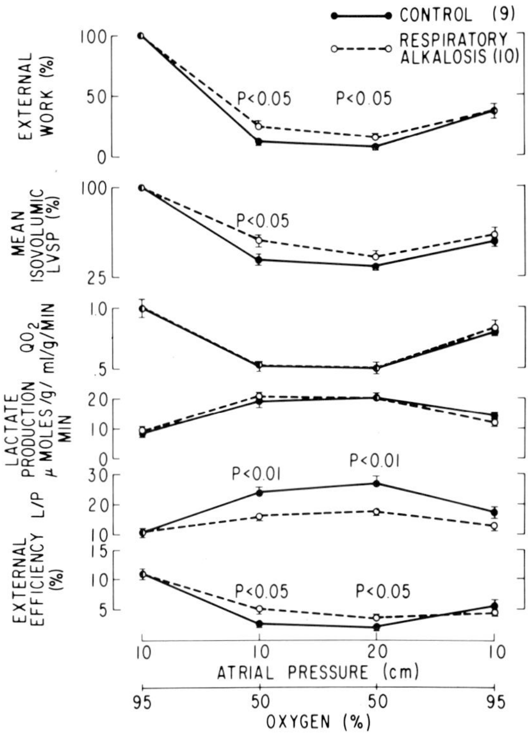

Figure 8. Calculated external work, mean isovolumic LV pressure, metabolic values, and external efficiency for experiments shown in Figure 7.

ANAEROBIC MITOCHONDRIAL ENERGY FORMATION

Penney and Cascarano (1970) suggested that Krebs cycle intermediates could provide energy for the hypoxic heart. This supposition was based upon the observation that mitochondria can form ATP anaerobically via fumarate-dependent oxidation of NADH (Sanadi and Fluharty, 1963) and by the oxidation of oxylacetate and of alpha-ketoglutarate in the cytochrome chain, thus producing energy in the absence of oxygen (Hunter, 1949).

Table 1 represents some data from the work of Penney and Cascarano (1970) in which rat hearts were perfused anaerobically in the presence of glucose alone; glucose, oxylacetate, and alpha-ketoglutarate; and glucose, fumarate, malate, and glutamate. The energy expenditure as indicated by the heart rate times blood pressure per gram was greater during anoxia in the presence of these Krebs cycle intermediates than with glucose alone. The ratio of energy expenditure per mole of glucose used or lactate produced was also higher when the tricarboxylic acid (TCA) cycle intermediates were present, suggesting that energy was contributed to the system by nonglycolytic mechanisms. However, the major factor in the increase in energy expenditure both at 10 min and at 70 min related to the increase in heart rate, whereas the aortic systolic pressure was either less in the presence of the Krebs cycle intermediates or the same as with glucose alone. Thus, a major effect of these intermediates may have been on the impulse-forming or conduction properties of the heart.

Table 2 shows the contents of some energy-related compounds in these hearts after 40 min of anaerobic perfusion. Glycogen levels were higher in the presence of the Krebs cycle intermediates, as were the levels of ATP and ADP. The hearts were entirely depleted of creatine phosphate. The hearts perfused with the Krebs cycle intermediates had higher levels of succinate than those perfused with glucose alone, and this suggests that mitochondrial metabolism was occurring during anoxia.

We attempted to repeat Cascarano's work in isolated perfused rat hearts with

Table 1. Effects of TCA cycle intermediates function of anaerobic hearts*

	Substrate (mM)		
	Gluc 20	Gluc, OAA, α-KG 20 14 14	Gluc, Fum, Mal, Glut 20 12.5 10 12.5
		10 min	
Heart rate (beats/min)	28±10	114±6	92±13
Aortic systolic pressure (mm Hg)	99±16	63±2	60±2
Energy expenditure (HR X Bp X K/g)	15±5	45±5	37±3
		70 min	
Heart rate (beats/min)	19±9	75±15	62±6
Aortic systolic pressure (mm Hg)	58±3	68±1	61±3
Energy expenditure (HR X Bp X K/g)	8.5±4.5	31.0±6.8	25.8±1.5

*Data from Penney and Cascarano (1970).

Table 2. Effects of TCA cycle intermediates on tissue metabolites of anaerobic hearts*

	Substrate (mM)		
	Gluc 20	Gluc, OAA, α-KG 20 14 14 Time: 40 min	Gluc, Fum, Mal, Glut 20 12.5 10 12.5
Glycogen (μmoles/g)	39.5±6.1	46.4±7.1	56.9±2.9
ATP (μmoles/g)	7.0±1.3	10.2±0.7	9.4±0.7
ADP (μmoles/g)	13.6±0.9	22.6±1.8	17.8±2.9
CP (μmoles/g)	0.0±0.0	0.0±0.0	0.2±0.1
Succinate (μmoles/g)	4.5	24.8	89.7

*Data from Penney and Cascarano (1970).

the identical compounds but with the heart rate kept constant at 330 beats/min by pacing (Scheuer and Stezoski, unpublished observations). We were not able to show any beneficial effects of these compounds on the mechanical activity of rat hearts during anoxia. Therefore, the physiological applicability of the Krebs cycle intermediates to improving anaerobic cardiac performance remains to be proven.

FREE FATTY ACIDS

Some workers have suggested that free fatty acids should be beneficial during moderate hypoxia, because 35% of the ATP formed from fatty acid metabolism is generated prior to the acetyl coenzyme A step (Olson, 1969).

Henderson and Sonnenblick (1972) reported that in the hypoxic papillary muscle, increasing free fatty acid levels were associated with a progressive depression of muscle function. The problem of interpreting these data in papillary muscles with profound hypoxia and translating them to the heart perfused through its coronary bed was illustrated by studies in the hypoxic (pO_2 150 mm Hg) isolated perfused rat heart (Apstein, Gmeiner, and Brachfeld, 1972). Studies were performed with 5.5 mM glucose present, with or without

0.4 mM palmitate. Hypoxia caused a marked reduction in heart rate, both in the presence and in the absence of palmitate; but in the hypoxic group, palmitate seemed to have a slightly beneficial effect on pressure development. However, when the product of heart rate and pressure is calculated, no over-all improvement in cardiac work would be seen in the hypoxic heart with palmitate. On the other hand, the type of deleterious effects seen in the isolated papillary muscle was not demonstrated by these studies.

The potential effect of free fatty acids in the ischemic heart must be considered, in view of the recent observations (Mjøs, 1971) that high levels of free fatty acid increased oxygen demands of the aerobic heart for any given level of external work.

In the ischemic dog heart (Mjøs, Kjekshus, and Lekven, 1974), free fatty acids were elevated by the infusion of norepinephrine. The increase in free fatty acids also caused a marked increase in ST segment elevations, but this could be reversed by inhibiting lipolysis with beta-pyridylcarbinol. The change in ST segment elevation paralleled the plasma free fatty acid levels. Thus, the norepinephrine per se was not responsible for the increased ischemic ST segment response but caused the response by elevating free fatty acids.

There are also recent data that indicate that physiological levels of free fatty acids increase the release of enzymes from the ischemic isolated perfused rat heart (de Leiris *et al.*, 1974).

Thus, contrary to the earlier suggestions that free fatty acids might be more efficient for ATP generation during ischemia or hypoxia, it now appears that free fatty acids are probably deleterious to the ischemic heart.

REFERENCES

APSTEIN, C. S., R. GMEINER, and N. BRACHFELD. 1972. Effect of palmitate on hypoxic myocardial metabolism and contractility, p. 136–146. *In* E. Bajusz and G. Rona (eds.), Recent Advances in Studies on Cardiac Structure and Metabolism, Vol. 1: Myocardiology. University Park Press, Baltimore.

de LEIRIS, J., L. H. OPIE, W. F. LUBBE, and O. BRICKNELL. 1974. Effect of substrate on enzyme release after coronary artery ligation in isolated rat heart, (This Book, page 00).

EVANS, C. L., A. C. DeGRAFF, T. KOSAKA, K. MacKENZIE, G. E. MURPHY, T. VACEK, D. H. WILLIAMS, and F. G. YOUNG. 1933. Utilization of blood sugar and lactate between heart-lung preparation. J. Physiol. 80: 21.

HENDERSON, A. H., and E. H. SONNENBLICK. 1972. Influence of free fatty acids on myocardial mechanics and oxygen consumption in rat papillary muscles and perfused hearts, under oxygenated and hypoxic conditions. *In* E. Bajusz and G. Rona (eds.), Recent Advances in Studies on Cardiac Structure and Metabolism, Vol. 1: Myocardiology, pp. 147–162. University Park Press, Baltimore.

HUNTER, F. E. 1949. Anaerobic phosphorylation due to a coupled oxidation-reduction between alpha ketoglutaric acid and oxalacetic acid. J. Biol. Chem. 177: 361.

KATZ, A. M., and H. H. HECHT. 1969. The early "pump" failure of the ischemic heart. Am. J. Med. 47: 497–502.

MJØS, O. D. 1971. Effect of free fatty acids on myocardial function and oxygen consumption in intact dogs. J. Clin. Invest. 50: 1386–1389.

MJØS, O. D., J. K. KJEKSHUS, and J. LEKVEN. 1974. Importance of free fatty acids as a determinant of myocardial oxygen consumption and myocardial ischemic injury during norepinephrine infusion in dogs. J. Clin. Invest. 53: 1290–1299.

MORGAN, H. E., M. J. HENDERSON, D. M. REGEN, and C. R. PARK. 1961. Regulation of glucose uptake in muscle. I. The effects of insulin and anoxia on glucose transport and phosphorylation in the isolated, perfused heart of normal rats. J. Biol. Chem. 36: 253–261.

NAKAMURA, Y., and A. SCHWARTZ. 1970. Possible control of intracellular calcium metabolism by (H^+): Sarcoplasmic reticulum of skeletal and cardiac muscle. Biochem. Biophys. Res. Comm. 41: 830–836.

NEELY, J. R., J. T. WHITMER, and M. J. ROVETTO. 1971. Effects of various degrees of whole heart ischemia on carbohydrate utilization in isolated working rat hearts. Abstract, 7th Annual Meeting of the International Study Group for Research in Cardiac Metabolism, Quebec, June 18th–21st.

OLSON, R. E. 1969. Metabolic interventions in the treatment of infarcting myocardium. Circulation 40: IV–195–201.

OPIE, L. H. 1971/72. Substrate utilization and glycolysis in the heart. Cardiology 56: 2–21.

PENNEY, D. G., and J. CASCARANO. 1970. Anaerobic rat heart. Effects of glucose and tricarboxylic acid-cycle metabolites on metabolism and physiological performance. Biochem. J. 118: 221–227.

SANADI, D. R., and A. L. FLUHARTY. 1963. On the mechanism of oxidative phosphorylation. VII. The energy-requiring reduction of pyridine nucleotide by succinate and the energy-yielding oxidation of reduced pyridine nucleotide by fumarate. Biochemistry 2: 523–528.

SCHEUER, J., and M. N. BERRY. 1967. Effect of alkalosis on glycolysis in the isolated rat heart. Am. J. Physiol. 213: 1143–1148.

SCHEUER, J., and S. W. STEZOSKI. 1970. Protective role of increased myocardial glycogen stores in cardiac anoxia in the rat. Circulat. Res. 27: 835–849.

SCHEUER, J., and S. W. STEZOSKI. 1972a. Effect of physical training on the mechanical and metabolic response of the rat heart to hypoxia. Circulat. Res. 30: 418–429.

SCHEUER, J., and S. W. STEZOSKI. 1972b. The effect of alkalosis upon the mechanical and metabolic response of the rat heart to hypoxia. J. Mol. Cell. Cardiol. 4: 599–610.

SMITHEN, C., J. CHRISTODOULOU, and N. BRACHFELD. 1974. Protective role of increased myocardial glycogen stores induced by propranolol. Abstract, 7th Annual Meeting of the International Study Group for Research in Cardiac Metabolism, Quebec, June 18th–21st.

Mailing address:
James Scheuer, M.D.,
Division of Cardiology, Montefiore Hospital and Medical Center,
111 East 210th Street, Bronx, New York 10467 (USA).

Recent Advances in Studies on
Cardiac Structure and Metabolism Volume 10
The Metabolism of Contraction
Edited by Paul-Emile Roy and George Rona

MYOCARDIAL CELL ELECTROLYTES AND WATER DURING ACUTE ISCHEMIA[1]

TIMOTHY J. REGAN, BUNYAD HAIDER,
S. SULTAN AHMED, and HENRY A. OLDEWURTEL

Department of Medicine, College of Medicine and Dentistry
New Jersey Medical School, Newark, New Jersey, USA

SUMMARY

This study was intended to define the early early electrolyte and water abnormalities of ischemic myocardium in the intact anesthetized dog. We have defined the appropriate circumstances for using ^{51}Cr-ethyelendiamineletraacetate as an extracellular marker during ischemia and have discerned no change in this space at a time when tissue water increments were observed, by 15 min of ischemia. Calculated on the basis of cell dry weight, the cell sodium increment exceeded potassium loss in this early period of ischemia. This is consistent with the view that interference with energy metabolism reduces the pumping of sodium from the cell. The enhanced entry of sodium is associated with a gain of water, initiating events that effect irreversibility.

INTRODUCTION

The problem of tissue swelling is important to a consideration of the pathogenesis of cardiac cell injury during acute ischemia. Although the proximate cause of irreversible change in cardiac cells is not known, it has been postulated that tissue swelling with subsequent distortion of cell membranes and resultant alteration of enzymatic activities may be a major factor (Leaf, 1973). The current study was undertaken to define the time course of the changes in cation and water distribution in ischemic tissue during the early period before irreversibility is established.

METHOD

Male mongrel dogs weighing between 21 and 26 kg were anesthetized 18 hours postprandial with morphine sulfate, 3 mg/kg, and pentobarbital; they were

[1]Supported in part by Contract NIH-72-2970 under the Myocardial Infarction Program, National Heart and Lung Institute, National Institutes of Health, Department of Health, Education and Welfare.

studied without opening the chest. After insertion of an endotracheal tube, respiration was regulated with a Harvard respiratory pump, so as to maintain arterial oxygen saturation and pH in the normal range. Catheters were placed in the aorta, left ventricle, and coronary sinus for blood sampling and pressure determinations. A balloon tip catheter with a patent distal lumen was inserted into the anterior descending artery under fluoroscopy. The distal lumen permitted measurement of peripheral coronary pressure before and after complete occlusion by balloon inflation. The catheters were initially filled with dilute heparin, and patency was maintained subsequently by slow saline infusions or intermittent flushes during the experiment.

The precordial leads as well as standard lead I were monitored for evidence of injury potential during ischemia. Ventricular and aortic pressures were measured by connecting the 50 cm Goodale-Lubin catheters, No. 8F, directly to Statham strain gauge transducers. Intermittent photographic recordings were made on a multichannel oscilloscope recorder.

The measurement of extracellular space was done with ^{51}Cr EDTA, a marker which has been described as a reliable method of measuring extracellular space (Lassen and Trap-Jensen, 1970). In our own laboratory, we have compared sucrose and ^{51}Cr EDTA during pulse injections and found the calculated extracellular space in myocardium to be essentially the same with both indicators (Effros *et al.*, 1974). For the purpose of this study, we have used a sustained infusion of the isotope. Initial experiments in normals indicated that a longer infusion period of 75 min measured a slightly larger space at 20% of tissue weight, compared to 17% after a 20-min infusion. This phenomenon was previously observed with sucrose (Page and Page, 1968) and was attributed to greater penetration of the sarcotubular system with time by the extracellular marker.

In studies of ischemia, we have used the longer term infusion, because equilibrium between the vascular and interstitial space was more readily sustained as judged by the constancy of serial venous plasma samples taken from the cardiac vein. These were taken at 1-min intervals in the last 10 min of the infusion and agreement within 5% was required for an acceptable study. At the conclusion of each study, the chest was rapidly opened and the heart was cold-arrested with iced Ringer's lactate. A transmural sample was rapidly obtained and included most of the ischemic area excluding the margin, and was separated into outer, middle, and inner thirds. The epicardium was carefully trimmed to remove perivascular adipose tissue. The samples were homogenized in distilled water, and potassium and sodium were determined on an AutoAnalyzer system with flame attachment. A portion was taken for analysis of ^{51}Cr EDTA in the Nuclear-Chicago scintillation counter. Assays of plasma cations and isotope were also performed. Separate samples were taken to determine wet and dry weights by heating in an oven at 100°C for 18 hours.

Three groups of animals were studied: a control group in which no ischemia was produced, a second group in which the ischemic duration was 15 min, and a third in which ischemia lasted 60 min.

RESULTS

Despite the fact that a substantial rise of coronary venous potassium is observed in the coronary venous plasma in the early period of ischemia (Regan *et al.*, 1967), changes in tissue concentrations of electrolytes expressed per gram wet weight are less evident after 15 min of ischemia. Sodium content is not altered and there is, in the inner and midventricular wall, an approximate 10% reduction of potassium concentrations (Fig. 1). This is associated with a significant increment of tissue water in these two layers of ischemic myocardium (Fig. 2). Calculation of the extracellular space, according to Manery (1954), revealed no significant change in this compartment. However, the distribution of water was altered, in that cell water was enhanced (Fig. 3). The slightly larger extracellular water space in normals was not significant.

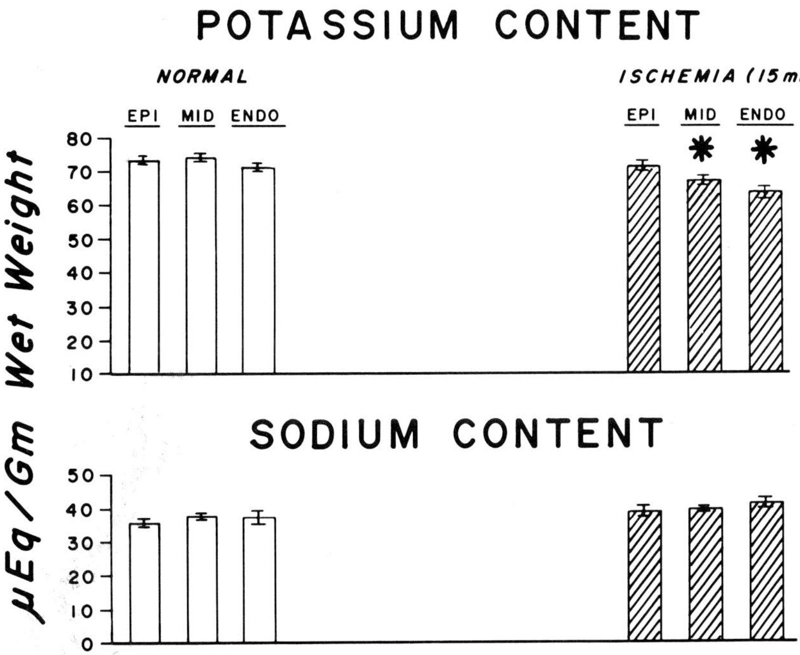

Figure 1. A significant transmural gradient of potassium was seen after 15 min of ischemia without an evident change in sodium. EPI refers to outer third layer of left ventricular myocardium, MID is middle third, and ENDO refers to inner third; * indicates $p<0.02$.

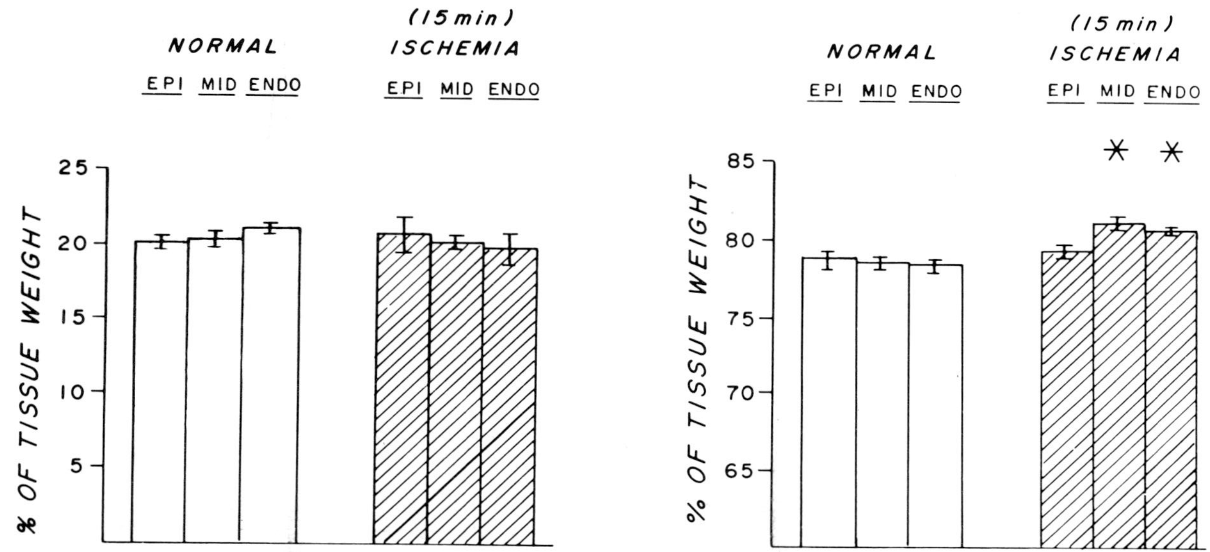

Figure 2. Panel on left depicts ^{51}Cr EDTA space; *right-hand panel* represents tissue water, which increased significantly in the inner layers after 15 min of ischemia.

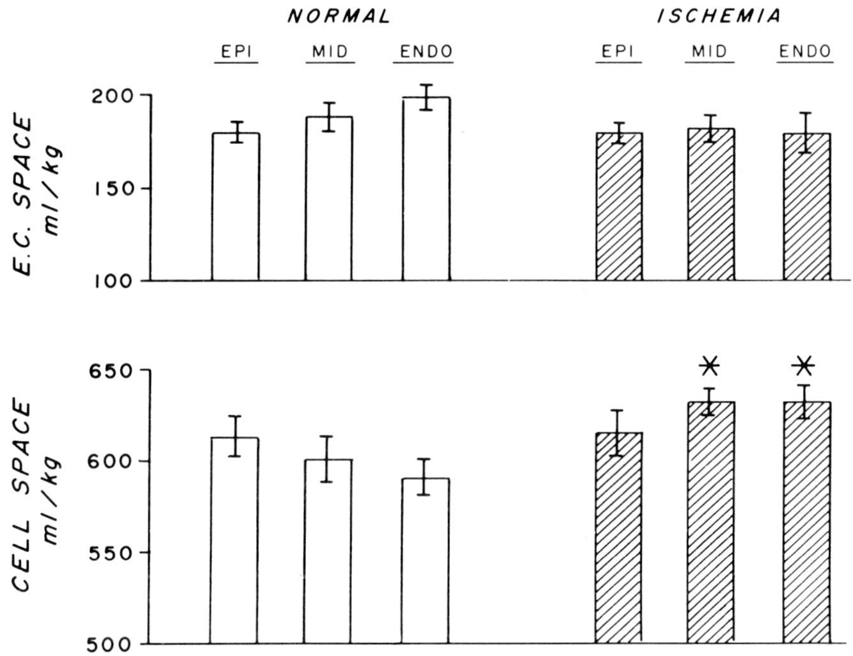

Figure 3. Calculation of water distribution indicates that increment is limited to cell space. E.C. = extracellular.

Calculation of the content of cations per kilogram of cell water revealed a more substantial reduction of potassium concentration, most evident in the inner and middle layers at 15 min of ischemia (Fig. 4). The cell sodium concentration was slightly, but not significantly, changed. By 60 min, however, when potassium content had diminished even further, there was a significant increment of cell sodium concentration despite further intake of water into the cell.

When the concentration of cations at 15 min were expressed as mEq/kg cell dry weight, there was a significant increase of sodium content in the endocardium of the ischemic area to 35.7 ± 2.0 mEq/kg cell dry weight, compared to 21.4 ± 1.1 in controls. On the other hand, potassium in the normal controls was 170 ± 1.3 mEq/kg dry cell weight and in the ischemic endocardium 166 ± 1.2 mEq/kg dry cell weight.

DISCUSSION

We have previously observed that, quite early after the onset of occlusion of the anterior descending vessel, coronary venous concentrations of potassium in the

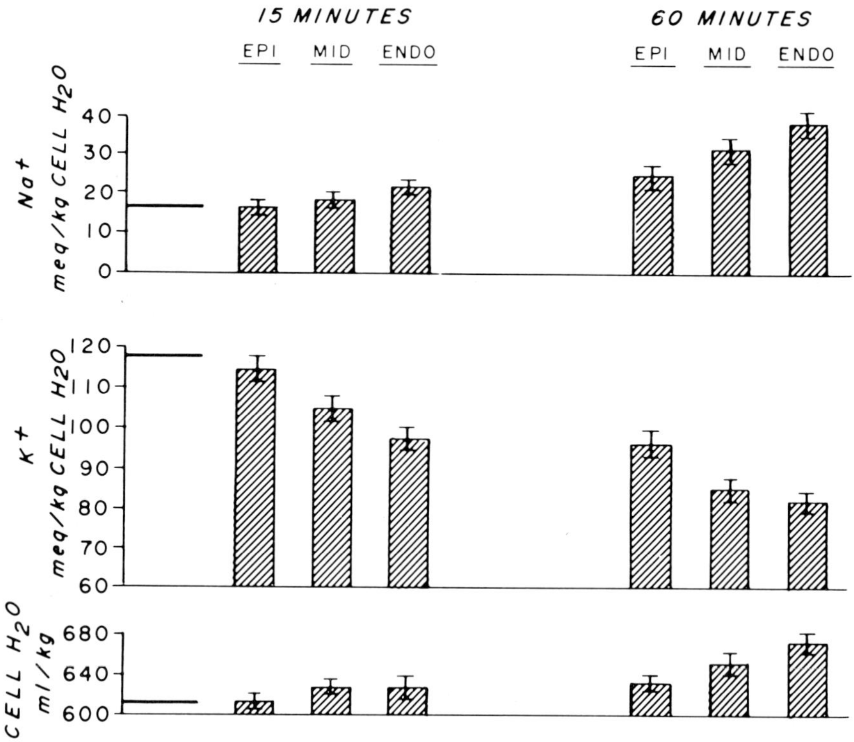

Figure 4. After 15 min of ischemia, there was a significant reduction of potassium in the middle and inner layers of the ischemic myocardium, while the sodium concentrations were not significantly altered. After 60 min of ischemia, there was progression of the decline in cell potassium concentration, partly accounted for by the additional increment of cell water. At this time, there was a significant accumulation of sodium in the ischemic cells. The *horizontal lines on the left* represent average values in normals.

plasma from the great cardiac vein become significantly elevated over simultaneous arterial samples (Regan *et al.,* 1967). This was assumed to represent cation leakage from cardiac cells. The data on tissue electrolyte analysis of this current study showed that the potassium leak was derived largely from the inner and middle layers of the ischemic myocardium. These data are in contrast to findings obtained in a study of transmural samples, in which no significant change in potassium ion concentration was observed (Whalen *et al.,* 1974). Assuming that coronary venous drainage was not impeded by the ligation procedure, the regional nature of the initial change of potassium ion content may have accounted for this lack of change in transmural samples.

Thus, it would appear that a decrease in energy production in the inner layers of myocardium results in liberation of potassium into the extracellular space, with a consequent reduction of the transmural gradient for this cation. This may be the major factor in the development of the injury potential as a representation of the reduced resting potential of the cell. A reduction of the transcellular ratio achieved by regional hyperkalemia during infusion of KCl into the anterior descending artery of animals without ischemia supports this view (Regan *et al.*, 1967). However, since the injury potential occurs within seconds after complete occlusion of the anterior descending vessel, it has yet to be established that K^+ leak occurs sufficiently early to account for the injury potential. The interrelationship of sodium, potassium, and water movement in the initial minutes of ischemia needs to be better defined to decide this issue.

The initial consequences of interfering with energy production during ischemia have been viewed as related to a reduced capacity to pump sodium from the cell (Leaf, 1973). Thus, cell osmolality would be enhanced, eliciting water movement from the interstitial space. This process may be intensified by the simultaneous increase in glycolytic intermediates, secondary to accelerated glycogenolysis known to occur with acute ischemia. The greater positive charge in the cell interior may facilitate loss of K^+ to the interstitium. However, it is conceivable that, before actual leakage of this cation occurs, a reduction of K^+ concentration in the cell by virtue of water gain may effectively reduce the transcellular gradient and thus diminish the resting potential.

The fact that tissue swelling is confined to the cellular space as defined in these studies is supported by histological studies during ischemia (Whalen *et al.*, 1974) as well as after reflow (Kloner *et al.*, 1974). Whether the cell edema can be reversed by increases in extracellular osmolality and whether this effect may normalize cation distribution as well as the pathophysiology of the cell remains to be determined.

Acknowledgments We appreciate the expert technical assistance of Mrs. B. Jenkins and Mrs. R. Nowak and the secretarial services rendered by Mrs. A. Binetti.

REFERENCES

EFFROS, R. M., B. HAIDER, P. ETTINGER, H. OLDEWURTEL, S. AHMED, and T. REGAN. 1974. In vivo response of myocardial cell pH to changes of plasma pH and ischemia. Fed. Proc. 33: 396.

KLONER, R. A., C. E. GANOTE, D. A. WHALEN, Jr., and R. B. JENNINGS. 1974. Effect of a transient period of ischemia on myocardial cells. II. Fine structure during the first few minutes of reflow. Am. J. Path. 74 (3): 399–422.

LASSEN, N. A., and J. TRAP-JENSEN. 1970. Capillary permeability for smaller hydrophilic tracers in exercising skeletal muscle in normal man and in patients with long-term diabetes mellitus. *In* C. Crone and N. A. Lassen (eds.), Capillary Permeability, pp. 135–152. Academic Press, New York.

LEAF, A. 1973. Cell swelling: A factor in ischemic tissue injury. Circulation 48 (3): 455–458.

MANERY, J. F. 1954. Water and electrolyte metabolism. Physiol. Rev. 34: 344–417.

PAGE, E., and E. G. PAGE. 1968. Distribution of ions and water between tissue compartments in the perfused left ventricle of the rat heart. Circulat. Res. 22: 435–446.

REGAN, T. J., M. A. HARMAN, P. H. LEHAN, W. M. BURKE, and H. A. OLDEWURTEL. 1967. Ventricular arrhythmias and K^+ transfer during myocardial ischemia and intervention with procaine amide, insulin, or glucose solution. J. Clin. Invest. 46: 1657–1668.

WHALEN, Jr., D. A., D. G. HAMILTON, C. E. GANOTE, and R. B. JENNINGS. 1974. Effect of a transient period of ischemia on myocardial cells. I. Effects on cell volume regulation. Am. J. Path. 74 (3): 381–398.

Mailing address:
Timothy J. Regan, M.D.,
College of Medicine and Denistry-New Jersey Medical School,
100 Bergen Street, Newark, New Jersey 07103 (USA).

Recent Advances in Studies on
Cardiac Structure and Metabolism Volume 10
The Metabolism of Contraction
Edited by Paul-Emile Roy and George Rona

REGIONAL MYOCARDIAL METABOLISM DURING SUBTOTAL AND TOTAL CORONARY ARTERY OCCLUSION[1]

LAWRENCE D. BAKER, OSCAR H. L. BING,
and JOSEPH V. MESSER[2]

Tufts and Harvard Medical Units, Boston City Hospital, and
the Department of Medicine, Harvard Medical School, Boston, Massachusetts, USA

SUMMARY

Regional myocardial metabolism was examined by means of coronary venous sampling in dogs having acute nonsimultaneous occlusions of the left anterior descending (LAD) and circumflex (CIRC) coronary arteries. Regional coronary venous metabolic sampling was not found to separate ischemic from nonischemic myocardium in LAD occlusion and did so imperfectly in isolated CIRC occlusion.

Myocardial metabolic consequences of subtotal and total coronary artery occlusion were compared in dogs during LAD and CIRC occlusion and in human subjects with subtotal and total anterior coronary artery occlusion. In spite of ECG evidence of greater myocardial ischemia during total coronary artery occlusion in the dog, transmyocardial metabolic changes were more consistent during subtotal occlusion. Human subjects stressed during cardiac catheterization exhibited a similar apparent paradox. This suggests variable coronary venous washout from ischemic myocardium during total coronary artery occlusion and, in humans, may also reflect the presence of scar tissue. Caution is advised when interpreting myocardial metabolic information in the presence of total coronary artery occlusion, since underestimation of the extent of myocardial ischemia is more likely than with an occlusion of lesser degree.

INTRODUCTION

Measurement of zonal myocardial metabolism has been advocated in the evaluation of human coronary artery disease (Herman, Elliott, and Gorlin, 1967). Such measurements, if reliable, would be of value in assessing the effectiveness of the collateral circulation distal to occlusive coronary lesions, in the investigation of the influence of drugs on ischemic regions of the heart, and in the critical evaluation of surgical revascularization procedures (Dwyer, 1972; Bemis *et al.*,

---

[1] Supported, in part, by a General Research Support Grant 57-209, Boston City Hospital, and Grants HL05986 and HL10539 from the National Heart and Lung Institute, National Institutes of Health.

[2] Present address: Section of Cardio-Respiratory Diseases, Presbyterian-St. Luke's Hospital, 1753 W. Congress Parkway, Chicago, Illinois 60612, USA.

1973). General difficulties related to interpretation of data derived from coronary venous metabolic sampling have been reviewed (Olson and Opie, 1972). The present study reports data bearing on two problems associated with regional coronary venous sampling: nonspecificity of coronary venous drainage and coronary venous sampling from a region whose nutrient artery has become totally occluded.

Specificity of regional coronary venous sampling in human coronary artery disease was claimed by Herman *et al.* (1967). However, such specificity has not been found by others working with experimental animals (Bartelstone, Schultz, and Wang, 1968; Owen *et al.*, 1970; Opie, 1971). A recent study utilizing balloon occlusion to separate anterior and posterior coronary venous effluent during left anterior descending coronary artery (LAD) occlusion has confirmed the presence of ischemic metabolic end products in the nonoccluded segment of myocardium (Meerbaum *et al.*, 1974). The present study examines the specificity of coronary venous metabolic sampling during separate occlusions of the LAD and circumflex artery (CIRC).

Specificity and sensitivity of coronary venous sampling may be altered by the degree of coronary artery occlusion. Investigators have suggested that washout of ischemic metabolites is reduced by total coronary artery occlusion (Sharma *et al.*, 1970). Indeed, a recent animal study noted that myocardial lactate production occurred following reperfusion of a totally occluded LAD (Lang *et al.*, 1974). In the present study, myocardial metabolic consequences of subtotal and total coronary artery occlusion were examined in both the open-chest dog and human subjects.

MATERIALS AND METHODS

Canine studies

Hemodynamics and regional myocardial metabolism were studied in 12 open-chest anesthetized dogs. Although advantages of the closed-chest model have been enumerated (Ferguson, Shadle, and Gregg, 1953; Hood, McCarthy, and Lown, 1967), thoracotomy was required in the present study for precise placement of the coronary venous catheters and uniform subtotal coronary artery occlusion. Comparison of open- and closed-chest models revealed no significant difference in coronary sinus flow characteristics (Stein *et al.*, 1969). Morphine-chloralose anesthesia was maintained at a level sufficient to prevent unconscious movement. Animals were supported with a Harvard pump respirator supplied with 40% oxygen. Lead II of the electrocardiogram was continuously monitored during each study and the amplitude setting was not changed. Dogs exhibiting serious arrhythmias or hypotension at any time were eliminated from the study.

Two no. 7 Goodale-Lubin catheters were inserted into the coronary sinus under fluoroscopic control, and final positioning was performed by direct

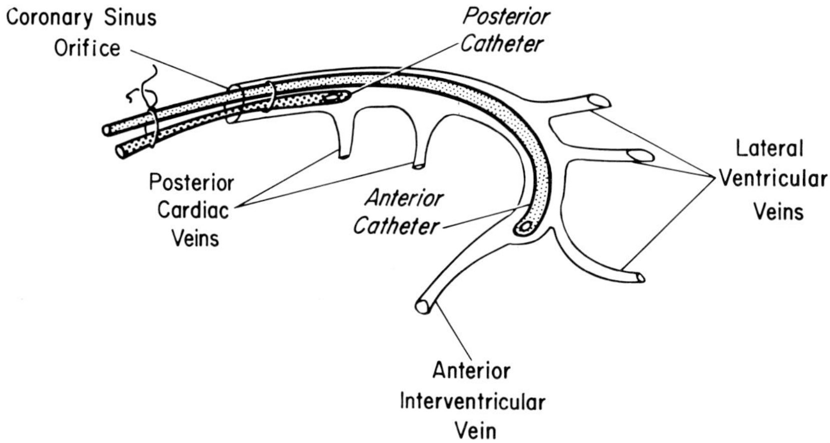

Figure 1. Coronary venous catheter position in the canine study. The anterior catheter is positioned upstream and the posterior catheter downstream from the coronary sinus entrances of all lateral left ventricular veins.

manipulation (Fig. 1). Catheter location was confirmed by blood gas analysis. One coronary venous catheter was advanced to the anterior interventricular vein. In order to avoid contamination with blood from the right atrium, the second catheter tip was placed 1.5 cm inside the coronary sinus orifice (Olsson and Gregg, 1965; Koberstein and Klocke, 1968). The posterior catheter was positioned downstream from the most posterior point of entrance of the lateral left ventricular veins.

No. 8 NIH catheters were advanced to the left ventricular chamber and proximal aorta. Transient subtotal and total occlusions of the LAD and CIRC were carefully performed using a silk ligature. Subtotal occlusion was produced by first including a wire obturator 0.8 mm in diameter within the ligature, and then removing the obturator after the tie had been made. The degree of coronary artery occlusion required to produce detectable anaerobic metabolism in this preparation without additional cardiac stress was observed to be about 85% in preliminary studies. Previous acute dog studies have indicated that, even during stress, LAD occlusion of less than 85% may not provoke detectable myocardial metabolic abnormalities (Pomposiello, Sharma, and Messer, 1970). The LAD was ligated distal to all branches appearing to nourish lateral left ventricle. Similarly, CIRC occlusion occurred distal to all anterior tributaries. The sequence of transient vessel occlusion was alternated throughout the study.

Metabolic and hemodynamic variables were measured in the control unoccluded state and promptly after 3 min of vessel occlusion. It has been demonstrated that metabolic evidence of ischemia appears soon after coronary artery occlusion (Opie, *et al.,* 1972; Chiong, West, and Parker, 1974), and may be near maximum 5 min after the onset of ischemia (Meerbaum *et al.,* 1974). After measurements, the ligature was released and the animal allowed to recover. An earlier study using this preparation has shown that this duration of coronary artery occlusion is compatible with complete metabolic and hemodynamic recovery (Baker *et al.,* 1969). During metabolic sampling, anterior and posterior coronary venous and aortic blood was drawn simultaneously and at the same rate from each site. At the end of each experiment, the coronary venous catheters were sewn in place and the heart removed. The anteroseptal distribution of the LAD and posterolateral distribution of the CIRC were confirmed by direct inspection, dye injection, and dissection. The coronary venous system was opened and a detailed map of the relationship of the catheters to the venous pattern was drawn.

Human studies

In this series, 22 human subjects with angiographically documented anterior coronary artery occlusion of 90% or greater were stressed with ergometer exercise, isoproterenol infusion, or pacing-induced tachycardia (PIT); 9 subjects were stressed twice. Subjects were selected according to the extent of anterior coronary artery occlusion and the completeness of data. A Goodale-Lubin catheter was positioned for anterior coronary venous sampling; catheter position was confirmed prior to venous sampling in each experimental state.

Simultaneous arterial and coronary venous blood samples were drawn in all subjects during rest, in 8 during bicycle ergometer exercise, in 14 during isoproterenol infusion, and in 9 during PIT. Subjects were exercised at a work load that could be sustained for 10 min without angina, and exercise was performed for at least 5 min before metabolic sampling. Isoproterenol was infused at 1–4 μg/min and titrated to achieve heart rates approximately 50% above resting levels. Samples were drawn 5 min after an increased heart rate, and increased rate of left ventricular pressure generation was observed. PIT was accomplished by means of a bipolar catheter electrode placed in the coronary sinus, except for 1 subject who was paced from the right atrium. Heart rate was increased an average of 49% above the resting mean. Of the 21 sets of observations in subjects with subtotal anterior coronary artery occlusion, 6 took place during exercise, 9 during isoproterenol infusion, and 6 during PIT. Of the 10 sets of observations in subjects with total occlusion, 2 occurred with exercise, 5 with isoproterenol infusion, and 3 with PIT. If a subject developed angina, measurements were obtained immediately and stress was reduced to subanginal levels before further sampling was performed.

General

Hemodynamic measurements were recorded on an Electronics for Medicine DR-8 recorder (Electronics for Medicine, Inc., 30 Virginia Road, White Plains,

New York). The zero for pressure measurements was set at the level of the mid-right atrium. Oxygen analyses of arterial and coronary venous blood were determined in duplicate on the Van Slyke apparatus, and pH, pCO_2 and pO_2 were measured using the electrometer, model 123, and the polarographic analyzer, model 125A (Instrumentation Laboratories, Inc., Boston, Mass.). Blood lactate and pyruvate concentrations were determined enzymatically by methods modified from those of Horn and Bruns (1956), and Redetzki, Bloedorn, and Bansi (1956), respectively. Lactate and pyruvate values were expressed in millimoles per liter of blood water. Duplicate arteriovenous sampling pairs were drawn in dogs and triplicate sampling pairs were drawn in humans. Duplicate determinations were performed on each sample. Duplicate lactate samples agreed within 5% and duplicate pyruvate samples within 9%.

Calculations

Pressure-time per minute was calculated as follows:

$$\text{PTM (mm Hg-sec/min)} = LV_{sm} \times \text{sep} \times \text{HR}$$

where LV_{sm} = left ventricular systolic mean pressure, sep = systolic ejection period, per beat.

Left ventricular dp/dt (mm Hg/sec) was measured directly from the tracing of the first derivative of the intracavitary left ventricular pressure.

Percent myocardial lactate extraction was calculated using the following formula:

$$\text{Percent myocardial lactate extraction} = \left(\frac{L_a - L_{cv}}{L_a} \times 100\right)$$

where L_a = arterial lactate concentration, L_{cv} = coronary venous lactate concentration.

Employment of the coronary venous lactate/pyruvate ratio has been restricted to states in which arterial lactate/pyruvate ratio did not change (Neill, 1968, 1969). In order to enlarge its usefulness in the present study, the coronary venous lactate/pyruvate ratio was corrected for changes in the arterial lactate/pyruvate ratio during the "stress" state by adding decrements and by subtracting increments in arterial lactate/pyruvate ratio.

Percent myocardial oxygen extraction was calculated from the following formula (Messer *et al.*, 1962):

$$\text{Percent myocardial oxygen extraction} = \left(\frac{O_{2a} - O_{2cv}}{O_{2a}} \times 100\right)$$

where O_{2a} = arterial oxygen content, O_{2cv} = coronary venous oxygen content.

Student's *t*-test was used to determine the significance of the paired data

(Snedecor and Cochran, 1967). Mean values are expressed, with standard error of the mean.

RESULTS

Canine studies

Hemodynamics Hemodynamic measurements were obtained during subtotal and total LAD and CIRC occlusion. Five variables were analyzed to assess changes in myocardial performance: heart rate, aortic systolic mean pressure, left ventricular end-diastolic pressure, left ventricular dp/dt, and pressure-time per minute. Significant depression of function occurred only in aortic mean systolic pressure during total CIRC occlusion (103 ± 6 mm Hg to 94 ± 7 mm Hg, $p<0.01$), in left ventricular end-diastolic pressure during total CIRC occlusion (12 ± 1 mm Hg to 18 ± 2 mm Hg, $p<0.05$), and in pressure-time per minute during subtotal LAD occlusion (2,767 ± 192 mm Hg-sec/min to 2,181 ± 256 mm Hg-sec/min, $p<0.01$). There was no evidence that irreversible hemodynamic deterioration occurred in any animal during transient coronary artery occlusion.

Coronary sinus pressure averaged 13 ± 1 mm Hg at rest and did not change during coronary artery occlusion. Coronary sinus pressure exceeded 12 mm Hg in one-half the data sets, possibly resulting in shunting of some left ventricular venous return away from the coronary sinus (Scharf, Bromberger-Barnea, and Permutt, 1971).

Electrocardiography The effect of subtotal and total coronary artery occlusion on S-T segment deviation in lead II of the ECG is shown in Figure 2. LAD occlusion produced reversible depression, while CIRC occlusion resulted in elevations of the S-T segment. In seven of eight paired sets of data, total

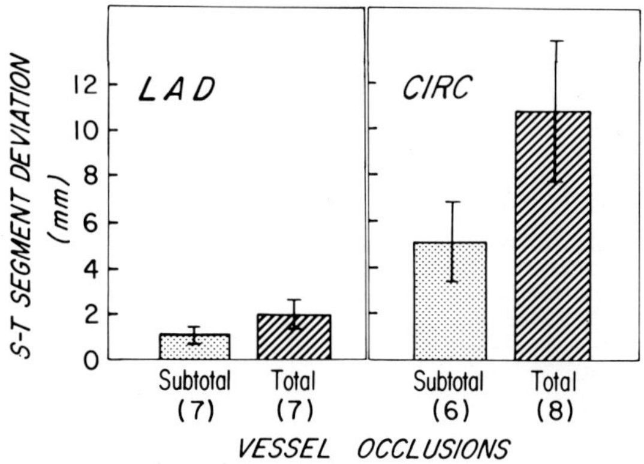

Figure 2. S-T segment deviation in lead II during coronary artery occlusion. Total coronary artery occlusion tended to produce greater S-T segment deviation than subtotal occlusion.

coronary artery occlusion produced more S-T deviation than the corresponding subtotal occlusion. During LAD occlusion in one dog, subtotal occlusion resulted in no S-T change, while total occlusion produced a small unexplained upward deflection of 0.5 mm.

Myocardial metabolism Mean resting arterial lactate was 3.06±0.22 mmole/liter rising 8% to 3.31 ± 0.24 mmole/liter during coronary artery occlusion. Arterial pO_2 was 91±10 mm Hg before occlusion, falling slightly to 82±9 with coronary artery occlusion.

The specificity of coronary venous metabolic sampling was investigated and the results are summarized (see Figs. 3 and 4). LAD occlusion produced metabolic evidence of ischemia to the same extent at anterior and posterior sampling sites. There was no difference between subtotal and total LAD occlusion in this regard. All four metabolic variables were significantly altered. Only coronary venous pO_2 indicated significantly greater ischemia at the anterior sampling site (Fig. 3*b*).

During CIRC occlusion (Fig. 4), all metabolic parameters again indicated significant ischemia. In contrast to findings during LAD occlusion, percent lactate extraction and coronary venous lactate/pyruvate ratio indicated significantly more ischemia at the posterior sampling site. Subgroup analysis showed this difference to occur during subtotal rather than total CIRC occlusion. It is noteworthy that metabolic evidence of ischemia appeared at the anterior sampling site after ligation of the artery supplying the lateral and posterior left ventricle.

The effects of subtotal and total coronary artery occlusion are shown in Figure 5. *Anterior* coronary venous sampling data were utilized in the comparison of the metabolic effects of subtotal and total *LAD* occlusion, and *posterior* coronary venous sampling data were used to analyze subtotal and total *CIRC* occlusion. Although greater ECG ST-T abnormalities were observed during total vessel occlusion, mean metabolic changes were similar during subtotal and total coronary occlusion. However, metabolic alterations achieved consistently greater statistical significance during subtotal occlusion (Fig. 5).

Human studies

Myocardial lactate metabolism data in human subjects with subtotal or total anterior coronary artery occlusion during exercise, isoproterenol infusion, or PIT is shown in Figure 6. Mean arterial lactate concentration rose during stress from 0.80±0.05 mmole/liter to 1.19±0.14 mmole/liter. Myocardial oxygen extraction is an unreliable indicator of human myocardial ischemia during isoproterenol infusion (Cohen *et al.*, 1966; Gorlin, 1972) and was excluded from the human data. Subjects with subtotal coronary occlusion exhibited significant shifts in percent myocardial lactate extraction and coronary venous lactate/pyruvate ratio during stress. On the other hand, stress produced widely varying patterns in

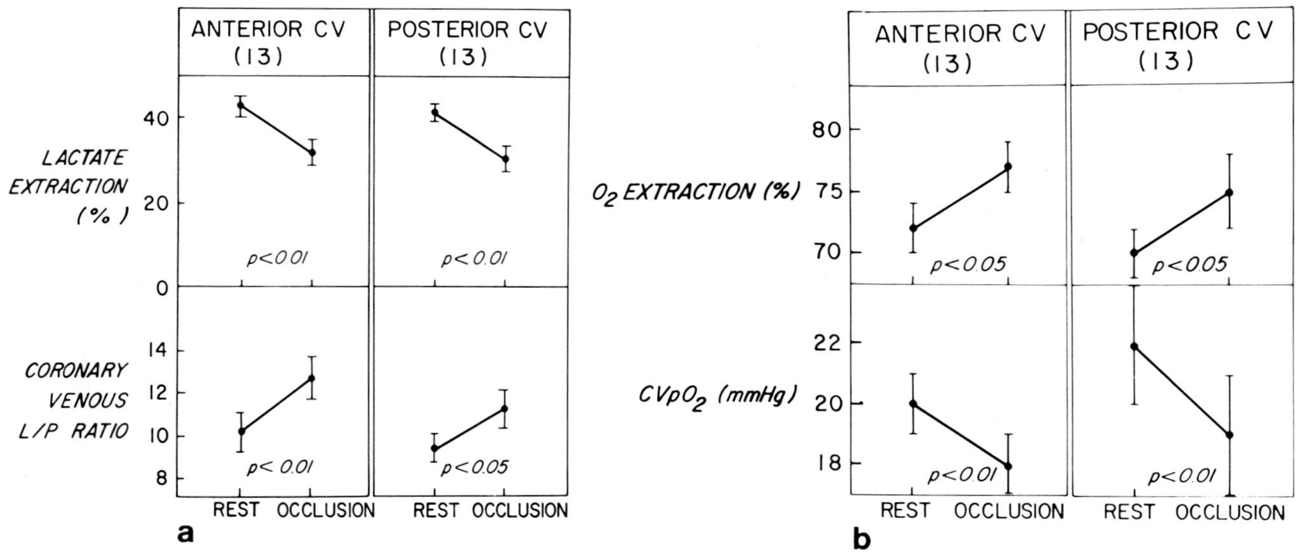

Figure 3. Measurement of myocardial lactate extraction, coronary venous lactate/pyruvate ratio, myocardial oxygen extraction, and coronary venous pO_2 during LAD occlusion. All metabolic parameters indicate the presence of myocardial ischemia at anterior and posterior sampling sites.

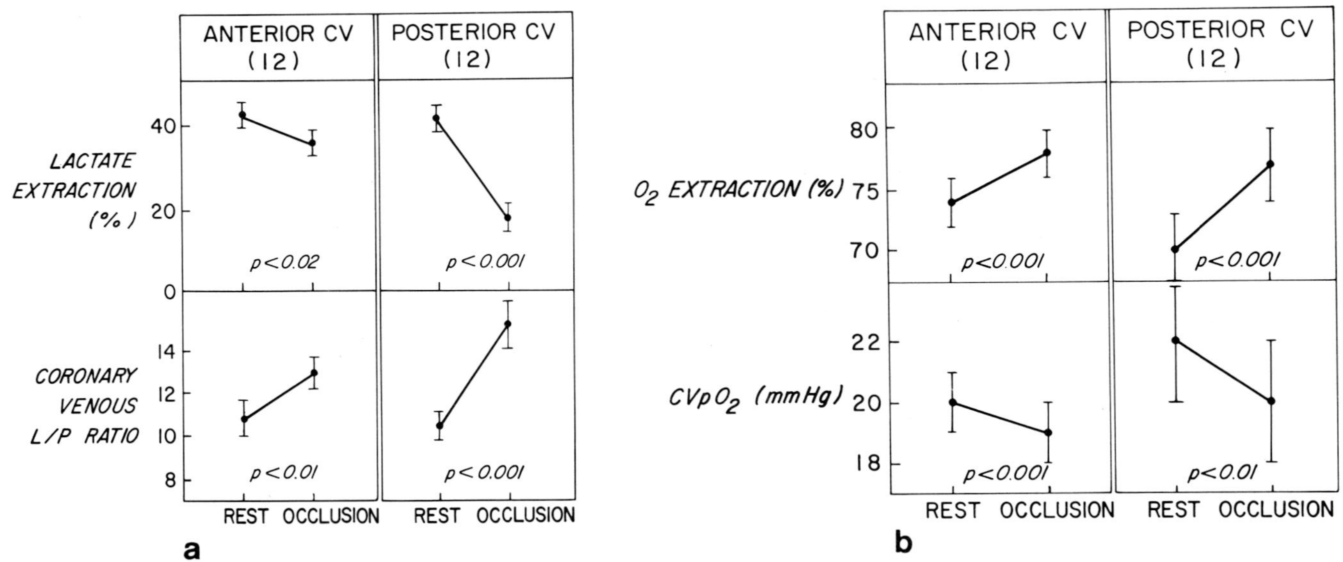

Figure 4. Measurement of myocardial lactate extraction, coronary venous lactate/pyruvate ratio, myocardial oxygen extraction, and coronary venous pO_2 during CIRC occlusion. Although all metabolic parameters indicate the presence of ischemia at anterior and posterior sampling sites, significantly greater changes occurred in lactate metabolism at the posterior sampling site.

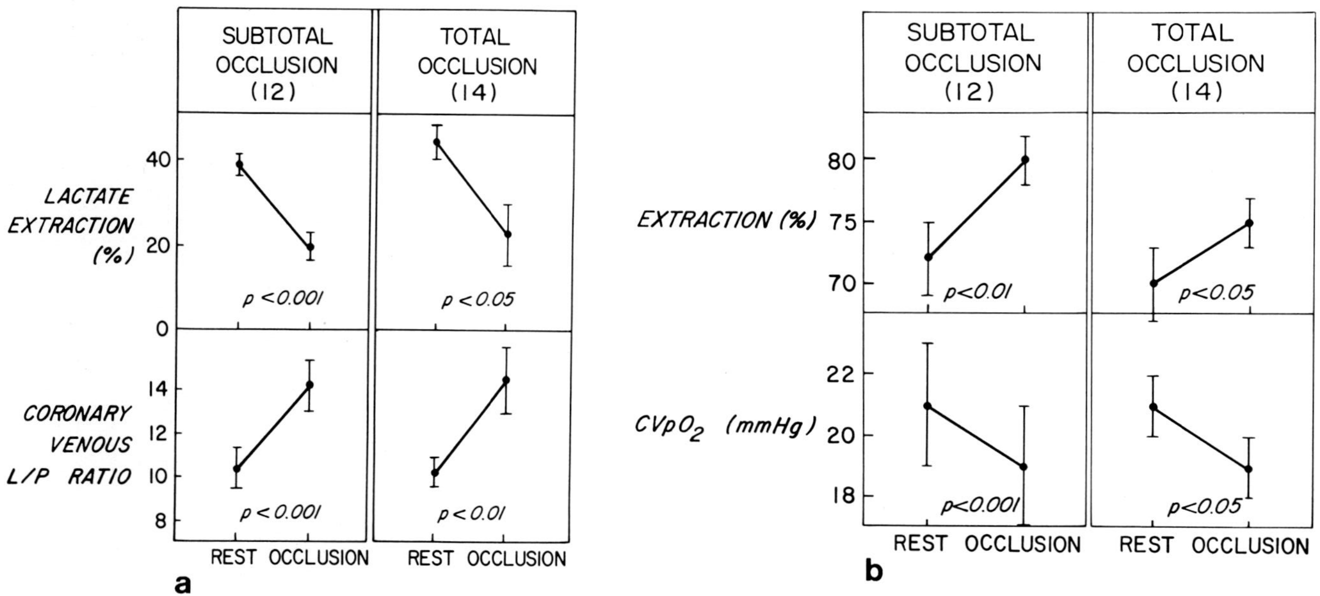

Figure 5. Effects of subtotal and total coronary artery occlusion on myocardial lactate extraction, coronary venous lactate/pyruvate ratio, myocardial oxygen extraction, and coronary venous pO_2. All metabolic parameters exhibit greater alteration during subtotal occlusion.

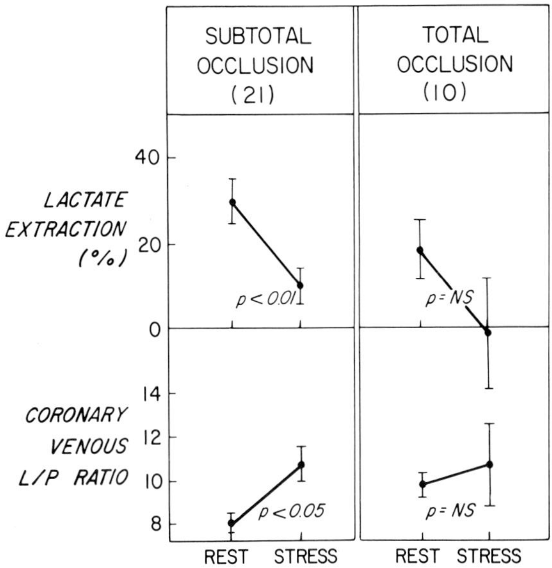

Figure 6. Effects of exercise, isoproterenol infusion, and PIT on myocardial lactate extraction and coronary venous lactate/pyruvate ratio in human subjects with subtotal and total anterior coronary artery occlusion. Greater metabolic changes occur in subjects with subtotal coronary artery occlusion.

subjects with total coronary occlusion, resulting in statistically insignificant changes in these two variables.

DISCUSSION

Specificity of coronary venous metabolic sampling

The LAD is reported to have an anteroseptal distribution in dogs and man (Rayford *et al.,* 1959; James, 1961). However, it has been noted that LAD ligation in dogs approximately 2.2 cm from the mouth of the left main coronary artery caused anterolateral infarction (Hood *et al.,* 1967). In the present study, LAD ligation was performed distal to all branches appearing to supply lateral left ventricle, and the CIRC was ligated distal to branches supplying anterior myocardium. The amount of myocardium made ischemic was probably not large, but discrete anteroseptal or posterolateral localization of the ischemia was assured.

Canine coronary venous anatomy is similar to that in humans (Rayford *et al.,* 1959). Hood (1968) showed that 83% of the radicals draining into the human coronary sinus originate from the left ventricle. In addition, 96% of the small veins arising from the left ventricle or interventricular septum empty into the coronary sinus (Hood, 1968). However, left ventricular coronary venous drainage is not as straightforward as it may seem. Significant and variable anastomoses occur between right and left ventricular veins in dogs and in man (Gregg and Shipley, 1947; Gregg, 1950; James, 1961; Pomposiello *et al.*, 1970). The partition between coronary sinus and anterior cardiac venous drainage, possibly not constant under stable resting conditions (Katz, Jochim, and Bohning, 1938; Katz, Jochim, and Weinstein, 1938; Gregg, Shipley, and Bidder, 1943), has been demonstrated to shift in dogs during various interventions. Left ventricular failure (Bartelstone *et al.*, 1968), increased coronary sinus pressure (Bartelstone *et al.*, 1968; Scharf *et al.*, 1971), and perhaps the mere presence of a coronary sinus catheter (James, 1971) may shift the flow partition in favor of anterior cardiac venous drainage to the right heart. Epinephrine and induced tachycardia also affect the venous partition (Lafontant, Feinberg, and Katz, 1962; Scharf *et al.*, 1971). The latter observations are pertinent to human myocardial metabolic evaluation utilizing catecholamine infusion and PIT. There is some evidence that coronary sinus flow may reverse for a brief period during each cardiac cycle (Stein *et al.*, 1969). Furthermore, the recent demonstration of canine veno-venous shunting between the anterior interventricular vein and the coronary sinus (Corday, 1974) will be of importance if such shunting is shown to be significant.

In the present study, four metabolic variables all indicated the presence of myocardial anaerobic metabolism at both anterior and posterior sampling sites. Diminished myocardial lactate extraction is a sensitive and relatively specific indicator of myocardial ischemia (Gorlin, 1972; Opie *et al.*, 1972, 1973; Levken, Mjos, and Kjekshus, 1973). Determination of coronary venous lactate/pyruvate ratios are attended by practical difficulties but may add greater specificity to metabolic determination of myocardial ischemia (Opie *et al.*, 1973). Coronary venous lactate/pyruvate ratios were corrected for changes in arterial ratios in an attempt to correct for noncardiac changes in lactate metabolism. Some observations have indicated that myocardial oxygen extraction is an unreliable detector of myocardial ischemia (Baker *et al.*, 1968; Neill, 1968). Other studies have found this variable to be useful (Messer *et al.*, 1962; Levken *et al.*,, 1973; Meerbaum *et al.*, 1974). Myocardial oxygen metabolism was monitored in the present study, although it is perhaps a less reliable indicator of ischemia than myocardial lactate metabolism.

The present study confirms earlier work done in this laboratory (Baker *et al.*, 1968) and by subsequent investigators (Owen *et al.*, 1970; Opie, 1971; Meerbaum *et al.*, 1974) showing that, during LAD occlusion, posterior coronary venous drainage contains metabolic evidence of myocardial ischemia. Changes detected in posterior coronary venous samples were of the same magnitude as those from anterior samples. Subgroup analysis revealed no difference in the relative influence of subtotal and total LAD occlusion data. Progressive confluence of coronary venous drainage should theoretically dilute the ischemic anterior effuent, resulting in smaller posterior alterations. This pattern was not seen in the present study, suggesting that mechanisms other than contamination of posterior samples with anterior effluent might be involved. It has been proposed that the increased performance by the "nonoccluded" segment of myocardium (Hood, 1970; Lewartowski, Sedek, and Gawin, 1970) is responsible for the appearance of anaerobic metabolism (Kaltenbach and Jennings, 1960; Gudbjarnason, Puri, and Mathes, 1971; Gudbjarnason, Ravens, and Mathes, 1972; Meerbaum *et al.,* 1974). One study (Meerbaum *et al.,* 1974) separated coronary venous drainage by inflating a coronary sinus balloon during sampling after total LAD occlusion. No coronary sinus pressures were reported, however, and it is possible that such balloon occlusion directed anterior drainage through venous interconnections to the posterior coronary sinus.

During CIRC occlusion, all measured metabolic variables again indicated the development of myocardial ischemia, as determined from anterior and posterior sites. The appearance of evidence of anaerobic metabolism at the anterior sampling site with CIRC occlusion similarly suggests ischemia of anterior myocardium not directly affected by the occlusion, or veno-venous shunting from posterior veins to the anterior interventricular vein. There is no reason to suspect reversal of coronary sinus blood flow and contamination of the anterior sampling site with retrograde flow of posterior effluent.

In contrast to LAD occlusion, CIRC occlusion resulted in significantly more pronounced changes at the posterior venous sampling site. It is possible that the absence of direct contamination allowed better metabolic separation of ischemic and nonischemic myocardium. Subgroup analysis in this case clearly showed that subtotal, rather than total, coronary artery occlusion was primarily responsible for the significant differentiation.

Regional coronary venous metabolic sampling does not separate ischemic from nonischemic myocardium in LAD occlusion but may do so in isolated CIRC occlusion. Although the compensatory increase in the performance of nonoccluded myocardium may give rise to anaerobic metabolism in this segment, further evaluation of the significance of coronary veno-venous shunting is desirable.

Metabolic consequences of subtotal versus total coronary artery occlusion

The S-T segment deviation of the epicardial electrocardiogram has been used to quantitate myocardial ischemia (Case, Rosell, and Crampton, 1966; Scheuer and Brachfeld, 1966; Redwood, Smith, and Epstein, 1972). In the present study, the S-T deviation of standard lead II of the electrocardiogram was monitored. Although scatter in the data prevented demonstration of a statistically significant difference between subtotal and total occlusion, S-T segment deviation during total occlusion tended to exceed that resulting from subtotal occlusion (Fig 2).

In spite of a tendency for the ECG to indicate greater myocardial ischemia during total coronary artery occlusion, all four metabolic parameters exhibited greater alteration during subtotal occlusion. No difference could be demonstrated in the relative effects of LAD vs. CIRC manipulation in this regard. It has been suggested that total coronary artery occlusion may prevent venous washout of metabolic end products in the acute phase of myocardial infarction (Sharma *et al.,* 1970). Reperfusion of the totally occluded LAD reveals an immediate increase in detectable ischemic metabolites, providing strong support for this concept (Lang *et al.,* 1974). No direct comparison of the effects of subtotal and total coronary artery occlusion was attempted in either of these previous studies.

A similar metabolic phenomenon was demonstrated in human subjects with subtotal or total anterior coronary artery occlusion who were stressed during cardiac catheterization. The distribution of the different stresses (exercise, isoproterenol infusion, PIT) was nearly the same in the subtotal and total occlusion groups. Myocardial oxygen extraction and coronary venous pO_2 are not of value in detecting myocardial ischemia during isoproterenol infusion (Cohen *et al.,* 1966; Gorlin, 1972), and were not utilized in this analysis. During stress, subjects with subtotal coronary artery occlusion exhibited a significant decrease in lactate extraction and increase in coronary venous lactate/pyruvate ratio. Subjects with total anterior coronary artery occlusion exhibited smaller resting values of lactate extraction. Lactate production occurred in this group during stress, but data scatter prevented changes in lactate extraction and coronary venous lactate/pyruvate ratio from achieving statistical significance.

Although poor venous washout in the presence of total coronary artery occlusion may interfere with the detection of ischemia, it is also possible that much of the myocardium once nourished by the occluded artery may be inert scar, incapable of yielding ischemic metabolic products during stress. Employing more severe degrees of stress in subjects with total coronary artery occlusion would probably be unfruitful, since those subjects experiencing angina during catheterization laboratory stress showed no increased tendency toward consistent myocardial metabolic abnormalities (Baker *et al.,* 1968).

In both dog and human studies of myocardial ischemia, myocardial metabolic changes were more consistent during subtotal than total coronary artery occlusion. In the dog, this was so despite ECG data suggesting greater ischemia during total occlusion. These observations suggest variable venous washout from ischemic myocardium during total occlusion and, in humans, may also reflect the presence of scar tissue. Caution is advised when interpreting myocardial metabolic information in the presence of total coronary artery occlusion, since underestimation of the extent of myocardial ischemia appears more likely than with lesser degrees of occlusion.

REFERENCES

BAKER, L. D., V. S. MATHUR, S. J. LESHIN, and J. V. MESSER. 1968. The significance of myocardial "Excess Lactate" in assessing human coronary arterial disease (Abstract). Circulation 38: VI–37.

BAKER, L. D., J. C. POMPOSIELLO, G. V. R. K. SHARMA, J. A. SNOW, and J. V. MESSER. 1969. Evaluation of regional coronary venous sampling in localized coronary artery obstruction (Abstract). Clin. Res. 17: 228.

BARTELSTONE, H. J., Y. SCHULTZ, and H. H. WANG. 1968. Partition of coronary drainage from the canine right heart (Abstract). Fed. Proc. 27: 631.

BEMIS, C. E., R. GORLIN, H. G. KEMP, and M. V. HERMAN. 1973. Progression of coronary artery disease. A clinical arteriographic study. Circulation 47: 455–464.

CASE, R. B., H. A. ROSELL, and R. S. CRAMPTON. 1966. Relation of S-T depression to metabolic and hemodynamic events. Cardiologia 48: 32–38.

CHIONG, M. A., R. WEST, and J. O. PARKER. 1974. Myocardial balance of inorganic phosphate and enzymes in man. Effects of tachycardia and ischemia. Circulation 49: 283–290.

COHEN, L. S., W. C. ELLIOTT, E. L. ROLETT, and R. GORLIN. 1966. Coronary heart disease. Clinical, cinearteriographic, and metabolic correlations. Am. J. Cardiol. 17: 153–168.

CORDAY, E., T. W. LANG, S. MEERBAUM, H. GOLD, S. HIROSE, S. RUBINS, and M. DALMASTRO. 1974. Closed chest model of intracoronary occlusion for study of regional cardiac function. Am. J. Cardiol. 33: 49–59.

DWYER, E. M. 1972. Regional myocardial blood flow. Circulation 45: 4–7.

FERGUSON, T. B., O. W. SHADLE, and D. W. GREGG. 1953. Effect of blood and saline infusion on ventricular end diastole pressure, stroke work, stroke volume and cardiac output in the open and closed chest dog. Circulat. Res. 1: 62–68.

GORLIN, R. 1972. Assessment of hypoxia in the human heart. Cardiology 57: 24–34.

GREGG, D. E. 1950. Coronary Circulation in Health and Disease, pp 21–36. Lea and Febiger, Philadelphia.

GREGG, D. E., and R. E. SHIPLEY. 1947. Studies of the venous drainage of the heart. Am. J. Physiol. 151: 13–25.

GREGG, D. E., R. E. SHIPLEY, and T. G. BIDDER. 1943. The anterior cardiac veins. Their functional importance in the venous drainage of the right heart. Am. J. Physiol. 139: 732–741.

GUDBJARNASON, S., P. S. PURI, and P. MATHES. 1971. Biochemical changes in noninfarcted heart muscle following myocardial infarction. J. Molec. Cell. Cardiol. 2: 253–276.

GUDBJARNASON, S., K. G. RAVENS, and P. MATHES. 1972. Metabolic changes in infarcted and noninfarcted myocardium during the post infarction period. *In* E. Bajusz and G. Rona (eds.), Recent Advances in Studies in Cardiac Structure and Metabolism, Vol. 1: Myocardiology, pp. 439–446. University Park Press, Baltimore.

HERMAN, M. V., W. C. ELLIOTT, and R. GORLIN. 1967. An electrocardiographic, anatomic, and metabolic study of zonal myocardial ischemia in coronary artery disease. Circulation 35: 834–846.

HOOD, W. B. 1968. Regional venous drainage of the human heart. Brit. Heart J. 30: 105–109.

HOOD, W. B. 1970. Experimental myocardial infarction. III. Recovery of left ventricular function in the healing phase. Contribution of increased fiber shortening in noninfarcted myocardium. Am. Heart J. 79: 531–538.

HOOD, W. B., B. McCARTHY, and B. LOWN. 1967. Myocardial infarction following coronary ligation in dogs. Hemodynamic effects of isoproterenol and acetylstrophanthidin. Circulat. Res. 21: 191–199.

HORN, H. D., und F. H. BRUNS. 1956. Quantitative bestimmung von L̆(+)-Milchsäure mit Milchsäuredehydrogenase. Biochim. biophys. Acta (Amst.) 21: 378–388.

JAMES, T. N. 1961. Anatomy of the Coronary Arteries, pp. 162–202. Paul B. Hoeber, Inc., New York.

JAMES, T. N. 1971. Functional anatomy of the coronary blood vessels. *In* H. L. Conn and O. Horwitz (eds.), Cardiac and Vascular Diseases, Vol. 1, pp. 39–40. Lea and Febiger, Philadelphia.

KALTENBACH, J. P., and R. B. JENNINGS. 1960. Metabolism of ischemic cardiac muscle. Circulat. Res. 8: 207–213.

KATZ, L. N., K. JOCHIM, and A. BOHNING. 1938. The effects of the extravascular support of the ventricles on the flow in the coronary vessels. Am. J. Physiol. 122: 236–251.

KATZ, L. N., K. JOCHIM, and W. WEINSTEIN. 1938. The distribution of the coronary blood flow. Am. J. Physiol. 122: 252–261.

KOBERSTEIN, R. C., and F. J. KLOCKE. 1968. Right atrial admixture in coronary venous blood. Fed. Proc. 27: 631.

LAFONTANT, R. R., H. FEINBERG, and L. N. KATZ. 1962. Partition of coronary flow and cardiac oxygen extraction between coronary sinus and other coronary drainage channels. Circulat. Res. 11: 686–698.

LANG, T. W., E. CORDAY, H. GOLD, S. MEERBAUM, S. RUBINS, C. COSTANTINI, S. HIROSE, J. OSHER, and V. ROSEN. 1974. Consequences of reperfusion after coronary occlusion. Am. J. Cardiol. 33: 69–81.

LEKVEN, J., O. D. MJOS, and J. KJEKSHUS. 1973. Compensatory mechanisms during graded myocardial ischemia. Am. J. Cardiol. 31: 467–473.

LEWARTOWSKI, B., G. SEDEK, and E. GAWIN. 1970. The increase of contractility of the intact portions of the myocardium during acute focal ischemia of the left ventricle in the dog. Kardiol. Pol. 13: 37–46.

MEERBAUM, S., T. W. LANG, E. CORDAY, S. RUBINS, S. HIROSE, C. COSTANTINI, H. GOLD, and M. DALMASTRO. 1974. Progressive alterations of cardiac hemodynamic and regional metabolic function after acute coronary occlusion. Am. J. Cardiol. 33: 60–68.

MESSER, J. V., R. J. WAGMAN, H. J. LEVINE, W. A. NEILL, N. KRASNOW, and R. GORLIN. 1962. Patterns of human myocardial oxygen extraction during rest and exercise. J. Clin. Invest. 4: 725–742.

NEILL, W. A. 1968. Myocardial hypoxia and anaerobic metabolism in coronary heart disease. Am. J. Cardiol. 22: 507–515.

NEILL, W. A. 1969. Effects of arterial hypoxemia and hyperoxia on oxygen availability for myocardial metabolism. Am. J. Cardiol. 24: 166–171.

OLSSON, R. A., and D. E. GREGG. 1965. Metabolic responses during myocardial reactive hyperemia in the unanesthetized dog. Am. J. Physiol. 208: 231–236.

OLSON, R. E., and L. H. OPIE. 1972. Report of the session on controversial methodological aspects. Cardiology 57: 102–117.

OPIE, L. H. 1971. Acute metabolic response in myocardial infarction. Brit. Heart J. 33: 129–137.

OPIE, L. H., P. OWEN, M. THOMAS, and R. SAMSON. 1973. Coronary sinus lactate measurements in assessment of myocardial ischemia. Comparison with changes in lactate/pyruvate and beta-hydroxybutyrate/acetoacetate ratios and with release of hydrogen, phosphate and potassium ions from the heart. Am. J. Cardiol. 32: 295–305.

OPIE, L. H., M. THOMAS, P. OWEN, and G. SHULMAN. 1972. Increased coronary venous inorganic phosphate concentrations during experimental myocardial ischemia. Am. J. Cardiol. 30: 503–513.

OWEN, P., M. THOMAS, V. YOUNG, and L. H. OPIE. 1970. Comparison between metabolic changes in local venous and coronary sinus blood after acute experimental coronary artery occlusion. Am. J. Cardiol. 25: 562–570.

POMPOSIELLO, J. C., G. V. R. K. SHARMA, and J. V. MESSER. 1970. Effects of graded coronary narrowing on canine myocardial metabolism (Abstract). Circulation 42: III–189.

RAYFORD, C. R., E. M. KHOURI, F. B. LEWIS, and D. E. GREGG. 1959. Evaluation of use of left coronary artery inflow and O_2 content of coronary sinus blood as a measure of left ventricular metabolism. J. Appl. Physiol. 14: 817–823.

REDETZKI, H., H. BLOEDORN, und H. W. BANSI. 1956. Einfluss der Thioctinsaure auf den brenztraubensaure und alpha-ketoglutarsaure Blutspiegel bei Leverkranken. Klin. Wschr. 34: 845–854.

REDWOOD, D. R., E. R. SMITH, and S. E. EPSTEIN. 1972. Coronary artery occlusion in the conscious dog. Effects of alterations in heart rate and arterial pressure on the degree of myocardial ischemia. Circulation 46: 323–332.

SCHARF, S. M., B. BROMBERGER-BARNEA, and S. PERMUTT. 1971. Distribution of coronary flow. J. Appl. Physiol. 30: 657–662.

SCHEUER, J., and N. BRACHFELD. 1966. Coronary insufficiency: Relations between hemodynamic, electrical and biochemical parameters. Circulat. Res. 18: 178–189.

SHARMA, G. V. R. K., R. KUMAR, F. MOLOKHIA, A. N. INAMDAR, W. B. HOOD, W. H. ABELMANN, and J. V. MESSER. 1970. Myocardial metabolism in acute and healing myocardial infarction: Evaluation of regional coronary venous sampling. Chest 58: 290–291.

SNEDECOR, G. W., and W. G. COCHRAN. 1967. Statistical Methods, 6th ed. Iowa State University Press, Ames, Iowa.

STEIN, P. D., H. S. BADEER, W. H. SHUETTE, and J. F. GLASER. 1969. Pulsatile aspects of coronary sinus blood flow in closed-chest dogs. Am. Heart J. 78: 331–337.

Mailing address:
Lawrence D. Baker, M.D.,
97 Red Gate Lane,
Reading, Massachusetts 01867 (USA).

Recent Advances in Studies on
Cardiac Structure and Metabolism Volume 10
The Metabolism of Contraction
Edited by Paul-Emile Roy and George Rona

EFFECTS OF PROLONGED CORONARY HYPOTENSION ON MYOCARDIAL SUBSTRATE UTILIZATION

JOHN J. SPITZER, L. J. GREENFIELD,
and L. B. HINSHAW

Department of Physiology,
Louisiana State University Medical Center, New Orleans, Louisiana,
and Veterans Administration Hospital and
Departments of Physiology and Biophysics and Surgery,
University of Oklahoma, Health Science Center,
Oklahoma City, Oklahoma, USA

SUMMARY

Myocardial substrate utilization was investigated following experimental coronary hypotension in isolated *in vitro* perfused dog hearts. Control observations were made at the beginning and at the end of the experiments with afterload of 100 mm Hg. Between the two controls, the afterload was lowered to 50 mm Hg and kept at that level for 4 hours. The observations during the hypotensive period were compared to the two controls. Lactate uptake and the percentage contribution of the lactate to myocardial CO_2 production were diminished during the hypotensive period. Myocardial free fatty acid (FFA) uptake and oxidation did not change significantly. Glucose uptake and myocardial respiratory quotient (RQ) were decreased. Thus, prolonged coronary hypotension caused a relative preference of FFA oxidation and diminished lactate and glucose uptake by the myocardium. It is of interest to note that these changes in substrate utilization were quite different from those observed during experimentally produced hemorrhagic hypotension or endotoxic shock.

INTRODUCTION

Alterations of myocardial substrate utilization have been previously studied employing various experimental shock models (Spitzer and Spitzer, 1972; Spitzer *et al.,* 1974). The shock-induced metabolic changes consisted of decreased myocardial free fatty acid (FFA) oxidation and increased utilization of lactate by the myocardium. These alterations were similar, regardless of the method used to produce the experimental shock. Therefore, it was thought to be of interest to dissect the various contributing factors and their relative importance in bringing about these metabolic changes.

Since coronary hypotension often accompanies shock, the present study was undertaken with the following factors in mind: a) to determine the effects of prolonged coronary hypotension on myocardial substrate utilization; b) to ascertain whether coronary hypotension causes demonstrable cardiac failure; and c) if failure does occur, would it be accompanied by characteristic alterations of myocardial lactate uptake or output?

MATERIALS AND METHODS

In 10 experiments, a large donor dog perfused an isolated dog heart according to a previously described technique (Hinshaw *et al.,* 1972). Arterial and coronary venous blood samples were taken simultaneously across the perfused heart. The afterload of the heart was adjusted to 100 mm Hg for the initial and the end control observations. Afterload was reduced to 50 mm Hg for 4 hours between the two control periods constituting the experimental portion of the study. A constant infusion of 1-^{14}C-palmitate was administered to study FFA oxidation (Spitzer and Spitzer, 1972).

RESULTS AND CONCLUSIONS

Table 1 shows the effects of prolonged coronary hypotension on myocardial blood flow and O_2 consumption. Both decreased considerably during hypotension and returned to initial values during the second control period. Heart rate was also decreased during hypotension. The arterial pH did not change throughout the experiments.

Changes of arterial FFA, lactate, and glucose and also myocardial uptake of these metabolites are shown in Table 2. Although arterial FFA concentration increased slightly in the course of the experiment, neither uptake nor oxidation of these metabolites changed significantly. It is of interest to note that myocardial FFA oxidation showed a similar positive correlation with arterial FFA

Table 1. Effect of prolonged coronary hypotension on myocardial metabolism (N=10)*

	1st Control	4-hour Hypotension (Change)	2nd Control
Mean arterial blood pressure (mm Hg)	100	−50	100
Heart rate (beats/min)	148±6	−17±2†	138±10
Coronary blood flow (ml/min)	45±5	−12±6‡	50±11
Arterial pH	7.46±0.04	−0.01±0.02	7.43±0.03
O_2 consumption (μmoles/min)	194±13	−56±8†	195±14

*Values expressed as mean ±SEM.
$^{\dagger}p<0.01$.
$^{\ddagger}p<0.05$.

Table 2. Effect of prolonged coronary hypotension on myocardial metabolism (N=10)*

	1st Control	4-hour Hypotension (Change)	2nd Control
Arterial FFA (μmole/ml)	0.486±0.043	+0.166±0.053†	0.732±0.094
FFA uptake (μmole/min)	7.4±2.2	−3.5±2.1	5.2±1.2
FFA oxidation (μmole/min)	2.3±0.6	+0.5±1.1	4.1±1.3
Arterial lactate (μmoles/ml)	3.29±.31	+0.61±0.34	4.37±0.45
Lactate uptake (μmole/min)	38.9±3.7	−20.1±4.1†	31.6±6.3
Arterial glucose (μmoles/ml)	6.71±0.65	−2.09±0.74†	4.81±0.56
Glucose uptake (μmoles/min)	15.0±4.8	−9.8±5.1‡	9.7±2.9
Myocardial RQ	0.95±0.01	−0.06±0.02†	0.87±0.04

*Values expressed as mean±SEM.
†$p<0.01$.
‡$p<0.05$.

concentration during hypotension (y = 10.6 X −3.5; r = 0.83) as during the two control periods (y = 9.6 X − 2.7; r = 0.79). In spite of an increasing trend in the arterial lactate concentration in the course of the experiments, myocardial lactate uptake decreased during the hypotensive period. Glucose uptake was also less during the hypotensive period. Thus, myocardial substrate utilization indicated a relative shift away from carbohydrates and more toward lipids, as indicated also by the decrease of myocardial RQ during hypotension.

Half of the 10 perfused hearts exhibited failure following hypotension, as indicated by the significantly elevated left ventricular end-diastolic pressure both at 50- and 100-mm Hg afterloads. This is seen in Table 3. It can also be observed that the failing hearts showed significantly lower efficiency than those that exhibited no signs of failure.

A positive lactate balance (removal of lactate by the myocardium) is a hallmark of the normal myocardium. During severe myocardial hypoxia, the myocardium no longer removes, but produces lactate (Scheuer and Brachfeld, 1966). Therefore, it is of interest to compare myocardial lactate uptake between the two groups of hearts, those that failed and those that exhibited no sign of

Table 3. Left ventricular end-diastolic pressure and cardiac efficiency in failing and nonfailing hearts*

	Failing (N=5)		Nonfailing (N=5)	
	Control	Following hypotension	Control	Following hypotension
Left ventricular				
End-diastolic pressure (mm Hg)				
At afterload of 50 mm Hg	2.0±0.5	13.7±3.9	2.6±0.8	4.5±0.8
At afterload of 100 mm Hg	3.2±0.9	17.5±4.5	3.5±0.5	4.8±0.8
Efficiency (%)				
At afterload of 50 mm Hg	5.1±0.1	1.0±0.4	5.2±0.3	5.3±0.3
At afterload of 100 mm Hg	7.2±0.4	5.9±0.3	7.7±0.4	7.5±0.5

*Values expressed as mean ± SEM.

Table 4. Arterial lactate concentration and myocardial uptake of lactate in failing and nonfailing hearts*

	Failing (N=5)		Nonfailing (N=5)	
	Control	During hypotension	Control	During hypotension
Arterial lactate (μmoles/ml)	3.35±0.41	3.72±0.51	3.23±0.52	4.09±0.51
Lactate uptake (μmoles/min)	38.0±6.6	20.6±5.3	39.4±4.3	16.6±3.4

*Values expressed as mean±SEM.

failure. It is shown in Table 4 that the net myocardial uptake of the two groups was not significantly different either under control conditions or during hypotension.

Thus, prolonged hypotension caused reversible alterations in myocardial substrate utilization, which consisted of decreased uptake of lactate and glucose and lowered myocardial RQ. These changes were quite the opposite of those observed following severe hemorrhage or endotoxin administration. Half of the perfused hearts exhibited increased left ventricular end-diastolic pressure and decreased cardiac efficiency following hypotension. In spite of these signs of failure, however, the hearts continued to show a net removal of lactate from the blood at a rate which was not different from that found in the nonfailing hearts.

Acknowledgments These investigations were supported by Grant HL 15037 from the National Institutes of Health and by Contract N00014-72-A-0317 from the Office of Naval Research.

REFERENCES

HINSHAW, L. B., L. J. GREENFIELD, S. E. OWEN, L. T. ARCHER, and C. A. GUENTER. 1972. Cardiac response to circulating factors in endotoxin shock. Am. J. Physiol. 222: 1047–1053.

SCHEUER, J., and N. BRACHFELD. 1966. Coronary insufficiency: Relations between hemodynamic, electrical, and biochemical parameters. Circulat. Res. 18: 178–189.

SPITZER, J. J., A. A. BECHTEL, L. T. ARCHER, M. R. BLACK, and L. B. HINSHAW. 1974. Myocardial substrate utilization in dogs following endotoxin administration. Am. J. Physiol. 227: 132–136.

SPITZER, J. J., and J. A. SPITZER. 1972. Myocardial metabolism in dogs during hemorrhagic shock. Am. J. Physiol. 222: 101–105.

Mailing address:
John J. Spitzer, M.D.,
Professor and Head, Department of Physiology,
L.S.U. Medical Center,
1542 Tulane Avenue,
New Orleans, Louisiana 70112 (USA).

Recent Advances in Studies on
Cardiac Structure and Metabolism Volume 10
The Metabolism of Contraction
Edited by Paul-Emile Roy and George Rona

NICOTINAMIDE ADENINE DINUCLEOTIDE DEGRADATION IN INFARCTED CARDIAC MUSCLE

ROSARIO NUNEZ, EDMUNDO CALVA,
and MEYTHA MARSH

Department of Biochemistry, Centro de Investigacion y de Estudios
Avanzados del Instituto Politécnico Nacional,
México D.F., México

SUMMARY

Nicotinamide adenine dinucleotides (NAD, $NADH_2$, NADP, and $NADPH_2$) levels decrease in myocardial dog tissue after the ligature of a coronary artery branch. The activity of a glycohydrolytic enzyme acting on NAD and releasing nicotinamide in an equivalent amount was of the same order of magnitude in infarcted tissue, irrespective of the time elapsed after the coronary artery occlusion, as it was in normal tissue. Most of the NAD contained in normal heart muscle was hydrolyzed as soon as the tissue was disrupted in a homogenizer, whereas no hydroylsis occurred when the whole fragment was incubated for 1 hour. The enzymatic activity was found mainly in a membranous fraction separated at 17,000 × *g* by differential centrifugation. Acid phosphatase, K^+-activated phosphatase, and NA^+-K^+-ATPase specific activities were greater in this fraction. It is suggested that the structural disorganization of the heart elicited either *in vitro* or during the infarction process determines the conditions for a reaction between the enzyme which is localized in the membranes and the NAD which is mainly in the cytosol.

INTRODUCTION

In previous work (Calva *et al.*, 1966; Nuñez *et al.*, 197), we showed that mitochondria isolated from infarcted heart muscle and incubated with β-hydroxybutyrate exhibit a low oxygen consumption rate, whereas succinate is oxidized at a normal rate. Furthermore, the P/O values measured with either of these two substrates decrease progressively, and the nicotinamide adenine dinucleotides[1] content diminishes in both the whole tissue and the isolated mitochondria. We suggested that the function of the respiratory chain linked to the oxidation of β-hydroxybutyrate may have been impaired at the coenzyme level as a consequence of the loss of the nicotinamide adenine dinucleotides. We

[1] Abbreviations used are: NAD, nicotinamide adenine dinucleotide; $NADH_2$, reduced NAD; NADP, nicotinamide adenine dinucleotide phosphate; $NADPH_2$, reduced NADP; NMN, nicotinamide mononucleotide.

proposed three possible explanations for the loss: a) a leakage of the tissue nicotinamide coenzymes toward the bloodstream; b) an enzymatic breakdown of these compounds in the myocardial tissue itself; and c) a defect in their synthesis. With regard to the first of these possible explanations, no significant changes were observed in the level and the ratio of NAD and $NADH_2$ in the coronary sinus blood taken after ligation of the coronary artery branch (Nuñez *et al.*, 1974). However, an enzyme which hydrolyzes NAD added to blood was detected in the dog white cells (unpublished results). Hence, we concluded it was possible that the myocardial nicotinamide adenine dinucleotides are released from the infarcted tissue and then hydrolyzed in the blood. To investigate the second possible explanation, an enzymatic activity in dog heart homogenates was measured. It was found that NAD, $NADH_2$, NADP, $NADPH_2$, and NMN were hydrolyzed between the pryidine and ribose moieties (unpublished results). It was also observed that most of the NAD contained in normal heart muscle was decomposed as soon as a tissue fragment was disrupted in a homogenizer, whereas no hydrolysis occurred when the whole fragment was incubated at 37°C for 1 hour.

The present report deals with the NAD glycohydrolase activity in infarcted myocardial tissue and with the subcellular distribution of this enzyme in normal cardiac tissue.

METHODS

Normal hearts were obtained from eight mongrel dogs anesthetized with intravenous pentobarbital or sacrificed by electrocution. Infarcted hearts were prepared in four anesthetized dogs by ligature of the left descending coronary artery after the pericardial sac was opened. Before excising the heart, the extent of infarcted tissue was measured by the electrocardiographic changes appearing in the unipolar epicardial records (Calva *et al.*, 1965). Animals were kept anesthetized by intraperitoneal injections of pentobarbital at intervals and received 250 ml of saline solution followed by a 10% glucose solution (60–70 ml/hour) through the femoral vein. Subsequently, two of the four animals were sacrificed 2 hours after occlusion, and the other two at 24 hours after occlusion.

Both normal and infarcted hearts were immersed immediately in ice-cold 0.25 M sucrose solution and perfused through both left and right coronary arteries with the sucrose solution.

Tissue homogenates

Left ventricle fragments were blotted with filter paper, and a tissue suspension was prepared with a glass Potter-Elvehjem homogenizer at a ratio of 0.5 g to 20 ml of ice-cold 0.25 M sucrose solution.

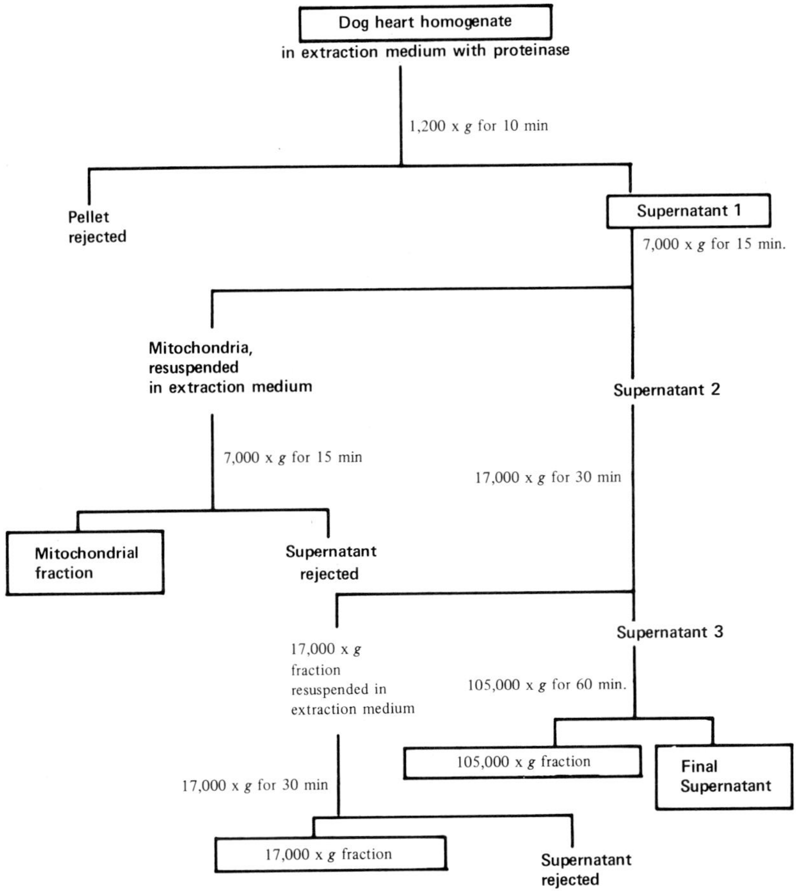

Figure 1. Flow sheet for preparation of subcellular fractions from dog heart homogenate. NAD glycohydrolase activity was measured in those fractions enclosed in a square. Relative centrifugal forces are average X *g* values.

Tissue fractionation

The homogenate was prepared, as described elsewhere (Calva *et al.*, 1965), in an extraction medium containing around 2,250 units of *B. subtilis* proteinase for each 2.5 g of myocardial left ventricle. Subcellular fractions were obtained by differential centrifugation (Fig. 1).

Measurement of NAD glycohydrolase activity

This activity was estimated as follows: 0.50 ml of 0.10 M sodium phosphate buffer, pH 7.4, and 1.0–1.75 ml of tissue fraction suspended in extraction medium were preincubated for 3 min at 37°C; 0.25 ml of 0.01 M NAD solution was then added. A sample of 1 ml of this mixture was withdrawn immediately after mixing and emptied into 2 ml of ice-cold 8% perchloric acid solution. After 15 min of incubation, a second sample was similarly treated. The NAD was measured directly in the acid extract with lithium lactate and lactate dehydrogenase, the nicotinamide was measured according to the technique of Strohecker and Henning (1966) after the pH of the tissue extract was adjusted to 4.6–5.7 by adding 0.5 M disodium hydrogen phosphate.

Normal hearts

Cylindrical pieces of the left ventricle weighing around 1g were incubated at 37°C. One piece was homogenized immediately in ice-cold 10% perchloric acid solution and the others at the end of 15, 30, and 60 min of incubation.

Homogenates were prepared in ice-cold sucrose solution (1 g in 3 ml) and incubated at 37°C. Proteins were precipitated by the addition of cold percholoric acid at 0, 5, 15, 30, and 60 min of incubation.

Infarcted hearts

Cylindrical fragments of the left ventricle were taken from the anterior wall (infarcted tissue) and from the posterior wall (control tissue), selected on the basis of the electrocardiographic records. The same procedures used on cylindrical pieces of normal heart were also performed on infarcted hearts, but only at time zero and after 30 min of incubation.

Assay of other enzymatic activities

The following were used as marker enzymes: cytochrome *c* oxidase (Wharton and Tzagoloff, 1967), Na^+-K^+-Mg^{++}-ATPase and Mg^{++}-ATPase (Kidwai *et al.*, 1971b), K^+-activated *p*-nitrophenylphosphatase (Kidwai, Radcliffe, and Daniel, 1971a), and acid *p*-nitrophenylphosphatase (Ostrowski and Tsugita, 1961).

RESULTS

Stability of NAD and nicotinamide in the cardiac tissue

The tissue contents of NAD and nicotinamide were well preserved in the cylindrical fragments of normal heart. In contrast, a remarkable decrease of NAD and a stoichiometric increase in nicotinamide occurred during incubation of the homogenate (Table 1).

In the infarcted cardiac tissue excised 2 hours after the coronary occlusion, the initial content of NAD was lower than that found in the control tissue whereas the level of nicotinamide appeared higher. The sum of the levels of the two compounds was similar in both tissues, and the increment of nicotinamide was

Table 1. Stability of NAD and nicotinamide in fragments and homogenates prepared from normal dog heart*

Incubation at 37°C (min)	NAD (nmoles/g wet wt.)	Nicotinamide (nmoles/g wet wt.)
Cylindrical fragment		
0	349	112
15	339	158
30	308	178
60	328	155
Tissue homogenate		
0	450	130
5	13	565
15	33	541
30	14	618
60	0	626

*Values presented are from 1 animal. They are representative of a typical pattern found in all 8 animals in the normal group.

Table 2. Content and stability of NAD and nicotinamide in fragments of dog hearts with experimental myocardial infarction*

Dog	Incubation at 37°C (min)	Control Tissue		Infarcted Tissue	
		NAD	Nicotinamide	NAD	Nicotinamide
2 hr after occlusion					
1	0	351	105	245	189
	30	322	172	121	381
2	0	410	101	341	175
	30	303	193	123	305
24 hr after occlusion					
3	0	438	97	41	34
	30	340	204	25	37
4	0	401	114	94	86
	30	306	204	87	100

*Figures (nmoles/g wet wt.) are average values from separated analysis of 2 tissue samples.

almost equivalent to the loss of NAD. Furthermore, the amount of NAD originally present in the infarcted tissue decreased from 50 to 68%, and the level of nicotinamide increased stoichiometrically after 30 min of incubation at 37°C (Table 2). On the other hand, in the infarcted cardiac tissue removed 24 hours after the arterial occlusion, the levels of NAD and nicotinamide were very low and did not change after incubation at 37°C (Table 2).

The control tissue taken from hearts with coronary artery branch occluded behaved as slightly disorganized tissue in that part of the endogenous NAD was hydrolyzed during incubation of a fragment of this tissue, whereas the changes observed in normal cardiac tissue under the same conditions were smaller (Table 2).

NAD glycohydrolase activity in control and infarcted cardiac tissue

The NADase activity measured in homogenates was of the same order of magnitude in both the control and the infarcted cardiac tissue, irrespective of the time elapsed after the occlusion of the artery. The amount of nicotinamide released during the reaction was approximately equivalent to the NAD loss (Table 3).

Table 3. NAD glycohydrolase activity in homogenates from dog hearts with experimental infarction

Dog	Cardiac Tissue Sample	Enzyme Activity (μmoles/g wet wt., 15 min)	
		NAD Consumed	Nicotinamide Produced
2 hr after coronary occlusion			
1	Control	15	17
	Infarcted	11	15
2	Control	9.4	12
	Infarcted	9.6	13
24 hr after coronary occlusion			
3	Control	12	10
	Infarcted	12	12
4	Control	11	9.8
	Infarcted	12	11

Table 4. Distribution of NAD glycohydrolase and marker enzymes in subcellular fractions from dog heart*

	Specific Activities (units/mg of protein)				
Fraction	NADase	Acid phosphatase	K^+-activated phosphatase	Na^+-K^+-Mg^{++} ATPase	Na^+-K^+ ATPase
Homogenate	43	64	62	2.9	0.9
Supernatant 1	19	64	46	1.5	0
Mitochondria	38	77	36	7.6	0.3
17,000 X *g*	515	327	266	27.3	7.0
105,000 X *g*	151	165	106	8.8	2.9

*Units: NADase, nmoles NAD hydrolyzed/mg, 15 min; acid phosphatase, nmoles *p*-nitrophenol hydrolyzed/mg, 15 min; K^+-activated phosphatase, nmoles *p*-nitrophenol hydrolylyzed/mg, 10 min; ATPase, μmoles P_i/mg, hour.

Distribution of the NADase in the subcellular fractions of dog cardiac tissue

Specific activities in subcellular fractions for NAD glycohydrolase and for marker enzymes are given in Table 4. No activity of NADase was found in the final supernatant. The values for cytochrome *c* oxidase (change of absorbance at 550 nm/mg, min) were 18.6 in the mitochondrial fraction and 2.0 in the 17,000 X *g* fraction. This last fraction contained 44% of the total NADase activity, and around 27% appeared in the mitochondrial fractions. The original activity was almost recovered (80% in all the subcellular fractions separated from supernatant 1 (Table 5).

DISCUSSION

The unfavorable conditions, namely anoxia and/or ischemia, under which a fragment of normal heart was maintained *in vitro* for 1 hour, were not noxious enough to stimulate the degradation of the endogenous NAD, whereas the disorganization of the cardiac tissue appeared to determine the conditions for a reaction between an enzyme and NAD contained in the tissue.

We concluded that the enzyme hydrolyzes the NAD at the *N*-glycosyl bond between the pyridine and the ribose moieties since, both in the homogenate from normal hearts and in the fragment of infarcted tissue excised 2 hours after the arterial occlusion, the increment of nicotinamide was equivalent to the loss of NAD. Consequently, this enzyme seems to be analogous to that detected previously in hearts from other animal species (Spaulding and Graham, 1947;

Table 5. Recovery of NAD glycohydrolase in subcellular fractions from dog heart

Fraction	Total Volume (ml)	Units/ml	Total Kilounits	Distribution (%)
Homogenate	665	343	228	–
Supernatant 1	526	58	30.5	100
Light mitochondria	21	253	5.3	17.3
Heavy mitochondria	10	300	3.0	9.8
17,000 × *g*	10	1340	13.4	44
105,000 × *g*	7.5	378	2.8	9
Final supernatant	476	0	0	0

Govier and Jetter, 1948; Hulsmans, 1961; Waravdekar and Griffin, 1964; Grigorovich, 1970) and known as NAD glycohydrolase.

It is possible that the nicotinamide liberated by the NADase activity could be converted to other compounds and/or to escape toward the bloodstream. This would explain the very low levels of this compound found in the infarcted tissue 24 hours after occlusion.

The NADase activity appears to be localized mainly in a membranous fraction which we are trying to characterize. This 17,000 × *g* fraction was vesicular when examined with the electron microscope (Fig. 2) and disclosed the maximal specific activities of those enzymes which have been found in the plasma membrane fraction prepared from hearts (Kidwai *et al.*, 1971b) or myometrium of rats (Kidwai *et al.*, 1971a). Moreover, the NADase seems to be tightly bound the membranes, since the activity measured in the greatly disorganized infarcted tissue (Calva *et al.*, 1969) was as high as in the normal myocardium. In contrast to this finding, a significant rise in the activity of the hexosemonophosphate shunt enzymes and a rapid decline in the oxidative and glycolytic enzymes in the infarcted cardiac tissue have been reported (Gudbjarnason *et al.*, 1967).

The fact that the NAD glycohydrolase does not hydrolyze the endogenous NAD in the normal cardiac tissue could be explained by the topological separation between the NAD and the NADase. On the other hand, it has been established that the NADase has low activity on enzyme-bound NAD, and it has been assumed that most of the nicotinamide nucleotides exist in the cell bound to protein (Kaplan, 1966). Consequently, it is possible that, during the infarction process or the *in vitro* disruption of the cardiac tissue, the enzyme-bound NAD is converted to unbound coenzyme. Finally, although alterations in lysosome stability have been proposed to be among the earliest structural modifications to occur during the ischemic myocardial injury (Ricciuti, 1972), further work is necessary to elucidate the role of these cell structures in the hydrolysis of the NAD.

In a previous work, we postulated (Calva *et al.*, 1966) that the blockade of the respiratory chain observed in mitochondria from infarcted cardiac tissue could be due to the loss of mitochondrial NAD. We can now add that it is possible this loss may be due to the activity of the NADase as a result of the tissue disintegration. It has not been determined whether such activity is also present

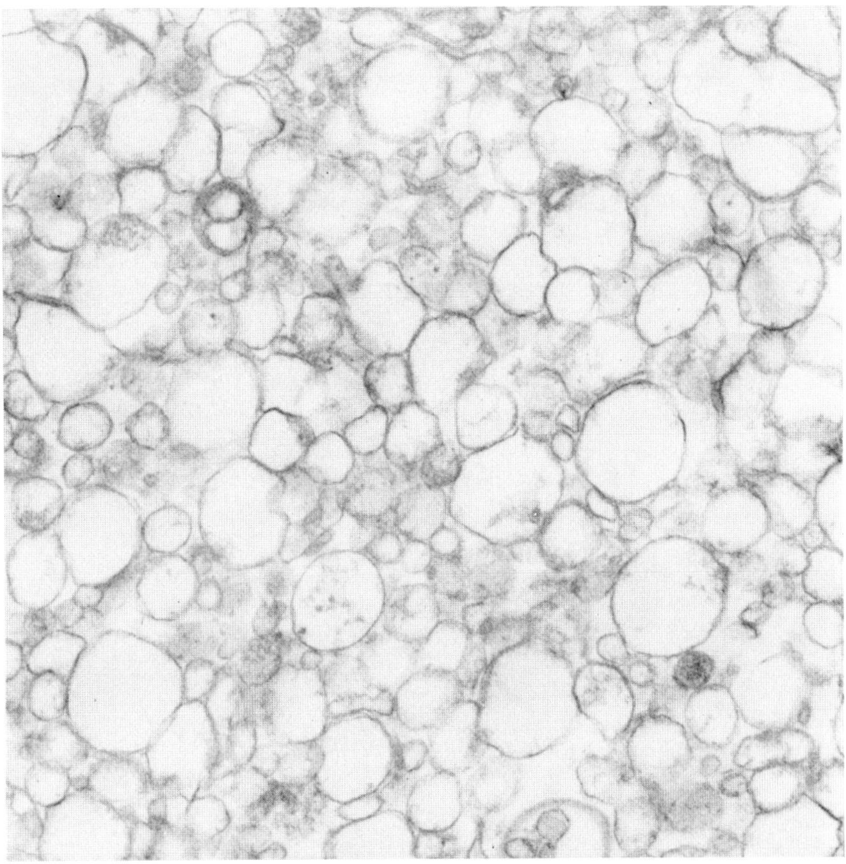

Figure 2. Electron micrograph of thin section prepared from the 17,000 × *g* fraction of dog heart homogenates. × 48,000. Courtesy of M. González del Pliego, Laboratory of electron microscopy, Inc.

in the cardiac mitochondria or whether it is due to a membranous contamination of the mitochondrial fraction.

REFERENCES

CALVA, E., A. MUJICA, A. BISTENI, and D. SODI-PALLARES. 1965. Oxidative phosphorylation in cardiac infarct. Effect of glucose-KCl-insulin solution. Am. J. Physiol. 209: 371–375.

CALVA, E., A. MUJICA, R. NUÑEZ, K. AOKI, A. BISTENI, and D. SODI-PALLARES. 1966. Mitochondrial biochemical changes and glucose-KCl-insulin solution in cardiac infarct. Am. J. Physiol. 211: 71–76.

CALVA, E., A. TRILLO, R. NUÑEZ, K. AOKI, and D. ARIZA. 1969. Relaciones entre los cambios bioquimicos y ultraestructurales en mitocondrias del corazón con infarto experimental. Arch. Inst. Cardiol. Méx. 39: 696–712.

GOVIER, W. M., and N. S. JETTER. 1948. The effect of alfatocopheryl phosphate on diphosphopyridine nucleotidase. Science 107: 146–147.

GRIGOROVICH, YU. A. 1970. Kordiamin effect on NAD-glycohydrolase in rabbit heart muscle. Vop. Med. Khim. 16: 90–94.

GUDBJARNASON, S., C. H. COWAN, W. BRAASCH, and R. BING. 1967. Changes in enzyme pattern of infarcted heart muscle during tissue repair. Cardiologia 51: 148–159.

HULSMANS, H. A. M. 1961. "Microsomes" in heart-muscle homogenates. Biochim. Biophys. Acta 54: 1–14.

KAPLAN, N. O. 1966. Regulation of enzyme and coenzyme levels. *In* N. O. KAPLAN and E. P. KENNEDY (eds.), Current Aspects of Biochemical Energetics, pp. 447–457. Academic Press, New York.

KIDWAI, A. M., M. A. RADCLIFFE, G. DUCHON, and E. E. DANIEL. 1971b. Isolation of zation of plasma membrane from rat myometrium. Biochim. Biophys. Acta 233: 538–549.

KIDWAI, A. M., M. A. RADCLIFFE, G. DUCHON, and E. E. DNAIEL. 1971b. Isolation of plasma membrane from cardiac muscle. Biochem. Biophys. Res. Commun. 45: 901–910.

NUÑEZ, R., E. CALVA, E. BRIONES, and F. LOPEZ-SORIANO. 1974. Nicotinamide coenzymes in heart and coronary blood during myocardial infarction. Am. J. Physiol. 226: 73–76.

OSTROWSKI, W., and A. TSUGITA. 1961. Purification of acid phosphomonoesterase from the human prostate gland. Arch. Biochem. Biophys. 94: 68–78.

RICCIUTI, M. A. 1972. Myocardial lysosome stability in the early stages of acute ischemic injury. Am. J. Cardiol. 30: 492–497.

SPAULDING, M. E., and W. D. GRAHAM. 1947. The enzymatic degradation of cozymase and the inhibitory action of alfa-tocopheryl phosphate. J. Biol. Chem. 170: 711–718.

STROHECKER, R., and H. M. HENNING. 1966. Vitamin assay. Tested methods, pp. 195–198 Verlag Chemie. GMBH, Weinheim/Bergstr.

WARAVDEKAR, V. S., and C. C. GRIFFIN. 1964. DPNase activity in livers from young, adult and tumor bearing mice. Exp. Cell Res. 33: 450–458.

WHARTON, D. C., and A. TZAGOLOFF. 1967. Cytochrome oxidase from beef heart mitochondria. *In* S. P. Colowick and N. O. Kaplan (eds.), Methods in Enzymology, Vol. X, pp. 245–250. Academic Press, New York.

Mailing address:
Edmundo Calva, M.D., Ph.D.,
Department of Biochemistry,
Centro de Investigación y de Estudios Avanzados
del Instituto Politécnico Nacional,
Apartado Postal 14–740, México 14, D.F.
(Mexico).

Recent Advances in Studies on
Cardiac Structure and Metabolism Volume 10
The Metabolism of Contraction
Edited by Paul-Emile Roy and George Rona

DISTRIBUTION OF GLYCOGEN AND LIPIDS IN THE ISCHEMIC CANINE LEFT VENTRICLE: BIOCHEMICAL AND LIGHT AND ELECTRON MICROSCOPIC CORRELATES

M. F. CRASS III and P. R. STERRETT

Departments of Physiology and Anatomy,
Texas Tech University School of Medicine, Lubbock, Texas, USA

SUMMARY

Transmural gradients in the contents of glycogen, triglycerides, and phospholipids were studied in the dog left ventricle with and without 4 hours of left circumflex artery ligation. In control dogs (patent circumflex), an increasing outer to inner glycogen gradient was observed. After 4 hours of ischemia, subepicardial glycogen content was similar to control values, but midventricular and subendocardial stores were nearly depleted, thus indicating a reversal of the control gradient. These changes were seen biochemically as well as by light and electron microscopic techniques. In the present studies, conclusive evidence for a transmural gradient in tissue triglyceride content was not obtained in either control or ligated dogs. However, an approximate three-fold increase in midventricular and subendocardial triglycerides was observed in ischemic hearts. Regional contents of total tissue phospholipids appeared to be similar across the left ventricular wall in both normal and ischemic left ventricles.

INTRODUCTION

Studies of regional coronary blood flow have shown that a reduction in coronary perfusion pressure results in a more marked underperfusion of the subendocardial myocardium than the subepicardial region (Salisbury, Cross, and Rieben, 1963; Moir and DeBra, 1967). Data obtained on normal dog hearts have suggested that transmural gradients in intramyocardial tissue pressure (Kirk and Honig, 1964) and tissue oxygen tension (Honig, Kirk, and Myers, 1967) exist in the left ventricular wall, thus rendering the subendocardial region more vulnerable to ischemic damage and necrosis. However, some recent regional blood flow studies do not entirely support the concept of subendocardial underperfusion in the normal canine left ventricle (Cobb, Bache, and Greenfield, 1974).

We have recently become interested in the possibility that, after total and/or subtotal occlusion of a branch of the left coronary artery, nonuniform metabolic responses occur across the left ventricular wall. In this context, the observations of Griggs, Tchokoev, and Chen (1972) showing differences in high energy phosphate levels and lactate to pyruvate ratios are supportive of this.

Glycogen depletion and probable lipid accumulation are characteristics of myocardial ischemia. However, relatively little has been reported concerning possible transmural differences in these processes or, in addition, correlations made with structural and ultrastructural changes. In the present study, we examined these factors in the acutely ischemic canine left ventricle.

MATERIALS AND METHODS

Mongrel dogs weighing 15–25 kg were given free access to food and water. The dogs were anesthetized with i.v. sodium pentobarbital and ventilated with room air, using a positive pressure respirator. Arterial pO_2, pCO_2 and pH were monitored and maintained within physiological range. Bilateral thoracotomy was performed, and the circumflex branch of the left coronary artery was isolated. The circumflex was ligated or left patent for 4 hours. A catheter was placed in the aorta via the left femoral artery for aortic pressures. Ventricular peak systolic and end-diastolic pressures were obtained with a catheter placed in the left ventricle via the right carotid artery. Pressures were recorded, using Statham P23Db transducers with an Electronics for Medicine model IR-4C recorder. Continuous electrocardiographic monitoring using limb leads was performed.

After the 4-hour experimental period, transmural biopsies were taken through the posterolateral wall of the left ventricle in the region of the posterior papillary muscle. The biopsy device consisted of a 3/8-inch electric hand drill with a 20 mm (ID) stainless steel boring attachment. Tissue plugs were 1) frozen intact in liquid nitrogen for chemical determination of glycogen (Good, Kramer, and Somogyi, 1933) and tissue extraction (Crass *et al.,* 1971) and determination of triglycerides (Van Handel and Zilversmit, 1957) and phospholipids (King, 1932); 2) sliced and immersed in (–) 60°C isopentane for light microscopy; or 3) processed by *in situ* coronary artery perfusion fixation for electron microscopy. For all analyses, the outer 2 mm were cut for subepicardial determinations. The middle 2 mm of the tissue plug were taken for midventricular analyses and, for the subendocardial region, the inner 2 mm including endocardium were sectioned.

Glycogen was cytochemically analyzed with light microscopy, using a modified PAS stain and counterstained with hematoxylin. Diatase of malt (0.1%) in phosphate buffer (pH 6.0) was used for confirmation of glycogen.

For electron microscopic analysis, left circumflex artery perfusion-washout with an oxygenated (95% O_2–5% CO_2) Krebs-Henseleit bicarbonate buffer containing 5.5 mM glucose and 4% albumin preceded perfusion with Karnovsky's fixative. Following dehydration with a graded series of alcohols and propylene oxide, the tissues were embedded in Epon 812 (Electron Microscopy Sciences). Ultrathin sections were mounted on 200-mesh rhodium-plated copper grids, stained with uranyl acetate and lead citrate, and examined in a Zeiss 9S-2 electron microscope.

RESULTS AND DISCUSSION

Cardiac performance

Heart rate, left ventricular peak systolic and end-diastolic pressures, and electrocardiogram were monitored at 10-min intervals throughout the 4-hour experimental period. Through the first 2 hours and 40 min, rate and pressures were stable. After 2½ hours, a transient decrease in left ventricular peak systolic pressure was observed in both control and ischemic dogs. No explanation is readily apparent for this hypodynamic interval. Left ventricular pressure development in ischemic hearts was comparable to controls throughout the 4-hour period. Left ventricular end-diastolic pressure was elevated slightly (mean of 3–5 mm Hg) in the ischemic group, the increase being statistically significant ($p<0.05$). Marked electrocardiographic changes were observed in the ischemic group 15 min after ligation, with the frequency of premature ventricular contractions increasing with time.

Transmural differences in glycogen content in the normal and ischemic canine left ventricle

Figure 1 shows subepicardial, midventricular, and subendocardial glycogen content after 4 hours with or without occlusion of the left circumflex artery. An increasing outer to inner gradient was observed in control hearts after 4 hours. The difference between subepicardial and midventricular glycogen contents was not significant ($p>0.05$). However, subendocardial glycogen content was greater ($p<0.01$) than that of the middle and outer regions. The finding of higher glycogen levels in the subendocardial region was reported previously by Jedeikin (1964) for a variety of species, including dogs. After 4 hours of ischemia, on the other hand, the gradient was reversed. In ischemic hearts, the outer region glycogen content was similar to that of controls; however, midmyocardial and subendocardial stores were virtually depleted.

The most striking functional changes we noted in these studies were abnormalities in the electrocardiogram. In this context, Jedeikin (1964) reported that conduction system tissues in nonischemic hearts contained the highest concentrations of glycogen of any region of the heart. Furthermore, it is possible that Purkinje cells may be more dependent upon glycolysis for ATP production. With this in mind, we employed light and electron microscopic techniques to examine control and 4-hour ischemic dog hearts for transmural distribution of glycogen, with a particular interest in the Purkinje cells of the subendocardium. Figures 2 to 4 represent photomicrographs of transmural tissue sections. Figure 2*A* shows the subendocardium containing Purkinje cells and adjacent inner myocardium from a control dog. The Purkinje cells contained a dense and homogenous distribution of glycogen granules. Adjacent inner myocardial cells also contained considerable amounts of glycogen, the perinuclear areas demonstrating the heaviest concentrations. The same region from an ischemic dog after 4 hours of

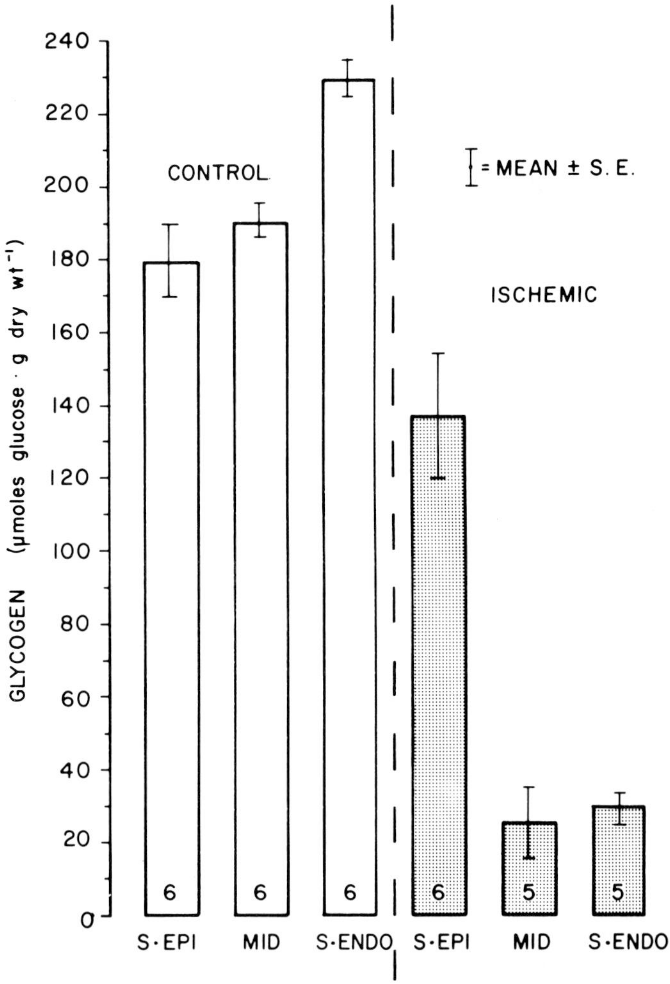

Figure 1. Transmural gradients in tissue glycogen content of canine left ventricle after 4 hours with or without left circumflex artery ligation. S·EPI = subepicardial region; MID = midventricular region; S·ENDO = subendocardial region.

ligation is shown in Figure 2*B*. Glycogen content of the Purkinje cells was depleted considerably, although most cells demonstrated scanty amounts of glycogen. The innermost myocardial cells showed almost total depletion. The midmyocardium from control dogs (Fig. 3*A*) contained a homogenous distribution of glycogen in all muscle cells, heavier concentrations being located in perinuclear areas. Ischemic midmyocardium showed essentially total absence of glycogen (Fig. 3*B*). Figure 4*A* illustrates the homogenous pattern of glycogen

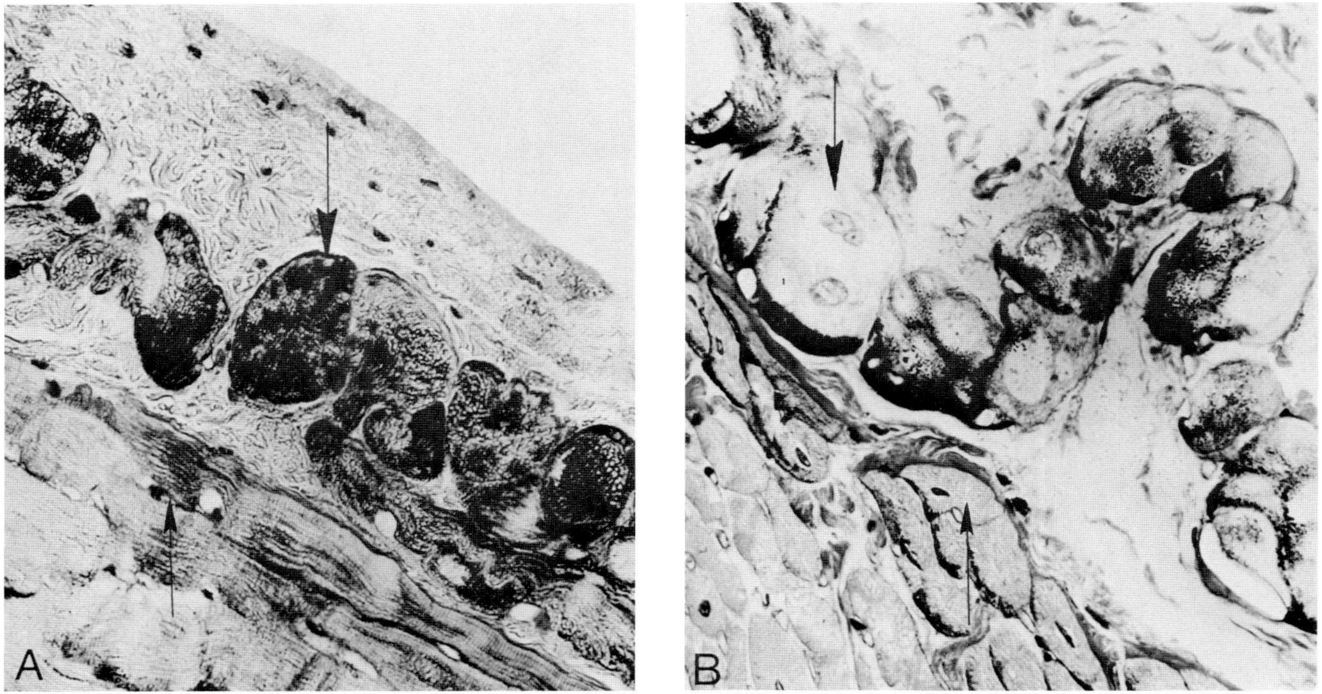

Figure 2. A: control subendocardial region, showing dense and homogenous distribution of glycogen (*large arrow*) in Purkinje cells of subendocardium. Adjacent inner myocardial cells also contain large amounts of glycogen (*small arrow*). × 60. *B:* ischemica subendocardial region, demonstrating substantial glycogen depletion in Purkinje cells (*large arrow*) as well as in adjacent myocardial cells (*small arrow*). × 60.

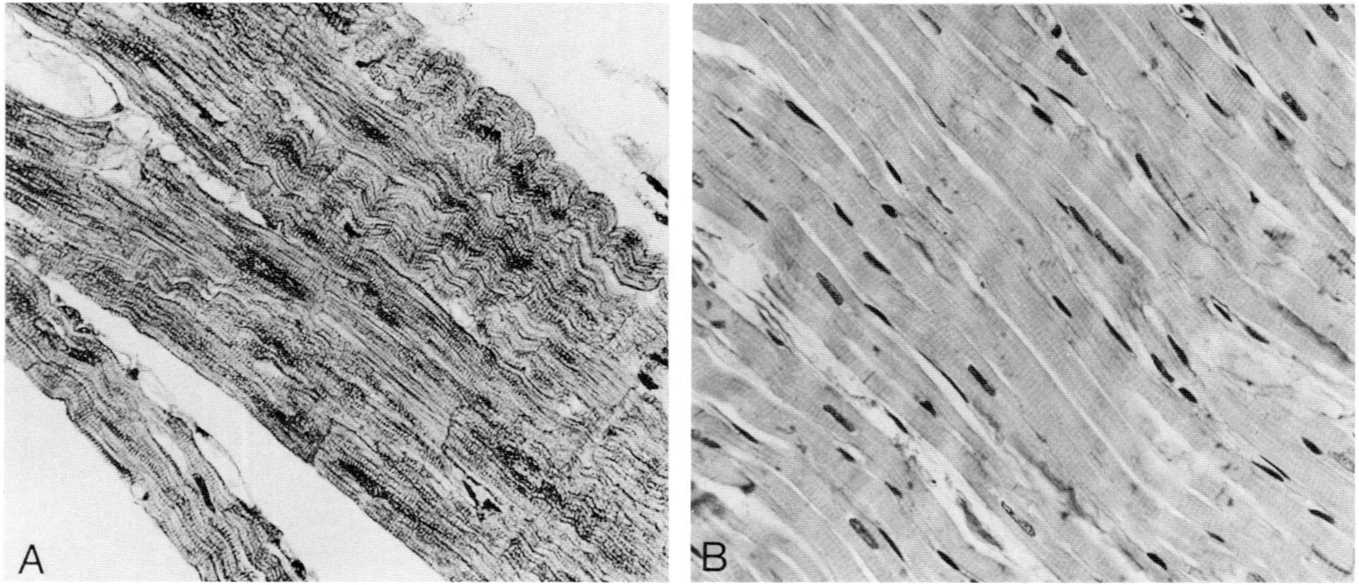

Figure 3. A: control midmyocardium, showing large amounts of glycogen contained within all cells. × 60. *B:* ischemic midmyocardium, demonstrating essentially total absence of glycogen. × 60.

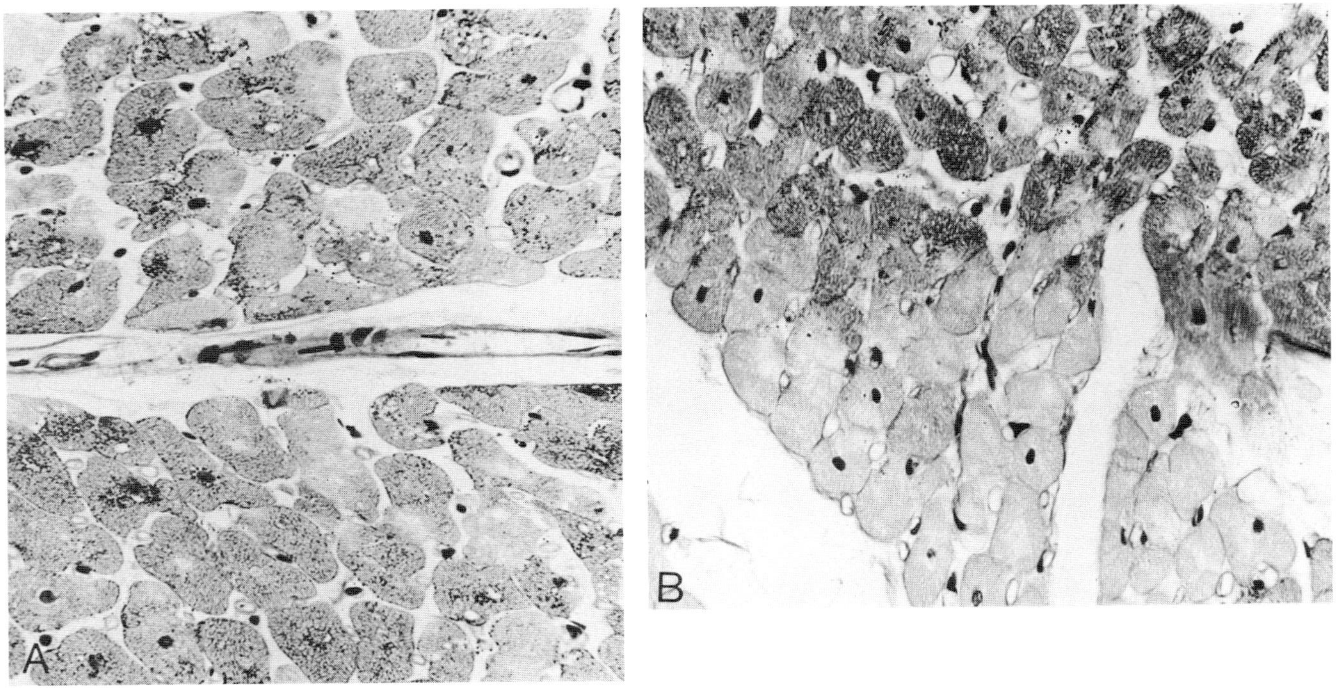

Figure 4. A: Photomicrograph of control subepicardial region (cross-section), demonstrating homogenous pattern of glycogen distribution. × 60. *B:* subepicardium of left ventricle after 4 hours of ischemia, showing patches of cells demonstrating both large and scanty amounts of glycogen. × 60.

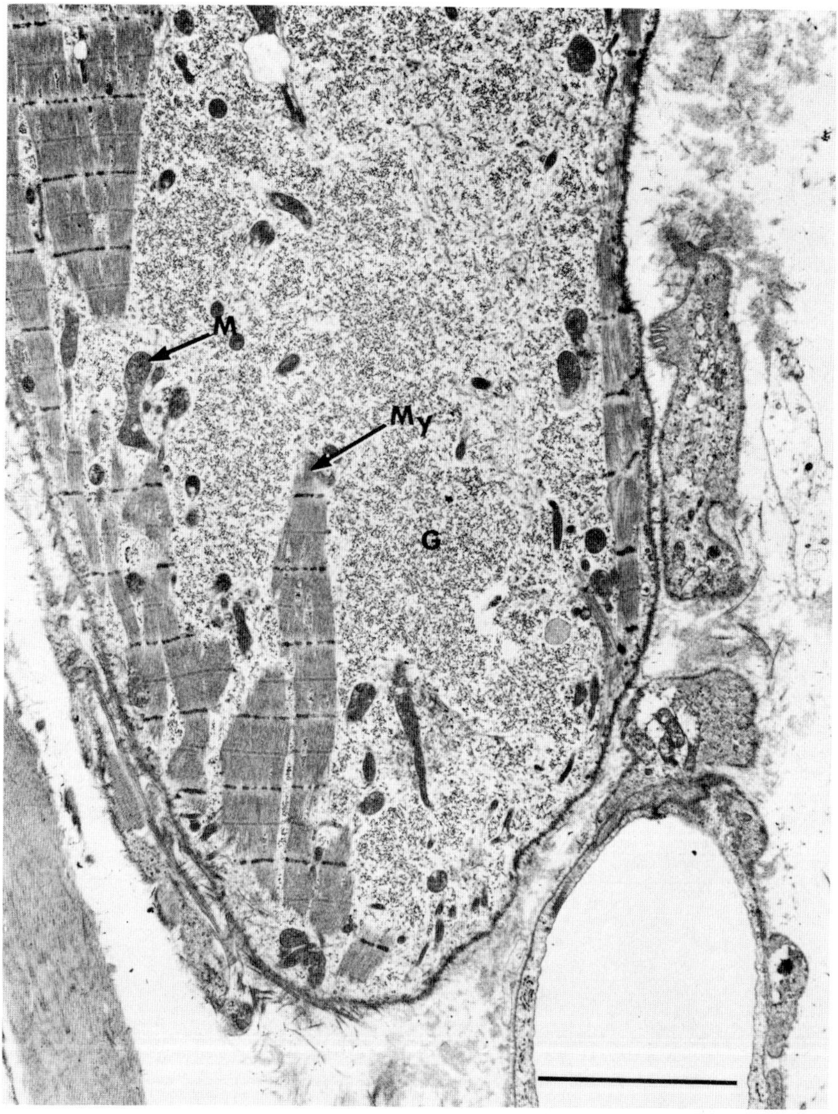

Figure 5. A and B. (*A*) Electron micrograph of Purkinje cells in control subendocardium depicting glycogen (G)-rich cytoplasm, pleomorphic mitochondria (M), and randomly dispersed myofibrils (My) within cell. Scale = 5μ. × 6, 337.

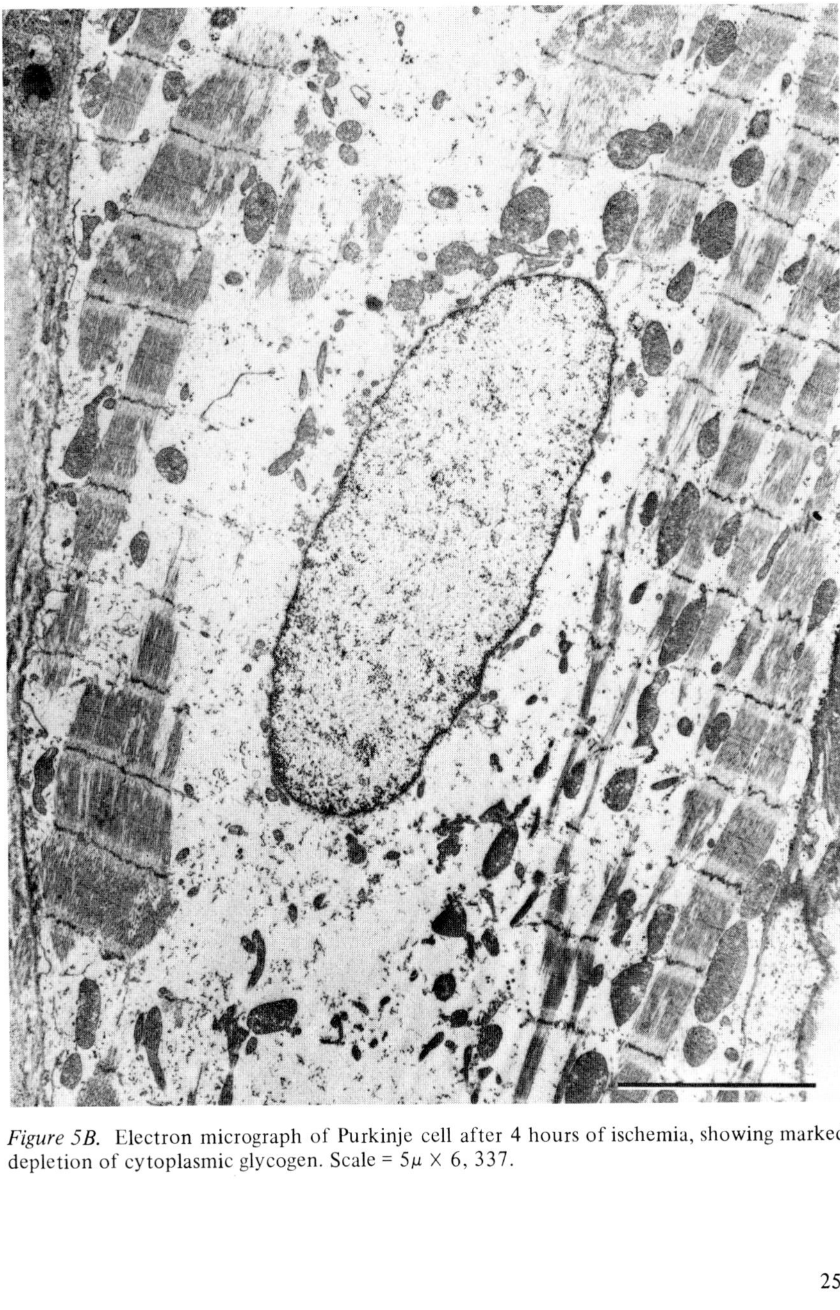

Figure 5B. Electron micrograph of Purkinje cell after 4 hours of ischemia, showing marked depletion of cytoplasmic glycogen. Scale = 5μ × 6, 337.

distribution in the control subepicardial region. After 4 hours of ischemia (Fig. 4*B*), a few patches or groups of muscle fibers showed some decrease in glycogen granules; however, most cells contained abundant quantities. These photomicorographs confirmed the respective increasing outer to inner and decreasing outer to inner glycogen gradients observed chemically in normal and ischemic dog left ventricles, respectively. It would appear safe to state that glycogen mobilization in ischemia-induced cellular hypoxia is most marked in the mid- and subendocardial regions, the latter expressed chiefly in Purkinje cell glycogen content. It would appear that compensatory or autoregulatory changes in blood flow accounted for restoration and maintenance of subepicardial glycogen with no evidence for similar compensatory events in the middle and inner zones. In apparent agreement with Jedeikin's observations, Purkinje cell glycogen in control dogs accounted for essentially all of the subendocardial stores. In ischemic dogs, it would be attractive to postulate that the pronounced conductivity disturbances we observed could reflect secondarily the depletion of endogenous substrate in these cells.

We extended our study in an attempt to characterize possible ultrastructural changes in the Purkinje cells of the subendocardium. In these studies, hearts were arrested *in situ* by perfusion of the left circumflex artery with Karnovsky's fixative following washout with bicarbonate buffer. Figure 5*A* shows a Purkinje cell in the subendocardium from a control dog. The characteristic appearance of glycogen-rich cytoplasm, pleomorphic mitochondria, and randomly dispersed myofibrils can be seen. A Purkinje cell from a 4-hour ischemic dog is shown in Figure 5*B*. Though the dispersed myofibrils looked relatively normal and mitochondria did not appear to contain dense granules, a marked depletion of glycogen was evident.

Transmural differences in lipid content in the normal and ischemic canine left ventricle

The uptake and intermediary metabolism of exogenous free fatty acids and the mobilization and synthesis of endogenous lipids by the ischemic myocardium has not been studied as thoroughly. In the present experiments, we measured regional total triglycerides chemically (Fig. 6). In control hearts, there was an apparent difference in midventricular and subendocardial triglycerides, but this was not statistically significant ($p > 0.05$). Similarly, after 4 hours of ligation, there was an apparent increase in subendocardial triglycerides relative to the middle region, but this difference again was not statistically significant ($p > 0.10$). However, triglyceride content of ischemic hearts was increased approximately three-fold in both regions relative to controls. Epicardial perivascular triglycerides prevented a valid estimation of tissue fat in the outer region; hence, values are not shown. Longer periods of ischemia could possibly differentiate neutral lipid accumulation in the subendocardial area if that region demonstrates rela-

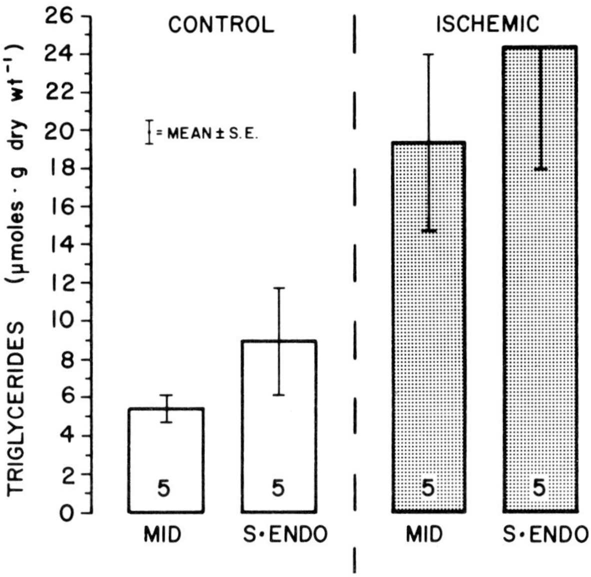

Figure 6. Transmural gradients in total tissue triglyceride content of canine left ventricle after 4 hours with or without left circumflex artery ligation. MID = midventricular region; S•ENDO = subendocardial region.

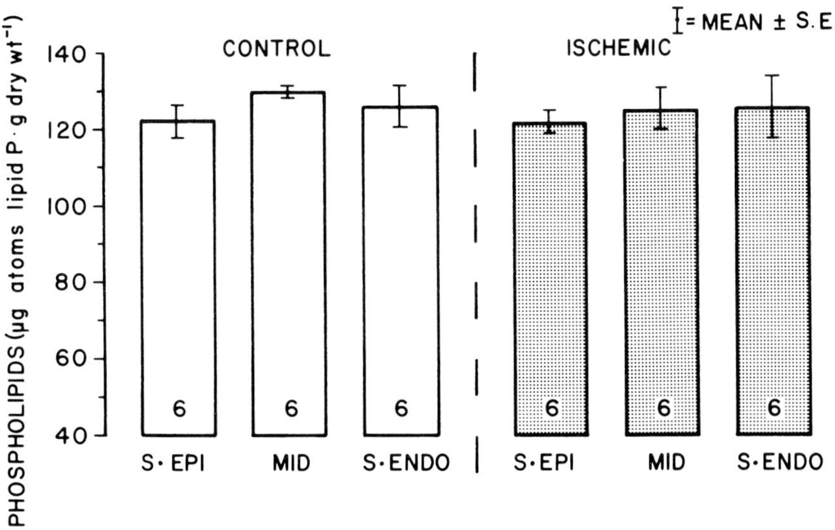

Figure 7. Transmural gradients in total tissue phospholipid content of canine left ventricle after 4 hours with or without left circumflex artery ligation. S•EPI = subepicardial region; MID = midventricular region; S•ENDO = subendocardial region.

tively greater underperfusion. It is conceivable that transmural gradients for tissue α-glycerol phosphate production as well as ischemia-induced alterations in carnitine palmityltransferase activity (and thus impaired mitochondrial fatty acid oxidation) exist under the present conditions. It has been shown that marked defects occur in mitochondrial carnitine-mediated long chain fatty acid oxidation after 1 day of canine left circumflex artery ligation (Schwartz *et al.*, 1973). If further studies confirm these events, conditions would favor a greater tendency toward fatty infiltration of the inner myocardium and subendocardium. Figure 7 shows the content of total phospholipids in the different transmural regions. Values for lipid phosphorus were virtually identical in the three regions in both control and ischemic dogs. Though by no means conclusive, the data suggested that the esterified lipid changes observed under the stated ischemic conditions may be limited to neutral fat. Studies are currently in progress to correlate changes in chemically determined triglycerides with light and electron microscopic tissue localization.

Acknowledgments The authors gratefully acknowledge the technical assistance of Galen M. Pieper, Sherrie Mankovsky, and Marsh Rossi.

REFERENCES

COBB, F. R., R. J. BACHE, and J. C. GREENFIELD, Jr. 1974. Regional myocardial blood flow in awake dogs. J. Clin. Invest. 53: 1618.

CRASS, M. F., III, E. S. McCASKILL, J. C. SHIPP, and V. K. MURTHY. 1971. Metabolism of endogenous lipids in cardiac muscle: effect of pressure development. Am. J. Physiol. 220: 428.

GOOD, C. A., H. KRAMER, and M. SOMOGYI. 1933. The determination of glycogen. J. Biol. Chem. 100: 485.

GRIGGS, D. M., Jr., V. V. TCHOKOEV, and C. C. CHEN. 1972. Transmural differences in ventricular tissue substrate levels due to coronary constriction. Am. J. Physiol. 222: 705.

HONIG, C. R., E. S. KIRK, and W. W. MYERS. 1967. Transmural distributions of blood flow. oxygen tension, and metabolism in myocardium: mechanism and adaptations. *In* International Symposium on the Coronary Circulation and Energetics of the Myocardium, pp. 31–46. Karger, Basel and New York.

JEDEIKIN, L. A. 1964. Regional distribution of glycogen and phosphorylase in the ventricles of the heart. Circulat. Res. 14: 202.

KING, E. J. 1932. The colorimetric determination of phosphorus. Biochem. J. 26: 292.

KIRK, E. S., and C. R. HONIG. 1964. An experimental and theoretical analysis of myocardial tissue pressure. Am. J. Physiol. 207: 361.

MOIR, T. W., and D. W. DeBRA. 1967. Effect of left ventricular hypertension, ischemia, and vasoactive drugs on the myocardial distribution of coronary flow. Circulat. Res. 21: 65.

SALISBURY, P. F., C. E. CROSS, and P. A. RIEBEN. 1963. Acute ischemia of inner layers of ventricular wall. Am. Heart J. 66: 650.

SCHWARTZ, A., J. M. WOOD, J. C. ALLEN, E. P. BORNET, M. L. ENTMAN, M. A. GOLDSTEIN, L. A. SORDAHL, and M. SUZUKI. 1973. Biochemical and morphologic correlates of cardiac ischemia. Am. J. Cardiol. 32: 46.

VAN HANDEL, E., and D. B. ZILVERSMIT. 1957. Micromethod for the direct determination of serum triglycerides. J. Lab. Clin. Med. 50: 152.

Mailing address:
Dr. M. F. Crass, III,
Department of Physiology,
Texas Tech University School of Medicine,
P.O. Box 4569,
Lubbock, Texas 79409 (USA).

Recent Advances in Studies on
Cardiac Structure and Metabolism Volume 10
The Metabolism of Contraction
Edited by Paul-Emile Roy and George Rona

EFFECTS OF CORONARY ARTERY LIGATION AND RELEASE AND PERFUSION WITH GLUCOSE ON CARDIAC ENERGY METABOLISM IN DOGS

ANIS I. OBEID, ROBERT H. EICH,
S. MOOKHERJEE, and ROBERT SPEAR

Department of Medicine, State University and Veterans Admistration Hospitals of the Upstate Medical Center, Syracuse, New York, USA

SUMMARY

Adenosine triphosphate (ATP), glycogen, and inorganic phosphorus (P_i) were assayed in myocardial tissues obtained from beating dog hearts subjected to the following procedures: Group 1. Ligation of anterior descending coronary artery (AD) for 60 min. Group 2. Perfusion of the AD with 20 mM glucose in Ringer's solution containing 1 mU regular insulin/ml, for 60 min. Group 3. Ligation of the AD for 60 min, followed by release of the ligature with re-establishment of blood flow for 30 min. Group 4. Perfusion of the AD with 20 mM glucose in Ringer's solution containing 1 mU/ml regular insulin for 60 min, followed by re-establishment of blood flow for 30 min.

Samples from normal myocardium (N zone) and from the area supplied by the AD (I zone) were simultaneously obtained and the level of metabolites compared.

In all four groups, ATP was significantly lower in I than N zone. However, ATP values were lower in N zone in group 1 as compared with the other groups ($p<0.05$) and higher in I zone in group 3 versus groups 1 and 2 ($p<0.05$). Glycogen was lower in I than N zones to a similar degree in all the groups. However, it was higher in the I zone in group 3 than in group 1 ($p<0.05$). P_i was significantly higher in I zone versus N zone only in group 1 ($p<0.05$).

These data suggest that, in the beating dog heart, ATP and glycogen preservation in a myocardial segment rendered ischemic for 1 hour is best achieved by re-establishing circulation. Glucose infusion into the ischemic segment did not contribute significantly to the ATP stores in that segment but may have exerted a protective effect on the nonischemic zones, possibly by providing a high glucose level in the circulating blood. This protection was equally well achieved by release of the ligature with or without prior glucose infusion. The increase in P_i in the ischemic zone in the dogs with coronary ligation only is probably related to accumulation of the ion under ischemic and hypoxic conditions.

INTRODUCTION

The dependence of myocardial cellular survival and function on aerobic metabolism is well recognized. However, under certain conditions, myocardial tissue can derive some of its energy requirements through anaerobic pathways (Opie, 1971/1972). In man as well as in experimental animals, myocardial infarction consists of an inhomogeneous mass with varying degrees of muscle involvement by the ischemic process (Sobel and Shell, 1973). Reversibility of myocardial ischemia in the zones that have been partially but reversibly damaged has been of major concern over the past few years (Maroko *et al.,* 1971; Maroko and Braunwald, 1973; Brachfeld, 1974).

Previous work on rat hearts has shown that, *in vitro,* protection of myocardial structure and function for a period of time is possible during anoxia if glucose is supplied in the perfusing medium (Weissler, 1968). In dog studies, release of coronary artery ligature after 1–3 hours of continuous ligation is associated with recovery of some ischemic areas that would have otherwise undergone necrosis (Maroko *et al.,* 1972). On the other hand, when blood flow is re-established after 3 hours or more of continuous ligation, it may lead to adverse effects including hemorrhage and possible extension of the infarcted area (Jarmakani *et al.,* 1974; Kane *et al.,* 1974). In dogs also, glucose and insulin administered intravenously during experimental coronary artery ligation result in reduction of myocardial damage as determined by creatine phosphokinase (CPK) content of the muscle and ST segment mapping (Kjekshus and Sobel, 1970; Maroko and Libby, 1972). While CPK content of the ischemic muscle and ST segment displacement have been recently used to assess the extent of myocardial damage in dogs undergoing coronary artery ligation, release, and glucose infusion, the behavior of myocardial high energy stores with these experimental interventions is less well delineated.

The following experiments were conducted in order to assess the effects of coronary artery ligation, release, and glucose infusion on the level of energy stores in the cardiac muscle in dogs. Glucose was infused intra-arterially into the distal descending coronary artery, perfusing a myocardial segment comprising approximately 10% of the left ventricular mass of a beating dog heart.

MATERIALS AND METHODS

A total of 20 mongrel dogs weighing 13–25 kg were anesthetized with sodium pentobarbital, 30 mg/kg, and respiration was maintained with a Harvard pump. A median sternotomy was performed, and the heart was suspended in a pericardial cradle. Catheters were inserted retrogradely into the left ventricle and ascending aorta for pressure monitoring and arterial sampling and, in the coronary sinus, for venous sampling. The anterior descending coronary artery (AD) segment distal to the first major diagonal branch was cannulated with a No. 16 or No. 18 plastic needle, the tip of the needle being placed about 0.5 cm distal to the puncture site in the arterial wall. A ligature was firmly placed around the artery just distal to the cannulation puncture site, preventing blood flow through the artery around the cannula. Another plastic catheter placed in the left carotid artery was connected through a stopcock with the cannula in the AD, allowing bypass of blood from the carotid artery as the sole supply to the distribution of the AD. This bypass was then used for experimental maneuvers such as coronary ligation by turning off the stopcock, release by turning it on, or infusion of metabolites. The color and pattern of contraction of the myocardial segment supplied by this bypass procedure appeared to be normal and similar to the rest of the myocardium as long as the bypass of arterial blood was maintained. Heparin was administered at the rate of 5,000 units intravenously every hour.

The dogs were then divided into four groups: Group 1 (five dogs): the AD was effectively ligated for 60 min by interrupting the carotid artery AD bypass. Group 2 (four dogs): the blood flow through the AD was interrupted, and the AD was perfused with 10–15 ml/min of 20 mmoles/ml of dextrose in Ringer's solution containing regular insulin, 1 mU/ml. Group 3 (six dogs): the AD blood flow was interrupted for 60 min, similar to group 1. At the end of 60 min, however, blood flow was re-established through the carotid artery AD bypass and maintained for 30 min. Group 4 (five dogs): the AD was perfused with glucose and insulin, similar to group 2 for 60 min. At the end of 60 min, blood flow was re-established through the carotid artery AD bypass for 30 min.

The temperature of the perfusing solutions was maintained at 37°C, and the pH was adjusted to physiological range in all experiments. Arterial and coronary venous pO_2, pH, pCO_2, and glucose were monitored at the beginning and the end of the experiment as well as at repeated intervals in the course of the experiment. Hemodynamic measurements were recorded at similar intervals.

At the end of each experiment, tissues from the normally perfused myocardium (N zone) and from the area of distribution of the anterior descending coronary artery (I zone) were obtained by a bone rongeur cooled in liquid nitrogen. Two simultaneous samples containing approximately 0.5 g of cardiac tissue were obtained and were rapidly frozen in liquid nitrogen, the sampling procedure taking place in about 10 sec. The tissues were stored at −90°C and later assayed in duplicate for the metabolites to be mentioned below.

Myocardial tissue segments were pulverized in a mortar in liquid nitrogen. A portion of the frozen powdered tissue was transferred to a pestle homogenizer containing a weighed amount of ice-cold 6% perchloric acid. This was then reweighed and homogenized, centrifuged, and decanted. The pellet was resuspended in perchloric acid, and the procedure was repeated. The combined supernatants were rapidly frozen in liquid nitrogen and stored at −90°C, for subsequent assay of adenosine triphosphate (ATP) and inorganic phosphorus (P_i).

A second portion of the frozen powdered tissue was transferred to a weighed amount of precooled 30% potassium hydroxide saturated with sodium sulfate, for glycogen assay. After digestion, the glycogen was precipitated by addition of ethanol and hydrolyzed to glucose with concentrated sulfuric acid and assayed by the method of Lo *et al.* (1970).

Table 1. Metabolic changes in control and perfused myocardial segments*

	Group 1		Group 2		Group 3		Group 4	
	N	I	N	I	N	I	N	I
ATP	3.3∓0.6†	1.5∓0.3	5.1∓0.5	1.8∓0.6	5.4∓0.3	3.4∓0.6†	5.1∓0.3	2.8∓0.6
Glycogen	8.2∓0.6	3.1∓0.3†	8.0∓1.2	3.7∓0.2	8.6∓0.3	5.4∓0.4†	9.0∓0.9	5.5∓1.4
P_i	5.7∓0.6	7.0∓0.7†	5.7∓0.2	5.1∓0.5	6.0∓0.3	5.1∓0.3†	4.2∓0.5	4.0∓0.5

*Tissue contents in μmoles/g wet tissue ± SEM. N refers to concentration of metabolites in normally perfused myocardium and I in experimental myocardial segment.
†Statistical significance at 0.05 level or less.

Table 2. Differences in concentration of metabolites between N and I zones in the various groups expressed in percentage, (N − I)/N X 100*

	Groups			
	1	2	3	4
ATP	49 ∓ 11	65 ∓ 10	38 ∓ 9	45 ∓ 12
Glycogen	60 ∓ 2†	51 ∓ 7	37 ∓ 4†	38 ∓ 14
P_i	−25 ∓ 3†	10 ∓ 7	14 ∓ 5	4 ∓ 6

*For abbreviations, see Table 1.
†$p \leqslant 0.05$.

ATP was determined by the method of Lamprecht and Trautschold (1963), and P_i by a modification of the method of Dryer *et al.* (1957). All results were expressed in μmoles/g wet tissue.

RESULTS

The metabolic data are presented in Tables 1 and 2.

Adenosine triphosphate

The ATP levels in the N zone were 3.3 ± 0.6, 5.1 ± 0.5, 5.4 ± 0.3, and 5.1 ± 0.3 μmoles/g for groups 1–4, respectively. Thus, the ATP concentration in the N zone in group 1 was significantly lower than the ATP concentration in the N zone of the other groups ($p<0.05$). For the I zone, the ATP levels were 1.5 ± 0.3, 1.8 ± 0.6, 3.4 ± 0.6, and 2.8 ± 0.6 μmoles/g for groups 1–4, respectively. The levels in group 1 were significantly lower than in group 3, with $p<0.05$. The rest showed no statistically significant differences, although it appeared that the level in group 2 was lower than in group 3 or 4. These data are presented graphically in Figure 1.

Within the same group, ATP content in the I zone was consistently lower than the ATP content in the N zone, $p<0.01$. When the differences (N − I) are expressed as percentage of the concentration in the N zone (N − I)/N, ATP content in the I zone was found to be reduced by 49, 65, 38, and 45% for groups 1–4, respectively. The smallest change was in group 3, but the differences between it and the other groups did not achieve statistical significance.

Glycogen

The levels of glycogen for the N zone were 8.2 ± 0.6, 8.0 ± 1.2, 8.6 ± 0.3, and 9.0 ± 0.9 μmoles/g for groups 1–4, respectively, with no statistically significant

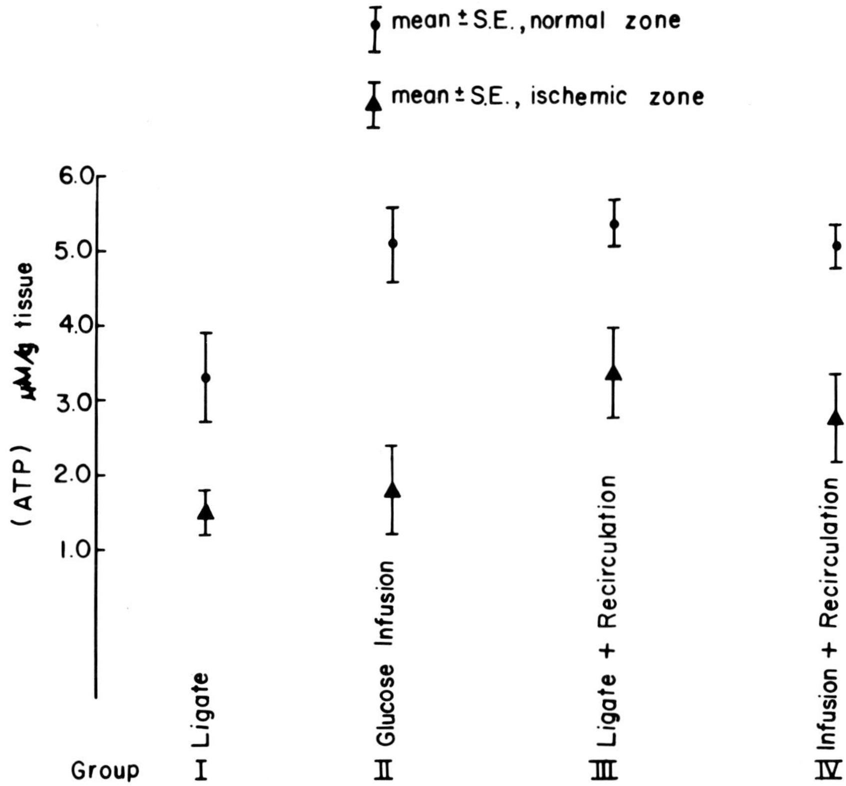

Figure 1. ATP levels in N (*closed circle*) and I (*closed triangle*) zones in four groups of dogs. Differences between N and I zones are significant in every group ($p<0.01$). Differences between I zones in groups 1 and 3 are significant ($p<0.05$).

differences between the groups. For the I zone, the values were 3.1 ± 0.3, 3.7 ± 0.2, 5.4 ± 0.4, and 5.5 ± 1.4 μmoles/g for groups 1–4, respectively. Statistically significant differences were found between groups 1 and 3, with $p<0.05$. The level in group 4 was also higher than that in group 1 and 2, but statistical significance was not achieved. These data are graphically depicted in Figure 2.

Within the same group, glycogen content in the I zone was significantly lower than that in the N zone in all the groups ($p<0.01$). The percentage changes (N − I)/N were 60, 51, 37, and 38% for groups 1–4, respectively, the change being significantly greater in group 1 than group 3 ($p<0.01$). Despite similar values in groups 3 and 4, the wider scatter in group 4 precluded statistical significance when compared with group 1.

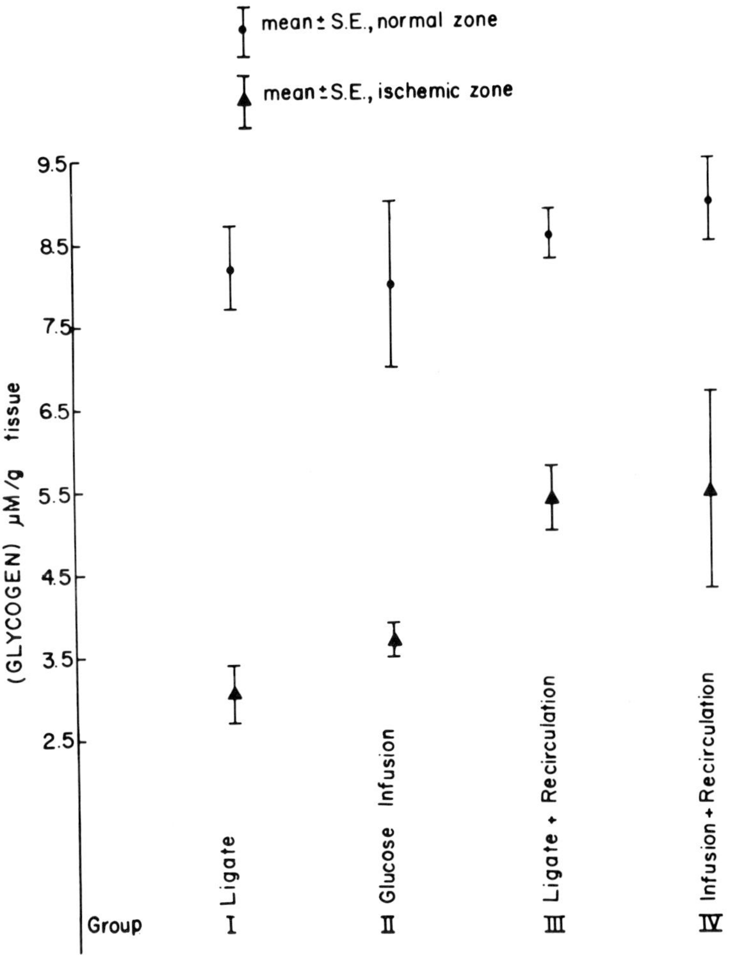

Figure 2. Glycogen concentration in N (*closed circle*) and I (*closed triangle*) in four groups of dogs. Differences between N and I zones are significant in every group ($p<0.01$). Differences between I zones in groups 1 and 3 are significant ($p<0.05$).

Inorganic phosphorus

The values for P_i in the N zone were 5.7 ± 0.6, 5.7 ± 0.2, 6.0 ± 0.3, and 4.2 ± 0.5 μmoles/g for groups 1–4, respectively, with no statistically significant differences between the various groups. For the I zone, the values were 7.0 ± 0.7, 5.1 ± 0.5, 5.1 ± 0.3, and 4.0 ± 0.5, group 1 being higher than the rest of the groups, with p <0.05. Furthermore, the content of P_i was significantly higher in the I zone than

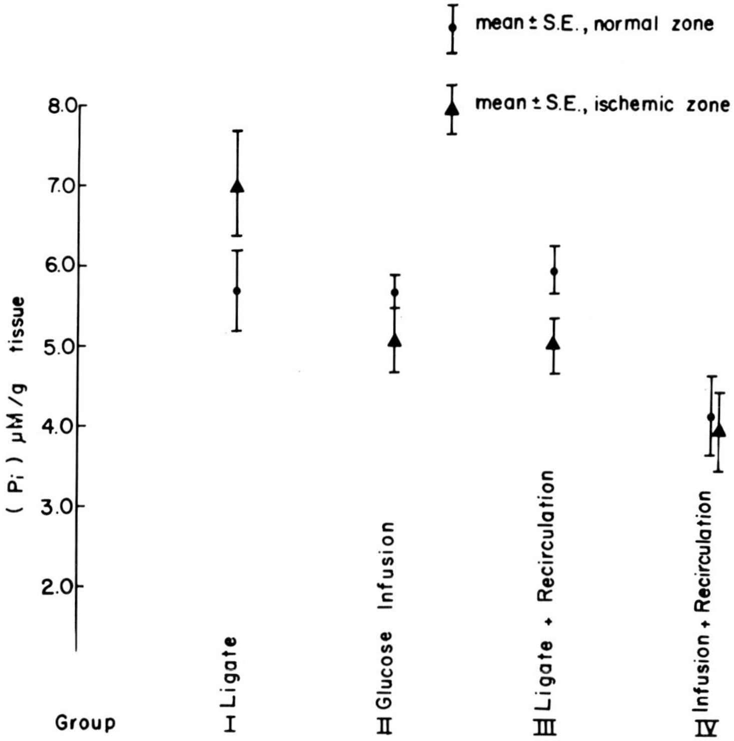

Figure 3. Concentration of P_i in N (*closed circle*) and I (*closed triangle*) zones in four groups of dogs. I zone was significantly higher than N zone in group 1 ($p<0.05$), and higher than I zones of other groups ($p<0.05$).

the N zone in group 1 only ($p<0.01$). In the rest of the groups, the concentration of P_i was slightly but consistently lower in the I than the N zones (Fig. 3). (N − I)/N for P_i were −25 ± 3, 10 ± 7, 14 ±7, and 4.6 ± 6% for groups 1–4, respectively, group 1 being significantly different from the rest of the groups ($p<0.05$).

Hemodynamic changes

The hemodynamic parameters showed no significant differences in heart rate, systemic blood pressure, or left ventricular end-diastolic pressure between the groups and within the same group during the course of the experiments. This is probably due to the small segment of myocardium that was involved with the experimental intervention. No major arrhythmias were encountered with interruption of blood flow to the small myocardial segment. However, with reestablishment of blood flow and during perfusion with glucose, there was a period of ventricular irritability of about 2-min duration, starting immediately after the institution of the intervention and progressing in some animals to ventricular fibrillation. The animals that developed ventricular fibrillation were eliminated from the study.

DISCUSSION

Among the interventions that help preserve anoxic myocardium, perfusion with glucose and re-establishment of blood flow were tested in this study. There is some controversy as to the duration of anoxia tolerated by the myocardium before incurring permanent damage. The initial findings of Jennings and coworkers (1960) indicated that, after 20 min of total ischemia, muscle necrosis begins. The preparation they used was the posterior papillary muscle in the dog, with end arterial blood supply. Other areas in the dog heart are not supplied by end arteries; therefore, when the main source of blood flow to such areas is interrupted, a nonuniform pattern of ischemic involvement takes place. This may be the main explanation for the discrepency in duration of ischemia tolerated by segments of dog myocardium reported by various investigators. We chose 1 hour of ischemia or anoxia, since it is well within the contested 3-hour limit and much beyond the 20 min reported by Jennings.

Our data show that reperfusion after 1 hour of continuous ischemia to the area of distribution of the distal AD in the beating dog heart is associated with evidence of considerable recovery of ATP concentration and glycogen content as compared with muscle that was subjected to similar ischemic process but was not reperfused. It is conceivable that ATP and glycogen content would have shown further recovery if reperfusion had been permitted for longer than 30 min.

An attempt to substitute coronary blood flow by perfusing the distribution of the distal AD by anoxic solutions containing glucose and insulin was then made. Washout of metabolites as well as provision of high concentration of substrate should theoretically stimulate the glycolytic pathways and help preserve energy stores in the anoxic myocardium (Opie, 1971/1972). In our experiments, a segment of the beating dog heart generating systemic pressures was perfused with glucose and insulin for 1 hour. Under these conditions, however, neither ATP stores nor glycogen concentration was tangibly improved by this intervention. While ATP depletion can be explained by the ineffectiveness of glycolysis to sustain myocardial tissue stores, except possibly in arrested or hypothermic states (Kubler and Spieckermann, 1970), glycogen depletion is not as readily explained. It appears that despite overwhelming the system with exogenous glucose to the extent of significantly raising systemic blood sugar levels, phosphorylase activation was not inhibited and endogenous glycogen was broken down. Neely, Whitfield, and Morgan, 1970) demonstrated glycogen-sparing effects of glucose during anoxia only at low levels of ventricular pressure development in the rat heart. In the turtle heart, Reeves (1963) could not demonstrate significant glycogen sparing when glucose was included in the anoxic solutions used to perfuse the heart. In other studies, arrested dog hearts pretreated with glucose and insulin were found to maintain higher glycogen content at various intervals of anoxia than dog hearts treated similarly but without premedication with glucose (Conn, Wood, and Morales, 1959). In the present study, no demonstrable glycogen-sparing effects were shown with glucose and insulin when perfusion was carried out for 1 hour.

On the other hand, while exogenous glucose did not contribute significantly to glycogen or ATP levels in the anoxic myocardial segment, it may have contributed to ATP levels in the normally perfused myocardium. Depression of ATP in the normal myocardium during coronary occlusion in our experiments is in agreement with recent data of Gudbjarnason, Puri, and Mathes (1972); its presumed improvement by glucose (spilling into the systemic circulation) is also in agreement with experiments using intravenous glucose during coronary occlusion. The other beneficial effect of glucose perfusion in our study is lowering of P_i in the anoxic segment, probably by washout of the metabolite generated during anoxia (Opie *et al.*, 1972).

In group 4, blood flow was re-established after 1 hour of glucose perfusion. The data from these dogs showed wider scatter and the effects were less dramatic than when reperfusion was instituted following 1 hour of total ischemia. The significance of this observation is uncertain, since the techniques of sampling were identical in all the groups. However, it is possible that, in dogs in which collateral circulation was well developed, the anoxic process was mild and glucose may have helped the partially anoxic muscle to survive well enough that, upon reperfusion with blood, there was complete recovery of energy stores. For example, in one such dog, the ATP levels at the end of the experiment were 5.4 μmoles/g in the I zone and 5.1 μmoles/g in the N zone, and the concentration of glycogen was 11.5 μmoes/g in the I zone and 12.0 μmoles/g in the N zone. On the other hand, in dogs with poor collateral circulation and severe myocardial anoxia, glucose may have had deleterious effects jeoparidizing whatever recovery was possible upon reperfusion with blood. The deleterious effects of glucose under these circumstances could be due to direct toxic properties of high concentration of intracellular glucose. Alternatively, they could be due to the inability of the anoxic cell to carry glycolysis from the early stages where ATP is consumed to the later stages where ATP is generated.

REFERENCES

BRACHFELD, N. 1974. Ischemic myocardial metabolism and cell necrosis. Bull. N.Y. Acad. Med. 50: 261–293.

CONN, H. L., Jr., J. C. WOOD, and G. S. MORALES. 1959. Rate of change in myocardial glycogen and lactic acid following arrest of coronary circulation. Circulat. Res. 7: 721–727.

DRYER, R. L., *et al.* 1957. the Determination of phosporous and phosphatase with *N*-phenyl-*p*-phenylenediamine. J. Biol. Chem. 225: 177.

GUDBJARNASON, S., P. S. PURI, and P. MATHES. 1971. Biochemical changes–Non-infarced heart muscle following myocardial infarction. J. Molec. Cell. Cardiol. 2: 253–276.

JARMAKANI, J. M. M., and J. L. COX, *et al.* 1974. Coronary hemodynamics, and myocardial perfusion after reestablishing coronary flow in experimental myocardial infarction. Am. J. Cardiol. 33: 146 (Abstr.).

JENNINGS, R. B., H. M. SOMMERS, and G. A. SMYTH *et al.* 1960. Myocardial necrosis induced by temporary occlusion of a coronary artery in dogs. Arch. Pathol. 70: 68–78.

KANE, J., and M. MURPHY *et al.* 1974 Does restoration of blood flow preserve acutely ischemic myocardium? Am. J. Cardiol. 33: 146. (Abstr.).

KJEKSHUS, J. K., and B. E. SOBEL. 1970. Depressed myocardial creatine phosphokinase activity following experimental myocardial infarction in rabbit. Circulat. Res. 27: 403–414.

KUBLER, W., and P. G. SPIECKERMANN. 1970. Regulation of glycolysis in the ischemic and the anoxic myocardium. J. Molec. Cell. Card. 1: 351–377.

LAMPRECHT, W., and I. TRAUTSCHOLD. 1963. Methods of Enzymatic Analyis. Academic Press, New York.

LO, S., . 1970. Determination of glycogen in small tissue samples. J. Appl. Physiol. 28: 234–236.

MAROKO, P. R., and J. K. KJEKSHUS *et al.* 1971. Factors influencing infarct size following experimental coronary artery occlusions. Circulat. 43: 67–82.

MAROKO, P. R., and E. BRAUNWALD. 1973. Modification of myocardial infarction size after coronary occlusion. Ann. Int. Med. 79: 720–733.

MAROKO, P. R., and P. LIBBY *et al.* 1972. Coronary artery reperfusion. I. Early effects on local myocardial function and the extent of myocardial necrosis. J. Clin. Invest. 51: 2710–2716.

MAROKO, P. R., and P. LIBBY. 1972. Precordial S-T segment evaluation mapping: An atraumatic method for assessing alterations in the extent of myocardial ischemic injury. The effects of pharmacologic and hemodynamic interventions. Am. J. Cardiol. 29: 223–230.

NEELY, J. R., C. F. WHITFIELD, and H. E. MORGAN. 1970. Regulation of glycogenolysis in hearts: Effects of pressure development, glucose, and FFA. Am. J. Physiol. 219: 1083–1088.

OPIE, L. H. 1971/1972. Substrate utilization and glycolysis in the heart. Cardiology 56: 2.

OPIE, L. H., and M. THOMAS *et al.* 1972. Increased coronary venous inorganic phosphate concentrations during experimental myocardial ischemia. Am. J. Cardiol. 30: 503–513.

REEVES, R. B. 1963. Control of glycogen utilization and glucose uptake in the anaerobic turtle heart. Am. J. Physiol. 205: 23–29.

SOBEL, B. E., and W. E. SHELL. 1973. Jeopardized, blighted and necrotic myocardium. Circulation 47: 215–216.

WEISSLER, A. M., and F. A. KRUGER *et al.* 1968. Role of anaerobic metabolism in the preservation of functional capacity and structure of anoxic myocardium. J. Clin. Invest. 47: 403–416.

Mailing address:
Anis I. Obeid, M.D., F.A.C.C.,
Chief, Cardiology Section,
Veterans Administration Hospital,
800 Irving Avenue, Syracuse, New York 13210 (USA).

Recent Advances in Studies on
Cardiac Structure and Metabolism Volume 10
The Metabolism of Contraction
Edited by Paul-Emile Roy and George Rona

ISCHEMIC DAMAGE AND METABOLISM DURING ELECTIVE CARDIAC ARREST

DAVID J. HEARSE and DAVID A. STEWART

Department of Biochemistry, Imperial College of Science and Technology, Kensington, London, Great Britain

SUMMARY

Isolated perfused working rat hearts were subjected to elective cardiac arrest for 20 or 30 min. Various methods of arrest were studied, either singly or in combination and with or without coronary perfusion. The functional recovery of the heart following the termination of arrest was found to be related to the concentration of ATP and creatine phosphate in the myocardium at the end of the period of arrest. In turn, these concentrations were dependent upon the method used to induce arrest. Normothermic ischemic arrest led to a marked reduction in high energy phosphates and a poor functional recovery. In contrast, coronary perfusion with hypothermic solutions or solutions containing high concentrations of potassium, induced arrest without depleting ATP or creatine phosphate. These procedures conferred considerable protection on the myocardium and thus permitted good recoveries. The energy status and recovery associated with ischemic arrest could be improved by combining the ischemia with hypothermia or potassium arrest. The latter, while increasing recovery significantly, still failed to afford complete protection to the myocardium. Potassium chloride gave greater protection than potassium citrate. When topical hypothermia was combined with ischemia, a time and temperature relationship was demonstrated but effective protection could only be obtained with severe topical hypothermia over a relatively short time period. The results stress the importance of maintaining high energy phosphates during arrest, and this requires the provision of a continuous supply of oxygen and nutrient, which may perhaps be best achieved by ensuring continuous and adequate coronary perfusion.

INTRODUCTION

Elective cardiac arrest (cardioplegia) in combination with cardiopulmonary bypass is used extensively during open-heart surgery. Under experimental conditions, cardiac arrest may be induced by one of a variety of procedures, including ischemia following aortic cross-clamping, electrically induced ventricular fibrillation, topical hypothermia, hypothermic perfusion, or the introduction of high concentrations of potassium into the coronary system. Each cardioplegic procedure has a different biochemical basis, and, therefore, each is likely to have different metabolic and functional consequences. Despite several detailed experimental (Bretschneider, 1964; Arnold and Lochner, 1965; Isselhard, Merguet, and

Aengenvoort, 1965; Lochner and Dudziak, 1965; Lochner, Arnold, and Muller-Ruchholtz, 1968; Iyengar *et al.,* 1973) and clinical (Griepp, Stinson, and Shumway, 1973; Isom *et al.,* 1973) studies of the biochemical and functional consequences of various methods of elective cardiac arrest, considerable controversy (Bloodwell *et al.,* 1969; Brantigan *et al.,* 1972; Griepp *et al.,* 1973; Isom *et al.,* 1973) still exists over the merits of various cardioplegic procedures.

Several cardioplegic procedures involve total cessation of coronary perfusion and consequent tissue ischemia. From a theoretical viewpoint and from a knowledge of the rapid and deleterious metabolic changes which follow coronary artery occlusion (Braasch *et al.,* 1968; Opie, 1971; Sobel, 1972), ischemic arrest may be expected to render the tissue susceptible to damage. Despite this, the importance of utilizing cardioplegic methods which avoid tissue ischemia, by maintaining coronary perfusion, is often questioned (Bloodwell *et al.,* 1969; Cooley, Reul, and Wukasch, 1972; Sapsford *et al.,* 1973).

We have therefore undertaken a series of studies with a rat heart model of cardiac bypass in which we have assessed to what extent ischemic arrest impairs the functional recovery of the heart after the termination of arrest. The possibility of improving cardiac recovery by combining ischemic arrest with hypothermia or potassium was also investigated. In addition, we have attempted to relate the cardioplegic method and the recovery of the heart to changes in the cellular energy metabolism, with particular reference to ATP and creatine phosphate.

MATERIALS AND METHODS

Male rats (280–320 g body weight) of the Sprague-Dawley strain were lightly anaesthetized with diethyl ether, and heparin (200 IU) was administered intravenously. The mounting procedure and perfusion apparatus were based on modifications (Hearse and Chain, 1972) of that described by Neely *et al.* (1967) and Chain, Mansford, and Opie (1969).

The perfusion apparatus was designed so that it could be used for two modes of perfusion which could be readily interconverted: 1) Nonworking Langendorff preparation: Here, hearts were perfused via the aorta as described by Langendorff (1895), with a perfusion pressure of 65 cm water. This mode of perfusion was used for an initial washout period and also for periods of elective cardiac arrest where coronary perfusion was required. 2) Working system: In this preparation, hearts were perfused via the left atrium as described by Neely *et al.* (1969), at an atrial perfusion pressure of 20 cm water. The left ventricle spontaneously ejected 40–50 ml of perfusate per minute against a hydrostatic pressure of 100 cm of water. The aortic output and coronary flow could be pooled and recirculated. This preparation was used for the control working period prior to elective cardiac arrest and for the recovery period following arrest.

It was therefore possible to simulate cardiac bypass with coronary perfusion (with or without elective cardiac arrest) by converting the preparation from an atrially perfused working heart to a Langendorff preparation (with or without cardiac arrest). In experiments requiring ischemic arrest (no coronary perfusion), the cannulae leading to the aorta and the left atrium were clamped. In all experiments, during the control and recovery periods the hearts were maintained at 37°C. During experimental periods, the hearts were maintained at 37°C or the desired degree of hypothermia by using a water-jacketed chamber at the same temperature.

Krebs-Henseleit bicarbonate buffer (Krebs and Henseleit, 1973; Umbreit, Burris, and Stauffer, 1964), pH 7.4, plus glucose, 11.1 mM, equilibrated with 95% O_2:5% CO_2 was the standard perfusion fluid. Immediately after mounting, the heart was perfused for a 5-min washout period by a nonrecirculating Langendorff-type perfusion. The preparation was then converted to a working heart system for a 15-min period. During this time, the stability of the preparation could be confirmed and control values for aortic flow rate and other parameters could be established. As an index of external work, aortic flow rate against a hydrostatic pressure of 100 cm of water could be monitored. At the end of the control period, each preparation was subjected to cardiac arrest under a variety of conditions. At the end of the experimental period, arrest was terminated and the hearts were converted back to a working system. The recovery of aortic flow was monitored over a 15-min period and was expressed as a percentage of the control value.

In a parallel series of experiments, hearts were not allowed to recover but, at the end of the period of arrest, were freeze clamped (Wollenberger, Ristau, and Schoffa, 1960) and the ventricular tissue was taken for the analysis (Hearse and Chain, 1972) of water content, ATP, and creatine phosphate concentrations.

RESULTS

Nonischemic controls

For comparative purposes, a series of nonischemic controls were obtained.

Normothermic coronary perfusion These hearts (N=6) were perfused aerobically for 30 min without cardioplegia. The results (Fig. 1) show that, after a 30-min Langendorff perfusion, when converted back to an atrially perfused working preparation all hearts recovered to 100% of the control aortic output in less than 1 min. At the end of the 30-min period of aerobic perfusion, the concentration of ATP and creatine phosphate (Table 1) in the hearts (N=8) was 26.2 ± 1.7 μmoles/g dry weight and 20.5 ± 1.6 μmoles/g dry weight, respectively. These are concentrations that one would associate with normal healthy myocardial tissue.

Normothermic coronary perfusion with potassium arrest To show the effect

Table 1. Myocardial concentration of ATP and creatine phosphate*

Condition during Experimental Period	Duration (min)	ATP(μmoles/g dry wt.)	Creatine Phosphate (μmoles/g dry wt.)
Normothermic coronary perfusion without cardioplegia	30	26.2±1.7	20.5± 1.6
Normothermic potassium chloride arrest with coronary perfusion	30	20.8±1.2	35.5±2.6
Normothermic potassium citrate arrest with coronary perfusion	30	24.5±1.4	35.7±1.9
Hypothermic (4°C) coronary perfusion	30	20.5±1.0	25.8±1.8
Normothermic ischemic arrest	20	13.5±0.8	4.3±0.4
Normothermic ischemic arrest	30	5.3±0.9	2.8±0.4
Normothermic potassium chloride arrest without coronary perfusion	30	11.1±4.2	9.4±2.1
Normothermic potassium citrate arrest without coronary perfusion	30	9.8±3.5	9.0±1.9
Ischemic arrest with hypothermia (4°C)	30	24.9±1.4	18.1±0.8
Ischemic arrest with hypothermia (15°C)	30	23.8±1.0	7.8±0.5
Ischemic arrest with hypothermia (21°C)	30	20.0±1.0	5.6±0.3
Ischemic arrest with hypothermia (30°C)	30	15.2±0.9	3.5±0.3
Ischemic arrest with hypothermia (4°C)	20	27.8±1.7	21.9±0.9
Ischemic arrest with hypothermia (15°C)	20	25.1±0.7	8.6±0.8
Ischemic arrest with hypothermia (21°C)	20	21.9±2.0	6.3±0.9
Ischemic arrest with hypothermia (30°C)	20	16.2±0.7	5.0±0.4

*Each value represents mean for 8 hearts and SEM values are indicated.

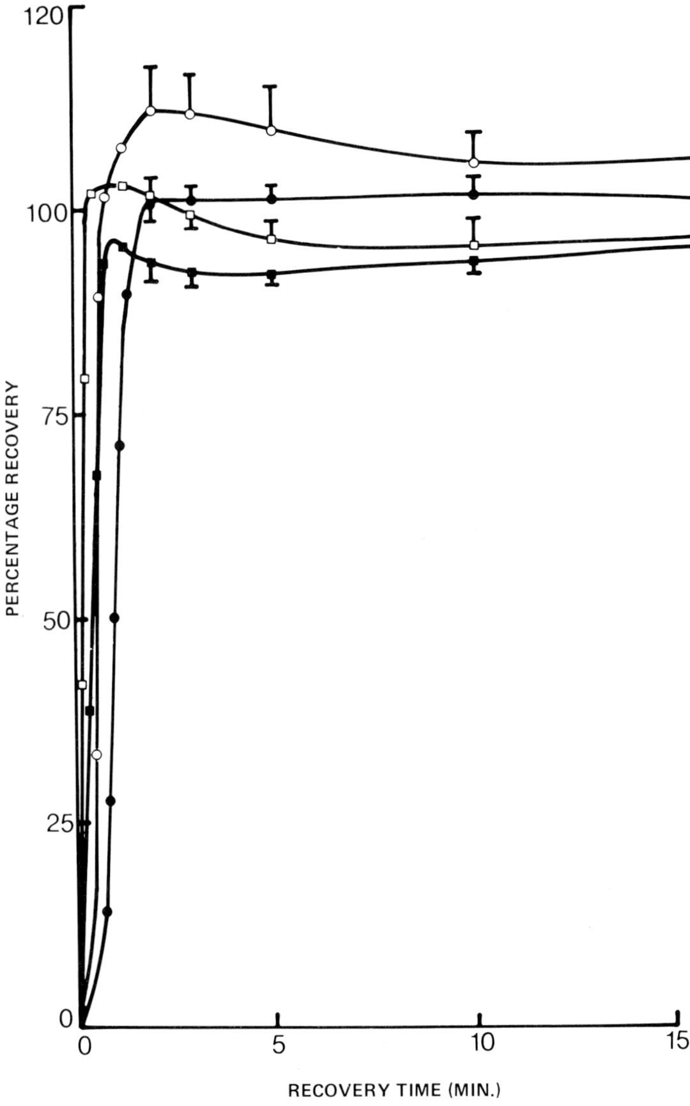

Figure 1. Percentage recovery of aortic flow rate in isolated working rat heart after: □, 30 min of normothermic coronary perfusion; ○, 30 min of normothermic potassium chloride arrest with coronary perfusion; ●, 30 min of normothermic potassium citrate arrest with coronary perfusion; ■, 30 min of hypothermic (4°C) coronary perfusion. Each *point* represents the mean for six hearts, and *bars* represent SEM.

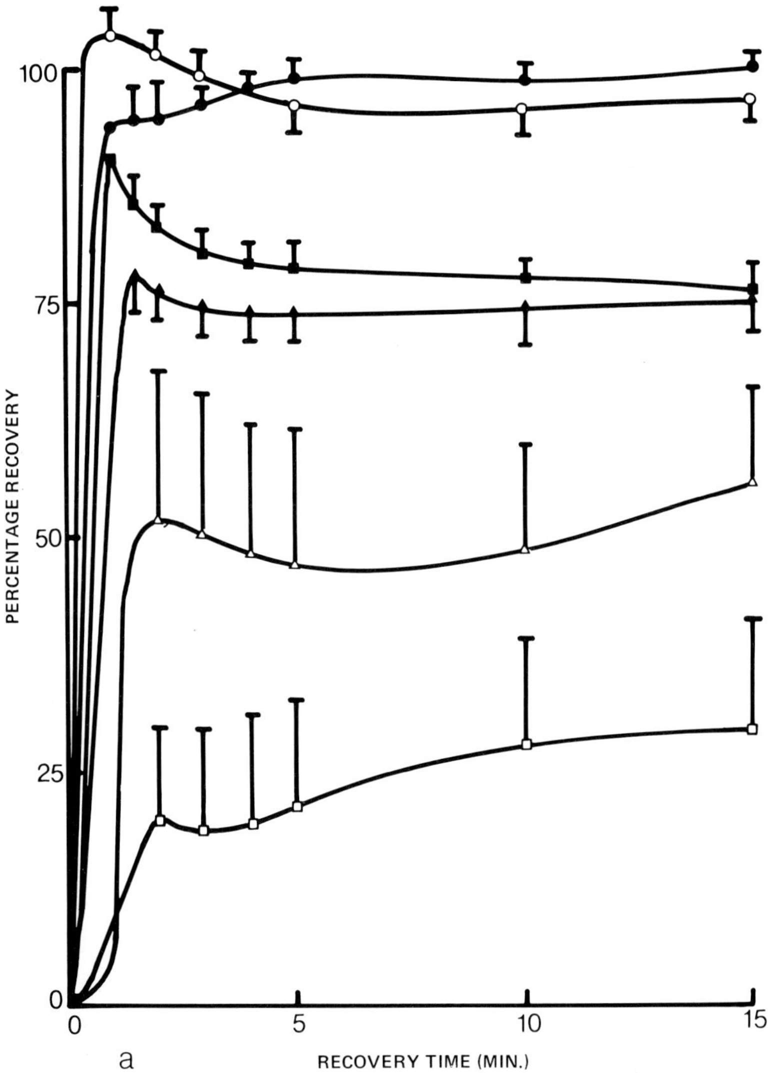

Figure 2. Percentage recovery of aortic flow rate in isolated working rat heart (*a*) after 20 min of arrest; (*b, facing page*) after 30 min of arrest. ○, normothermic coronary perfusion without arrest (control); ●, ischemic arrest with topical hypothermia at 4°C; ■, ischemic arrest with topical hypothermia at 15°C; ▲, ischemic arrest with topical hypothermia at 21°C; △, ischemic arrest with topical hypothermia at 30°C; □, ischemic arrest with 37°C normothermia. Each *point* represents mean for six hearts, and *bars* represent SEM.

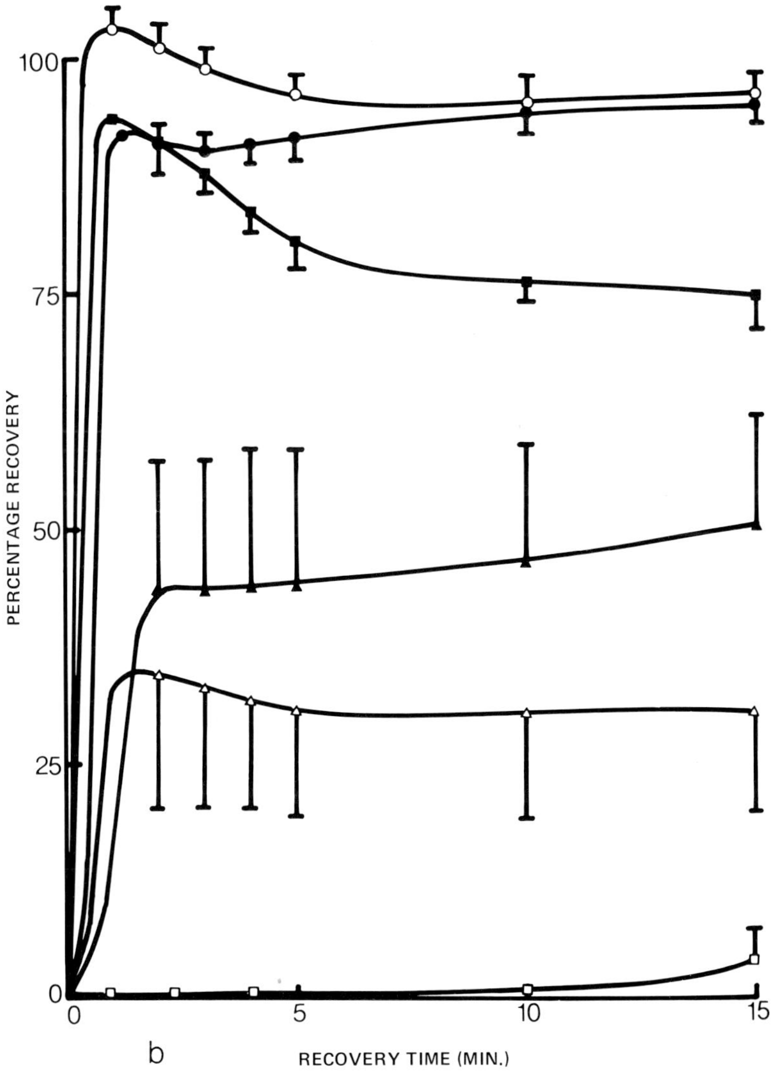
PERCENTAGE RECOVERY
100
75
50
25
0
0
5
10
15
b
RECOVERY TIME (MIN.)

of potassium arrest in nonischemic myocardium, hearts (N=8) were arrested by aerobic Langendorff perfusion (30 min) with buffer containing a high concentration of KCl (K^+ 16.0 mM, Na^+ correspondingly reduced). At the end of this period, the concentrations of ATP and creatine phosphate in the myocardium (Table 1) were 20.8 ± 1.2 and 35.5 ± 2.6 μmoles/g dry weight, respectively. As with the nonarrested control hearts, the recovery profile (Fig. 1) for these hearts (N=6) reflected the high concentration of ATP and creatine phosphate; it reached 100% recovery in approximately 1 min and was maintained at this or a higher level for the duration of the experiment.

In a parallel series of studies, hearts (N=8) were arrested with potassium citrate (16.0 mM). These hearts also recovered well (Fig. 1), and high levels of ATP and creatine phosphate (Table 1) were found in the myocardium at the end of the period of arrest.

Hypothermic coronary perfusion and arrest Hearts (N=8) were hypothermically arrested for 30 min by aerobic perfusion at 4°C. At the end of this period, the myocardial concentrations of ATP and creatine phosphate (Table 1) were 20.5 ± 1.0 and 25.8 ± 1.8 μmoles/g dry weight, respectively. Again, these levels were reflected in the recovery profiles (Fig. 1) of the hearts (N=6).

Ischemic arrest

Normothermic ischemic arrest Coronary perfusion was halted for a period of either 20 min (8 hearts) or 30 min (8 hearts) and hearts were maintained at 37°C. After 20 min of normothermic ischemia, the concentrations of ATP and creatine phosphate had declined to 13.5 ± 0.8 and 4.3 ± 0.4 μmoles/g dry weight, respectively, and, after 30 min, had declined further to 5.3 ± 0.9 and 2.8 ± 0.4 μmoles/g dry weight (Table 1). In comparison with the normothermically perfused control hearts (Fig. 1), the recovery profile (Fig. 2*a*) after 20 min of ischemia was greatly reduced. Recovery was variable, slow, and incomplete, a result which is in agreement with the observed low concentrations of myocardial high energy phosphates. When ischemic arrest was extended to 30 min (Fig. 2*b*), these concentrations were even lower, and this resulted in a further reduction in recovery.

In comparison with the cardiac recovery after potassium arrest or hypothermic arrest (Fig. 1), ischemic arrest greatly impairs cardiac recovery. We therefore investigated the degree of protection which could be afforded to the heart during ischemic arrest by combining it with either topical hypothermia or potassium arrest.

Ischemic arrest with graded topical hypothermia Hearts were subjected to 20 or 30 min of ischemic arrest at either 4, 15, 21, 30, or 37°C. The recoveries (normothermic) and high energy phosphate values are illustrated in Figure 2 and Table 1, respectively. The results illustrate a marked temperature/time dependence, with very good recoveries occurring with severe hypothermia (4°C) and a

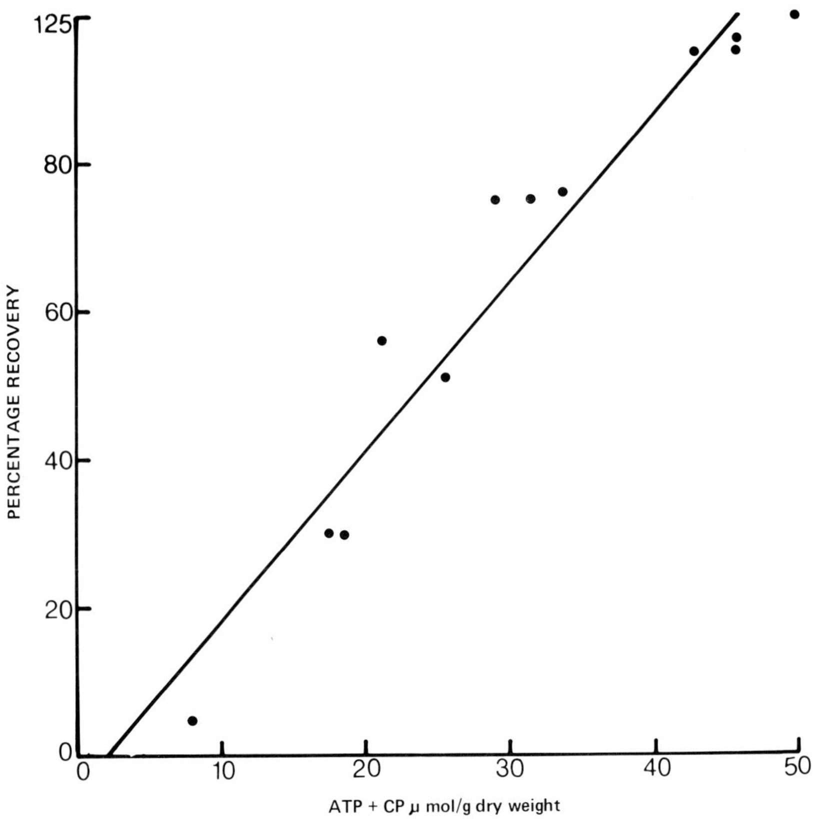

Figure 3. Correlation (r=0.96, $p \leqslant 0.001$) between recovery of aortic flow rate in isolated working rat heart after a period of hypothermic arrest, and total concentration (μmoles/g dry weight) of ATP and creatine phosphate in myocardium at end of period of arrest.

rapidly deteriorating recovery when the temperature was greater than 15–20°C. The high energy phosphate values reflect the recoveries, and Figure 3 shows the correlation (r=0.96, $p \leqslant 0.001$) between the final recovery and the total concentration of ATP and creatine phosphate in all the above experiments which involved ischemic arrest.

Normothermic ischemia with potassium arrest Hearts (N=8) were arrested by perfusing the coronaries with high potassium chloride (16.0 mM) buffer for 1 min; coronary perfusion was then terminated and the arrested hearts were maintained at 37°C for 30 min. At the end of this period, the concentrations of ATP and creatine phosphate in the myocardium (Table 1) were 11.1 ± 4.2 and 9.4 ± 2.1 μmoles/g dry weight, respectively. The recovery of these hearts (Fig. 4), while poor and only reaching 40% after 15 min was considerably better than

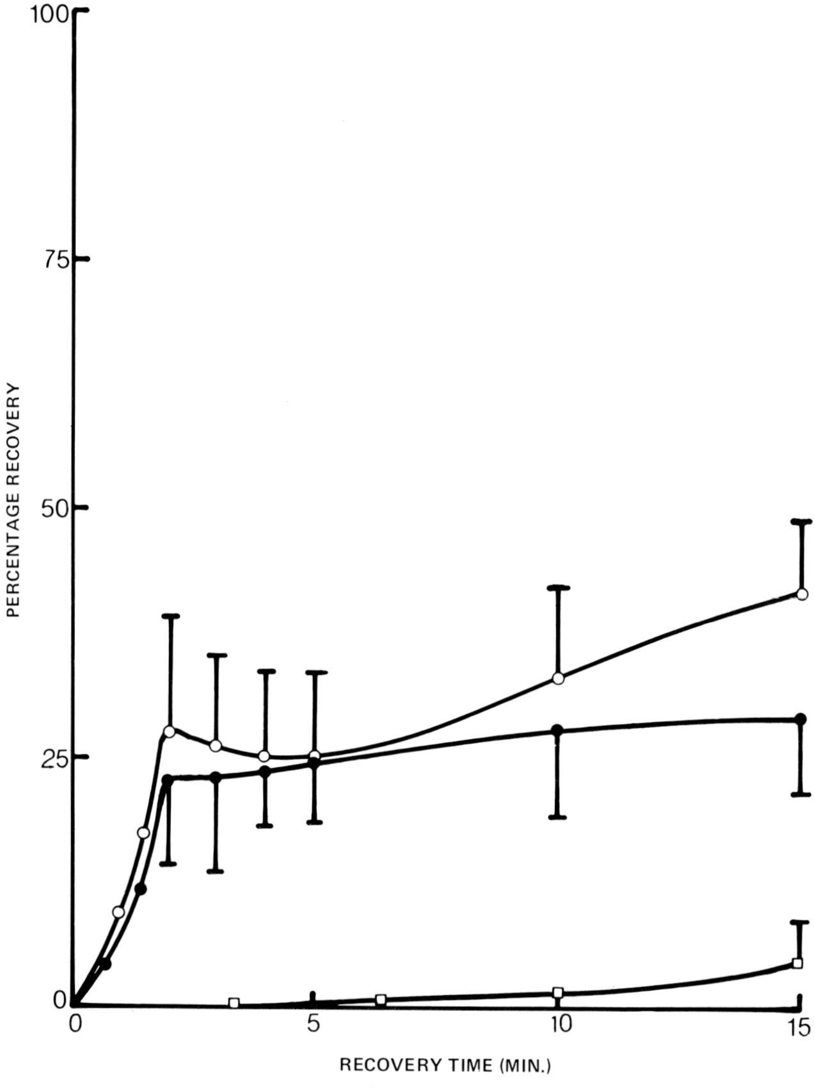

Figure 4. Percentage recovery of aortic flow rate in isolated working rat heart after: □, 30 min of normothermic ischemic arrest; ○, 30 min of normothermic ischemia combined with potassium chloride arrest; •, 30 min of normothermic ischemia combined with potassium citrate arrest. Each *point* represents mean of six hearts, and *bars* represent SEM.

those hearts (Fig. 2*b*) subjected to 30 min of normothermic ischemia alone. The same was found to be true for the high energy phosphate levels (Table 1).

While these results would indicate a protective effect for the potassium ion, it seemed important to consider the contribution of the anion. In many early studies of potassium arrest, potassium citrate has been used. However, under circumstances when high concentrations of potassium citrate coexist with tissue ischemia, it is possible to argue that the citrate may exacerbate damage by reducing anaerobic ATP production by inhibition (Randle, Denton, and England, 1968) of the glycolytic pathway. The above study was therefore repeated with potassium chloride replaced by potassium citrate (16.0 mM). The recovery profiles and high energy phosphate values are shown in Figure 4 and Table 1 and do indicate that the use of citrate does reduce cardiac recovery somewhat.

DISCUSSION

The results of the studies described in this paper illustrate how the cardioplegic method affects the survival and subsequent recovery of the heart. In each instance, the recovery was closely related to the level of ATP and creatine phosphate in the myocardium at the end of the period of arrest. Thus, methods such as normothermic ischemic arrest which greatly reduce cellular energy reserves are detrimental to the survival and recovery of the tissue. In contrast, procedures such as potassium arrest with normothermic coronary perfusion or hypothermic coronary perfusion induce arrest without depleting ATP or creatine phosphate and thus permit good recoveries.

The reduction of high energy phosphates and the poor cardiac recovery associated with normothermic ischemic arrest can be markedly altered by the use of topical hypothermia. Our results have shown a time-dependent relationship between the degree of hypothermia and the recovery of the hearts. This finding paralleled a similar relationship between the level of high energy phosphates and the recovery of the hearts. Such results would be consistent with the fact that, in addition to inducing cardiac arrest, hypothermia has the advantage of conferring considerable protection on the myocardium by greatly reducing the metabolic rate. Bretschneider (1964) has reviewed the effect of normothermic and hypothermic arrest upon the survival and recovery time of the human heart and has also shown that protection increases with increasing degrees of hypothermia.

High levels of ATP and creatine phosphate and cardiac recovery to near control values can only be obtained with arrest by hypothermic coronary perfusion or ischemia with severe (4°C) topical hypothermia. In contrast, recoveries to above control values can be obtained following normothermic arrest by continuous perfusion of the coronaries with 16 mM K^+. It is interesting to note that, in addition to the recovery exceeding 100%, the level of creatine phosphate in the myocardium at the end of arrest also exceeds the control value. This finding may reflect a much reduced cellular energy requirement during arrest; a similar elevation of creatine phosphate during aerobic potassium arrest has been reported by Lochner *et al.* (1968). Our studies have shown that the combination of potassium arrest with ischemia does afford a measure of protection to the ischemic myocardium, resulting in improved recovery and improved ATP and creatine phosphate levels. This may possibly be due to the more rapid induction of arrest by potassium and the consequent conservation of cellular energy supplies. However, under the conditions of this study, poor and variable recoveries were still obtained, and potassium chloride afforded a greater measure of protection than potassium citrate, possibly because of the inhibition of anaerobic glycolysis by citrate.

In conclusion and from the viewpoint of myocardial protection during cardioplegia, these studies would stress the importance of maintaining levels of cellular high energy phosphates. This condition requires the provision of a continuous supply of oxygen and substrate, which is best achieved by ensuring continuous and adequate coronary perfusion. Under ischemic conditions, this does not occur, and effective protection can only be obtained in conjunction with severe topical hypothermia.

Acknowledgments This work was supported with the aid of a grant from the British Heart Foundation. The authors thank M. V. Braimbridge for his advice and discussion.

REFERENCES

ARNOLD, G., und W. LOCHNER. 1965. Die Temperaturabhängigkeit des Sauerstoffverbrauches stillgestellter, Kunstlich perfundierter Warmblüterherzen zwischen 34° und 4°C. Pflügers Arch. 284: 169–175.

BLOODWELL, R. D., J. N. KIDD, G. L. HALLMAN, W. J. BURDETTE, M. J. McMURTNEY, and D. A. COOLEY. 1969. Cardiac valve replacement without coronary perfusion: Clinical and laboratory observations. *In* L. A. Brewer (ed.), Prosthetic Heart Valves, pp. 397–410. Charles C Thomas, Springfield, Ill.

BRAASCH, W., S. GUDBJARNASON, P. S. PURI, K. G. RAVENS, and R. F. BING. 1968. Early changes in energy metabolism in the myocardium following acute coronary artery occlusion in anaesthetized dogs. Circulat. Res. 23: 429–438.

BRANTIGAN, J. W., A. M. PERNA, T. J. GARDNER, and V. L. GOTT. 1972. Intramyocardial gas tensions in the canine heart during anoxic cardiac arrest. Surg. Gynecol. Obstet. 134: 67–72.

BRETSCHNEIDER, H. J. 1964. Überlebenszeit und Wiederbelebungszeit des Herzens bei Normo- und Hypothermie. Verh. Dtsch. Ges. Kreislaufforsch. 30: 11–34.

CHAIN, E. B., K. R. L. MANSFORD, and L. H. OPIE. 1969. Effects of insulin on the pattern of glucose metabolism in the perfused working and Langendorff heart of normal and insulin deficient rats. Biochem. J. 115: 537–546.

COOLEY, D. A., G. J. REUL, and D. A. WUKASCH. 1972. Ischemic contracture of the heart: "Stone Heart". Am. J. Cardiol. 29: 575–577.

GRIEPP, R. B., E. B. STINSON, and N. E. SHUMWAY. 1973. Profound local hypothermia for myocardial portection during open heart surgery. J. Thorac. Cardiovasc. Surg. 66: 731–739.

HEARSE, D. J., and E. B. CHAIN. 1972. The role of glucose in the survival and recovery of the anoxic isolated perfused rat heart. Biochem. J. 128: 1125–1133.

ISOM, O. W., N. D. KUTIN, E. A. FALK, and F. C. SPENCER. 1973. Patterns of myocardial metabolism during cardiopulmonary bypass and coronary perfusion. J. Thorac. Cardiovasc Surg. 66: 705–721.

ISSELHARD, W., H. MERGUET, und J. AENGENVOORT. 1965. Vergleich des Herzstoffweshsels bei verschiedenen Methoden des Künstlichen Herzstillstandes. Pflügers Arch. 286: 336–371.

IYENGAR, S. R. K., S. RAMCHAND, E. J. P. CHARRETTE, C. K. S. IYENGAR, and R. B. LYNN. 1973. Anoxic cardiac arrest: An experimental and clinical study of its effects. J. Thorac. Cardiovasc. Surg. 66: 722–730.

KREBS, H. A., und K. HENSELEIT. 1932. Untersuchungen uber die Harnstoffbildung im Tierkörper. Hoppe Seyler's Z. Physiol. Chem. 210: 33–66.

LANGENDORFF, O. 1895. Untersuchungen, am uberlebenden Saugerthierherzen. Pflügers Arch. 61: 291–332.

LOCHNER, W., G. ARNOLD, and E. R. MULLER-RUCHHOLTZ. 1968. Metabolism of the artifically arrested heart and of the gas perfused heart. Am. J. Cardiol. 22: 299–311.

LOCHNER, W., und R. DUDZIAK. 1965. Skilstandsumsatz und Ruheumsatz des Herzens. Pflügers Arch. 285: 169–177.

NEELY, J. R., H. LEIBERMEISTER, E. J. BATTERSBY, and H. E. MORGAN. 1967. Effect of pressure development on oxygen consumption by isolated rat heart. Am. J. Physiol. 212: 804–814.

OPIE, L. H. 1971. Acute metabolic response in myocardial infarction. Brit. Heart J. 33: 129–137.

RANDLE, P. J., R. M. DENTON, and P. J. ENGLAND. 1968. *In* T. W. Goodwin (ed.), The Metabolic Role of Citrate, pp. 87–103. Academic Press, London.

SAPSFORD, R., E. BLACKSTONE, J. KIRKLIN, R. KARP, N. KOUCHOUKOS, A. PACIFICO, and C. ROE. 1973. Coronary perfusion vs. ischemic arrest for aortic valve replacement: a prospective, randomized study. Suppl. IV to Circulation VII and VIII (Suppl.): 73.

SOBEL, B. E. 1972. Biochemical and morphological changes in infarcting myocardium. Hosp. Pract. pp. 59–71.

UMBREIT, W. W., R. H. BURRIS, and J. F. STAUFFER. 1964. Manometric Techniques, p. 132. Burgess, Minneapolis.

WOLLENBERGER, A., O. RISTAU, und G. SCHOFFA. 1960. Eine einfache Technik der extremschnellen Abkuhlung grosserer Gewebstucke, Pflügers Arch. 270: 399–412.

Mailing address:
Dr. D. J. Hearse,
Department of Biochemistry,
Imperial College,
Kensington, London SW7 (Great Britain).

Recent Advances in Studies on
Cardiac Structure and Metabolism Volume 10
The Metabolism of Contraction
Edited by Paul-Emile Roy and George Rona

EFFECT OF SUBSTRATE ON ENZYME RELEASE AFTER CORONARY ARTERY LIGATION IN ISOLATED RAT HEART[1]

J. de LEIRIS,[2] L. H. OPIE,
W. F. LUBBE, and O. BRICKNELL[3]

Ischaemic Heart Disease Laboratory, Department of Medicine,
University of Cape Town and Groote Schuur Hospital,
Cape Town, South Africa

SUMMARY

In the present work, we tested the hypothesis that the development of myocardial infarction in isolated perfused working rat hearts could be influenced by the nature of the substrate reaching the infarcting zone. We measured cardiac output and rates of release of lactate dehydrogenase (LDH) after coronary artery ligation in hearts perfused with different substrates: glucose; palmitate, oleate, linoleate bound to albumin; palmitate-albumin plus glucose and/or insulin. Cardiac output after ligation was lowest with palmitate and highest with glucose. LDH release was 5–10 times greater with FFA than with glucose, but was significantly reduced by the addition of glucose or insulin to the palmitate-albumin solution. Our results argue for an important role for substrate effects on the metabolic outcome of experimental myocardial infarction.

INTRODUCTION

It has been suggested (Oliver, 1972; Opie, 1972) that the outcome of myocardial ischaemia is dependent on the nature of the substrate reaching the ischemic zone. We tested the hypothesis that provision of glucose might be beneficial and provision of free fatty acids (FFA) harmful for the survival of ischaemic infarcting tissue, by perfusing isolated working rat hearts after coronary artery ligation, using media containing glucose or glucose and albumin, or a long chain fatty acid and albumin in different FFA:albumin molar ratios, or albumin-bound fatty acid together with glucose and insulin.

---

[1] Full text submitted for publication in Journal of Molecular and Cellular Cardiology.

[2] Present address: Laboratoire de Physiologie Comparée et de Physiologie Cellulaire, Université Paris XI, Orsay, France.

[3] Supported by the Ethical Drug Foundation Association.

METHODS

Male rats of the Sprague-Dawley strain, fed *ad libitum,* were used in these experiments. Following anaesthetization with ether and heparinization (100 I.U. injected into the femoral vein), the heart was rapidly excised according to the technique of Neely *et al.* (1967), arrested by plunging into ice-cold perfusion fluid, mounted by the aorta on a cannula, and perfused in a retrograde manner by the Langendorff (1895) technique in a nonrecirculating fashion at 65 cm H_2O pressure. The left atrium was then cannulated to allow atrial perfusion ("working heart" technique) according to the method of Neely *et al.* (1967) as modified by Opie, Mansford, and Owen (1971).

During the working heart period, preparations were perfused via the left atrium at an atrial pressure of 10 cm H_2O.

In nonligated hearts (control experiments) perfused with glucose as substrate, the left ventricle spontaneously ejected 40–50 ml of perfusate per min against a hydrostatic pressure of 100 cm of water. The aortic output and coronary flow were pooled and recirculated.

The left coronary artery was stitched with silicone-treated silk 5-0 by means of a 3/8 circle taper needle (Davis and Jeck, Pearl River, New York). The ligature was placed about 1 mm below the root of the aorta at the left bottom edge of the left atrium, as described by Bajusz (1963) and by Kannengiesser, Lubbe, and Opie (1975). Substrates used included: glucose (11 mM); glucose (11 mM) and albumin (0.45 mM); palmitate (0.5–1.5 mM) bound to albumin (0.1–0.45 mM); linoleate (0.5 mM) bound to albumin (0.1 mM); oleate (0.5 mM) bound of albumin (0.1 mM); and palmitate-albumin (0.5 mM:0.1mM) together with glucose (11 mM) and insulin (2 mμ/ml).

Release of enzymes is thought to be related to loss of integrity of the cell membrane. Therefore, to check the extent of celular alterations, we measured lactate dehydrogenase (LDH) activity in perfusate by the method of Wroblewski and La Due (1955).

RESULTS

Cardiac output after coronary artery ligation was lowest with palmitate (28–38% of initial value) and highest with glucose (61%). LDH release was 5–10 times greater with palmitate-albumin as substrate than with glucose or glucose-albumin. Increasing the molar ratio of palmitate to albumin from 1:1 to 5:1 doubled the rate of release of LDH. The addition of glucose and insulin to palmitate-albumin decreased release of LDH by two-thirds. LDH release was halved when palmitate was replaced by equimolar linoleate or oleate.

CONCLUSION

We conclude that the extent of myocardial cell damage resulting from coronary ligation is greatly increased by the presence of FFA (especially palmitate) in the perfusing fluid and decreased by the additional presence of glucose and insulin. Our results argue for an important role for substrate effects on the metabolic outcome of experimental myocardial infarction.

Acknowledgments We acknowledge with pleasure the support of the MRC of South Africa, the Chris Barnard Fund, and N A T O.

REFERENCES

BAJUSZ, E. 1963. Conditioning Factors for Cardiac Necrosis, p. 250. Karger, Basel.
KANNENGIESSER, G. J., W. F. LUBBE, and L. H. OPIE. 1975. Experimental myocardial infarction with left ventricular failure in the isolated perfused rat heart. J. Molec. Cell. Cardiol. 7: 135–151.
LANGENDORFF, O. 1895. Untersuchungen am über lebender Saügertierherzen. Pflügers Arch. ges Physiol. 61: 251–332.
NEELY, J. R., H. LIEBERMEISTER, E. J. BATTERSBY, and H. E. MORGAN. 1967. Effect of pressure development on oxygen consumption by isolated rat heart. Am. J. Physiol. 212: 804–814.
OLIVER, M. F. 1972. Metabolic response during impending myocardial infarction. II-Clinical implications. Circulation 45: 491–500.
OPIE, L. H. 1972. Metabolic response during impending myocardial infarction. I-Relevance of studies of glucose and fatty acids metabolism in animals. Circulation 45: 483–490.
OPIE, L. H., K. R. L. MANSFORD, and P. OWEN. 1971. Effect of increased heart work on glycolysis and adenine nucleotides in the perfused heart of normal and diabetic rats. Biochem. J. 124: 475–480.
WROBLEWSKI, F., and J. S. La DUE. 1955. Lactic dehydrogenase activity in blood. Proc. Soc. Expt. Biol. Med. 90: 210–213.

Mailing address:
Dr. J. de Leiris,
Laboratoire de Physiologie Comparée,
Bâtiment 443 -Centre Universitaire, Orsay 91405 (France).

Recent Advances in Studies on
Cardiac Structure and Metabolism Volume 10
The Metabolism of Contraction
Edited by Paul-Emile Roy and George Rona

REGIONAL MYOCARDIAL METABOLISM FOLLOWING ACUTE CORONARY ARTERY OCCLUSION

TZU-WANG LANG, SAMUEL MEERBAUM, ELIOT CORDAY, COSTANTINO COSTANTINI, STEVEN RUBINS, and MANUEL CORDOBA

Cedars-Sinai Medical Center, Los Angeles, California, USA

SUMMARY

Regional measurements of the consequences of coronary artery occlusions indicate myocardial derangements in both the coronary occluded and nonoccluded myocardium. The proximal left anterior descending coronary artery (LAD) was acutely occluded in 34 closed-chest anesthetized dogs through inflation of an intracoronary balloon. Temporary separation of venous compartments was achieved by means of a balloon catheter placed within the great cardiac vein, and simultaneous blood sampling from both occluded and nonoccluded segments was carried out.

Acute coronary occlusion resulted in early cardiac dysfunction which persisted during the subsequent 3 hours of LAD occlusion. Oxygen extraction increased 6–8% in both occluded and nonoccluded zones. Lactate extraction in the nonoccluded segment decreased significantly from 34.4 to 23.7% ($P < 0.05$). In the occluded zone, lactate extraction dropped from 35.0 to 7.8% ($P < 0.05$). Of the 34 dogs studied, 17 exhibited lactate production in the occluded segment, while 7 dogs also had lactate production in the nonoccluded region. Significant postocclusion myocardial potassium efflux was noted in both zones.

INTRODUCTION

The elucidation of mechanisms and assessment of factors which contribute to cardiac dysfunction during regional ischemia and infarction requires truly regional myocardial measurements. Knowledge of myocardial metabolism is important, since it reflects the degree of dysfunction and may also be a precursor of further alterations in function. The mutual relationship between alterations or derangements within the ischemic and nonischemic segments of the left ventricle, as well as their influence on global ventricular function, need to be clarified if proper interpretation of the consequences of ischemia and of the effectiveness of interventions is to be arrived at.

Most of the recent attempts to measure regional myocardial metabolism have been in open-chest preparations (Owen *et al.*, 1970; Obeid *et al.*, 1972; Opie *et al.*, 1973). Moreover, there has been only limited study of metabolism alterations in the nonischemic zone of the left ventricle during regional coronary occlusion (Gudbjarnason, Puri, and Mathes, 1971; Mathes and Grudbjarnason, 1971). This nonoccluded zone may be jeopardized by altered myocardial oxygen demands and supply due to systemic and cardiac compensatory events. Simultaneous measurements in both the coronary occluded and nonoccluded regions are required to understand their individual alterations and their contribution to global cardiac function. Only then can there be a rational choice and evaluation of treatment of cardiac dysfunction following ischemic events.

The experimental model described in this paper was designed to allow simultaneous myocardial metabolic determinations in both occluded and nonoccluded regions in the closed-chest preparation (Corday *et al.*, 1974b). In addition, the model features intracoronary balloon occlusion at selected sites of the coronary arteries. Regional metabolic measurements in this model have been routinely performed in several experimental programs. Similar study of regional ischemic zone metabolism is currently being applied in clinical evaluation (Chatterjee, 1975).

METHODS

The new experimental close-chest intracoronary balloon occlusion model was previously described in detail (Corday *et al.*, 1974b). A double lumen balloon catheter is inserted via the right carotid artery into the left anterior descending coronary artery (LAD), with the balloon at the catheter tip placed at the desired (proximal LAD) site. The center lumen of this catheter may be used for sampling, measurements, and regional interventions distal to the balloon occlusion. Occlusion is accomplished through inflation of the balloon by means of hypaque solution, which facilitates fluoroscopic monitoring of balloon location and extent of inflation.

A second double lumen balloon catheter is inserted via the right jugular vein into the coronary sinus (CS) and is then advanced into the great cardiac vein (GCV). Its balloon is placed under fluoroscopic control at the desired site in the GCV, while a specially designed tip with its center lumen protruding into the anterior interventricular vein (AIV) is located alongside the LAD distal to its balloon occlusion. A separate CS catheter is employed for CS blood sampling and thermodilution determinations of CS flow. Brief inflation of the GCV balloon results in separation of venous compartments and permits simultaneous blood sampling from the ischemic (AIV) and nonischemic (CS) zone drainages. Normally, the AIV venous drainage of the LAD perfused zone flows through GCV and mixes in the CS with the venous drainage from the myocardium perfused by the left circumflex coronary artery (LCF).

The venogram shown in Figure 1 illustrates the temporary separation of the venous compartment during an LAD occlusion. Prior to inflation of the GCV balloon, contrast medium injected into the GCV and AIV drains normally into the CS (Fig. 1A). When the GCV balloon is inflated, there is a high degree of isolation of the GCV from the CS, and separation of the ischemic from the nonischemic myocardial venous drainages (Fig. 1B).

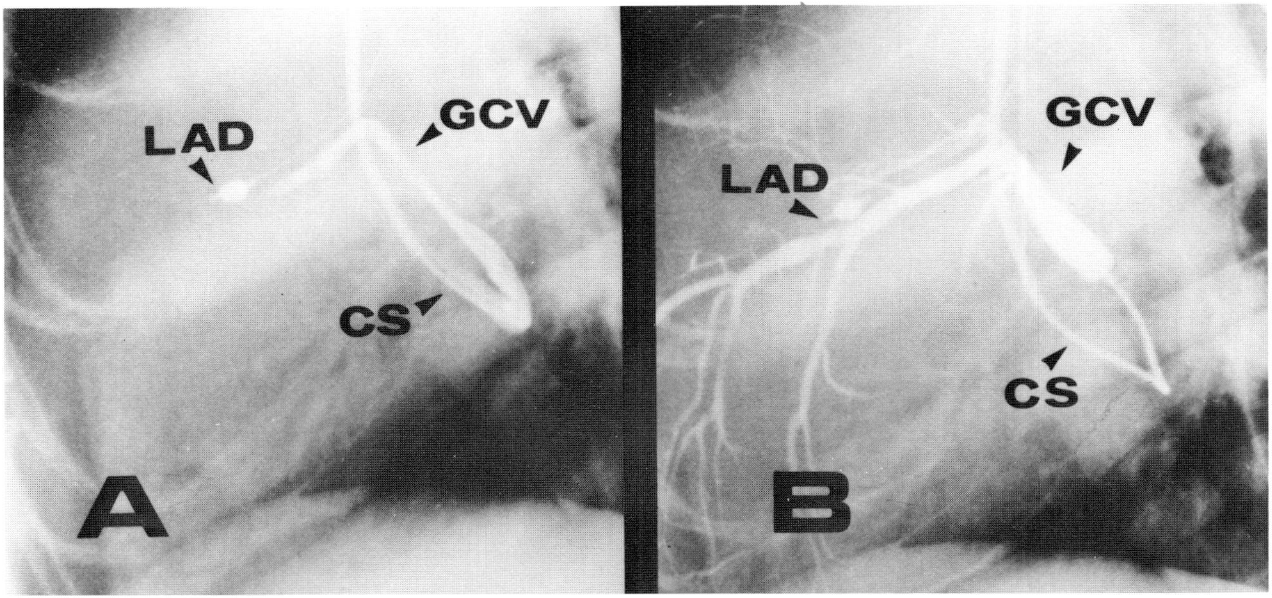

Figure 1. Venogram of great cardiac vein (GCV) and anterior interventricular coronary vein (AIV). *A:* Contrast material was injected into the GCV through the center lumen of the catheter with its balloon deflated. Note the contrast material regurgitation into the coronary sinus (CS), consistent with normal drainage of venous blood from the AIV and GCV. The intracoronary balloon is inflated in the proximal left anterior descending coronary artery. *B:* Proximal intracoronary balloon LAD occlusion in the same dog. The GCV balloon was temporarily inflated and contrast material was injected through the center lumen of the GCV catheter. Note there was no regurgitation of contrast material into the CS. Also note that there was no evidence of shunting from AIV to the CS.

Measurements were carried out in 34 normotensive closed-chest dogs weighing 30–40 kg, anesthetized with morphine (2 mg/kg intramuscularly) and pentobarbital (15 mg/kg intravenously). A Harvard respirator provided artificial respiration via tracheal intubation. Heparinization was carried out initially (5 mg/kg) and supplemented throughout the experiment (3 mg/kg/hour). Fluid loss was estimated and replaced by saline infusion at a rate of 100 cc/hour. Cardiac output and coronary sinus flow were measured by the thermodilution technique (Ganz *et al.*, 1971a, b). Left ventricular pressure was measured with a tip transducer catheter (BioTech Instruments). Maximal LV dP/dt was determined by electrical differentiation. A Statham P23Db pressure transducer was employed for measurement of aortic root pressure.

Metabolic determinations included myocardial lactate and potassium balances. Lactate extraction (or production) in the coronary occluded and nonoccluded segments were computed from aortic, CS, and AIV blood sample analyses as [(A-V)/A]×100%. Chemical analysis was based on the Hans-Jurgens-Hohorst technique. Myocardial potassium balance in these segments was measured by photometry and calculated as (A-V) in mEq/l. Regional oxygen extraction was also determined. Blood gases (pO_2 and pCO_2) and pH were monitored throughout the experiments.

A separate series of 5 dogs was used for baseline studies. The sham preparation consisted of full instrumentation, manipulations, and measurements identical to the primary 34-dog series, except that the LAD remained unoccluded with the uniflated balloon in the coronary artery. The total period of these experiments corresponded to the 30-minute control and 3-hour occlusion period of the primary series, with measurements at regular intervals.

Control hemodynamic and metabolic data were measured during a 30-minute stable period prior to acute occlusion of the proximal LAD. These measurements were repeated at 5 minutes, 30 minutes, and 3 hours after the coronary occlusion. Simultaneous metabolic determinations were obtained from the coronary occluded and nonoccluded segments of the left ventricle.

Hemodynamic recordings at papers speed of 50 mm/sec were obtained on the Electronics for Medicine unit (Model DR-16). Statistical results were computed as mean ± standard error of mean (SEM). Significance was established through paired t-tests.

Following each of the experiments, the heart was excised for gross analysis and histopathologic investigations. Sections of the heart were stained with hematoxylin-eosin for light microscopic study.

RESULTS

Sham series (5 dogs)

During the 3½ hour period of measurements, the 5 dogs exhibited a rise in heart rate (105 to 132 beats/min), a minor decrease in peak systolic pressure (150 to

146 mm Hg), and small fluctuations in cardiac output and systemic vascular resistance. Mild fluctuations were also noted in myocardial lactate and potassium balance, but there was no significant difference between the measurements in the LAD or LCF perfused regions.

Primary series (34 dogs)

Hemodynamics The statistical results (mean ± SEM) are presented in bar chart form (Fig. 2) in terms of percent change from preocclusion control measurements.

Immediately following acute occlusion of the proximal LAD, there was a slight rise in heart rate (8%), a significant though small drop in peak systolic pressure (-5%), a significant rise in left ventricular end-diastolic pressure (LVEDP) (34%) from a preocclusion control level of 7 mm Hg, and a significant drop in max LV dP/dt (-20%).

A full set of hemodynamic measurements was performed at 30 minute occlusions. Few arrhythmias were noted, and those which occurred were limited to the first 30 minutes after occlusion. The previous parameters were essentially unchanged. Cardiac output was significantly depressed (-20%), systemic vascular resistance significantly increased (20%), and stroke work significantly lowered relative to preocclusion control (-28%). Measurements of coronary sinus flow 30 minutes after occlusion indicated a reduction of 9%.

Repeat measurements of hemodynamics at 3 hours of LAD occlusion indicated: heart rate rose significantly (20% relative to control); peak systolic pressure did not change (-6%); LVEDP rose (50%); max LV dP/dt remained unaltered compared with the 30-minute measurements (−19%); cardiac output reflected a slight compensation (17%) as did systemic vascular resistance (15%). The stroke work was further lowered (-34%), and the reduced coronary sinus flow did not change.

In summary, global cardiac function, measured by means of hemodynamic parameters, appeared significantly depressed in the earliest stages of the LAD occlusion. With the exception of slight compensation and fluctuations, cardiac function remained significantly deranged at 3 hours occlusion.

Metabolism (Fig. 2)

Lactate Balance Control measurements in the nonoccluded segments of the left ventricle indicated a lactate extraction of 34.4%. Acute LAD occlusion caused an immediate and significant decrease of this balance to 23.5%, and this altered metabolism persisted to the end of the 3-hour experimental occlusion. Review of individual dogs indicated that, of the 34 animals, 5 exhibited normal postocclusion lactate extraction in the nonoccluded segment (>20%), but 22 had some degree of lactate derangement (<20% but >0%), and 7 dogs showed lactate production in this usually assumed noninvolved region of the myocardium.

In the occluded zone, lactate balance was severely and immediately deranged

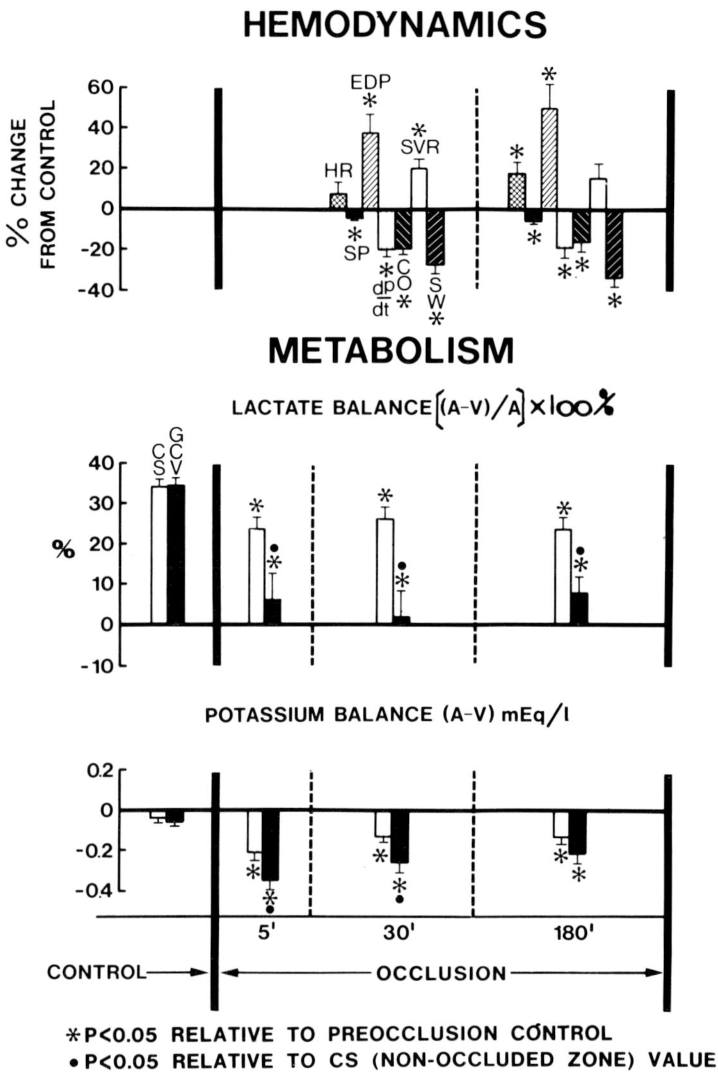

Figure 2. Statistics of hemodynamics and regional metabolism during 3-hour occlusion in 34 dogs. Hemodynamics are expressed as percent change from control, lactate balance as [(A-V)/A] × 100% for both the coronary occluded and nonoccluded myocardium and potassium balance as A-V in mEq/l. Significance ($p<0.05$) is indicated relative to preocclusion control (*), and also relative to nonoccluded zone metabolism (•). Control hemodynamics (mean ± SEM) were as follows: heart rate (HR, beat/min) 109 ± 5.6; peak systolic pressure (SP, mm Hg) 141 ± 3; left ventricular end-diastolic pressure (EDP, mm Hg) 7 ± 0.6; max LV dP/dt (dP/dt, mm Hg/sec) 2840 ± 135; cardiac output (CO, l/min) 2.7 ± 0.18; systemic vascular resistance (SVR, dynes sec cm^{-5}) 3740 ± 0.18; stroke work (SW, g·m/beat) 42 ± 3. Note that hemodynamic function was depressed shortly after the acute LAD

following LAD occlusion. Measurements 5 minutes after occlusion indicated a drop from a control extraction of mean 35% to 6.1%. At 30 minutes, there was lactate production in this region (-1%). After 3-hour occlusion, lactate balance showed a +7.8% extraction, still significantly deranged. Analysis of individual dogs showed that 17 of the 34 had occluded zone lactate production during the occlusion, and the remaining 17 had deranged lactate metabolism (<20%).

The differences between lactate balances in the coronary occluded and non-occlusion, and the remaining 17 had deranged lactate metabolism (<20%).

Potassium Balance The nonoccluded segment exhibited significant potassium loss at 5 minutes of acute proximal LAD occlusion (-0.21 mEq/l). Subsequently, the value was -0.13 mEq/l at both the 30-minute and 3-hour occlusion measurements. At 5 minutes after LAD occlusion, the coronary occluded segment showed a significant and higher myocardial potassium loss (-0.35 mEq/l). This loss was -0.26 mEq/l and -0.22 mEq/l at 30 minutes and 3 hours occlusion, respectively. During the 3-hour occlusion period, 17 of the 34 dogs had significant potassium loss from the occluded zone and, in 12, there were instances of significant potassium loss from the nonoccluded region.

Oxygen Extraction Control oxygen extractions were 57% in both occluded and nonoccluded zones. During the early period of LAD occlusion, there was a 6–8% increase in oxygen extraction relative to control in both regions. This increase in oxygen extraction persisted to the end of the 3-hour experiment.

Individual Experiments Figure 3 presents hemodynamic and metabolic data in an individual animal. Except for a decrease in heart rate and return of LVEDP to near control at 3-hour occlusion, the general hemodynamic trends were maintained. This graph is indicative of the variability and fluctuation of data which must be expected in a population of animals. It is noteworthy that, in this experiment, pronounced lactate production occurred in both occluded and nonoccluded segments of the left ventricle within 5 minutes of the acute LAD occlusion.

Histopathology The anterior left ventricular tissue exhibited mild peripheral vascular congestion, but nuclei and fibers appeared intact. There was no inflammatory response. Occasional increase in myocardial fiber granularity and margination of leukocytes were noted. Infarct size was grossly estimated at 20% of the left ventricle. In some dogs which exhibited significantly depressed lactate balance in the nonoccluded myocardium during the 3-hour LAD occlusion,

---

occlusion. This depression persisted over 3 hours of occlusion. Nonoccluded zone metabolism is based on blood sampling from the CS. Occluded region metabolism is based on blood sampling from the GCV. Note that the occluded region exhibited significantly depressed lactate and potassium metabolism and that the metabolic derangement persisted throughout the 3 hours of occlusion. The nonoccluded region exhibited a significant but lesser derangement of lactate balance and potassium loss from the myocardium. Throughout the occlusion period, there was a significant difference between metabolism in the coronary occluded and nonoccluded regions.

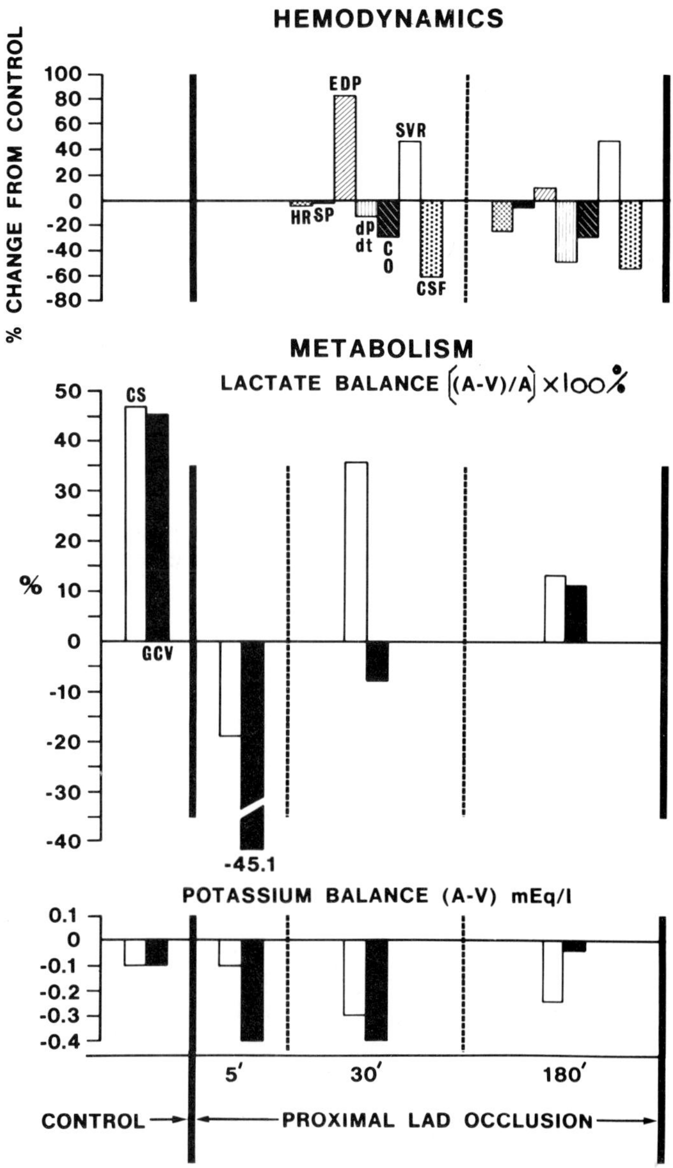

Figure 3. This bar graph indicates hemodynamics and metabolism in an individual dog. Control hemodynamics were: HR 118; SP 131; EDP 6; dP/dt 3400; CO 2.83; SVR 2835; coronary sinus flow (CSF, ml/min) 203. Note that, shortly after occlusion, there was a sharp rise in EDP, a drop in CO, an increase in SVR, and a 50% reduction in CSF. At 3 hours of occlusion, EDP returned to near control, but the other hemodynamic derangements persisted. Immediately after acute LAD occlusion, both the coronary occluded and non-occluded segments of the left ventricle exhibited lactate production. During the subsequent

patchy or focal infarctions were noted in the nonoccluded segment of the left ventricle.

DISCUSSION

This series of experiments provided further evidence of the suitability of the intracoronary balloon occlusion model for comprehensive studies of regional myocardial function. The simultaneous hemodynamic and regional myocardial metabolic data from the coronary occluded and nonoccluded regions of the left ventricle provide a useful description of the progressive alterations following acute occlusion of the proximal LAD in closed-chest dogs. The acute LAD occlusion causes an early hemodynamic dysfunction and metabolic derangements which tend to persist throughout the 3-hour occlusive period. Previous evidence (Chatterjee, 1975; Corday *et al.*, 1974b) indicated that there is only minor postocclusion increase of LAD pressure distal to the occlusion during the first few hours, and this is taken to mean that there is no significant change in coronary collateral flow. The 34 dogs reported remained normotensive, but another 5 dogs of the total series (39 dogs) developed hypotension and were classified as being in cardiogenic shock (both peak systolic pressure and cardiac output reduced more than 30%). The progressive alterations in cardiac function in these dogs will be described in a separate communication (Corday *et al.*, 1974a).

The severe metabolic derangements, i.e., lactate production and myocardial potassium loss, in the ischemic zone were expected and were particularly pronounced shortly after LAD occlusion. Surprisingly, the nonoccluded region also exhibited significant lactate derangements, occasional lactate production and potassium loss, even though these metabolic derangements were significantly less pronounced than in the occluded region. Careful checks of the coronary anatomy, including venographic study of veno-venous shunting as well as consideration of the experimental model, sampling procedures, and functional pressure differentials, would tend to preclude major contamination of samples from the nonoccluded zone, while allowing for somewhat greater regional metabolic derangements in the local ischemic region than is indicated by the measurements. Recent studies dealt with reduction in ATP and CP content in the noninfarcted myocardium during coronary occlusion (Gudbjarnason *et al.*, 1971; Mathes and Gudbjarnason, 1971). In a recent investigation, we have found that

---

3 hours of occlusion, lactate balance improved in both zones but remained depressed. At 5 and 30 minutes following occlusion, a major difference was noted between the lactate balances of the two zones. There was a major decrease in myocardial perfusion in the occluded zone during the early occlusion period. At 30 minutes after occlusion, both regions exhibited potassium loss. At 3 hours of occlusion, there was a return of potassium balance to near control.

the metabolic derangements in the nonoccluded myocardium were occasionally associated with patchy and focal necroses in this zone (Corday *et al.*, 1975; Lang *et al.*, 1974).

Regional measurements of myocardial forces and lengths in open-chest preparations (Meerbaum *et al.*, 1974; Wyatt *et al.*, 1974) indicate that a coronary occlusion may cause the nonoccluded portions of the left ventricle to hypercontract or hypocontract in compensation for the sharply reduced contractile function of the occluded region. Furthermore, LAD occlusion was almost invariably associated with significantly increased systemic vascular resistance. Increased impedance to left ventricular ejection and hpercontraction of the nonoccluded myocardium could result in regionally elevated oxygen demands (Driscoll and Eckstein, 1967). Coronary occlusion may also interfere with and diminish the overlapping coronary branch circulation in the adjacent nonoccluded myocardium.

Oxygen extraction following LAD occlusion was noted in both occluded and nonoccluded segments. It is believed that the occluded region would tend to increase oxygen extraction to compensate for its greatly diminished blood flow. In the nonoccluded zone, higher oxygen extraction could be related to the increased oxygen requirements due to the factors discussed above.

The presence of persisting and significant metabolic derangements in the nonoccluded myocardium can have important consequences to the global cardiac function and should therefore be assessed and treated, if necessary. These derangements may in some instances accelerate the development of shock (Corday *et al.*, 1974a). Treatment of a dysfunctioning nonoccluded segment is not only essential to maintain global function, but would also benefit the ischemic zone by maintaining systemic and coronary perfusion.

REFERENCES

CHATTERJEE, K., J. M. MATLOFF, H. J. C. SWAN, W. GANZ, V. S. KAUSHIK, P. MAGNUSSON, M. M. HANIS, and J. S. FORRESTER. 1975. Abnormal regional metabolism and mechanical function in patients with ischemic heart disease: Improvement after successful regional revascularization by aorto-coronary bypass. Circulation (in press).

CORDAY, E., T. W. LANG, and S. MEERBAUM. 1974a. Mechanisms of cardiogenic shock following coronary occlusion: hemodynamic and metabolic precursors. (In preparation).

CORDAY, E., L. KAPLAN, S. MEERBAUM, J. BRASCH, C. COSTANTINI, T. W. LANG, H. GOLD, S. RUBINS, and J. OSHER. 1975. Consequences of coronary artery occlusion on remote myocardium: Effects of occlusion and reperfusion. Am. J. Cardiol. (in press).

CORDAY, E., T. W. LANG, S. MEERBAUM, H. GOLD, S. HIROSE, S. RUBINS, and M. DALMASTRO. 1974. A closed chest intracoronary occlusion model for the study of regional cardiac function. Am. J. Cardiol. 33: 49–59.

DRISCOL, T. E., and R. W. ECKSTEIN. 1967. Coronary inflow and outflow responses to coronary artery occlusion. Circulat. Res. 20: 485–495.

GANZ, W., R. DONOSO, H. S. MARCUS, J. S. FORRESTER, and H. J. C. SWAN. 1971a. A new technique for measurement of cardiac output by thermodilution in man. Am. J. Cardiol. 27: 392–396.

GANZ, W., K. TAMURA, H. S. MARCUS, R. DONOSO, S. YOSHIDA, and H. J. C. SWAN. 1971b. Measurement of coronary sinus blood flow in continuous thermodilution in man. Circulation 44: 181–195.

GUDBJARNASON, S., P. S. PURI, and P. MATHES. 1971. Biochemical changes in non-infarcted heart muscle following myocardial infarction. J. Mol. Cell. Cardiol. 2: 253–276.

LANG, T. W., E. CORDAY, H. GOLD, S. MEERBAUM, S. RUBINS, C. COSTANTINI, S. HIROSE, J. OSHER, and V. ROSEN. 1974. Consequences of reperfusion after coronary occlusion. Am J. Cardiol. 33: 69–81.

MATHES, P., and S. GUDBJARNASON. 1971. Changes in norepinephrine stores in the canine heart following experimental myocardial infarction. Am. Heart J. 81: 211–219.

MEERBAUM, S., J. OSHER, T. W. LANG, G. LEWIS, C. FELDSTEIN, E. DURAN, and E. CORDAY. 1974. Miniature transducer study of local myocardial mechanics following coronary occlusion. Circulation 49–50 (Suppl. III): 14.

OBEID, A., H. SMULYAN, R. GILBERT, and R. H. EICH. 1972. Regional metabolic changes in the myocardium following coronary artery ligation in dogs. Am. Heart J. 83: 189–196.

OPIE, L. H., P. OWEN, M. THOMAS, and R. SAMSON. 1973. Coronary sinus lactate measurements in assessment of myocardial ischemia. Comparison with changes in lactate/pyruvate and beta-hydroxybutyrate/acetoacetate ratios and with release of hydrogen, phosphate and potassium ions from the heart. Am. J. Cardiol. 32: 295–305.

OWEN, P., M. THOMAS, V. YOUNG, and L. OPIE. 1970. Comparison between metabolic changes in local venous and coronary sinus blood after actue experimental coronary artery occlusion. Am. J. Cardiol. 25: 562–570.

WYATT, H. L., P. L. da LUZ, J. FORRESTER, G. DIAMOND, R. CHAGRASULIS, and H. J. C. SWAN. 1974. Depression of function in nonischemic myocardium after coronary occlusion. Circulation, 49–50 (Suppl. III): 119.

Mailing address:
Eliot Corday, M.D.,
Cedars-Sinai Medical Research Institute,
4751 Fountain Avenue, Los Angeles, California 90029 (USA).

Recent Advances in Studies on
Cardiac Structure and Metabolism Volume 10
The Metabolism of Contraction
Edited by Paul-Emile Roy and George Rona

MYOCARDIAL ENZYME RELEASE FROM ISCHEMIC ISOLATED PERFUSED WORKING RAT HEART

ANDERS P. WALDENSTRÖM and ÅKE C. HJALMARSON

Department of Medicine I, Sahlgren's Hospital,
University of Göteborg, Göteborg, Sweden

SUMMARY

Isolated working rat hearts were made ischemic for 5, 10, and 30 minutes respectively. After the ischemic period, all hearts were perfused in a retrograde nonworking way for 30 minutes. During the 5 first minutes of ischemia, there was a marked fall of cardiac output and coronary flow. A significant release of GOT was seen and this was more marked after longer periods of ischemia. Addition of adrenaline to the perfusate increased the enzyme release. Pacing at 400/minute, high preload, high afterload, acidosis, or alkalosis did not alter enzyme release. Glycogen, ATP and CrP levels were depressed at the end of the ischemic period, but were seen to be rising again during the retrograde perfusion. This study indicates that myocardial tissue may release enzymes without being irreversibly damaged.

INTRODUCTION

Several techniques have been used to induce experimental myocardial infarction. One of the most widely used methods involves occlusion of a coronary artery of dog hearts *in situ* (Case, Nasser, and Crampton, 1969; Kjekshus and Sobel, 1970; Karlsson, Templeton, and Willerson, 1973). This method, however, entails several technical problems. When studying myocardial metabolism under ischemic conditions, the isolated perfused heart has many advantages. Until recently, such models have only been used when studying the anoxic state (De Leiris *et al.*, 1969; Butterworth *et al.*, 1970; Hearse, Humphrey, and Chain, 1973).

The aortic ball valve as described by Neely *et al.* (1973) provides an excellent means of inducing whole heart ischemia. This method allows control of coronary flow, independent of the mechanical performance of the heart. Some slight modifications were made in this model as shown in Figure 1. The one-way aortic valve is connected with the aorta by cannulation. The left ventricle ejects the perfusate during systole against a hydrostatic pressure head of 80 cm H_2O. The valve prevents back-flow in the aortic tube and reduces the diastolic perfusion of the coronaries. The one-way valve is bypassed by a tube which is opened during

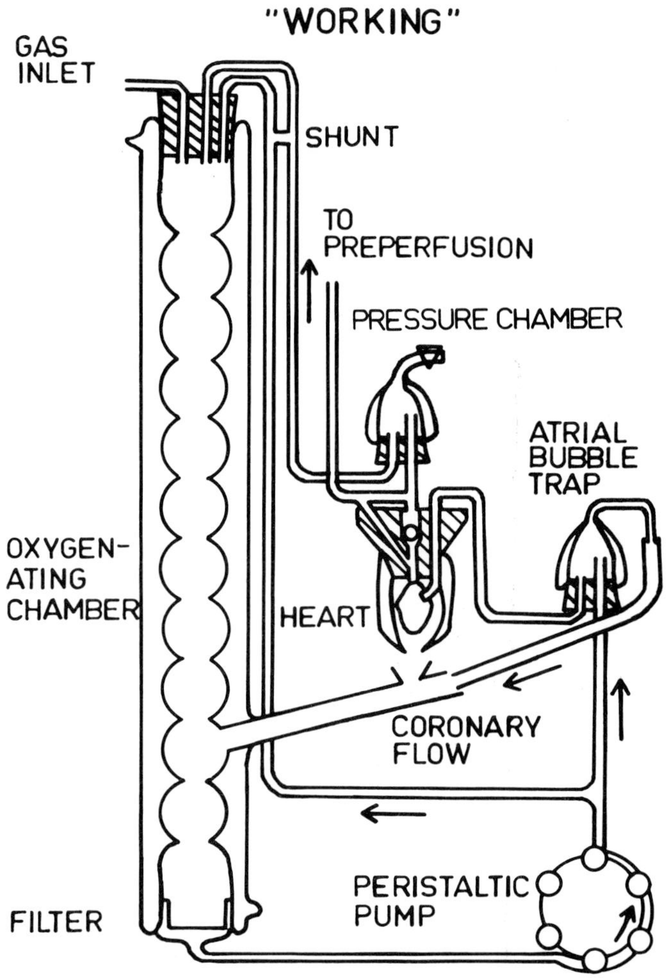

Figure 1. Perfusion apparatus as described by Neely *et al.* (1973), modified by connecting a shunt between overflow tube from pump and aortic tube. Shunt was open only during retrograde reperfusion.

control, and pre- and post-ischemia perfusions. The left atrium is cannulated and provided with the recirculating perfusion fluid at an adjustable filling pressure. The present investigation was done to study the release of GOT, CPK, and LDH from the isolated rat heart and to correlate this release to mechanical performance and heart content of glycogen, lactate, ATP, and creatine phosphate (CrP).

MATERIALS AND METHODS

Male Wistar rats (350–450 g) were heparinized (2.5 mg) 30 minutes prior to Nembutal anaesthesia (60 mg/kg b.w.). The hearts were quickly excised and transferred to ice-chilled saline. The aorta and left atrium were cannulated and the hearts were perfused with 90 ml of recirculating Krebs-Henseleit bicarbonate buffer with 14 mM of glucose added. The perfusion fluid was continuously gassed with 95% O_2 - 5% CO_2 equilibrated with water at 37°C. All perfusions were carried out at 37°C. Preload was constantly held at 10 cm H_2O and afterload at 80 cm H_2O. The hearts were paced at 300 beats/minutes. Prior to ischemia, the hearts were perfused in a retrograde nonworking way and thereafter for 10 minutes in the anterograde working way. Statistical significance was considered if $p < 0.05$. Each value represents mean ± S.E.M. for 6 hearts. Ischemia was induced for 5, 10, and 30 minutes by closing the one-way valve bypass tube. After the ischemic period, all hearts were perfused in the retrograde nonworking way at a constant perfusion pressure of 80 cm H_2O. This was made possible by opening a shunt from the pump overflow tube to the aortic tube. This shunt was opened just sufficiently to provide fluid in the aortic tube to the top of the oxygenating chamber. A sample of the perfusate was taken at 10, 20, and 30 minutes after the ischemic period. At the end of the perfusion, all hearts were frozen using a Wollenberger clamp cooled to the temperature of liquid nitrogen.

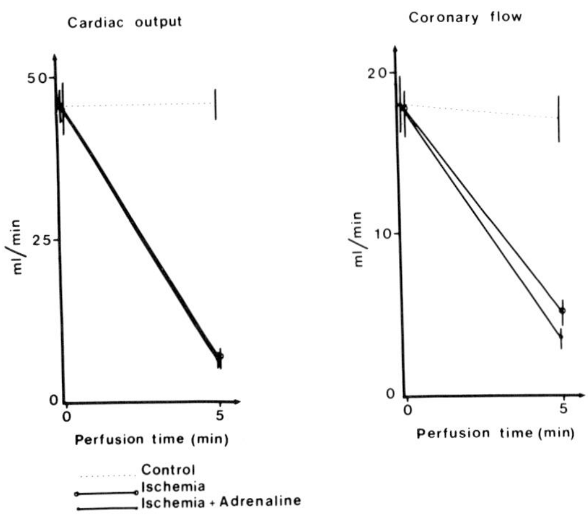

Figure 2. Mechanical performance of controls, ischemic hearts, and ischemic hearts with adrenaline (10^{-6}M) added. Each heart was perfused for 10 minutes as a Langendorff preparation for washout. They were then switched over for 10 minutes of working hearts before start of ischemia. Controls were continued as working hearts for 5 minutes instead of ischemia. Adrenaline was added to 1 group at start of ischemia.

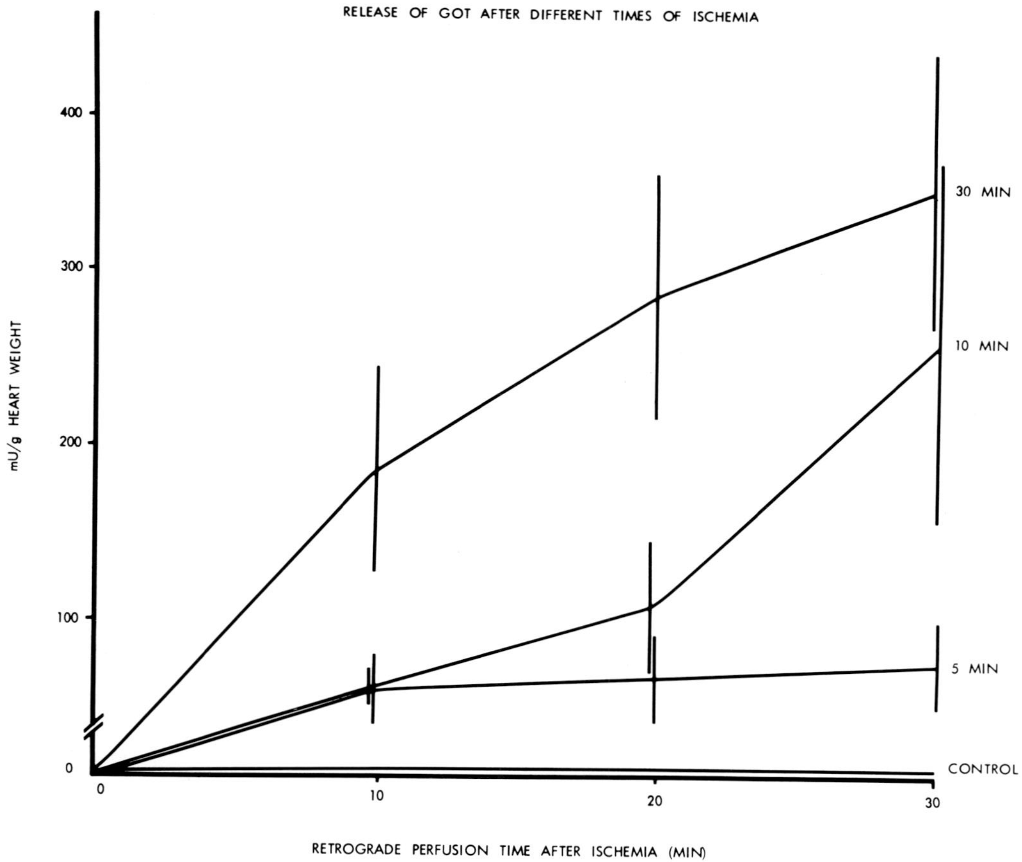

Figure 3. Enzyme release after 5, 10, and 30 minutes of ischemia, measured 10, 20, and 30 minutes after end of ischemic period. Significant release was seen at all sampling times compared to controls.

RESULTS

Hemodynamics

Cardiac output and coronary flow decreased rapidly during the first 5 minutes of ischemia, and then more slowly. During the subsequent reperfusion for 30 minutes all hearts made ischemic for 5 minutes developed normal cardiac output and coronary flow. The hearts made ischemic for 10 minutes recovered their functions almost completely while, after 30 minutes of ischemia, all hearts were irreversibly damaged. Addition of adrenaline (10^{-6}M) to the perfusate did not alter these hemodynamic parameters during the 5 minutes of ischemia (Fig. 2). However, such hearts did not recover as well as the controls during reperfusion. The control hearts maintained their flow rates at a constant level during the period of investigation.

Enzyme release

Release of GOT was measured at 10, 20, and 30 minutes after ischemic periods lasting for 5, 10, and 30 minutes (Fig. 3). The control hearts were perfused in the working way for 40 minutes and then switched over to retrograde perfusion for 30 minutes. There was no enzyme release from these hearts; the longer the ischemic period, the more enzyme was released. A continuous release of GOT took place during the whole 30-minute period of retrograde perfusion. It was found necessary to reperfuse retrogradely at a perfusion pressure of 80 cm H_2O to elevate coronary flow and washout enzymes. During ischemic periods of 5–30 minutes there was a small and variable release of enzymes. More significant release of enzymes was obtained at the end of ischemia by washing through the coronaries with 5 ml of perfusate 5 times, using a syringe, while the heart was arrested in ice-chilled buffer. However, the retrograde perfusion, as previously stated, gave rise to the most reliable release of enzymes. CPK and LDH were released in proportion to GOT, while only very small amounts of GPT were obtained in the perfusate. Addition of adrenaline (10^{-6}M) to the perfusate of hearts made ischemic for 5 minutes resulted in a marked increase in the release of GOT, which was comparable to that of hearts made ischemic for 30 minutes without addition of adrenaline (Fig. 4). Other conditions which were suspected of being unfavourable during ischemia (Maroko and Braunwald, 1973) were investigated as shown in Table 1. There was, however, no significant difference in the release of GOT as compared to 5 minutes of control ischemia.

Metabolism

From Figure 5, it can be seen that the glycogen content of the heart was markedly depressed after the ischemic periods. However, reperfusion of the coronaries for 30 minutes normalized the glycogen levels, with the exception of the hearts that were ichemic for 30 minutes. Also, a rise was seen here, but the reperfusion

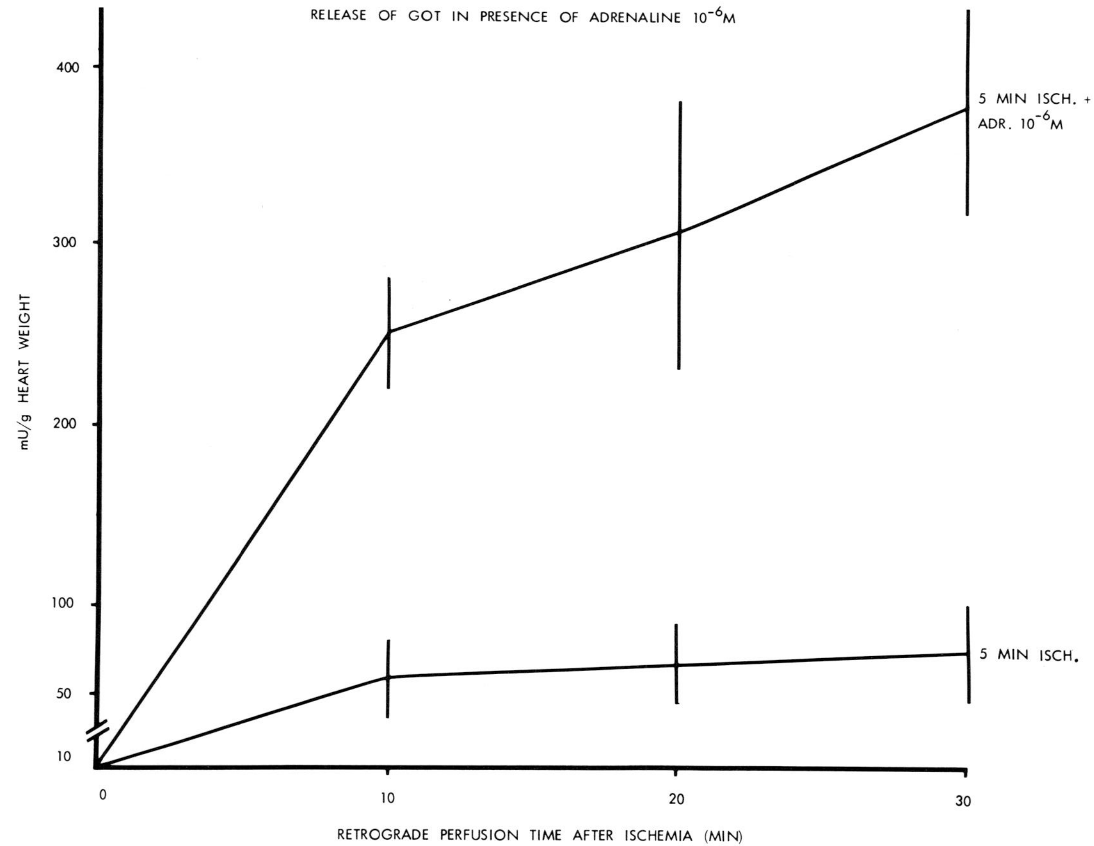

Figure 4. Comparison of enzyme release between hearts ischemic for 5 minutes and when adrenaline (10^{-6}M) was added to ischemic hearts.

Table 1. Release of GOT (as mU/g heart weight) under different conditions during ischemia*

Conditions during 5 minutes of ischemia	Retrograde perfusion time after ischemia		
	10 minutes	20 minutes	30 minutes
Pacing 400/minute	32 ± 5	58 ± 15	49 ± 17
High preload (25 cm H_2O)	32 ± 7	53 ± 12	53 ± 11
High afterload (aorta clamped)	41 ± 14	61 ± 21	70 ± 32
Acidosis (pH 7.0)	15 ± 5	17 ± 3	18 ± 2
Alkalosis (pH 7.8)	27 ± 7	24 ± 6	26 ± 4

*Each value represents the mean ± S.E.M. of 6 hearts.

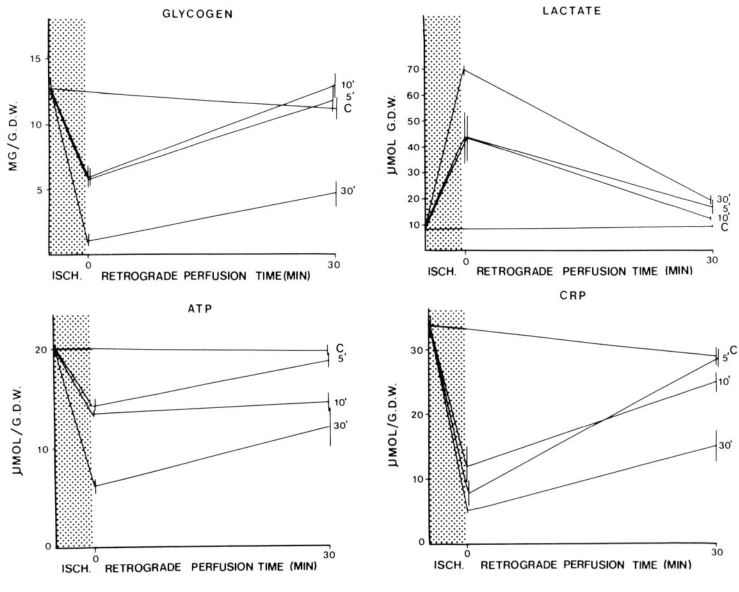

Figure 5. Tissue levels of glycogen, lactate, ATP, and creatine phosphate (CrP) in controls and hearts ischemic for 5, 10, and 30 minutes. Dotted area represents various periods of ischemia. All hearts were perfused retrogradely for 30 minutes after end of ischemic period.

time may not have been long enough to normalize the glycogen content. A high tissue lactate content at the end of ischemia rapidly fell towards control levels. This fall may be due to metabolism of lactate and/or release to the perfusate. During the periods of ischemia, ATP and CrP levels fell markedly. The high-energy phosphates of the hearts made ischemic for 5 minutes were completely normalized during the reperfusion period. A rise in ATP and CrP was also found after 10 and 30 minutes of ischemia.

CONCLUSION

The results of this study demonstrate leakage of enzymes (GOT, CPK, and LDH) from isolated rat hearts after various periods of ischemia. The enzyme release from the hearts seems to be a function of time and the severity of ischemia, the length of reperfusion, and the rate of coronary flow after ischemia. If an adequate coronary flow is not reestablished after ischemia, enzymes may not be washed out; reestablishment of contraction may also be important. Studies of anoxic perfused hearts (Hearse *et al.,* 1973) showed a longer time lag before enzymes could be detected. This may depend on a lower demand for high energy phosphates, as the hearts were not working. Neely *et al.* (1973) showed that the isolated rat heart was irreversibly damaged when ATP was depleted 70% during the ischemia, a finding that was confirmed in this study. These severely damaged hearts had a marked release of enzymes during reperfusion following the ischemic period. Interestingly, there was a significant release of enzymes in hearts made ischemic for 5 minutes although, in these hearts, cardiac function and tissue levels of glycogen, lactate, ATP and CrP completely normalized during the reperfusion. This indicates that cell death might not be obligatory to induce enzyme release. Addition of adrenaline to the perfusate increased the enzyme release and aggravated the cardiac damage. This finding is in agreement with the observation that interventions that increase myocardial oxygen consumption increase myocardial ischemic damage (Maroko and Braunwald, 1973). However, in this study, pacing-induced tachycardia, high preload, and high afterload did not change myocardial enzyme release or damage. Unlike the dog heart *in situ* (Maroko and Braunwald, 1973), the isolated rat heart is not influenced by sympathetic tone or circulating catecholamines. The ischemic isolated rat heart has proved suitable for studies of factors of importance for ischemic cell damage and enzyme release.

Acknowledgments The present study was supported by grants from the Swedish National Association against Heart and Chest diseases, the Swedish Medical Research Council (B74-13X-2529-05), and Göteborg Medical Society.

REFERENCES

BUTTERWORTH, P. J., D. T. PLUMMER, H. ROCHMAN, and W. V. WEBER. 1970. Factors affecting the release of anzymes from the isolated perfused rat heart. J. Physiol., London 209:3P.

CASE, R. B., M. G. NASSER, and R. S. CRAMPTON. 1969. Biochemical aspects of early myocardial ischemia. Am. J. Cardiol. 24: 766–775.

HEARSE, D. J., S. M. HUMPHREY, and E. B. CHAIN. 1973. Abrupt reoxygenation of the anoxic potassium-arrested perfused rat heart: A study of myocardial enzyme release. J. Mol. Cell Cardiol. 5: 395–407.

KARLSSON, J., G. H. TEMPLETON, and J. T. WILLERSON. 1973. Relationship between epicardial S-T segment changes and myocardial metabolism during acute coronary insufficiency. Circulat. Res. 32: 725–730.

KJEKSHUS, J. K., and B. E. SOBEL. 1970. Depressed myocardial creatine phosphokinase activity following experimental myocardial infarction in rabbit. Circulat. Res. 27: 403–414.

DE LEIRIS, J., D. BRETON, D. FEUVRAY, and E. CORABOEUF. 1969. Lacticodehydrogenase release from perfused rat heart under the effect of abnormal media. Arch. Internat. Physiol. Biochim. 77: 749–762.

MAROKO, P. R., and E. BRAUNWALD. 1973. Modification of infarct size after coronary occlusion. Ann. Int. Med. 79: 720–733.

NEELY, J. R., M. J. ROVETTO, J. T. WHITMER, and H. E. MORGAN. 1973. Effects of ischemia on ventricular function and metabolism in the isolated working rat heart. Am. J. Physiol. 225: 651–658.

Mailing address:
Dr. Anders Waldenström,
Department of Medicine 1, Sahlgren's Hospital,
S-413 45 Göteborg (Sweden).

Exogenous and Endogenous Factors Influencing Ischemic Damage

Recent Advances in Studies on
Cardiac Structure and Metabolism Volume 10
The Metabolism of Contraction
Edited by Paul-Emile Roy and George Rona

EFFECT OF A THROMBOLYTIC AGENT (UROKINASE) ON NECROSIS AFTER ACUTE MYOCARDIAL INFARCTION[1]

W. T. HERMENS,[2] S. A. G. J. WITTEVEEN,
L. HOLLAAR, and H. C. HEMKER[2]

SUMMARY

Using a two-compartment model with continuous input function, the rate and extent of myocardial damage were calculated from plasma enzyme levels measured in patients suffering from acute myocardial infarction (AMI). Results were compared for four different enzymes. As regards the course of necrosis, two different hypotheses were tested and it was shown that the necrotic process has to be considered as a multiplicative process rather than as a sudden disintegration of cells due to anoxia. Comparison of results obtained in a group of 5 patients treated with urokinase and in an untreated control group of 10 patients showed that the necrotic phase as defined by the enzyme release from the infarcted area was significantly shortened.

INTRODUCTION

The importance of determining plasma enzyme levels in patients suffering from acute myocardial infarction (AMI) has been well established. However, these data are mainly used in a qualitative way for diagnostic or prognostic reasons. As the routine laboratory determinations of these enzymes are sufficiently accurate to allow a thorough analysis of enzyme levels as a function of time, much information is lost that way. In the last few years, we have developed a method of obtaining a quantitative anlysis of myocardial damage after AMI (Hemker *et al.*, 1972; Witteveen, 1972). In the present paper, this method is used to investigate the effect of urokinase on the necrotic process after AMI.

MATERIALS AND METHODS

The experimental group consisted of 15 patients with AMI, of whom 5 received urokinase. Patients with cardiogenic shock or severe heart failure were not

---

[1] Supported by the Dutch Heart Foundation.
[2] Laboratory of Cardiobiochemistry, University Hospital, Leiden, The Netherlands
Present address: Medical Faculty Maastricht, Maastricht, The Netherlands.

included in the investigation. For further clinical details see the publication by Witteveen (1972).

Blood was drawn from an indwelling catheter as soon as possible after admission to hospital and, subsequently, every 4 hours for 2 or 3 days. At the end of this period, progressively fewer samples were taken until plasma levels returned to normal.

Samples consisted of 10 ml of blood collected in tubes containing 0.2 ml of 20% sodium citrate to prevent clotting and were immediately centrifuged, initially for 10 min at 900 *g* to remove erythrocytes and leukocytes, and subsequently for 20 min at 40,000 *g* to remove thrombocytes. The plasma was then stored at $-20°C$ until enzyme determinations were carried out.

The following enzymes were studied: glutamic oxalacetic transaminase (GOT; E.C. 2.6.1.1), creatine phosphokinase (CPK; E.C. 2.7.3.2), phosphohexose isomerase (PHI; E.C. 5.3.1.9), lactate dehydrogenase (LDH, E.C. 1.1.1.27), and α-hydroxybutyrate dehydrogenase (α-HBDH; E.C. 1.1.1.27). The last two enzymes are identical but are determined in a different assay, so that comparison of results permits an estimation of over-all experimental error.

Enzyme determinations were done spectrophotometrically. For GOT, LDH, and α-HBDH commercially available test kits were used (Boehringer). Recently, Boehringer introduced so-called optimized test kits. If these are used, the results presented below should be adjusted. CPK was determined according to Rosalki, (1967), and PHI according to Weber and Wegmann (1968). Enzyme activities were expressed as I.U./liter, in which 1 I.U. catalyzes the turnover of 1 μmole of substrate per minute.

Urokinase, a nonantigenic thrombolytic agent acting through activation of plasminogen, was made available by Hoffmann La Roche. The drug was administered as soon as possible after admission to hospital. An initial loading dose of 7,200 units/kg body weight was given in 10 min, after which an infusion of 3,600 units/hour/kg body weight was given for 18 hours.

Heart tissue was obtained at thoracic surgery. To determine the enzyme activities in the normal myocardium, the tissue was kept in physiological saline at 0°C and homogenized within 10–15 min (Turrax). After sedimentation of the remaining particles, the activities were measured in the supernatant (Table 2). To study the shed-out of enzymes *in vitro,* tissue was kept at 37°C for up to 100 hours in plasma under anoxic conditions and gentle shaking; the activity appearing in the plasma was measured. After this period, the tissue was homogenized to obtain the residual tissue activity (Table 1). The molecular weights of the human enzymes were determined by column chromatography with DEAE Sephadex G-200 of plasma containing high activities. To obtain a standard, albumin (M = 69,000) and ceruloplasmin (M = 160,000) were also determined by quantitative immunoelectrophoresis (Laurell).

The mathematical analysis of the patient data is based upon the two-compartment model presented in Figure 1, where V_1 is the vascular volume, V_2 the extravascular volume, $F(t)$ the enzyme release as a function of time, k the sum of first order clearance constants, and D the diffusion constant. This model is based upon the following considerations. The observed (double) exponential clearance from the plasma indicates that there is only first order elimination proportional to the enzyme concentration.

The assumption of the absence of extravascular elimination is based upon the *in vitro* experiments showing no appreciable denaturation. This suggests that the enzyme breakdown is circulation-dependent. Insofar as the enzymes from the necrotic cells do not diffuse directly into the blood stream, we assumed that, in view of the relatively direct transport of lymph from the myocardium to the vascular compartment by way of the thoracic duct, the input $F(t)$ is only into V_1. A more detailed discussion of the implicit assumptions of this model will be published elsewhere (Hermens *et al.*). It can be shown (Witteveen, 1972) that the plasma concentration of enzyme as a function of time $X(t)$ in this model is given by the double exponential function

$$X(t) = C_1(t)e^{-a_1 t} + C_2(t)e^{-a_2 t} + X_n \tag{1}$$

$(a_1 > a_2 > 0)$, where X_n is the normal steady state enzyme concentration, and the constants a_1 and a_2 depend on V_1, V_2, k, and D. The time-dependent coefficients in Eq. (1) are given by:

$$C_1(t) = [(a_1 - a_1 a_2/k)/(a_1 - a_2)] \int_0^t f(\tau)e^{a_1\tau}d\tau \tag{2}$$

$$C_2(t) = [(a_1 a_2/k - a_2)/(a_1 - a_2)] \int_0^t f(\tau)e^{a_2\tau}d\tau \tag{3}$$

where $f(t) = F(t)/V_1 - kX_n$ = extra enzyme release (i.e., due to infarction) per liter vascular volume.

As soon as the enzyme release from the infarcted area is completed, i.e., $f(t) = 0$, $C_1(t)$ and $C_2(t)$ become constants and a logarithmic plot gives a curve with slopes a_1 and a_2 as shown in Figure 1.

The vascular volume V_1 was calculated from body weight and hematocrit. The remaining parameters, k, V_2, and permeability constant P (which is related to the diffusion constant D by $P = DS/\delta$, where S and δ are the total interface area and the characteristic thickness of the membrane between both compartments), were obtained by curve-fitting techniques (Hermens *et al.*, to be published). However, an assumption has to be made regarding the release function $f(t)$.

Two different hypotheses concerning $f(t)$ were proposed as showin in Figure 2. In case A, we assume that necrosis is a multiplicative process; i.e., local degradation of cells has an amplifying effect on the necrosis of adjacent tissue, for instance, because of the release of cell contents, e.g., potassium. It can be shown

(Aitchison and Brown, 1963) that, under these circumstances, $f(t)$ will be a log normal function given by

$$f(t) = \frac{Q}{\sigma t\sqrt{2\pi}} \exp\left(-0.5(Int - \mu)^2/\sigma^2\right) \tag{4}$$

This is a three-parameter function in which Q is the total extra release of enzyme per liter vascular volume.

In case B, we assume that, after a period of ischemia, the majority of cells become permeable in a relatively short lapse of time (for instance, in a few minutes). This results in a sudden maximal release of diffusing enzymes with an exponential fall-off which is described by the following function:

$$f(t) = Qb\exp(-b(t-t_p)) \qquad t \geqslant t_p \tag{5}$$

which is again a three-parameter function with total release Q.

RESULTS

As reported earlier (Hemker *et al.* 1972), the biphasic disappearance shown in Figure 1 is only observed for PHI, meaning that the plasma clearance curves for CPK, GOT, and LDH (α-HBDH) are monoexponential. This phenomenon is caused by the fact that the enzyme release from the infarcted tissue continues for an appreciable period of time (about 34 hours for patients treated with urokinase and about 48 hours for untreated patients; see Figure 5). Owing to this tailing release, the fast phase of the clearance curve will not be observed in

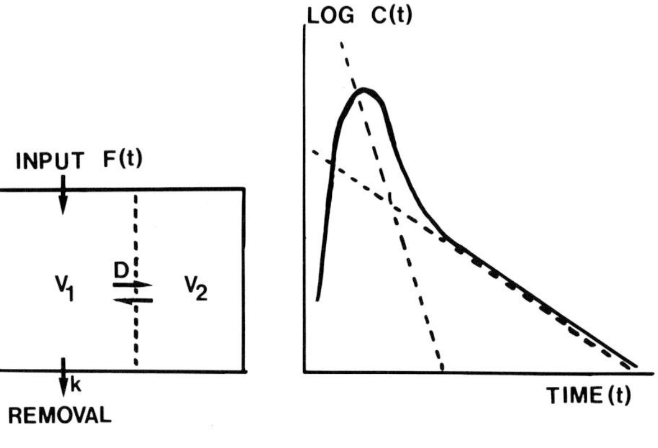

Figure 1. Two-compartment model.

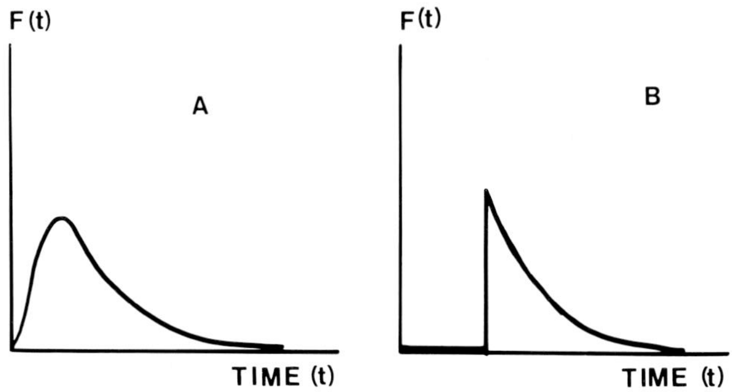

Figure 2. Two hypotheses regarding necrosis; A=multiplicative necrosis, B=sudden necrosis.

those cases where the exchange by diffusion (P) is fast compared to the clearance (k).

PHI is the only enzyme which is metabolized fast enough to show a biphasic disappearance of those patients with a relatively short period of enzyme release (mostly urokinase-treated patients).

The results of curve fitting for PHI showed that, in all cases, the assumption of multiplicative necrosis gave a better fit than the assumption of sudden necrosis; for more details, see the work of Hermens *et al.* (to be published). The best fit parameter values obtained for P and V_2/V_1 are (for PHI):

$$P = 0.15 \pm 0.1 \text{ (liter/hour)}$$

$$V_2/V_1 = 0.30 \pm 0.12 \qquad (6)$$

For the molecular weights, the following values were found:

LDH (α–HBDH)	M = 145,000
PHI	M = 98,000
CPK	M = 82,000
GOT	M = 80,000

All values are ± 5,000.

As the diffusion constants of proteins are not very sensitive to molecular weight (Schultze and Heremans, 1966), it is improbable that the values for P and V_2/V_1 will differ much from those found for PHI. Values (6) were therefore used in a routine computer program for all enzymes. In this program, the slow phase of the clearance curve is used to calculate k. Subsequently, the extravascular concentrations, the release function $f(t)$, and the total release are calculated from the plasma concentrations. The results are shown in Figure 3.

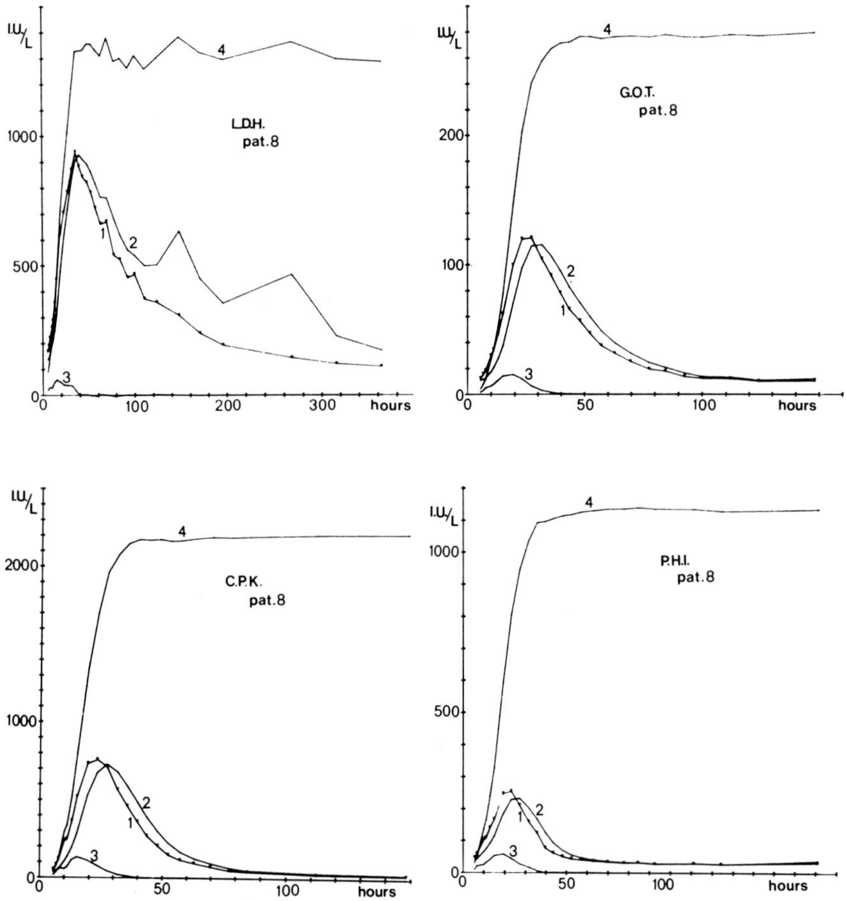

Figure 3. Patient 8. *1* = measured plasma concentrations. *2* = calculated extravascular concentrations. *3* = calculated enzyme release/hour. *4* = calculated total enzyme release.

Figure 4 shows the correlations between the calculated total release Q for the different enzymes compared to LDH. As the clearance for LDH (α-HBDH) is slow (cf. Table 3), this enzyme is the least sensitive to the irregularities in the release and was therefore chosen as reference. The systematic error in Q_{LDH} corresponding to the uncertainty in P and V_2/V_1 is estimated at 10–15%.

The correlation for CPK is not satisfactory, possibly attributable to the fact that this is the only enzyme that showed an unpredictable denaturation during in vitro studies (Witteveen, 1972). This phenomenon could invalidate our assumption of absence of extravascular elimination for this enzyme.

The results of four *in vitro* experiments of anoxic incubation of tissue are

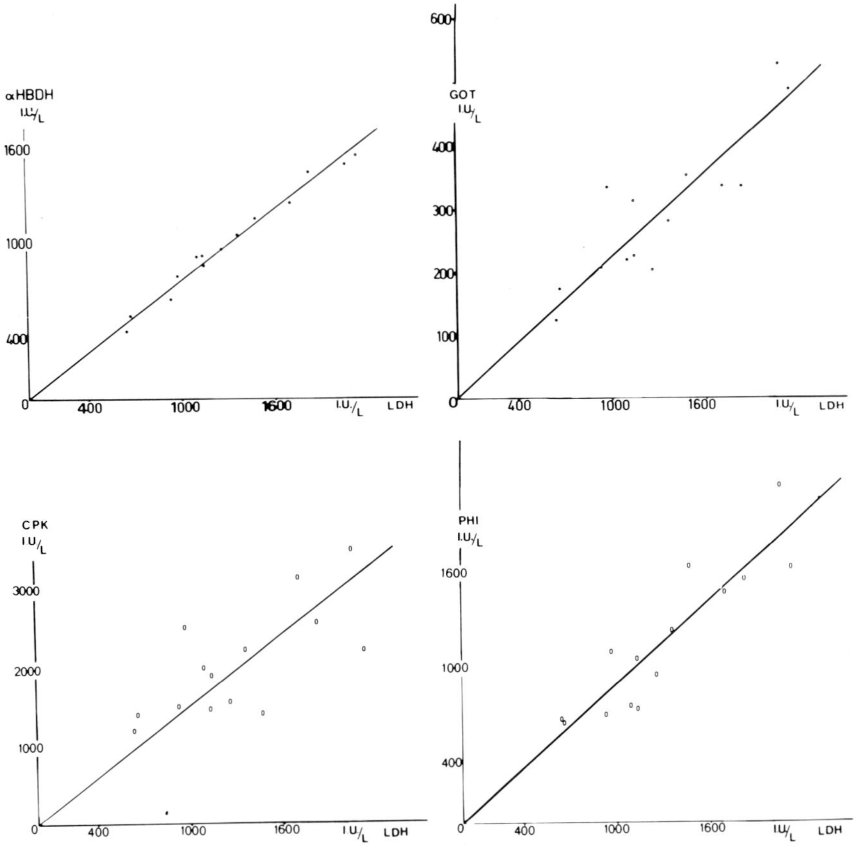

Figure 4. Correlations between calculated total enzyme release. Best fit straight line through the origin for α-HBDH: $y = 0.74x$; GOT: $y=0.22x$; CPK: $y=1.50x$; PHI: $y=0.89x$. 0.89 x.

shown in Table 1. As seen from Figure 4 and Table 1, the calculated total amounts of enzyme released from the infarcted area in patients exhibit approximately the same ratio as the amounts released *in vitro.* However, if we compare these *in vitro* values with the activities measured in fresh tissue (Table 2), we find that, for LDH (α-HBDH) and PHI, the original activities are recovered but, for GOT and CPK, roughly one-third of the activity has disappeared by a local process which apparently (also) proceeds *in vitro.*

Table 3 shows the calculated clearance constants k, total enzyme release Q, and equivalent amount of heart tissue A in grams for all patients except patient 15; in this patient, the levels were too low to permit calculation of LDH, GOT, and PHI.

Table 1. In vitro enzyme release from human heart tissue incubated in plasma*

	LDH		α-HBDH		GOT		CPK		PHI	
Experiment	A	B	A	B	A	B	A	B	A	B
1	122	21	99	19	43	22	261	71	112	19
2	136	23	87	16	23	28	88	37	84	19
3	127	33	87	21	35	37	126	36	120	28
4	89	20	64	15	32	42	164	96	106	19
Mean:	118	24	84	18	34	32	160	60	105	21

* Activities expressed in I.U./g wet tissue. A, activity in plasma; B, residual tissue activity after 100 hours' incubation.

The amounts A were calculated for LDH by multiplying Q_{LDH} by 1.20 to account for the residual tissue activity (see Table 1) and by the vascular volume, and dividing by 128 (see Table 2). For the other enzymes, the calculated Q was first transformed into the corresponding Q_{LDH} by means of the best fit equations given in Figure 4, followed by the same procedure. The effect of urokinase is shown in two ways. Table 4 gives the time of maximal plasma activity in hours after the onset of severe symptoms. As can be seen from this table, there is a considerable shortening for patients treated with urokinase. The same phenomenon is observed in Figure 5, where the calculated enzyme release $f(t)$ in I.U./liter/hours has been averaged for each group of patients.

DISCUSSION

The effect of urokinase in patients presenting electrocardiographic evidence of necrosis, such as elevation of the ST-segment, has been revealed in the multicentered European Urokinase Trial in myocardial infarction. It could be demon-

Table 2. Human heart tissue enzyme activity*

	LDH	α-HBDH	GOT	CPK	PHI
Mean	128	101	96	370	143
S.D.	15.4	11.1	18.2	37.0	14.3

*Activities expressed in I.U./g wet tissue (results of 15 determinations).

Table 3. Calculated parameters*

Patient	LDH			α-HBDH			GOT			CPK			PHI		
	k	*Q*	*A*	*k*	*Q*	*A*	*k*	*Q*	*A*	*k*	*Q*	*A*	*k*	*Q*	*A*
1 φ	0.0126	2,113	58	0.0125	1,533	55	0.0576	485	60	0.0849	2,205	40	0.1803	1,624	50
2 φ	0.0134	2,042	52	0.0113	1,483	50	0.0576	523	60	0.0705	3,473	59	0.0759	2,137	61
3	0.0157	1,699	45	0.0164	1,240	43	0.0513	334	40	0.0736	3,123	55	0.1652	1,463	43
4	0.0194	1,253	44	0.0202	957	44	0.0742	204	33	0.0602	1,559	36	0.2903	948	37
5	0.0131	1,819	39	0.0133	1,423	40	0.0546	334	33	0.0661	2,562	37	0.1536	1,548	37
6	0.0139	1,469	34	0.0167	1,145	35	0.0405	353	38	0.0684	1,402	22	0.1577	1,630	43
7	0.0182	1,131	32	0.0181	917	34	0.0670	315	41	0.0711	1,468	28	0.1714	1,050	33
8	0.0157	1,355	36	0.0189	1,038	36	0.0715	280	34	0.1037	2,217	39	0.2436	1,236	37
9	0.0129	1,090	38	0.0134	906	30	0.0478	221	26	0.0615	1,998	34	0.1625	749	21
10 φ	0.0102	969	27	0.0130	785	29	0.1321	335	43	0.1128	2,505	47	0.1714	1,095	35
11 φ	0.0082	1,134	29	0.0086	855	29	0.0428	225	26	0.0558	1,895	32	0.1054	730	21
12 φ	0.0135	928	22	0.0122	634	20	0.0580	207	22	0.0662	1,498	23	0.1611	697	18
13	0.0274	664	17	0.0243	535	18	0.0781	176	20	0.0692	1,396	24	0.1675	648	18
14	0.0124	639	17	0.0121	442	15	0.0439	127	15	0.0754	1,193	21	0.1890	665	20
15	–	–	–	–	–	–		–	–	0.0413	255	4	–	–	–

*k, first order clearance constant $(\text{hours})^{-1}$; Q, total enzyme release in I.U./liter vascular volume; A, equivalent amount of heart tissue (g) (see text); and φ, patients treated with urokinase.

Table 4. Time of maximal plasma enzyme activity*

	LDH	α-HBDH	GOT	CPK	PHI
A	37.7 (6.3)	39.1 (7.0)	27.6 (4.9)	22.9 (5.4)	22.2 (5.3)
B	25.7 (6.3)	25.7 (6.3)	19.6 (4.2)	17.3 (3.6)	16.8 (2.8)

*Values are expressed as means, with S.D. shown in parentheses. A, mean value for control patients (N = 10); B, mean value for patients treated with urokinase (N = 5). Difference between A and B is significant at a 0.02 level for LDH, α-HBDH, and GOT, and at a 0.05 level for CPK and PHI (Wilcoxon).

strated that the electrocardiogram of patients treated with urokinase showed a more rapid normalization of the ST-segment as compared to the control group (Burkart *et al.*, 1973).

As yet, no important therapeutic advantages of thrombolytic agents have been shown for AMI. Clinical trials on urokinase and another thrombolytic agent, streptokinase, revealed no appreciable effect on mortality in patients treated in coronary care units. In some earlier studies (Verstraete, 1973) in which patients were not treated in coronary care units, a favorable effect was found in trials with streptokinase. Whether this effect implies that there have been fewer fatal arrhythmias in the treated group, a finding which could be explained, for example, by a shorter period of tissue necrosis and, thus, a shorter period of electrical instability, remains to be demonstrated.

The two largest infarctions were found in urokinase-treated patients, with the result that this group showed a somewhat larger average enzyme release than did the control group. This effect is not significant in a rank test (see Table 4) and

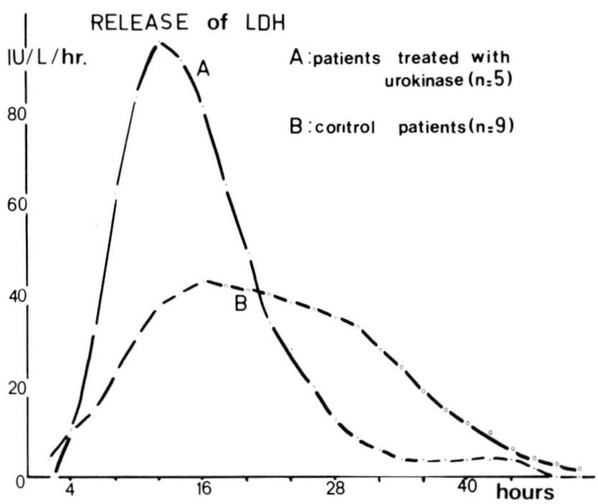

Figure 5. Averaged release functions *f(t)* for LDH.

might have been caused by the fact that the urokinase treatment was given to early recognized and hospitalized patients, the group that tends to present the larger infarctions. However, it remains to be answered whether urokinase, apart from accelerating the enzyme release, could also enhance it.

As regards the quantitation of the amount of necrotic tissue, it should be noted that, although we account for the amount of enzyme that is not released *in vitro* (20% for LDH) and although the release *in vitro* shows the same ratio for the amounts of different enzymes as the release *in vivo,* there might be an over-all underestimation due to local *in vivo* breakdown of tissue, e.g., by inflammatory processes.

Experiments in which CPK release was studied after experimental AMI in dogs showed that less than 30% of total tissue CPK appeared in the circulation (Shell, Kjekshus, and Sobel, 1971). However, the time course of plasma enzyme activity in dogs differs considerably from that in man; therefore, caution should be used before applying these data in studies of infarction in man.

REFERENCES

AITCHISON, J., and J. A. C. BROWN. 1963. *In* The Lognormal Distribution. ,Cambridge.

BURKART, F., F. DUCKERT, P. W. STRAUB, P. G. FRICK, W. SCHWEIZER, und F. KOLLER. 1973. Die fibrinolytische Therapie beim akuten Myokardinfarkt. Schweiz Med. Wschr. 103: 1814.

HEMKER, H. C., S. A. G. J. WITTEVEEN, W. Th. HERMENS, and L. HOLLAAR. 1972. Quantitation of the size of a myocardial infarction by determination of plasma enzymes. *In* H. A. Snellen, H. C. Hemker, P. G. Hugenholtz, and J. H. van BEMMEL (eds.), Quantitation in Cardiology. Leiden University Press, Leiden.

HERMENS, W. Th., S. A. G. J. WITTEVEEN, L. HOLLAAR, and H. C. HEMKER. Study of myocardial necrosis after acute myocardial infarction in man by means of plasma enzyme levels. (To be published).

ROSALKI, S. B. 1967. An improved procedure for serum creatine phosphokinase determination. J. Lab. Clin. Med. 69: 696.

SCHULTZE, H. E., and J. F. HEREMANS. 1966. *In* Molecular Biology of Human Proteins, Vol. 1, p. 176. Elsevier Publishing Company, Amsterdam.

SHELL, W. E., J. K. KJEKSHUS, and B. E. SOBEL. 1971. Quantitative assessment of the extent of myocardial infarction in the conscious dog by means of analysis of serial changes in serum creatine phosphokinase activity. J. Clin. Invest. 50: 2614.

VERSTRAETE, M. 1973. The present status of thrombolytic agents (Review article). Drugs 5: 353.

WEBER, H., und T. WEGMANN. 1968. *In* Atlas der klinischen Enzymologie (Methoden). Georg Thieme Verlag, Stuttgart.

WITTEVEEN, S. A. G. J. 1972. Assessment of the extent of a myocardial infarction on the basis of plasma enzyme levels. Thesis, Leiden.

Mailing address:
Dr. S. A. G. J. Witteveen,
Department of Cardiology, University Hospital,
Rijnsburgerweg 10, Leiden (The Netherlands).

Recent Advances in Studies on
Cardiac Structure and Metabolism Volume 10
The Metabolism of Contraction
Edited by Paul-Emile Roy and George Rona

AN IMPROVED ANIMAL MODEL FOR STUDYING THE EFFECT OF DRUGS ON MYOCARDIAL METABOLISM DURING ISCHEMIA[1]

A. H. M. JAGENEAU, W. VAN GERVEN, R. KRUGER,
H. VAN BELLE, and R. S. RENEMAN

Cardiovascular Department, Janssen Pharmaceutica,
Beerse, Belgium

SUMMARY

An improved dog model to study the effect of drugs on myocardial metabolism during ischemia is described. A reproducible degree of ischemia could be obtained by partial occlusion of the anterior descending branch of the left coronary artery (LAD), using an inflatable cuff with a micrometer. The possibility of inducing the stenosis twice in the same animal has the advantage that the animal can be used as its own control. The reproducibility of the degree of ischemia was demonstrated by the nonsignificant differences in local venous lactate, inorganic phosphate, and glucose concentrations after the first and second stenosis.

The mean pressure difference over the stenosis was used to express the degree of coronary artery narrowing. In this model, one does not have to rely on the collateral circulation in collecting local venous blood. Moreover, it is very likely that this blood is obtained from the most pronounced ischemic area, which was localized with radioactive microspheres. At this degree of stenosis, left ventricular function was not affected too much, as was demonstrated by the slight changes in dP/dt max, and systolic and diastolic aortic pressure after induction of the stenosis.

The usefulness of our model to evaluate the activity of drugs is demonstrated by the effect of fentanyl, a potent morphine-like analgesic, on the poststenotic local venous lactate and inorganic phosphate concentrations.

INTRODUCTION

Much valuable information on metabolic changes in experimental myocardial ischemia has been obtained from a model described by Owen *et al.* (1970). In this model, a side branch of the anterior descending branch of the left coronary artery was ligated and the venous blood, taken from one of the concomitant veins of the ligated side branch, was analyzed for metabolic changes. A disadvantage of this model is that metabolic products from the ischemic area can only

[1] Supported by a grant from I.W.O.N.L.

be obtained through blood coming from collateral areas, so that the local venous blood is mixed with blood from less- or even nonischemic tissue. This means that the concentration of metabolic products in the local venous blood will be low when the blood is obtained from small ischemic areas with marked collateral circulation and high when the blood is obtained from large ischemic areas with slight collateral circulation. Moreover, the ischemic area belonging to the ligated artery does not necessarily have to drain through the concomitant veins.

The varying concentrations of metabolic products in the local venous blood, due to anatomical variations, make the model of Owen *et al.* less suitable to evaluate the effect of drugs on myocardial metabolism during ischemia, when comparison from animal to animal is required. Yet the spontaneous decrease in lactate (Owen *et al.*, 1970; personal observations) and inorganic phosphate (Opie *et al.*, 1972), 10–15 min after ligation of the coronary artery branch, makes it difficult to study the effect of drugs on the metabolism of ischemic myocardium in the same animal.

Therefore, it was desirable to design an animal model for myocardial ischemia, in which the effect of drugs could be tested. In the technique described here, a reproducible degree of ischemia could be obtained by partial occlusion of the anterior descending branch of the left coronary artery (LAD). The possibility of applying the stenosis twice in the same animal allowed us to use the animal as its own control.

MATERIALS AND METHODS

The experiments were performed on mongrel dogs of either sex and unknown age, ranging in weight from 22 to 30 kg. The animals were premedicated with Hypnorm (1 ml/kg body weight) as described by Marsboom *et al.* (1964). Anesthesia was induced with sodium pentobarbitol (10 mg/kg body weight) and, after endotracheal intubation, was maintained with oxygen-nitrous oxide. Ventilation was kept constant during the experiments with a positive-pressure respirator (Bird).

The ECG was derived from the limb leads. The chest was opened through the left fifth intercostal space and the pericardium was incised over the antero-lateral aspect of the heart. Distal to the diagonal branch, an electromagnetic flow probe (Schaper, Rens, and Jageneau, 1965), and an inflatable cuff (Jageneau and Rens, 1969), which was also used for zero flow determination, were placed on the LAD from proximally to peripherally, respectively. The electromagnetic flow probe was connected to a 600-Hz sine wave electromagnetic flowmeter (Transflow). Ascending aortic pressure was measured through the femoral artery with a polyethylene catheter connnected to a pressure transducer (Telco). The pressure in the LAD distal to the site of stenosis was measured through a small anterolateral branch (Meer and Reneman, 1972) with a polyethylene catheter (P.E. 50 Clay Adams) connected to a pressure transducer (Telco). The length and diameter of the catheters for aortic and coronary artery pressure measurements were so selected that no difference in delay could be detected between both measuring systems. The difference between mean aortic and mean coronary pressure was determined as described previously (Spencer and Edmunds, 1968). Left ventricular pressure was measured through the left brachial artery with a catheter-tip manometer (Millar) and its maximal first derivative (dP/dt max) was determined with an analogue differentiator of our own design (Schaper, Lewi, and Jageneau, 1965). The frequency response of the differentiator was 100 Hz-3 dB. Recordings of the hemodynamic variables were made continuously on a multichannel Schwarzer recorder.

The cuff (Fig. 1, *A* and *B*) on the LAD was through Silastic tubing connected to a micrometer (Fig. 1*C*). The whole system was filled with distilled water so that the cuff could be inflated carefully until the desired degree of stenosis, i.e., a mean pressure difference between the ascending aorta and the LAD of 40–70 mm Hg, was reached. This range of pressure differences was selected, since pilot experiments had shown that lactate production consistently occurred at this degree of coronary artery stenosis. The cuff system can also be inflated with a servomotor pump, for instance, controlled by the pressure difference over the stenosis, using an autoregulating feedback system (Fig. 1*D*).

Arterial blood samples were obtained through the catheter used for aortic blood pressure measurements. Local venous blood samples were obtained through a polyethylene catheter (P.E. 60; Clay Adams) inserted into the great cardiac vein with a Seldinger technique. The tip of the catheter was placed halfway between the site of stenosis and the apex, being approximately the site of maximal ischemia (see below). In some experiments, the local venous sampling catheter was inserted at this level into one of the branches of the great cardiac vein. All blood samples were immediately deproteinized with cold perchloric acid and analyzed, using automated techniques for lactate (Apstein, Puchner, and Brachfeld, 1970), phosphate (Van Belle, 1970), and glucose (Weilinger).

After a recorded control period, the cuff was inflated until the desired degree of stenosis was reached and was kept inflated at this level for a period of 15 min. Then the cuff was deflated during 30 min. After this stabilization period, a second stenosis similar to the first one was applied. In both procedures, arterial and venous blood samples were taken 10 min prior to, just before, and usually 5, 10, and 15 min after the stenosis was induced.

Ten experiments were performed without constant heart rate. In 10 other experiments, bipolar stimulation electrodes were sutured to the right ventricle. In these experiments, heart rate was kept constant at 140 beats/min with a

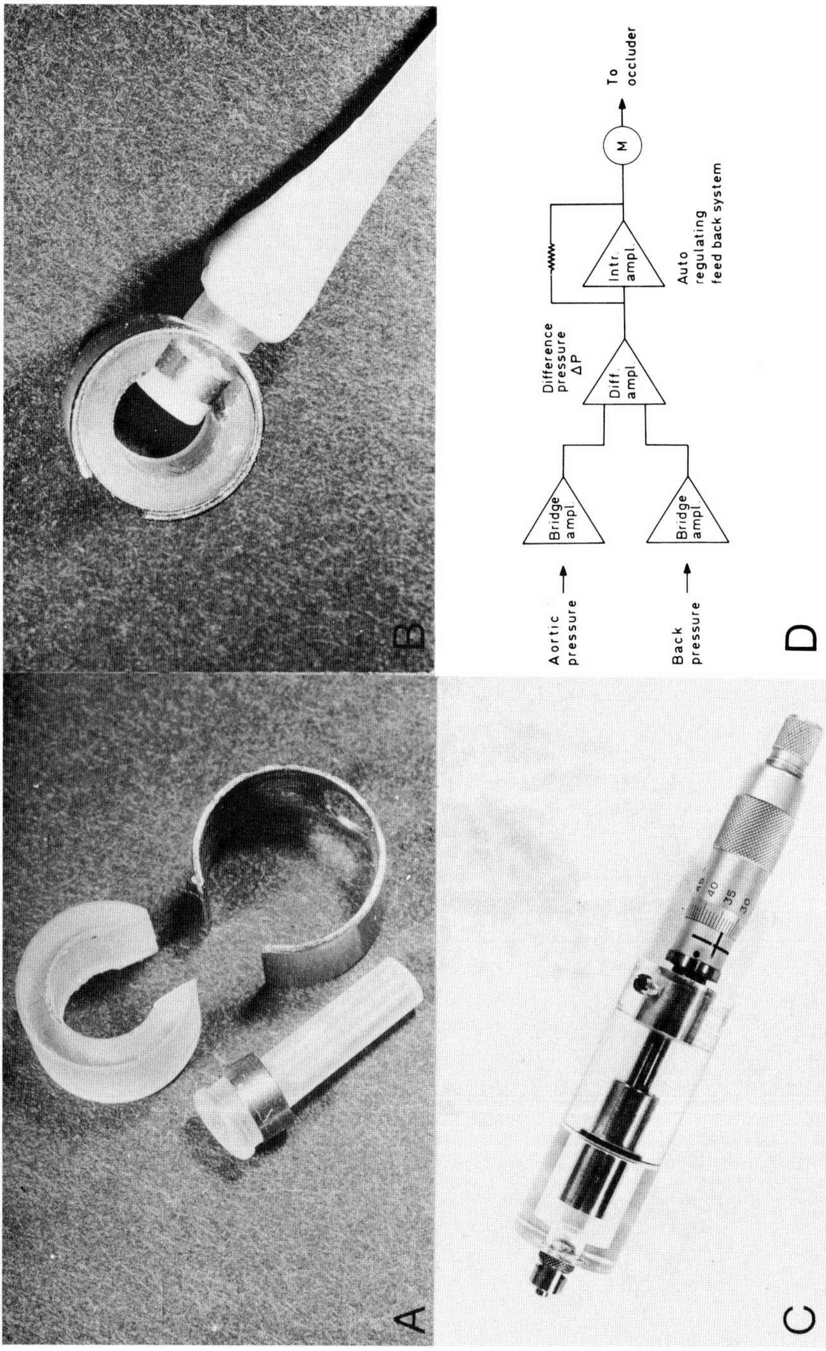
A
B
C
D
Aortic pressure
Back pressure
Bridge ampl.
Bridge ampl.
Difference pressure ΔP
Diff. ampl.
Intr. ampl.
M
To occluder
Auto regulating feed back system

stimulator of our own design. In 5 other experiments, fentanyl, a potent morphine-like analgesic (Janssen *et al.*, 1963), was injected intravenously, 5–10 min before the second stenosis.

To localize the site of maximal ischemia, radioactive microspheres were injected into the left atrium before and after the first stenosis in 6 additional experiments. The microspheres were 15 μ in diameter and suspended in 10% dextran and 0.5% Tween 80. They were labeled with either ^{125}I, ^{141}Ce, ^{85}Sr, or ^{95}Nb. With each injection, 1.8–2.3 million microspheres were administered to guarantee an adequate number of spheres in the tissue samples. The order of injection of the various isotopes was randomized as much as possible. Reference samples (20 ml) were obtained through the right brachial artery with a Harvard suction pump over a period of 1 min. After the animals had been killed with an overdose of the anesthetic, the hearts were removed, rinsed, and fixed in 4% formaldehyde during 1–3 days. After this period, endo- and epicardial samples were taken along the LAD: 1 cm proximal (1) and distal (2) to the stenosis site, halfway between the site of stenosis and apex (3), and at the distal 1/4 of this distance (5); along the left marginal branch: at the level of 3 (4) and 5 (7); halfway between 5 and 7 (6). The endocardial and epicardial samples both enclosed 50% of the wall thickness. The tissue and blood samples were placed in vials and counted by a modification (Schaper *et al.*,, 1973) of the method of Rudolph and Heymann (1967).

RESULTS

Only slight changes in heart rate, left ventricular dP/dt max, and systolic and diastolic aortic pressure could be detected after induction of both the first and the second stenosis in the experiments without controlled heart rate. The maximal increase in mean heart rate after the first stenosis was 15 beats/min (11%) and, after the second, 4 beats/min (3%). The maximal decrease in mean dP/dt max, and mean systolic and mean diastolic aortic pressure after the first and the second stenosis was, respectively, 320 (11%) and 100 mm Hg/sec (4%), 5 (4.8%) and 2 mm Hg (2%), and 4.5 (6%) and 1 mm Hg (1.5%). In these experiments, the average mean pressure difference between the ascending aorta and the coronary artery varied between 52.5 and 54.5 mm Hg during the first stenosis and between 49 and 53 mm Hg during the second stenosis. Under these circumstances, mean coronary blood flow varied, respectively, between 33 and 36% and between 32 and 37% of the control value. At these degrees of coronary artery stenosis, reproducible increases in local venous lactate and inorganic

---

Figure 1 (facing page). Inflatable cuff (*A* and *B*), syringe micrometer (*C*), and the electronic circuit of the autoregulating feedback system for automatic control of the degree of inflation of the cuff (*D*), M = servomotor.

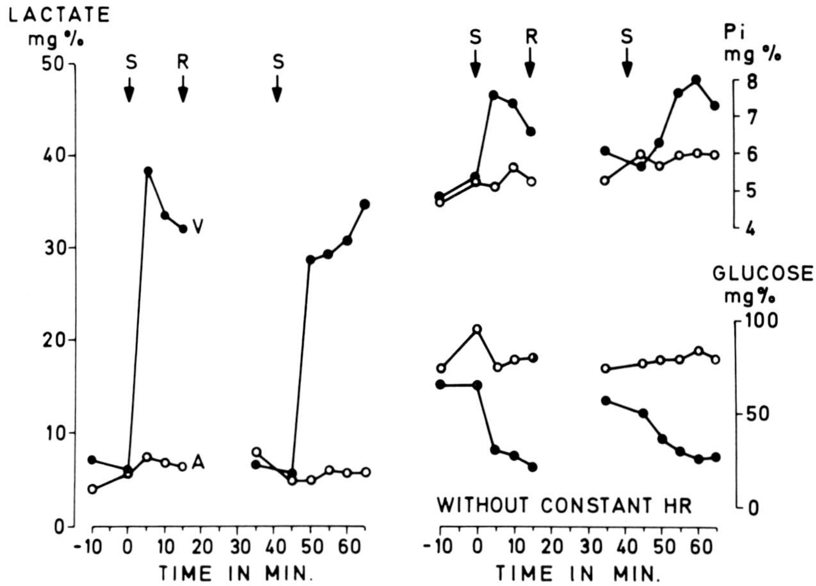

Figure 2. Arterial (A) and local venous (V) lactate, inorganic phosphate (Pi), and glucose concentrations, before and after the first and second coronary artery stenosis (S). R = release of stenosis. HR = heart rate.

phosphate values and reproducible decreases in local venous glucose value were found (Fig. 2).

When heart rate was kept constant, the changes in dP/dt max and systolic and diastolic aortic pressure were also minimal. The maximal decrease in mean dP/dt max, and mean systolic and mean diastolic aortic pressure after the first and second stenosis was, respectively, 270 (15%) and 160 mm Hg/sec (10%), 7 (7.1%) and 8.5 mm Hg (9%), and 4 (5.4%) and 7.5 mm Hg (10.5%). At constant heart rate, the average mean pressure difference over the sentosis varied between 51.5 and 53 mm Hg during the first stenosis and between 46 and 47 mm Hg during the second. Under these circumstances, mean coronary blood flow varied between 41 and 48% and between 41 and 43% of the control value, respectively. Also at constant heart rate, the increases in local venous lactate and inorganic phosphate values and the decreases in local venous glucose value were reproducible at these degrees of conronary artery narrowing (Fig. 3).

In the experiments in which fentanyl was given, the average mean pressure difference between the ascending aorta and the coronary artery varied between 58 and 59.5 mm Hg during the first stenosis and between 51 and 54 mm Hg during the second. Heart rate and dP/dt max fell after the administration of fentanyl, so that during the second stenosis the mean heart rate and mean dP/dt max were, respectively, 26–40 beats/min and 480–600 mm Hg/sec lower than during the first stenosis. No changes could be detected in systolic and diastolic aortic pressure after the first and second stenosis. dP/dt max slightly decreased after the first stenosis (maximal decrease in mean dP/dt max: 120 mm Hg/sec = 4.9%), but not after the second stenosis. After fentanyl, 25 μg/kg i.v., a comparable degree of coronary artery stenosis caused a significantly smaller increase in local venous lactate value, (and in some experiments, in inorganic phosphate value, also), than before drug administration, while the decrease in local venous glucose concentration was not much affected by the drug (Fig. 4).

The radioactivity per gram of tissue, expressed as percentage of the activity per ml of the reference sample, before and after stenosis is shown in Table 1. The most pronounced fall in radioactivity was seen in areas 3, 5, and 6, i.e., along or close by the LAD, halfway between and at the distal 1/4 of the distance between the site of stenosis and the apex. In areas 5 and 6, the decrease in radioactivity was more pronounced in the endocardial than in the epicardial samples. The activity in the collateral areas 1, 4, and 7 did not change significantly.

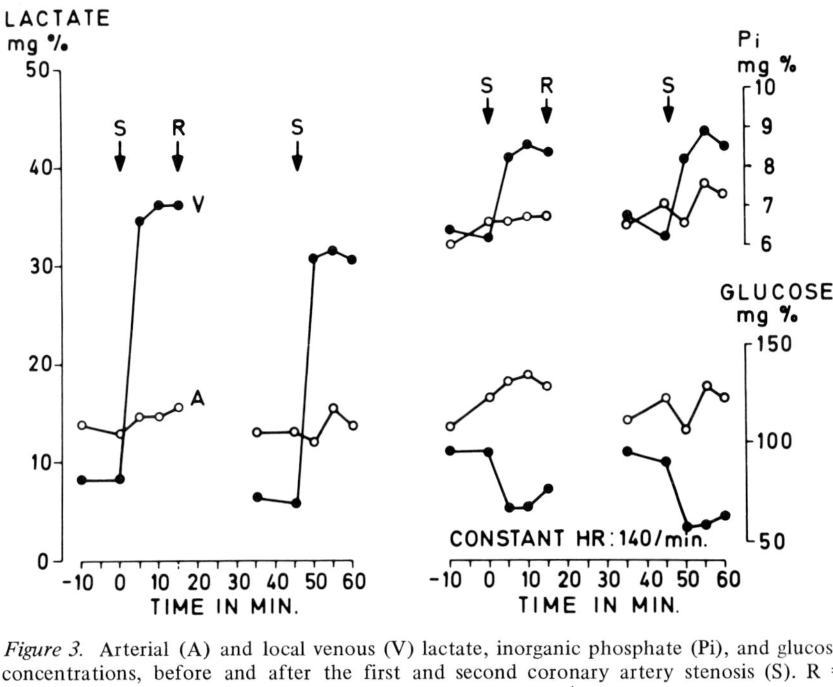

Figure 3. Arterial (A) and local venous (V) lactate, inorganic phosphate (Pi), and glucose concentrations, before and after the first and second coronary artery stenosis (S). R = release of stenosis. Heart rate (HR) kept constant at 140 beats/min.

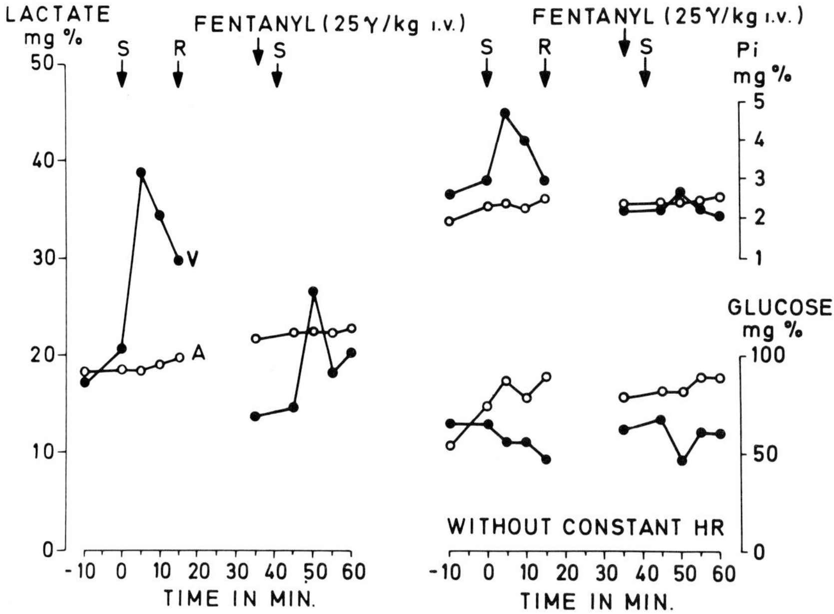

Figure 4. Arterial (A) and local venous (V) lactate, inorganic phosphate (Pi), and glucose concentrations, before and after the first and second coronary artery stenosis (S). R = release of stenosis. HR = heart rate. In this experiment, fentanyl was injected before the second stenosis was induced.

DISCUSSION

In the present paper, we have described an improved animal model for studying the effect of drugs on myocardial metabolism during ischemia. A reproducible degree of ischemia could be obtained by partial occlusion of the LAD, using an inflatable cuff and a micrometer. The possibility of applying the stenosis twice in the same animal allows us to use the animal as its own control. The reproducibility of the degree of ischemia was demonstrated by the nonsignificant differences in local venous lactate, inorganic phosphate, and glucose concentrations after the first and second stenosis. The fact that heart rate can be held constant in this model makes it possible to differentiate between chronotropic and other effects on ischemic myocardial metabolism.

Stenosis of a major coronary artery was preferred as against total occlusion of a small side-branch of the LAD (Owen *et al.*, 1970; Opie *et al.*, 1972) to obtain a large, but still perfused, ischemic area, so that one does not have to rely on collateral circulation in collecting local venous blood. Moreover, partial occlusion of a coronary artery does not affect collateral circulation, as demonstrated

Table 1. Radioactivity/g tissue, expressed as percentage of activity/ml of reference sample, both before and after coronary artery stenosis

Sample*		Before Stenosis		After Stenosis	
		Median	Range	Median	Range
1.	Endo	71.5	64.1–187.7	73.35	47.0–151.8
	Epi	49.95	41.7–231.9	51.3	40.6–188.5
2.	Endo	71.5	49.5–174.3	53.3	37.5–157.4
	Epi	48.9	36.2–161.2	43.95	29.3–160.5
3.	Endo	68.75	45.7–135.4	29.15	5.1– 60.0
	Epi	47.9	26.5–105.2	23.95	10.2– 54.4
4.	Endo	84.75	59.6–183.9	100.05	74.2–151.6
	Epi	91.1	68.8–230.7	84.55	76.5–214.2
5.	Endo	60.4	45.5–164.2	10.75	6.8– 55.3
	Epi	52.8	35.2–127.3	33.6	14.2– 77.1
6.	Endo	60.4	15.3– 82.1	10.35	1.7– 60.3
	Epi	64.05	31.5– 96.8	27.6	12.8– 69.8
7.	Endo	95.1	76.9–150.8	95.6	57.8–153.8
	Epi	89.7	67.3–186.6	105.7	78.0–167.1

*See method section for localization of these samples in myocardium.

by the insignificant changes in radioactivity in the collateral areas after stenosis of the LAD, whereas total occlusion of the LAD results in improved circulation in the same collateral areas (Reneman *et al.,* 1974). However, a large ischemic area makes it more likely that the local venous blood obtained is draining the most pronounced ischemic area, an assumption which was supported by the results of the radioactive microsphere experiments. These experiments, after all, showed that we were sampling blood in the most pronounced ischemic area. That this technique does not affect left ventricular function too much is demonstrated by the slight changes in heart rate, dP/dt max, and systolic and diastolic aortic pressure after induction of the stenosis.

The mean pressure difference over the stenosis rather than the decrease in mean coronary blood flow was used to express the degree of coronary artery narrowing since, at these low flows, the determination of mean coronary blood flow is rather inaccurate (Meer and Reneman, 1972). Although the pressure difference over the stenosis can be measured easily and accurately, the necessary occlusion of a small side branch of the LAD causes a small infarct, which is a drawback of our model. Because of this infarcted area, the local venous lactate concentration in the control samples will probably be higher than it should be, as is indicated by the fact that, in some experiments, this concentration even slightly exceeded the arterial lactate concentration.

In our first approach, we confined ourselves to the determination of lactate, inorganic phosphate, and glucose, although any other biochemical parameter can be determined in this model. Lactate and inorganic phosphate concentrations were determined, since it was shown by Owen *et al.* (1970), and Opie *et al.* (1972), respectively, that these concentrations increased significantly in the local venous blood during myocardial ischemia. Glucose concentrations were measured because of the importance of glucose as a substrate, especially for the ischemic myocardium.

The stenosis was not usually maintained longer than 15 min, since the lactate and inorganic phosphate concentrations in the local venous blood normally decrease spontaneously after 10–15 min of ischemia (Owen *et al.*, 1970; Opie *et al.*, 1972; personal observations). The stabilization period of 30 min between two succeeding stenoses seems to be sufficient, since the biochemical control values before the second stenosis do not differ very much from those before the first stenosis.

The effect of fentanyl, a potent morphine-like analgesic, on local venous lactate and inorganic phosphate concentrations demonstrates that our model can be used to evaluate the action of drugs. The finding that after the administration of fentanyl a comparable degree of coronary artery stenosis causes a significantly smaller increase in local venous lactate value and, in some experiments, in inorganic phosphate value, also, while the decrease in local venous glucose concentration is not very much affected by the drug, might indicate that fentanyl decreases the oxygen demand of the left ventricle, for instance, by a decrease in heart rate. However, further experiments are necessary to assess these findings at their true value.

Acknowledgments The authors are greatly indebted to Mr. J. Lambregts, Mr. P. Beirnaert, and Mrs. Martha Raeves-Van Reusel for their technical assistance, and to Mr. L. Pardoel for his help in analyzing the blood samples.

REFERENCES

APSTEIN, C. S., E. PUCHNER, and N. BRACHFELD. 1970. Improved automated lactate determination. Analyt. Biochem. 38: 20–34.

JAGENEAU, A. H. M., and W. RENS. 1969. A simple pneumatic cuff for occlusion of small arteries in dogs, pigs and sheep. Pflügers Arch. Ges. Physiol. 310: 182–184.

JANSSEN, P. A. J., C. J. E. NIEMEGEERS, and J. G. H. DONY. 1963. The inhibitory effect of fentanyl and other morphine-like analgesics on the warm water induced tail withdrawal reflex in rats. Arzneimittel-Forschung 13: 502–507.

MARSBOOM, R. A., D. VERSTRAETE, D. THIENPONT, and D. MATTHEEUWS. 1964. The use of haloanisone and fentanyl for neuroleptanalgesia in dogs. Brit. Vet. J. 120: 466–468.

MEER, v.d., J. J., and R. S. RENEMAN. 1972. An improved technique to induce a standardized functional stenosis of a coronary artery. Europ. Surg. Res. 4: 407–418.

OPIE, L. H., M. THOMAS, P. OWEN, and G. SHULMAN. 1972. Increased coronary venous inorganic phosphate concentrations during experimental myocardial ischemia. Am. J. Cardiol. 30: 503–513.

OWEN, P., M. THOMAS, V. YOUNG, and L. OPIE. 1970. Comparison between metabolic changes in local venous and coronary sinus blood after acute experimental coronary arterial occlusion. Am. J. Cardiol. 25: 562–570.

RENEMAN, R. S., A. H. M. JAGENEAU, W. VAN GERVEN, and L. STYNEN. 1974. The effect of counterpulsation (CP) on regional myocardial blood flow (MBF) after occlusion of the l. ant. desc. cor. art. (LAD). Europ. J. Clin. Invest. (In press).

RUDOLPH, A. M., and M. A. HEYMANN. 1967. The circulation of the fetus in utero. Methods for studying distribution of blood flow, cardiac output and organ blood flow. Circulat. Res. 21: 163–184.

SCHAPER, W., P. LEWI, W. GLAMENG, and L. GIJPEN. 1973. Myocardial steal produced by coronary vasodilatation in chronic coronary artery occlusion. Basic Res. Cardiol. 68: 3–20.

SCHAPER, W. K. A., P. LEWI, and A. H. M. JAGENEAU. 1965. The determinants of the rate of change of the left ventricular pressure (dP/dt). Arch. Kreislaufforschung 46: 27–41.

SCHAPER, W. K. A., W. RENS, und A. H. M. JAGENEAU. 1965. Konstruktion und EigenschaftenneuerMessköpfe zur electromagnetischen Regisstrierung der Blutströmung. Pflügers Arch. 283: 242–258.

SPENCER, M. P., and L. H. EDMUNDS. 1968. Evaluation of operative left ventricular outflow tract lesions with a fluid impedance plot. Circulation 37: 912–921.

VAN BELLE, H. 1970. New and sensitive reaction for automatic determination of inorganic phosphate and its application to serum. Analyt. Biochem. 33: 132–142.

WEILINGER, H. G.O.D. Period Methods. Boehringe GmbH, Mannheim.

Mailing address:
A. Jageneau, Cardiovascular Department,
Janssen Research Foundation,
Koninklijke Laan, 17, 2340 Beerse (Belgium)

Recent Advances in Studies on
Cardiac Structure and Metabolism Volume 10
The Metabolism of Contraction
Edited by Paul-Emile Roy and George Rona

FACTORS INFLUENCING TOLERANCE OF CARDIAC MUSCLE TO HYPOXIA[1]

OSCAR H. L. BING, CARL S. APSTEIN,[2] and WESLEY W. BROOKS

The Thorndike Laboratory, Harvard Medical Unit, Boston City Hospital, the Department of Medicine, Harvard Medical School, and the Boston University Medical Service, Boston City Hospital and the Boston University School of Medicine, Boston, Massachusetts, USA

SUMMARY

The effects of isoproterenol, glucose, and pH on the responses of isolated rat cardiac muscle to hypoxia (95% N_2, 5% CO_2) were examined while the muscles were contracting isometrically 12 times a minute at 28°C.

In the presence of 5.5 mM glucose, 10^{-5} M isoproterenol and alkaline pH (7.8) improved the performance of cardiac muscle during early hypoxia. This was followed by a premature decline in developed tension, and contracture appeared. Recovery of function following reoxygenation with 95% O_2 and 5% CO_2 after a 60-min period of hypoxia was poor. Acid pH (6.8) resulted in an early decline of mechanical activity during hypoxia; but contracture did not appear, and full recovery of developed tension was seen upon reoxygenation after 60 min of hypoxia.

When 22 mM glucose was used as substrate, the early responses to hypoxia were not altered; but late performance was improved, contracture did not appear, and full recovery after 60 min of hypoxia was seen. If additional glucose was added to the bath after 30 min of hypoxia (concentration 22 mM), little effect on developed tension was evident; but contracture diminished and recovery after 60 min of hypoxia was improved. Addition of 22 mM glucose allowed isoproterenol to exert its inotropic effect in the absence of late deleterious changes.

The data support the concept that factors tending to increase the utilization of limited stores of anaerobic substrate during hypoxia facilitate deterioration. By increasing exogenous glucose, the support of inotropic activity without late adverse effects appears possible, and recovery is improved upon reoxygenation.

INTRODUCTION

Mammalian cardiac muscle function is exquisitely sensitive to oxygen. In its absence, mechanical performance declines abruptly and persisting function and subsequent integrity of myocardial tissue are compromised. A number of factors

---

[1] Supported, in part, by Massachusetts Heart Association Grant 940, Boston City Hospital General Research Support Grant 5S01 RR-5569, and Grant HL10539 from the National Heart and Lung Institute, National Institutes of Health.

[2] Present address: Cardiology Unit, Boston City Hospital, 818 Harrison Avenue, Boston, Massachusetts, USA.

influencing the viability of ischemic heart muscle have been examined (Maroko *et al.*, 1971). While important practical information has been obtained, some of the techniques used for these studies are relatively imprecise. Further, since myocardial ischemia is a complex entity, it is not clear which factors are responsible for observed changes. The present report summarizes recent experiments dealing with the effects of hypoxia on cardiac muscle under carefully controlled conditions. These data help to define a component of the ischemic process and hopefully provide a basis for further studies in the intact animal.

METHODS

Following decapitation, the hearts of male Charles River CD^R rats weighing 150–250 g were rapidly removed and placed in oxygenated Krebs-Henseleit solution (Krebs and Henseleit, 1932) at 30°C. Trabeculae carneae and papillary muscles were carefully dissected from the left ventricle, mounted between two spring clips, and placed in a chamber containing Krebs-Henseleit solution with 5.5 mM glucose. The solution was bubbled through with 95% O_2 and 5% CO_2 (pH 7.4), and the temperature was kept at 28°C by a Lauda model K-2 circulating pump. The lower spring clip attached to the muscle preparation was connected to a Statham G7B-0.75-350 force transducer by a hollow stainless steel section of 30-gauge needle tubing, which passed through a mercury seal at the bottom of the muscle chamber. The upper clip was connected by a thin gold chain to a lever arm, above which a micrometer stop was mounted for the adjustment of muscle length. The preparations were stimulated 12 times/min by a Grass model S88 stimulator which delivered 7-msec square wave pulses through parallel platinum electrodes at voltages which were approximately 10% greater than the minimum required to produce a maximal mechanical response.

After a 30-min period during which the preparations were permitted to shorten while carrying light loads, the muscles were carefully stretched to the apices of their length-tension curves. Preparations contracted isometrically at this length for a 15-min period. Hypoxia was induced by bubbling the solution through with 95% N_2 and 5% CO_2. During reoxygenation, the original 95% O_2, 5% CO_2 gas mixture was used. In all experiments, hypoxia was present for 60 min. During or prior to hypoxia, glucose, mannitol, or isoproterenol was added. Reoxygenation took place in the original bath solution containing 5.5 mM glucose at pH 7.4 without added agents such as isoproterenol or mannitol.

The *in vitro* length of each muscle preparation was measured at the apex of its length-tension curve, using a Gaertner cathetometer and telescope. At the end of each experiment, the muscle between the two clips was blotted and weighed, and the cross-sectional area of the muscle was calculated, assuming cylindrical uniformity and a specific gravity of 1.000. Tension was normalized for muscle cross-sectional area, and absolute values of tension were calculated. Data were also expressed as a percent of prehypoxia control values. Mechanical events were obtained at a speed of 0.25 mm/sec on a Sanborn model 7700 recorder. Irritable and poor preparations were discarded. Peak isometric tension and contracture tension were continuously recorded. The latter was defined as an increase in resting tension above control values. Isometric tension and contracture development were followed during a 60-min period of hypoxia and a subsequent 15- to 30-min period of reoxygenation.

RESULTS

Interventions prior to hypoxia

Effect of pH The pH of the bath solution was changed from 7.4 to 6.8 or 7.8 by the addition of 1 N HCl or 1 N NaOH. Associated with this change in pH, tension declined to 93.7 and 90% of control values in the case of acid and base addition, respectively. After 15 min of stable contraction at the new pH, hypoxia was initiated. After 15 min of hypoxia, developed tension values at pH 6.8, 7.4, and 7.8 were 13 ± 2, 26 ± 2, and 39 ± 3% of prehypoxia control values, respectively (Fig. 1). Developed tension at pH 7.4 was significantly different from that at pH 6.8 (p<0.001) and pH 7.8 (p<0.01). During late hypoxia, however, these differences were no longer present and, after 60 min, approximately 10% of prehypoxia tension was developed by all the pH groups.

Contracture during hypoxia was most marked at pH 7.8; at 60 minutes, contracture tension was 2.5±0.3 g or 44% of prehypoxia developed tension. A lesser degree of contracture was seen at pH 7.4, while no contracture developed at any time during hypoxia at pH 6.8 (Fig. 1). Despite the more rapid decline in mechanical activity seen at acid pH, recovery of developed tension after 15 min

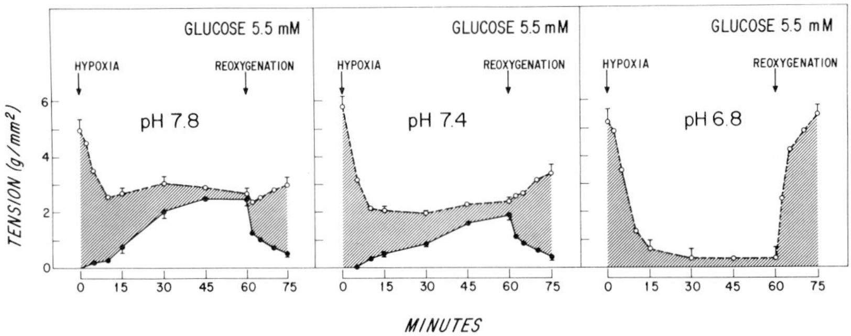

Figure 1. Effects of pH on mechanical performance of isolated rat left ventricular papillary muscle preparations during 60 min of hypoxia (95% N_2, 5% CO_2) and 15 min of reoxygenation (95% O_2, 5% CO_2). *Closed circles* indicate contracture tension and *open circles* total tension developed by muscle preparation. *Shaded area between lines connecting open and closed circles* indicate values of active (developed) tension. *Brackets* at intervals during hypoxia and reoxygenation indicate ± 1 SEM. At pH 7.8, enhanced tension development is seen during early hypoxia with subsequent late deterioration and poor recovery on reoxygenation. At pH 6.8, less active tension is developed during early hypoxia, but contracture does not appear and recovery of developed tension upon reoxygenation is prompt.

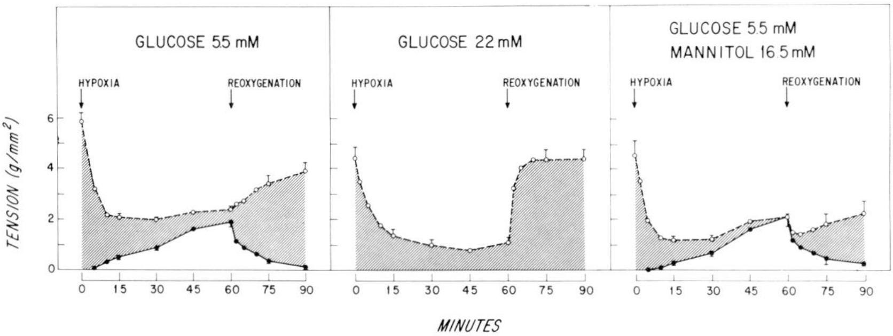

Figure 2. Effects of glucose and mannitol added prior to hypoxia on mechanical performance during subsequent hypoxia and reoxygenation. Format is same as in Figure 1. High glucose concentrations prevent contracture development during hypoxia and improve recovery upon reoxygenation. These effects are due to metabolic and not osmotic effects of glucose, since mannitol administration does not lead to similar changes (see text).

of reoxygenation was most complete at pH 6.8. Delayed recovery of both developed and contracture tension was seen at neutral or alkaline pH. Developed tension values at pH 6.8, 7.4, and 7.8 after 15 min of reoxygenation were 104±3, 56±8, and 50±2% of prehypoxia values, respectively.

Effect of Glucose No significant change in mechanical activity was observed when the glucose concentration of the bath was increased four-fold to a bath concentration of 22 mM. Following the initiation of hypoxia, a rapid decline in isometric tension took place and, after 15 min, 27±2 and 30±2% of prehypoxia tension was developed in the presence of 5.5 and 22 mM glucose, respectively. During late hypoxia, considerable differences are observed (Fig. 2, *left* and *center*). Preparations in the bath containing 5.5 mM glucose developed 8±2% of prehypoxia tension, while 1.9±0.2 g of contracture tension appeared. In those muscles bathed in 22 mM glucose, developed tension stabilized at 24±3% of prehypoxia values, and no contracture developed at any time during the 60-min period of hypoxia. Upon reoxygenation, developed tension increased promptly in those preparations bathed in 22 mM glucose, while the decline in contracture and recovery of isometric tension was delayed in preparations bathed in 5.5 mM glucose solution. After 30 min of reoxygenation, developed tension was 63±4 and 100±5% of prehypoxia values in preparations bathed in 5.5 and 22 mM glucose during hypoxia, respectively. Osmolality equivalent to 22 mM glucose was produced by adding 5.5 mM glucose and 16.5 mM mannitol to the bath. Tension development at 60 min had fallen to 1±1% of prehypoxia values, and 2.1±0.3 g of contracture tension had appeared at this time (Fig. 2, *right*). After 30 min of reoxygenation, isometric tension recovered to 41±7% of prehypoxia values. Improved performance during hypoxia was thus due to the metabolic and not osmotic effects of glucose.

Effect of Isoproterenol When isoproterenol (10^{-5} M) was added to the bath under oxygenated conditions, no significant change in isometric tension developed by rat ventricular muscle was observed. A clearly positive inotropic effect, however, was manifested by a significant increase in the rate of tension development (57±5 to 68±9 g/mm^2/sec) and an abbreviation of the time to peak tension (165±8 to 127±3 msec). After 15 min of hypoxia, isometric tension had declined less rapidly in muscles exposed to isoproterenol (Fig. 3); 42±3 and 27±2% of prehypoxia tension was present in the isoproterenol- and non-isoproterenol-treated groups, respectively. By 60 min of hypoxia, however, preparations exposed to isoproterenol demonstrated severe contracture and no longer developed active tension. After 30 min of reoxygenation, contracture tension had decreased to prehypoxia control values and developed tension recovered to 29±4 and 63±4% of prehypoxia control values in isoproterenol- and non-isoproterenol-treated preparations, respectively.

Effect of High Glucose and Isoproterenol Combined With 22 mM glucose in the oxygenated bath solution, the inotropic effect of isoproterenol (10^{-5} M) did not differ from that seen in 5.5 mM glucose. During early hypoxia, improved performance similar to that observed in the presence of 5.5 mM glucose was again seen. After this time, however, considerable differences in the response of the isoproterenol-treated muscle preparations were observed (Fig. 4). In the bath containing 22 mM glucose, developed tension did not decline to zero; but at 60 min, 37±5% of prehypoxia developed tension was present. In contrast to the severe contracture seen when the bath solution contained 5.5 mM glucose, no contracture was seen with isoproterenol during hypoxia in the presence of 22

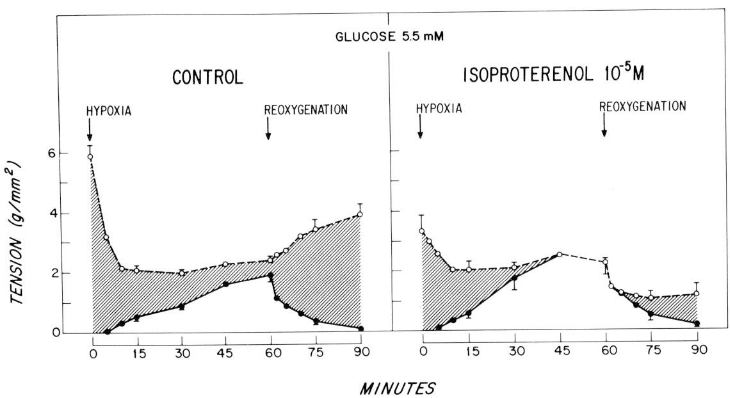

Figure 3. Isoproterenol added to bath solution prior to hypoxia improves performance of muscle preparation during early hypoxia. Late deleterious effects are manifested by enhanced contracture, loss of developed tension, and poor recovery upon reoxygenation. Format is same as in Figure 1.

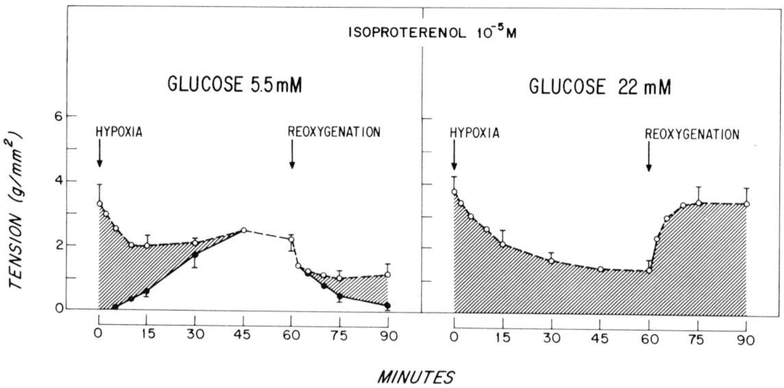

Figure 4. When isoproterenol and high glucose concentrations are combined prior to hypoxia (*right*), improved performance is seen during hypoxia in absence of late adverse effects. Format is same as in Figure 1.

mM glucose. Upon reoxygenation, recovery of developed tension was prompt when 22 mM glucose was present during hypoxia, while it was delayed in the presence of 5.5 mM glucose. After 30 min of reoxygenation, developed tension was 29±4 and 92± 4 of prehypoxia control values in preparations exposed to 5.5 and 22 mM glucose in addition to isoproterenol during hypoxia.

Interventions during hypoxia

Effect of Glucose After 30 min of hypoxia, the concentration of glucose in the bath was increased from 5.5 to 22 mM. While little change in developed

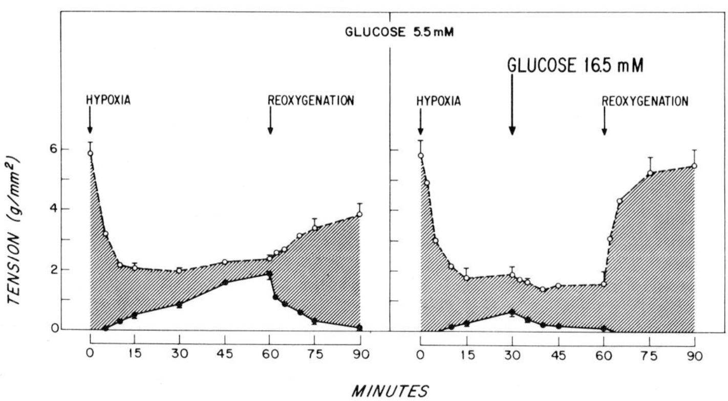

Figure 5. Effects of adding high glucose concentrations *during* hypoxia (*right*). Although there is little improvement of developed tension, contracture diminishes during hypoxia and recovery of developed tension is improved upon reoxygenation. Format is same as in Figure 1.

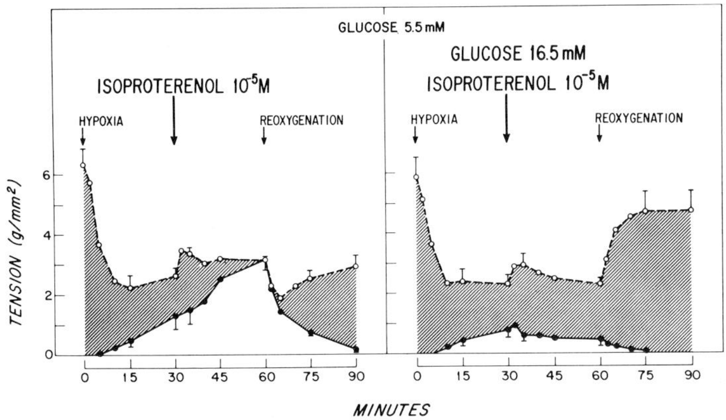

Figure 6. When isoproterenol is added to muscle bath *during* hypoxia (*left*), transient improvement in developed tension is seen. This is followed, however, by more severe contracture and poor recovery upon reoxygenation. When isoproterenol at 30 min of hypoxia is combined with addition of a high glucose concentration (*right*), improved performance is seen without an increase in contracture or impaired recovery of developed tension upon reoxygenation. Format is same as in Figure 1.

tension was observed during the subsequent 30 min, isometric tension did not continue to fall gradually as in the presence of 5.5 mM glucose (Fig. 5). More striking was the change in contracture tension, which decreased from 0.7±0.2 to 0.2±0.1 g/mm^2 in the presence of 22 mM glucose. In 5.5 mM glucose, contracture tension increased from 0.9±0.1 g/mm^2 at 30 min of hypoxia to 1.9±0.2 g/mm^2 at 60 min. After 30 min of reoxygenation, developed tension increased to 95±7% of prehypoxia values when increased glucose concentrations were added to the bath at 30 min of hypoxia. Developed tension increased to 63±4% of prehypoxia values after 30 min of reoxygenation in preparations where 5.5 mM glucose was present in the bath for the 60-min period of hypoxia.

Effect of Isoproterenol Isoproterenol calculated to yield a bath concentration of 10^{-5} M was added after 30 min of hypoxia; 5.5 mM glucose was present in the bath. A transient increase in developed tension was observed (Fig. 6, *left*), followed by a decline, so that isometric tension had fallen to zero by 60 min of hypoxia. While developed tension declined, contracture tension increased and was 3.2±0.4 g/mm^2 by 60 min of hypoxia. In non-isoproterenol-treated preparations, developed tension at 60 min of hypoxia was 8 ± 1% of prehypoxia values, while contracture tension had increased to 1.9±0.2 g/mm^2. After 30 min of reoxygenation, contracture had disappeared in muscles treated with isoproterenol during hypoxia. Developed tension recovered to 43±5% of prehypoxia values as compared to 63±4% in preparations not exposed to isoproterenol.

Effect of Glucose and Isoproterenol At 30 min of hypoxia, both isoproterenol (10^{-5} M) and glucose (bath concentration 22 mM) were added to the muscle chamber. As in the case of isoproterenol alone, a transient increase in developed tension took place (Fig. 6, *right*). This was not followed, however, by a decline in developed tension and, after 60 min of hypoxia, isometric tension was 31±3% of prehypoxia values. In contrast to the effects of isoproterenol alone, there was a slight decrease in contracture tension from 0.8±0.3 to 0.5±0.2 g/mm^2 at 30 and 60 min of hypoxia, respectively. Upon reoxygenation, contracture tension fell to zero by 15 min in preparations treated with isoproterenol and 22 mM glucose during hypoxia, and developed tension increased to 80±6% of prehypoxia values by 30 min.

DISCUSSION

In this study, the effects of hypoxia on the function of isolated cardiac muscle have been examined in three ways. First, the course of isometric tension development during hypoxia has been evaluated; second, the development of contracture has been followed; and third, the recovery of developed tension during reoxygenation after a 60-min period of hypoxia has been examined.

A fall in isometric tension during hypoxia is not necessarily associated with deleterious alterations of the other parameters. In fact, depressed tension development during hypoxia frequently appears to prevent or delay contracture and to enhance recovery of function upon reoxygenation. On the other hand, the stimulation of mechanical activity during the course of hypoxia may be associated with more severe contracture development and poor recovery upon reoxygenation. The decline in developed tension associated with severe contracture appears to reflect the exhaustion of energy stores. Isometric tension depression of this type is seen during hypoxia following stimulation by isoproterenol and the presence of alkalosis during hypoxia. In both instances, this type of loss of function was preceded by enhanced isometric tension development during the period of hypoxia (Bing, Brooks, and Messer, 1972a, 1973). These observations are consistent with the concept that the relatively limited stores of anaerobic substrate are more rapidly depleted by interventions which enhance the activity of hypoxic heart muscle. In support of this idea is the observation that contracture development during hypoxia and poor recovery of developed tension upon reoxygenation following isoproterenol addition can be prevented by simultaneously increasing the glucose concentration of the bath (Fig. 6).

Depression of mechanical activity during hypoxia which is not a result of exhaustion, on the other hand, appears to have a protective effect upon cardiac muscle. Contracture development is delayed and recovery of isometric tension is improved upon reoxygenation. This was demonstrated by the additional depressant effect of an acid pH on mechanical function during hypoxia (Bing *et al.*, 1973). Similar protection has been observed following a decrease (Ekblom and

Table 1. Developed tension*

	Bath Concentration			Prehypoxia Control (%)				
				Hypoxia				
N	Glucose (mM)	Isoproterenol (M)	Control (g/mm^2)	15 min	30 min	35 min	60 min	Reoxygenation, 90 min
				Additions prior to hypoxia				
11	5.5	—	5.9±0.6	27±2	19±2		8±1	63±4
5	22	—	4.3±0.6	31±2	22±3		24±3	100±5
6	5.5	10^{-5}	3.6±0.6	42±3	10±3		0	29±4
6	22	10^{-5}	4.2±0.5	57±5	43±5		37±5	92±4
				Additions at 30 minutes of hypoxia				
5	22	—	5.8±0.6	26±3	21±3	21±3	25±6	95±7
6	5.5	10^{-5}	6.3±1.1	28±2	21±2	30±4	0	43±5
6	22	10^{-5}	5.8±0.7	33±4	26±4	40±4	31±3	80±6

*Effects of glucose and isoproterenol on isometric tension during 60 min of hypoxia (95% N_2, 5% CO_2) and 30 min of reoxygenation (95% O_2, 5% CO_2). Additions to the bath are made either prior to hypoxia (top) or at 30 min of hypoxia (bottom). Values indicate means ± 1 SEM.

Table 2. Contracture tension (g/mm^2)*

	Bath Concentration		Hypoxia			Reoxygenation
N	Glucose (mM)	Isoproterenol (M)	15 min	30 min	60 min	90 min
			Additions prior to hypoxia			
11	5.5	–	0.5±0.1	0.9±0.1	1.9±0.2	0.1±0.1
5	22	–	0.0	0.0	0.0	0.0
6	5.5	10^{-5}	0.6±0.2	1.8±0.5	2.3±0.4	0.2±0.1
	22	10^{-5}	0.0	0.0	0.0	0.0
			Additions at 30 minutes of hypoxia			
5	22	–	0.3±0.1	0.7±0.2	0.2±0.1	0.0
6	5.5	10^{-5}	0.5±0.2	1.3±0.4	3.2±0.4	0.2±0.1
6	22	10^{-5}	0.5±0.2	0.8±0.3	0.5±0.2	0.0

*Effects of glucose and isoproterenol on contracture tension during 60 min of hypoxia and 30 min of reoxygenation. Format is otherwise the same as in Table 1.

Bing, 1974) or cessation (unpublished observations) of stimulation rate. This principal is applied clinically in open-heart surgery, where mechanical activity is arrested for protracted periods by various means.

The protective effect of acidosis during hypoxia seems at first to be a paradoxical concept, since the deleterious effects of an acid pH, particularly during hypoxia, have been emphasized (Downing, Talner, and Gardner, 1966; Gelet, Altschuld, and Weissler, 1969). While an acid pH may facilitate hydrolytic breakdown of tissue or have deleterious effects on cardiac rhythm, a protective effect on subsequent function, although possibly transient, is demonstrated. It is recognized that pH influences the activity of several glycolytic enzymes. An acid pH, by inhibiting glycolysis, may conserve carbohydrate stores. Another mechanism by which a low pH may preserve energy stores is by depression of contractile activity, a major energy-consuming action of the heart (Katz, 1973).

If these observations in isolated cardiac muscle can be extended to the intact animal during asphyxia or ischemia, a number of hypotheses may be ventured. Such a mechanism may provide transient protection to aquatic diving species, where a considerable decline in pH and pO_2 has been demonstrated during prolonged submersion (Robin *et al.*, 1964). If protection of the myocardium is provided by a fall in pH in association with acute occlusive coronary disease, it may be reasonable to re-evaluate the use of alkali in patients with myocardial infarction. In the patient with ischemic cardiac arrest during open-heart surgery, the acidosis associated with ischemia may similarly have a protective effect. In patients with the so-called "stone-heart" (Cooley, Reul, and Wukasch, 1972; Katz and Tada, 1972)—a state of apparent irreversible contracture occasionally seen in patients undergoing open-heart surgery–it might be suggested that an insufficiently acid pH may play a contributory pathophysiological role. This entity is seen in patients with severe cardiac hypertrophy and exhaustion of glycogen stores. In addition to inadequate reserves of anaerobic substrates, it may not be possible to generate a sufficiently acid pH to help prevent contracture. Similarly, it may be possible to prolong cardiac surgery without deleterious effects by a consideration of cardiac pH.

During hypoxia, cardiac muscle must depend on glycolytic activity to maintain activity and preserve viability. In comparison to such species as the turtle, the intrinsic performance of mammalian cardiac muscle during hypoxia is diminished (Bing *et al.,* 1972a). Tolerance of cardiac muscle to hypoxia in general appears to primarily reflect differences in glycolytic activity (Wu and Chang, 1947; Scheuer and Stezoski, 1970). If glycolysis is blocked by iodoacetate, the difference in tolerance of rat and turtle heart muscle to hypoxia is no longer present, and tension development declines precipitously in both species (Bing *et al.,* 1972a). Dependence of mechanical activity on glycolysis is not present under oxygenated conditions, as performance is not altered by iodoacetate when the muscle bath is well oxygenated.

Significant improvement in the tolerance of mammalian ventricular muscle to hypoxia can be provided by exogenous glucose (Winbury, 1956; Yang, 1963; Weissler *et al.,* 1968). This effect is due to the metabolic action of glucose and not the increased osmolality of the bath solution, since enhanced tolerance is not seen with non-metabolizable carbohydrates. Improved tension development due to exogenous glucose appears late during the period of hypoxia, and the increase in isometric tension is relatively small. Much more striking is the decrease or absence of contracture development during hypoxia and the improved performance upon reoxygenation. When the administration of isoproterenol and 22 mM glucose are combined, it also appears possible to significantly improve developed tension during hypoxia in the absence of late adverse effects (Apstein, Bing, and Levine, 1974). These salutory influences of exogenous glucose occur when glucose is added to the bath during as well as prior to hypoxia. While the late deleterious effects of hypoxia seem to be importantly influenced by the presence of anaerobic substrate, early changes in performance following hypoxia are less clearly understood.

This study serves to illustrate some of the multifactorial influences on the tolerance of cardiac muscle to hypoxia. In the present investigation of pH, glucose, and isoproterenol, the effects of muscle performance and substrate availability are interrelated. Other areas to be explored include the evaluation of methods which enhance the diffusion of substrates or preserve the integrity of intracellular organelles. In evaluating the effects of any intervention on the tolerance of cardiac muscle to hypoxia, this investigation emphasizes the importance of evaluating performance parameters in addition to those analogous to isometric tension development. Important information is obtained by observing contracture development during hypoxia and recovery of mechanical activity following reoxygenation.

REFERENCES

APSTEIN, C. S., O. H. L. BING, and H. J. LEVINE. 1974. Sustained inotropism in hypoxic cardiac muscle. Clin. Res. 22: 258A.

BING, O. H. L., W. W. BROOKS, A. N. INAMDAR, and J. V. MESSER. 1972a. Tolerance of isolated heart muscle to hypoxia: turtle vs. rat. Am. J. Physiol. 223: 1481–1485.

BING, O. H. L., W. W. BROOKS, and J. V. MESSER. 1972b. Effects of isoproterenol on heart muscle performance during myocardial hypoxia. J. Molec. Cell. Cardiol. 4: 319–328.

BING, O. H. L., W. W. BROOKS, and J. V. MESSER. 1973. Heart muscle viability following hypoxia: Protective effect of acidosis. Science 180: 1297–1298.

COOLEY, D. A., G. J. REUL, and D. C. WUKASCH. 1972. Ischemic contracture of the heart: "stone heart." Am. J. Cardiol. 29: 575–577.

DOWNING, S. E., N. S. TALNER, and T. H. GARDNER. 1966. Influence of hypoxemia and acidemia on left ventricular function. Am. J. Physiol. 210: 1327–1333.

EKBLOM, B., and O. H. L. BING. 1974. Tolerance of isolated cardiac muscle to hypoxia: Force-frequency interrelationships. (In preparation).

GELET, T. R., R. A. ALTSCHULD, and A. M. WEISSLER. 1969. Effect of acidosis on the performance and metabolism of the anoxic heart. Circulation 40 (Suppl. 4): 60–69.

KATZ, A. M. 1973. Effects of ischemia on the contractile processes of heart muscle. Am. J. Cardiol. 32: 456–460.

KATZ, A. M., and M. TADA. 1972. The "stone heart": A challenge to the biochemist. Am. J. Cardiol. 29: 578–580.

KREBS, H. A., und K. HENSELEIT. 1932. Untersuchungen über die Harnstoffbildung im Tierkörper Hoppe-Seyler's Z. Physiol. Chem. 210: 33–66.

MAROKO, P. R., J. K. KJEKSHUS, B. E. SOBEL, T. WATANABE, J. W. COVELL, J. R. ROSS, JR., and E. BRAUNWALD. 1971. Factors influencing infarct size following experimental coronary artery occlusions. Circulation 43: 67–82.

ROBIN, E. D., J. W. VESTER, H. V. MURDAUGH, Jr., and J. E. MILLER. 1964. Prolonged anaerobiasis in a vertebrate: Anaerobic metabolism in the freshwater turtle. J. Cell. Comp. Physiol. 63: 287–297.

SCHEUER, J., and S. W. STEZOSKI. 1970. Protective role of increased myocardial glycogen stores in cardiac anoxia in the rat. Circulat. Res. 27: 835–849.

WEISSLER, A. M., F. A. KRUGER, N. BABA, D. C. SCARPELLI, R. F. LEIGHTON, and J. K. GALLIMORE. 1968. Role of anaerobic metabolism in the preservation of functional capacity and structure of anoxic myocardium. J. Clin. Invest. 47: 403–416.

WINBURY, M. M. 1956. Influence of glucose on contractile activity of papillary muscle during and after anoxia. Am. J. Physiol. 187: 135–138.

WU, J. J., and I. CHANG. 1947. Glycolytic activity of the hearts of vertebrates. Quart. J. Exp. Physiol. 34: 91–95.

YANG, W. C. 1963. Anaerobic functional activity of isolated rabbit atria. Am. J. Physiol. 205: 781–784.

Mailing address:
Oscar H. L. Bing, M.D.,
Cardiovascular Unit, Beth Israel Hospital,
330 Brookline Avenue, Boston, Massachusetts 02215 (USA).

Recent Advances in Studies on
Cardiac Structure and Metabolism Volume 10
The Metabolism of Contraction
Edited by Paul-Emile Roy and George Rona

EFFECTS OF pH AND pCO_2 ON PERFORMANCE OF ISCHEMIC MYOCARDIUM[1]

MYRON L. WEISFELDT, RICHARD L. BISHOP, and H. LEON GREENE

Peter Belfer Laboratory for Myocardial Research, Cardiovascular Division and Department of Medicine, The Johns Hopkins University School of Medicine and Hospital, Baltimore, Maryland, USA

SUMMARY

Contractile performance of ischemic feline myocardium was evaluated under conditions of selective changes in perfusate pH and pCO_2. A substantial increase in myocardial performance was noted when the pCO_2 was lowered at constant pH, and depression of performance was noted when the pCO_2 was increased at constant pH. Perfusate acidosis at constant pCO_2 resulted in depression of performance only after 20 min of exposure. Alkalosis did not increase performance and decreased performance transiently during mild ischemia. These studies suggest that performance of myocardium during ischemia is closely related to tissue pCO_2 and is minimally related to the level of extracellular pH.

INTRODUCTION

Hypoxic cardiac muscle exposed to perfusate alkalosis (Case, 1971/1972) produced by increasing perfusate buffering capacity demonstrates improved performance, while acidosis (Downing, Talner, and Gardner, 1966) produced by decreasing buffering capacity results in depressed performance. These and other studies (Katz and Hecht, 1969; Katz, 1973; Safer and Davies, 1973) suggest that intracellular acidosis may lead to the depression of performance in either hypoxic or ischemic myocardium. However, intracellular acidosis during ischemia (Effros *et al.*, 1973) appears to result not only from increased lactate and thus hydrogen ion production (Case, 1971/1972; Opie *et al.*, 1973), but also from decreased removal of metabolic products. This is reflected in the higher tissue (Flaherty *et al.*, 1974) and coronary venous (Opie *et al.*, 1973; Marshall, Parratt, and Ledingham, 1974) pCO_2 during ischemia. The present studies compare the relative effects of selective changes in perfusate pH and pCO_2 on the performance of ischemic myocardium.

[1] Supported by Contract PH 43 NHLI 67-1444 under Myocardial Infarction Research Unit Program, National Heart and Lung Institute, National Institutes of Health, Department of Health, Education, and Welfare.

METHODS

An isolated, isovolumic, non-blood-perfused, feline left ventricular preparation was used for these studies. Cats were anesthetized with 40 mg/kg of pentobarbitol intraperitoneally. The heart was removed and suspended from a column which perfused the coronary arteries with Krebs-Ringers bicarbonate (Weisfeldt and Shock, 1970) to which 16 mM glucose and 1% dextran (molecular weight 70,000) were added. The perfusate was pumped through a 14-μ Millipore filter to a height equivalent to 81 mm Hg pressure and kept constant by an overflow system. Ischemia was induced by adjusting the height of the column to either 33 mm Hg (mild ischemia) or 22 mm Hg (moderate ischemia). The temperature was maintained at 35°C, and the perfusate was not recirculated. The fluid was equilibrated with 90% oxygen, 5% CO_2, 5% N_2, except as indicated. The hearts were bathed in 5–10 ml of effluent at 35°C.

A vent was placed in the left ventricle to drain Thebesian flow. A button was sewn into the mitral valve orifice after left atriotomy. A flange from the button occluded the aortic outflow tract. A 15-gauge needle, 1.5 cm in length, transversed the button and then entered a saline-filled condom attached to the button. The volume of fluid was adjusted to produce an end-diastolic pressure in the isovolumic left ventricle of 5–13 mm Hg. Left ventricular pressures were measured with a Statham P23Db pressure transducer. Left ventricular pressure, dp/dt, and the stimulus artifact were recorded continuously, using a Brush Mark 200 recorder (Gould, Inc., Cleveland, Ohio). Peak isovolumic left ventricular pressure and maximal dp/dt were utilized as indices of performance. The hearts were paced via clip electrodes attached to the right ventricular outflow tract with a square wave pulse of 3-msec duration and a voltage 20% above threshold at a rate of 102 beats/min. Coronary flow was measured as the volume of effluent from the bath under steady state conditions. Electrolyte and coronary venous pH and pCO_2 measurements were made on fluid withdrawn from the pulmonary artery after ligation of the venae cavae.

The pH of the perfusate was increased from 7.41 to 7.80 at constant pCO_2 by an equimolar increase in the sodium bicarbonate and decrease in sodium chloride content. Likewise, the pH was lowered to 7.10 by an equimolar increase in sodium chloride and decrease in sodium bicarbonate. The pCO_2 of the perfusate was lowered from 35 to 16 mm Hg or raised to 60 mm Hg at constant pH and pO_2 by changing the equilibrating gas from 90% O_2, 5% CO_2, 5% N_2 to 90% O_2, 2.5% CO_2, 7.5% N_2, or 90% O_2, 7.5% CO_2, 2.5% N_2 and by equimolar decreases or increases in sodium bicarbonate and chloride content as appropriate. The final value for the pH and pCO_2 of all solutions was checked by direct measurement. Thus, under all conditions the osmolality, sodium concentration, and concentration of all constituents of the perfusate except bicarbonate and chloride were constant. After 10 min of perfusion at 81 mm Hg, all hearts were made ischemic for 10 min prior to alterations in perfusate pH or pCO_2. Changes in perfusate were made rapidly by switching to a second reservoir. In control hearts, there was no change in the perfusate pH or pCO_2.

At the conclusion of each experiment, the heart was arrested with 10% KCl, and the column height was changed from 81 to 22 or 33 mm Hg. At ventricular pressures of 5–10 mm Hg or 100–120 mm Hg, a fall in pressure of no more than 3 mm Hg was observed when the column height was changed. Thus, there was no significant balloon herniation from the ventricle into the aorta, and perfusion pressure was not reflected in the measured intracavitary pressure.

RESULTS

Peak left ventricular pressure and end-diastolic pressure from eight control hearts without ischemia are shown in Figure 1. The initial mean peak left ventricular pressure was recorded immediately after completion of the preparation. Peak left ventricular pressure remained above the initial value for over 90 min. Experiments were performed over a 50-min period, beginning at 20 min after completion of the preparation. Mean left ventricular end-diastolic pressure began at 6.2±0.8 mm Hg and at the end of 90 min it was 8.6±1.8 mm Hg.

The results of eight hearts subjected to mild ischemia are shown in Figure 2. After 10 min, the perfusion pressure was changed to 33 mm Hg, and performance declined rapidly. After 30 min of ischemia, perfusion pressure was increased, and full recovery occurred. A continuous recording from a single heart subjected to

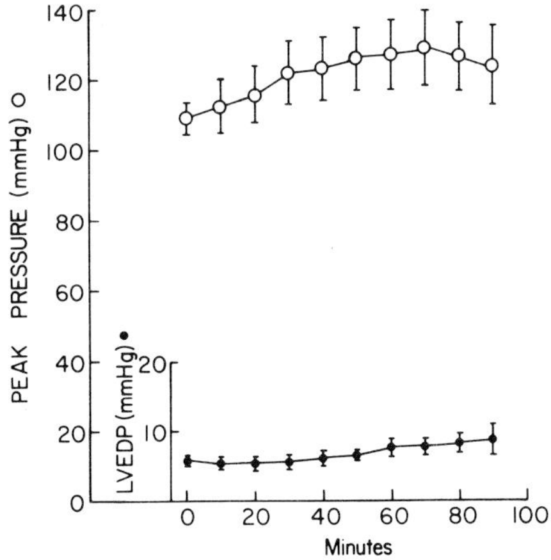

Figure 1. Peak isovolumic left ventricular pressure and end-diastolic pressure, beginning immediately following completion of preparation in eight hearts not subjected to ischemia. *Bars* indicate SEM.

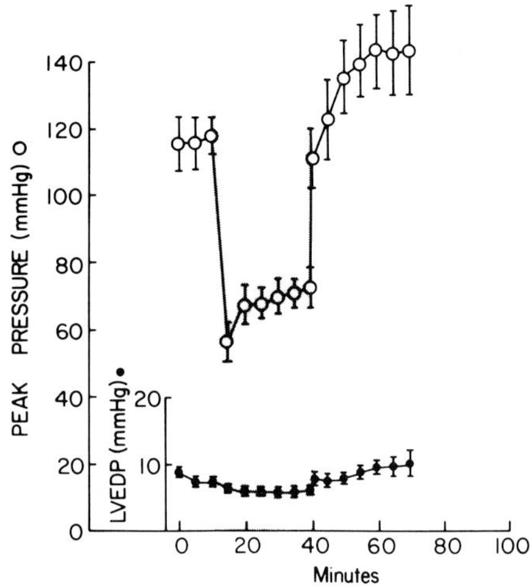

Figure 2. Peak isovolumic left ventricular pressure and end-diastolic pressure in eight hearts subjected to 30-min period of mild ischemia (perfusion pressure = 33 mm Hg).

moderate ischemia is shown in Figure 3. Following the reduction in perfusion pressure from 81 to 22 mm Hg, there was a rapid decline in performance as indicated by the fall in peak left ventricular pressure and dp/dt.

Coronary flow, 2.1±0.12 ml/min/g wet weight, fell to 1.08±0.08 ml/min/g in mild ischemia and to 0.88±0.07 ml/min/g in moderate ischemia. The control tissue lactate/pyruvate ratio was 15 in nonischemic heart muscle and increased five-fold during mild ischemia (Table 1). There was a significant potassium loss from the myocardium during both mild and moderate ischemia. Reactive hyperemia occurred on return of perfusion pressure to control levels. A transient decrease of perfusion pressure to zero for 30 sec during ischemia resulted in no reactive hyperemia on return to ischemic perfusion pressure.

Figure 4 shows the effect of varying pCO_2 and alkalosis on peak left ventricular pressure during mild ischemia. After 10 min of ischemia (zero time), peak pressure was 67.7±2.2 mm Hg and was not significantly different between groups. Left ventricular pressures are expressed as the percentage of this value at zero time. Alkalosis (pH = 7.80) at constant pCO_2 resulted in a significant transient decline in performance compared to the control group. There was no significant improvement in performance throughout the entire period of observation. In contrast, lowering of the pCO_2 at constant perfusate pH (7.48) resulted in a significant improvement in performance relative to the control group. Maximal improvement was achieved by 1.0 min after exposure to lowered pCO_2. Conversely, increasing the pCO_2 to 60 mm Hg at constant pH (7.47) depressed peak pressure relative to the control group. Similar results were obtained when

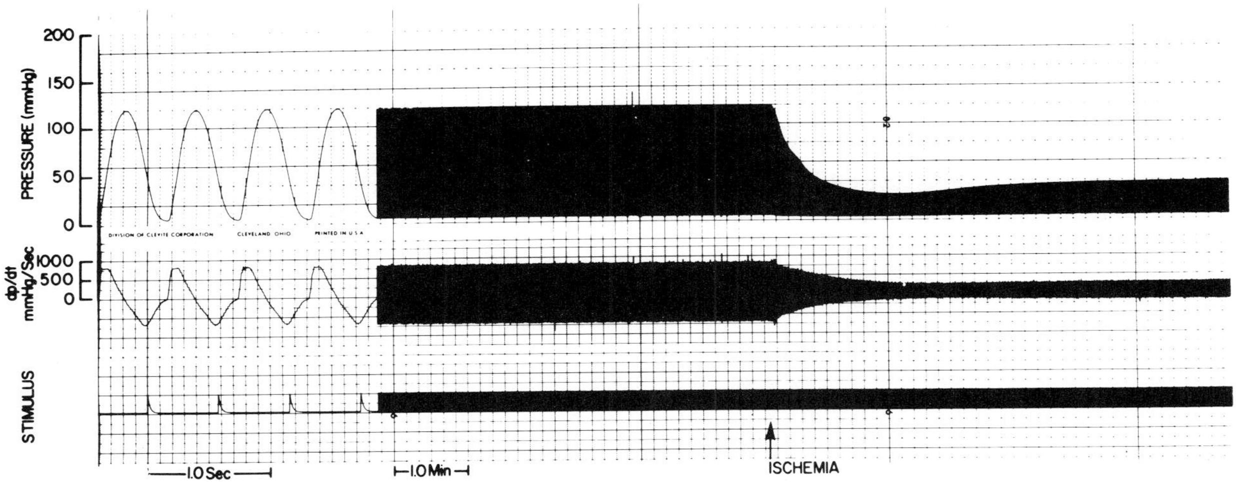

Figure 3. Tracing from heart subjected to moderate ischemia (perfusion pressure = 22 mm Hg). *Top tracing* shows isovolumic left ventricular pressure. *Middle tracing* is first derivative of isovolumic pressure. *Lower tracing* is stimulus artifact. Initial 6 min of ischemia are shown.

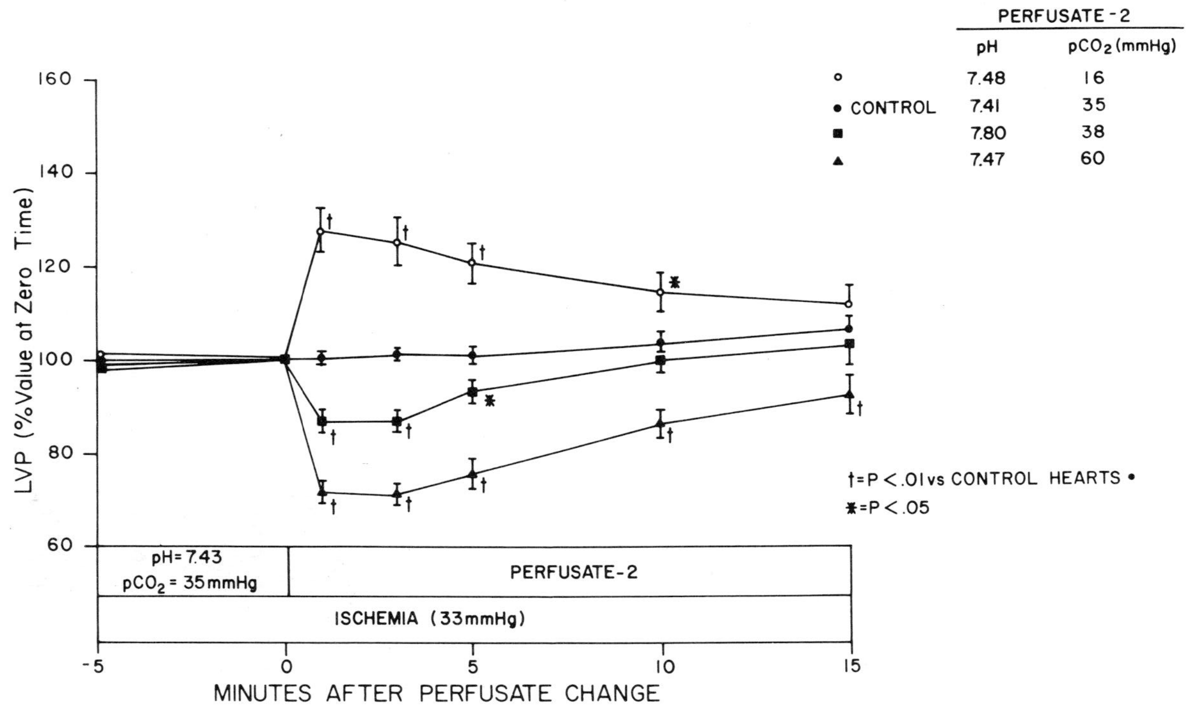

Figure 4. Peak isovolumic left ventricular pressure expressed as percentage of value at zero time. Zero time is at 10 min of perfusion under ischemic conditions (33 mm Hg) and at time of alteration of perfusate. In control hearts, perfusate was not altered. For discussion, see text.

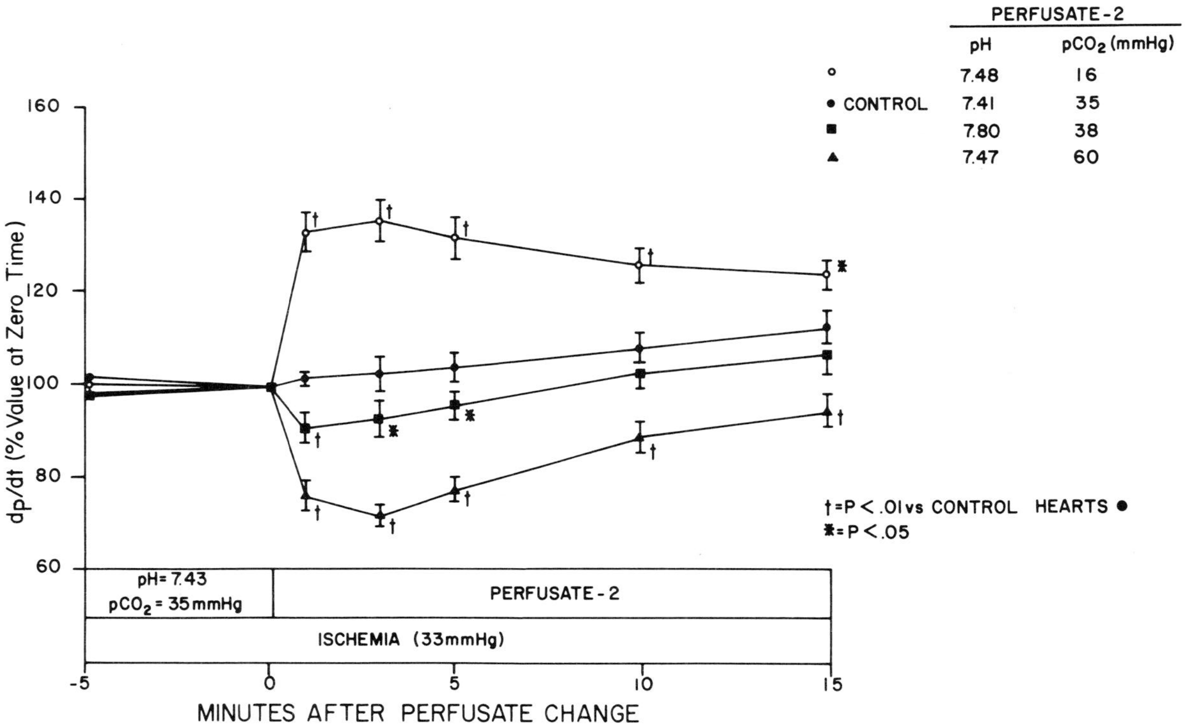

Figure 5. Isovolumic left ventricular dp/dt expressed as percentage of value at zero time (10 min of perfusion at 33 mm Hg).

Table 1. Tissue lactate and lactate/pyruvate ratio

	N	Lactate (μmoles/g)	Lactate/Pyruvate
Control	5	0.71±0.16	15.1±2.2
Mild ischemia	4	2.77±0.61	85.2±5.1
p		<0.02	<0.001

dp/dt was used as an index of performance (Fig. 5). Mean dp/dt prior to ischemia was 795±25 mm Hg/sec and 505.3±20 mm Hg/sec at time zero.

Since it might be argued that under more severely ischemic conditions a greater improvement in performance might be observed with alkalosis, alkalosis and acidosis were examined under conditions of moderate ischemia (22 mm Hg). The mean peak left ventricular pressure at zero time was 43.5±1.7 mm Hg; 6 hearts were studied in each group under these conditions. As seen in Figure 6, alkalosis (pH = 7.80) at constant pCO_2 (39 mm Hg) again did not result in a significant improvement in performance above control. Acidosis (pH = 7.10) at constant pCO_2 (37 mm Hg) did not depress performance significantly relative to control until more than 15 min of exposure. Even after 20 min of exposure to acidosis, the magnitude of depression of performance was relatively small: 6.6±0.2%.

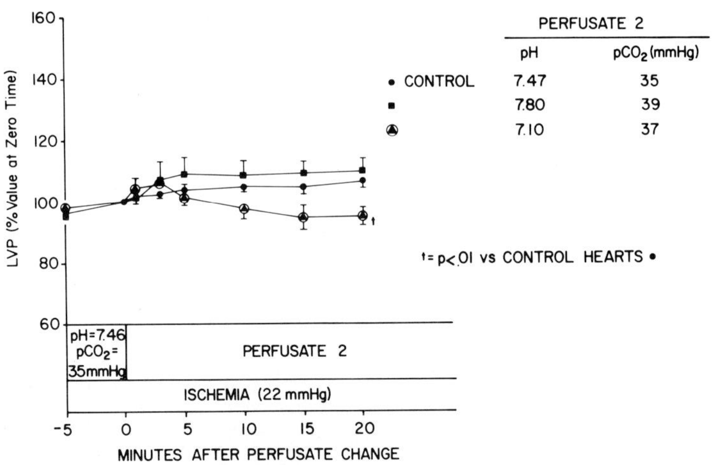

Figure 6. Peak isovolumic left ventricular pressure expressed as percentage of value at zero time (10 min of perfusion at 22 mm Hg).

DISCUSSION

The present studies demonstrate that alteration in perfusate pCO_2 independent of perfusate pH is associated with rapid, significant alterations in performance. They suggest that the early depression of myocardial performance during ischemia may be related to increased tissue pCO_2 owing, at least in part, to decreased CO_2 removal. In contrast to hypoxic models, alteration of extracellular pH per se had relatively little effect upon ischemic muscle performance.

To alter pCO_2 at constant pH with constant osmolality and sodium concentration, the perfusate bicarbonate content was altered. Thus, the hearts exposed to low pCO_2 were also exposed to low buffer capacity. This was reflected in a lower coronary venous pH. Since a lower buffer capacity would tend to depress performance (Fig. 6), the increase in performance with lower pCO_2 occurred despite a lower perfusate buffer capacity. Conversely, increased pCO_2 depressed performance despite increased buffer capacity.

It has been suggested that intracellular acidosis as a result of ischemia inhibits lactate production (Rovetto, Whitmer, and Neely, 1973). Thus , exposure to a lower pCO_2 may initially increase lactate production substantially as a result of a higher intracellular pH. This increase in lactate production may then lead to a subsequently lower intracellular pH. Such changes in intracellular pH may explain the relative decrease in performance toward the end of the period of low pCO_2. This is now the subject of further investigation.

REFERENCES

CASE, R. B. 1971/1972. Ion alterations during myocardial ischemia. Cardiology 56: 245–262.

EFFROS, R. M., P. O. ETTINGER, K. MAROLD, B. HAIDER, H. A. OLDEWURTEL, and T. J. REGAN. 1973. Myocardial cell acidosis during acute ischemia. J. Clin. Invest. 52: 25a. (Abstr.).

FLAHERTY, J. T., S. F. KHURI, J. B. O'RIORDAN, B. PITT, and V. L. GOTT. 1974. Quantitation of myocardial ischemia by intramyocardial carbon dioxide, oxygen, and electrograms. (Abstract). Fed. Proc. 33: 395.

DOWNING, S. E., N. S. TALNER, and T. H. GARDNER. 1966. Influences of hypoxemia and acidemia on left ventricular function. Am. J. Physiol. 210: 1327–1334.

KATZ, A. M. 1973. Effects of ischemia on the contractile processes of heart muscle. Am. J. Cardiol. 32: 456–460.

KATZ, A. M., and H. H. HECHT. 1969. The early "pump" failure of the ischemic heart. Am. J. Med. 47: 497–502.

MARSHALL, R. J., J. R. PARRATT, and I. McA. LEDINGHAM. 1974. Changes in blood flow and oxygen consumption in normal and ischaemic regions of the myocardium following acute coronary artery ligation. Cardiovasc. Res. 8: 204–215.

OPIE, L. H., P. OWEN, M. THOMAS, and R. SAMSON. 1973. Coronary sinus lactate measurements in assessment of myocardial ischemia: Comparison with changes in lactate/pyruvate and beta-hydroxybutyrate/acetoacetate ratios and with release of hydrogen, phosphate and potassium ions from the heart. Am. J. Cardiol. 32: 295–305.

ROVETTO, M. J., J. T. WHITMER, and J. R. NEELY. 1973. Comparison of the effects of anoxia and whole heart ischemia on carbohydrate utilization in isolated working rat hearts. Circulat. Res. 32: 699–711.

SAFER, B., and R. E. DAVIES. 1973. Effects of [H+] on physiological and biochemical function of heart. (Abstract). Fed. Proc. 32: 449.

SCHEUER, J., and S. W. STEZOSKI. 1972. The effect of alkalosis upon the mechanical and metabolic response of the rat heart to hypoxia. J. Molec. Cell. Cardiol. 4: 599–610.

WEISFELDT, M. L., and N. W. SHOCK. 1970. Effect of perfusion pressure on coronary flow and oxygen usage of nonworking heart. Am. J. Physiol. 218: 95–101.

Mailing address:
Myron L. Weisfeldt, M.D.,
508 Blalock Building, The Johns Hopkins Hospital,
601 N. Broadway, Baltimore, Maryland 21205 (USA).

Recent Advances in Studies on
Cardiac Structure and Metabolism Volume 10
The Metabolism of Contraction
Edited by Paul-Emile Roy and George Rona
Copyright 1975 University Park Press Baltimore

ROLE OF ADENINE NUCLEOTIDE TRANSLOCASE IN METABOLIC CHANGE CAUSED BY ISCHEMIA

AUSTIN L. SHUG, JOSEPH R. KOKE,
JOHN D. FOLTS, and NEVILLE BITTAR

Veterans Administrations Metabolic Research Laboratory, and the Department of Medicine,
University of Wisconsin, Madison, Wisconsin, USA

SUMMARY

Inhibition of adenine nucleotide translocase by elevated levels of long chain acyl-CoA esters has been shown to occur during the onset of ischemia in experiments conducted on dogs. Other findings indicate that, as a consequence of translocase inhibition, the production of mitochondrial creatine phosphate was abolished and, in this manner, respiration was slowed to state 4 or an ischemic-like condition. A variety of biochemical, hemodynamic, and ultrastructural evidence further suggests that this inhibition of adenine nucleotide transport in and out of the heart mitochondria may be the initial and key disturbance which "triggers" the more drastic metabolic changes known to occur as the degree of ischemia becomes more severe. The mitochondrial "damage" caused by long chain acyl-CoA ester inhibition of adenine nucleotide translocase appears to be reversible by carnitine.

INTRODUCTIONS

Adenine nucleotide translocase, an enzyme localized on the mitochondrial inner membrane (Lehninger, 1972), catalyzes exclusively a molecule-for-molecule exchange of ADP and ATP. In this manner, the enzyme confers the specificity of oxidative phosphorylation upon adenine nucleotides (Stadtman, 1955) and controls the ATP/(ADP + P_i) or phosphate potential of the cell (McLean, Gumaa, and Greenbaum, 1971). Adenine nucleotide translocase, therefore, links the energy metabolism of the mitochondrial compartment with that of the cytosol.

Control of this important enzyme activity has been shown to be related directly to the relative cellular concentrations of long chain acyl-CoA esters and carnitine (McLean *et al.*, 1971; Lerner *et al.*, 1972; Shug and Shrago, 1973). These and other studies (Shug *et al.*, 1971; Harris, Farmer, and Ozawa, 1972; Shrago *et al.*, 1972) have established that free fatty acids do inhibit adenine nucleotide translocase, and they have indicated that the presence of ADP in the acyl-CoA molecule may account for the specific and competitive inhibition of adenine nucleotide translocase by long chain acyl-CoA esters (Shrago *et al.*, 1974).

It is now well established that fatty acids are the preferred substrate in well oxygenated hearts (Neely, Rovetto, and Oram, 1972) and that, during ischemia, the myocardium is forced to oxidize glucose and to store the fatty acids as triglycerides (Scheuer and Brachfeld, 1966). This special relationship between lipid and carbohydrate in myocardial metabolism suggested that long chain acyl CoA esters could accumulate in the heart muscle during ischemia, and it was consequently shown *in vitro* that acyl CoA esters did, indeed, inhibit adenine nucleotide translocation in heart mitochondria (Shug and Shrago, 1973). Recent experiments (Bittar *et al.,* 1974) from this laboratory have shown *in vivo* that long chain acyl CoA levels increase in the ischemic dog heart myocardium, and that adenine nucleotide translocase activity is diminished in mitochondria isolated from the ischemic myocardium.

In the present investigation, we have shown accumulation of acyl CoA esters during the early stages of ischemia and demonstrated significant inhibition of adenine nucleotide translocase *in vivo* in the ischemic dog heart as early as 1 min after coronary artery ligation. In another series of experiments, we have found that, in heart mitochondria, long chain acyl CoA esters prevent the coupling of creatine phosphorylation to electron transport by inhibition of adenine nucleotide translocase. The inhibition was reversed by carnitine. This important relationship between creatine kinase and adenine nucleotide translocase has been reported to be the basis for the mechanism by which heart mitochondria maintain a high (state 3) and constant rate of respiration (Jacobus and Lehninger, 1973).

Further support for the hypothesis that long chain acyl CoA inhibition of adenine nucleotide translocation may be the key metabolic disturbance during the onset of ischemia was obtained by the perfusion of atractyloside into the dog myocardium. Atractyloside is a plant glycoside that is generally considered to be specific for the inhibition of adenine nucleotide translocase (Lehninger, 1972). Infusion of this inhibitor directly into the anterior descending artery produced electrocardiographic changes typical of ischemia and resulted in the corresponding inhibition of only adenine nucleotide translocase.

METHODS

Samples of myocardial tissue for study were obtained from mongrel dogs weighing between 20 and 30 kg. Anesthesia was produced using 3 mg/kg morphine sulfate subcutaneously, followed in 1 hour by intravenous pentobarbital (30 mg/kg). Ventilation was maintained by means of a Harvard animal respirator using room air.

Short segments of the anterior descending branch of the left coronary artery were dissected free, thus allowing blood flow to be measured using a Biotronix flowmeter and flow transducer. Epicardial electrodes were sutured to the anterior and lateral walls of the left ventricle. Normal dogs were those in whom no intervention was carried out. Ischemia was produced by ligating the anterior descending artery and, in a number of animals, distal segments of diagonal branches and circumflex artery were also ligated. At designated times, cardiac inflow and outflow was stopped by clamps, and portions of the left ventricle were excised rapidly for the preparation of mitochondria.

Other small muscle samples for study were removed by needle biopsy. Such samples were rapidly homogenized and assayed for adenine nucleotide translocase as described in Table 3.

Preparation of myocardial tissue samples for ultrastructural studies was accomplished by fixation of small bits of the desired area in 4% formaldehyde, 2% glutaraldehyde, 0.001 M $CaCl_2$—all in 0.1 M Na cacodylate at pH 7.2. After several hours, the tissue was washed, postfixed in 2% OsO_4, and processed routinely for electron microscopy. Isolated mitochondria were fixed by resuspension in isolation medium to which glutaraldehyde had been added to a final concentration of 3%. After 15–30 minutes, the mitochondria were centrifuged into a pellet and processed routinely for electron microscopy.

For mitochondrial preparation, samples of the excised left ventricle were homogenized in 0.25 M sucrose, 0.004 M Tris-HCl (pH 7.4), and 1.0 mM EGTA, and mitochondria were obtained by differential centrifugation according to the method of Henderson and Lardy (1970). Oxygen consumption and respiratory control ratios were measured polarographically as described by Estabrook (1967), with a Clark oxygen supplied with the Gilson oxygraph.

Translocation of adenine nucleotides was determined by the method of Wojtczak and Zaluska (1967) with a slight modification (McLean *et al.*, 1971). Protein was determined by the Biuret procedure (Gornall, Bardwill, and David, 1949), and acyl CoA levels were measured by standard methods (Stadtman, 1955, 1957; Williamson and Corkey, 1969). The studies on creatine kinase activity in beef heart mitochondria were made essentially as described by Jacobus and Lehninger (1973). In these studies, beef heart mitochondria (5 mg) were added to a test tube at room temperature containing 6 ml of the reaction mixture (210 mM sucrose; 3 mM $MgCl_2$; 0.1 mM EGTA; 5 mM K phosphate buffer, pH 7.4; 20 mM Tris; and 4 mM creatine). Malate and glutamate were used as substrates (5.8 mM each). Reaction was initiated by addition of 7 μmoles of ADP, and creatine phosphate was measured on 0.1-ml filtrates of the medium, using a coupled enzymatic assay involving creatine kinase, hexokinase, and glucose 6-phosphate dehydrogenase and resulting in reduction of NADP with an increase in absorption at 340 nm. This was varied by adding atractylate, oleoyl-CoA, and carnitine to the reaction mixture, and creatine phosphate was measured as in the control.

Table 1. Adenine nucleotide translocase activity in control and ischemic muscle

Period of Ischemia (min)	Treatment	No. of Arteries Clamped	DPM/Mitochondrial Pellet	
			Control	Ischemic
60	None	1	10,500 ± 825	5,900 ± 910
60	None	2	10,200 ± 780	6,000 ± 850
60	None	3	9,100 ± 550	5,300 ± 610
60	None	4	8,000 ± 470	2,500 ± 320
60	400 mg carnitine	3	10,600 ± 930	7,000 ± 790
30	None	2	12,700 ± 680	9,300 ± 620
30	None	3	13,050 ± 650	8,200 ± 615
30	None	4	9,500 ± 720	4,300 ± 590
15	None	3	14,800 ± 910	10,300 ± 630
30	Narrowed down		14,000 ± 480	8,100 ± 530
3	None	1	16,800 ± 670	10,050 ± 585
1*	None	1	23,000	12,100
Normal value 17,500 ± 760 /0.1 mg protein				

*Assayed directly in homogenized biopsy.

RESULTS AND DISCUSSION

Table 1 summarizes the effect of varying the time and severity of ischemia. The general effect of ligating the small branches (listed under number of arteries clamped) was to decrease the collateral flow and, as a result, the injury pattern was more severe. The hemodynamic and ultrastructural effects of these interventions were discussed more completely by Koke *et al.* (1974). It should be noted that, as the number of arteries clamped increased, the adenine nucleotide translocase activity decreases. The results shown in Table 1 represent an average value of from three to five animals, and it should be mentioned that, in most mitochondria tested, respiratory control values were over 20. When the "degree of ischemia" was very severe (60 min with three or four arteries clamped, and 30 min with four arteries clamped), respiratory control values dropped to about 10 or less, indicating inhibition of other respiratory mitochondrial systems. When carnitine was added (*in vitro*) to mitochondria from less severe ischemic areas, some reversal of translocase inhibition was observed, whereas the more severe ischemic injury, as described above, was not reversible by carnitine. We have not been able to demonstrate, so far, a beneficial effect of perfusing carnitine into the heart before ligating the artery, as indicated in Table 1. It can be seen that, when the lumen of a dog was reduced by about 80% by narrowing down the artery, significant inhibition of translocase was obtained, as was also true for only 3 min of ischemia. Adenine nucleotide translocase was assayed by the modified method described in Table 3 for the results given for only 1 min of ischemia, which may account for the larger amount of inhibition observed. In any case, the significant amount of translocase inhibition observed for such short periods of ischemia suggests that the acyl-CoA inhibition of adenine nucleotide translocase may be the primary event that can initiate the drastic metabolic changes that occur during the onset of ischemia (Neely and Morgan, 1974). The close correlation between the increase in the acyl-CoA concentration and the decrease in adenine nucleotide translocase activity as the time and severity of ischemia are increased is shown in Figure 1, and it is suggested that, as acyl-CoA concentrations increase, adenine nucleotide translocation is inhibited.

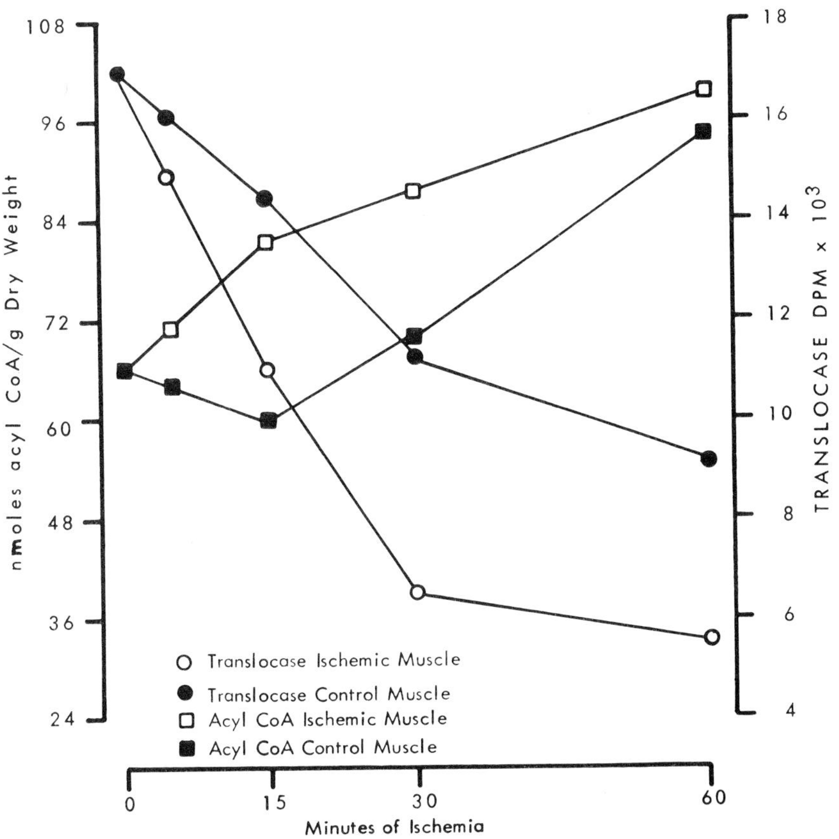

Figure 1. Comparison of amount of acyl-CoA esters and adenine nucleotide translocase activity in ischemic dog heart. Experimental details in text.

Table 2. Effect of long chain acyl-CoA esters on creatine phosphate formation by beef heart mitochondria

Additions	Concentration ($\times 10^{-6}$M)	Creatine Phosphate (nmole/ml filtrate)			
		1 min	2 min	5 min	10 min
Control		4	13	28	50
Atractyloside	1.66	2	5	13	24
	4.16	0	1	1	1
Oleoyl-CoA	0.80	0	7	19	29
	1.6	0	0	11	11
Oleoyl-CoA + carnitine	0.8 3.0	2	11	20	42
	1.6 3.0	0	3	14	29

The investigations of Jacobus and Lehninger (1973) have suggested that the phosphocreatine-creatine kinase system of heart mitochondria serves to buffer the rate of respiration at a high and constant rate, as well as to channel high energy phosphate to the contractile system through the joint action of the mitochondrial and sarcoplasmic isoenzymes of creatine kinase. The phosphorylation of creatine was found to occur just outside the atractyloside-sensitive ADP-ATP carrier system (adenine nucleotide translocase). These results suggested that the particular function of translocase in heart mitochondria was to provide a constant source of ATP for mitochondrial creatine kinase. In this manner, the end product of oxidative phosphorylation would be creatine phosphate, and a constant supply of ADP would always be present to stimulate respiration.

Table 2 shows the effect of acyl-CoA esters on creatine phosphate production by isolated beef heart mitochondria. As can be seen, oleoyl-CoA gives a pronounced inhibition of creatine phosphate production which is reversed by carnitine. These results suggest that, as the acyl-CoA concentration increased during ischemia, adenine nucleotide translocation is inhibited, thus cutting off the supply of ATP for creatine phosphate production and preventing ADP stimulation of respiration. The rapid drop in creatine phosphate formation during the onset of ischemia (Braasch *et al.*, 1968; Gudbjarnason, Mathes, and Ravens, 1970) tends to support these suggestions.

Because of the obvious technical problems, not only in the size of the sample, but in the amount of time necessary to prepare mitochondria for adenine nucleotide translocase, it has been very difficult to determine adenine nucleotide translocase activity in tissue exposed to short periods of ischemia. Attempts to quick-freeze (in liquid nitrogen) ischemic and nonischemic dog heart muscle and later to prepare mitochondria for respiratory and adenine nucleotide translocase studies have been unsuccessful. In such preparations, the respiratory control values, regardless of substrate, dropped well below 20, and adenine nucleotide translocase activities were from 30 to 90% less than those of mitochondria from nonfrozen tissues.

The rather unusual properties of adenine nucleotide translocase—the very high activity even at 0°C, the enormous affinity of mitochondria for ADP, a property not displayed by other intracellular organelles, and its specific sensitivity to inhibition by atractylate—all suggested that activity might be determined directly in a cell homogenate. A series of experiments were then undertaken using various amounts of drill biopsy samples from beef heart. A summary of the results is shown in Table 3. It can be seen that the activity per mg protein in the assay of adenine nucleotide translocase in the homogenate compared favorably with the value for isolated mitochondria and, in some cases, gave even slightly higher values. In assaying adenine nucleotide translocase in a homogenate after

Table 3. Assay of adenine nucleotide translocase directly in beef heart homogenate*

Preparation	Volume of Preparation (ml)	Mitochondrial Protein (ml)	Activity (dpm) ANT	ANT + Atractylate
Crude homogenate	0.1	0.013	3501	135
	0.2	0.026	6191	170
	0.4	0.052	8932	158
Washed mitochondria	0.05	0.02	2950	73
	0.1	0.04	6400	131
	0.2	0.08	9024	143

*Approximately 0.5 g (weigh exactly, drill biopsy) of tissue are homogenized in 10 ml of EGTA media (media used for mitochondrial preparation) on fast homogenizer. After homogenizing, 10 ml EGTA are added, and homogenate is mixed. Translocase assay is run on aliquots of this homogenate. Assay system contains 0.85 ml of translocase media, 20 λ ^{14}CADP = 53,000 dpm, and 0.05 ml of atractylate 10^{-2} which is used to terminate reaction. Assay is run 5 min in ice and is started by addition of homogenate. After reaction is stopped with atractylate, 7 ml of EGTA media is added and tubes mixed. Tubes are spun 10 min at 12,500 rpm (Sorvall); an aliquot of this supernatant is counted and tubes are decanted; 7 ml of EGTA are added and tubes are inverted to wash pellet. Tubes are respun 10 min at 12,500 rpm, decanted, and this pellet (mitochondria and nuclei) is solubilized with 0.2 ml of Soluene for 1–2 hours and counted. ANT = adenine nucleotide translocase.

Table 4. Effect of perfusion of atractyloside and 4-pentenoic acid on mitochondrial respiration

Treatment	Time (min)	Amount (g)	Respiratory Rate (μmole O_2 consumed/min)			
			Control area		Ischemic area	
			State 3	Uncoupled	State 3	Uncoupled
None	—	—	91	90	85	86
Atractyloside	2	3.2	83	84	31	89
	10	2.9	68	87	22	63
	60	1.3	89	89	83	80
4-Pentenoic acid	60	2.2	93	93	90	90

terminating the assay with atractylate, it was found essential to wash the mitochondrial-nuclear pellet once to remove nonspecifically absorbed ^{14}C-ADP. Virtually no ADP was taken up by the nucleus. Similar results were obtained with small biopsy samples from dog heart.

The advantages of estimating adenine nucleotide translocase activity in such a simple assay system are many. It is possible to test adenine nucleotide translocase activity in very small biopsy samples obtained after short times of ischemia. By increasing the amount of ^{14}C-ADP in the assay, it should be possible to assay adenine nucleotide translocase activity quickly in biopsy samples as small as 10 mg, or even less. This possibility may allow the determination of adenine nucleotide translocase activity in human biopsy material when it becomes available. In addition, it is possible to conduct temporal studies on the change in adenine nucleotide translocase during the course of ischemia in a single open-chest anesthetized dog as shown in Table 1. Such biopsies will also be obtained more exactly from specified areas of the ischemic tissue. In general, this rapid assay of adenine nucleotide translocase in small biopsy samples should make it possible to follow minute-to-minute changes in activity, following the production of myocardial ischemia or other hemodynamic interventions.

As has been previously mentioned (Lehninger, 1972), atractyloside is considered a specific inhibitor of adenine nucleotide translocation. Because of this, in an attempt to employ adenine nucleotide translocase inhibitors as models for clinical heart disease, a series of experiments have been made in which relatively large amounts of atractyloside have been infused into the artery of dog hearts. Table 4 shows the effects of varying the amount, time, and flow rate of atractylate perfusion of the dog heart and the respiratory rate. Mitochondria were isolated from the affected control areas of the left ventricle as previously described (Methods) and the respiratory rates (state 3 and uncoupled) were determined. Atractylate was infused directly into the anterior descending artery in the amounts, times, and flow rates indicated in Figure 2 and Table 4. The enormous inhibition of state 3 respiration and its reversal by uncoupler seen for the higher doses of atractylate are typical of translocase inhibition. Direct adenine nucleotide translocase assay (see Methods) indicated more than 95% inhibition of ^{14}C-ADP translocation. However, when a lesser amount (1.3 g in 60 min) was infused, there was virtually no effect on mitochondrial respiration, and only marginal translocation inhibition. Infusion of 4-pentenoic acid, a compound thought to inhibit carnitine transferase (Senior, Robson, and Sherratt, 1968), proved to be technically difficult and without appreciable effect on either hemodynamic or biochemical findings.

Figure 2 depicts the hemodynamic findings after infusion of 2.9 g of atractyloside in 20 ml of 0.1 M Tris buffer (pH 7.4) at a rate of 2 cc/min. (Results also shown in Table 4.) Within 1 min of the onset of infusion, injury potentials were noted in the epicardial lead attached to the area of left ventricle supplied

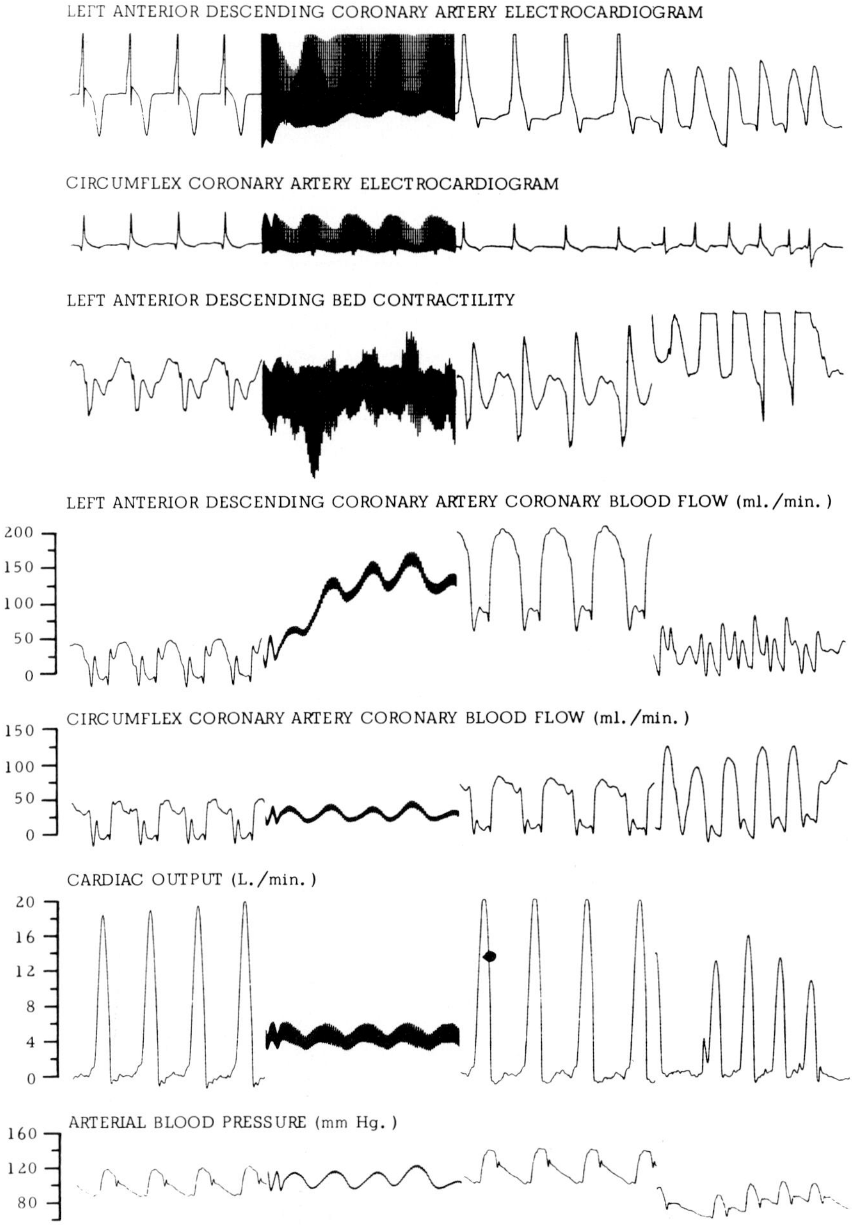

Figure 2. Effects of perfusion of atractyloside on various Hemodynamic parameters. Atractylate (2.9g) in 20 cc of Tris buffer was infused directly into anterior descending artery at an infusion rate of 1.94 cc min for 10 min. Other information in text.

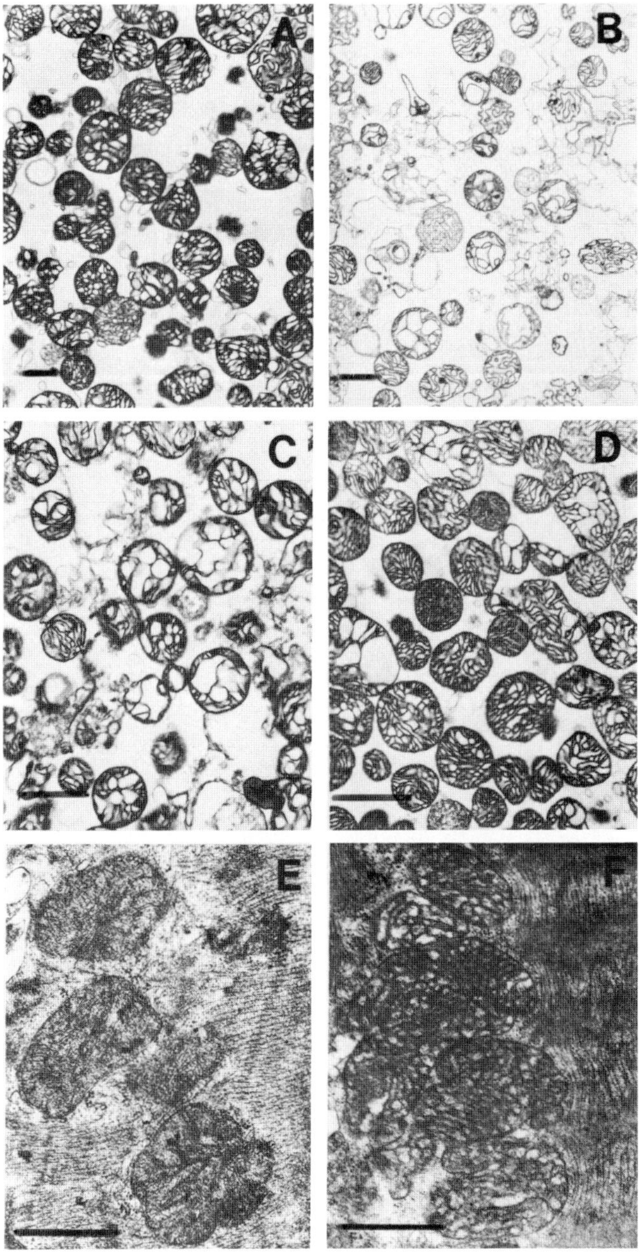

Figure 3. Effects of adenine nucleotide translocase inhibitors on ultrastructure of heart mitochondria. Calibration lines represent 1.0 μm. *A:* beef heart mitochondria incubated with substrate (7.5 μmoles of pyruvate + 7.5 μmoles of malate) and ADP (0.5 μmoles). *B:* beef heart mitochondria, as in *A* + 2 × 10^{-5} M atractyloside. *C:* beef heart mitochondria, as in *A* + 2 × 10^{-6} M oleoyl-CoA. *D:* beef heart mitochondria, as in *C* + 2 μmoles of carnitine. *E:* dog heart mitochondria, *in situ* normal. *F:* dog heart mitochondria, *in situ* after infusion of 2.9 g of atractyloside for 10 min. Other details of experiments in text. Magnification values: A, × 5,000 μm; B, × 7,000 μm; C, × 10,000 μm; D, × 10,000 μm; E, × 14,000 μm; and F, × 15,000 μm.

by the left anterior descending artery. Left anterior descending flow rose from a mean value of 30 cc/min to 150 cc/min. Cardiac output and blood pressure also rose. By 5 min, however, ventricular tachycardia was noted to arise from the left anterior descending bed. This resulted in a decline in flow to control value, while cardiac output and blood pressure fell to below control levels.

In a series of studies on the adenine nucleotide-induced contraction of the inner mitochondrial membrane, Stoner and Sirak (1973a, b) have shown that the inner membrane of isolated bovine heart mitochondria undergoes pronounced contraction upon incubation with substrate and ADP (state 3 respiration). In addition, they have shown that atractyloside specifically inhibits the ADP-catalyzed contraction, because of inhibition of adenine nucleotide translocase. The inhibition of contraction by atractyloside resulted in an increase in intra-cristal space and a decrease in the optical density of the mitochondrial suspension or, in other words, a change in mitochondial configuration. In Figure 3, we show similar effects of ADP and ADP + atractyloside on beef heart mitochondria. In addition, we have found that oleoly-CoA has an identical effect with atractylate (presumably because of inhibition of adenine nucleotide translocase), and that carnitine reverses the effect of oleoyl-CoA. In Figure 3, we also show the ultrastructure of normal dog heart mitochondria *in situ* as previously described (Koke *et al.*, 1974), as well as the effect of infusing 2.9 g of atractylate for 10 min (see also Table 4 and Fig. 2). The results indicate a very early ultrastructural change in the mitochondrial membrane, which is caused specifically by inhibition of adenine nucleotide translocase.

These experiments suggest that the initial event that occurs after the onset of ischemia is the inhibition of adenine nucleotide translocase by long chain fatty acid acyl-CoA, thus interrupting the vital link between adenine nucleotide translocase and mitochondrial creatine kinase. This disturbance then triggers the well documented metabolic changes that occur with continuing ischemia.

Acknowledgments This work was supported in part by grants from the Wisconsin Heart Association and the Graduate School of the University of Wisconsin and by National Institute of Health Postgraduate Training Grant HL 5364.

REFERENCES

BITTAR, N., A. L. SHUG, J. R. KOKE, J. D. FOLTS, and E. S. SHRAGO. 1973. Inhibited adenine nucleotide translocation in mitochondria isolated from ischemic myocardium. *In* Abstracts, Myocardial Cell Damage, The 6th Annual Meeting of the International Study Group for Research in Cardiac Metabolism. Freiburg, Germany, September 25–28, 1973. (In press)

BRAASCH, W., S. GUDJARNASON, P. S. PURI, K. G. RAVENS, and R. J. BING. 1968. Early changes in energy metabolism in the myocardium following acute coronary artery occlusion in anesthetized dogs. Circulat. Res. 23: 429–438.

ESTABROOK, R. W. 1967. Mitochondrial respiratory control and the polarographic measurement of ADP:O ratios. Meth. Enzymol. 10: 42.

GORNALL, A. G., C. J. BARDWILL, and M. M. DAVID. 1949. Determination of serum proteins by means of the biuret reaction. J. Biol. Chem. 177: 751–766.

GUDBJARNASON, S., P. MATHES, and K. G. RAVENS. 1970. Functional compartmentation of ATP and creatine phosphate in heart muscle. J. Mol. Cell. Cardiol. 1: 325–339.

HARRIS, R. A., B. FARMER, and T. OZAWA. 1972. Inhibition of the mitochondrial adenine nucleotide transport system by oleoyl CoA. Arch. Biochem. Biophys. 150: 199.

HENDERSON, P. J. F., and H. A. LARDY. 1970. Bongkrekic acid. An inhibitor of the adenine nucleotide translocase of mitochondria. J. Biol. Chem. 245: 1319–1326.

JACOBUS, W. E., and A. L. LEHNINGER. 1973. Creatine kinase of rat heart mitochondria. Coupling of creatine phosphorylation to electron transport. J. Biol. Chem. 248: 4803–4810.

KOKE, J. R., A. L. SHUG, J. D. FOLTS, and N. BITTAR. 1974. Ultrastructural changes in myocardial cells subjected to short periods of ischemia. *In* Abstracts, 7th Annual Meeting of the International Study Group for Research in Cardiac Metabolism, Quebec, June 18th–21st. University Park Press, Baltimore.

LEHNINGER, A. L. 1972. Biochemistry, pp. 400–401. Worth Publishers, New York.

LERNER, E., A. L. SHUG, C. ELSON, and E. SHRAGO. 1972. Reversible inhibition of adenine nucleotide translocation by long chain fatty acyl coenzyme A esters in liver mitochondria of diabetic and hibernating animals. J. Biol. Chem. 247: 1513–1519.

McLEAN, P., K. A. GUMAA, and A. L. GREENBAUM. 1971. Long chain acyl CoAs, adenine nucleotide translocase and the coordination of the redox states of the cytosolic and mitochondrial compartments. FEBS Letters 17: 345–350.

NEELY, J. R., M. J. ROVETTO, and J. F. ORAM. 1972. Myocardial utilization of carbohydrate and lipids. Progr. Cardiovasc. Dis. 15: 289–329.

NEELY, J. R., and H. E. MORGAN. 1974. Relationship between carbohydrate and lipid metabolism and the energy balance of heart muscle. Ann. Rev. Physiol. 36: 413.

SCHEUER, J., and N. BRACHFELD. 1966. Myocardial uptake and fractional distribution of palmitate-1-C^{14} by the ischemic dog heart. Metabolism 15: 945–954.

SENIOR, A. E., B. ROBSON, and E. S. A. SHERRATT. 1968. Biochemical effects of the hypoglycaemic compound pent-4-enoic acid and related non-hypoglycaemic fatty acids. Biochem. J. 110: 511–519.

SHRAGO, E., A. SHUG, C. ELSON, and E. LERNER. 1972. Regulation of the translocation of adenine nucleotides across the inner mitochondrial membrane by long chain acyl CoA esters. *In* M. A. Mehlman and R. W. Hanson (eds.), Symposium on Metabolic Regulation, pp. 165–182. Academic Press, New York and London.

SHRAGO, E., A. L. SHUG, C. ELSON, T. SPENNETTA, and C. CROSBY. 1974. Long chain fatty acyl CoA regulation of metabolite transport in rat and guinea pig liver mitochondria. J. Biol. Chem. (In Press).

SHUG, A., E. LERNER, C. ELSON, and E. SHRAGO. 1971. The inhibition of adenine nucleotide translocase activity by oleoyl CoA and its reversal in rat liver mitochondria. Biochem. Biophys. Res. Commun. 43: 557–563.

SHUG, A. L., and E. SHRAGO. 1973. A proposed mechanism for fatty acid effects on energy metabolism of the heart. J. Lab. Clin. Med. 81: 214–218.

SOUVERIJN, J. H. M., P. J. WEIJERS, G. S. P. GROOT, and A. KEMP, Jr. 1970. The adenine nucleotide translocator and the nucleotide specificity of oxidative phosphorylation. Biochim. Biophys. Acta 223: 31–35.

STADTMAN, E. R. 1955. Phosphotransacetylase from *Clostridium kluyreri*. *In* S. P. Colowick and N. O. Kaplan (eds.), Methods of Enzymology, Vol. 1, pp. 596–599. Academic Press, New York.

STADTMAN, E. R. 1957. Preparations and assay of acetyl phosphate. *In* S. P. Colowick and N. O. Kaplan (eds.), Methods of Enzymology, Vol. 3, pp. 230–231. Academic Press, New York.

STONER, C. D., and H. D. SIRAK. 1973a. Adenine nucleotide-induced contraction of the inner mitochondrial membrane. I. General characterization. J. Cell. Biol. 56: 51–64.

STONER, C. D., and H. D. SIRAK. 1973b. Adenine nucleotide-induced contraction of the inner mitochondrial membrane. II. Effect of bongkrekic acid. J. Cell. Biol. 56: 65–73.

WILLIAMSON, J. R., and B. E. CORKEY. 1969. Assays of intermediates of the citric acid cycle and related compounds by fluorometric enzyme methods. *In* J. M. Lowenstein (ed.) and S. P. Colowick and N. O. Kaplan (general eds.), Methods of Enzymology, Vol. 13, pp. 436–440. Academic Press, New York.

WOJTCZAK, L., and H. ZALUSKA. 1967. The inhibition of translocation of adenine nucleotides through mitochondrial membrane by oleate. Biochem. Biophys. Res. Commun. 28: 76.

Mailing address:
Austin L. Shug, Ph.D.,
Metabolic Research Laboratory, Veterans Administration Hospital,
2500 Overlook Terrace, Madison, Wisconsin 53705 (USA).

Recent Advances in Studies on
Cardiac Structure and Metabolism Volume 10
The Metabolism of Contraction
Edited by Paul-Emile Roy and George Rona

HYPOXIA-INDUCED PROSTAGLANDIN RELEASE FROM RABBIT HEART

ÅKE WENNMALM

Department of Physiology, Karolinska Institutet, Stockholm, Sweden

SUMMARY

Rabbit hearts were perfused at a pressure of 60 cm H_2O with 37°C Tyrode solution aerated with 5% CO_2 in O_2. The effluent from the heart was collected in four consecutive 10-min periods. The samples were analyzed for prostaglandins of the E series (PGE), using thin layer chromatography for identification and assay on the superfused rat stomach strip for quantitative estimation. Three different series of experiments were performed, the hearts in each series during the second effluent collection period being perfused under conditions of either hypoxia (Tyrode solution aerated with 5% O_2 and 5% CO_2 in N_2), absence of glucose in the Tyrode solution, or hypotension/low perfusion flow (perfusion pressure lowered to 30 cm H_2O). In the series where glucose was omitted or hypotension/low perfusion flow was induced during the second effluent collection period, the outflow of PGE decreased exponentially throughout the four periods. In the series where hypoxia was maintained during the second effluent collection period, a marked increase in the outflow of PGE was noted after the end of hypoxia.

Prostaglandins of the E series are powerful vasodilators, and the PG released may, by inducing coronary vasodilation, counteract the hypoxia. However, it has been shown that, in human plasma, conversion of arachidonic acid to PGE is parallelled by platelet aggregation. Therefore, if PG synthesis is also stimulated by hypoxia in man, myocardial ischemia must be considered as a possible mechanism for platelet aggregation and subsequent coronary thrombosis.

INTRODUCTION

Prostaglandin release from tissues can be evoked by different types of stimuli. Not only autonomous nerve stimulation or vasoactive drugs elicit prostaglandin liberation (Gilmore, Vane, and Wyllie, 1968; Wennmalm, 1971; Junstad and Wennmalm, 1973; Wennmalm and Junstad, 1973), but also mechanical influences and ischemia (McGiff *et al.*, 1970; Gryglewski and Vane, 1972; Jaffe *et al.*, 1972). Increased prostaglandin synthesis in the kidney is consistently observed following renal ischemia and has been given different physiological and pathophysiological interpretations (McGiff *et al.*, 1970; Jaffe *et al.*, 1972; Lonigro *et al.*, 1973). In the ischemic rabbit heart, release of prostaglandin-like substances has been reported to occur only occasionally (Minkes, Douglas, and Needleman, 1973).

Since ischemia involves several different alterations in the perfusion conditions, the adequate stimulus–induced by ischemia–for prostaglandin (PG) synthesis and release seems unclear. Thus, we found it of interest to study the liberation of PG from the isolated perfused rabbit heart under three conditions (referred to as "ischemia equivalents") which mimic some of the physical alterations induced by ischemia, namely hypoxia, hypoglycemia, and hypotension.

METHODS

Rabbit Heart Perfusion

Rabbits of mixed strains and sexes were used for the study. The weight of the animals varied from 1.2 to 2.4 kg. After a blow on the head, the animal was exsanguinated by cutting the left carotid artery. The heart was rapidly removed and transferred to the perfusion apparatus, where it was perfused according to Langendorff with Tyrode solution of the following composition (in mM): NaCl, 136.9; KCl, 2.7; $CaCl_2$, 1.8; $MgCl_2$, 1.0; NaHCO, 11.9; NaH_2PO_4, 0.4; glucose, 5.6. The solution was continuously aerated with 5% CO_2 in O_2. The pH of the solution was 7.4–7.5. The perfusion pressure was 60 cm H_2O, and the temperature was kept at 37°C. The apex of the heart was connected to a strain gauge transducer, and heart rate and contractile force were recorded on a Grass model 5 D polygraph. The experiments were started after the heart had been spontaneously beating in the perfusion apparatus for 10–15 min. During the experiments, the effluent from the organ was collected over consecutive 10-min periods for 40 min. The effluent was immediately acidified to pH 3 and chilled.

The perfusion apparatus (Dinckelacker & Co, Mainz) consisted of two shanks, 55 cm in height, supplied with Tyrode solution from separate reservoirs. In the shanks, the solution was continuously heated and aerated by a separately thermostated circulation pump (Lauda Therm) and gas tube (AGA, Stockholm) systems. The shanks of the perfusion apparatus were joined together at the lower end with a stopcock, which allowed instant changing of the perfusion from either of the shanks. The system described thus permitted solutions at different pressures and of different composition to be kept in the shanks. The modified Tyrode solution, the prostaglandin-releasing capacity of which was to be investigated, was regularly used for perfusion during the second perfusate collection period (10–20 min). It was modified with regard to its content of oxygen (pO_2 decreased from 720 to 40 mm Hg) or glucose (glucose decreased from 5.6 to 0 mM) and as regards the perfusion pressure (lowered from 60 to 30 cm H_2O).

Analysis of Prostaglandins

Lipids in the chilled and acidified effluent from the heart were extracted twice with equal amounts of ethyl acetate. After washing of the organic phase to neutral reaction, it was evaporated to dryness. The residue was dissolved in 2 ml of water. The biological activity of the dissolved residue was tested against known amounts of PGE_2 on the superfused rat stomach strip. The assay organ was superfused by Tyrode solution of the composition given above, to which was added phentolamine (7×10^{-7} M), propranolol (8×10^{-7} M), atropine (10^{-7} M), methysergide (6×10^{-7} M), and diphenhydramine (7×10^{-7} M), in order to block activity in the strip due to the presence in the lipid extract of noradrenaline, acetylcholine, serotinin, and histamine.

Usually, 1/10 of the lipid extract from the effluent was tested on the stomach strip. Since the assay organ was sensitive to prostaglandins of the E series in amounts of less than 1 ng, outflow rates of prostaglandin-like substances in the effluent from the heart of less than 1 ng/min could be detected. Each lipid extract was tested twice, and the intraindividual variation in the samples on repeated assays was usually less than 15%. The recovery of the prostaglandin extraction procedure was checked by the use of internal standards, and ranged between 65 and 80%. The values given for outflows of prostaglandins are not corrected for losses during the extraction procedure.

RESULTS

The spontaneous beating rate of the hearts in the perfusion apparatus ranged between 105 and 165/min. It usually declined somewhat during the course of the experiments, as did the contractile force. The coronary flow ranged between 25 and 45 ml/min at the beginning of the experiments and was also found to be moderately decreased toward the end of the perfusion. The initial spontaneous outflow of prostaglandins in the perfusate ranged between 0.9 and 4.2 ng/min. The outflow varied insignificantly between consecutive periods of effluent collection, provided the heart was allowed to beat spontaneously. In most experiments, a weak tendency of the outflow of prostaglandins to decline during the course of the experiments was observed.

Perfusion of the heart with a *hypoxic Tyrode solution* (5% O_2, corresponding to a pO_2 of 40 mm Hg) induced a marked coronary vasodilation, as evidenced by an increase in the flow rate from 29 to 45 ml/min. The heart rate could not be recorded, since ventricular arrythmia was induced by the hypoxia in most of the experiments. The contractile force was markedly decreased, by 75–100% (mean 81 ± 4%). Following discontinuation of the hypoxia, regular supraventricular rhythm was restored in most cases. The contractile force also returned toward the prehypoxic value in most cases; complete recovery was, however, not achieved. During hypoxia, the liberation of prostaglandins was low. Immediately following discontinuation of perfusion with low pO_2, a significant ($p < 0.05$) increase in the outflow of prostaglandins was observed (Fig. 1). Thin layer chromatography revealed that the biological activity was due to the presence of

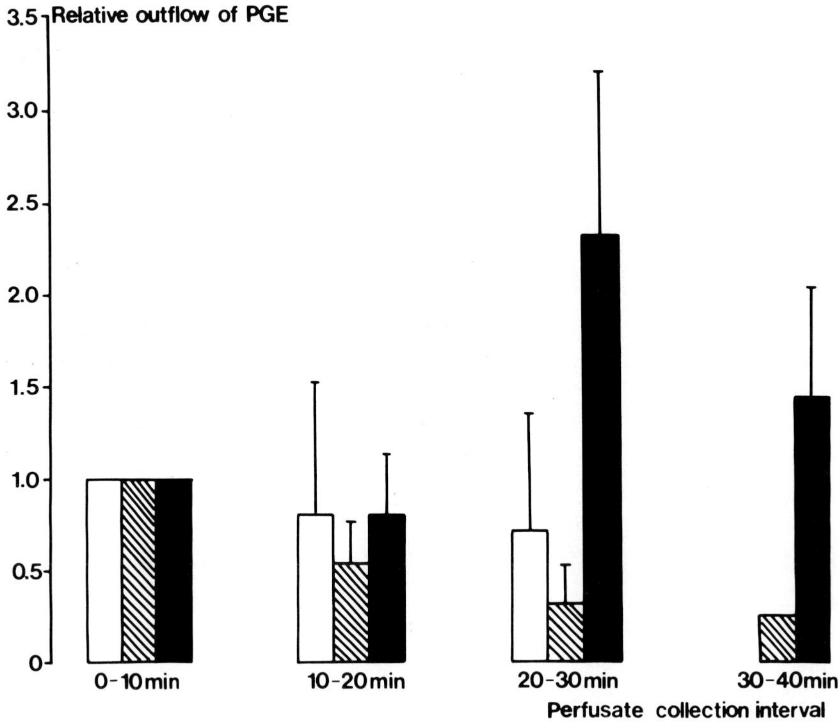

Figure 1. Perfused rabbit heart. Columns demonstrate relative outflow of PGE during four consecutive 10-min periods, expressed in relation to outflow during first 10-min period. Three different series were performed. *1, white columns:* low perfusion pressure (30 cm H_2O) during 10–20 min; normal perfusion pressure during the preceding and following periods. *2, striped columns:* absence of glucose during 10– 20 min; normal glucose concentration (5.6 mM) during preceding and following periods. *3, black columns:* hypoxia (pO_2 = 40 mm Hg) during 10–20 min; normal oxygen pressure (pO_2 = 720 mm Hg) during preceding and following periods.

PGE, mainly PGE_2, in the lipid extract of the effluent. The outflow rate of prostaglandins remained elevated during the last perfusate collection period (30–40 min).

Absence of glucose did not significantly affect the mechanical performance of the heart; the beating frequency remained unchanged and the contractile force displayed only a minor change (−6 ± 8 %). The coronary flow showed an elevation during the period of low glucose (from 26 to 37 ml/min). The outflow of prostaglandins decreased continuously during the four perfusate collection periods in this series, without any change during the low glucose period (Fig. 1).

During *hypotension* (perfusion at 30 cm H_2O during 10–20 min), the heart rate was diminished from 107 ± 2 to 82 ± 7 min. The contractile force was slightly elevated, by 12 ± 6 %. The outflow of prostaglandins was unaffected (Fig. 1).

DISCUSSION

The aim of the present study was to analyze the influence of some variations in the perfusion conditions, induced by ischemia, on the formation and liberation of prostaglandins from the rabbit perfused heart. The method used was to separately study the effect of hypoxia, absence of glucose, and low perfusion pressure/flow on the appearance of prostaglandins in the effluent from the organ.

Hypoxia was constantly found to cause liberation of prostaglandins from the heart. Recent reports in the literature have shown divergent results on this point. Kent *et al.* (1973) found that anoxia caused prostaglandin release from dog hearts and suggested that endogenous prostaglandins are important in the regulation of the coronary blood flow. Similar results were obtained by Block and Vane (1973) after induction of anoxia in the rabbit heart. However, Minkes *et al.* (1973) reported that 5–8 min of ischemia only occasionally elicited release of prostaglandin-like substances from the perfused rabbit heart. Since ischemia has earlier been shown to effect prostaglandin liberation from other tissues and species (McGiff *et al.*, 1970; Jaffe *et al.*, 1972; Lonigro *et al.*, 1973), it seems probable that ischemia is in fact a constant stimulus for prostaglandin formation and release. The current data suggest that the proper stimulus for prostaglandin release is the hypoxia induced during ischemia. Other changes in the perfusion conditions induced by ischemia, e.g., absence of glucose and low perfusion pressure, did not elicit prostaglandin release in the present experiments and seem to be insignificant in this respect.

If hypoxia, induced via ischemia, is a constant stimulus for the release of prostaglandins from tissues, it seems logical, on the basis of their activity as vasodilators, that endogenously formed prostaglandins play an important role in the development of reactive hyperemia. Recent results in our laboratory have supported this hypothesis by showing that inhibition of prostaglandin synthesis in humans is followed by a decrease, but not complete inhibition, of reactive muscle hyperemia in man (Kilbom and Wennmalm, 1974).

It is well known that human blood platelets are capable of forming and releasing prostaglandins (Smith and Willis, 1970). It has also been shown that this formation occurs in parallel with blood clotting as well as platelet aggregation (Minkes *et al.*, 1973). Recent results have suggested that the prostaglandins released are not responsible per se for the aggregation; this would more likely be induced by an intermediate product in the synthesis of prostaglandins from arachidonic acid, probably the cyclic endoperoxide (Silver *et al.*, 1973).

Since hypoxia is a stimulus for the formation and release of prostaglandins, the possibility must be considered that ischemia, when stimulating prostaglandin formation, also enhances or induces platelet aggregation and thrombosis. The consequence of ischemia stimulation on the synthesis of prostaglandins in the myocardium might thus be dual: dilation of the coronary vessels but also enhancement or induction of platelet aggregation.

In conclusion, the experiments presented show that rabbit myocardial hypoxia elicits prostaglandin release from the organ. The prostaglandins released might be beneficial in that they cause vasodilation and thereby counteract the hypoxia. However, since conversion of arachidonic acid to prostaglandins involves the synthesis of a compound capable of inducing platelet aggregation, the possibility must be considered that hypoxia developing during myocardial ischemia elicits or enhances the formation of thrombi in the coronary vessels.

Acknowledgment This study was supported by Swedish Medical Research Council Project 4341-01.

REFERENCES

BLOCK, A. J., and J. R. VANE. 1973. Release of prostaglandins (PGs) from rabbit isolated hearts. Naunyn-Schmiedeberg's Arch. Pharmak. Suppl. 279, 19.

GILMORE, N., J. R. VANE, and J. H. WYLLIE. 1968. Prostaglandins released by the spleen. Nature (Lond). 218: 1135–1140.

GRYGLEWSKI, R., and J. R. VANE. 1972. The release of prostaglandins and rabbit aorta contracting substance (RCS) from rabbit spleen and its antagonism by anti-inflammatory drugs. Brit. J. Pharmacol. 45: 37–47.

JAFFE, B. M., C. W. PARKER, G. R. MARSHALL, and P. NEEDLEMAN. 1972. Renal concentrations of prostaglandin E in acute and chronic renal ischemia. Biochem. Biophys. Res. Comm. 49: 799–805.

JUNSTAD, M., and Å. WENNMALM. 1973. On the release of prostaglandin E_2 from the rabbit heart following infusion of noradrenaline. Acta Physiol. Scand. 87: 573–574.

KENT, K. M., R. W. ALEXANDER, J. J. PISANO, H. R. KEISER, and T. COOPER. 1973. Prostaglandin dependent coronary vasodilator responses. Physiologist 16: 361.

KILBOM, Å., and Å. WENNMALM. 1974. Prostaglandins and post-ischemic muscular vasodilation: Effect of indomethacin on forearm blood flow after ischemia. IRCS 2: 1077.

LONIGRO, A. J., H. D. ISTKOVITZ, K. CROWSHAW, and J. C. McGIFF. 1973. Dependency of renal blood flow on prostaglandin synthesis in the dog. Circulat. Res. 32: 712–717.

McGIFF, J. C., K. CROWSHAW, N. A. TERRAGNO, A. J. LONIGRO, J. C. STRAND, M. A. WILLIAMSON, J. B. LEE, and K. K. F. NG. 1970. Prostaglandin-like substances appearing in canine renal venous blood during renal ischemia. Circulat. Res. 27: 765–782.

MINKES, M. S., J. R. DOUGLAS, Jr. and P. NEEDLEMAN. 1973. Prostaglandin release by the isolated perfused rabbit heart. Prostaglandins 3: 439–445.

SILVER, M. J., J. B. SMITH, C. INGERMAN, and J. J. KOCSIS. 1972. Human blood prostaglandins: Formation during clotting. Prostaglandins 1: 429–436.

SILVER, M. J., J. B. SMITH, C. INGERMAN, and J. J. KOCSIS. 1973. Arachidonic acid-induced human platelet aggregation and prostaglandin formation. Prostaglandins 4: 863–875.

SMITH, J. B., and A. L. WILLIS. 1970. Formation and release of prostaglandins in response to thrombin. Brit. J. Pharmacol. 40: 545P.

WENNMALM, Å. 1971. Studies on mechanisms controlling the secretion of neurotransmitters in the rabbit heart. Acta Physiol. Scand. 82 (Suppl. 365).

WENNMALM, Å., and M. JUNSTAD. 1973. Endogenous prostaglandin mediated inhibition of parasympathetic neurotransmission in the rabbit heart? Acta Physiol. Scand. 90 (Suppl. 396).

Mailing address:
Dr. Åke Wennmalm,
Fysiologiska Institutionen, Karolinska Institutet,
S-104 01 Stockholm (Sweden).

Recent Advances in Studies on
Cardiac Structure and Metabolism Volume 10
The Metabolism of Contraction
Edited by Paul-Emile Roy and George Rona

INFLUENCE OF THE ADRENAL MEDULLA ON CARDIAC RHYTHM DISTURBANCES FOLLOWING ACUTE CORONARY ARTERY OCCLUSION

GERALD J. KELLIHER, CARLOS WIDMER,
and JAY ROBERTS

Department of Pharmacology, Medical College of Pennsylvania,
Philadelphia, Pennsylvania, USA

SUMMARY

The present study was designed to examine the relationship between epinephrine secreted from the adrenal medulla and acute occlusion of the left anterior descending (LAD) coronary artery in α-chloralose-anesthetized cats. Occlusion of the LAD artery produced arrhythmia in all of the animals, with 39% developing ventricular fibrillation (VF). Epinephrine content in the adrenal venous blood increased 5-fold immediately after occlusion and was associated with the development of arrhythmia and VF. The adrenal venous epinephrine levels increased 8-fold after occlusion in the animals that developed VF while, in those that did not develop VF, the epinephrine levels increased only 1.5-fold. Bilateral ligation of the adrenal veins 1 hour prior to occlusion of the LAD artery did not alter the pattern of development of arrhythmia but prevented the occurrence of VF. Pretreatment of animals with either practolol or sotalol 15 min prior to LAD artery occlusion also did not affect the occurrence of arrhythmia but prevented the incidence of VF. The data indicate that acute occlusion of the LAD coronary artery produces a sharp rise in epinephrine from the adrenal medulla and that epinephrine is a major factor in producing VF but does not seem to be involved to any great extent in the production of less ominous ventricular arrhythmias.

INTRODUCTION

Myocardial catecholamine levels and the release of these substances have been examined following coronary artery occlusion. There have been several reports that plasma norepinephrine levels and the urinary excretion of catecholamines are elevated in animals and humans following coronary artery occlusion (Richardson, 1963; Valori, Thomas, and Shillingford, 1967; Klein *et al.*, 1968; Lewis *et al.*, 1972). Mathes and Gudbjarnason (1970) reported that myocardial norepinephrine levels fell rapidly following coronary occlusion in both infarcted and normal cardiac tissue and that norepinephrine levels in the infarcted tissue were entirely depleted within 48 hours. Indeed, Ceremuzynski, Staszewska-Barczak, and Herbaczynska-Cedro (1969) found that acute occlusion of the anterior descending branch of the left coronary artery in dogs led to the development of ventricular arrhythmia and that the arrhythmia was associated with an increased release of catecholamines into the blood stream. Furthermore, these workers found that, in 86% of the animals studied, epinephrine was the catecholamine secreted. They also found that, in dogs which failed to secrete catecholamines and showed no ectopic activity after coronary ligation, intravenous infusion of epinephrine provoked ventricular arrhythmia. Since epinephrine in the doses used does not induce arrhythmia in dogs without coronary artery ligation, the development of ventricular arrhythmias in these circumstances appears to be related to the combination of coronary artery occlusion and increased plasma catecholamine levels. Correlation of the severity of ventricular arrhythmias with an increase in norepinephrine and epinephrine excretion in patients with acute myocardial infarctions has been reported by Jewitt *et al.* (1969). More recently, Lubonsky and Oganov (1972) have shown that there is a marked increase in norepinephrine and epinephrine levels in both the plasma and urine of patients with acute myocardial infarction. Furthermore, Lewis *et al.* (1972), Valori *et al.* (1967), and Januszewicz *et al.* (1968) have shown that a high incidence of arrhythmia is associated with high plasma levels of epinephrine. Since the adrenal medulla is the primary source of endogenous epinephrine, these findings suggest that catecholamine secretion from the adrenal gland plays an important part in the sequelae following coronary occlusion.

The present study was designed to determine if acute occlusion of the left anterior descending coronary artery produces an increase in the outflow of epinephrine from the adrenal medulla and whether this effect is associated with arrhythmia and death in the early stages after occlusion.

MATERIALS AND METHODS

Conditioned cats, unselected as to sex and ranging in weight from 1.3 kg to 5.1 kg, were anesthetized with α-chloralose, 70 mg/kg, intravenously. The trachea was cannulated to ventilate the animals artificially throughout the experiment with room air. The animals were respired at a rate of 18 cycles/min. The total air delivered by each inspiratory movement was adjusted by varying the volume per stroke in such a way as to barely overcome the spontaneous respiratory movements observed at the beginning of the experiment.

The right femoral artery and vein were cannulated for monitoring blood pressure and for drug administration, respectively. Lead II of the electrocardiogram (ECG) was monitored continuously and, along with the blood pressure, was recorded on a Grass polygraph. The animal's temperature was monitored by means of rectal probe connected to a Yellow Springs telethermometer; a heating pad was used to keep the temperature at 37°C.

Adrenal vein cannulation

In order to obtain samples of adrenal venous blood, an 8-cm longitudinal incision was made just caudal to the xiphoid appendix along the midventral line.

The hollow viscera were displaced with a piece of gauze sponge and the retroperitoneal leaflet was incised in order to expose the left adrenal gland. The animal was then given 0.5 ml/kg of sodium heparin (1,000 USP units/ml) through the femoral vein catheter. The left suprarenolumbar vein (adrenal vein) was tied at its entrance to the inferior vena cava. A polyethylene catheter (PE 90) was rapidly inserted through a needle into the adrenal vein and fixed in place with silk (000). No more than 3 mm of catheter were introduced into the vein. The PE 90 catheter was then guided through the right lateral abdominal wall by means of a spinal needle. This adrenal vein catheter was connected to another catheter (PE 160) which had previously been introduced into the femoral vein. A snug fit between the PE 90 and PE 160 catheters allowed us to maintain blood flow through the left adrenal gland between samples and to recirculate the adrenal venous blood into the general circulation. The abdominal wall was then closed, using surgical clips. Adrenal vein blood samples of 2 ml were collected by disconnecting the catheters, clamping the femoral catheter, and allowing the adrenal vein effluent to flow through the catheter into the collecting tube for a period of approximately 2 min. The blood flow through the adrenal vein catheter was approximately 1.0 ml/min.

Centrifuge tubes were kept in ice for collecting blood samples. Immediately after collection, the blood samples were spun down at $80 \times g$ for 5 min, and the plasma, along with the appropriate internal standards for epinephrine, were immediately frozen in Dry Ice. The samples were assayed for epinephrine the following morning, using the fluorometric technique of Anton and Sayre (1962).

A control adrenal venous blood sample was taken 15 min after a ligature was placed beneath the left anterior descending (LAD) coronary artery, immediately distal to the circumflex artery. This ligature was securely tied at the moment when occlusion was produced. A second adrenal blood sample was taken immediately after the coronary occlusion, and at 15-min intervals thereafter until the end of the experiment.

A control series of experiments was carried out in which the coronary surgery was carried out up to the point where the ligature was placed beneath the LAD coronary artery, but no occlusion was performed (see below). The animals were then followed for 2 hours monitoring ECG, blood pressure, and temperature, with adrenal blood sampling for epinephrine determination every 15 min. At the end of 2 hours, these animals were exsanguinated to determine the effects of hypotension on adrenal epinephrine output.

Coronary occlusion

The left chest wall was incised along the fifth rib. The muscles were cut in between ties, using umbilical ligatures. The fifth rib was cut, and the chest was opened. The pericardium was incised in order to expose the heart. The medial border of the incised pericardium was temporarily attached to the chest wall and the left atrial appendage was lifted from the heart, using a silk tie (000). This procedure exposed the LAD artery, which was then dissected to place a ligature (00 silk) beneath it, distal to the circumflex artery.

A piece of saline-wet gauze was placed in the chest wound; 15 min later, occlusion of the artery was produced by securely tying the ligature.

Adrenal vein ligation

To produce ligation of both adrenal veins, the glands were exposed through the same surgical approach as described above. The adrenal vein was then tied as it approaches and leaves the adrenal gland. Coronary occlusion was produced 1 hour after the second adrenal vein was ligated. To make this series of experiments comparable to those in which cannulation of the adrenal vein was performed, heparin (500 units) was administered to the animals.

A control series of experiments was also carried out, in which both adrenal veins were tied as described above and the coronary surgery was performed to the point where the ligature was placed beneath the LAD coronary artery but the artery was not occluded. The animals were followed for 4 hours without coronary occlusion, monitoring ECG, blood pressure, and temperature; at the end of the observation period the animals were sacrificed.

Beta-blockade

To evaluate the effects of beta-receptor blockade, another series of experiments were performed in which the left adrenal vein was cannulated and coronary occlusion was produced. The adrenal veins were not ligated in these experiments. At 15 min before occlusion of the LAD coronary artery, practolol (4 mg/kg) was given intravenously to the animals at the rate of 0.5 mg/kg/min. Since practolol has been reported to possess a neural depressant action in addition to producing beta-blockade (Kelliher and Roberts, 1974), we also used sotalol, which has been reported not to exert effects other than beta-blockade (Somani and Lum, 1965; Kelliher and Roberts, 1974). The drug was administered in a dose of 2 mg/kg, under the same conditions as those used for practolol.

The beta-blockers were prepared in saline just prior to administration, and the dose is expressed in terms of the hydrochloride salt. The degree of receptor blockade was determined in each cat by noting the response to isoproterenol before and after practolol. This was done during the period before ligation of the coronary artery.

The mean ± standard error are presented in the text. Student's *t*-test was used to determine the significance of differences for epinephrine values; Chi square analysis was used to test the significance of incidence of fibrillation; p values of 0.05 or less were considered significant.

RESULTS

Time to death and incidence of fibrillation

Controls In 4 animals in the control series in which a ligature was placed under the LAD but not tied, there were no instances of arrhythmia and ventricular fibrillation (VF) did not occur. All of the animals survived the 4-hour observation period and were sacrificed at that time. Acute occlusion of the LAD branch of the left coronary artery in 23 cats resulted in ventricular arrhythmia in 96% (22/23) of the animals. Coronary occlusion was followed within 1 min by sinus bradycardia in about 70% (16/23) of the animals (Table 1). Shortly thereafter (within 2 min), ventricular arrhythmia developed with brief episodes of ventricular tachycardia. In 39% (9/23) of the animals, ventricular arrhythmia was followed by VF; this usually occurred within 11 min after the tie was applied. The ECG from a typical experiment is depicted in Figure 1. Of the animals that did not fibrillate, 43% (10/23) developed very low blood pressure with a "shocklike" state and died at an average time of 64 min after occlusion. The remaining 4 animals were sacrificed about 2 hours after occlusion, at which time their heart rate and blood pressure were similar to preocclusion values and they were in normal sinus rhythm with occasional premature ventricular complexes (PVC).

Adrenal vein ligation Bilateral ligation of the adrenal veins was produced in 7 cats 1 hour prior to coronary occlusion. Acute occlusion of the LAD coronary artery produced a pattern of arrhythmic activity similar to that in the control series (Fig. 1). Sinus bradycardia occurred in less than 1 min in about 43% (3/7) of the animals after occlusion and was followed within 6 min by ventricular arrhythmia in 100% (7/7) of the animals (Table 1). The most striking finding in this series of experiments was that VF did not occur in any of the animals. Over one-half (57%; 4/7) of the animals developed very low blood pressures and died in an average time of 138 min. The remaining animals (3/7) were sacrificed 4–5 hours after coronary occlusion; at the time of sacrifice, the blood pressure and heart rate were similar to control values and there was normal sinus rhythm with an occasional PVC.

Beta-blockade In the practolol-treated group, sinus bradycardia was recorded in 20% (1/5) of the animals in less than 1 min, with ventricular beats occurring in all of the animals within 2 min after occlusion (Table 1). VF did not occur in this group; however, 40% (2/5) of the animals developed low blood pressure and died at about 44 min after coronary occlusion. The remaining 60% (3/5) of the animals were sacrificed 2 hours after occlusion; the blood pressure and heart rate were similar to control values and there was normal sinus rhythm with an occasional PVC.

Sinus bradycardia was absent in the sotalol-treated group, but ventricular beats occurred within about 2 min and were present in 100% (5/5) of the animals

Table 1. Incidence and time of occurrence of arrhythmia after coronary occlusion

Experimental group	Bradycardia*		Premature Ventricular Complex		Ventricular Fibrillation	
	Occurrence (%)	Onset after occlusion (min ± SE)	Occurrence (%)	Onset after occlusion (min ± SE)	Occurrence (%)	Onset after occlusion (min ± SE)
Adrenal veins, intact	69.6 (16/23)	0.8±0.1	96 (22/23)	2.4±0.5	39.1 (9/23)	11.0±3.9
Adrenal veins, ligated	42.8 (3/7)	0.7±0.1	100(7/7)	6.2±2.6	0	–
Practolol-pretreated	20.0 (1/5)†	0.6	100 (5/5)	2.1±0.2	0	–
Sotalol-pretreated	0 (0/5)†	–	100 (5/5)	2.3±0.3	0	–

*Sinus bradycardia = 20% drop in preocclusion heart rate.
†Significantly different from adrenal veins intact group.

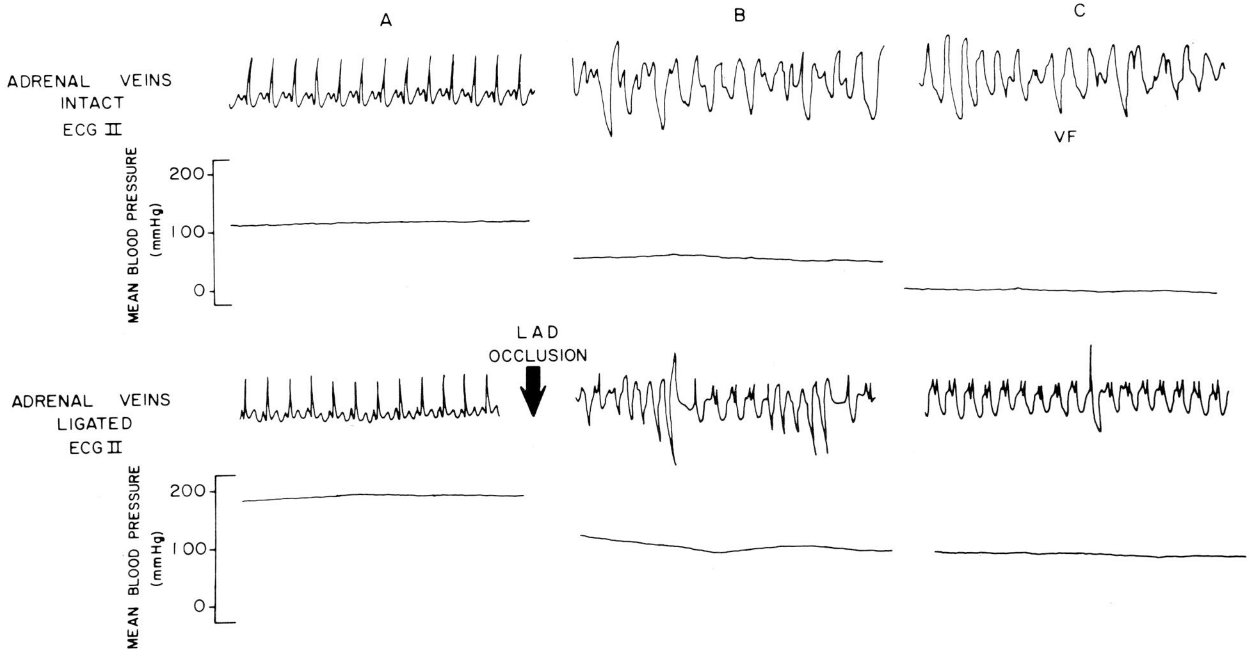

Figure 1. Effect of coronary artery occlusion on cardiac rhythm and blood pressure. *Top and bottom panels* are tracings from two separate experiments in α-chloralose anesthetized cats. *Top panels* are from an experiment in which adrenal veins were left intact, while *bottom panels* are from an experiment in which both adrenal veins were ligated 1 hour before coronary artery occlusion. In *panel A* is electrocardiogram (lead II) and mean arterial blood pressure before occlusion of left anterior descending (LAD) coronary artery in each experiment. Acute occlusion of LAD coronary artery was produced between *panels A and B* and is indicated by *arrow. Panel B* was recorded 8 min after coronary occlusion and shows ventricular arrhythmias in both cats. *Top half of panel C* was recorded 10 min after occlusion, when the animal was in ventricular fibrillation (VF). *Bottom half of panel C* shows ECG and blood pressure 1 hour after occlusion, when the animal was in normal sinus rhythm with an occasional premature ventricular complex.

(Table 1). Nevertheless, VF did not occur in any of the animals in this group. In 40% (2/5), low blood pressure developed and these animals died in about 78 min; the remaining 60% (3/5) were sacrificed 2 hours after occlusion, with a blood pressure close to control values and in normal sinus rhythm with an occasional PVC.

Adrenal vein epinephrine

Controls In a control series of experiments using four animals, adrenal vein epinephrine samples were determined before and within 2 min after a ligature was placed under the LAD artery but not tied. These values were 0.06 ± 0.03 and 0.03 ± 0.01 μg/ml, respectively, demonstrating that in the control situation adrenal vein epinephrine does not increase because of manipulation of the artery ($p>0.05$; Fig. 2.). To simulate the low blood pressure encountered in fibrillation, rapid hemmorhage of the animal was performed at the time of sacrifice; this produced levels of epinephrine in the adrenal effluent of 0.46 ± 0.09 μg/ml ($p<0.05$).

When the LAD artery was occluded in a series of 17 animals, adrenal venous epinephrine increased five-fold from 0.04 ± 0.02 μg/ml before ligation to 0.22 ± 0.14 μg/ml within 2 min after ligation (Fig. 2). The sharp rise in epinephrine output from the adrenal medulla was closely associated with the onset of ventricular arrhythmia and VF. This is in marked contrast to the epinephrine value obtained in cats without occlusion (0.03 ± 0.01 μg/ml), suggesting that acute occlusion of the LAD artery produces a marked rise in epinephrine output from the adrenal medulla and that the epinephrine is involved in the production of fibrillation.

Since all animals did not fibrillate, a comparison of the epinephrine values in those which did or did not was made. The epinephrine values after occlusion in the animals that fibrillated were increased 8-fold over controls, while those in animals that did not fibrillate were increased only 1.5-fold. In the animals that developed VF, the adrenal venous blood epinephrine was 0.088 ± 0.068 μg/ml before occlusion and increased to 0.747 ± 0.426 μg/ml after occlusion, while those without VF had epinephrine values of 0.022 ± 0.003 and 0.031 ± 0.007 μg/ml before and after occlusion, respectively ($p<0.005$; Fig. 3).

Adrenal vein ligation No blood samples were obtained from these animals, since both adrenal veins had been ligated prior to coronary artery occlusion.

Beta-blockade Pretreatment with practolol 15 min prior to LAD artery occlusion appeared to diminish the response of the adrenal medulla to coronary occlusion. Adrenal venous blood epinephrine content after practolol but prior to occlusion averaged 0.03 ± 0.01 μg/ml in three cats and rose after occlusion to 0.06 ± 0.001 μg/ml (Fig. 2). In three animals of the sotalol-treated group, epinephrine values also approximately doubled after occlusion, rising from 0.026 to 0.056 μg/ml (Fig. 2). It should be noted that the values for epinephrine after

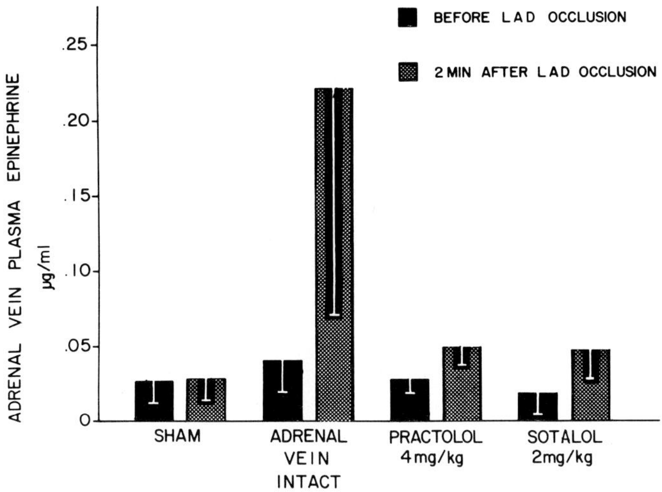

Figure 2. Effect of coronary artery occlusion on adrenal venous blood epinephrine content. Adrenal venous blood samples were analyzed for epinephrine and are expressed on the ordinate in µg/ml plasma. ***Solid bars*** represent mean epinephrine values before occlusion of left anterior descending (LAD) coronary artery, while ***stippled bars*** represent mean epinephrine values obtained 2 min after occlusion. ***Vertical lines within bars*** represent standard error of mean. In sham series (N=3), all surgery was done except that the coronary artery was not occluded. Practolol (N=3) or sotalol (N=3) was administered intravenously 15 min prior to LAD artery occlusion. Epinephrine level after occlusion in the adrenal vein intact series (N=17) is not significantly different from epinephrine level after occlusion in the sham series (p>0.05).

occlusion in the practolol- and sotalol-treated groups are considerably less than the comparable values in groups not treated with the beta-blockers (0.22 ± 0.14 µg/ml).

DISCUSSION

The purpose of the present study was to examine the importance of epinephrine release from the adrenal medulla in the cardiac rhythm disturbances that are produced by acute occlusion of the anterior descending branch of the left coronary artery. The results indicate that epinephrine may be a major determinant in the production of fatal rhythm disorders, i.e., VF, but is not involved in less ominous arrhythmias such as premature ventricular beats and ventricular tachycardia. These findings have far reaching implications, since it was previously thought that PVC are harbingers of VF (Lown, Kleinand and Hershberg, 1969; Moss *et al.,* 1972) and that antiarrhythmic measures should be aimed at reducing PVC to lessen the chances of developing VF. The present results suggest that PVC *per se* need not be suppressed in order to protect against VF and that

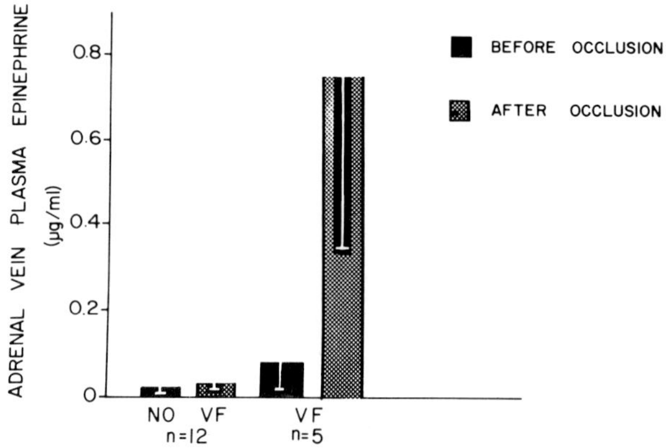

Figure 3. Correlation of adrenal venous blood epinphrine levels with ventricular fibrillation (VF) after coronary artery occlusion. Epinphrine was determined in blood samples from adrenal vein before and immediately after acute occlusion of anterior descending branch of left coronary artery. *Ordinate* is epinephrine content of blood expressed in µg/ml. *Solid bars* are mean values obtained from animals before occlusion of artery, while *stippled bars* are mean values obtained from animals 2 min after coronary artery occlusion. *Vertical line within each bar* represents standard error of mean. Number of animals in each experiment is depicted by n. Animals that developed VF had a significantly higher epinephrine value after occlusion than animals that did not develop VF ($p<0.005$).

the two phenomena are not necessarily related. Our studies indicate that VF after acute coronary occlusion is closely associated with the secretion of epinephrine from the adrenal medulla.

There are many reports in the literature demonstrating an elevation of serum and urinary catecholamine levels after myocardial infarction (Richardson, 1963; Valori *et al.*, 1967; Klein *et al.*, 1968; Lewis *et al.*, 1972). Most investigators have attempted to relate the increase in catecholamine level to the presence and severity of the arrhythmia. Although some of these studies showed a positive correlation (Jewitt *et al.*, 1969; Lubonsky and Oganov, 1972), other investigators reported no correlation (Videbaek, Christensen, and Sterndorff, 1972). We have found that, although pretreatment with 6-hydroxydopamine or bilateral adrenalectomy alone does not prevent the development of ventricular rhythm disorders, such measures substantially reduce the incidence of VF (Kelliher and Roberts, unpublished observations). It appears from our data and those of others that the question of whether a correlation exists between catecholamine activity and cardiac rhythm disorders seems to depend on the type of arrhythmia. In this regard, the less ominous types of arrhythmia show little relationship with the level of catecholamine activity, while the more serious rhythm disorders may be a direct result of the increase in catecholamine activity. Other evidence suggesting an involvement of catecholamines in the more ominous types of arrhythmia comes from the adrenal vein ligation experiments. The results of these experiments showed that epinephrine may be involved in the production of VF after myocardial infarction but not in the genesis of ventricular arrhythmia. Furthermore, the results with practolol and sotalol suggest that epinephrine is producing VF through an action on beta receptors, since both of these agents are effective beta-blocking agents and both agents prevented the occurrence of VF.

The results of the present study show that pretreatment with either practolol or sotalol reduced the secretion of epinephrine from the adrenal medulla and it is possible that this action rather than the beta-receptor blocking action was responsible for the antifibrillatory effect of these agents. There is recent evidence to support the finding that these agents can affect the release of catecholamine from the adrenal medulla. In this regard, Proakis *et al.* (1974) have reported that sotalol can reduce the release of catecholamine from the adrenal medulla that is produced *in vitro* by acetylcholine or potassium. In addition, our own results have demonstrated that practolol but not sotalol diminished adrenergic nervous discharge in postganglionic fibers of the cardiac accelerator nerve in doses similar to those employed in the present study (Kelliher and Roberts, 1973). Thus, in the present study, it is possible that sotalol—through an action to diminish the effects of the transmitter acetylcholine to produce release of epinephrine from the adrenal medulla—and practolol—by an action to diminish neural discharge in the nerves innervating the adrenal medulla—prevented the sudden surge of epinephrine that occurs after acute coronary artery occlusion and thus prevented VF.

It should be emphasized that neither practolol nor sotalol was effective in preventing ventricular arrhythmia. Previous reports have presented conflicting evidence of the effectiveness of beta-blockers as antiarrhythmic agents after myocardial infarction. Lucchesi, Whitsitt, and Stickney (1967) and Somani and his colleagues (1965, 1968) reported that they were ineffective against ventricular arrhythmias after myocardial infarction, while Pentecose and Austen (1966), Schmid and Hanna (1967), Fearson (1967), Ceremuzynski *et al.* (1969), Lemberg, Castellanos, and Arcebal (1970), and Bocage, Otero, and Harris (1972) have reported that the agents were effective against these arrhythmias. It is possible that these reports are not in conflict with each other, since the discrepancies in results may be due to the severity of the arrhythmia being evaluated. Thus, beta-blocking agents appear to be very effective in preventing the more ominous rhythm disorders such as VF, while they are not very effective against less serious arrhythmias.

Epinephrine from the adrenal medulla may not be solely responsible for the occurrence of VF after acute myocardial infarction. Harris, Estandia, and Tillotson (1951) as well as Ebert, Allgood, and Sabiston (1970) have demonstrated that surgical removal of the adrenergic innervation to the heart is effective in reducing but not eliminating the development of VF after occlusion of the LAD coronary artery. Since myocardial infarction by itself lowers the threshold to fibrillation and since catecholamines have a similar effect, it is not surprising that the combination of these two interventions produces a high incidence of VF. Most authors have investigated the role of adrenergic innervation to the heart in this interaction, while the present study has examined the influence of epinephrine from the adrenal gland. Our results indicate that epinephrine from the adrenal medulla is mainly involved, since complete abolition of VF may be achieved by measures which eliminated secretions from the adrenal medulla.

In the clinical or other experimental situation, it is uncertain which source of catecholamine is most important in producing VF. It is probable that both the adrenergic nervous system and the adrenal medulla play a role. In any case, the clinical implication of our results is that blockade of the action of catecholamines on the heart would at least prevent the development of lethal rhythm disorders (VF). This will allow the patient sufficient time to be transported to the coronary care unit where other antiarrhythmic and therapeutic measures can be instituted.

The most widely used agent in the treatment of ventricular arrhythmias produced by acute myocardial infarction is lidocaine (Lown *et al.*, 1967; Lown and Vassaux, 1968). Recently, however, there have been a growing number of reports suggesting that this agent is of only limited usefulness in the early stages after myocardial infarction (Chopra *et al.*, 1971; Gamble and Cohn, 1972; Beiser *et al.*, 1972). The ineffectiveness of this agent in suppressing arrhythmia in the acute phase of myocardial infarction may be due to a negligible effect against the high level of adrenergic nervous activity that results immediately after myocardial infarction (Harris *et al.*, 1951), while the lack of effectiveness against VF may be due to the drug's failure to block or influence the increased secretion of epinephrine from the adrenal gland. Indeed, Allen *et al.* (1974) have reported recently that lidocaine is ineffective against ventricular rhythm disorders induced by epinephrine in dogs with occlusion of the anterior descending branch of the left coronary artery. Since it is mainly the direct effects of lidocaine on the heart which explain its antiarrhythmic activity, we would anticipate that it would be more effective at a later time interval after myocardial infarction when adrenergic nervous activity and catecholamine secreted from the adrenal medulla are less involved in the genesis of the rhythm disorder. This situation may be the case in the coronary care unit, since the patient usually does not reach the coronary care unit until several hours after the initial episode of infarction (McNeilly and Pemberton, 1968). In this circumstance, lidocaine would then be administered against a background of adrenergic activity that is much less than that prevailing in the 1st hour.

Acknowledgments The original investigations reported in this manuscript were supported in part by United States Public Health Service Grant HL 16283 and by the Whitehall Foundation. The authors wish to express their appreciation to Robert Trent for technical assistance and to Mrs. Ruth Adams for her help in preparing this manuscript.

REFERENCES

ALLEN, J. D., R. G. G. JAMES, B. T. McNAMEE, R. G. SHANKS, and J. F. PANTRIDGE. 1974. Adrenaline-induced lignocaine-resistant dysrhythmias in experimental myocardial infarction. Am. Heart J. 87: 21–25.

ANTON, A., and D. F. SAYRE. 1962. A study of the factors affecting the aluminum oxide trihydroxyindole procedure for the analysis of catecholamines. J. Pharmacol. Exp. Therap. 138: 360–375.

BEISER, G. D., D. R. ROSING, J. V. TALANO, R. B. KARSH, N. H. MILLER, and S. E. EPSTEIN. 1972. Treatment of ventricular arrhythmias occurring during acute coronary artery occlusion in conscious dogs. Am. J. Cardiol. 29: 253.

BOCAGE, A. J., H. OTERO, and A. S. HARRIS. 1972. Effectiveness of beta adrenergic blockade in prevention of ventricular arrhythmias after coronary occlusion. 5th Int. Congr. Pharmacol., p. 24.

CEREMUZYNSKI, L., J. STASZEWSKA-BARCZAK, and K. HERBACZYNSKA-CEDRO. 1969. Cardiac rhythm disturbances and the release of catecholamine after acute coronary occlusion in dogs. Cardiovasc. Res. 3: 190–197.

CHOPRA, M. P., U. THADANI, R. W. PORTAL, and C. P. ABER. 1971. Lignocaine therapy for ventricular ectopic activity after acute myocardial infarction: a double-blind trial. Brit. Med. J. 3: 668–670.

EBERT, P. A., R. J. ALLGOOD, and D. C. SABISTON. 1968. The antiarrhythmic effects of cardiac denervation. Ann. Surg. 168: 728.

FEARSON, R. D. 1967. Propranolol in the prevention of ventricular fibrillation due to experimental coronary artery occlusion. Am. J. Cardiol. 20: 222–228.

GAMBLE, O. W., and K. COHN. 1972. Effect of propranolol, procainamide, and lidocaine on ventricular automaticity and re-entry in experimental myocardial infarction. Circulation 46: 498–506.

HARRIS, A. S., A. ESTANDIA, and R. F. TILLOTSON. 1951. Ventricular ectopic rhythms and ventricular fibrillation following cardiac sympathectomy and coronary occlusion. Am. J. Physiol. 165: 505–512.

JANUSZEWICZ, W., M. SZNAJDERMAN, B. WOCIAL, and J. PREIBISZ. 1968. Urinary excretion of free norepinephrine and free epinephrine in patients with acute myocardial infarction in relation to its clinical course. Am. Heart J. 76: 345.

JEWITT, D. E., C. J. MERCER, D. REID, M. THOMAS, C. VALORI, and J. SHILLINGFORD. 1969. Free noradrenaline and adrenaline excretion in relation to the development of cardiac arrhythmias and heart-failure in patients with acute myocardial infarction. Lancet *i*: 635–641.

KELLIHER, G. J., and J. ROBERTS. 1973. A comparison of the effects of sotalol and of the isomers of practolol on adrenergic nervous activity. Life Sciences 13: 1231–1243.

KELLIHER, G. J., and J. ROBERTS. 1974. A study of the antiarrhythmic action of certain beta-blocking agents. Am. Heart J. 87(4): 458–467.

KLEIN, R. F., W. G. TROYER, H. G. THOMPSON, M. D. BOGDONOFF, and A. G. WALLACE. 1968. Catecholamine excretion in myocardial infarction. Arch. Intern. Med. 00: 122–476.

LEMBERG, L., A. CASTELLANOS, and A. ARCEBAL. 1970. The use of propranolol in arrhythmias complicating acute myocardial infarction. Am. Heart J. 80: 479–487.

LEWIS, R. P., H. BOUDOULAS, W. F. FORESTER, and A. M. WEISSLER. 1972. Shortening of electromechanical systole as a manifestation of excessive adrenergic stimulation in acute myocardial infarction. Circulation 46: 856–892.

LOWN, B., A. M. FAKHRO, W. B. HOOD, Jr., 1967. The coronary care unit. New perspectives and directions. J. Am. Med. Ass. 199: 188–198.
LOWN, B., M. D. KLEINAND, and P. I. HERSHBERG. 1969. Coronary and precoronary care. Am. J. Med. 46: 705–724.
LOWN, B., and C. VASSAUX. 1968. Lidocaine in acute myocardial infarction. Am. Heart J. 76: 586.
LUBONSKY, P. E., and R. G. OGANOV. 1972. Blood plasma catecholamines and their urinary excretion in patients with acute myocardial infarction. Am. Heart J. 83: 182–188.
LUCCHESI, B. R., L. WHITSITT, and J. L. STICKNEY. 1967. Antiarrhythmic effects of beta adrenergic blocking agents. Ann. N.Y. Acad. Sci. 139: 940–951.
MATHES, P., and S. GUDBJARNASON. 1970. Uptake and subcellular distribution of norepinephrine in the canine heart following experimental myocardial infarction. Proc. Soc. Exp. Biol. Med. 135: 504–508.
McNEILLY, R. H., and J. PEMBERTON. 1968. Duration of last attack in 998 fatal cases of coronary artery disease and its relation to possible cardiac resuscitation. Brit. Med. J. 3: 139–142.
MOSS, A. J., S. GOLDSTEIN, W. GREENE, and J. De CAMILLA. 0000. Prehospital precursors of ventricular arrhythmias in acute myocardial infarction. Arch. Intern. Med. 129: 756–762.
PENTECOSE, B. L., and W. G. AUSTEN. 1966. Beta-adrenergic blockade in experimental myocardial infarction. Am. Heart J. 72: 790–796.
PROAKIS, A. G., J. L. BOROWITZ, M. R. McDOUGAL, and G. R. SPRATTO. 1974. Interaction between acetylcholine and sotalol (MJ-1999) in adrenal medulla. Proc. Soc. Exp. Biol. Med. 145: 1368–1371.
RICHARDSON, J. A. 1963. Circulating levels of catecholamines in acute myocardial infarction and angina pectoris. Progr. Cardiovasc. Dis. 6: 56–62.
SCHMID, J. R., and C. HANNA. 1967. A comparison of the antiarrhythmic actions of two new synthetic compounds, iproveratril and MJ-1999 with quinidine and pronethalol. J. Pharmacol. Exp. Therap. 156: 331–338.
SOMANI, P., and B. K. LUM. 1965. The antiarrhythmic actions of beta adrenergic blocking agents. J. Pharmacol. Exp. Therap. 147: 194–204.
SOMANI, P., and P. L. WATSON. 1968. Antiarrhythmic activity of the dextro and levo-rotatory isomers of 4-(2-isopropylamino-1-hydroxyethyl) methanesulfonanilide (MJ-1999). J. Pharmacol. Exp. Therap. 164: 317–325.
VALORI, C., M. THOMAS, and J. P. SHILLINGFORD. 1967. Urinary excretion of free noradrenaline and adrenaline following acute myocardial infarction. Lancet *i*: 127–130.
VIDEBAEK, J., N. J. CHRISTENSEN, and B. STERNDORFF. 1972. Serial determination of plasma catecholamines in myocardial infarction. Circulation 46: 846–855.

Mailing address:
Gerald J. Kelliher,
Department of Pharmacology,
Medical College of Pennsylvania,
3300 Henry Avenue, Philadelphia, Pennsylvania 19129 (USA).

Recent Advances in Studies on
Cardiac Structure and Metabolism Volume 10
The Metabolism of Contraction
Edited by Paul-Emile Roy and George Rona

AN ENZYME MECHANISM IN EXPLANATION OF PAIN IN ANGINA PECTORIS

KURT A. OSTER and DONALD J. ROSS

Department of Biology, Fairfield University,
Fairfield, Connecticut, USA

SUMMARY

The effect of catecholamines, represented by epinephrine and norepinephrine, on the activity of phospholipase A_2 from bee venom was studied. It was shown that the hydrolysis of *l*-α-lecithin to lysolecithin and a fatty acid was considerably activated by preincubation of the lecithin with the biogenic amine. On the other hand, addition of nitroglycerin or propranolol to the enzyme solution considerably curtailed activation by the catecholamines. The pharmacological effect of the split products of *l*-α-lecithin, free fatty acids (FFA), and lysolecithin in the nascent state on the myocardial cell membrane might be more plausible than the commonly accepted theory that the FFA derive from lipolysis of remote fat deposits. Certain arrhythmias and ion imbalances might be caused by catecholamine activation of phospholipase A_2. Of great pharmacological interest is the observation that this activation is inhibited by a beta-adrenergic blocking agent without the presence of cyclic adenosine monophosphate as messenger. The reaction may serve as a model for the study of the pharmacological influence of nitroglycerin and propranolol on angina pectoris.

INTRODUCTION

The cause of the pain experienced in angina pectoris is unknown. An attempt had been made to attribute the pain sensation experienced during angina pectoris attacks to the action of myocardial phospholipase A_2 (E.C.3.1.1.4) on cell membrane-associated phospholipids (Oster, 1968). The authors, recalling the known algesic effect of subcutaneously injected bee venom, which consists mainly of phospholipases, have postulated that a similar enzyme, situated in the myocardial cell, might also effect pain when triggered by such stimuli of heart action as emotion, mild exertion, heavy meals, and nonspecific stimuli which might occur during sleep.

In a published schematic approach to the metabolism of plasmalogen (Oster, 1972), the enzyme phospholipase A_2 was shown to be activated by Ca^{++} ions and catecholamines. At that time, literature describing calcium activation of the enzyme had been reviewed by Meldrum (1965). However, the postulated activation of phospholipase A_2 by catecholamines had never been described before the schematics were published. Feeling that such a significant clue to the origin of pain in heart muscle merited a thorough inquiry, we set out to investigate this postulated catecholamine activation of phospholipase A_2, hoping that such a study might also provide additional insight into the pharmacological effects of two categories of antianginal drugs, nitrites (nitroglycerin) and the beta-receptor blockers (propranolol).

The effects on lipid metabolism of injected catecholamines have thus far been tested in the intact animal or the open-chest animal preparation (Opie, 1972). In order to learn more about the basic mechanism, we preferred to use the pure enzyme in bee venom as a cell-free model, thus eliminating such extraneous influences as foreign proteins and fats, morphological structures, and the presence of other amine-inactivating enzymes.

It is understood that this approach may not be entirely applicable to the biochemical metabolism of the mammalian myocardium. Nevertheless, as will be shown in the "Discussion," we know that the cyclic AMP-mediated, lipolytic action of injected catecholamines liberates large amounts of free fatty acids (FFA) in the test animal. Combined with plasma protein, this reaction may well mask the pharmacological effects of the FFA arising from the action of phospholipase A_2, situated in the myocardial cell, on its substrate, the phospholipids. We thought that these cardiogenic FFA in their nascent state might possess more pharmacological significance than those which are transported through the circulatory system after liberation. Many investigators have reported conflicting results from FFA obtained through lipolytic action of the catecholamines (Opie, 1972). Our *in vitro* test method has also enabled us to vary the amounts of catecholamines within their physiological range without fear that other metabolic systems may trap them and make them unavailable.

METHOD

Enzymatic assay of phospholipase A_2 derived from bee venom

Reaction: Lecithin + $H_2O \xrightarrow[\text{lipase } A_2]{\text{Phospho-}}$ Lysolecithin + an unsaturated fatty acid ion.

Unit definition: 1 unit will cleave 1 μmole of *l*-α-lecithin to lysolecithin and a fatty acid in 1 min at 37°C at pH 6.6.

Method of assy: Colorimetric at 570 mμ, pH 6.6, 37°C, based primarily on the method of Magee and Thompson (1960) as modified by Dawson (1963). This measures the decrease of lecithin and is based on the ability of carboxylic ester groups in intact lipids to react with hydroxylamine in alkaline solution to yield hydroxamic acids. The hydroxamic acids react with ferric chloride to give colored ferric hydroxamic acid complexes.

Reagents	
Buffer:	0.5 M Tris, pH 6.6
Substrate:	20 mg/ml lecithin in 0.01 M calcium chloride
Activator:	1.5% sodium deoxycholate
Other:	3:1 v/v ethanol:ethyl ether
	2 M hydroxylamine HCl
	14% NaOH
	3 N HCl
	10% anhydrous ferric chloride in 0.1 N HCl
Enzyme:	Diluted in 50% glycerol to give 2 units/ml
Reaction mixture:	1) 0.7 ml lecithin emulsion, 20 mg/ml
	2) 0.7 ml 0.5 M buffer, pH 6.6
	3) 0.7 ml 1.5% sodium deoxycholate (warm to 37°C)
	4) 0.1 ml diluted enzyme (approximately 0.2 unit)
Physiological concentrations	
Activators of phospholipase A_2:	Epinephrine, 0.01 μg/100 ml
	Norephinephrine, 0.05 μg/100 ml, added to reaction mixture 5 min before enzyme
Inhibitors of activated phospholipase A_2:	Nitroglycerin, 2.4×10^{-5} mg/ml
	Propranolol, 1.0 μg/ml, added with enzyme

RESULTS

Findings in the catecholamine-activated cleavage of phosphatidylcholine by phospholipase A_2 and its partial curtailment by nitroglycerin and propranolol are depicted in Figures 1–3 (a more detailed description of the results will be published elsewhere).

Figure 1 shows the considerable activation of phospholipase A_2 by epinephrine (0.01 μg/100 ml) and the influence of nitroglycerin (2.4×10^{-5} mg/ml) on this effect. In 5 min, 8.44 μmoles of *l*-α-lecithin were hydrolyzed. This value was reduced to 6.65 μmoles by the action of nitroglycerin.

As is seen in Figure 2, the norepinephrine activation of phospholipase A_2 in 5 min to 9.63 μmoles (an increase over the normal 6.14 μmoles) was significantly

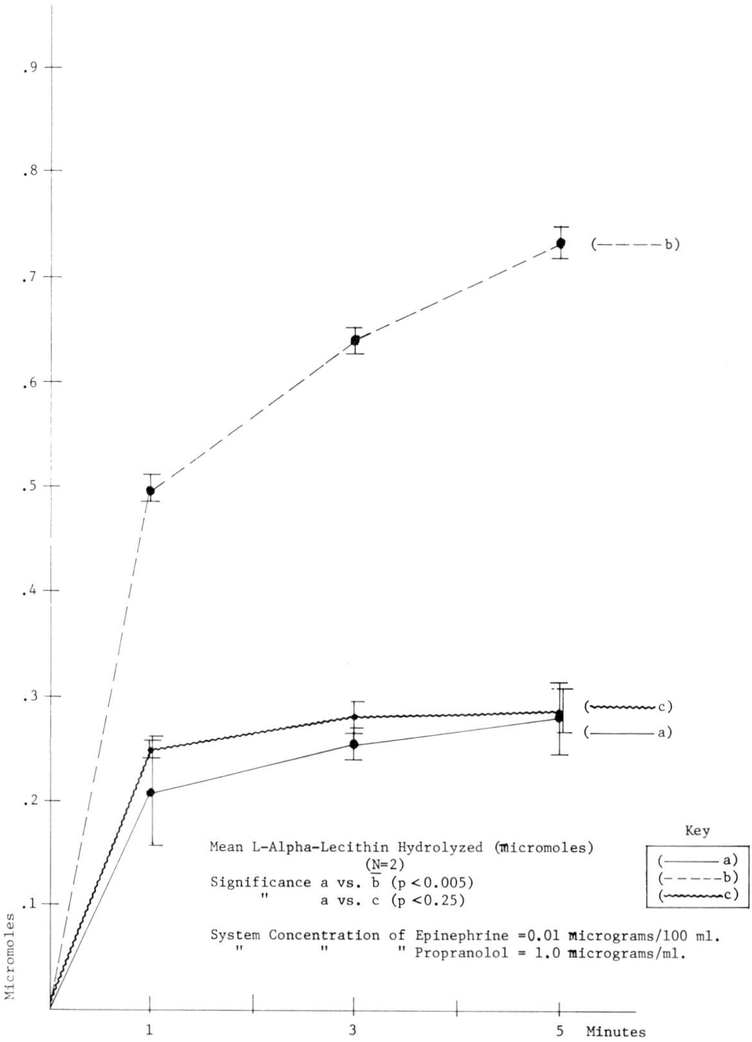

Figure 3. Epinephrine-activated phospholipase A_2 with propranolol deactivation.

lesions and prenecrotic myocardial infarction. The presence of xanthine oxidase in these locations was described (Ross, Ptaszynski, and Oster, 1973). Xanthine oxidase can act on the aldehydic moiety of plasmalogen only if there is first a cleavage of the fatty acid from the ester bond. Phospholipase A_2 seems to require the presence of only one fatty ester linkage adjacent to the alcohol-phosphate bond, and the carbon atom to which this fatty acid is attached must have a precise stereochemical configuration (Gurr and James, 1971).

In a previous publication, it was hypothesized that the known algesic effect of bee venom might be due to its high content of phospholipases, in addition to the presence of polypeptides, apamine, and mellitin (Oster, 1968). One might explain the nature of the pain-producing mechanism by assuming that algesia is caused by: a) the enzyme itself, b) a product of the enzyme action, c) a lack of enzyme substrate, or d) a possible activator of the enzyme. Phospholipases have been found in myocardial cells, either in mitochondria or in lysosomes. We have postulated that the pain experienced in angina pectoris might be associated with the presence of phospholipase A_2 in the myocardium and its algesic activity via one of the four suggested pathways. Use was made of the known pharmacological action of nitroglycerin, i.e., that of possibly aborting attacks of chest pain when administered sublingually. A direct influence of nitroglycerin on the enzyme itself was not found.

It is known that the onset of chest pain is a crescendo phenomenon; i.e., pain is not produced immediately following physical activity or emotional upheaval, but it does occur some time later. We concluded, therefore, that the activation of an eventual pain mechanism could not be a simple pain threshold diminution but, rather, a slow stimulation of a pain-producing metabolic process, which should be influenced by nitroglycerin and/or propranolol in dosages equivalent to those found to be effective *in vivo.* These postulates have found an explanation in our model system.

Oliver (1972) reviewed the evidence supporting local release of catecholamines in the myocardium triggered by ischemia. We have examined the influence of certain biogenic amines on an enzyme situated in the myocardial cell and found a gradual activation of phospholipase A_2 by epinephrine at an optimum 5-min contact with the lecithin substrate and a specific inhibition of this activation by both nitroglycerin and propranolol when added to the enzyme itself. It is of great interest and scientific relevance that this activation of enzyme (phospholipase A_2) by hormone (epinephrine) took place in a cell-free medium without the messenger action of cyclic AMP. This direct action, if paralleled *in vivo,* might have physiological significance for the understanding of other hormone action on cellular enzymes.

The split products of phosphatidyl choline, either a fatty acid (unsaturated or saturated) and/or lysolecithin, display individual pharmacological qualities. Cardiac dysrhythmia has been ascribed to the release of fatty acids from fat depots through lipolysis by catecholamine activation of cyclic AMP. We offer as an explanation the concept of fatty acids released by catecholamine activation of phospholipase A_2 acting on phospholipids. We believe that since fatty acids liberated in such a way are present *in status nascens,* not bound to a carrier protein, their dysrhythmic effect is more direct than that of fatty acids carried in the blood stream from remote reservoirs (Opie, 1972).

Lysolecithins are very active pharmacologically, releasing potassium, inorganic

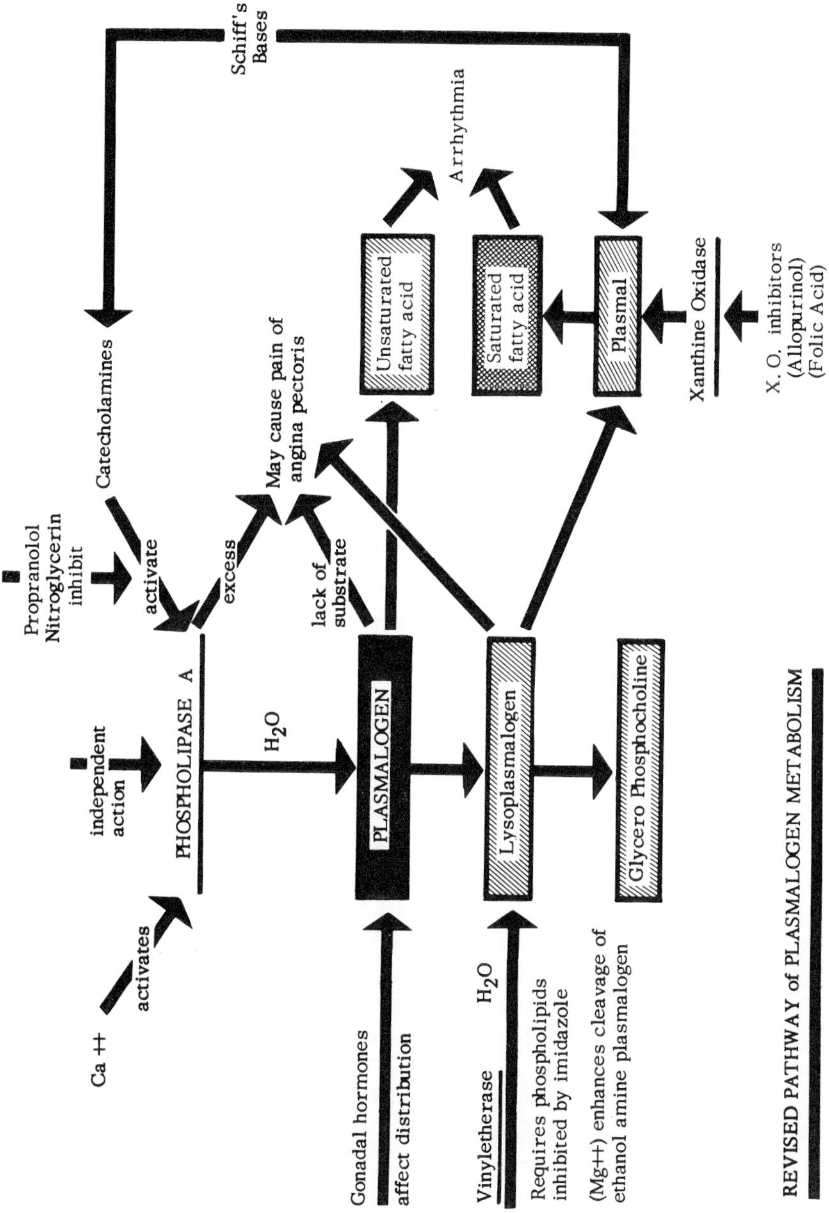

Figure 4. Revised pathway of plasmalogen metabolism.

phosphates, and soluble enzymes from muscle cells. The mechanism of action of these substances lies in their detergent qualities, which render the cell membrane more permeable, creating a difference in permeability for many electrolytes (Khairallah and Page, 1960).

Although this study has not yet fully explained what causes the actual pain of angina pectoris, we have created a model which is easily applied to further studies and to investigational manipulation *in vitro* applied to situations which might parallel the *in vivo* condition without interference from extraneous influences.

The fact that lysoplasmalogen exerts a much lesser detergent effect on the cell membrane than lysolecithin (Piantadosi and Snyder, 1970) is considered by us as further confirmation of the prenecrotic plasmalogen depletion concept. When plasmalogen is depleted by the influence of ectopic xanthine oxidase from bovine milk, this protective influence of lysoplasmalogen becomes dissipated. Lysolecithin, arising from the catecholamine activation of phospholipase A_2, can then exert its full detergent force on the cell membrane. This observation is in excellent agreement with previously predicted pathways.

The newly acquired knowledge about the activation of phospholipase A_2 by catecholamine and the influences of nitroglycerin and propranolol have been incorporated into a revised scheme of plasmalogen metabolism (Fig. 4).

Acknowledgment The authors acknowledge with gratitude the technical assistance of Mr. S. Delco.

REFERENCES

DAWSON, R. M. C. 1963. On the mechanism of action of phospholipase A. Biochem. J. 88: 114.

GURR, M. I., and A. T. JAMES. 1971. Lipid Biochemistry, pp. 140–141. Cornell University Press, Ithaca.

KHAIRALLAH, P. A., and I. H. PAGE. 1960. A vasopressor lipid in incubated plasma. Am. J. Physiol. 199(2): 341–345.

MAGEE, W. L., and R. H. S. THOMPSON. 1960. The estimation of phospholipase A activity in aqueous systems. Biochem. J. 77: 526–534.

MELDRUM, B. S. 1965. The actions of snake venoms on nerve and muscle. The pharmacology of phospholipase A and of polypeptide toxins. Pharmacol. Rev. 17: 393–445.

OLIVER, M. F. 1972. Metabolic response during impending myocardial infarction. II. Clinical implications. Circulation 45: 491–500.

OPIE, L. H. 1972. Metabolic response during impending myocardial infarction. I. Relevance of studies of glucose and fatty acid metabolism in animals. Circulation 45: 483–490.

OSTER, K. A. 1968. Treatment of angina pectoris. Cardiol. Digest (original article) 3: 29–34.

OSTER, K. A. 1972. Role of plasmalogen in heart disease. *In* E. Bajusz and G. Rona (eds.), Recent Advances in Studies on Cardiac Structure and Metabolism, Vol. 1: Myocardiology, pp. 803–813. University Park Press, Baltimore.

PIANTADOSI, C., and F. SNYDER. 1970. Plasmalogens and related derivatives: their chemistry and metabolism. J. Pharmacol. Sci. 59: 283–297.

ROSS, D. J., M. PTASZYNSKI, and K. A. OSTER. 1973. The presence of ectopic xanthine oxidase in atherosclerotic plaques and myocardial tissues. Proc. Soc. Exp. Biol. Med. 144: 523–526.

Mailing address:
Kurt A. Oster, M.D.,
881 Lafayette Boulevard,
Bridgeport, Connecticut 06604 (USA).

Recent Advances in Studies on
Cardiac Structure and Metabolism Volume 10
The Metabolism of Contraction
Edited by Paul-Emile Roy and George Rona

EFFECT OF DEEP HYPOTHERMIA, LIMITED CARDIOPULMONARY BYPASS, AND TOTAL ARREST ON GROWING PUPPIES[1]

THEODORE L. FOLKERTH, WILLIAM W. ANGELL, RICHARD G. FOSBURG, and JAMES H. OURY

Department of Thoracic-Cardiovascular Surgery and the Clinical Investigation Center, Naval Regional Medical Center, San Diego, California, USA

SUMMARY

The advantages of a bloodless field and total cardiac relaxation have popularized the technique of deep hypothermia and total circulatory arrest for the correction of complex congenital cardiac defects in infancy. There is, however, a significant potential for cerebral and pulmonary complications. Presently, the most common technique is that of using a combination of surface cooling and cardiopulmonary bypass cooling and rewarming. Normal neurological development has been claimed with the present technique of hypothermia at 20°C and total circulatory arrest for periods up to an hour; however, there are reports of seizure activity in the early postoperative period. There is also a disturbing incidence of respiratory insufficiency and, occasionally, hemorrhagic pulmonary edema.

This study, using growing puppies and subjecting them to deep hypothermia and total circulatory arrest for varying periods of time, disclosed that animals subjected to 60 min of circulatory arrest recovered neurologically; however, there were histological changes of anoxia in the brain. Animals subjected to 30 min of total circulatory arrest were normal neurologically and there was no histological evidence of anoxic damage to brain tissue. Puppies that were continuously on cardiopulmonary bypass had no significant pulmonary changes caused by increasing the inspired oxygen tension in the ventilator; however, striking changes were noted when limited cardiopulmonary bypass was employed for core cooling and total circulatory arrest combined with pulmonary ventilation with 100% oxygen. We conclude from this experimental study that the use of surface cooling and core cooling with subsequent total circulatory arrest at 20°C is a safe procedure, providing the period of time of cardiac arrest is kept around 30 min. We also conclude that the alveolar oxygen tension should be maintained at the lowest level possible during the interval of circulatory arrest to avoid the apparent rapid onset of post-traumatic pulmonary insufficiency.

---

[1] Supported in part by the Bureau of Medicine and Surgery, Department of the Navy, Clinical Investigation Control Center Work Unit 4-16-254; the American Heart Association Grant 72-901; and the National Institutes of Health Grant HL 14845-02, Palo Alto Veterans Administration Hospital, Division of Cardiovascular Surgery, Palo Alto, California. The opinions or assertions contained herein are those of the authors and are not to be construed as official or as reflecting the views of the Navy Department.

[2] Present address: Western Heart Associates Medical Group, Inc., Santa Clara Valley Medical Center, San Jose, California, USA.

INTRODUCTION

Deep hypothermia and total circulatory arrest have greatly facilitated operative correction of complex congenital cardiovascular defects in infancy. The advantages of a bloodless field and total cardiac relaxation have popularized this technique for repair of many complex congenital defects (Barratt-Boyes *et al.*, 1972; Barratt-Boyes and Neutze, 1973; Subramanian and Wagner, 1973). A potential for significant cerebral and pulmonary complications is borne out by the experience in many cardiovascular centers (Rittenhouse *et al.*, 1974). In 1962, Björk and Hultquist reported significant brain damage in a high percentage of patients undergoing open-heart surgery under profound hypothermia. The technique was not popular until Mohri and Hikasa and their co-workers used a combination of surface cooling and cardiopulmonary bypass cooling and rewarming, which significantly aided in cardiac resuscitation. Barratt-Boyes and Neutze (1973); Subramanian and Wagner (1973); and Mohri *et al.* (1968), have further refined the technique to the point of common usage in these centers. There are disturbing reports of brief episodes of seizure activity in the early postoperative period (Barratt-Boyes and Neutze, 1973). Subsequent normal neurological development is claimed with the present technique of deep hypothermia (20°C) and total circulatory arrest for periods up to an hour (Barratt-Boyes and Neutze, 1973; Subramanian and Wagner, 1973). These reports are confirmed by similar observations of our own, both clinically and in experimental cardiovascular procedures in young animals. We were also disturbed by the high incidence of pulmonary problems with respiratory insufficiency and, occasionally, hemorrhagic pulmonary edema. This study was undertaken to help define the parameters responsible for maintaining normal neurological and pulmonary function with deep hypothermia and total circulatory arrest.

METHOD

Two groups of animals were used. The first 33 puppies, weighing 5 kg or less, underwent ice surface cooling to 28°C, followed by limited cardiopulmonary bypass for core cooling to 18–20°C. In 16 animals, a Travenol heat exchanger and a Bentley infant bubble oxygenator were used and, in 17, a disk oxygenator was used. The cooling gradient was not allowed to exceed 11°C. Animals were exsanguinated during total circulatory arrest for periods of 25–90 min at 20°C. The right ventricular outflow was reconstructed. Following suture of the ventriculotomy, the puppies were rewarmed by cardiopulmonary bypass to 36°C. Based on observations made during this experiment, a second group of 24 puppies (Table 1) was sutdied to determine whether extracorporeal circulation at 20°C or total circulatory arrest at 20°C was more deleterious to the lungs and, secondly, how the inspired oxygen tension interacted with these variables in

Table 1. Pulmonary effects of deep hypothermia*

Group	Procedure	No.	Fi O_2 (%)
A	Cardiopulmonary bypass	6	20
B		4	100
C	Circulatory arrest	6	20
D		6	100

*Puppies at 20°C for 60 min.

causing pulmonary damage. Left atrial pressure was continuously monitored and not allowed to exceed 15 cm of water pressure. The technique of cardiopulmonary bypass and surface and core cooling was the same in all the groups. Lung samples from the right lungs not traumatized by surgery were taken at the end of bypass and at 12 and 18 hours after bypass. The animals were then sacrificed, lung water was measured, and the lung samples were fixed in formalin and studied under light microscopy, using hematoxylin and eosin stain. Arterial pO_2 and pCO_2, arterial and venous pH, and percentage saturation of hemoglobin determinations were made.

RESULTS

There were 16 long term survivors among the 33 puppies undergoing right ventricular outflow reconstruction with deep hypothermia. They were sacrificed at 2–9 months postoperatively, with complete autopsy being performed. Those puppies arrested for less than 45 min recovered completely, with no clinical evidence of neurological impairment. The majority of puppies cooled for over 60 min had obvious neurological impairment, with many never regaining spontaneous respiration. Those few animals surviving 60 min and the group arrested for 45–60 min had consistent neuropathological findings, pyknotic changes in the nuclei, and dropping out of ganglion cells in the hippocampus or Ammon's horn; this is seen in Figure 1. Figure 2 shows Purkinje cell dropout in the cerebellum. While this was a consistent finding in all animals with total circulatory arrest for periods of time greater than 45 min, no animals with a 30-min period of total circulatory arrest had histological evidence of anoxia.

Results of the lung study showed that the paO_2 was maintained above 70 mm mercury in animals ventilated with room air, and above 400 mm mercury in puppies ventilated with 100% oxygen. During cardiopulmonary bypass, paO_2 was consistently in excess of 400 mm mercury. The mean arterial-alveolar gradient in the room air group was 54 mm mercury, and the value was 481 mm

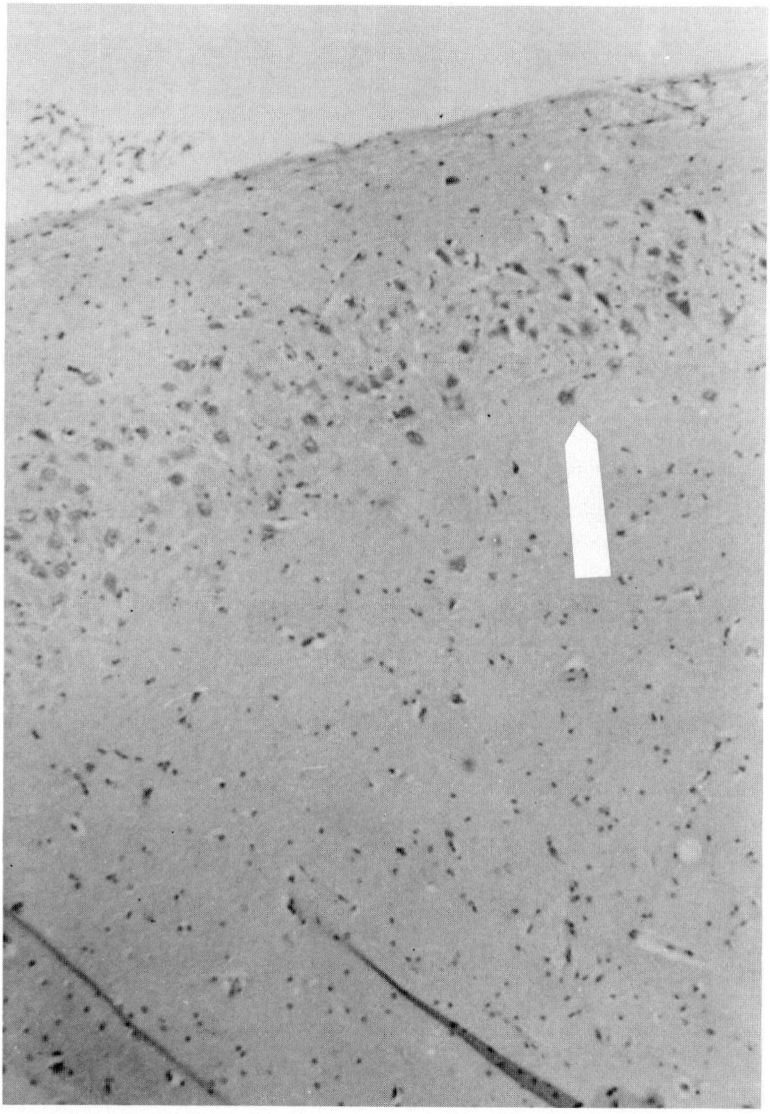

Figure 1. Ammon's horn. Note dark pyknotic nuclei (*arrow*) and reduction in pyrimidal cells. H and E. × 100.

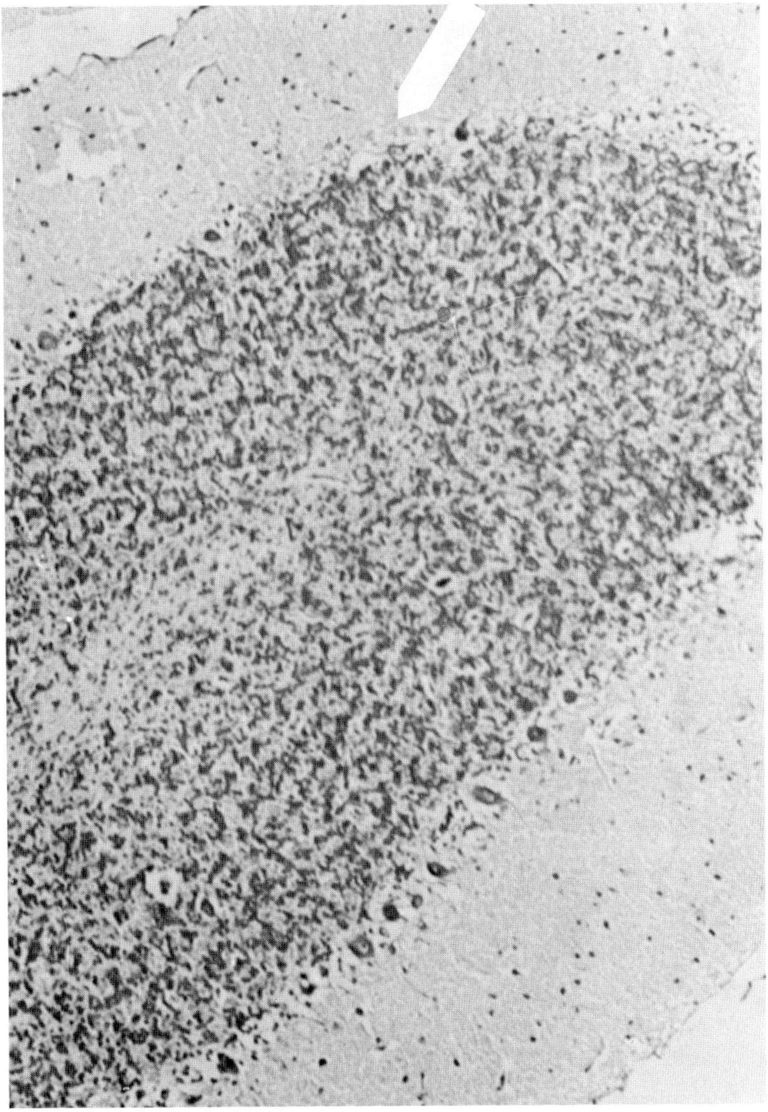

Figure 2. Cerebellum. Note dropping out of Purkinje cells (*arrow*) between molecular and granular layers. Nissl's stain. × 100.

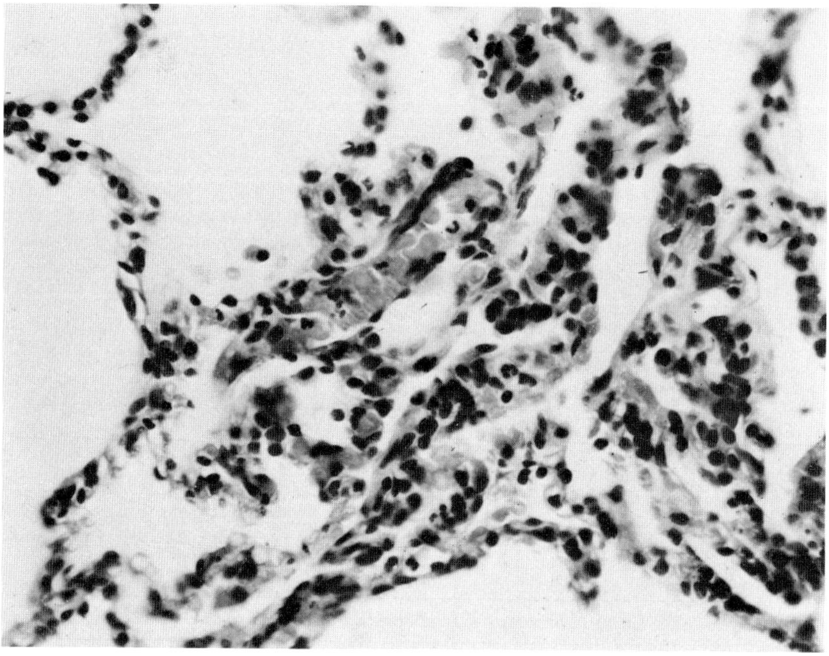

Figure 3. Tissue sample taken 12–18 hours after bypass from bypass-100% oxygen group. Note vascular congestion (*upper center*) and alveolar hemorrhage (*lower center*). × 125.

mercury in the 100% oxygen group. Oxygen saturation showed a shift to the left in the oxyhemoglobin dissociation curve, consistent with the effects of temperature and alkalosis. The respective mean cardiopulmonary bypass times for groups A and B were 109 and 125 min, and for groups C and D they were 70 and 73 min.

Animals undergoing bypass and those ventilated with room air showed minimal changes in the samples taken immediately following cardiopulmonary bypass. Of 16 puppies in groups A, B, and C, 7 had normal lung morphology; 5 showed mild changes; and the remaining 4 had mild to moderate changes, notably alveolar edema and hemorrhage. These changes are shown in Figure 3. The later samples taken 12–18 hours after bypass showed greater morphological changes, with congestion and hemorrhage or more severe changes. The significant difference was in group D (circulatory arrest-100% oxygen at 20°C), which showed profound changes in the samples taken immediately after cardiopulmonary bypass. These changes are shown in Figures 4–6. None of the lungs in these instances were normal; 2 puppies showed mild changes of vascular congestion, alveolar hemorrhage, and edema, and the remaining 4 showed moderate to

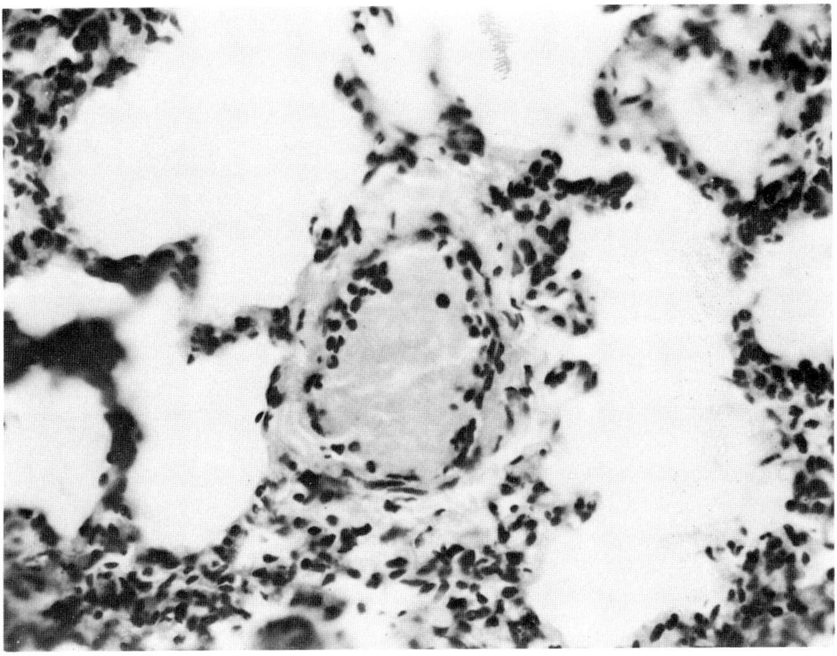

Figure 4. Vascular congestion and polymorphonuclear leukocyte margination in sample taken at end of bypass from arrest-100% oxygen group. × 125.

severe changes. The late samples from this group showed further degeneration of pulmonary microscopic morphology.

DISCUSSION

With the resurgence in the use of deep hypothermia, limited cardiopulmonary bypass, and total circulatory arrest for the correction of complex congenital heart disease in the newborn and in infants, it is imperative that in depth analysis be undertaken of morphological and functional changes in the brain in particular, and in other organ systems in general.

Using the technique of surface cooling, limited cardiopulmonary bypass for core cooling, and total circulatory arrest for variable periods of time, the clinical setting currently being used for the correction of infant complex congenital heart disease was thus simulated. Various investigators, including Subramanian and Wagner (1973) and Barratt-Boyes *et al.* (1972), have reported their recent results using deep hypothermia and circulatory arrest and pointed out that there is an approximate 60-min period of time of total circulatory arrest at 20°C

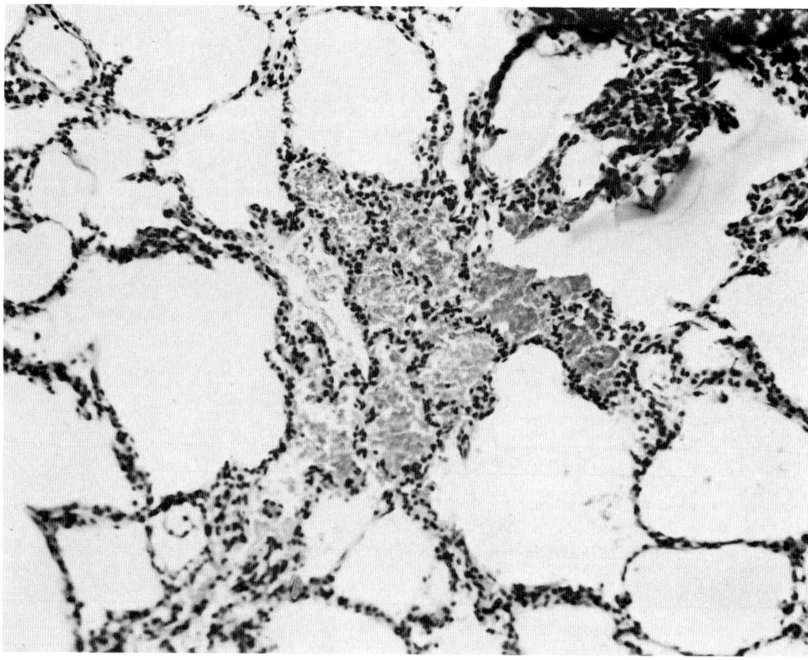

Figure 5. Alveolar hemorrhage in sample taken at end of bypass from arrest-100% oxygen group. × 50.

which is well tolerated by the infant brain. There are, however, other reports (Barratt-Boyes and Neutze, 1973) of seizure activity and abnormal neurological behavior in the early postoperative period in infants who have been subjected to arrest over this same interval of time. There are no long term follow-up studies on any of these infants, although they are described as neurologically normal with normal electroencephalograms in the early postoperative period (Subramanian and Wagner, 1973). It would seem, however, that a more sensitive index of significant neurological sequelae may be required. It is well known that ischemic damage to the central nervous system tissue is irreversible and it may be that the amount of ischemic or anoxic damage in these young infants may not be clinically apparent at a young age and only become manifest as they mature.

In the present study, the animals that were subjected to total circulatory arrest at 20°C for up to 60 min were neurologically normal; however, there was a rather significant incidence of histological changes in the brain, characterized by pyknotic changes in the nuclei and dropping out of ganglion cells in the hippocampus, and Purkinje cell dropout in the cerebellum. This would indicate that animals neurologically normal from a clinical standpoint had suffered a significant anoxic insult to brain tissue. Some factors responsible for avoidance

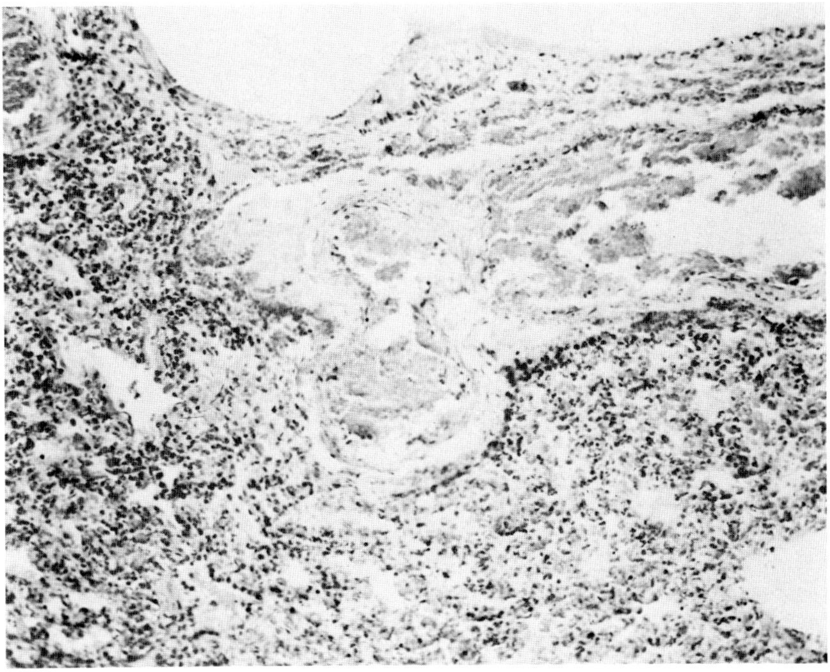

Figure 6. Tissue sample taken 12–18 hours after bypass from arrest-100% oxygen group, showing profoundly disrupted pulmonary architecture. × 40.

of anoxic changes in the brain tissue include the avoidance of a marked cooling gradient during the core cooling phase of the procedure as shown by Almond *et al.* (1964). During the core rewarming phase, the warming gradient must be kept below the level at which gas bubbles can form because of the rapidly warming blood in the extracorporeal unit. These experiments indicate a difference between 30, 45, and 60 min of total circulatory arrest at 20°C. If the interval of arrest is kept at approximately 30 min, there is nearly universal complete recovery from a neurological standpoint and no histological evidence of anoxic damage to brain tissue. Pulmonary insufficiency is a problem second only to the neurological changes that occur following hypothermic cardiopulmonary bypass and limited circulatory arrest. Morphological alterations that occur in the lungs are nonspecific in that they occur in other experimental preparations (Folkerth *et al.,* 1970) and consist primarily of interalveolar edema and hemorrhage, and the later occurrence of polymorphonuclear leukocyte infiltration into the interstitial and alveolar spaces. This progresses to the classic picture of congestive atelectasis that has been characterized as post-traumatic pulmonary insufficiency (Wilson, 1972). In the present study, animals that were continuously on cardiopulmonary bypass had no significant pulmonary changes evoked by increasing the inspired oxygen tension in the ventilator. The striking changes were noted when limited cardiopulmonary bypass was employed for core cooling and then total circulatory arrest combined with pulmonary ventilation with 100% oxygen. It is theorized that the total circulatory arrest group with high inspired oxygen tension had persistent exposure of the alveolus to high oxygen tension without circulatory capillary blood to prevent oxygen toxicity. Cardiopulmonary bypass may have a protective effect by persistent circulation at least through the bronchial system, limiting the concentration and the time exposure of the alveolus to this high concentration of oxygen. Ventilation with lower oxygen concentrations during total arrest has the same beneficial effect.

In summary, it would appear that the use of surface cooling and core cooling with subsequent total circulatory arrest at 20°C is a safe procedure, provided the period of time of cardiac arrest is not allowed to exceed 30 min. We also conclude that the alveolar oxygen tension should be maintained at the lowest level possible during the interval of circulatory arrest, to avoid the apparent rapid onset of post-traumatic pulmonary insufficiency. Clinically, while longer periods of cardiac arrest at this temperature are apparently well tolerated, animal studies indicate that there are significant histological changes in the central nervous system tissue. We are presently using this technique with total circulatory arrest for approximately 30 min and the low flow perfusion for longer intervals. Palliative shunting operations are used in patients where a total correction would necessitate a much longer period of circulatory complex congenital heart disease that would not respond to palliative shunting operations and in whom one would have to accept the inherent risks of this procedure in order to attempt total repair.

REFERENCES

ALMOND, C. H., J. C. JONES, H. M. SNYDER, S. M. GRANT, and B. W. MEYER. 1964. Cooling gradients and brain damage with deep hypothermia. J. Thorac. Cardiovasc. Surg. 48: 890–897.

BARRATT-BOYES, B. G., and J. M. NEUTZE. 1973. Primary repair of tetralogy of fallot in infancy using profound hypothermia with circulatory arrest and limited cardiopulmonary bypass: A comparison with conventional two stage management. Ann. Surg. 178: 406–411.

BARRATT-BOYES, B. G., J. M. NEUTZE, E. R. SEELYE, and M. SIMPSON. 1972. Complete correction of cardiovascular malformations in the first year of life. Progr. Cardiovasc. Dis. 15: 229–253.

BJÖRK, V. O., and G. HULTQUIST. 1962. Contraindications to profound hypothermia in open heart surgery. J. Thorac. Cardiovasc. Surg. 44: 1–13.

FOLKERTH, T. L., C. J. CARRICO, E. A. WAHRENBROCK, and M. J. TRUMMER. 1970. Role of the atrial myocardium in the pathogenesis of the "shock lung." Arch. Surg. 101: 416–420.

MOHRI, H., R. W. BARNES, L. C. WINTERSCHEID, D. H. DILLARD, and K. A. MERENDINO. 1968. Challenge of prolonged suspended animation: A method of surface-induced deep hypothermia. Ann. Surg. 168: 779–787.

RITTENHOUSE, E. A., H. MOHRI, D. H. DILLARD, and K. A. MERENDINO. 1974. Deep hypothermia in cardiovascular surgery. Ann. Thorac. Surg. 17: 63–98.

SUBRAMANIAN, S., and H. WAGNER. 1973. Correction of transposition of the great arteries in infants under surface-induced deep hypothermia. Ann. Thorac. Surg. 16: 391–401.

WILSON, J. W. 1972. Treatment or prevention of pulmonary cellular damage with pharmacologic doses of corticosteroid. Surg. Gynecol. Obstet. 134: 576–681.

Mailing address:
Cdr. Theodore L. Folkerth,
Naval Regional Medical Center,
San Diego, California 92134 (USA).

Recent Advances in Studies on
Cardiac Structure and Metabolism Volume 10
The Metabolism of Contraction
Edited by Paul-Emile Roy and George Rona

CONSEQUENCES OF REPERFUSION FOLLOWING ACUTE CORONARY OCCLUSION

SAMUEL MEERBAUM, TZU-WANG LANG,
ELIOT CORDAY, SHIGERU HIROSE,
COSTANTINO COSTANTINI, and HERBERT GOLD

Cedars-Sinai Medical Center, Los Angeles, California, USA

SUMMARY

The sequential effects of reperfusion following acute intracoronary balloon occlusion were investigated in closed-chest anesthetized dogs by simultaneous measurement of hemodynamics and regional metabolism and histopathologic study. Simultaneous regional determinations of myocardial metabolic function were carried out in coronary-occluded and nonoccluded segments of the left ventricle. A 3-hour occlusion of the proximal left anterior descending coronary artery (LAD) resulted in significant decrease in cardiac function. Initial reperfusion was marked by major washout of metabolites and development of frequently serious arrhythmias. Reperfusion following 3 hours of LAD occlusion led to further deterioration in hemodynamics and metabolism; 30% of the dogs developed shock. Lactate extraction decreased further in the nonoccluded region and reached production levels in the occluded zone. Potassium loss increased in both zones. Oxygen extraction remained unchanged. Pathophysiologic examination of tissue specimens from the reperfused zone revealed microthrombi, patchy necrosis, fragmentation of myofibrils, vascular congestions, and hemorrhages.

INTRODUCTION

Clinical reports of coronary revascularization following acute myocardial infarction differ in their evaluation of the effectiveness of this surgical procedure (Favaloro *et al.*, 1971; Scanlon *et al.*, 1971). Significantly increased survival rate and improved left ventricular function have been reported by some (Chatterjee *et al.*, 1973), while others find no change in function and a high risk when performing the bypass operation during acutely evolving myocardial ischemic injury (Dawson *et al.*, 1974; Hammermeister, Kennedy, and Hamilton, 1974). It is generally agreed that the timing of the revascularization can be a critical factor and mechanisms of myocardial viability and criteria of reperfusion require further definition.

Recent experimental studies of reperfusion after 3–5 hours of acute coronary occlusion revealed additional hemorrhagic infarction in about 50% of the cases

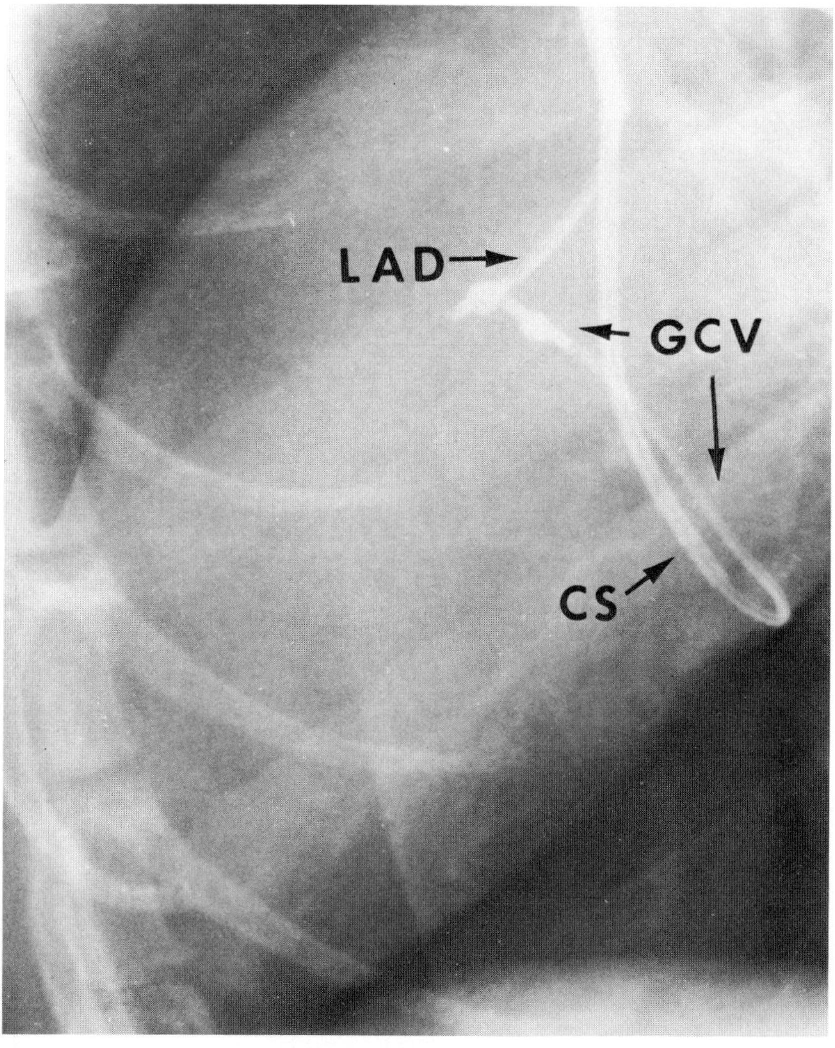

Figure 1. Angiographic demonstration of coronary balloon occlusion in closed-chest model. Left anterior oblique view of relative position of occlusion balloon within left anterior descending coronary artery (LAD) and within great cardiac vein (GCV). Note proximity of GCV balloon to LAD occlusion, thus insuring regional blood sampling from ischemic zone through central lumen of the GCV catheter. CS, coronary sinus.

(Breshnahan, Shell, and Ross, 1972). However, other investigations indicate that, after 3 hours of occlusion, reperfusion tends to result in substantial recovery of cardiac function to near normal (Ginks *et al.,* 1972). Difficulties are encountered in interpreting and correlating the experimental results, because of differing modes and sites of coronary occlusion, preexisting coronary collaterals, physiologic artifacts due to open-chest preparations, and frequent inadequacy of end-points used in evaluating the effects and effectiveness of reperfusion.

The present study was aimed at simultaneous and detailed closed-chest monitoring of global and regional left ventricular function in the ischemic and nonischemic segments during a 3 hour intracoronary occlusion and subsequent 5 hours of reperfusion. These measurements, coupled with histologic evaluation, would provide further insights into the process of reperfusion.

EXPERIMENTAL MODEL

Our closed-chest experimental model featuring intracoronary balloon occlusion of coronary arteries has been fully described and its application reported in studies of coronary occlusion and reperfusion (Corday *et al.,* 1974). Regional metabolism in the coronary occluded and nonoccluded zones of the left ventricle are based on simultaneous blood sampling from the coronary sinus and the anterior interventricular cardiac vein (AIV), performed during brief periods of great cardiac vein balloon separation of the coronary venous compartments. Figure 1 is a left oblique angiographic view demonstrating the model and the proximity of the LAD and great cardiac vein balloons.

METHODS

The experimental preparation and procedures have been described in a preceding paper of the Symposium (Lang *et al.,* 1975). Acute occlusion of the proximal LAD was performed in 7 closed-chest dogs through inflation of an intracoronary balloon. Deflation of the balloon and removal of the intracoronary catheter into the aorta provided total reperfusion of the coronary artery.

Cardiac function was measured in a preocclusion control period and at regular intervals during 3 hours' balloon occlusion of the proximal LAD and throughout 5 hours of reperfusion. After completion of the experiment, the hearts were excised and subjected to gross examination as well as to histopathologic study.

RESULTS

Hemodynamics and regional metabolic measurements in the coronary occluded and nonoccluded zones of the left ventricle are shown in Figure 2. The statistical

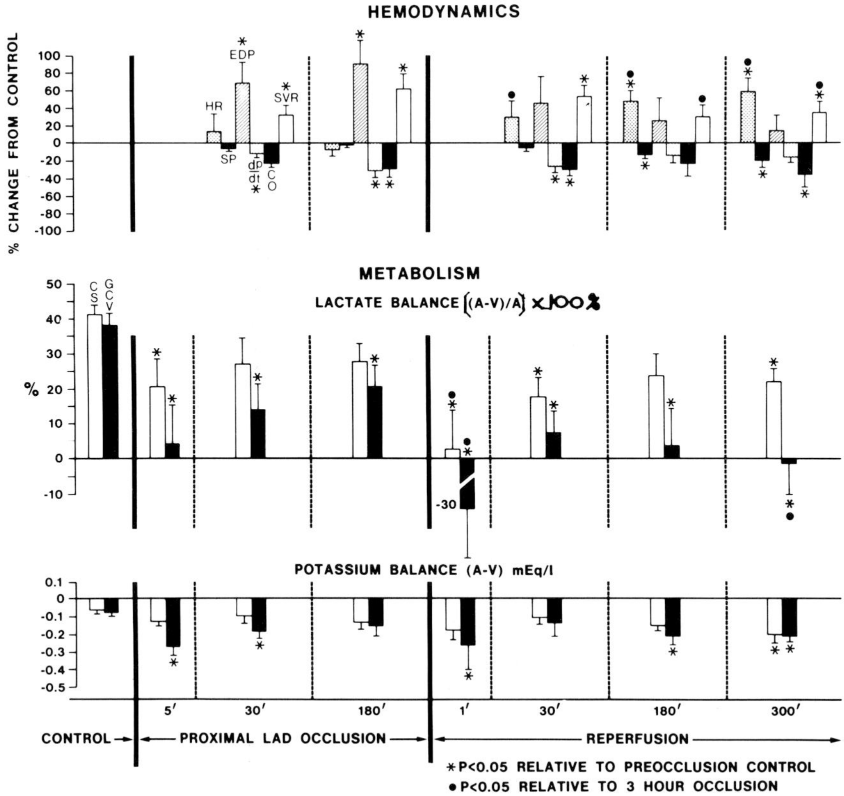

Figure 2. Statistical data of hemodynamics and metabolism during 3-hour coronary occlusion and 3 and 5 hours of reperfusion. Hemodynamics are expressed as percent change from control, lactate balance as [(A-V)/A] × 100% in coronary occluded (GCV) and nonoccluded (CS) segments of left ventricle, potassium balance as A-V in mEq/l in the two zones. Significance ($P<0.05$) is indicated relative to control (*) and 3-hour occlusion (•). Control hemodynamics (mean ± SEM) were as follows: heart rate (HR; beats/min) 109 ± 9; peak systolic pressure (SP; mm Hg) 136 ± 8; left ventricular end-diastolic pressure (EDP; mm Hg) 6 ± 1.5; max LV dP/dt (dP/dt; mm Hg/sec) 2740 ± 260; cardiac output (CO; l/min) 3.3 ± 0.3; systemic vascular resistance (SVR; dynes sec cm $^{-5}$) 2720 ± 150. Hemodynamic data indicated significant cardiac dysfunction during 3 hours of acute LAD occlusion. Shortly after reperfusion, there was a significant rise of heart rate, and cardiac function remained depressed. At 3 hours' reperfusion, a further deterioration of hemodynamics was noted. Shock developed in 1 animal at 3 hours of reperfusion and in another just prior to 5 hours of reperfusion. Lactate balance was significantly depressed shortly after LAD occlusion, in both coronary occluded and nonoccluded regions. Lactate balance improved subsequently, but was still deranged at 3 hours of occlusion. Immediately upon reperfusion, there was a major metabolite washout from myocardium. During 5 hours of reperfusion, metabolism in occluded region was significantly depressed, and lactate balance in nonoccluded zone was also deranged. Significant myocardial potassium loss occurred in occluded region shortly after occlusion. A lesser potassium loss persisted in both regions throughout 3 hours of occlusion. There was a washout of potassium immediately after reperfusion. Significant myocardial potassium loss persisted throughout 5-hour reperfusion.

data are presented as mean ± standard error of mean (SEM). Significance ($P < 0.05$) is indicated relative to control and relative to 3 hours' occlusion.

Hemodynamics

Hemodynamics are presented in Figure 2 as percent change from control. During the 3-hour acute occlusion of the LAD, heart rate dropped slightly below the preocclusion control level. Immediately after reperfusion, the heart rate rose 29% and reached 39% and 57% above control at 3 and 5 hours of reperfusion, respectively. Few arrhythmias were noted during the occlusion period, but reperfusion was frequently associated with persistent arrhythmias.

During the coronary occlusion, peak systolic pressure dropped slightly (6%). Reperfusion resulted in a further decrease in this pressure to 10 and 21% below control at the end of 3 and 5 hours. At the latter time, a substantial reduction in peak systolic pressure had occurred in 3 of the 7 dogs, with 2 reaching shock levels (defined as a 30% reduction below control in both the systolic pressure and cardiac output).

Left ventricular end-diastolic pressure rose significantly (89%) during the 3 hours of occlusion. Subsequent reperfusion lowered LVEDP in the direction of control (17 and 6% above control at 3 hours and 5 hours' reperfusion).

Maximal LV dP/dt dropped 33% below control during the LAD occlusion. Following 3 hours of reperfusion, max LV dP/dt improved but was still 24% below the control level. Mean LV dP/dt was 17% below control at 5 hours' reperfusion.

The 3-hour LAD occlusion caused a 30% decrease in cardiac output. After reperfusion, cardiac output appeared to recuperate at first; but it subsequently deteriorated (−27% at 3 hours) and, at 5 hours of reperfusion, it was 36% below control, with 2 animals developing shock.

Systemic vascular resistance rose 63% during occlusion and dropped to 40 and 33% above control as a result of 3 and 5 hours of reperfusion.

Stroke work decreased with occlusion and dropped further to 49% and 64% below control at the end of the 3- and 5-hour reperfusion periods, partly due to associated increases in heart rate.

Measurements of coronary sinus flow in 5 dogs indicated a 36% reduction below control at 3 hours of occlusion. During reperfusion, coronary flow rose but remained 15 and 11% below the preocclusion control at the end of 3 and 5 hours of reperfusion, respectively.

Metabolism

Regional measurements of metabolism included percent oxygen extraction (A-V difference), lactate balance (A-V difference/arterial × 100%), and myocardial potassium loss (A-V difference) in mEq/l (Fig. 2).

Coronary occluded region The oxygen extraction of the coronary occluded segment increased about 5% during the 3 hours of LAD occlusion, and this slightly elevated oxygen extraction was maintained during the subsequent 5 hours of reperfusion.

The ischemic zone exhibited a sharp derangement in lactate balance changing from extraction (38.7%) toward production (4.3%) during the earliest period of LAD occlusion. After 3 hours of occlusion, lactate balance in the occluded zone returned to 20% extraction. Immediately upon reperfusion, there was evidence of a major washout of accumulated lactates. At 3 hours' reperfusion, myocardial lactate balance in this region was 3.7%, and at 5 hours the previously occluded zone exhibited lactate production (-2.9%).

The myocardial potassium balance, which was near zero in control, decreased sharply to a mean of -0.28 mEq/l during the early period of LAD occlusion. This represents a significant potassium loss from the coronary occluded region. Potassium loss persisted and was -0.23 mEq/l at the end of the occlusion period. Reperfusion led to initial washout of potassium. Subsequently, at 3 and 5 hours of reperfusion, the previously occluded segment of the myocardium exhibited a persisting loss of potassium (-0.22 mEq/l).

Nonoccluded region of the left ventricle Regional oxygen extraction in the nonoccluded segment of the left ventricle was increased by about 4% during the occlusion and remained at this level during reperfusion.

In the nonoccluded zone, lactate balance showed a derangement during the initial period of coronary occlusion, from 41% to 20.9%. This change was significant relative to control, but also significantly less pronounced than in the coronary occluded region. At the end of the 3-hour occlusion period, lactate balance in the nonoccluded segment was 28.1%. After 3 and 5 hours of reperfusion, the nonoccluded zone indicated 23.7 and 22.2% lactate extraction.

The potassium loss during the earliest period of coronary occlusion was slight (-0.13 mEq/l) in the nonoccluded myocardium and persisted at this level during the subsequent 3-hour occlusion. At 3 hours' reperfusion, myocardial potassium loss was -0.16 mEq/l and, at 5 hours of reperfusion, this loss increased to -0.21 mEq/l.

Individual case

Figure 3 illustrates the hemodynamics and regional metabolism in an individual dog. In this case, it was noted that reperfusion was associated with marked arrhythmias and marked metabolic derangements relative to the 3-hour occlusion state. This metabolic dysfunction was measured in both the coronary occluded and nonoccluded regions of the left ventricle.

Histopathology

Myocardial infarction, largely of a transmural character, was found in the excised heart, in the anterior left ventricular wall and near the apex of the left

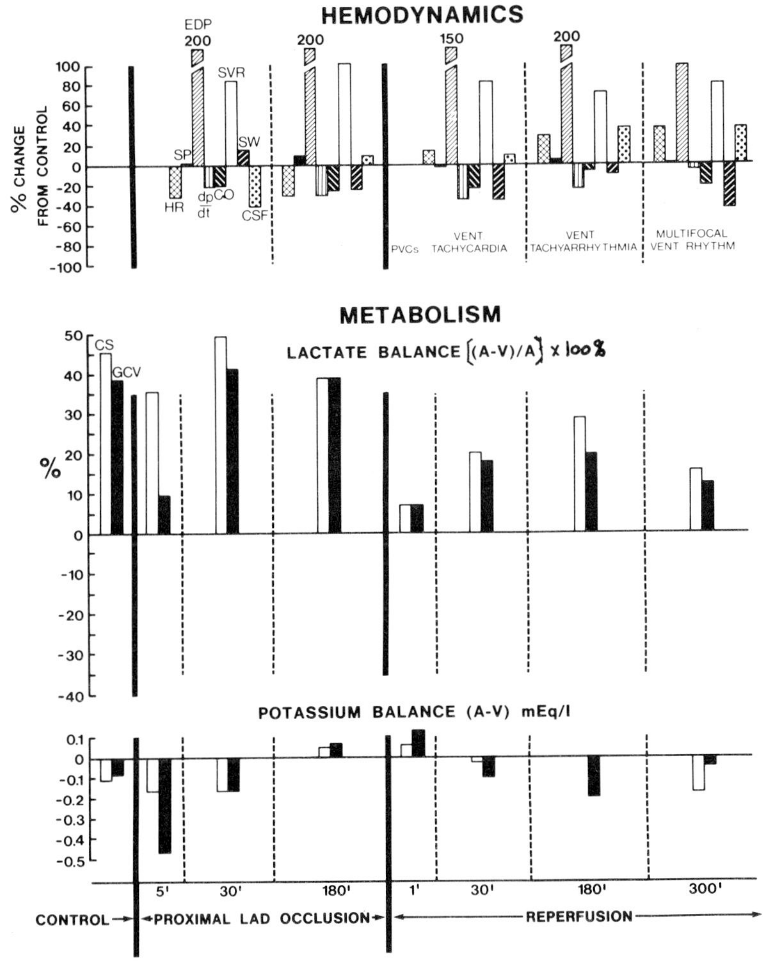

Figure 3. Individual case hemodynamics and metabolism during LAD occlusion and reperfusion. Control hemodynamics were HR 141; SP 139, EDP 2; dP/dt 2685; CO 2.76; SVR 2500; SW 30; coronary sinus flow (CSF ml/min) 175. Shortly after LAD occlusion there was a drop of HR, a sharp rise in EDP, a major increase in SVR, and a 40% reduction in CSF. Cardiac dysfunction persisted throughout 3-hour occlusion. Following reperfusion, HR increased and arrhythmias developed. After 5 hours of reperfusion, hemodynamic indices showed continued depression of cardiac function. Lactate metabolism in nonoccluded region remained unchanged during coronary occlusion. In occluded zone, lactate balance was depressed only immediately after occlusion and returned subsequently to near control. Major derangements of lactate balance were noted in both occluded and nonoccluded regions during 5-hour reperfusion period. Potassium balance was significantly depressed in occluded zone shortly after occlusion. Remaining period of occlusion and reperfusion exhibited minor myocardial potassium loss.

ventricle. Some signs of infarction were also observed in the posterior portion of the left ventricle, particularly in those dogs which developed significant hypotension. There was evidence of microthrombi in the areas of infarction. Further observations included cellular necrosis and fragmentation of myofibrils, intramyocardial vascular congestion (hyperemia), and hemorrhages accompanying the infarction.

DISCUSSION

The period of myocardial viability following an acute LAD coronary artery occlusion in the dog and cat has been variously reported as 20 minutes to several hours (Fisher and Edwards, 1963; Jennings *et al.*, 1963; Hamer, 1969; Maroko *et al.*, 1972; O'Brien *et al.*, 1972; Jennings and Reimer, 1973; Symes, Arnold, and Blurdell, 1973). Reperfusion after 45 minutes of occlusion has often been associated with some irreversible myocardial damage (Blumgart, Gilligan, and Schlesinger, 1941; Krug, 1965; Bolooki *et al.*, 1968; Krug, 1970; Torok *et al.*, 1971). However, there is equally convincing evidence (Ginks *et al.*, 1972; McNamara *et al.*, 1973; Puri, 1973) that reperfusion after up to 4 hours of proximal LAD occlusion may eventually bring about a substantial return of contractile function of the left ventricle. The exact site and size of the coronary artery chosen for the occlusion are clearly of primary importance. The role played in this connection by other factors, particularly the collateral circulation, in determining the size and distribution of the zone of infarction have yet to be clarified. There have been extensive investigations of reperfusion using histopathologic, histochemical, and enzymatic analysis (Krug, 1972). These studies generally concluded that irreversible damage was done to the ischemic myocardium by 75–90 minutes of coronary artery occlusion. Reperfusion after longer periods of coronary occlusion was occasionally associated with damage to the microcirculation, increased tissue hydrogen ion content, reduced glycogen, and intercellular and interstitial edema (Krug, 1972).

Several investigators (Sewell, Koth, and Higgins, 1955; Arnulf and Martin, 1973; Verrier, Corbalan, and Bown, 1974) have noted that arrhythmic and other functional derangements tend to supervene in the early phase of reperfusion. These effects had to be verified to examine the influence of the specific preparation and to assess their contribution to the reperfusion effectiveness and eventual survival. The reported hemorrhagic damage associated with reperfusion (Breshnahan *et al.*, 1972) also requires clarification as to its genesis and possible relationship to observed damaged small intramyocardial coronary vessels. It has been pointed out that much of the ischemic damage may be of a patchy character, leading to infarction chiefly at the subendocardial level at the center of the ischemic zone, and that the remainder of the underperfused myocardium may indeed be recoverable by reperfusion.

In viewing the results of experimental reperfusion and clinical revascularization, it appears important to relate the reperfusion effectiveness to the nature and extent of myocardial damage and functional derangements caused by the preceding acute infarction. An appropriate understanding of the effects of reperfusion following coronary occlusions requires comprehensive measurements and comparisons with preocclusion control, and such data are frequently unavailable in man. Hence, the mechanisms and characteristic effects of reperfusion are best resolved through well-controlled experimental study. The closed-chest model we have devised appears to be suitable in terms of physiologically valid simulation of coronary occlusion and reperfusion, with simultaneous measurements obtained from the occluded and nonoccluded segments of the left ventricle. Open-chest models may be employed, but these are often associated with physiologic artifacts, including compromised cardiac function, which may preclude true assessment of reperfusion effects.

The presented data indicate that 5 hours of reperfusion failed to correct the left ventricular dysfunction caused by 3 hours of acute intracoronary occlusion of the proximal LAD and resulted in further hemodynamic and metabolic myocardial derangements; 2 of the 7 dogs developed cardiac shock after 3 hours of reperfusion. Reperfusion arrhythmias were frequent and persistent. Hemorrhagic myocardial damage noted in histopatholigic study is ascribed to the reperfusion. Some improvements in hemodynamics are apparent in the initial period of reperfusion, but these are negated during the remainder of the 5-hour reperfusion. Major metabolite washout and arrhythmias in early reperfusion may contribute to functional derangements which, when untreated, could lead to shock and ventricular fibrillation.

We have recently carried out another study to determine whether extended duration reperfusion, after 3 hours' proximal LAD occlusion, would bring about a partial or total return to normal cardiac function (Constantini *et al.*, 1975). The same closed-chest intracoronary balloon occlusion model was employed, and hemodynamic as well as regional metabolic measurements were performed following 1 hour as well as after 7 days of reperfusion. Thorough histopatholigic analysis was performed on 8 reperfused hearts, and also on 4 additional control dogs, in which LAD occlusion was allowed to persist for 7 days.

In the latter 4 dogs, the extent of infarction (based on light microscopy with H and E stains) ranged from 20 to 25 volume % of the left ventricle, and in most instances ventriculography on the seventh day indicated severe hypokinesis of the anterior left ventricular segment. In contrast, the 7-days reperfused group, while corroborating the previously observed functional derangements in early reperfusion, exhibited a significantly lower mean infarction size, approximately 12% of the left ventricle. Of the 8 dogs reperfused for 7 days, 2 exhibited larger infarcts (15 and 29%). Metabolic and hemodynamic measurements on the seventh day indicated a significant improvement relative to the 3-hour occlusion, even though in 7 of the 8 dogs there was a residual segmental LV hypokinesis. These recent experiments lead to the tentative conclusion that 7-day or longer periods of reperfusion after a 3-hour LAD occlusion may, on the average, bring about a return of cardiac function toward control levels and significantly decrease the size of infarction.

REFERENCES

ARNULF, G., et C. L. MARTIN. 1973. Fibrillation ventriculaire au cours du retablissement circulatoire dan les arteres coronaires. Arch. Mal. Coeur. 66: 869–876.

BLUMGART, H. L., D. R. GILLIGAN, and M. J. SCHLESINGER. 1941. Experimental studies on the effect of temporary occlusion of coronary arteries. II. The production of myocardial infarction. Am. Heart J. 22: 374–389.

BOLOOKI, H., J. J. ROOKS, C. E. VIERA, B. SMITH, K. MOBIN-UDDIN, C. R. LOMBARDO, and J. R. JUDE. 1968. Comparison of the effect of temporary or permanent myocardial ischemia on cardiac function and pathology. J. Thorac. Cardiovasc. Surg. 56: 590–598.

BRESHNAHAN, G. F., W. F. SHELL, and J. ROSS, Jr. 1972. Deleterious effects of reperfusion in evolving myocardial infarction. Circulation 45–46(suppl II): 13.

CHATTERJEE, K., H. SWAN, H. SUSTAITA, J. MATLOFF, and W. PARMLEY. 1973. Impending infarction and direct myocardial revascularization, pp. 326–370. *In* H. J. C. Swan and E. Corday (eds.), Myocardial infarction. Williams and Wilkins, Baltimore.

CORDAY, E., T. W. LANG, S. MEERBAUM, H. GOLD, S. HIROSE, S. RUBINS, and M. DALMASTRO. 1974. A closed chest intracoronary occlusion model for the study of regional cardiac function. Am. J. Cardiol. 33: 49–59.

COSTANTINI, C., E. CORDAY, T. W. LANG, S. MEERBAUM, J. BRASCH, L. KAPLAN, S. RUBINS, H. GOLD, and J. OSHER. 1975. Revascularization after 3 hours of coronary occlusion: Effects on regional cardiac metabolic function and infarct size. Am. J. Cardiol. (In press)

DAWSON, J. T., R. J. HALL, G. L. HALLMAN, and D. A. COOLEY. 1974. Mortality in patients undergoing coronary artery bypass surgery after myocardial infarction. Am. J. Cardiol. (In press)

FAVALORO, R. G., D. B. EFFLER, C. CHEANVECHAI, R. A. QUINT, and F. M. SONES. 1971. Acute coronary insufficiency (impending myocardial infarction) and myocardial infarction. Surgical treatment by saphenous vein graft technique. Am. J. Cardiol. 28: 598–613.

FISHER, S. M., and W. S. EDWARDS. 1963. Tissue necrosis after temporary artery occlusion. Am. Surg. 29: 617–619.

GINKS, W. R., H. D. SYBERS, P. R. MAROKO, J. W. COVELL, B. E. SOBEL, and J. ROSS, Jr. 1972. Coronary artery reperfusion. II. Reduction of myocardial infarct size at one week after the coronary occlusion. J. Clin. Invest. 51: 2717–2723.

HAMER, J. 1969. Discussion. *In* N. Brachfield, Maintenance of cell viability. Circulation 39–40(suppl IV): 212–219.

HAMMERMEISTER, K. E., J. W. KENNEDY, and G. W. HAMILTON. 1974. Aortocoronary saphenous-vein bypass. Failure of successful grafting to improve resting left ventricular function in chronic angina. New Engl. J. Med. 290: 186–192.

JENNINGS, R. B., J. P. KATENBACH, H. M. SOMMERS, G. F. BAHR, and W. B. WATERMAN. 1963. Studies of the dying myocardial cell, pp. 189–205. *In* T. N. James and J. W. Keyes (eds.), The etiology of myocardial infarction. Little Brown, Boston.

JENNINGS, R. B., and K. A. REIMER. 1973. The fate of the ischemic myocardial cell, pp 13–25. *In* H. J. C. Swan and E. Corday (eds.), Myocardial infarction. Williams and Wilkins, Baltimore.

KRUG, A. 1965. Untersuchungen am Katzenherzen nach temporaerer Unterbindund einer Koronararaterie. Ein Beitrag zum Problem der Wiederbelebung mit Defect des Herzmuskels. Z. Kreislaufforsch. 54: 510–516.
KRUG, A. 1970. The extent of ischemic damage in the myocardium of the cat after permanent and temporary coronary occlusion. J. Thorac. Cardiovasc. Surg. 60: 242–247.
KRUG, A. 1972. Myokardennekrosen nach Permanentem oder Temporarein Koronararterienverschluss mit und ohne Herzentlastung durch Extrakorporale Zirkulation. Arch Kreislaufforsch. 67: 326–388.
LANG, T. W., S. MEERBAUM, E. CORDAY, C. COSTANTINI, S. RUBINS, and M. CORDOBA. 1975. Regional myocardial metabolism following acute coronary artery occlusion. (This Book, page 000).
MAROKO, P. R., P. LIBBY, W. R. GINKS, C. M. BLOOR, W. E. SHELL, B. E. SOBEL, and J. ROSS, Jr. 1972. Coronary artery reperfusion. I. Early effects on local myocardial function and the extent of myocardial necrosis. J. Clin. Invest. 51: 2710–2716.
McNAMARA, J. J., J. R. SOETER, G. T. SUEHIRO, and A. L. MORGAN. 1973. Myocardial viability after transient ischemia. Am. J. Cardiol. 31: 146.
O'BRIEN, C. M., M. CARROLL, P. T. O'ROURKE, E. L. RHODES, O. GAG, M. M. KIRSH, J. D. MORRIS, and H. E. SLOAN. 1972. The reversibility of acute ischemic injury to the myocardium by restoration of coronary flow. J. Thorac. Cardiovasc. Surg. 64: 840–846.
PURI, P. S. 1973. Coronary reperfusion. Correlation between recovery of contractility and high energy phosphates. Physiologist 16: 426.
SCANLON, P. J., R. NEMICKAS, J. R. TOBIN, W. ANDERSON, A. MONTOYA, and R. PIFARRE. 1971. Myocardial revascularization during acute phase of myocardial infarction. J. Am. Med. Ass. 218: 207–000.
SEWELL, W. H., D. R. KOTH, and C. E. HIGGINS. 1955. Ventricular fibrillation in dogs after sudden return of flow to the coronary artery. Surgery 38: 1050–1053.
SYMES, J. R., I. M. F. ARNOLD, and P. E. BLURDELL. 1973. Early revascularization of the acute myocardial infarction: the critical time factor. Canad. J. Surg. 16: 275–283.
TOROK, B., I. TOTH, G. TEMES, E. ROTH, and T. SZMOLENSKY. 1971. Mortality after restoration of circulation following temporary coronary ligation. Acta Chir. Acad. Sci. Hung. 12: 57–64.
VERRIER, R., R. CORBALAN, and B. BOWN. 1974. Analysis of vulnerability changes during acute coronary occlusion and release. Am. J. Cardiol. 33: 174.

Mailing address:
Eliot Corday, M.D.,
Cedars-Sinai Medical Center, Research Institute,
4751 Fountain Avenue, Los Angeles, California 90029 (USA).

Recent Advances in Studies on
Cardiac Structure and Metabolism Volume 10
The Metabolism of Contraction
Edited by Paul-Emile Roy and George Rona

REDUCTION IN INFARCT SIZE FOLLOWING EXPERIMENTAL CORONARY OCCLUSION BY ADMINISTRATION OF VERAPAMIL[1]

B. N. SINGH, H. J. SMITH,
and R. M. NORRIS

Departments of Cardiology and Coronary Care, Green Lane Hospital,
and Department of Medicine, University of Auckland School of Medicine,
Auckland, New Zealand

SUMMARY

The effects of verapamil, a calcium antagonist with antiarrhythmic and anti-anginal actions, were studied on the size of infarct in open-chest anesthetized dogs with left anterior descending coronary artery occlusion. Hemodynamic functions were monitored by standard techniques, and infarct size was assessed by epicardial electrocardiography. Myocardial blood flow was measured by radioactive microspheres, and regional metabolism was studied by measuring lactate concentration in the venous drainage of the ischemic myocardium. Verapamil produced a significant reduction in ST segment elevation in the epicardial electrocardiogram and prevented the hemodynamic deterioration seen in the control animals in which ST segment elevation persisted. The drug had no effect on either total collateral blood flow or on the relative distribution of flow to the inner and outer halves of the ventricular wall. No significant differences were found between the levels of coronary venous lactate in the control animals and those in the verapamil-treated group. The results raise the possibility that the apparent protective effect of verapamil in myocardial ischemia may be due to its electrophysiological action on the cardiac membrane in relation to its calcium-antagonistic properties in excitable tissues.

INTRODUCTION

In 1968, Kaumann and Aramendia reported that verapamil (Isoptin; Cordilox; Knoll AG: Fig. 1a), a potent smooth muscle relaxant and coronary vasodilator (Haas and Hartfelder, 1962), provided almost complete protection against ventricular fibrillation due to experimental coronary occlusion in dogs. They made the observation that, whereas animals which received the beta-adrenergic blocking drug, sotalol (MJ1999), died within 24 hours of coronary occlusion, those

[1] Supported in part by grants from the National Heart Foundation and the Medical Research Council of New Zealand.

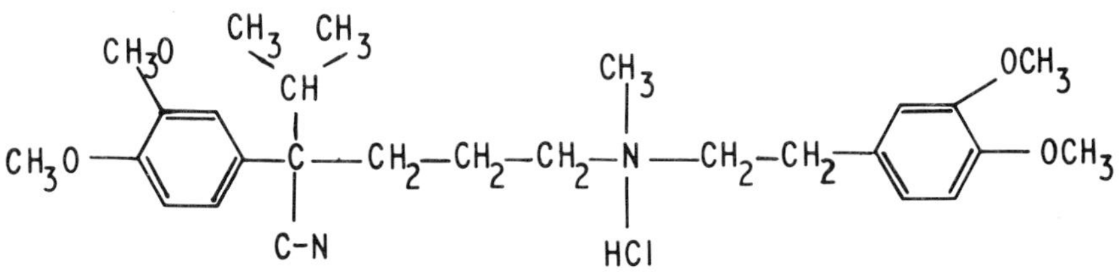

IPROVERATRIL (VERAPAMIL)

Figure 1a. Structural formula of verapamil.

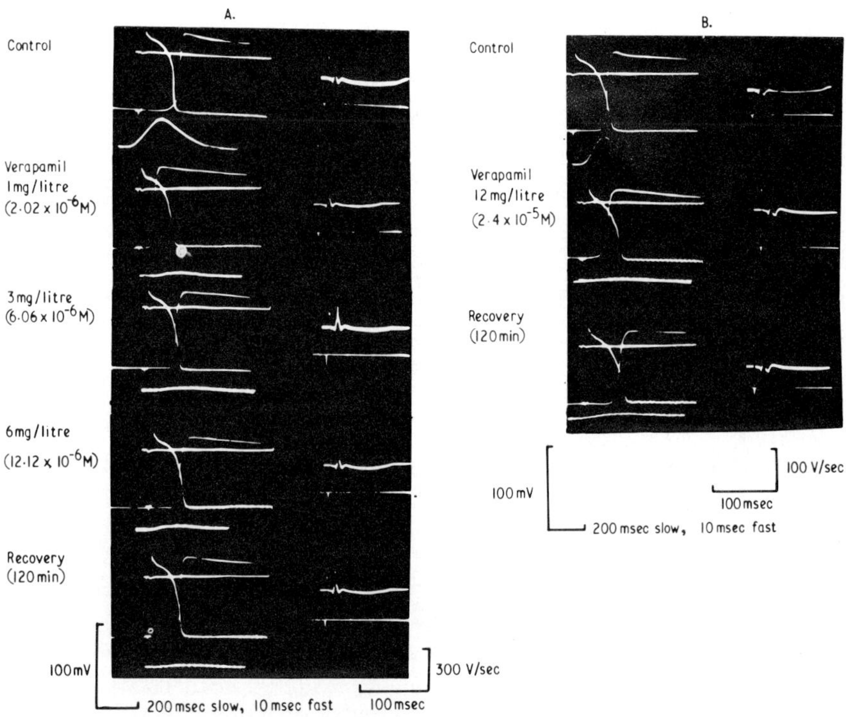

Figure 1b. Effect of verapamil on ventricular intracellular potentials. On left in each frame, horizontal line indicates zero potential; superimposed traces depict intracellular potentials at slow and fast sweep speeds. Bottom trace shows isometric tension. On right in each frame, upper trace shows surface potential; maximal rate of rise of action potential is indicated by depth of differential signal in lower trace. Verapamil has a markedly depressant effect on isometric contraction in concentrations in which it does not affect maximal rate of rise of action potential. (From Singh, 1971)

pretreated with verapamil before coronary ligation survived for many months. The authors suggested that the coronary dilatation induced by verapamil contributed to the hemodynamic recovery with prolonged survival after coronary artery occlusion.

The antiarrhythmic actions of verapamil under experimental conditions (Melville, Shister, and Huq, 1964; Schmid and Hanna, 1967; Singh, 1971; Singh and Vaughan Williams, 1972) have now been confirmed in the clinic (Heng et al., 1975; Schamroth, Krikler, and Garrett, 1972). Electrophysiological studies using microelectrodes in isolated heart muscle preparations (Melville *et al.,* 1964; Singh and Vaughan Williams 1972; Cranefield, Aronson, and Wit, 1974; Zipes and Fischer, 1974) have shown a striking difference (Fig. 1b) between the antiarrhythmic actions of verapamil and those of other conventional agents (Singh and Hauswirth, 1974). Such studies, however, have not suggested a mechanism whereby the drug might act to prolong survival in animals subjected to coronary artery ligation. We have attempted to investigate the possibility that verapamil might act by limiting infarct size, the experimental model used being the open-chest anesthetized dog with ligation of the left anterior descending coronary artery.

METHODS

Studies were performed in mongrel dogs of either sex (18 to 42 kg) anesthetized with pentobarbitone, 30 mg/kg during induction and 5 mg/kg/hour thereafter as an intravenous infusion (Vaughan Williams, Bagwell, and Singh, 1973). Respiration was maintained through a cuffed endotracheal tube and a Harvard respirator with 30% oxygen in air, initial adjustment being made according to the values of arterial blood gases which were monitored intermittently. Blood was rendered incoagulable by intravenous heparin.

Polyethylene catheters were placed in an external jugular vein and femoral artery for intravenous injections and for the recording of systemic arterial pressure. Through a lateral thoracotomy in the fifth left intercostal space, the heart was exposed and suspended in a pericardial cradle. Wide-bore catheters were placed in the main pulmonary artery and in the left atrium and, together with the catheter in the femoral artery, were connected to Statham P23Db pressure transducers with a reference level at mid-left ventricular cavity. All outputs were displayed and recorded on a Sanborn 6-channel model 350 recorder, one channel of which was connected to lead II of the electrocardiogram to monitor heart rate. This channel was later used for recording epicardial electrocardiograms.

Cardiac output was measured by the indicator dilution technique, indocyanine green (2.5 mg) being injected into the main pulmonary artery and the blood withdrawn from the femoral artery through a Gilford densitometer. Cardiac

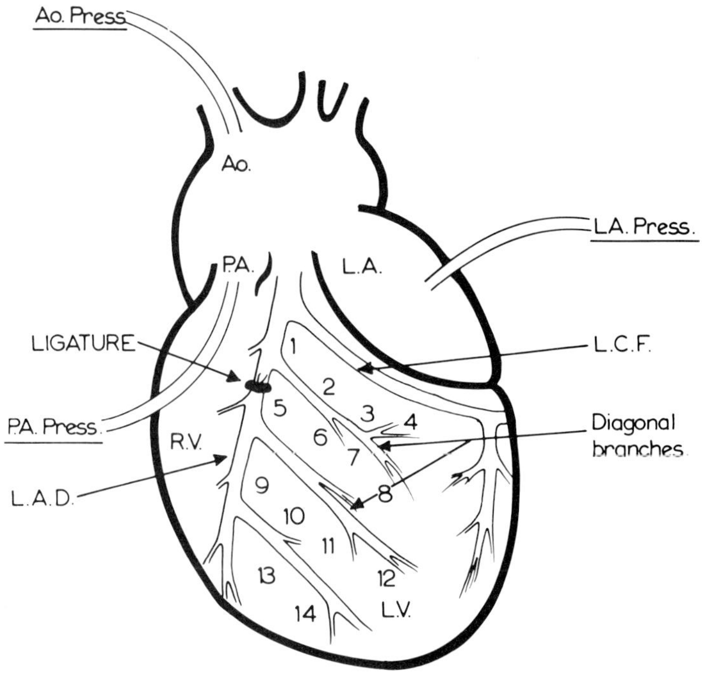

Figure 2. Schematic representation of experimental preparation used to study effects of verapamil on hemodynamic functions and epicardial electrocardiogram, following left anterior descending (LAD) artery occlusion. Locations of numbered sites from which electrocardiographic records were taken and occlusive LAD tie are shown in center of diagram. Arterial catheter was introduced into the aorta (Ao) through right common carotid artery in neck. LA = left atrium, LCF = left circumflex artery, PA = pulmonary artery, RV = right ventricle, LV = left ventricle.

index (ml/min/kg), stroke volume index (ml/kg), systemic resistance index (units-kg), and stroke work index (g-m/kg/beat) were calculated from the measured hemodynamic variables.

Left anterior descending (LAD) artery was ligated in two stages (Harris, 1950) at a level 2 cm from the left mainstem bifurcation. The experimental preparation used is shown schematically in Figure 2. Epicardial electrocardiograms were recorded after the method of Maroko and his associates (1972), with three modifications (Smith *et al.,* 1975b). These were the use of a steel-ball electrode (Neutze, Kerr, and Whitlock, 1973), 14 predetermined electrode-recording positions over the anterior left ventricular wall, and measurement of the ST segment shift at 0.06 seconds after the end of the S wave (Reid, Pelides, and Shillingford, 1971).

Three series of experiments were performed:

1. In 14 animals, hemodynamic and ECG data were obtained before ligation of

the LAD artery, 15 minutes after LAD occlusion, and serial measurements were made thereafter for periods up to 2 hours.

2. In 9 animals, hemodynamic and ECG recordings were taken before and at 15 minutes after ligation of the LAD artery. Verapamil was then given as a bolus injection of 0.2 mg/kg intravenously over a period of 2 minutes. This was followed by a constant infusion of the drug for 2 hours at a rate of 0.005 mg/kg/min, a dose schedule which has been considered clinically relevant (Ryden and Saetre, 1971). Hemodynamic functions and ECG maps were recorded 10 minutes after the bolus injection of verapamil and, as in the control group, serial measurements were made for a period of 2 hours.

3. In 8 animals, the effects of verapamil on myocardial blood flow (MBF) and lactate production by the ischemic myocardium were investigated. In this series, MBF and levels of lactate in the vein draining the infarct zone in 9 untreated animals were compared with the results obtained in 8 animals given intravenous verapamil in the manner described above. These studies, which were carried out in collaboration with P. J. Hurley M.D. and M. B. John B.Sc. (Department of Nuclear Medicine, Auckland Hospital), will be reported in detail elsewhere (Smith *et al.*, 1975c). They will be mentioned here only briefly. MBFs were determined by injection into the left atrium of radioactively-labelled carbonized microspheres, 15 plus-minus 5 μ diameter, labelled with gamma-emitting nuclides, ^{141}Ce, ^{85}Sr, and ^{169}Yb (Becker, Fortuin, and Pitt, 1971). The technique of sampling venous blood from the infarct zone has been presented elsewhere (Smith *et al.*, 1975a); venous lactate concentrations were measured by the method of Ellis, Cain, and Williams (1963).

RESULTS

Effects of LAD artery occlusion on hemodynamic variables and epicardial ST segment

The mean data for hemodynamic variables and the epicardial ST segment shifts before and after coronary occlusion in the control animals, as well as in those which subsequently received verapamil, are shown in Table 1. There were no significant differences between the two groups with respect to heart rate, cardiac index (CI), mean arterial pressure (MAP), or mean left atrial pressure (MLAP). The mean ΣST of 0.03 ± 0.01 mV in the verapamil group (before coronary occlusion) was, however, significantly different ($p<0.01$) from the mean value of 6.3 ± 1.5 mV for ΣST in the control series. The reason for this is not clear, but the difference does not appear to bear on the conclusions subsequently drawn with respect to the effects of verapamil on the ST segment alterations since, 15 minutes after LAD occlusion, the mean ΣST of 50.3 ± 9.4 mV for the control series was not significantly different from 58.0 ± 9.7 mV in the group of animals which were given verapamil subsequently.

Table 1. Hemodynamic and epicardial *ST* segment changes following LAD occlusion in anesthetized dogs

		Heart Rate (beats/min)	Cardiac Index (ml/min/kg)	Mean Arterial Pressure (mm Hg)	Mean Left Atrial Pressure (mm Hg)	Σ ST (mV)
Series A	precoronary occlusion	141.1 ± 6.5	126.4 ± 16.7	105.0 ± 4.0	5.9 ± .07	6.4 ± 1.5
n = 14	15 minutes post-occlusion	149.1 ± 5.4	119.2 ± 10.5	99.4 ± 4.7	6.6 ± 1.1	50.3 ± 9.4
Series B[a]	precoronary occlusion	125.7 ± 11.7	129.4 ± 9.4	103.9 ± 4.8	4.9 ± 0.8	0.3 ± 0.01
n = 9	15 minutes post-occlusion	137.4 ± 8.4	101.1 ± 12.0	99.8 ± 5.4	7.1 ± 0.7	58.0 ± 9.7

[a]At 15 minutes after coronary occlusion, the animals in series B were given verapamil 0.2 mg/kg intravenously as a bolus, followed by 0.005 mg/kg/min infusion for 2 hours. Subsequent changes which occurred with time in both series are shown in Figures 2 and 3. Values are means ± S.E.M.'s; n refers to the number of animals from which the data were obtained.

Following LAD occlusion, and before verapamil was administered, there was a trivial increase in heart rate with a mild reduction in MAP in both series (Table 1). The increases in MLAP were significant in the control (5.9 ± 0.7 to 6.6 ± 1.1 mm Hg; $p<0.05$) as well as in the verapamil series (4.9 - 0.1 to 7.1 ± 0.7 mm Hg; $p<0.05$). The decrease in cardiac output was, however, significant only in the verapamil series (129.4 ± 9.4 to 101.1 ± 12.0 ml/min/kg; $p<0.01$).

Effects of verapamil on hemodynamic variables and epicardial ST segment after LAD occlusion

The time course of changes in the measured and derived hemodynamic variables and the ST segment alterations in both the control and treated groups of animals is shown in Figures 3 and 4.

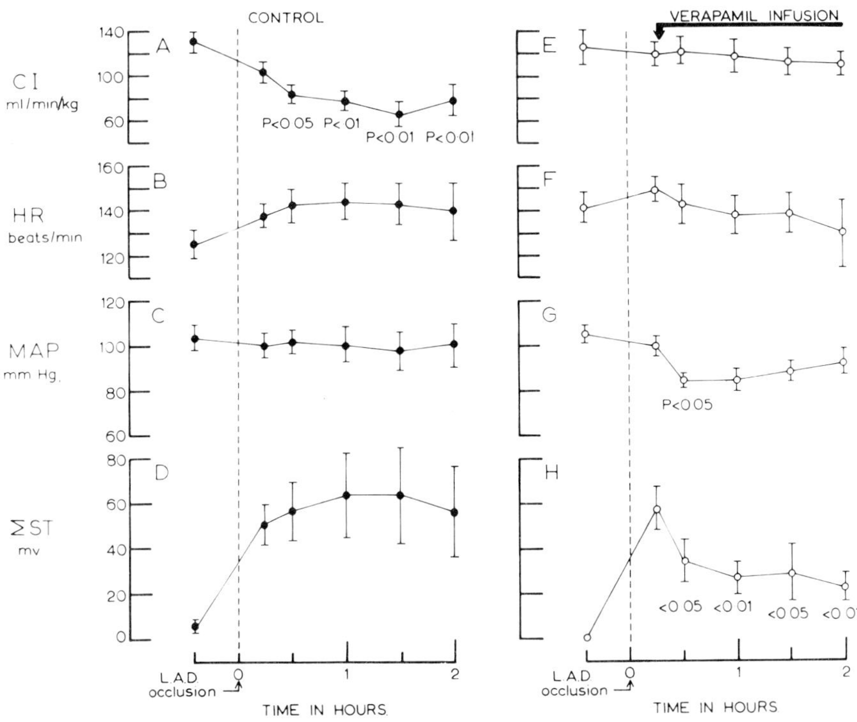

Figure 3. Effects of verapamil on cardiac index (CI), heart rate (HR), mean arterial pressure (MAP) and summated ST segment elevation (ΣST), in open-chest anesthetized dogs with LAD occlusion. Panels A, B, C, and D demonstrate changes with time in control group and panels E, F, G and H show those occurring in verapamil-treated group. In both groups, each level of significance of change in a particular variable indicated refers to difference from value at 15 minutes after LAD occlusion. In control group, there is progressive decrease in CI, and ΣST remains persistently elevated. In verapamil-treated group, cardiac index is maintained, and ΣST is significantly reduced.

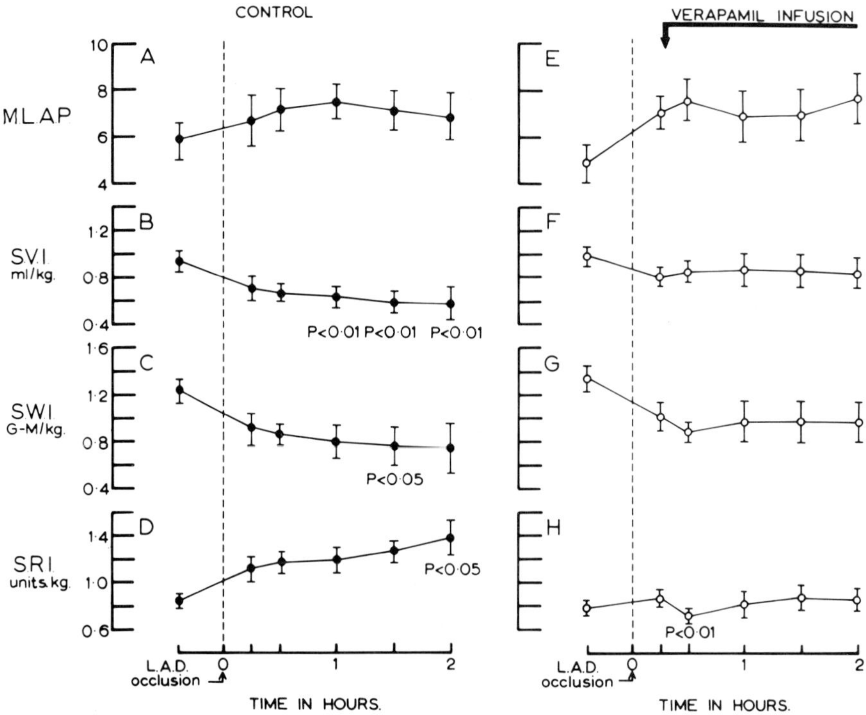

Figure 4. Effects of verapamil on mean left atrial pressure (MLAP), stroke volume index (SVI), stroke work index (SWI), and systemic resistance index (SRI), in anesthetized dogs with LAD occlusion. Panels A, B, C, and D demonstrate changes with time in control group and panels E, F, G, and H show those occurring in the verapamil-treated group. As in Figure 2, in both groups, each level of significance of change in a particular variable indicated refers to difference from value at 15 min *after* LAD occlusion. Note progressive increase in SRI with fall in SVI in control but not verapamil-treated group.

Cardiac index, heart rate, and stroke volume In the control group, CI declined with time from 119.2 ± 10.5 ml/min/kg 15 minutes after LAD occlusion, to reach a level of 79.1 ± 14.3 ml/min/kg at 2 hours (Fig. 3; $p<0.01$). In animals given verapamil, CI remained relatively constant after the initial reduction following coronary occlusion (Fig. 3). There was no significant change in CI either 15 minutes after the intravenous bolus injection of verapamil, or during the subsequent continuous infusion of the drug.

Since the changes in heart rate were not significant in either group (Fig. 3), the alterations in stroke volume index paralleled those in CI (Fig. 4).

Mean arterial pressure and mean left atrial pressure Within 2 minutes of the bolus dose of verapamil, the MAP fell from 99.4 ± 4.7 to 67.5 ± 3.4 mm Hg. The maximum fall in blood pressure of 30 ± 3.5 mm Hg at 2 minutes was significant

($p<0.01$). The recovery was, however, rapid and by 15 minutes the mean level of MAP for the whole group was 84.8 ± 3.8 mm Hg.

There was no significant change in MLAP following the bolus injection of verapamil. A continuous recording of the initial time course of change in MAP and MLAP is shown in Figure 5. The subsequent changes in MAP for the control animals (Fig. 3) and for those given verapamil (Figure 3) were not significant, although in the latter animals the levels tended to be lower. The changes in MLAP for both groups are shown in Figure 4.

Stroke work index (SWI) LAD produced a significant reduction ($p<0.01$) in the SWI in both series of animals at 15 minutes. Thereafter, minor changes in SWI occurred in the control as well as the verapamil group; these changes were not statistically significant (Figure 4.)

Systemic resistance index (SRI) At 15 minutes after LAD occlusion, SRI was 1.21 ± 0.11 units-kg in the control group and 0.87 ± 0.07 units-kg in the verapamil series. When verapamil was given SRI fell transiently from 0.87 ± 0.07 units-kg to 0.77 ± 0.05 units-kg ($p<0.01$). The subsequent values for SRI were, however, not significantly different from the level before the drug was given (Figure 4). In contrast, the SRI in the control group showed a progressive rise from 1.21 ± 0.15 units-kg at 15 minutes to 1.34 ± 0.15 units-kg at 2 hours ($p<0.05$; Figure 4).

Epicardial ST segment The administration of verapamil after LAD occlusion followed by a constant infusion of the drug reduced the ΣST by 39% (from 58.0

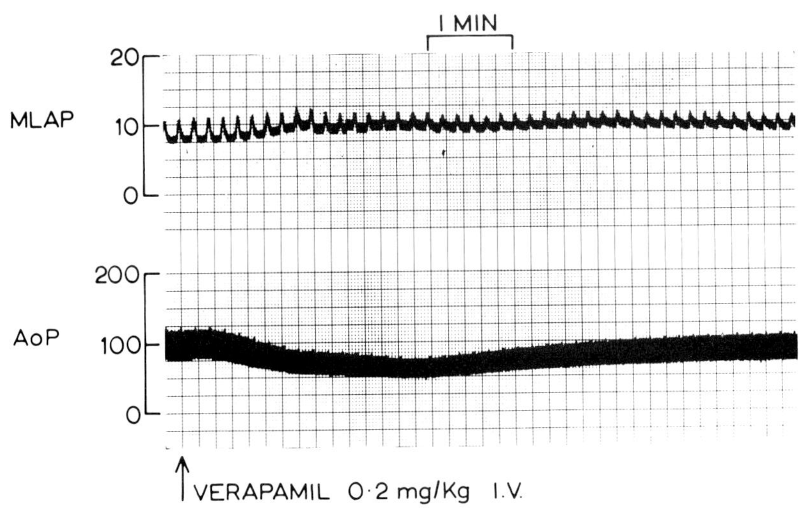

Figure 5. Effect of verapamil 0.2 mg/kg intravenously on aortic blood pressure (AOP) and mean left atrial pressure (MLAP). Note effect on MLAP is trivial; arterial pressure is significantly reduced by drug, but almost complete recovery occurs in 15 minutes.

Table 2. Effects of verapamil on myocardial blood flow (*MBF*)[a] after left anterior descending (*LAD*) coronary artery occlusion in anesthetized dogs

	A. Normally perfused myocardium		B. Ischemic myocardium	
Time after coronary occlusion	Control group n = 9	Treated group n = 8	Control group n = 9	Treated group n = 8
15 min	122.0 ± 11.8	109.0 ± 5.4	11.6 ± 1.4	9.6 ± 2.3
60 min	102.2 ± 7.4	125.0 ± 12.7[b]	18.1 ± 2.9	12.5 ± 2.4[b]
240 min	144.9 ± 11.7	119.0 ± 8.3[b]	21.1 ± 3.8	17.8 ± 2.5[b]

[a]*MBF* has been expressed as ml/min/100g of myocardium. Verapamil was given *after* LAD occlusion (see text). Values shown are means ± S.E.M.'s; n refers to the number of animals from which data were obtained.
[b]After verapamil.

Table 3. Effects of verapamil on endocardial/epicardial ratios of flow after left anterior descending coronary artery occlusion in anesthetized dogs[a]

Time after coronary occlusion	A. Normally perfused myocardium		B. Ischemic myocardium	
	Control group n = 9	Treated group n = 8	Control group n = 9	Treated group n = 8
15 min	1.08 ± 0.07	1.08 ± 0.05	0.52 ± 0.10	0.47 ± 0.08
60 min	1.17 ± 0.07	0.96 ± 0.05[b]	0.60 ± 0.08	0.63 ± 0.09[b]
240 min	1.12 ± 0.06	1.04 ± 0.07[b]	0.53 ± 0.12	0.43 ± 0.12[b]

[a]Endocardial/epicardial flow ratios were computed from the radio-activity counts taken from the pieces of left ventricular wall subdivided into inner and outer halves.

[b]After Verapimil

± 9.4 mV to 34.9 ± 9.7 mV; $p<0.05$) at 15 minutes and by 57% at 1 hour (to 27.7 ± 7.0 mV; Fig. 3). In contrast, in the control series, the mean ΣST of 50.3 ± 9.4 mV at 15 minutes after LAD occlusion showed no significant changes throughout the entire period of observation (Figure 3). A detailed analysis of the effects of verapamil on epicardial electrograms after coronary occlusion is reported elsewhere (Smith *et al.*, 1975b).

Effects of verapamil on myocardial blood flow

The effects of verapamil on MBF as measured by the distribution of microspheres in dogs with LAD occlusion compared to the changes found in an untreated group are presented in Table 2. In the areas of normally perfused myocardium, verapamil increased MBF from 109.0 ± 5.4 to 125.0 ± 12.7 ml/min/100 g at 1 hour ($p<0.01$) but the increase was less significant at 2 hours ($p<0.05$).

In the myocardium distal to the occluded LAD artery, MBF was approximately 10% of that in the normally perfused areas. In the group of animals given verapamil, the MBF in the ischemic area increased by 25% at 1 hour and by 85% at 2 hours. In the control group, MBF increased by 55% at 1 hour and by 82% at 2 hours. The differences between the MBF in the two groups were significant at 1 hour ($p<0.01$) but not at 2 hours. It is significant, however, that, compared to the controls, verapamil *did not increase* MBF to the ischemic area.

The changes in MBF were also analysed in terms of distribution of the flow to the inner and outer halves of the left ventricular wall. After LAD occlusion, a disproportionate degree of endocardial ischemia, as reflected by an approximately 50% fall in the endocardial/epicardial flow ratios, was found. Verapamil had no significant effect on the endocardial/epicardial ratios. The results are presented in Table 3.

Effects of verapamil on lactate production by the ischemic myocardium

The time course of changes in veno-arterial (v-a) lactate concentrations expressed as a percentage of the value at 15 minutes after LAD occlusion for the control group as well as for the group of animals given verapamil is shown in Figure 6. No significant differences were found between the two groups during the entire period of observation.

DISCUSSION

These results show that epicardial ST segment elevation over areas of ischemic myocardium can be reduced by verapamil when it is given 15 minutes after LAD occlusion in open-chest anesthetized dogs. This apparently salutary action of the drug was accompanied by beneficial effects on cardiovascular functions, but was not associated with an increase in myocardial blood flow to the ischemic area or

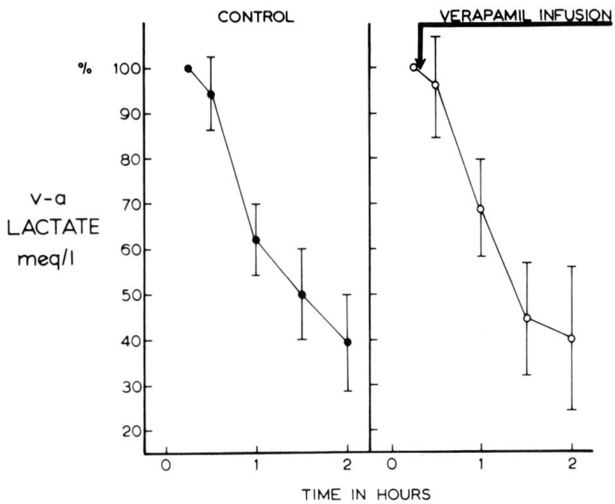

Figure 6. Effect of verapamil on veno-arterial (v-a) differences in lactate following coronary occlusion. *Ordinate:* V-a difference in lactate concentration is expressed as a percentage of value found 15 minutes after coronary ligation. *Abscissa:* Time in hours after coronary ligation. The changes shown for control group are similar to those found in animals given verapamil. The bars in graphs indicate standard errors of means.

a reduction in the levels of lactate in the veins draining the infarcting myocardium. The precise mechanism of the drug's apparent protective action in myocardial ischemia is thus not clear from the present studies. However, the results are of importance and may be of clinical relevance, in view of the recent interest in the possibility of reducing the size of myocardial infarcts by pharmacological and metabolic interventions (Maroko *et al.*, 1971, 1972b; Libby *et al.*, 1973; Maroko and Braunwald, 1973).

The alteration in ST segment is now increasingly used as an index of acute ischemic damage in clinical as well as experimental infarction (Maroko and Braunwald, 1973; Reid *et al.*, 1974). The extent of cellular damage is known to vary with the degree of reduction in myocardial creatine phosphokinase (CPK) activity following experimental coronary occlusion (Maroko *et al.*, 1972a; Maroko and Braunwald, 1973). Furthermore, it has been found that the degree of ST segment elevation 15 minutes after coronary ligation reliably predicts cellular integrity as reflected in the severity of myocardial CPK depletion and ultrastructural changes 24 hours later (Maroko *et al.*, 1972a; Maroko and Braunwald, 1973). Metabolic studies have also emphasized the significance of ST segment elevation as an indication of myocardial ischemia. For example, Karlsson, Templeton, and Willerson (1973) demonstrated that, during myocardial ischemia, epicardial sites of ST segment elevation were areas which had high levels of tissue lactate with pronounced depletion of ATP and creatine-phosphate. Thus, the reduction by verapamil of the epicardial ST segment elevation following coronary occlusion demonstrated here, may represent an improvement in the severity and extent of myocardial damage, even though the effects of the drug on myocardial enzyme histochemistry or structural alterations have not yet been studied.

The pharmacological properties of verapamil are of particular interest when considering the possible mechanisms through which the drug might act in myocardial ischemia. Verapamil is a potent smooth muscle relaxanat (Golenhofen and Lammel, 1972) and a weak nonspecific sympathetic antagonist (Nayler *et al.*, 1968). Clinically, it has been used in the management of angina (Sandler, Clayton, and Thornicroft, 1968; Livesley *et al.*, 1973) and cardiac arrhythmias (Heng *et al.*, 1975; Schamroth *et al.*, 1972). The details of the hemodynamic and electrophysiological effects of verapamil are therefore especially relevant.

In doses up to 0.25 mg/kg intravenously, hemodynamic changes due to verapamil in dogs without coronary occlusion could be accounted for by peripheral vasodilatation (Ross and Jorgensen, 1967; Nayler *et al.*, 1968). In patients with heart disease, 0.15 mg/kg of the drug given intravenously produced minor increases in cardiac output and left ventricular end diastolic pressure, with a slight reduction in LVdp/dt max associated with significant but transient decreases in arterial pressure and systemic resistance (Singh, Roche, and Agnew, 1975).

The electrophysiological properties of verapamil differ markedly from those of conventional antiarrhythmic drugs (Singh, 1971; Singh and Vaughan Williams, 1972; Cranefield *et al.*, 1974; Singh and Hauswirth, 1974; Zipes and Fischer, 1974), there being evidence to suggest that its primary action is a selective

The electrophysiological properties of verapamil differ markedly from those of conventional antiarrhythmic drugs (Singh, 1971; Singh and Vaughan Williams, 1972; Cranefield *et al.*, 1974; Singh and Hauswirth, 1974; Zipes and Fischer, 1974), there being evidence to suggest that its primary action is a selective inhibition of calcium ion movement across the membrane of excitable tissues (Fleckenstein, 1971). This action in the heart has been postulated to be responsible for its antiarrhythmic actions and for the depression of contractility (Singh and Vaughan Williams, 1972; Cranefield *et al.*, 1974). Kaumann and Aramendia (1968) emphasized the antiarrhythmic actions of verapamil following coronary occlusion in dogs; and it is of significance that late ventricular fibrillation occurred in none of the verapamil-treated animals in our experiments, whereas 3 of the 14 untreated animals developed ventricular fibrillation (Smith *et al.*, 1975b). Calcium antagonism in smooth muscle cells has also been thought to be the basis for the drug's peripheral and coronary vasodilator actions (Golenhofen and Lammel, 1972). Thus, the peripheral, coronary and myocardial effects of verapamil may all be of importance in accounting for the drug's protective action in myocardial ischemia.

Because verapamil is a coronary dilator, it might be expected to act by increasing collateral blood flow to the ischemic myocardium. Our studies with radioactive microspheres demonstrated that, while an increase in blood flow to the ischemic area did occur with time, it was not enhanced significantly by verapamil treatment. This finding was further supported by the observation that lactate production in the ischemic myocardium was also not influenced by verapamil.

In this context, it is of interest that the coronary dilator, dipyridamole, does not affect ST segment elevation following coronary occlusion (Bleifeld *et al.*, 1975), whereas nitroglycerin, which alters the distribution of blood flow in the ischemic myocardium (Fam and McGregor, 1968; Cohen *et al.*, 1973), have a beneficial effect on experimental infarct size (Smith *et al.*, 1973). Propranolol and nitroglycerin, both with anti-anginal actions as well as having salutary effects on ST segment alterations in experimental infarcts (Maroko et al., 1971; Smith et al., 1973), have been shown to increase the distribution of flow to the subendocardium without increasing total myocardial blood flow in ischemia (Becker *et al.*, 1971). No evidence for such a preferential increase in flow to the subendocarium by verapamil was found in our studies.

The action of nitroglycerin on ST segment elevation reported by Smith and his colleagues (1973) is relevant to the findings obtained with verapamil. They observed that the effect of nitroglycerin on the ST segment was particularly striking when the hypotensive response was prevented by alpha-adrenergic receptor stimulation by methoxamine. It would thus seem that a decrease in the afterload alone is unlikely to account for the changes in the ST segment elevation found after verapamil. A reduction in systemic arterial pressure by hemorrhage has been shown to aggravate ST segment alterations (Maroko *et al.*, 1971), probably due to impaired coronary perfusion. In our studies with verapamil, an improvement in ST segment elevation occurred, even during the hypotensive phase of the drug's action, suggesting that the salutary effects of the compound are likely to be due to its direct action on the myocardium. Moreover, in contrast to propranolol (Maroko *et al.*, 1971) and practolol (Libby *et al.*, 1973), both of which tend to affect the major determinants of myocardial oxygen consumption, verapamil had minor effects on heart rate, blood pressure (except transiently), and stroke work, suggesting that changes in the ST segments were not completely explained in terms of reduced cardiac work. If the peripheral effects of the drug do not play a major role in protecting the ischemic myocardium, it seems likely that, when the hypotensive effects of verapamil are prevented by the concomitant administration of methoxamine, a greater effect on ST segment might be achieved than with verapamil alone. Thus, the combination therapy of verapamil and methoxamine may have potential therapeutic applications in the early stages of clinical myocardial infarction.

The possibility that verapamil exerts its action in ischemia largely by a direct effect on cellular metabolism must also be considered. The fact that the electrocardiographic changes due to verapamil in our experiments were accompanied by favorable hemodynamic effects suggests that the actions of verapamil cannot be related simply to those on the cardiac membrane. In this context, it is pertinent to note that catecholamine-induced myocardial necrosis, which has been thought to be due to excessive intracellular calcium uptake by the ischemic myocardium, is prevented by pretreatment with verapamil (Fleckenstein, 1971). Experimental studies designed to investigate the effects of verapamil on electrophysiological and ultrastructural parameters as well as enzyme histochemistry are likely to provide further help in defining the mechanism of the apparent beneficial effects of the drug in experimental myocardial infarction.

Acknowledgments We wish to thank Drs E. A. Harris, R. M. L. Whitlock, and Sir Brian Barratt-Boyes for encouragement, help, and criticism, and Miss Heather D. Nisbet for expert technical assistance.

REFERENCES

BECKER, L. C., N. J. FORTUIN, and B. PITT. 1971. Effect of ischemia and anti-anginal drugs on the distribution of radioactive microspheres in the canine left ventricle. Circulat. Res. 28: 265–269.

BLEIFELD, W., W. WENDE, J. MEYER, und W. D. BUSSMAN. 1975. Einfluss einer Vaso-dilatation durch Dipyridamol ant die gross des akuten experimentellen Herzinfarktes. Naunyn Schmiederbergs Arch. Pharmakol. (In press).

COHEN, M. V., J. M. DOWNEY, E. H. SONNENBLICK, and E. S. KIRK. 1973. The effects of nitroglycerin on coronary collaterals and myocardial contractility. J. Clin. Invest. 52: 2836–2847.

CRANEFIELD, P. F., R. S. ARONSON, and A. L. WIT. 1974. Effect of verapamil on the normal action potential and on a calcium-dependent slow response of canine cardiac Purkinje fibers. Circulat. Res. 34: 203–213.

ELLIS, J. P., S. M. CAIN, and E. W. WILLIAMS. 1963. Rapid, accurate analysis of blood lactate. Technical Documentary Report No SAM/TDR/63/4G, USAF School of Aerospace Medicine, Texas.

FAM, W. M., and M. McGREGOR. 1968. Effect of nitroglycerin and dipyridamole on regional coronary resistance. Circulat. Res. 22: 649–658.

FLECKENSTEIN, A. 1971. Specific inhibitors and parameters of calcium action in the excitation-contraction coupling of heart muscle and their role in the prevention or production of myocardial lesions, pp. 135–188. P. Harris and L. Opie (eds.) *In* Calcium and the heart. Academic Press, London and New York.

GOLENHOFEN, K., and E. LAMMEL. 1972. Selective supression of some components of spontaneous activity in various types of smooth muscle by Iproveratril (Verapamil). Pfleugers Arch. 331: 233–243.

HAAS, H., und G. HARTFELDER. 1962. Alpha-Isopropyl-alpha-(N-methyl-N-homoveratryl)-gamma-amino-propyl)- 3-4 dimethoxyphenylacetonitril, eine substanz mit coronargefasserweiternden Eigenschaften. Arzneim. Forsch. 12: 549–558.

HARRIS, A. S. 1950. Delayed development of ventricular ectopic rhythms following experimental coronary occlusion. Circulation 1: 1318–1328.

HENG, M. K., B. N. SINGH, R. M. NORRIS, A. H. G. ROCHE, and C. J. MERCER. 1975. Immediate effects of verapamil on cardiac arrhythmias. Am. Heart J. (in Press).

KARLSSON, J., G. H. TEMPLETON, and J. T. WILLERSON. 1973. Relationship between epicardial ST segment changes and myocardial metabolism during acute coronary insufficiency. Circulat. Res 32: 725–730.

KAUMANN, A. J., and P. ARAMENDIA. 1968. Prevention of ventricular fibrillation induced by coronary ligation. J. Pharmacol. Exp. Ther. 164: 326–332.

LIBBY, P., P. R. MAROKO, J. W. COVELL, C. I. MALLOCK, J. ROSS, and E. BRAUNWALD. 1973. The effects of practolol on the extent of myocardial ischemic injury following experimental coronary occlusion and its effects on ventricular function in the normal and ischemic heart. Cardiovasc. Res. 7: 167–173.

LIVESLEY, B., P. F. CATLEY, R. C. CAMPBELL, and S. ORAM. 1973. Double-blind evaluation of verapamil, propranolol and isosorbide dinitrate against a placebo in the treatment of angina pectoris. Brit. Med. J. 1: 375–378.

MAROKO, P. R., and E. BRAUNWALD. 1973. Modification of myocardial infarct size after coronary occlusion. Ann. Intern. Med. 79: 720–733.

MAROKO, P. R., J. K. KJEKSHUS, B. E. SOBEL, T. WATANABE, J. W. COVELL, J. ROSS, and E. BRAUNWALD. 1971. Factors influencing infarct size following experimental coronary artery occlusions. Circulation 43: 67–82.

MAROKO, P. R., P. LIBBY, W. R. GINKS, W. E. SHELL, and J. ROSS Jr. 1972a. Coronary artery reperfusion I. Early effects on myocardial contractility and myocardial necrosis. J. Clin. Invest. 51: 2710–2176.

MAROKO, P. R., P. LIBBY, B. SOBEL, C. M. BLOOR, H. D. SYBERS, J. W. COVELL, and E. BRAUNWALD. 1972b. Effect of glucose insulin-potassium infusion on myocardial infarction following experimental coronary artery occlusion. Circulation 45: 1160–1175.

MELVILLE, K. J., H. E. SHISTER, and S. HUQ. 1964. Iproveratril: experimental data on coronary dilatation and anti-arrhythmic action. Canad. Med. Ass. J. 761–770.

NAYLER, W. G., I. McKINNES, J. B. SWANN, J. M. PRICE, V. CARSON, D. RACE, T. E. LOWE. 1968. Some effects of iproveratril (isoptin) on the cardiovascular system. J. Pharmacol. Exp. Ther. 161: 247–261.

NEUTZE, J. M., A. R. KERR, and R. M. L. WHITLOCK. 1973. Epicardial mapping in a variant of type A Wolff-Parkinson-White Syndrome. Circulation 48: 662–666.

REID, D. S., L. J. PELIDES, and J. P. SHILLINGFORD. 1971. Surface mapping of RS-T segment in acute myocardial infarction. Brit. Heart J. 33: 370–374.

REID, P. R., D. R. TAYLOR, D. T. KELLY, M. L. WEISFELDT, J. O. HUMPHRIES, R. ROSS, and B. PITT. 1974. Myocardial infarct extension detected by praecordial ST segment mapping. New Engl. J. Med. 290: 123–128.

ROSS, G., and C. R. JORGENSEN. 1967. Cardiovascular actions of Iproveratril. J. Pharmacol. Exp. Ther. 158: 504–509.

RYDEN, L., and H. SAETRE. 1971. The hemodynamic effect of verapamil. Europ. J. Clin. Pharmacol. 3: 153–157.

SANDLER, G., G. A. CLAYTON, and S. G. THORNICROFT. 1968. Clinical evaluation of verapamil in angina pectoris. Brit. Med. J. 3: 224–227.

SCHAMROTH, L., D. M. KRIKLER, and C. GARRETT. 1973. Immediate effects of intravenous verapamil in cardiac arrhythmias. Brit. Med. J. 1: 660–662.

SCHMID, J. R., and C. HANNA. 1967. A comparison of the antiarrhythmic actions of two new synthetic compounds, iproveratril and MJ 1999 and quinidine and pronethalol. J. Pharmacol. Exp. Ther. 156: 331–338.

SINGH, B. N. 1971. A study of the pharmacological actions of certain drugs and hormones with particular reference to cardiac muscle. D. Phil. Thesis, Oxford University, England.

SINGH, B. N., and O. HAUSWIRTH. 1974. Comparative mechanisms of action of antiarrhythmic drugs. Am. Heart J. 87: 367–382.

SINGH, B. N., A. H. G. ROCHE, and T. M. AGNEW. 1975. Immediate hemodynamic effects of intravenously administered verapamil in patients with cardiac disease. Submitted to Circulation.

SINGH, B. N., and E. M. VAUGHAN WILLIAMS. 1972. A fourth class of antidysrhythmic action? Effect of verapamil on ouabain toxicity, on atrial and ventricular intracellular potentials and on other features of cardiac function. Cardiovasc. Res. 6: 109–119.

SMITH, H. J., R. M. NORRIS, B. N. SINGH, and E. A. HARRIS. 1975a. Regional and temporal changes in coronary venous lactate concentration following experimental coronary occlusion. Submitted to Eur. J. Cardiol.

SMITH, E. R., D. R. REDWOOD, W. E. McCARRSEN, and S. E. EPSTEIN. 1973. Coronary artery occlusion in the conscious dog. Effects of alterations in arterial pressure produced by nitroglycerin, hemorrhage and alpha-adrenergic agonists on the degree of myocardial ischemia. Circulat. Res. 47: 51–57.

SMITH, H. J., B. N. SINGH, H. D. NISBET, and R. M. NORRIS. 1975b. The effects of verapamil on infarct size following experimental coronary occlusion. Cardiovasc. Res. (In press)

SMITH, H. J., B. N. SINGH, R. M. NORRIS, H. D. NISBET, M. B. OHNS, and P. J. HURLEY. 1975c. Effects of verapamil on patterns of myocardial blood flow and lactate production in relation to ST segment changes following experimental coronary occlusion. Submitted to Eur. J. Cardiol.

VAUGHAN WILLIAMS, E.M., E. E. BAGWELL, and B. N. SINGH. 1973. Cardio-specificity of beta-adrenoceptor blockade. Cardiovasc. Res. 7: 226–240.

ZIPES, D. R., and J. C. FISCHER. 1974. Effects of agents which inhibit the slow channel on sinus node automaticity and atrioventricular conduction in the dog. Circulat. Res. 34: 184–192.

Mailing Address:
Dr. B. N. Singh, Department of Medicine,
Auckland University School of Medicine,
Park Road, Auckland (New Zealand).

Coronary Vessel Changes Related to Cardiac Ischemia

Recent Advances in Studies on
Cardiac Structure and Metabolism Volume 10
The Metabolism of Contraction
Edited by Paul-Emile Roy and George Rona

EFFECTS OF ISCHEMIA ON ADENINE NUCLEOTIDE METABOLISM AND CORONARY FLOW DURING CARDIAC HYPERTROPHY

FRIEDRICH H. DEGENRING[1]

Department of Physiology, University of Virginia, School of Medicine, Charlottesville, Virginia, USA

SUMMARY

The effects of ischemia on myocardial adenine nucleotide metabolism and coronary flow during cardiac hypertrophy were studied in 140 rats and 20 guinea pigs, respectively. During increasing periods of ischemia, the initially lower ATP contents decreased significantly as did the initially elevated ADP levels, whereas AMP, adenosine, inosine, and hypoxanthine showed a continually rising elevation compared with the normal hearts. The sum of ATP, ADP, AMP, and their degradation products in the hypertrophied myocardial tissues started to decline after 5 min of ischemia. The remainder was found in the 0.9% NaCl solution in which the rat hearts were incubated, in the form of hypoxanthine, which was the largest fraction, followed by inosine and adenosine, which was the lowest fraction. In normal hearts, these changes occurred only after 60 min of ischemia. The coronary flow of the isolated guinea pig hearts increased significantly with decreasing content of the oxygen gas phase in the Krebs-Henseleit perfusion medium. These changes were more significant in normal than in hypertrophied hearts, despite the clear initial elevations of the coronary flow in these hearts at 95% oxygen saturated perfusion, as well as the essential increases of the adenosine content in the myocardial tissues and in the perfusates during the development of the hypoxemia. Consequently, these results significantly demonstrate the curtailed compensation possibilities of hypertrophied hearts for the maintenance of their functions during the development of ischemia in comparison with normal hearts, a factor obviously caused by the inefficient utilization of their energy supply even without ischemia.

INTRODUCTION

The rate of decrease in high energy phosphate stores during myocardial ischemia of dogs following coronary artery ligation observed by Braasch *et al.* (1968) led Gudbjarnason (1972) to conclude that myocardial energy transfer is inhibited because these ischemic muscles stop contracting at a relatively high ATP level. Furthermore, Gunning and Coleman (1972) showed an inefficient energy utilization (increased oxygen consumption—MVO_2—and decreased contractility) by

[1] Recipient of a Fortbildungsstipendium der Deutschen Forschungsgemeinschaft, Bonn - Bad Godesberg, Fed. Rep. Germany.

papillary cat muscles from hypertrophied right ventricles. Simultaneous studies of myocardial adenine nucleotide metabolism and the coronary flow also give an insight into the myocardial energy utilization, since they reveal the high energy phosphate levels and the oxygen supply of the hearts. Accordingly, the purpose of the present investigation was to correlate the adenine nucleotide metabolism and coronary flow of normal and of hypertrophied hearts, with and without ischemia.

METHODS

Compensated cardiac hypertrophy was produced in 140 rats either by daily swimming exercise, by treatment with daily intramuscular injections of 0.2–0.3 mg of DL-thyroxine for 7 weeks, or by surgical constrictions of the abdominal aorta just below the diaphragm for 9 days. The hearts of the animals were rapidly removed under pentobarbital anesthesia and frozen with aluminium tongs precooled in liquid nitrogen, either immediately or at various intervals (1.5 and 20 min) after the start of ischemia; this was produced by sealing the isolated hearts in glass beakers containing 1.0 ml of 0.9% NaCl solution, with gentle shaking in a water bath at 37°C. In addition, 20 guinea pig hearts were made hypertrophic by daily intramuscular injections of 0.3 mg of DL-thyroxine for 7 weeks.

After rapid removal from the animal, the hearts were immediately perfused with Krebs-Henseleit solutions equilibrated with 95% O_2 at 37°C (pH 7.4) at a pressure of 60 cm H_2O for 30 min. The hearts were paced at 180 impulses/min. After the initial 30-min period of perfusion, the hearts were perfused for 10 min with a solution equilibrated with either 95, 30, 20, 15, or 5% O_2. Immediately after the 10-min perfusion period, the hearts were frozen for subsequent enzymatic determination of the adenine nucleotides and their derivatives by spectrophotometry as described previously (Dobson, Rubio, and Berne, 1971).

RESULTS AND DISCUSSION

The changes in the myocardial purine contents were the same, irrespective of the method used to produce cardiac hypertrophy. Figure 1*A* illustrates the reduction of ATP in hypertrophied hearts and the further significant decline with increasing hypoxemia or ischemia in comparison with that of normal hearts subjected to the same duration of ischemia or hypoxemia. The faster depletion of ATP cannot be explained by the present data. Nevertheless, it could be caused by the elevated ATPase activity in hearts stressed by exercise (Hearn and Gollnick, 1961) and/or by lower creatine phosphate stores as observed by the author (unpublished results) and by Feinstein (1962); alternatively, it could be attributable to the apparently decreased synthetic pathways in the final stages of

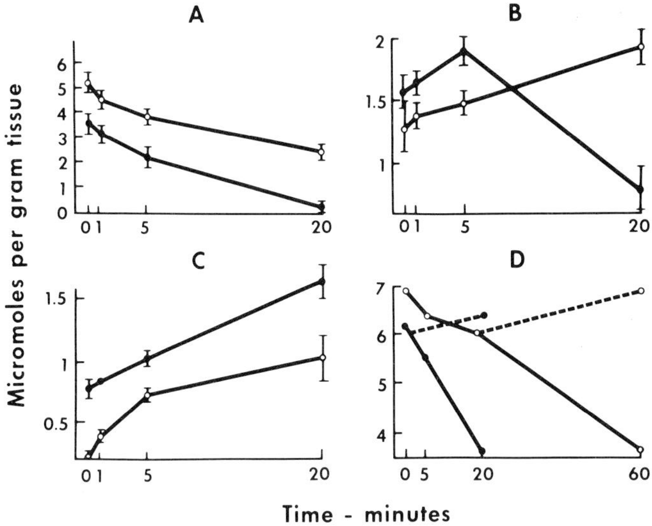

Figure 1. ATP (*A*), ADP (*B*), AMP (*C*), and sum of the purine contents with adenosine, inosine, and hypoxanthine (*D*), in normal (*open circles*) and hypertrophied myocardium (*closed circles*) at different periods of ischemia. *Ordinates:* mean contents ± standard deviation in μmoles/g tissue wet weight. *Abscissae:* period of ischemia, in minutes. *Dashed lines* in *D:* sum of purine contents in myocardial tissues and in fluid bathing hearts. Note different time scale in *D* compared with those in *A, B,* and *C.*

the cardiac hypertrophy. Figure 1, *B* and *C*, shows the changes of ADP and AMP accompanying the degradation of ATP during ischemia. ADP was at first slightly increased compared with the normal hearts. However, with increasing duration of ischemia, the ADP content of the hypertrophied hearts decreased significantly whereas the levels in the normal hearts were augmented. AMP showed a steady increase which was more significant in hypertrophied than in normal hearts with increasing periods of ischemia. The decrease of ATP during ischemia is accompanied not only by changes of ADP and AMP, but also by distinct increases of adenosine, inosine, and hypoxanthine in the myocardial tissues of the hypertrophied hearts.

Figure 1*D* summarizes the changes of ATP, ADP, AMP, adenine nucleosides, and hypoxanthine in normal and in hypertrophied hearts. The *solid lines* disclose an essential decrease of the sum of these purine contents in the myocardial tissues at increasing periods of ischemia. When the amounts of adenosine, inosine, and hypoxanthine which appeared in the 0.9% NaCl solution bathing the hearts after prolonged ischemic periods were added to the myocardial tissue contents, the total nucleotide and derivative pool tended to remain constant (*dashed lines*). It should be noted that the changes in the hypertrophied hearts after 20 min of ischemia correspond to those of the normal hearts subjected to a

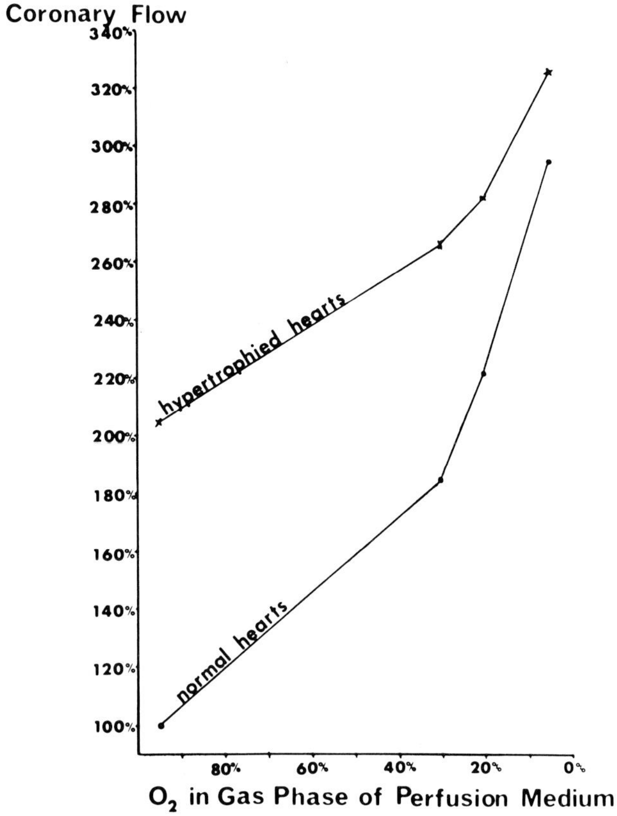

Figure 2. Coronary flows expressed as percentage increases, in normal and hypertrophied guinea pig hearts during development of hypoxemia. 100% = coronary flow of normal hearts at 95% oxygen in gas phase of Krebs-Henseleit perfusion medium.

three-fold greater period of ischemia. The appearance of adenosine, inosine, and hypoxanthine in the fluid bathing the hearts also suggests an impairment of metabolism of these compounds in the synthesis of the adenine nucleotides after prolonged ischemic periods. Since adenosine is incorporated into adenine nucleotides at a faster rate than inosine or hypoxanthine (Liu and Feinberg, 1971), the adenosine levels in the fluid were always the lowest, even after 20 min of ischemia.

Figure 2 illustrates the exponential increases of the coronary flow with decreasing content of the oxygen gas phase in the Krebs-Henseleit perfusion medium. The coronary flow of the hypertrophied hearts perfused with a 95% O_2-equilibrated solution was significantly elevated in comparison with that of

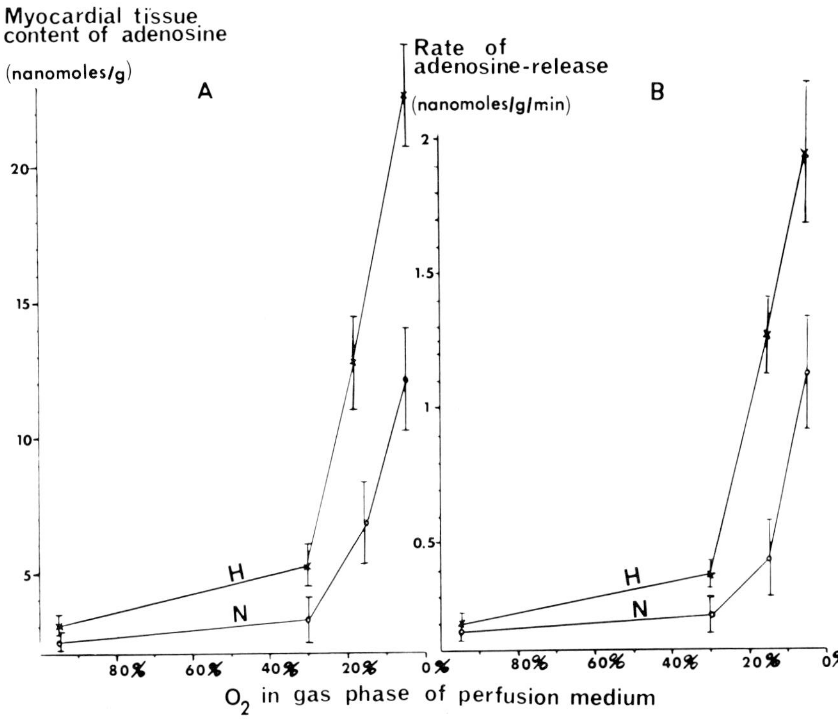

Figure 3. A: myocardial tissue content of adenosine in normal (N) and hypertrophied (H) hearts. *B:* rate of release of adenosine into perfusate during development of hypoxemia. *Ordinates:* mean values ± standard deviation in nanomoles/g, or nanomoles/g/min, respectively. *Abscissae:* percentage of oxygen in gas phase of perfusion medium.

normal hearts subjected to the same conditions. However, the flows tended to move to the same levels with increasing hypoxemia, because of the smaller coronary flow elevations in the hypertrophied hearts.

Figure 3*A* shows the increases in the myocardial tissue content of adenosine accompanying the coronary flow elevations, with respect to the decreasing oxygen equilibration of the perfusion medium, while Figure 3*B* illustrates the increasing rate of adenosine release with increasing hypoxemia. Figure 4 indicates that, during hypoxemia, the coronary flow of normal hearts increases more than that of hypertrophied hearts as related to the respective myocardial weights. In comparison with the hypertrophied myocardium, the elevations of the coronary flow of the normal heart per gram heart weight become even more significant during hypoxemia when they are related to the myocardial tissue content of adenosine. These observations suggest that the development of

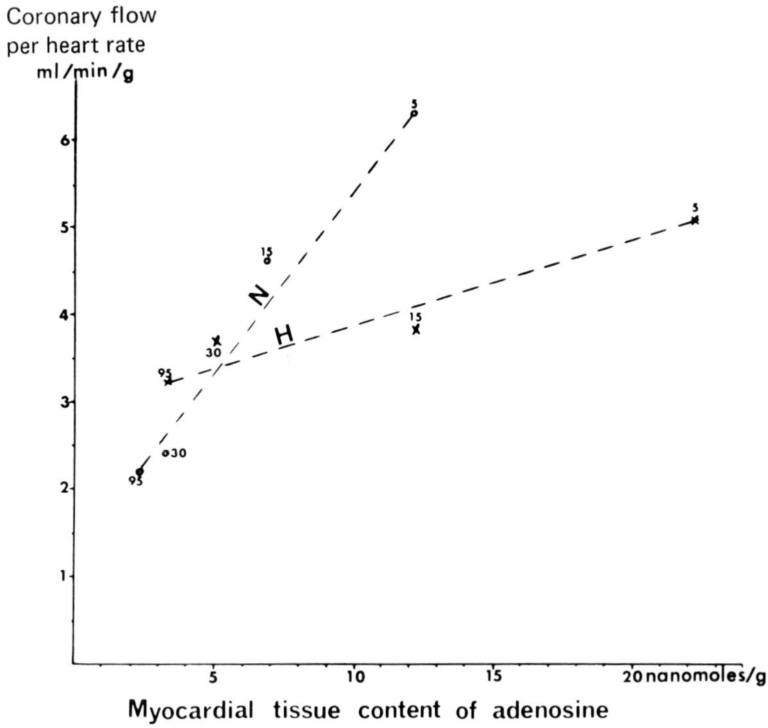

Figure 4. Coronary flows related to myocardial tissue levels of adenosine during development of hypoxemia. *Ordinates:* mean values of coronary flow/g myocardial tissue wet weight in ml/min/g. *Abscissae:* mean contents of adenosine/g myocardial tissue wet weight in nanomoles/g. *From left to right:* values after 10 min of perfusion with 95% (95), 30% (30), 15% (15), and 5% (5) O_2-equilibrated Krebs-Henseleit solutions.

ischemia reveals impaired functions of the hypertrophied myocardium significantly as to its moderate coronary flow elevations, despite an increased production of adenosine, which is most probably due to the predominant decline and degradation of the high energy phosphate stores. The increased intercapillary distances of the hypertrophied hearts (Linzbach, 1947), including the augmented muscle fiber diameters, cause an essential increase in the oxygen supply. However, this increased supply obviously very soon reaches its limits during the development of ischemia, because the coronary flow increase is then curtailed by the initially dilated coronary vessels, unlike in the normal hearts.

Acknowledgments I acknowledge with pleasure the helpful advice of Dr. Robert M. Berne and Dr. Rafael Rubio as well as the able assistance of Mr. Richard R. Curnish and Mrs. Luvenia E. Wigginton.

REFERENCES

BRAASCH, W., S. GUDBJARNASON, P. S. PURI, K. G. RAVENS, and R. J. BING. 1968. Early changes in energy metabolism in the myocardium following acute coronary artery occlusion in anesthetized dogs. Circulat. Res. 23: 429–438.

DOBSON, J. G., Jr., R. RUBIO, and R. M. BERNE. 1971. Role of adenine nucleotides, adenosine, and inorganic phosphate in the regulation of skeletal muscle blood flow. Circulat. Res. 29: 375–384.

FEINSTEIN, M. B. 1962. Effects of experimental congestive heart failure, Ouabain, and asphyxia on the high-energy phosphate and creatine content of the guinea pig heart. Circulat. Res. 10: 333–346.

GUDBJARNASON, S. 1972. Inhibition of energy transfer in ischemic heart muscle. *In* E. Bajusz and G. Rona (eds.), Recent Advances in Studies on Cardiac Structure and Metabolism, Vol. 1: Myocardiology, pp. 17–26. University Park Press, Baltimore.

GUNNING, J. F., and H. N. COLEMAN. 1972. The effects of hypertrophy in myocardial energy untilization. *In* E. Bajusz and G. Rona (eds.), Recent Advances in Studies on Cardiac Structure and Metabolism, Vol. 1: Myocardiology, pp. 190–199. University Park Press, Baltimore.

HEARN, G. R., and Ph.D. GOLLNICK. 1961. Effects of exercise on the adenosinetriphosphatase activity in skeletal and heart muscles of rats. Intern. Zschr.angew.Physiol. einschl.Arbeitsphysiol. 19: 23–26.

LINZBACH, A. 1947. Mikrometrische und histologische Analyse hypertropher menschlicher Herzen. Virchows Arch. Abt. A. 314: 534–594.

LIU, M. S., and H. FEINBERG. 1971. Incorporation of adenosine-8-^{14}C and inosine-8-^{14}C into rabbit heart adenine nucleotides. Am. J. Physiol. 220: 1242–1248.

Mailing Address:
Dr. F. H. Degenring,
Division of Cardiology, Department of Medicine,
University of Heidelberg,
58 Bergheimer Str., D-69 Heidelberg (Fed. Rep. Germany).

Recent Advances in Studies on
Cardiac Structure and Metabolism Volume 10
The Metabolsim of Contraction
Edited by Paul-Emile Roy and George Rona

DEMONSTRATION OF THE "NO-REFLOW" PHENOMENON IN THE DOG HEART AFTER TEMPORARY ISCHEMIA

ROBERT A. KLONER, CHARLES E. GANOTE,
ROBERT B. JENNINGS, and KEITH A. REIMER

Department of Pathology, Northwestern University Medical School, Chicago, Illinois, USA

SUMMARY

The effects of 40- or 90-min periods of temporary myocardial ischemia on the distribution of coronary flow and capillary structure were assessed in groups of mongrel dogs. Thioflavin S, a fluorescent dye which stains vascular endothelium when injected intravenously, was used to demonstrate the distribution of microvascular perfusion at 10 sec, 5 min, or 20 min following release of a 40- or 90-min circumflex coronary artery occlusion. Hearts which demonstrated perfusion defects were sampled for electron microscopy. Following 40 min of occlusion, thioflavin S was distributed uniformly throughout the myocardium. In contrast, following 90-min periods of coronary occlusion, perfusion defects always were present in the subendocardial half of the posterolateral left ventricular wall. Several morphological features in these areas of no reflow were observed by electron microscopy, including decreased endothelial pinocytotic vesicles, endothelial gaps and bleb formation, capillaries packed with erythrocytes, occasional intraluminal thrombi, and extravascular erythrocytes and fibrin. Myocardial cells showing severe injury always were seen within but also extending beyond the areas of poor perfusion. These results demonstrate that areas of no reflow occur following 90-min periods of ischemic injury in the dog, but that primary myocardial cell injury occurs during the ischemic period and not as a function of the "no-reflow" phenomenon.

INTRODUCTION

Perfusion defects following episodes of temporary ischemia have been observed in brain (Majno *et al.*, 1967; Chaing *et al.*, 1968), kidney (Summers and Jamison, 1971; Flores *et al.*, 1972), and skin (Willms-Kretschmer and Majno, 1969), and have been referred to as the "no-reflow" phenomenon. This phenomenon also has been observed in the cat heart, following 60–120 min of temporary coronary occlusion (Krug *et al.*, 1966), and has been suspected in canine myocardium (Willerson *et al.*, 1972). However, the etiology and timing of its development in the dog heart have not been established.

The aims of the experiments presented in this paper were to determine if a no-reflow phenomenon occurs in the dog heart following temporary coronary occlusion and, if so, to evaluate the distribution of coronary flow and structural changes associated with this entity. The results show that the no-reflow phenomenon does occur in inner portions of myocardium after 90 min but not after 40 min of ischemia and that it is associated with extensive capillary damage.

MATERIALS AND METHODS

Mongrel dogs of both sexes, weighing between 9 and 23 kg, were used. They were fasted overnight and were anesthetized with intravenous sodium pentobarbital (30 mg/kg). Ventilation was maintained by a cuffed endotracheal tube attached to a model 1063 Harvard respirator pump. Limb lead II of the electrocardiograph and peripheral arterial blood pressure obtained with a Statham (model P23) strain guage were monitored continuously. The posterior papillary muscle infarction preparation was used as described previously (Jennings, Wartman, and Zudyk, 1957). Briefly, a thoracotomy was performed via the fourth left intercostal space, the left lung was gently retracted, and the pericardium was opened widely to expose the heart. The circumflex branch of the left coronary artery was dissected free 1–2 cm from the aorta. A Goldblatt clamp was placed around the artery and was tightened following control electrocardiograph and blood pressure readings. Occlusion of this vessel produces a large area of ischemia on the posterolateral wall of the left ventricle. The posterior papillary muscle is an area of severe ischemia in this preparation and was the tissue studied for the presence of arterial reflow. Posterolateral wall ischemia was confirmed in the present experiments by the presence of cyanosis and ST segment elevation in leads II, III, and aVf.

Experimental design

1. In 2 dogs, sham operations without occlusion were performed.
2. In 4 dogs, 90 min of permanent occlusion was performed.
3. In 12 dogs, 40 min of temporary coronary occlusion was performed, followed by 10–12 sec (1 dog), 5 min (7 dogs), or 20 min (4 dogs) of coronary arterial reperfusion induced by quickly opening the Goldblatt clamp.
4. In 15 dogs, 90 min of temporary coronary occlusion was performed, followed by coronary arterial reperfusion for 10–12 sec (4 dogs), 5 min (7 dogs), or 20 min (4 dogs).

Evaluation of coronary flow distribution

The transmural distribution of coronary flow was determined by intravenous injection of 1 cc/kg body weight of a 4% solution of thioflavin S, a fluorescent vital stain for endothelium, 10 sec prior to excision of the heart. Following excision of the heart, the left ventricle was cut into 0.2-cm longitudinal slices which were examined for thioflavin S fluorescence under ultraviolet light through a Y2 barrier filter. Myocardium receiving significant arterial blood flow will fluoresce yellow-green under ultraviolet light, while myocardium receiving low or no flow is nonfluorescent (Kloner *et al.*, 1974a). The percentage of nonfluorescent tissue in the posterior papillary muscle and subjacent myocardium (PP) was quantitated from photographs of tissue slices as described previously (Kloner *et al.*, 1974a).

Light and electron microscopy

Tissue from dogs receiving 90 min of permanent or temporary ischemia was sampled for electron microscopy from both fluorescent and nonfluorescent portions of PP and also from the anterior wall of the left ventricle. Small blocks (about 0.1–0.3 mm thick) were cut with a clean razor blade, minced under cold phosphate-buffered 1% osmic acid, and fixed for 60 min at 5°C. They were dehydrated, embedded in Epon, sectioned, and examined according to techniques described previously (Kloner *et al.*, 1974c). These sections were compared with those sampled from a recently reported (Kloner *et al.*, 1974c) large series of dogs with 40 min of temporary ischemia and 2–20 min of reflow.

RESULTS

Coronary flow distribution

Nonischemic myocardium All slices from the left ventricle of sham-operated dogs demonstrated homogeneous yellow-green thioflavin S fluorescence. In all dogs with occlusions, the anterior interventricular septum and portions of the anterior left ventricular free wall also fluoresced homogeneously yellow-green and thus served as an internal control for each dog (Fig. 1*a*).

Permanent coronary occlusion In dogs with 90 min of permanent ischemia, the PP always had large areas of nonfluorescence, especially in the inner two-thirds of the myocardial wall. Some fluorescence usually was present, however, in the subepicardium with patchy extensions into the middle third of the myocardium (Fig. 1*b*). The percentage of nonfluorescent tissue in the PP after 90 min was 49 ± 10%, which was similar to the percentage nonfluorescence previously observed after 15 and 60 min of permanent ischemia (Kloner *et al.*, 1974a).

Temporary coronary occlusion In 11/12 dogs with 40 min of temporary occlusion and varying lengths of arterial reperfusion, the PP was homogeneously fluorescent, indicating that coronary reperfusion into the ischemically damaged tissue had occurred (Fig. 1*c*). In 1 dog which received 40 min of ischemia and 20 min of reperfusion, a small area at the tip of PP (4% of PP) was nonfluorescent, indicating that arterial blood reperfusion did not occur into this region.

In all dogs with 90 min of temporary occlusion and varying lengths of arterial blood reperfusion, nonfluorescent areas were present in the subendocardium and inner midmyocardium, indicating that arterial reperfusion into these areas was

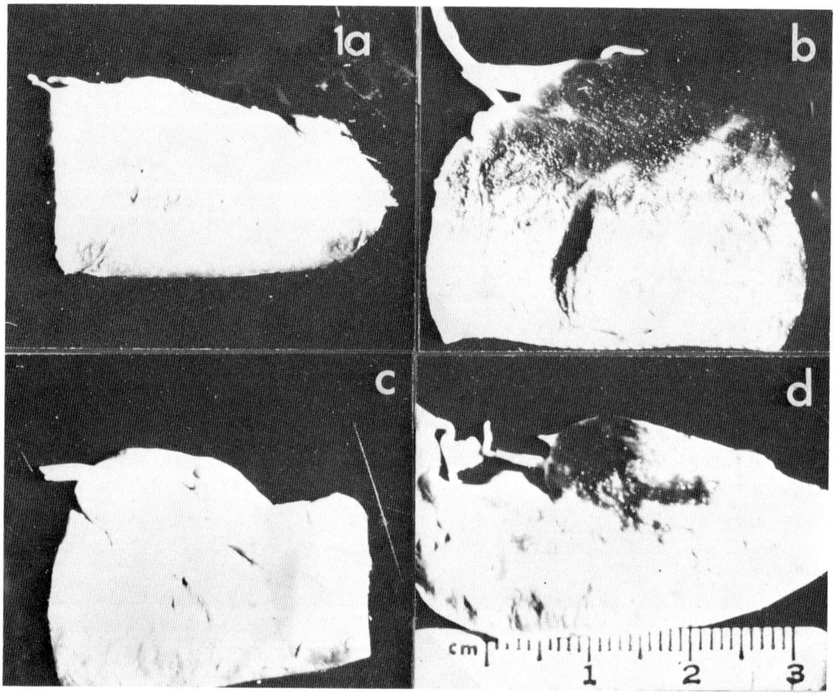

Figure 1. Distribution of coronary flow as determined by fluorescence. *a:* section of nonischemic control myocardium taken from anterior papillary muscle and subjacent myocardium. Myocardium exhibits homogeneous fluorescence under ultraviolet (UV) light. Endocardial surface is *toward top of picture* and epicardial surface is *toward bottom of picture. b:* posterior papillary muscle and subjacent myocardium (PP) permanently ischemic for 90 min under UV light. While subepicardial myocardium and midmyocardium are largely fluorescent, subendocardium and papillary muscle are largely nonfluorescent. *c:* PP temporarily ischemic for 40 min with 5 min of coronary reperfusion. Myocardium fluoresces homogeneously, suggesting that significant reflow occurred. *d:* PP temporarily ischemic for 90 min with 20 min of coronary reperfusion. Portions of posterior papillary muscle are nonfluorescent and, thus, represent areas of no reflow.

Table 1. Number of dogs exhibiting areas of no reflow

Procedure	No. of Dogs
40-min occlusion plus blood reperfusion	1/12
90-min occlusion plus blood reperfusion	15/15

poor (Table 1; Fig. 1*d*). These zones of "no reflow" represented 11 ± 2%, 12 ± 4%, and 10 ± 3% of the PP in dogs which received 10–12 sec, 5 min, and 20 min of coronary reperfusion, respectively. Small areas of nonfluorescence also were present in the inner myocardium of slices between the PP and the anterior papillary muscle (AP) and occasionally involved the tip of the AP.

Fine structure

Nonischemic myocardium Myocardial cells and capillaries from the anterior interventricular septum and anterior left ventricular free wall were intact. Capillary endothelium had numerous pinocytotic vesicles, and intraluminal projections of endothelium were short or absent (Fig. 2*a*).

Permanently ischemic myocardium Myocardial cells permanently ischemic for 90 min were morphologically similar to myocardium permanently ischemic for 40 and 60 min in previous studies (Kloner *et al.,* 1974c). I bands, nuclear chromatin clumping, and intermyofibrillar edema were prominent. Mitochondria were swollen and contained amorphous intramitochondrial dense bodies. Glycogen was absent. Although most capillaries were normal after 40 min of ischemia, they showed marked abnormalities when the period of ischemia was extended to 90 min. Endothelial pinocytotic visicles were absent or decreased in number, and nuclear chromatin clumping was prominent. In addition, large endothelial bleb-like protrusions and large, clear, membrane-bound vesicular structures were present in the lumen (Fig. 3*a*). The latter vesicles may have been swollen protrusions of endothelium, which either were detached from the endothelium or were attached at a site not included in the plane of section. Fibrin and platelet thrombi were observed in two of four dogs studied by electron microscopy.

Myocardium from areas of no reflow Myocardial cells from nonfluorescent regions of no reflow in hearts with temporary 90-min coronary occlusions showed similar changes to myocardium permanently ischemic for 90 min. Several additional changes were present in capillaries. With 10–12 sec of reflow, capillaries were often tightly packed with red blood cells, suggesting a downstream obstruction to flow, perhaps by the large endothelial protrusions or membrane-bound vesicles (Figs. 2*b* and 4*a*). Endothelial pinocytotic vesicles were sparse (Figs. 2*b* and 3*b*–5). In addition, with 5–20 min of reperfusion, capillaries also often had large gaps in the endothelium. These were commonly associated with intraluminal fibrin thrombi (Figs. 3*b* and 4*b*). Extravascular red blood cells (Fig. 3*b*) and fibrin also were common. Occasionally, capillaries appeared compressed by adjacent swollen myocardial cells (Fig. 5). These latter cells usually showed areas in which the sarcolemmal membrane was lifted from the myofilaments (Table 2).

Myocardium from areas adjacent to zones of no reflow In dogs with 90 min of temporary ischemia, myocardium from fluorescent areas adjacent to zones of

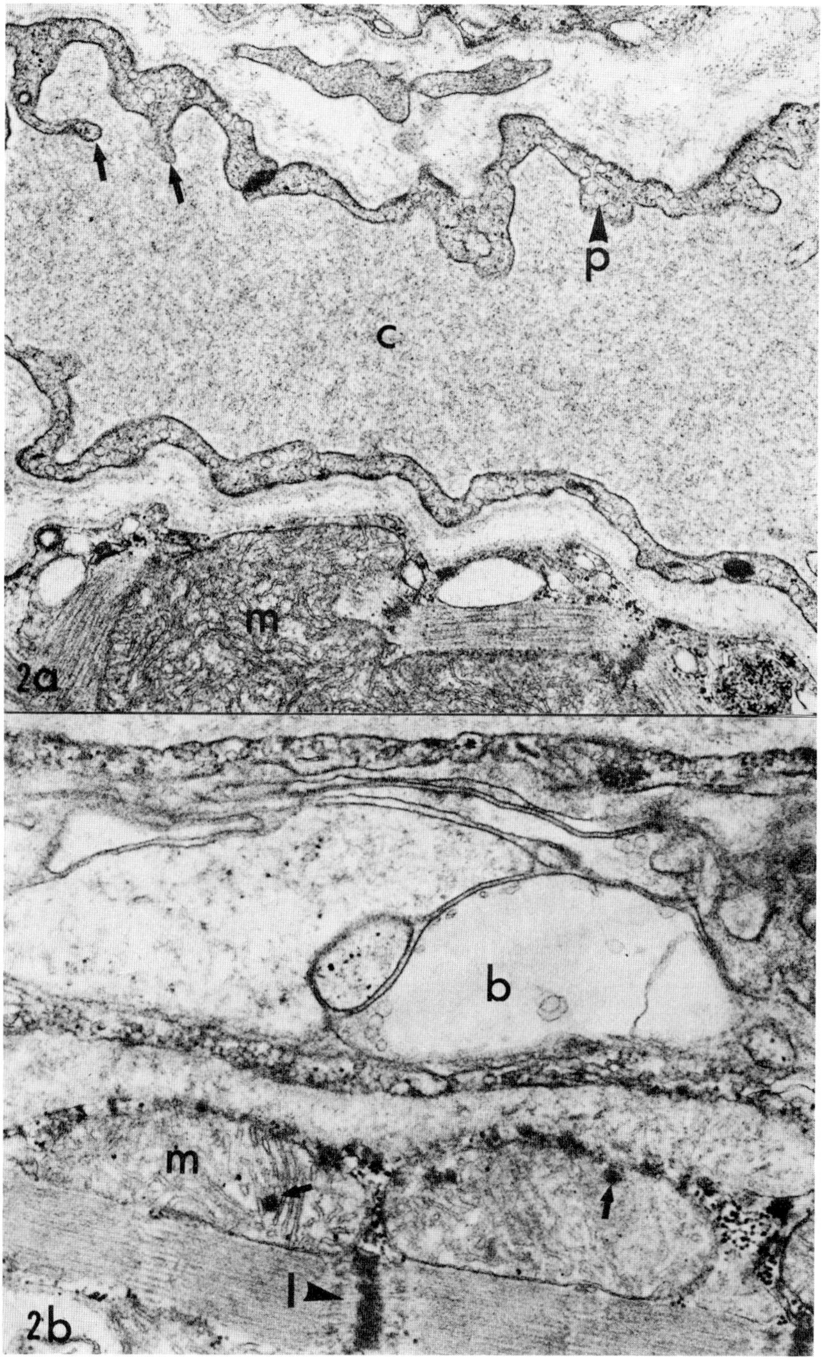

Figure 2. Electron microscopic observations following ischemia. *a:* nonischemic control myocardium. Capillary endothelium is thin, and pinocytotic vesicles (p) are numerous. Short endothelial folds are present (*arrows*). Mitochondrial cristae are tightly packed. c, capillary lumen; m, mitochondrion. × 24,000. *b:* PP temporarily ischemic for 90 min with

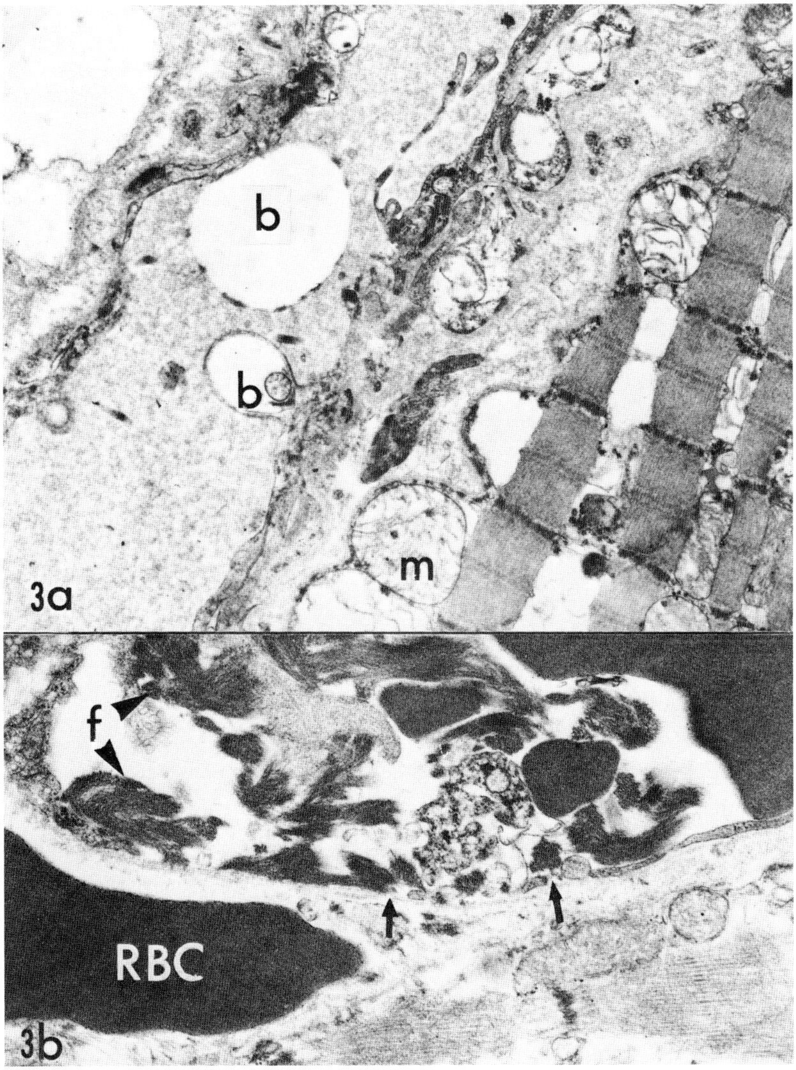

Figure 3. Electron microscopic observations following ischemia. *a:* PP permanently ischemic for 90 min. Two swollen endothelial blebs (b) appear to be budding off of the intraluminal surface of the capillary. The larger blebs shown in Figure 2b may have originated from such forms. Pinocytotic vesicles are absent. Intermyofibrillar edema is prominent and mitochondria are swollen with disrupted cristae and amorphous dense bodies. I bands are present. m, mitochondrion. × 12,950. *b:* PP temporarily ischemic for 90 min with 20 min of coronary blood reperfusion (from an area of no reflow). Gaps are present in capillary endothelium (*arrows*), with adjacent intraluminal fibrin tactoids (f). A red blood cell (RBC) is present in extravascular space. × 18,000.

10–12 sec of coronary arterial blood reperfusion (from an area of no reflow). Endothelial pinocytotic vesicles are absent. Capillary lumen is full of membrane-bound vesicular blebs (b), which appear capable of obstructing flow. Some of these could represent degranulated platelets. Mitochondria (m) contain amorphous matrix dense bodies (*arrows*) characteristic of irreversible myocardial ischemic injury. I, I band. × 32,000.

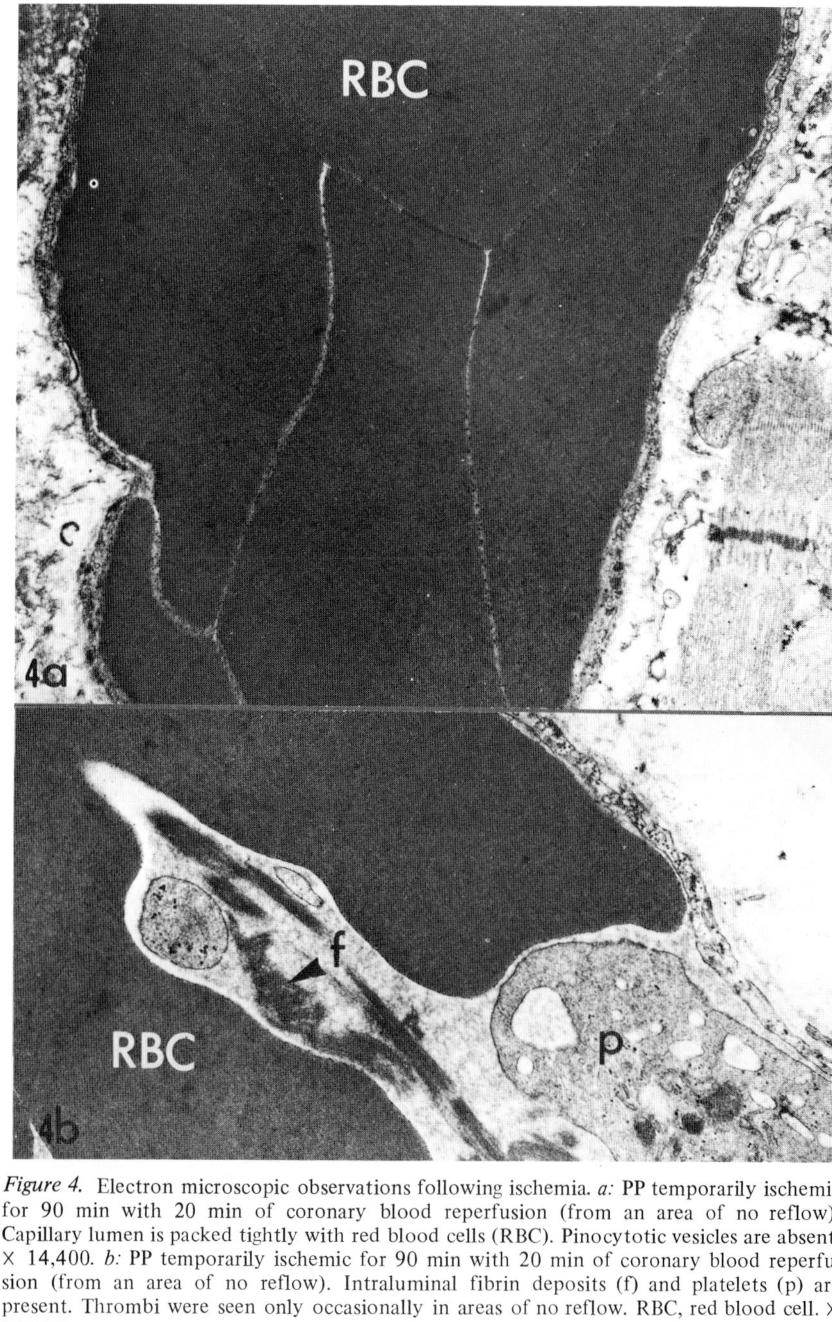

Figure 4. Electron microscopic observations following ischemia. *a:* PP temporarily ischemic for 90 min with 20 min of coronary blood reperfusion (from an area of no reflow). Capillary lumen is packed tightly with red blood cells (RBC). Pinocytotic vesicles are absent. × 14,400. *b:* PP temporarily ischemic for 90 min with 20 min of coronary blood reperfusion (from an area of no reflow). Intraluminal fibrin deposits (f) and platelets (p) are present. Thrombi were seen only occasionally in areas of no reflow. RBC, red blood cell. × 16,000.

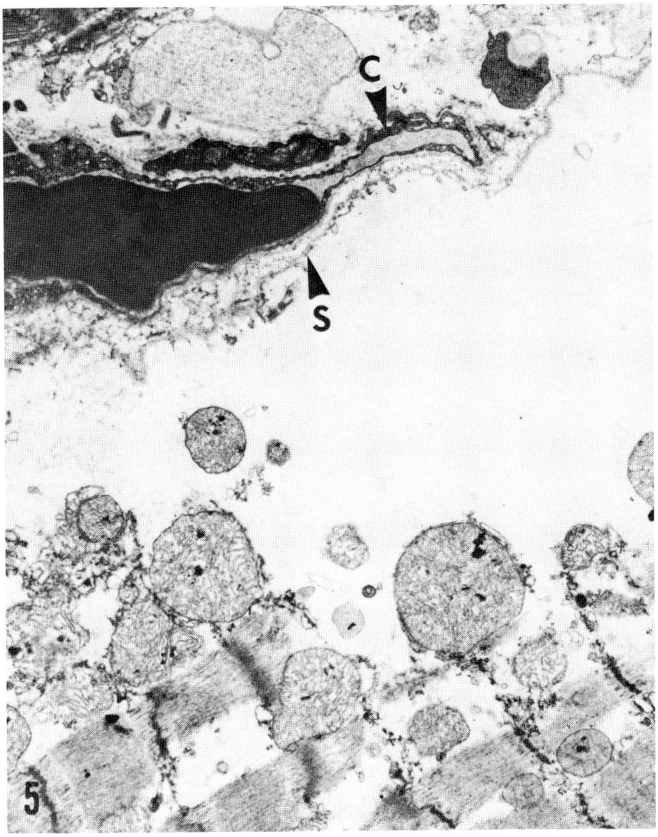

Figure 5. Electron microscopic observations following ischemia. PP temporarily ischemic for 90 min with 20 min of blood reperfusion (from an area of no reflow). Sarcolemmal membrane (s) is lifted off of myofilaments, forming a bleblike space which appears to be compressing a portion of an adjacent capillary (c). Intermyofibrillar edema and I bands are prominent. Mitochondria contain amorphous matrix densities. × 7,000.

no reflow showed changes similar to those seen with 40 min of temporary ischemia (Kloner *et al.*, 1974c). Myocardial cells were massively swollen and contained contraction bands. Mitochondria were swollen and contained amorphous dense bodies. However, granular $CaPO_4$ intramitochondrial dense bodies characteristically seen with 40 min of temporary ischemia were absent. Except for endothelial nuclear chromatin clumping and a decrease in the number of pinocytotic vesicles, the capillaries were similar to controls. Myocardial cells from fluorescent areas in the outer half of the ventricular wall showed only minimal changes, including occasional nuclear chromatin clumping and mitochondrial swelling. Capillaries appeared normal.

Table 2. Morphological changes in capillaries in areas of no reflow

Decreased endothelial pinocytotic vesicles
Nuclear chromatin clumping
Endothelial "blebs"
Endothelial gaps with extravascular erythrocytes and fibrin tactoids
Erythrocytes packed tightly together in capillary lumen
Occasional thrombi
Occasional capillary compression by swollen myocardial fibers

DISCUSSION AND SUMMARY

The results suggest that the no-reflow phenomenon is present in severely ischemic canine myocardium after 90 min of ischemia. Since areas of no reflow were detected as early as 10–12 sec following the release of the occlusion, the defect which prevented homogeneous reperfusion was probably present at the end of the 90-min episode of permanent ischemia. Capillary changes were closely associated with areas of no reflow and included absence of pinocytotic vesicles, endothelial bleb formation, red blood cell aggregation, endothelial gaps, and occasional thrombi. The close association between the no-reflow phenomenon and evidence of capillary injury suggests that direct vascular injury is the principal cause of the poor perfusion observed. However, capillary compression by adjacent swollen myocardial cells also may contribute to the depressed arterial reflow.

On the other hand, this study demonstrates that myocardium irreversibly injured by a 40-min coronary occlusion is uniformly perfused by blood containing thioflavin S after 10–12 sec, 5 min, or 20 min of blood reflow. The absolute rate of flow within previously ischemic areas is unknown, but the amount of thioflavin S staining was equal to that seen in control tissue. This observation with thioflavin S correlated with the ultrastructural demonstration that the microvasculature was largely intact following 40 min of ischemia with or without reflow. Increases in tissue water, Ca^{++}, and Na^{+} which occur in myocardium after 40 min of temporary coronary occlusion but which do not occur after 40 min of ischemia in the absence of reflow (Shen and Jennings, 1972a, b; Whalen *et al.*, 1974) also indicate that significant coronary blood reperfusion takes place following reflow at this time.

What is the significance of the no-reflow phenomenon? We believe that it probably plays no primary role in causing early myocardial cell ischemic injury, since: 1) irreversible myocardial cell damage occurs as early as 40 min after coronary occlusion in areas of severe ischemia, i.e., well before significant capillary damage or the no-reflow phenomenon is usually seen; 2) severe myocardial cell damage was seen, not only within, but also well beyond areas of poor perfusion. The no-reflow phenomenon does have important clinical implications. Damaged tissue in areas of no reflow is not in close contact with the circulation and should be unsalvageable by drug therapy or coronary bypass surgery. Also, it should heal more slowly than tissue in close contact with the arterial circulation. However, zones of no reflow following 90 min of ischemia are confined to the most severely ischemic portions of the inner myocardium. There is a large area of less severely ischemic myocardium present in the outer myocardial wall (Kloner *et al.*, 1974b) of the heart which is free of significant structural alterations in the microvasculature. Hence, the potential for myocardial salvage by restoration of flow still exists in these less severely ischemic zones.

Acknowledgments This work was supported in part by Grant HE 08729 and Contract NHLI-72-2984 from the National Institutes of Health and a grant from the Chicago and Illinois Heart Association. Dr. Kloner was a predoctoral fellow (National Institutes of Health Grant GM 00131) during the time of this study.

The authors are grateful for the technical assistance provided by C. Kawahara, J. L. Hines, B. Schumacher, R. L. Martinez, R. Godfrey, S. Browning, S. M. Liu, S. Safavi, B. Riebe, and S. Decker.

REFERENCES

CHAING, J., M. KOWADA, A. AMES, II, R. L. WRIGHT, and G. MAJNO. 1968. Cerebral ischemia III. Vascular changes. Am. J. Path. 52: 455–476.

FLORES, J. D. R., D. R. DiBONA, C. H. BECK, and A. LEAF. 1972. The role of cell swelling in ischemic renal damage and the protective effect of hypertonic solute. J. Clin. Invest. 51: 118–126.

JENNINGS, R. B., W. B. WARTMAN, and Z. E. ZUDYK. 1957. Production of an area of homogeneous myocardial infarction in the dog. Arch. Path. 63: 580–585.

KLONER, R. A., C. E. GANOTE, K. A. REIMER, and R. B. JENNINGS. 1974a. Coronary flow during acute myocardial ischemia. Fed. Proc. 33: 592.

KLONER, R. A., C. E. GANOTE, K. A. REIMER, and R. B. JENNINGS. 1975. Distribution of coronary arterial flow in acute myocardial ischemia. Arch. Path. 99: 86–94.

KLONER, R. A., C. E. GANOTE, D. WHALEN, and R. B. JENNINGS. 1974c. Effect of a transient period of ischemia on myocardial cells II. Fine structure during the first few minutes of reflow. Am. J. Path. 74: 399–422.

KRUG, A., W. M. ROCHEMONT, and G. KORB. 1966. Blood supply of the myocardium after temporary coronary occlusion. Circulat. Res. 19: 57–66.

MAJNO, G., A. AMES, II, J. CHAING, and R. L. WRIGHT. 1967. No reflow after cerebral ischemia. Lancet *ii:* 569.

SHEN, A. C., and R. B. JENNINGS. 1972a. Myocardial calcium and magnesium in acute ischemic injury. Am. J. Path. 67: 417–440.

SHEN, A. C., and R. B. JENNINGS. 1972b. Kinetics of calcium accumulation in acute myocardial ischemic injury. Am. J. Path. 67: 441–452.

SUMMERS, W. K., and R. L. JAMISON. 1971. The no reflow phenomenon in renal ischemia. Lab. Invest. 25: 635–643.

WHALEN, D. A., D. G. HAMILTON, C. E. GANOTE, and R. B. JENNINGS. 1974. Effect of a transient period of ischemia on myocardial cells I. Fine structure during the first few minutes of reflow. Am. J. Path. 74: 381–398.

WILLERSON, J. T., W. J. POWELL, Jr., T. E. GRUINEY, J. J. STARK, C. A. SANDERS, and A. LEAF. 1972. Improvement in myocardial function and coronary blood flow in ischemic myocardium after mannitol. J. Clin. Invest. 51: 2989–2998.

WILLMS-KRETSCHMER, K., and G. MAJNO. 1969. Ischemia of the skin. Electron microscopic study of vascular injury. Am. J. Path. 54: 327–353.

Mailing address:
Robert B. Jennings, M.D.,
Department of Pathology, Northwestern University Medical School,
303 East Chicago Avenue, Chicago, Illinois 60611 (USA).

Recent Advances in Studies on
Cardiac Structure and Metabolism Volume 10
The Metabolism of Contraction
Edited by Paul-Emile Roy and George Rona

CHANGES IN ISCHEMIC CORONARY VASODILATION PRODUCED BY BLOOD, PLASMA, AND SALINE INFUSIONS[1]

NEVILLE BITTAR, THOMAS J. PAULY,
and JOSEPH R. KOKE

Cardiovascular Research Laboratory, Department of Medicine,
University of Wisconsin Medical School, Madison, Wisconsin, USA

SUMMARY

Reactive hyperemia was diminished when blood of varying oxygen content was infused into the vascular bed distal to an occluded coronary artery. The degree of response appeared to be related to the oxygen content of the infused sample. However, plasma and saline infusions also caused a reduction in reactive hyperemia. It is therefore possible to conclude that the prevention of the fall in intravascular pressure plays an important role in the regulation of vascular smooth muscle tone in the dog heart.

INTRODUCTION

The mechanism responsible for control of coronary vascular resistance is unknown (Berne, 1964). It was Bayliss (1902) who first suggested that the change in tone of arteriolar smooth muscle is a direct response to stretch or to tension resulting from a pressure difference across the vessel wall. Thus, the autoregulation of blood flow in the kidney (Waugh and Shanks, 1960), intestine (Johnson, 1960), and skeletal muscle (Blair *et al.*, 1959) has now been attributed by some to such a myogenic mechanism. As pointed out by Berne (1964), the heart is a more difficult organ for the study of a myogenic response because the contracting myocardium adds an extravascular component to resistance so that a precise method for continuous measurement of such a component has not been developed. This paper describes experiments in dogs in which an attempt has been made to examine the possible contribution of a myogenic mechanism to the regulation of arteriolar smooth muscle tone. The results of these studies support the idea that such a mechanism exists and that it does play a role in the regulation of coronary vascular tone.

[1] This work was supported in part by a grant from the American Heart Association.

Table 1. Effects of infusions of arterial blood, femoral venous blood, coronary sinus blood, plasma, and saline on reactive hyperemia*

Experiment	Oxygen Delivery Debt (ml)	Oxygen Delivered by Infusion (ml)	Repayment of Delivery Debt (%)	Reduction in Reactive Hyperemia (%)
		Infusion of Arterial Blood		
1	1.52	1.52	100	100
2	1.60	1.60	100	100
3	1.87	1.87	100	97
4	2.80	2.80	100	100
6	1.68	1.68	100	95
7	0.95	0.95	100	95
10	1.99	1.99	100	100
11	1.02	1.02	100	100
12	1.30	1.30	100	100
		Infusion of Femoral Venous Blood		
2	1.60	1.32	88	79
3	1.87	1.65	89	80
4	2.80	1.96	70	57
7	0.95	0.75	79	70
11	1.02	0.84	82	75

		Infusion of Coronary Sinus Blood		
1	1.52	0.56	36	24
2	1.60	0.52	34	20
3	1.87	0.88	47	30
11	1.02	0.36	35	22
12	1.30	0.46	36	26
		Infusion of Plasma		
1	1.52	0.036	2.3	27
5	1.90	0.045	2.4	22
8	1.40	0.032	2.2	30
9	2.25	0.054	2.3	25
10	1.99	0.045	2.3	31
11	1.02	0.027	2.6	30
		Infusion of Saline		
7	0.95	0.023	2.4	19
11	1.02	0.027	2.6	20
12	1.30	0.032	2.4	26

*In each case, a volume equal to the flow debt was infused. Infusions were made during the period of occlusion via a cannulated branch of the anterior descending artery.

MATERIALS AND METHODS

The experiments were performed on 12 healthy mongrel dogs weighing between 15 and 25 kg. Anesthesia was produced by the administration of morphine sulfate (3 mg/kg) subcutanesouly, followed in 1 hour by chloralose (100 mg/kg). The animals were ventilated through a cuffed endotracheal tube with a Harvard respirator using room air. A left thoracotomy was performed through the fifth intercostal space, the pericardium incised, and the heart suspended in a pericardial cradle. A short segment of the anterior descending branch of the left coronary artery was dissected free from the surrounding tissue and an electromagnetic flow transducer of appropriate size was placed on the vessel. In some experiments, a second flow transducer was applied to the circumflex branch of the left coronary artery. Coronary blood flow was recorded with a two channel Biotronex model BL-610 pulsed logic electromagnetic flowmeter. The flow transducers were calibrated *in vitro,* using segments of canine coronary artery and pooled dog blood. A snare occluder constructed of polyethylene tubing was placed around the anterior descending artery immediately distal to the flow transducer. Arterial blood pressure was measured in the femoral artery with a Statham p-23Db strain gauge. Mean values for blood pressure and flow were obtained by electronic integration and recorded on a direct writing Sanborn Polyviso. For direct intracoronary infusion of blood, plasma, and saline, a small diagonal branch of the anterior descending artery, distal to flow transducer and occluder, was dissected free and cannulated with a 21-gauge Teflon catheter. The catheter was advanced so that its tip lay immediately adjacent to the lumen of the anterior descending artery. For sampling of coronary sinus blood, a catheter was manipulated via the external jugular vein into the coronary sinus. The femoral vein was cannulated for withdrawal of peripheral venous blood.

Control responses to 10- and 15-sec occlusions of the anterior descending artery were recorded. For each occlusion, a flow debt was calculated from the following formula: flow debt = control flow rate × duration of occlusion. A volume of arterial blood equal to the calculated flow debt was then obtained from the femoral artery. The occlusion of the anterior descending artery was then repeated and the arterial blood was infused via the cannulated side branch. The infusions were accomplished by hand injection at a near constant rate throughout the period of occlusion in an attempt to restore, as accurately as possible, the blood flow which would have occurred had the vessel not been occluded. After several responses to infusion of arterial blood had been recorded, infusions of femoral venous blood, coronary sinus blood, plasma, and 0.9% saline were performed. Plasma was obtained from freshly centrifuged venous blood drawn from the same dog. The saline solution was warmed to 37°C before infusion. In each case, control reactive hyperemia was recorded. Not all substances were infused in every experiment.

The oxygen content of arterial, venous, and coronary sinus blood was determined for each experiment, using the method of Van Slyke and Neill. The concentrations of oxygen dissolved in plasma and saline were calculated from the formula $CO_2 = \alpha O_2 \times pO_2$, where CO_2 is the concentration of oxygen in solution. αO_2 is the solubility coefficient at 37°C, and pO_2 is the partial pressure of oxygen. The amount of oxygen which was delivered by each of the infusions was determined from the expression

$$\text{Oxygen delivered} = (\text{concentration of oxygen in the sample}) \times (\text{volume of the sample})$$

For each period of coronary artery occlusion, a value termed oxygen delivery debt was calculated. This was simply the quantity of oxygen which would have been delivered if no occlusion had been performed or oxygen delivery debt = (flow debt) × (arterial oxygen content). The theoretical repayment of this oxygen delivery debt which was accomplished by each infusion was calculated from the expression

$$\text{Repayment of oxygen delivery debt} = \frac{\text{oxygen delivered by infusion} \times 100}{\text{oxygen delivery debt}}$$

RESULTS

When blood, plasma, or saline was infused into the anterior descending artery distal to the area of occlusion, the volume of reactive hyperemia was diminished. The relative magnitude of this diminution varied with the substance infused. Typically, when a volume of arterial blood equal to the flow debt was infused during the period of occlusion, reactive hyperemia was abolished as shown in Figure 1. In this experiment, the flow debt resulting from 10 sec of occlusion is 8 ml. Infusion of 8 ml of arterial blood during the period of occlusion abolished reactive hyperemia. Following the infusion of 8 ml of peripheral venous blood, hyperemia was seen, but it was less than that of the control. Infusions of the same volume of blood taken from the coronary sinus, plasma, or 0.9% saline resulted in a less marked, although definite attenuation of the response. The infusion of plasma or saline into the anterior descending vascular bed without arterial occlusion caused either a slight increase or no perceptible change in coronary blood flow.

The results obtained in 12 dogs are shown in Table 1. In each case, the flow debt is totally repaid. Because of the variation in oxygen content of the samples, a variable percentage of the oxygen delivery debt was repaid with each infusion. This difference was most marked for the plasma and saline samples, since the quantity of oxygen carried in solution was very small as compared to that

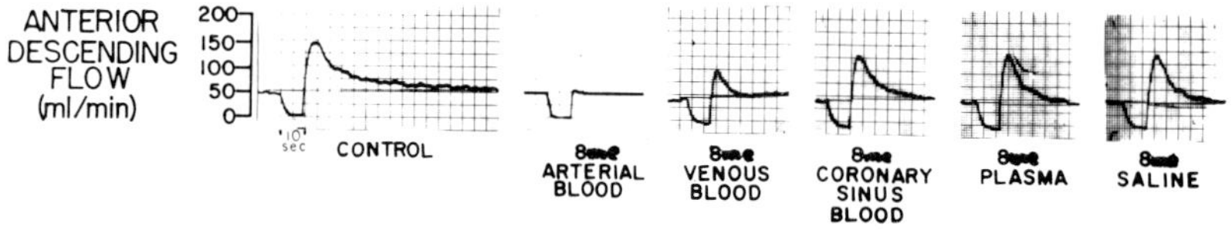

Figure 1. Effect of intracoronary infusion of equal volumes of arterial blood, systemic venous blood, coronary sinus blood, plasma, and saline on reactive hyperemia following a 10-sec occlusion is shown.

contained in samples of whole blood. Despite this relatively low oxygen delivery capacity of plasma and saline, infusion of these substances still resulted in a considerable diminution of reactive hyperemia. The reduction in reactive hyperemia flow following infusion of plasma and saline solutions was comparable to that produced by infusing coronary sinus blood (19–30%) (Fig. 2).

COMMENT

These experiments offer further support to the idea that the magnitude of the oxygen debt produced by occlusion of a coronary artery is an important determinant of the volume of reactive hyperemia. The infusion of a volume of

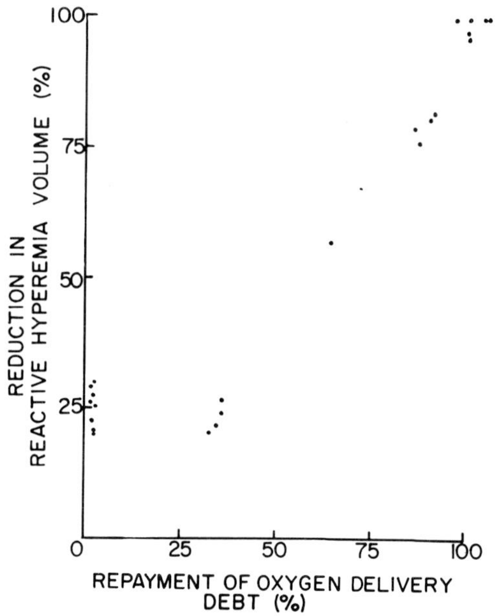

Figure 2. Relationships between the percentage of oxygen delivery debt repaid and the resultant reduction in reactive hyperemia are shown.

arterial blood equal to the flow debt resulted in total repayment of the oxygen delivery debt. That this effectively removed the stimulus for coronary vasodilation was shown by the absence of reactive hyperemia.

These observations do not contradict the work of Coffman and Gregg (1961), who showed that the oxygen debt was consistently overpaid during reactive hyperemia. In their experiments, myocardial oxygen consumption was noted to increase during the hyperemia period, indicating that the heart had in some way been stimulated to increase its oxygen uptake. Although the nature of this stimulus is unknown, it would appear to occur during the period of ischemia. In the present experiments, such a period of ischemia was generally not allowed to take place, as oxygenated blood was often infused distal to the point of occlusion. Thus, when fully oxygenated blood was infused, no actual deficit of either blood flow or oxygen could have occurred, and therefore a change in the tone of the coronary resistance vessels would not be anticipated. That this was the case is indicated by the fact that coronary blood flow returned to control levels immediately upon release of the occlusion (see Fig. 1).

The failure of an equal quantity of femoral venous blood to abolish reactive hyperemia suggests that this blood was deficient in some factor necessary for the maintenance of coronary vascular tone. Another possibility is that the venous blood contained a vasodilator substance which was not present in arterial blood. Although the latter possibility was not excluded, it seems likely that the lower oxygen content of the venous blood was a critical factor. Although less oxygen was carried by the venous sample, a sufficient gradient should have existed between myocardial tissue and blood to allow partial repayment of the oxygen debt.

The decreases in reactive hyperemia resulting from the infusion of coronary sinus blood, plasma, and saline were comparable in magnitude (20–31%). This was true despite a wide difference in the calculated repayment of the oxygen delivery debt (2.2–47.0%). Although coronary sinus blood contained a greater amount of oxygen, its effect was no better than that of plasma or saline. This suggests that an insufficient gradient of oxygen existed so that diffusion into myocardial cells did not take place.

It thus appears that some other factor, other than oxygen, must be implicated to explain the reduction in the volume of reactive hyperemia seen following infusion of coronary sinus blood, plasma, and saline. That some other blood-borne metabolite may account for this phenomenon is ruled out by the effectiveness of saline. Furthermore, neither plasma nor saline possessed significant vasoactive properties when injected into the nonoccluded coronary vascular bed. It is therefore likely that the infusion of these substances distal to the area of coronary artery occlusion prevented the fall in intravascular pressure in the distal vessels that would have activated a myogenic mechanism. Thus, one can conclude that a portion of the vasodilation which occurs during occlusion of a major coronary artery is a direct response of the vascular smooth muscle to the change in transmural pressure.

REFERENCES

BAYLISS, W. M. 1902. On the local reactions of the arterial wall to changes of internal pressure. J. Physiol. (London) 28: 220–231.

BERNE, R. M. 1964. Regulation of coronary blood flow. Physiol. Rev. 44: 1–29.

BLAIR, D. A., W. E. GLOVER, A. D. M. GREENFIELD, and I. C. RODDIE. 1959. The increase in tone in forearm resistance blood vessels exposed to increased transmural pressure. J. Physiol. (London) 149: 614–625.

COFFMAN, J. D., and D. E. GREGG. 1961. Oxygen metabolism and oxygen debt repayment after myocardial ischemia. Am. J. Physiol. 201: 881–887. 1961.

JOHNSON, P. C. 1960. Autoregulation of intestinal blood flow. Am. J. Physiol. 199: 311–318.

WAUGH, W. H., and R. G. SHANKS. 1960. Cause of genuine autoregulation of the renal circulation. Circulat. Res. 8: 871–888.

Mailing address:
Neville Bittar, M.D.,
Cardiovascular Research Laboratory
Room 523, 420 N. Charter Street, Madison, Wisconsin 53706 (USA).

Recent Advances in Studies on
Cardiac Structure and Metabolism Volume 10
The Metabolism of Contraction
Edited by Paul-Emile Roy and George Rona

TIME-DEPENDENT CHANGES OF COLLATERAL CORONARY REACTIVITY IN THE DOG

A. JUHÁSZ-NAGY and J. KUDÁSZ

Department of Cardiovascular Surgery, Semmelweis University
Medical School, Budapest, Hungary

SUMMARY

Vascular reactivity of the developing coronary collaterals has been investigated by studying local flow responses elicited by adrenergic and hypoxic stimuli in the ischemic foci of the canine heart. Tissue blood flow was measured by means of the heat clearance technique. The myocardial area explored was rendered ischemic by ligating the left anterior descending coronary artery 2–90 days prior to the experiment. This area is supplied exclusively by collateral channels. Adrenergic sensitivity was tested by i.v. administration of noradrenaline and isoproterenol (1 μg/kg), as well as by supramaximal electrical stimulation of the stellate ganglion. Metabolic sensitivity of the vessels was tested by brief periods of general hypoxia. During the first 3 weeks of the myocardial infarction, the collateral vascular response induced by each form of the adrenergic activation was considerably different from that observed in the normal coronary bed: instead of the conventional beta-type dilation, noradrenaline and stellate stimulation induced constriction in the ischemic zone whereas isoproterenol, having only pure beta-stimulating effects proved to be ineffective. Following the 3rd postoperative week, a gradually increasing sensitivity to beta-dilator influences appeared on the collateral vessels. However, this transformation of the adrenergic effect was not accompanied by a similar restoration of the local vascular propensity for metabolic (hypoxic) dilation. The results suggest that time courses of the changes of the adrenergic and metabolic vascular reactivity in the collateral network, although probably interrelated, do not parallel each other.

INTRODUCTION

The collateral blood supply to the ischemic foci holds a prominent place in the outcome of myocardial infarction. After coronary obstruction, the collateral network enlarges enormously as a consequence of vigorous growth processes. Very little is known, however, of the adrenergic and metabolic sensitivity of the developing collaterals, and we are unable, therefore, to compare their reactivity to the behavior of the normal coronary vessels. The aim of the present investigation was to clear up some aspects of the latter problem.

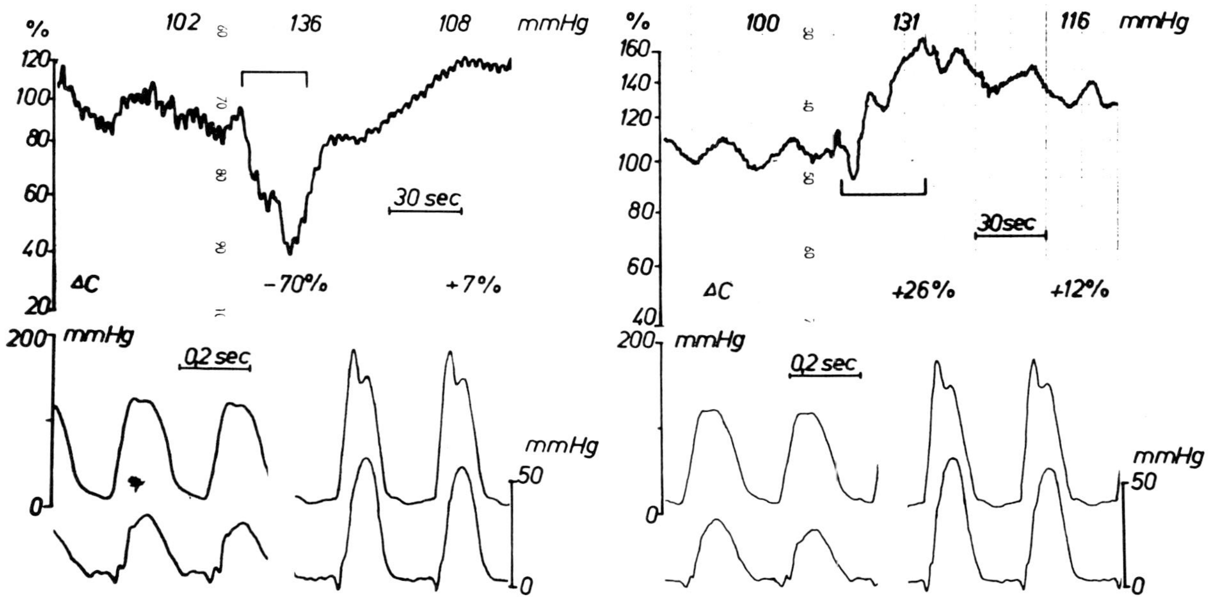

Figure 2. Effects of stellate ganglion stimulation on blood flow in the collateral-dependent area (*left*) and in the normal myocardium (*right*). Dog, 18 days after coronary occlusion. *Upper tracing:* local blood flow. *Lower tracings:* left and right intraventricular pressure curves registered before and during stimulation. Figures above the flow curve refer to mean arterial blood pressure (mm Hg). Figures below the flow curve refer to changes of calculated local vascular conductance (% Δ C).

Table 1. Hemodynamic effects of noradrenaline (1 μg/kg body weight) on the blood supply in the ischemic myocardial zone*

	Local Blood Flow (%)		Mean Blood Pressure (mm Hg)		Vascular Conductance (%)	
	Control	Change	Control	Change	Control	Change
A. 2–20 days after coronary ligation ($N = 10$)	100 ± 0	−7 ± 10	90 ± 6	+49 ± 5†	100 ± 0	−39 ± 7†
B. 29–90 days after coronary ligation ($N = 11$)	100 ± 0	+56 ± 19‡	111 ± 11	+49 ± 5†	100 ± 0	+2 ± 9
p values between groups A and B	–	<0.02	>0.10	>0.9	–	<0.01

*Values expressed as mean ± SEM.
†$p < 0.001$.
‡$p < 0.02$.

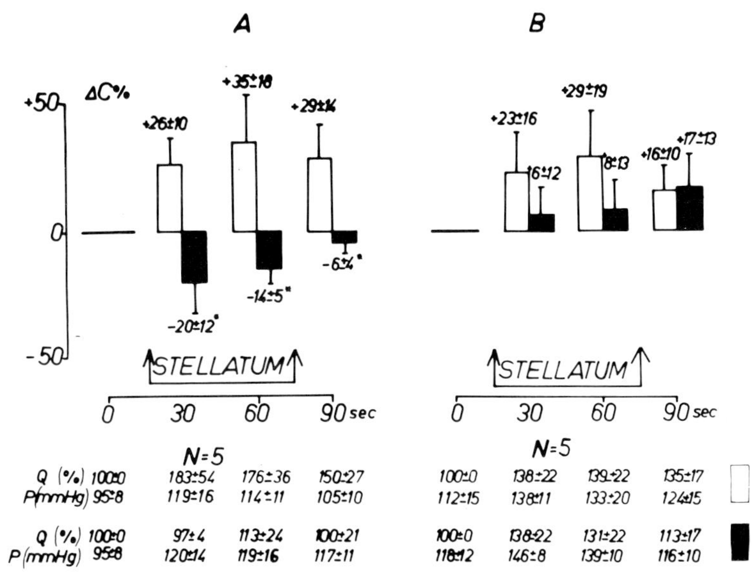

Figure 3. Effects of stellate stimulation on vascular conductance in the collateral dependent area (*black columns*) and in the normal myocardium (*white columns*). *Below:* values of local blood flow (Q) and mean arterial blood pressure (P). *A:* 5–18 days after coronary occlusion. *B:* 28–80 days after coronary occlusion. Mean values ± SEM are indicated. Asterisks refer to significance of differences between the collateral and normal coronary responses. (p<0.05).

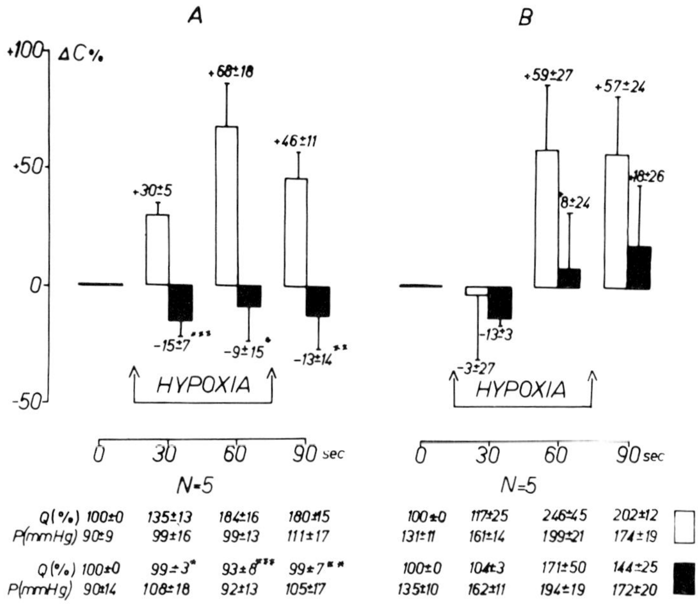

Figure 4. Effects of 60-sec asphyxia on local vascular conductance. *A:* 5–18 days after coronary occlusion. *B:* 28–80 days after coronary occlusion. * $p<0.05$; ** $p<0.01$; *** $p<0.001$; other symbols as in Figure 3.

Figure 3, indicate significant differences between the vascular conductance changes of the normal and the collateral vessels in the subacute group (*A*) whereas, in the chronic group (*B*), these changes did not significantly differ from each other.

At the same time, arterial hypoxia failed to elicit metabolic vascular dilation within the infarcted zone, whereas it caused a considerable flow increase in the normal myocardium supplied by the circumflex artery (Fig. 4). This finding seems to be in complete accord with the defective beta-adrenergic sensitivity of the collaterals during the first (acute) phase of infarction (Fig. 3, *A*). However, collateral metabolic adaptability was still highly inadequate during the second phase of infarction, where the "constrictor" character of the adrenergic effect was replaced by a more or less normal vascular dilation of the beta type. As can be seen in Figure 5, an otherwise normal hemodynamic response to general hypoxia can be associated with a defective metabolic adaptive capacity of the collateral vessels in the "healing" phase of infarction.

DISCUSSION

In previous studies, we showed that constrictor (alpha-adrenergic) sympathetic impulses reach the normal canine coronaries through specific vasomotor fibres

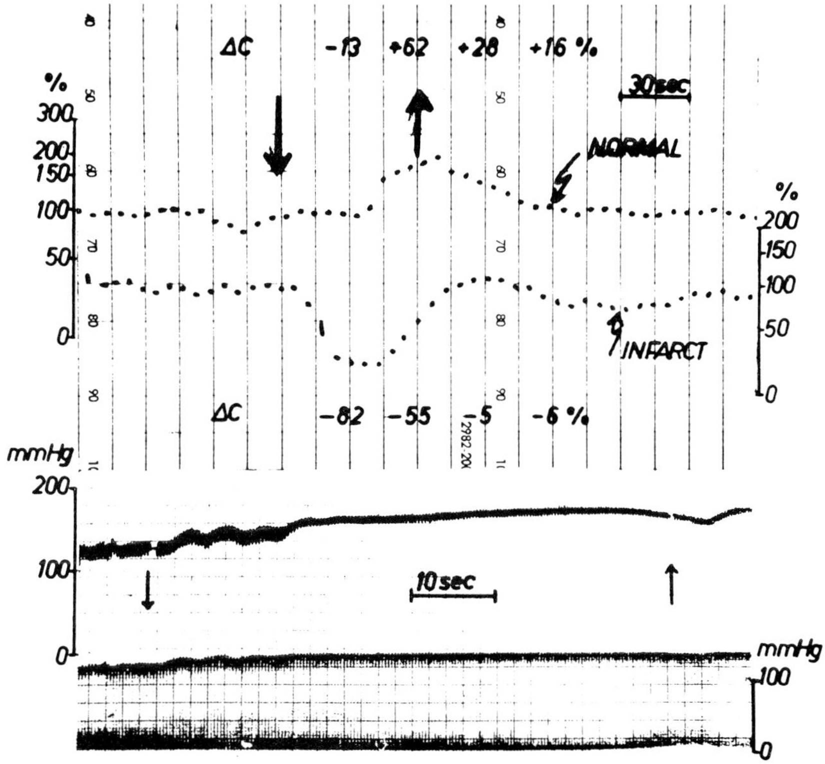

Figure 5. Effects of asphyxia (*between arrows*), 33 days after coronary occlusion. *Upper tracings:* flow curves in the collateral-dependent and normal areas registered simultaneously with the aid of a channel selector. *Lower tracings:* arterial blood pressure and left ventricular pressure. Figures refer to changes in vascular conductance.

(Juhász-Nagy and Szentiványi 1961). At the same time it is true that the alpha-constrictor response is more difficult to elicit than the beta-type coronary dilation. Strong stimulation of the stellate or administration of large doses of catecholamines also increases cardiac activity, and the direct action of the smypathetic mediator on the coronary smooth muscle is invariably masked by the concomitant and overriding dilator effects secondary to the augmented myocardial metabolism.

According to the present data, the mode of action of the sympathetic mediator on the developing collaterals is considerably different from the mechanism outlined above. In the initial phase of cardiac infarction, flow-diminishing adrenergic influences, possibly of alpha type, affect the coronary collaterals to an extent not evident with the normal vessels. The exact mechanism of this "reversed" adrenergic vascular response remains to be elucidated. It was demonstrated earlier by Kunos and Szentiványi (1968) that the adrenergic mechanism in the myocardium is sensitive to metabolic changes: the receptor quality of the same adrenergic response proved to be beta at higher metabolic rates and alpha at lower rates. It may be assumed, therefore, that the decreased dilator responsiveness of the developing collaterals reflects the low metabolic rate of the damaged myocardial cells. In the subsequent ("healing") phase of the infarction, a gradually increasing sensitivity to beta-dilator influences appears on the coronary collaterals. This transformation of the adrenergic effect is, however, unparalleled by a similar restoration of the true metabolic sensitivity of the same vessels. The metabolic insensitivity of the developed collateral channels seems to be an analogous phenomenon, with "exhaustion" of the metabolic adaptive capacity occurring in the normal coronary circulation after undue utilization of the adaptive vascular reserve; the phenomenon was described earlier by us (Juhász-Nagy and Sz ordoray, 1972; Kudász and Juhász-Nagy, 1973). The characteristic feature of this pathological state is the selective loss of the vascular propensity for hypoxic adaptation. As far as collateral blood supply is concerned, the latter aspect of the coronary reactivity remained similarly defective, even in the healing phase of the infarction, depriving the collateral-dependent myocardial areas of the most important physiological protective mechanism against tissue hypoxia.

REFERENCES

JUHÁSZ-NAGY, A., and M. SZENTIVÁNYI. 1961. Separation of cardioaccelerator and coronary vasomotor fibers in the dog. Am. J. Physiol. 200: 125–129.

JUHÁSZ-NAGY, A., M. SZENTIVÁNYI, and G. GRÓSZ. 1974. Effect of adrenergic activation on collateral coronary blood flow. Jap. Heart J. 15: 289–298.

JUHÁSZ-NAGY, A., and P. SZODORAY. 1972. Effect of transitory myocardial ischaemia on coronary adaptability and on heat production of the left ventricle. Acta physiol. Acad. Sci. Hung. 42: 275–285.

KUDÁSZ, J., and A. JUHÁSZ-NAGY. 1973. Human myocardial blood flow during cardiac operations. J. Cardiovasc. Surg. 13: 568–575.

KUNOS, G., and M. SZENTIVÁNYI. 1968. Evidence favouring the existence of a single adrenergic receptor. Nature 183: 1077–1078.

Mailing address:
Dr. A. Juhász-Nagy,
Dept. of Cardiovascular Surgery,
Semmelweis University Medical School,
H-1122 Budapest, Városmajor u.68 (Hungary).

Recent Advances in Studies on
Cardiac Structure and Metabolism Volume 10
The Metabolism of Contraction
Edited by Paul-Emile Roy and George Rona

HEMODYNAMICS AND METABOLIC DETERMINANTS OF DISTRIBUTION OF MYOCARDIAL FLOW

JOHN H. KENNEDY, RAYMOND LIMET,
N. H. C. HWANG, JUDY BARNETT,
LEONARD SCHWABE, and ALLAN FLASTERSTEIN

Taub Laboratories for Mechanical Circulatory Support,
Cora and Webb Mading Department of Surgery,
Baylor College of Medicine; and the Department of Civil Engineering,
University of Houston, Houston, Texas, USA

SUMMARY

A computer simulation of left coronary outflow response to branch occlusion was carried out, using physiological data derived from canine left coronary artery generated on an IBM 360/44 in order to provide multidimensional analysis of the problem of the distribution of flow within the myocardium. During coronary occlusion in this computer model, distal impedance varied little during the occlusion of one or the other arterial branch. This observation was taken as a boundary condition in the simulation exercise. Some of the important factors in the intramyocardial distribution of flow are discussed.

INTRODUCTION

The redistribution of available cardiac output to the coronary circulation in normal and in diseased states both in animals (Essex *et al.*, 1939) and in man (Bing *et al.*, 1949) has been known for 35 years. Limited to some degree by technology, studies of the distribution of perfusion within myocardium have awaited the development of more recent techniques and instrumentation.

A number of experimental and clinical studies have emphasized the importance of the multiple factors, both local (Domenech, Douglas, and Talesnik, 1968; Bing *et al.*, 1969; Hood, 1971; Kubler and Spieckermann, 1970; Schaper, 1971; Tomoda *et al.*, 1971; Walters, Hundeshagen, and Borst, 1971; Buckberg *et al.*, 1971a; Moir, 1972a; Smith *et al.*, 1972; Fisher *et al.*, 1973; Leaf, 1973) and systemic, (Martin *et al.*, 1969; Marshall and Parratt, 1973a; Vance, Brown, and Smith, 1973), which influence the distribution of net myocardial perfusion within the myocardium and have emphasized the role of these factors in this genesis of subendocardial ischemic necrosis (Buckberg *et al.*, 1971, 1972a; Kudasz and Juhasz-Nagy, 1972). In a previous report from this laboratory (Kennedy *et al.*, 1966), the demonstration of coronary arteriovenous shunts by injection of glass microspheres (53μ–210μ in diameter) was reported. Increased oxygen content in blood samples from the coronary sinus was noted in animals in which left ventricular failure had been induced by the administration of levarterenol, 0.01 mg/kg/min for 30 min, as well as following serial ligation of branches of the circumflex and anterior descending coronary artery. Further corroboration of this concept has been advanced by Marshall and Parratt (1973b) and others (Bruley *et al.*, 1973; Jorgensen *et al.*, 1973; Mark *et al.*, 1973), who have described markedly elevated oxygen tension (and decreased flow measured by ^{133}Xe) of blood from the infarcted muscle in greyhounds, following left anterior descending coronary artery ligation and dipyridimole administration. These authors concluded that the results could be explained by the opening up of vessels that did not take part in tissue exchange. The significance of myocardial perfusion and alteration in the gradient in metabolite concentration or the dilution of injected tracer substances must be reconsidered in the light of the possibility of alterations in intramyocardial distribution of perfusion.

The present mathematical study was undertaken in order to assess the possible role of one factor, coronary vascular impedance, in the distribution of myocardial perfusion; changes in myocardial perfusion when the left coronary artery is alternately occluded in a theoretical mathematical model were solved on a computer.

MATERIALS AND METHODS

In order to provide multidimensional analysis of the problem of the distribution of flow within myocardium, a computer simulation of left coronary outflow response to branch occlusion was carried out, using the canine left coronary artery as a function of time (Fig. 1), generated on an IBM 360/44 based on available physiological data. Similarly, flow in the two immediate branches of the left coronary artery, the anterior descending and circumflex coronary arteries were generated (Fig. 1). These were used with the input function for calculation of coronary outflow response to occlusion of one of the branches.

RESULTS

As shown in the lower four sections of Figure 1, based on the recorded pressure differences between the two ends of each of the branches of the above-mentioned flow rates in a computer analogy, both being functions of time, the input impedance of each of the branches were calculated. From examination of Driscol and Eckstein's (1967) experimental data, it is noted that the distal impedance varies very little during the occlusion of one or the other arterial branch. This observation was taken as a boundary condition in our simulation exercise.

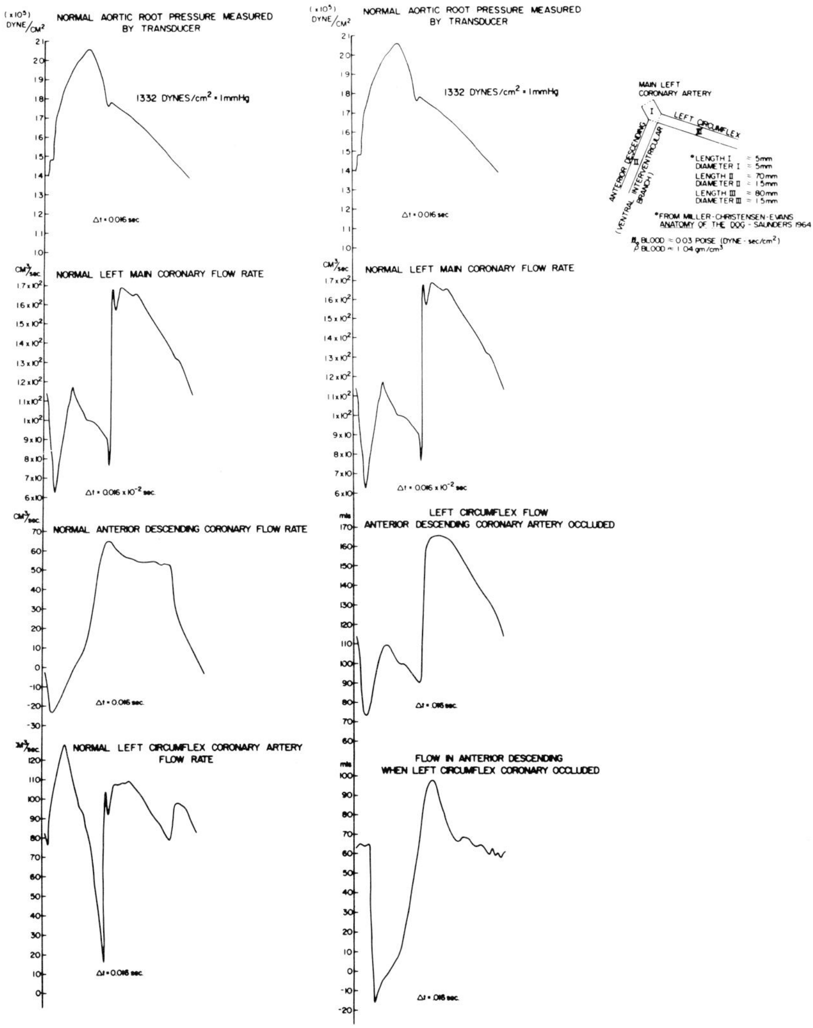

Figure 1. Computer simulation, left coronary flow.

DISCUSSION

This limited theoretical study is consistent with what is already known; there are many difficult-to-control factors which influence the distribution of myocardial perfusion, including both systemic and local metabolic factors. Computer analogy enables one to isolate several of these variables, however. In the previously published integral solution (Chao and Hwang, 1971), two of the hydraulic transmission-line equations were used to predict the flows in these three segments, using coupled differential equations and solving by the continuous sampling model program. Pressure as a function of time and position could therefore be determined. Changes in the geometry and parameters of the system, such as with considerable atherosclerosis and a saphenous vein bypass graft, could be simulated with reasonable accuracy. Further studies of such systems using actual data from controlled experiments would enable the investigator to obtain dynamic lumped parameters for the constants described above. This is of obvious value, since measurements of Young's, Modulus' and Poison's ratio under other than living and healthy conditions can only be used as reference points and cannot be regarded as absolutely correct for living systems.

Recruitment of the data necessary for a more specific study of the distribution of myocardial perfusion by computer analogy may prove difficult, however. As shown in Table 1, which summarizes the factors in myocardial perfusion to an area of relative ischemia, noradrenaline (Eckenhoff, 1947; Kennedy, 1966; Marshall and Parratt, 1973), collateralization (including vein bypass graft) (Hood, 1971; Schaper, 1971; Moir, 1972; Smith *et al.*, 1972), local ischemia (Tomoda *et al.*, 1971), increased diastolic flow as in the case of intra-aortic balloon counter pulsation (Lin, Galysh, and Patel, 1972; Weber and Janicki, 1973; Schenk, 1974), carotid body (vagal) stimulation (Moir, 1972b; Uchida and Ueda, 1972; Limet, Fourny, and Kennedy, 1974), and adaptation to altitude (Meerson, Gomzakov, and Shimkovich, 1973) all may increase myocardial perfusion to an area of relative ischemia. Adaptation to high altitude, is of particular interest. Meerson *et al.* (1973), conditioned rats to an ambient environment

Table 1. Factors in myocardial perfusion to area of relative ischemia*

Increase	Decrease
Noradrenaline	Increased intramyocardial pressure
Adaptation to altitude	Cell swelling
Collateralization (including VBPG)	LV tension, duration
Ischemia, local	Coronary occlusion
Increased diastopic flow (IABC)	Increased ventricular volume
Carotid body (vagal) stimulation	Hypocapnia
	CPBP, oligemia

*LV = left ventricle; CPBP = cardiopulmonary bypass; IABC = intraaortic balloon counterpulsation.

Table 2.

Animal species–identif. No., sex BABOON 105 F
Start time (TOCP device turned on)* 11:48 AM 5/21/1974
Death of animal (time & date) 14: 0 PM 5/21/1974

Regional blood flow in % of injection site blood flow (BFP) & BFP divided by the region weight in % of body weight at death(FDP)

Injection time	−0.30			1.25
Region name				
Weight(%body weight)		BFP	FDP	
Right Ventricle				
0.1600	0.166	1.04	1.307	8.17
Septum endocardium				
0.1052	0.190	1.81	0.124	1.18
Septum myocardium				
0.0526	0.059	1.13	0.041	0.77
Left ventricle epicardium				
0.1340	0.162	1.21	1.770	13.21
Left ventricle myocardium				
0.1000	0.187	1.87	2.104	21.04
Left ventricle endocardium				
0.0860	0.173	2.01	1.411	16.40
Left ventricle papillary				
0.0180	0.037	2.08	0.212	11.79
Sum of BFP	0.975		6.968	
Sum of % wts.				
0.6558				

similar to that at 6,000 m over a 40-day period in a pressure vessel and challenged both these animals and controls with left coronary artery ligation. Conditioning resulted in a 5-fold lower mortality and a reduction of the area of ischemic necrosis by 35%, as well as minimizing the disturbances in myocardial contractile force 4-fold. There was evidence of increased capacity for oxidative ATP resynthesis in the myocardium. While conditioning to high altitude is known to increase the release of oxygen by the red cells and to increase the concentration of the metabolite diphosphoglycerate in the erythrocytes (Leaf, 1973; Marchetti *et al.,* 1973), the latter determination was not made in these experiments. Of the many factors which decrease myocardial perfusion to an area of relative oligemia, those of major importance are: increase in intramyocardial pressure due to cell swelling (Leaf, 1973) or left ventricular tension or duration (Buckberg, 1971; 1972a, b; Baird, Goldbach, and de la Rocha, 1972); coronary occlusion (Bing *et al.,* 1969; Fisher *et al.,* 1973); increase in ventricular volume (Martin *et al.*, 1969); hypocapnea (Vance *et al.*, 1973); and cardiopulmonary bypass, particularly when oligemia (Baird *et al.*, 1972) is present. The last is of particular interest when one recognizes the magnitude of the use of cardiopulmonary bypass to permit contemporary cardiac surgery, 16,000 such operations having been done in calendar year 1969 (Glenn, 1972). In a series of pilot experiments from this laboratory in baboons subjected to hypothermia, perfusion, cooling, and 30-min circulatory arrest followed by perfusion rewarming, there was redistribution of blood flow within the left ventricular myocardium (Table 2). While individual variations were present among animals at the same net coronary flow, epicardial flow increased 5-fold and papillary muscle flow increased 13-fold over control values. After rewarming, total coronary flow increased 3.8 times over control values. Endocardial flow rose slightly, while flow to the central zone and epicardium increased to 20–30 times the control value. These determinations were made using radioactive microspheres, after the method of Rudolph and Heyman (1967).

REFERENCES

BAIRD, R. J., M. M. GOLDBACH, and A. de la ROCHA. 1972. Intramyocardial pressure; the persistence of its transmural gradient in the empty heart and its relationship to myocardial oxygen consumption. J. Thorac. Cardiovasc. Surg. 64: 635–646.

BING, R. J., S. GUDBJARNASON, H. TSCHOPP, and W. BRAASH. 1969. Molecular changes in myocardial infarction in heart muscle. New York Acad. Sci. 156: 583–593. January, 1969.

BING, R. J., M. D. HAMMOND, J. C. HANDLESMAN, S. R. POWERS, F. C. SPENCER, J. E. ECKENHOFF, W. T. GOODALE, J. H. HAFKENSHIEL, and S. S. KETY. 1949. The measurement of coronary blood flow, oxygen consumption and efficiency of the left ventricle in man. Am. Heart J. 38: 1–24.

BRULEY, D. F., D. H. HUNT, H. I. BICHER, and M. H. KNISELY. 1973. Simulating myocardial oxygen dynamics. Adv. Exp. Med. Biol. 37: 859–866.

BUCKBERG, G. D., J. B. ARCHIE, D. E. FIXLER, and J. I. E. HOFFMAN. 1971. Experimental subendocardial ischemia during left ventricular hypertension. Surg. Forum 00: 124–125.

BUCKBERG, G. D., D. E. FIXLER, J. P. ARCHIE, and J. T. HOFFMAN. 1972a. Experimental subendocardial ischemia in dogs with normal coronary arteries. Circulat. Res. 30: 67–81.

BUCKBERG, D. G., V. POWERS, D. E. PAGLIA, D. G. MULDER, and J. V. MALONEY. 1972b. Subendocardial ischemia after cardiopulmonary bypass. J. Thorac. Cardiovasc. Surg. 64: 669–684.

CATCHPOLE, P., D. V. HACKEL, and F. A. SIMEONE. 1955. Coronary peripheral blood flow: experimental hemorrhagic hypotension treated with norepinephrine. An. Surg. 1423: 372–381.

CHAO, J. C., and N. H. C. HWANG. 1971. An intregal form of the hydraulic transmission line theory as applied to circulatory systems. *In* Proceedings, 9th Annual Symposium of Biomathematics and Computer Science in the Life Sciences, Houston, Texas, March.

CHIU, C. J., W. A. MERSEREAU, and H. J. SCOTT. 1972. Subendocardial hemorrhagic necrosis. The role of direct mechanical trauma on the endocardium. J. Thorac. Cardiovasc. Surg. 64: 66–75.

DOMENECH, R., R. DOUGLAS, and J. TALESNIK. 1968. Regulating mechanisms of coronary blood flow and their relation to myocardial performance. Arch. Biol. Med. Exp. (Santiago) 5: 101–104.

DRISCOL, T. E., and R. W. ECKSTEIN. 1967. Coronary inflow and outflow responses to coronary artery occlusion. Circulat. Res. 20: 485–495.

ECKENHOFF, J. E. 1947. Cardiac oxygen metabolism in control of coronary circulation. Am. J. Physiol. 149: 634–647.

ESSEX, H. E., J. F. HERRICK, E. J. BALDES, and F. C. MANN. 1939. Influence of exercise on blood pressure, pulse rate and coronary blood flow of the dog. Am. J. Physiol. 125: 614–623.

FISHER, W., D. HEIMBACH, I. M. LEDINGHAM, and R. J. MARSHALL. 1973. Blood flow and oxygen extraction in ischemic and normal regions of the myocardium following acute coronary artery ligation. J. Physiol. (Lond.) 230: 15–17.

GLENN, W. W. L. 1972. Presidential address; American Heart Association.

GUDBJARNASON, S., P. MATHES, and K. G. RAVENS. 1970. Functional compartmentalization of ATP and creatine phosphate in the heart muscle. J. Mol. Cell. Cardiol. 1: 325–339.

HOOD, W. B., Jr. 1971. Pathophysiology of ischemic heart disease. Progr. Cardiovasc. Dis. 14: 297–320.

JORGENSEN, C. R., K. WANG, Y. WANG, F. L. GOBEL, R. R. NELSON, and H. TAYLOR. 1973. Effects of propranolol on myocardial oxygen consumption and its hemodynamic correlates during upright exercise. Circulation 48: 1173–1182.

KENNEDY, J. H., O. HAIDERER, J. D. REID, and N. BAILAS. 1966. Experimental heart failure; importance of precapillary coronary arteriovenous shunts. Arch. Surg. 93: 759–766.

KHOURI, A., D. GREGG, and R. A. RAYFORD. 1965. Effective exercise on cardiac output, left coronary flow and myocardial metabolism in the unanesthetized dog. Circul. Res. 17: 427–437.

KLUGE, T. 1971. The cardiac lymph circulation. Nord. Med. 86: 1325–1332.

KUBLER, W., and P. G. SPIECKERMANN. 1970. Regulation of glycolysis in the ischemic and the anoxic myocardium. J. Mol. Cell. Cardiol. 1: 351–377.

KUDASZ, J., and A. JUHASZ-NAGY. 1972. Human myocardial blood flow during cardiac operations. J. Cardiovasc. Surg. 13: 568–575.

LEAF, A. 1973. Cell swelling. A factor in ischemic tissue injury. Circulation 48: 455–458.

LIMET, R., J. FOURNY, and J. H. KENNEDY. 1974. Effects of the carotid sinus reflex upon coronary vessel tone. J. Surg. Res. (In press).

LIN, C. Y., F. T. GALYSH, and A. S. PATEL. 1972. Coronary blood flow responses to intraaortic balloon pumping in dogs treated with cardioactive drugs. J. Ass. Adv. Med. Instrum. 6: 343–347.

MARCHETTI, G. V., L. MERLO, V. NOSEDA, and O. VISIOLI. 1973. Myocardial blood flow in experimental cardiac hypertrophy in dogs. Cardiovasc. Res. 7: 519–527.

MARK, A. L., H. E. MAYER, P. G. SCHMID, D. D. HEISTAD, and F. M. ABBOUD. 1973. Adrenergic control of the peripheral circulation in cardiomyopathic hamsters with heart failure. Circulat. Res. 33: 74–81.

MARSHALL, R. J., and J. R. PARRATT. 1973a. The effect of noradrenalin on blood flow and oxygen consumption in normal and ischemic areas of myocardium. Am. Heart J. 86: 653–662.

MARSHALL, R. J., and J. R. PARRAT. 1973b. Efforts of dipyridamole on blood flow and oxygen handling in the acutely ischemic and normal canine myocardium. Brit. J. Pharmacol. 49: 391–399.

MARTIN, A. M., Jr., D. B. HACKLE, M. L. ENTMAN, M. P. CAPP, and M. S. SPACH. 1969. Mechanisms in the development of myocardial lesions and hemorrhagic shock. New York Acad. Sci. 156: 79–90.

MEERSON, F. Z., O. A. GOMZAKOV, and M. V. SHIMKOVICH. 1973. Adaptation of high altitude hypoxia as a factor preventing development of a myocardial ischemic necrosis. Am. J. Cardiol. 00: 30–34.

MOIR, T. W. 1972a. Coronary vascular adjustments to acute myocardial ischemia. Arch. Intern. Med. 129: 799–807.

MOIR, T. W. 1972b. Subendocardial distribution of coronary blood flow and the effect of antianginal drugs. Circulat. Res. 30: 621.

ROWELL, L. B. 1974. Human cardiovascular adjustments to exercise and thermal stress. Physiol. Rev. 54: 75–159.

RUDOLPH, A. M., and M. A. HEYMAN. 1967. Methods for studying distribution of blood flow. Circulat. Res. 21: 163–184.

SCHAPER, W. 1971. Pathophysiology of coronary circulation. Progr. Cardiovasc. Dis. 14: 275–296.

SCHENK, W. G., Jr. 1974. Counterpulsation; Symposium on mechanical supportive circulation. Tenad. Med. Ass. J. (In press).

SMITH, S. C., Jr., R. GORLIN, M. V. HERMAN, W. J. TAYLOR, and J. J. COLLINS, Jr. 1972. Myocardial blood flow in man; effects of coronary collateral circulation and coronary artery bypass surgery. J. Clin. Invest. 51: 2556–2565.

TOMODA, H., W. W. POMLEY, S. FUJIMURA, and J. M. MATLOFF. 1971. Effects of ischemia and reoxygenation on regional myocardial performance of the dog. Am. J. Physiol. 221: 1718–1721.

TYBERG, J. V., W. PARMLEY, and E. H. SONNENBLICK. 1969. In vitro studies of myocardial asynchrony and regional hypoxia. Circulat. Res. 25: 569–579.

UCHIDA, Y., and A. H. UEDA. 1972. Non-uniform blood flow through the ischemic myocardium induced by stellate ganglion stimulation. Jap. Circulat. J. 36: 673.

VANCE, J. P., D. M. BROWN, and G. SMITH. 1973. The effects of hypocapnea on myocardial blood flow and metabolism. Brit. J. Anesth. 45: 455–463.

WALTER, P., H. HUNDESHAGEN, and H. G. BORST. 1971. Treatment of acute myocardial infarction by transmural blood supply from the ventricular cavity. Europ. Surg. Res. 3: 130–138.

WEBER, K. T., and J. S. JANICKI. 1973. Coronary collateral flow in intra-aortic balloon counter pulsation. Trans. Am. Soc. Artif. Organs. 19: 395–401.

Mailing address:
Dr. John H. Kennedy,
Cora and Webb Mading Department of Surgery,
Baylor College of Medicine, Houston, Texas 77025 (USA).

Investigative Methods in Cardiac Ischemia

Recent Advances in Studies on
Cardiac Structure and Metabolism Volume 10
The Metabolism of Contraction
Edited by Paul-Emile Roy and George Rona

STUDIES OF REGIONAL MYOCARDIAL PERFUSION IN PATIENTS WITH CORONARY ATHEROSCLEROSIS, USING XENON-133 AND A MULTIPLE CRYSTAL SCINTILLATION CAMERA

P. J. CANNON, D. H. SCHMIDT,
M. B. WEISS, D. L. FOWLER,
W. J. CASARELLA, and K. ELLIS

Departments of Medicine and Radiology, College of Physicians and Surgeons,
Columbia University, New York, USA

SUMMARY

Regional myocardial perfusion was measured in 164 patients at coronary arteriography. Washout of xenon-133 from multiple areas of the heart was monitored with a multiple crystal scintillation camera after tracer injection into the left or right coronary artery. Rate constants of radioisotope clearance were computed by monoexponential analysis of the initial portion of each washout curve. Regional myocardial blood flow rates in ml/100 g · min were calculated by the Kety formula. The pattern of local myocardial perfusion rates was compared to the coronary arteriogram obtained during the same study.

In patients with normal coronary arteriograms, the average mean myocardial perfusion rate in the left ventricle (LV) significantly exceeded that of the right ventricle and the right atrial area. Mild heterogeneity of local myocardial flow rates in the LV was observed. In patients with essential hypertension, aortic stenosis, and aortic insufficiency, the average LV perfusion rates were similar to those of subjects with normal coronary arteriograms, pressures, and ventriculograms. The mean LV perfusion rates were significantly reduced in patients without coronary disease who had cardiomyopathy and cardiac failure.

Patients with less than 50% obstructions apparent on coronary arteriograms had normal or slight (insignificant) reductions of mean LV perfusion at rest. However, in 43 patients with greater than 50% obstructions of two branches of the left coronary, the average mean LV perfusion rate was significantly lower than that of the normal group. Increased heterogeneity of local perfusion rates in the LV was also observed in the patients with coronary disease. The mean perfusion rate in the left anterior descending (LAD) subregion was subnormal at rest. In 26 patients with significant LAD disease, there was no direct correlation between the resting perfusion rate in the LAD region and the arteriographic reading. In many individual patients with significant lesions, the perfusion rates in the LAD region were normal. The flow in the myocardial subregion distal to an isolated LAD occlusion of 50–95% was equal to that in the remainder of the ventricle in 7 patients, and it was reduced in 4 with total LAD occlusion.

When the myocardial O_2 consumption of patients was increased by right atrial pacing, the myocardial response index (ΔLV perfusion/(Δsystolic BP × heart rate)) observed in subjects with greater than 50% coronary obstructions was less than in those with no or insignificant disease. The mean regional flow response observed in myocardial areas distal to significant coronary obstructions which were well supplied by collateral vessels, although less than normal, was significantly greater than that observed in areas of disease which lacked a collateral supply. The data suggest that measurement of regional myocardial perfusion in patients with coronary disease may enhance the physiological interpretation of coronary obstructions and collateral blood vessels visualized by coronary arteriography.

INTRODUCTION

Current hypotheses suggest that coronary artery disease is associated with angina pectoris or myocardial dysfunction when the blood supply to a region of the myocardium is insufficient to meet the requirements of the tissue for oxygen (Friedberg, 1972). Because it is impossible to assess the balance between the metabolic demands of heart muscle and its blood supply using coronary arteriography alone, investigators in this laboratory have been working to develop a new method to estimate quantitatively the capillary blood flow in multiple areas of the myocardium. The initial experiments in animals used xenon-133 and a single crystal scintillation camera (Cannon, Haft, and Johnson, 1969b). The subsequent adaptation of the method for study of patients undergoing diagnostic coronary arteriography used a multiple crystal scintillation camera combined with computer processing of the multiple radioisotope washout curves recorded from different regions of the heart (Cannon *et al.*, 1969a; Cannon, Dell, and Dwyer, 1970, 1972a).

The purpose of this report is to present preliminary results of studies conducted on 164 patients at Columbia-Presbyterian Medical Center. The regional myocardial perfusion rates obtained with the radioisotope method were correlated with independent radiographic assessments of the degree of coronary artery occlusive disease in each patient.

METHODS

Cardiac catheterization and coronary arteriography were performed only on patients in whom the studies were clinically indicated. Usual indications were a history of angina pectoris, atypical cardiac pain, congestive heart failure of uncertain etiology, and evaluation for cardiac valve surgery. Informed consent for the isotopic measurements of myocardial blood flow was obtained from each patient. The investigative protocols were approved by the Human Investigation and Joint Radioisotope Committees of the Columbia-Presbyterian Medical Center. There were no complications related to the blood flow studies.

Coronary arteriography was performed by a modified Judkins (1967) technique. Images of contrast material injected into the coronary vessels were recorded on biplane cine and/or serial cut films. Subsequent to visualization of each artery, the measurements of regional myocardial perfusion using ^{133}Xe were performed. In most patients, biplane cineangiography was then performed

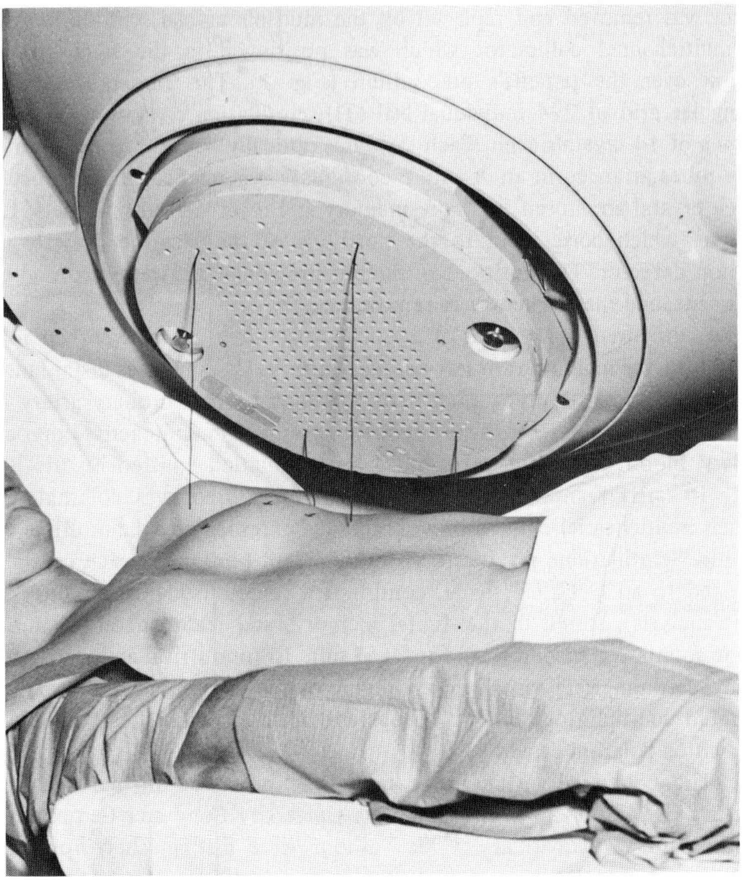

Figure 1. Multiple crystal scintillation camera. The head of the camera has been obliqued in the picture so that the multiple channel collimator and the 21 columns of 14 scintillation crystals can be seen. The patient is positioned in the LAO position.

while contrast material was injected into the left ventricle. Left ventricular end-diastolic volume and ejection fraction were calculated by a modification of the method of Dodge *et al.* (1966).

The measurements of regional myocardial perfusion were made by a method outlined briefly below which has been described in detail previously (Cannon *et al.,* 1972a; Dwyer, Dell, and Cannon, 1973). At the conclusion of diagnostic coronary arteriography, a left anterior oblique (LAO) cineangiogram was obtained with a series of radiopaque-radioactive markers positioned on the chest wall; 5–10 min were then allowed to elapse before measurements of myocardial blood flow were made. Without any change in the patient's position, the cine

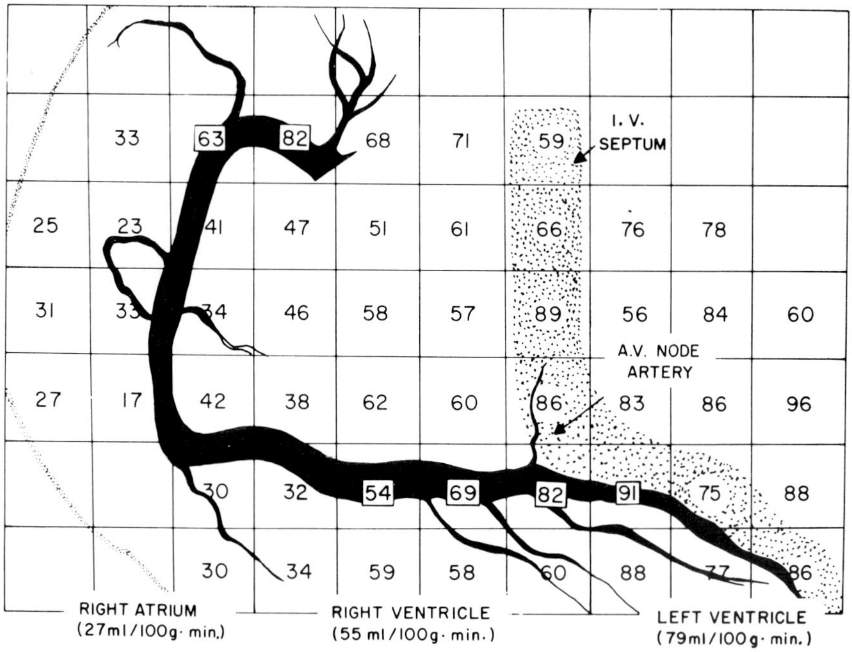

Figure 5. Normal right coronary artery (LAO view).

in this study, the mean LV perfusion rate was 68 ml/100 g · min. The local perfusion rates of LV subregions supplied by the anterior descending or the circumflex arteries were determined by averaging the perfusion rates of crystals in the appropriate areas of the perfusion pattern. In other studies, average perfusion rates proximal and distal to discrete coronary artery stenoses of patients with ischemic heart disease were estimated in a similar fashion.

Figure 5 shows the myocardial perfusion pattern obtained after radioactive xenon was injected into an arteriographically normal dominant right coronary artery. Myocardial regions were again distinguished by anatomical landmarks provided by the coronary arteriogram; the proximal right coronary artery separated the right ventricle from the right atrial area, and the AV node artery and the arteriographic "blush" in the interventricular septum separated the right ventricle from that portion of the inferior left ventricle supplied by the right coronary.

Measurements of regional perfusion in 25 patients with normal coronary arteriograms who were studied at Roosevelt Hospital (Cannon, Dell, and Dwyer, 1972b; Dwyer *et al.*, 1973) are summarized in Figure 6. In this group, the average mean myocardial blood flow rate in the left ventricle, 63 ml/100 g ·

min, significantly exceeded that of the right ventricle and that of the right atrial area.

Figure 7 summarizes preliminary observations made at Columbia-Presbyterian Medical Center of the average of the mean LV myocardial perfusion rates in groups of patients who were not in heart failure and who had normal coronary arteriograms (*black bars*) or arteriograms which revealed insignificant coronary artery disease (*shaded bars*). Insignificant coronary disease was defined arbitrarily as the presence of obstructions of less than 50% on the coronary angiograms which were not associated with prolonged transit time of contrast material. In patients with normal coronary arteriograms with essential hypertension, with aortic stenosis, or with aortic insufficiency, the averages of the mean LV perfusion rates were not statistically different from the average mean LV myocardial perfusion rate of 63 ml/100 g · min found in 11 patients with normal coronary arteriograms, normal intracardiac pressures, and normal ventriculograms. These observations in hypertensive and aortic valvular heart disease contrast markedly with the depressed average mean LV perfusion, 46 ml/100 g · min, found in 19 patients with normal coronary arteriograms who had LV

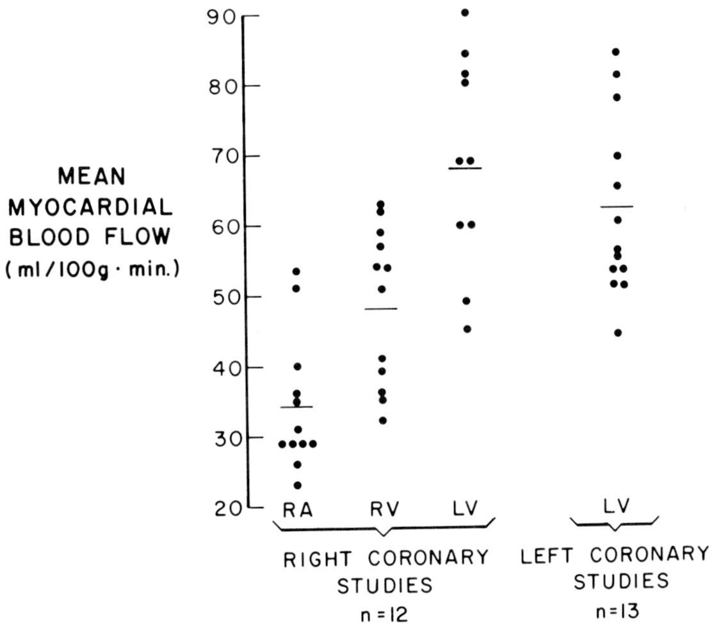

Figure 6. Regional myocardial perfusion in patients with normal coronary arteriograms. RA = right atrium; RV = right ventricle; LV = left ventricle.

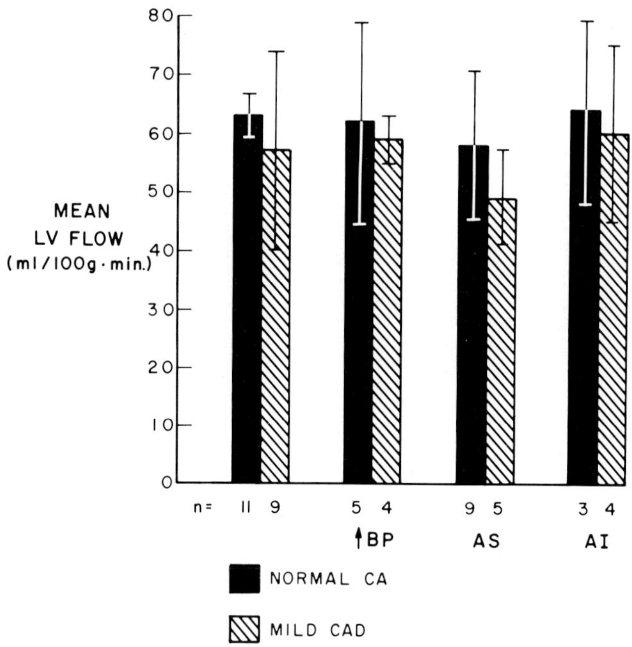

Figure 7. Effect of heart diseases and mild coronary artery disease on myocardial perfusion (N=50). ↑BP = hypertensive heart disease; AS = aortic stenosis; AI = aortic insufficiency.

hypertrophy and cardiac failure secondary to cardiomyopathy (Weiss *et al.*, 1972). In the four groups of patients (Fig. 7), the presence of insignificant coronary obstructions on the arteriograms was associated with slight average reductions of the mean LV perfusion rates which were not statistically significant.

Diffuse reductions of myocardial blood flow rates were observed at rest in patients with radiographically significant atherosclerosis of both branches of the left coronary artery. In 43 patients with greater than 50% obstructions of both branches of the left coronary artery (Fig. 8), the average of the mean LV myocardial perfusion rates of 44 ml/100 g · min was significantly below the value found in the studies of patients with normal arteriograms. In these patients with significant coronary disease, increased heterogeneity of myocardial perfusion was also found. The coefficient of variation of the local myocardial blood flow rates in the left ventricle was 22% in those with significant abnormal left coronary arteriograms and only 16% in those with normal arteriograms.

The myocardial perfusion rates present at rest in the left anterior descending (LAD) subregion of the left ventricle of 26 patients with radiographically significant LAD disease appear in Figure 9. Each of the patients had no disease

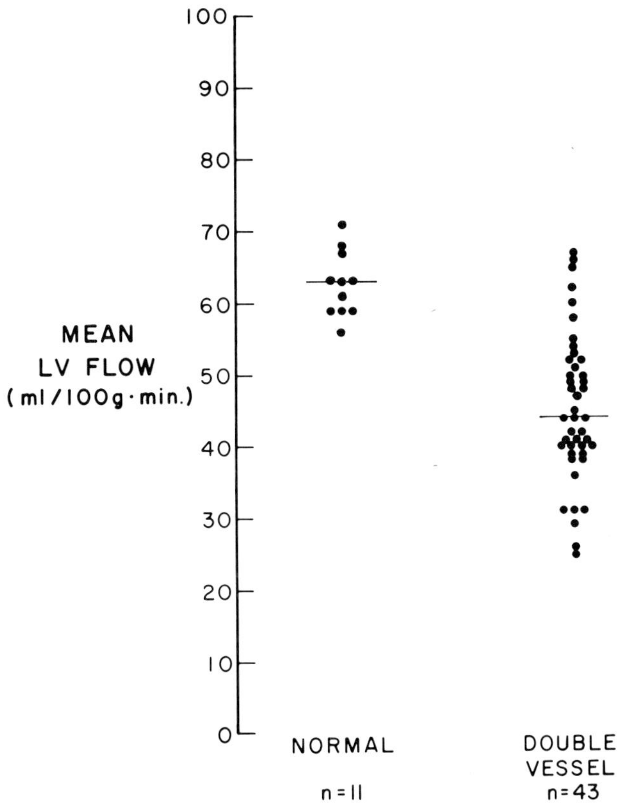

Figure 8. Effect of radiographically significant disease of two branches of the left coronary artery on mean LV perfusion.

or insignificant disease of the circumflex artery; 12 of the patients had significant right coronary obstructions. The mean LAD regional perfusion rate in the disease group was significantly lower than the mean perfusion rate found in the LAD region of the control group of patients. However, there was no correlation between the degree of coronary obstruction greater than 50% apparent on the coronary arteriograms and the degree to which resting LAD myocardial perfusion was reduced. In many of the patients with radiographically significant LAD disease, the resting myocardial perfusion in the region supplied by the diseased artery was normal. In 11 patients it was possible to examine regional perfusion in the myocardial area distal to a radiographically observed isolated LAD obstruction and to compare perfusion in this region with that of myocardium proximal to the lesion and in the remainder of the same ventricle. No localized reductions of regional ventricular perfusion distal to the obstruction were

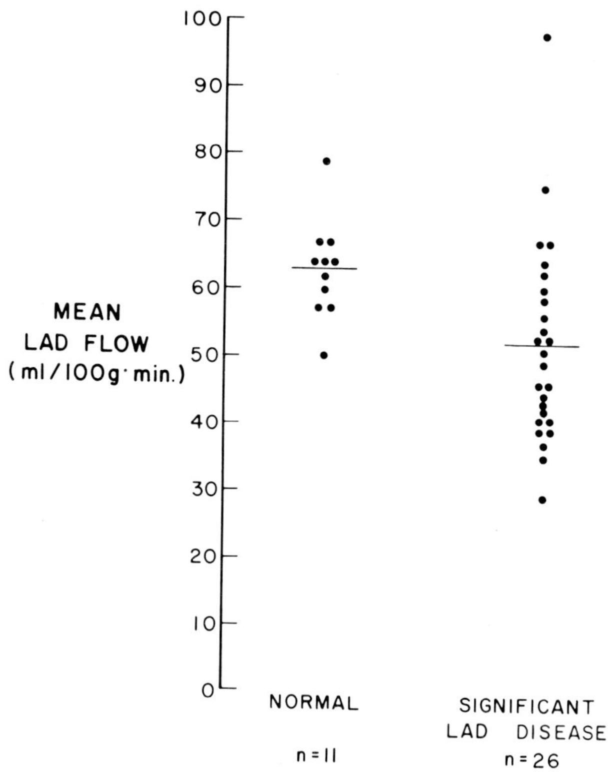

Figure 9. Myocardial perfusion in the left anterior descending subregion at rest; effect of significant LAD disease.

observed at rest, except in 4 patients with total LAD obstruction who either gave history of heart attack or had electrocardiographic evidence of anterior myocardial infarction.

To assess more readily the significance of a radiologically apparent coronary obstruction, studies of regional myocardial blood flow were performed before and during the time when myocardial oxygen consumption was increased by right atrial pacing (Schmidt *et al.*, 1973). Figure 10 shows the myocardial blood flow response to atrial pacing to a heart rate of 150 beats/min in a hypertensive patient with scleroderma who had normal coronary arteries. As pacing was performed, the "double product" of heart rate times systolic blood pressure (which is proportional to myocardial oxygen consumption) (Holmberg and Varnauskas, 1971) increased from 16,800 to 34,500. In response to the increased metabolic demand of the faster heart rate, there were diffuse increases in regional myocardial blood flow throughout the left ventricle.

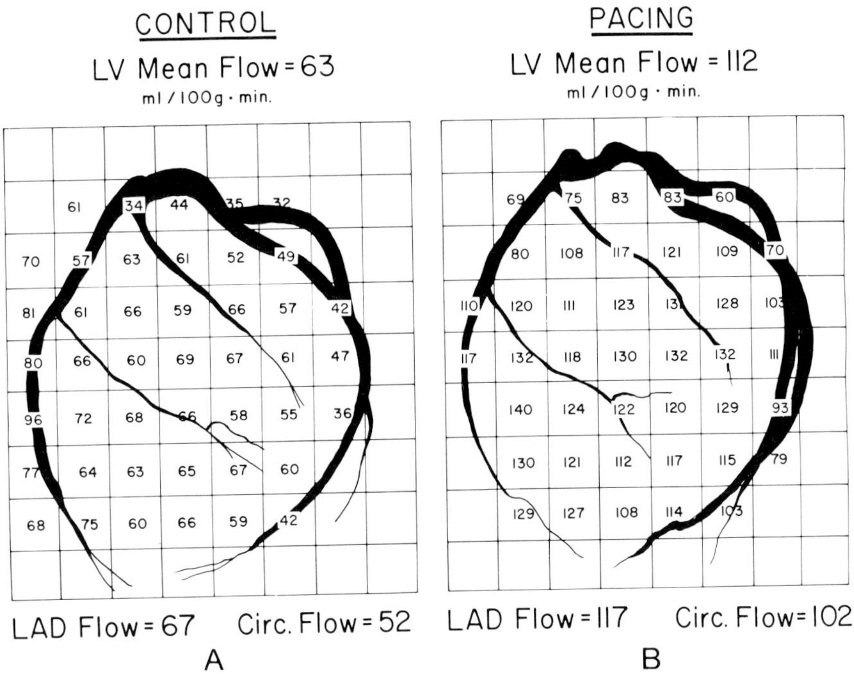

Figure 10. Myocardial flow response of a patient with a normal coronary arteriogram to right atrial pacing. *a:* study performed at rest. *b:* study performed during atrial pacing.

Figure 11 illustrates the inadequate regional flow response to atrial pacing observed in a patient with a totally occluded LAD artery. At rest (Fig. 11*a*), the myocardial perfusion rate in the LAD subregion was maintained at nearly the level of the remainder of the left ventricle by retrograde collateral vessels from the circumflex artery. The collateral vessels on this and other arteriograms were independently assessed by two radiologists. The collaterals were termed "inadequate or marginal" if the vessel supplied by the collaterals did not fill to a 1-mm diameter and if the passage of contrast material through the vessel was significantly delayed. The resting study (Fig. 11*a*) illustrates that the measurement technique is capable of measuring the rate of collateral blood flow in man. It illustrates, also, that collateral vessels may maintain resting perfusion distal to coronary obstructions at normal or nearly normal levels. Figure 11*b* shows the response of the patient to right atrial pacing. At a heart rate of 150, the patient developed angina pectoris. Even though the double product increased significantly during pacing, and the flow in the circumflex area rose from 48 to 71 ml/100 g · min, the myocardial blood flow in the LAD subregion supplied by

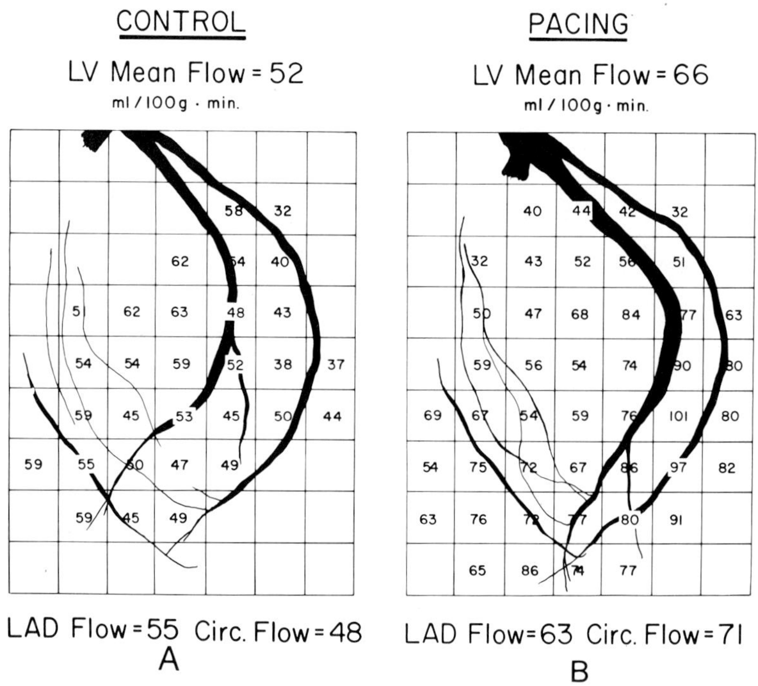

Figure 11. Myocardial flow response to pacing of a patient with an occluded left anterior descending artery. *a:* study performed at rest. *b:* study performed during atrial pacing.

collateral vessels rose only slightly. Total LV myocardial perfusion increased moderately during atrial pacing in this study.

The average responses of mean LV myocardial perfusion to right atrial pacing in 19 patients without radiographically significant coronary disease (group I), and in 18 patients with significant coronary disease, i.e., 50% stenosis of one or both left coronary arteries (group II), are summarized in Figure 12. In the two groups of patients, the double product of heart rate times systolic blood pressure

Table 1. Regional myocardial perfusion response to atrial pacing

	Myocardial Perfusion (ml/100 g·min)		
	Mean LV Total	LAD	LC
Rest	57	53	62
Paced	71	73	73

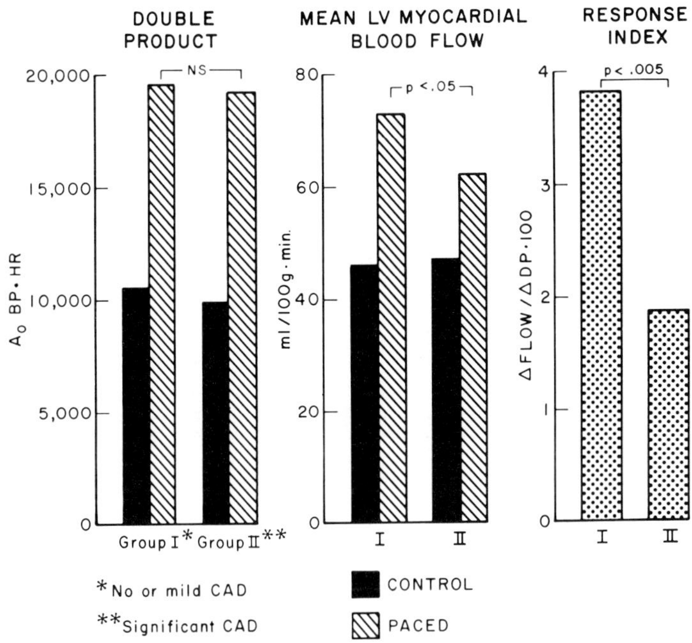

Figure 12. Responses to right atrial pacing in patients with and without significant coronary artery disease. Group I, $N = 19$; group II, $N = 18$.

rose equally in response to atrial pacing to a heart rate of 150 or to the onset of precordial pain. However, the response of the mean LV myocardial blood flow to pacing was significantly less ($p<0.05$) in the patients with radiographically significant coronary disease (group II). Coronary vascular resistance per 100 g myocardium fell 31% in group I patients and only 15% in group II patients, a difference that was also significant ($p<0.01$). A myocardial blood flow response index was calculated for each patient by dividing the increment in flow produced with pacing by the simultaneous increment in myocardial oxygen consumption (the double product). The rise in myocardial blood flow which occurred in response to the increased metabolic demand of atrial pacing was significantly reduced in the patients with radiographically significant single or double vessel left coronary disease.

Table 1 summarizes the flow responses to atrial pacing observed in a patient with 90% stenosis of the LAD artery. Extensive collateral blood vessels from the right coronary and the circumflex arteries to the region of LV supplied by the LAD artery were apparent on the patient's arteriogram. The distal LAD artery filled rapidly to a diameter of 2 mm by retrograde flow from the collateral vessels. In this patient, myocardial blood flow increased uniformly throughout

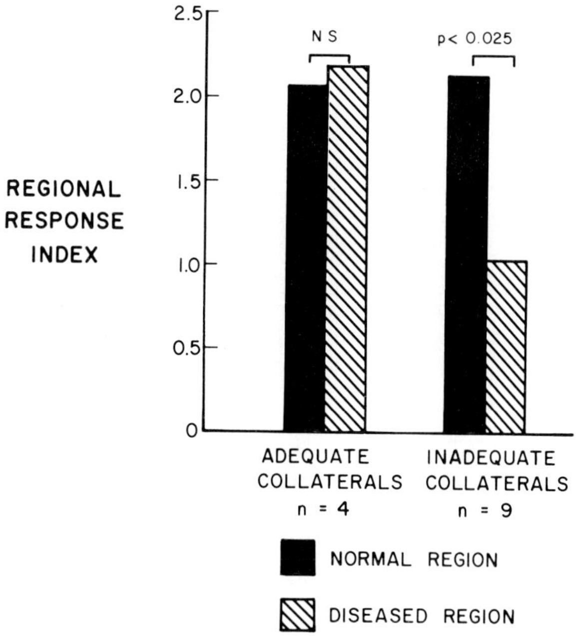

Figure 13. Influence of collateral flow on the regional flow response to atrial pacing.

the left ventricle in response to atrial pacing. The response index (Δ flow/ Δ double product) of the LAD area supplied by collateral vessels was equal to that of the noninvolved region of the left ventricle.

Figure 13 summarizes the influence of a collateral blood supply on the regional myocardial blood flow response to pacing in 13 patients. The flow response of a myocardial area distal to a coronary obstruction equalled that of the noninvolved area in the same ventricle if the region was adequately supplied by collateral vessels (i.e., if abundant collaterals were apparent on the coronary arteriograms). In the absence of adequate collaterals, the regional blood flow response of an area distal to a radiographically significant coronary obstruction was significantly reduced.

DISCUSSION

Methodological considerations

The present studies demonstrate that the new method to measure regional myocardial perfusion in intact man (Cannon *et al.*, 1972a) can be rapidly and safely performed at the time of coronary arteriography. The technique provides quantitative estimates of capillary perfusion in multiple regions of the heart from measurements carried out simultaneously after the radioisotope is injected into the coronary artery. The multiple detectors and the multichannel collimator subdivide the myocardium into separate regions from which individual ^{133}Xe washout curves are obtained. Thus, the uncertainties of attempting to resolve a single multiexponential ^{133}Xe, ^{85}Kr, or H_2 curve obtained in a patient with coronary disease to calculate regional flow data are avoided (Eckenhoff *et al.*, 1948; Ross *et al.*, 1964; Klein, Cohen, and Gorlin, 1965; Klocke *et al.*, 1968). Computer analysis of the multiple ^{133}Xe washout curves gives not only accurate estimates of isotope clearance from heart muscle but also statistical assessments of the accuracy of the measurements. The quantitative data provide more information concerning the circulation of the myocardium than that supplied by qualitative imaging of perfusion defects in the myocardium with radioactive biodegradable albumin microspheres (Ashburn *et al.*, 1971; Weller *et al.*, 1972). In addition, the use of a labeled inert gas, such as ^{133}Xe, avoids the streaming artifacts in the coronary circulation which may occur due to inadequate mixing when labeled albumin microspheres are injected directly into a coronary artery. Because it is possible to align the pattern of local myocardial perfusion rates with the coronary arteriograms obtained during the same study, one can determine regional perfusion proximal and distal to the obstructive lesions seen on the coronary arteriograms. Such localization of perfusion defects in relation to coronary anatomy is impossible with techniques which assess regional LAD perfusion by sampling H_2 or He in the great cardiac vein (Bunnell *et al.*, 1974).

It is important to emphasize, however, that the primary data obtained in the present studies are the rate constants of ^{133}Xe washout from areas of the heart viewed by the multiple detectors, which were calculated by monoexponential least squares analysis of the initial portions of each tracer curve. As such, each bit of primary data is a quantitative expression of the ability of the local circulation to clear a diffusible indicator and, by inference, to supply the tissue with diffusible nutrients. The expression of the primary data in terms of tissue blood flow must be interpreted with caution to the extent that it involves use of the Kety formula and, therefore, depends upon assumptions inherent in a monoexponential analysis of the curves and an assumed partition coefficient. The basic observations, nevertheless, retain a validity which is independent of the assumptions involved in expressing the data in terms of flow.

The method used to measure regional myocardial blood flow with ^{133}Xe and the scintillation camera in its present stage of development has certain limitations. These are discussed in more detail elsewhere (Cannon *et al.*, 1972a; Cannon, 1974). Currently, the method is unable to distinguish epicardial from endocardial blood flow; it is our belief, however, that the current form of data analysis yields a weighted average of the flows that are present in the subepicardial and subendocardial regions viewed by the same crystal (Cannon, 1974). Because the heart is three-dimensional, one or more myocardial surfaces may obviously be viewed by the same crystal in certain projections. For this reason, the present studies were performed with the patient in the LAO position, a projection which displays the regions supplied by the LAD artery to greatest advantage. With the advent of new, more sensitive and faster scintillation cameras, it is anticipated that collimation can be increased so that multiple flow measurements will be possible in different projections which display the regions distal to various coronary obstructions optimally and with minimal overlap of other areas of the myocardium supplied by the same vessel.

Regional myocardial perfusion

The data from the studies which indicate that subjects with normal coronary arteriograms have mild heterogeneity of local perfusion rates in the left ventricle (e.g., Fig. 4) are consistent with animal and human experiments performed at the time of thoracotomy in which ^{133}Xe was injected locally into different areas of the myocardium (Sullivan *et al.*, 1967; Brandi, Fam, and McGregor, 1968). Similarly, the observation that the myocardial blood flow rate in patients with normal coronary arteriograms is significantly higher in the left ventricle than in the right ventricle or right atrial area (Fig. 6) is consistent with observations made in experimental animals using radioactive ions or microspheres to measure the tissue perfusion (Love and Burch, 1957; Domenech *et al.*, 1969). The higher LV perfusion rate suggests that the pressure work of the left ventricle may require not only a greater muscle mass but also a larger blood supply per unit tissue than the right ventricle or the atria (Braunwald, 1969). The fact that blood flow per gram was reduced in the myocardium of patients with heart failure secondary to cardiomyopathy and was increased in patients without significant coronary artery obstructions during atrial pacing (Fig. 10) is evidence in man that the factors which contribute to myocardial oxygen consumption, such as heart rate, myocardial contractility, and peak wall stress, determine the resting myocardial blood flow (Braunwald, 1969). This observation has several implications for the interpretation of the regional blood flow changes which may result from coronary occlusive disease. The resting flow in noninvolved areas of the myocardium and also in ischemic regions where there are abnormalities of ventricular wall motion may both be determined, at least in part, by the performance characteristics of those respective portions of the ventricular wall. The need of techniques to distinguish situations when myocardial blood flow (either total or regional) is adequate for the demands of myocardial work from situations when blood flow is inadequate to meet the metabolic demands of the tissue (ischemia) is apparent.

Diffuse reductions of local myocardial blood flow rates have been observed at rest in patients with radiographically significant occlusive disease of one or both branches of the left coronary artery in this and a previous study (Cannon *et al.*, 1972b). The present studies of 43 patients at Columbia-Presbyterian Medical

Center which indicate that the average mean LV myocardial blood flow rate in patients with greater than 50% occlusion of two left coronary vessels was significantly lower than in 11 patients with normal coronary arteriograms confirm our earlier study of similar patients at Roosevelt Hospital (Cannon *et al.*, 1972b). The group of Klocke *et al.*, using H_2 or He washout combined with coronary sinus sampling to measure LV perfusion, has also recently confirmed these results (Bunnell *et al.*, 1974). Failure of investigators in the past to find different flow rates between groups of patients with normal and abnormal coronary arteriograms (Ross *et al.*, 1964; Klein *et al.*, 1965; Klocke *et al.*, 1968) may have resulted from the fact that localized areas of reduced perfusion were obscured by larger areas with normal flow when the measurements were made with single crystal detectors or coronary sinus sampling and only single N_2O, H_2, ^{133}Xe, or ^{85}Kr washout curves were obtained. In the present, as in previous studies (Sullivan *et al.*, 1967; Klocke *et al.*, 1968; Cannon *et al.*, 1972b) evidence for increased heterogeneity of local LV perfusion rates was found in the patients with significant coronary heart disease. In this group, the coefficient of variation of the local LV myocardial blood flow rates was increased.

In patients with radiographically insignificant coronary disease, total myocardial perfusion at rest was not reduced significantly below values found in subjects with normal coronary arteriograms and normal ventricular performance (Fig. 7). These data are consistent with the data of Smith *et al.* (1972) indicating that the resting myocardial perfusion rates in patients with less than 80% stenosis of a coronary artery were not significantly lower than normal distal to the lesion; the latter measurements were performed in the operating room, where myocardial flow estimates were made by monitoring ^{133}Xe washout after local injection of ^{133}Xe directly into the coronary arteries.

Previous studies (Cannon *et al.*, 1972b) and individual cases in the present series have indicated that this technique to estimate blood flow regionally in the heart with ^{133}Xe and a scintillation camera is capable of detecting reductions of perfusion in localized areas of the myocardium. The present data in 7 patients indicated that resting perfusion rates distal to localized LAD obstruction of 50–95% were not different from those in the remainder of the heart, whereas in 4 patients with total LAD occlusions, resting flows distal to the lesion were reduced. In a previous study of patients with transmural myocardial infarction, regional reductions of perfusion were found at rest in 10 patients with an anterior transmural infarction; reductions of flow to the right ventricle and inferior left ventricle were found in 10 others with transmural inferior infarctions (Dwyer *et al.*, 1973). Various factors may account for the relatively uniform resting regional flow distribution found in many patients with 50–95% lesions; among them are included: the degree of obstruction is not flow limiting at basal levels of myocardial work; the presence of collateral blood supply to the ischemic region; the presence of diffuse atherosclerosis elsewhere, which was not visualized on the angiograms; and the possibility that impaired performance of the remaining (apparently uninvolved) myocardium may require less blood flow.

The present studies also demonstrate that the new measurement technique can be used to assess collateral blood flow in man. They indicate (e.g., Figs. 9 and 11*a*) that myocardial perfusion in regions distal to 80–100% coronary artery occlusions can be maintained at normal levels at rest by a collateral blood supply. The observations that human collateral blood vessels can increase the myocardial nutrient flow to diseased regions in response to increased metabolic demands (Table 1, Fig. 13) are of interest, since they are consistent with several studies of collateral blood flow in animals with coronary occlusion who were given nitroglycerine or norepinephrine (Cohen *et al.*, 1973; Pass *et al.*, 1974). Additional studies of this type may resolve the current uncertainty as to whether or not radiologically visible collateral vessels have a functional significance (Gensini and Bruto da Costa, 1969; Sheldon, 1969; Helfant, Kemp, and Gorlin, 1970).

The preliminary studies with atrial pacing (Schmidt *et al.*, 1973) confirm several previous reports using pacing or exercise (Holmberg and Varnauskas, 1971; Knoebel *et al.*, 1971) which showed that average mean LV myocardial perfusion increased inadequately when related to an index of myocardial oxygen consumption in patients with radiographically significant coronary disease. In the patients with isolated disease of the LAD or circumflex artery, the regional myocardial blood flow responses varied greatly during pacing. Reductions in flow, no change in flow, and subnormal and normal responses in the diseased region were observed (Schmidt *et al.*, 1973). The variations appeared to depend upon the extent of coronary disease, the functional consequences of ischemia, and the adequacy of the collateral blood supply to the area. When abundant collateral vessels to a diseased area were visualized radiographically, the regional flow response to pacing was equal to that of the regions of myocardium with lesser coronary disease (although the responses of both regions in patients with significant coronary disease were less than in normal individuals). These findings in man are similar to observations in dogs in which collateral flow responses of regions distal to coronary occlusions were studied with radioactive ions or microspheres during nitroglycerine administration, pacing, or norepinephrine infusion (Cohen *et al.*, 1973; Pass *et al.*, 1974). The collateral flow response to increased demand in animal and human studies equalled the rest of the ventricle but was subnormal (Schmidt *et al.*, 1973; Pass *et al.*, 1974).

In conclusion, the present studies provide evidence for the hypothesis that angina pectoris results from an imbalance between coronary blood supply and the metabolic demands of the myocardium for oxygen. The regional myocardial flow studies of patients with radiographically significant coronary occlusive disease indicate that, even if resting regional perfusion is normal, blood flow in the area of myocardium distal to the coronary lesion may respond inadequately or not at all during atrial pacing, with the frequent consequence of induced chest pain. The studies also suggest that the physiological assessment of coronary atherosclerotic lesions and of collateral blood vessels in man may be enhanced by combining high grade coronary arteriography and measurements of regional myocardial perfusion.

Acknowledgments This work was supported by Grants HL 14148 and HL 14236 from the National Institute of Health. Dr. Cannon is recipient of Career Development Award HL-15031-10 from the United States Public Health Service.

REFERENCES

ASHBURN, W. L., E. BRAUNWALD, A. L. SIMON, K. L. PETERSON, and J. H. GAULT. 1971. Myocardial perfusion imaging with radioactive labelled particles injected directly into the coronary circulation of patients with coronary artery disease. Circulation 44: 851–865.

BASSINGTHWAIGHTE, J. B., T. STRANDELL, and D. E. DONALD. 1968. Estimation of coronary blood flow by washout of diffusible indicators. Circulat. Res. 23: 259–278.

BRANDI, G., W. M. FAM, and M. McGREGOR. 1968. Measurement of coronary flow in local areas of myocardium using Xenon^{133} . J. Appl. Physiol. 24: 446–450.

BRAUNWALD, E. 1969. The determinants of myocardial oxygen consumption. Thirteenth Bowditch Lecture. Physiologist 12: 65–93.

BUNNELL, I., F. KLOCKE, D. GREENE, D. ARANI, R. TANDON, R. OLIVEROS, and S. WITTENBERG. 1974. Effects of isolated left anterior descending coronary artery stenoses on coronary blood flow. Clin. Invest. 53: 13a (Abstract 48).

CANNON, P. J. 1975. The measurement of regional myocardial perfusion in patients with ischemic heart disease with ^{133}Xenon and a scintillation camera. *In* V. P. Kriss, R. N. Pierson, Jr., R. H. Jones, and W. J. MacIntyre (eds.), Quantitative Nuclear Cardiology, pp. 155–201. John Wiley and Sons, Inc., New York.

CANNON, P. J., R. B. DELL, and E. M. DWYER, Jr. 1970. Regional myocardial perfusion in man. J. Clin. Invest. 49: 16a.

CANNON, P. J., R. B. DELL, and E. M. DWYER, Jr. 1972a. Measurement of regional myocardial perfusion in man with ^{133}Xenon and a scintillation camera. J. Clin. Invest. 51: 964–977.

CANNON, P. J., R. B. DELL, and E. M. DWYER, Jr. 1972b. Regional myocardial perfusion rates in patients with coronary artery disease. J. Clin. Invest. 51: 978–994.

CANNON, P. J., R. B. DELL, E. M. DWYER, and P. M. JOHNSON. 1969a. A method for quantitative assessment of regional myocardial perfusion in the intact animal. Circulation 40 (Suppl. VL): III–56.

CANNON, P. J., J. I. HAFT, and P. M. JOHNSON. 1969b. Visual assessment of regional myocardial perfusion using radioactive xenon and scintillation photography. Circulation 40: 277–288.

COHEN, M. J., J. M. DOWNEY, E. N. SONNENBLICK, and E. S. KIRK. 1973. The effects of nitroglycerine on coronary collaterals and myocardial contractility. J. Clin. Invest. 52: 2836–2847.

CONN, H. L., Jr. 1961. Equilibrium distribution of radioxenon in tissue: Xenon-hemoglobin association curve. J. Appl. Physiol. 16: 1065–1070.

DODGE, H. T., H. SANDLER, W. A. BAXLEY, and R. R. HAWLEY. 1966. Usefulness and limitations of radiographic methods for determining left ventricular volume. Am. J. Cardiol. 18: 10–24.

DOMENECH, R. J., J. I. E. HOFFMAN, M. I. M. NOBLE, K. B. SAUNDERS, J. R. HENSON, and S. SUBIJANTO. 1969. Total and regional coronary blood flow measured by radioactive microspheres in conscious and anesthetized dogs. Circulat. Res. 25: 581–596.

DWYER, E. M., Jr., R. B. DELL, and P. J. CANNON. 1973. Regional myocardial blood flow in transmural myocardial infarction. Circulation 48: 924–935.

ECKENHOFF, J. E., J. H. HAFKENSCHIEL, M. H. HARMEL, W. T. GOODALE, M. LUBIN, R. J. BING, and S. S. KETY. 1948. Measurement of coronary blood flow by the nitrous oxide method. Am. J. Physiol. 152: 356–364.

FRIEDBERG, C. K. 1972. Some comments and reflections on changing interests and new developments in angina pectoris. Circulation 46: 1037–1047.

GENSINI, G. G., and B. C. BRUTO da COSTA. 1969. The coronary collateral circulation in living man. Am. J. Cardiol. 24: 393–400.

HELFANT, R. H., H. G. KEMP, and R. GORLIN. 1970. Coronary atherosclerosis, coronary collaterals, and their relation to cardiac function. Ann. Int. Med. 73: 189–193.

HERD, J. A., M. HOLLENBERG, G. D. THORBURN, H. KOPALD, and A. C. BARGER. 1962. Myocardial blood flow determined with Krypton^{85} in unanesthetized dogs. Am. J. Physiol. 203: 122–124.

HOLMBERG, S., and Varnauskas. 1971. Coronary circulation in pacing induced tachycardia. Acta. Med. Scand. 190: 481–490.

JAFFE, C. C., and K. ELLIS. 1974. Angiocardiographic quantitation of ventricular volume, shape and mass. Curr. Probl. Radiol. 4: 1–55.

JUDKINS, M. P. 1967. Selective coronary arteriography by the percutaneous femoral technique. Radiology 89: 815–824.

KETY, S. S. 1960. I. Blood-tissue exchange methods. Theory of blood-tissue exchange and its application to measurement of blood flow. Meth. Med. Res. 8: 223–227.

KLEIN, M. D., L. S. COHEN, and R. GORLIN. 1965. Krypton^{85} myocardial blood flow: Precordial scintillation versus coronary sinus sampling. Am. J. Physiol. 209: 705–710.

KLOCKE, F. J., R. C. KOBERSTEIN, D. E. PITTMAN, I. L. BUNNELL, D. G. GREENE, and R. D. ROSING. 1968. Effects of heterogeneous myocardial perfusion on coronary venous H_2 desaturation curves and calculations of coronary flow. J. Clin. Invest. 47: 2711–2724.

KNOEBEL, S. B., W. C. ELLIOTT, P. L. McHENRY, Jr., and R. ROSS. 1971. Myocardial blood flow in coronary artery disease. Correlation with severity of disease and treadmill exercise response. Am. J. Cardiol. 27: 51–58.

LOVE, W. D., and G. E. BURCH. 1957. Differences in the rate of Rb^{86} uptake by several regions of the myocardium of control dogs and dogs receiving l-norepinephrine or pitressin. J. Clin. Invest. 36: 479–484.

PASS, H. I., J. L. COX, H. N. O. WECHSLER, Jr., and D. C. SABISTON, Jr. 1974. Response of coronary collateral circulation to increased myocardial demands. Circulation 48(Suppl. IV): IV–92.

ROSS, R. S., K. UEDA, P. R. LICHTLEN, and J. R. REES. 1964. Measurement of myocardial blood flow in animals and man by selective injection of radioactive inert gas into the coronary arteries. Circulat. Res. 15: 28–41.

SCHMIDT, D. H., M. B. WEISS, W. CASARELLA, E. POLLOCK, and P. J. CANNON. 1973. The effect of atrial pacing on regional myocardial perfusion in coronary artery disease. Circulation 48(Suppl.): IV–64.

SHAW, D. J., A. PITT, and G. C. FRIESINGER. 1971. Autoradiographic study of the ^{133}Xenon disappearance and method for measurement of myocardial blood flow. Cardiovasc. Res. 6: 268–276.

SHELDON, W. C. 1969. Editorial on the significance of coronary collaterals. Am. J. Cardiol. 24: 303–304.

SMITH, S. C., Jr., R. GORLIN, M. V. HERMAN, W. J. TAYLOR, and J. J. COLLINS, Jr. 1972. Myocardial blood flow in man: effects of coronary collateral circulation and coronary artery bypass surgery. J. Clin. Invest. 51: 2556–2565.

SULLIVAN, J. J., W. J. TAYLOR, W. C. ELIOTT, and R. GORLIN. 1967. Regional myocardial blood flow. J. Clin. Invest. 46: 1402–1412.

WEISS, M., D. H. SCHMIDT, C. JAFFE, W. CASARELLA, K. ELLIS, and P. J. CANNON. 1972. Regional myocardial blood flow in patients with idiopathic myocardial hypertrophy. Circulation 46: II–235.

WELLER, D. A., R. J. ADOLPH, H. N. WELLMAN, R. G. CARROLL, and O. KIMM. 1972. Myocardial perfusion scintigraphy after intracoronary injection of ^{99m}Tc-labelled human albumin microspheres: Toxicity and efficacy for detecting myocardial infarction in dogs: Preliminary results in man. Circulation 46: 963–875.

Mailing address:
Dr. Paul J. Cannon,
Department of Medicine, College of Physicians and Surgeons,
Columbia University, 630 West 168th Street, New York, New York, 10032 (USA).

Recent Advances in Studies on
Cardiac Structure and Metabolism Volume 10
The Metabolism of Contraction
Edited by Paul-Emile Roy and George Rona

LOCAL MYOCARDIAL BLOOD FLOW MEASURED BY THE USE OF A NEEDLE-TYPE Pt-H_2 ELECTRODE

TOMIYASU KOYAMA

Research Institute of Applied Electricity, Hokkaido University, Sapporo, Japan

SUMMARY

In order to measure local myocardial blood flow repeatedly, a needle-type recessed Pt electrode sealed in a thin injection needle was constructed. By means of this electrode, the myocardial blood flow in the inner and outer portions of the anterior wall of the exposed left ventricle of anesthetized dogs was studied. Under control conditions, the blood flow in the inner portion (112.4 ml/100 ml/min) exceeded that in the outer portion (98.7 ml/100 ml/min). But the blood flow distribution expressed in terms of the blood flow ratio in both portions was readily varied by several pharmacological treatments, starvation, or prolongation of the experiment. Combining these observations with the results obtained in the decrease of the O_2 current in the coronary occlusion test, the following possibilities seemed probable: 1) The myocardial blood flow under normal conditions was so distributed that the inner layer could be supplied with an increased blood flow and more oxygen. 2) Such a balance between the local blood flow and O_2 demand can often be disturbed under various conditions, as was seen in the case of isoproterenol infusion in the present study.

INTRODUCTION

Modern techniques for the measurement of myocardial blood flow permit a repetition of only a few measurements in each experimental animal. Therefore, it seemed worthwhile to make an H_2 electrode applicable to repetitive measurements of blood flow. In the present study, a needle-type recessed Pt electrode was constructed and employed in the measurement of local blood flow in the inner and outer portions of left ventricular myocardium under various conditions.

METHODS

The structure of the electrode is shown in Figure 1. A Pt wire 100 μ in diameter was fused in a thin thermister glass rod; its tip was polished obliquely. The insulated Pt wire in turn was sealed in a thin injection needle by means of a resin, to facilitate its insertion into the beating myocardium. Two types of electrode tips, A and B, were prepared; type B probably causes less trauma to

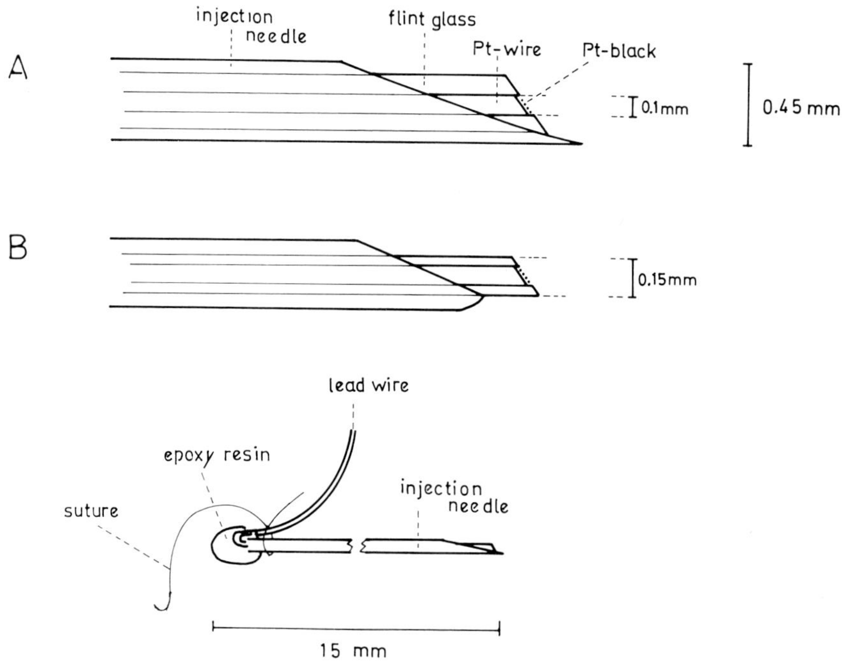

Figure 1. Schematic illustration of the needle-type recessed Pt-H_2 electrode.

the tissue than type A. But as the tip of the former type was fragile and since both types of electrode tips gave essentially the same results in *in vivo* measurements, the type A electrode was used in most measurements because of the solidity of the tip. To make the active electrode surface recessed from the surrounding sheath glass, the Pt surface was etched electrochemically, in a saturated $NaNO_2$ solution (Silver, 1965) and the Pt surface was covered with Pt-black.

Mongrel dogs anesthetized by intravenous injection of pentobarbital received a ventilation of 250 ml/kg/min from a positive pressure respirator. The heart was exposed by a thoracotomy at the 6th intercostal space. Two electrodes were inserted into the anterior wall of the left ventricle at a site 2–3 cm higher than the apex and 1–2 cm left of the anterior descending artery. One electrode was placed in the outer portion or outer layer of the myocardium at a depth of 2–3 mm and the other was fixed in the inner portion of the myocardium at a depth of 6–9 mm from the epicardial surface. The electrodes were fastened to the epicardial surface by a thin suture and the thorax was then closed. The H_2 in the aortic blood was recorded by a Pt-H_2 catheter electrode (Koyama and Marutani, 1971) placed in the aorta via an exposed femoral artery.

After the currents flowing through the electrodes became stable, the inspiration gas was changed from room air to a gas mixture containing H_2. The currents through the three electrodes increased, attaining a plateau in 3–6 min. The gas was then switched back to room air, and the H_2 in the myocardium was washed out. From these washout curves, the myocardial blood flow in the outer and inner layers was calculated separately according to the formula of Kety and Schmidt (1945), with the assumption that the mean tissue P_{H_2} was approximately equal to that in the coronary vein.

RESULTS

An example of H_2 current through the Pt electrodes is shown in Figure 2. The *upper and lower curves* indicate the P_{H_2} changes in the aorta and in the myocardium, respectively. When the H_2 content in the inspired gas mixture was increased stepwise, both the H_2 currents increased in proportion to the H_2 content in the inspired gas mixture. The calculated blood flow values agreed within a 5% error.

An example of actual successive recordings of H_2 currents is shown in Figure 3, where the *curves from top to bottom* indicate the H_2 currents in the aorta, and the outer and inner layers of the myocardium, respectively. In Figure 3*A*, the curves were recorded during artificial pacemaking at the rate of 216 beats/min. Blood flows in the outer layer $\dot{Q}_{out}$ and inner layer $\dot{Q}_{inn}$ were 139.9 and 108.9 ml/100 ml/min, respectively, and the ratio of blood flow of both layers, $\dot{Q}_{inn}/\dot{Q}_{out}$, was as low as 0.784. The pacemaker was then switched off and the heart rate became spontaneous; it fell to 132 beats/min, as shown in Figure 3*B*. The changes of the myocardial H_2 curves subsequently became slower. $\dot{Q}_{out}$ and $\dot{Q}_{inn}$ decreased to 105.0 and 95.1 ml/100 ml/min, respectively, but the ratio $\dot{Q}_{inn}/\dot{Q}_{out}$ increased to 0.906. In Figure 3*C*, the heart rate was again controlled at a rate of 170 beats/min. Both $\dot{Q}_{out}$ and $\dot{Q}_{inn}$ increased in comparison with the values obtained at the spontaneous heart rate, while the ratio $\dot{Q}_{inn}/\dot{Q}_{out}$ decreased to 0.876. In Figure 3*D*, norepinephrine was infused at the constant heart rate. Both H_2 curves became steeper, and the increase of $\dot{Q}_{inn}$ was larger than that of $\dot{Q}_{out}$. The ratio increased to 0.965.

Another example is shown in Figure 4. The curves in Figure 4*A* were recorded without any treatment. Carbocromen (Nitz and Pötzsch, 1963), a coronary vasodilatory substance, was then injected intravenously in a dose of 2 mg/kg, and registration of H_2 curves, shown in Figure 4*B*, was started 15 min later. The rise and fall of the two myocardial curves were clearly accelerated as compared with those in Figure 4*A*. In Figure 4*C* and *D*, the measurements were repeated over subsequent 15-min periods. Even 45 min after the injection, both $\dot{Q}_{out}$ and $\dot{Q}_{inn}$ were larger than the control values.

Mean values of $\dot{Q}_{out}$, $\dot{Q}_{inn}$, the ratio of $\dot{Q}_{inn}/\dot{Q}_{out}$, the heart rate, and the

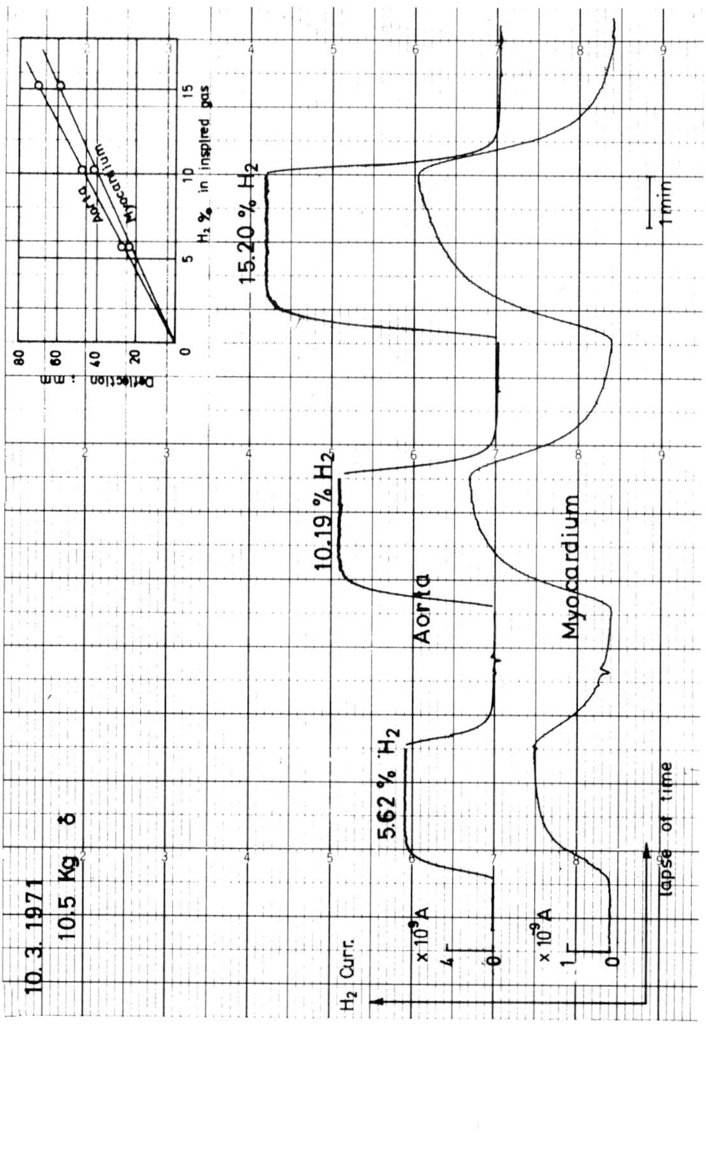

Figure 2. Relationship between H_2 content in the inspired gas mixture and H_2 currents obtained in the myocardium as well as in the aorta.

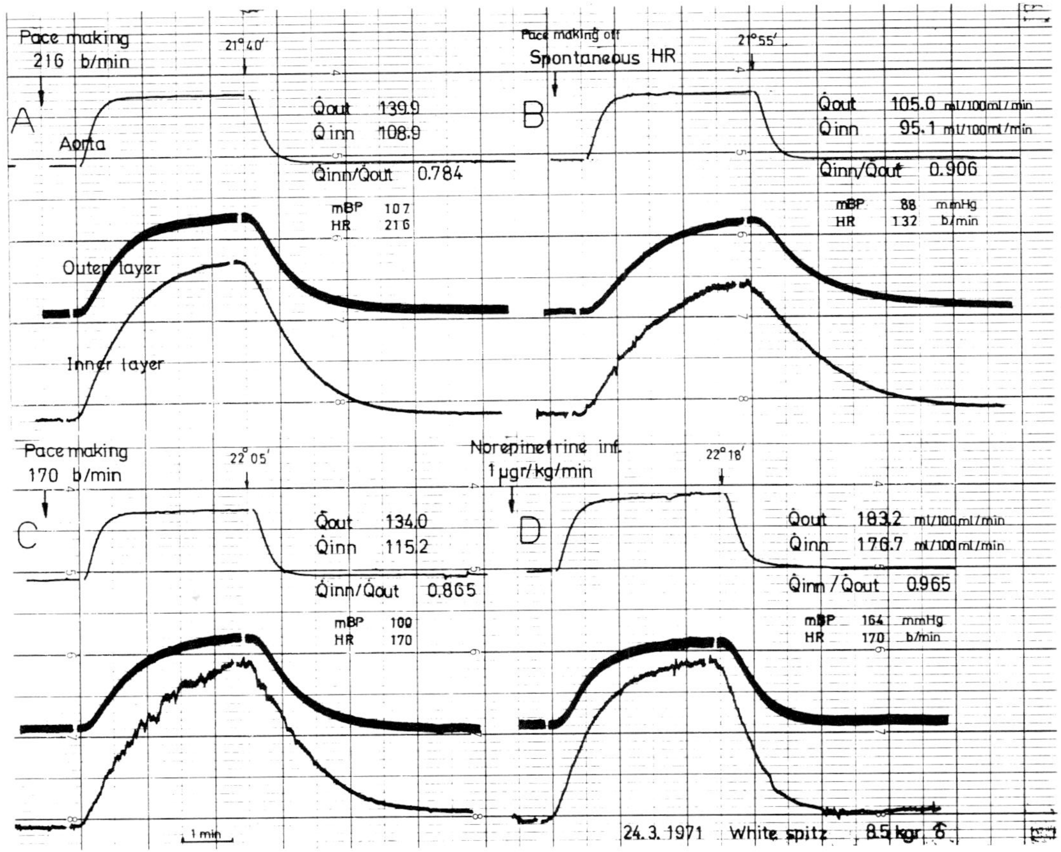

Figure 3. Examples of actual recordings at artificially controlled heart rates. *A:* atrial pacemaking at the rate of 216 beats/min. *B:* spontaneous heart rate of 132 beats/min. *C:* atrial pacemaking at the rate of 170 beats/min. *D:* norepinephrine infusion at the fixed heart rate of 170 beats/min. In each frame, the *curves from top to bottom* were obtained in the aorta, and the myocardial outer and inner portions, respectively.

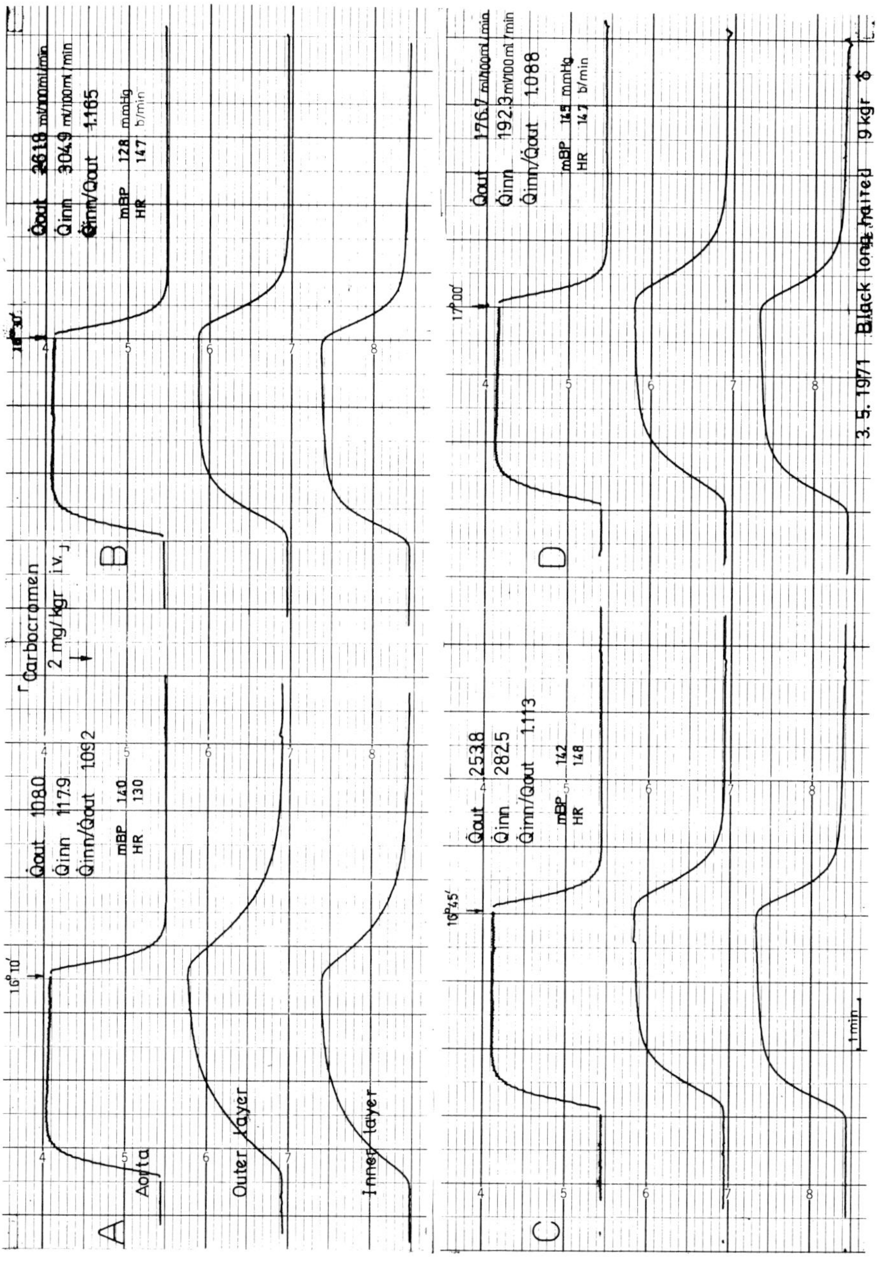
A
B
C
D
Aorta
Outer layer
Inner layer
Carbocromen
2 mg/kgr i.v.
1 min
3. 5. 1971 Black long haired 9 kgr ♂

mean aortic blood pressure are summarized in Figure 5. Isoproterenol infusion, in a dose of 0.3 μg/kg/min, caused a larger increase of $\dot{Q}_{out}$ than of $\dot{Q}_{inn}$. Intravenous propranolol injection, in a dose of 0.3 mg/kg, resulted in a decrease of both $\dot{Q}_{out}$ and $\dot{Q}_{inn}$, but the decrement was more marked in $\dot{Q}_{out}$. Norepinephrine infusion, in a dose of 1 μg/kg/min, caused a stronger increase of $\dot{Q}_{inn}$ than of $\dot{Q}_{out}$. This effect was clearly apparent, especially when a long time elapsed after measurement in each animal had started. Angiotensin infusion (N=9), in a dose of 0.2 μg/kg/min, caused no significant changes in myocardial blood flow. In hypoxemia induced by breathing 6.7% O_2 (N=9), both $\dot{Q}_{out}$ and $\dot{Q}_{inn}$ increased, but the ratio $\dot{Q}_{inn}/\dot{Q}_{out}$ decreased significantly. Furthermore, following carbocromen injection (N=9), $\dot{Q}_{out}$ and $\dot{Q}_{inn}$ increased significantly, with $\dot{Q}_{inn}$ showing a rather stronger increase than $\dot{Q}_{out}$.

The effects of heart rate on myocardial blood flow are shown in Figure 6, where the serial measurements are connected by straight lines. The series numbered from 1 to 5 were obtained in one dog, repeatedly, over a period of 7 hours. $\dot{Q}_{inn}$ increased only slightly in accordance with the increment of heart rate. Moreover, the blood flow values showed a gradual slight decrease on repetition of the serial measurements 1 to 4, while the arterial blood pressure at each given heart rate remained almost unchanged. On the other hand, $\dot{Q}_{out}$ showed a stronger dependency on the time lapse and the heart rate. In particular, it showed a tendency toward a gradual increase from the series 1 to 4. The last series (5*) was obtained after norepinephrine infusion. The $\dot{Q}_{inn}$ s of this series became larger than those in the preceding series, while $\dot{Q}_{out}$ s decreased. Accordingly, the ratio $\dot{Q}_{inn}/\dot{Q}_{out}$ showed a tendency to decrease in the course of serial measurements and, at the last serial measurement, it somewhat increased, as shown in Figure 7.

In order to observe the relationship between blood flow distribution and local O_2 consumption, the left coronary anterior descending artery was totally occluded and the decrease of the myocardial O_2 currents was recorded (Winbury, Howe, and Weiss, 1971). Examples of actual recordings are shown in Figure 8, where the *upper and lower curves* indicate the O_2 currents in outer and inner layers of left ventricular myocardium, respectively. Figure 8*A* was obtained under control conditions, with no pharmacological treatment except anesthesia. The O_2 current in the inner layer decreased slightly faster than in the outer layer at the occlusion. The averages and standard deviations at the period required for an 80% decrease in seven measurements in four dogs were 76 ± 36 and 41 ± 10 sec ($p<0.02$), respectively. Figure 8*B* shows the O_2 currents under isoproterenol infusion in the same animal. The O_2 current in the inner layer decreased more rapidly (28 sec for an 80% decrease) than under control conditons, while that in

Figure 4 (facing page). Examples of actual recordings, showing the long lasting effects of intravenous injection of carbocromen on myocardial blood flow.

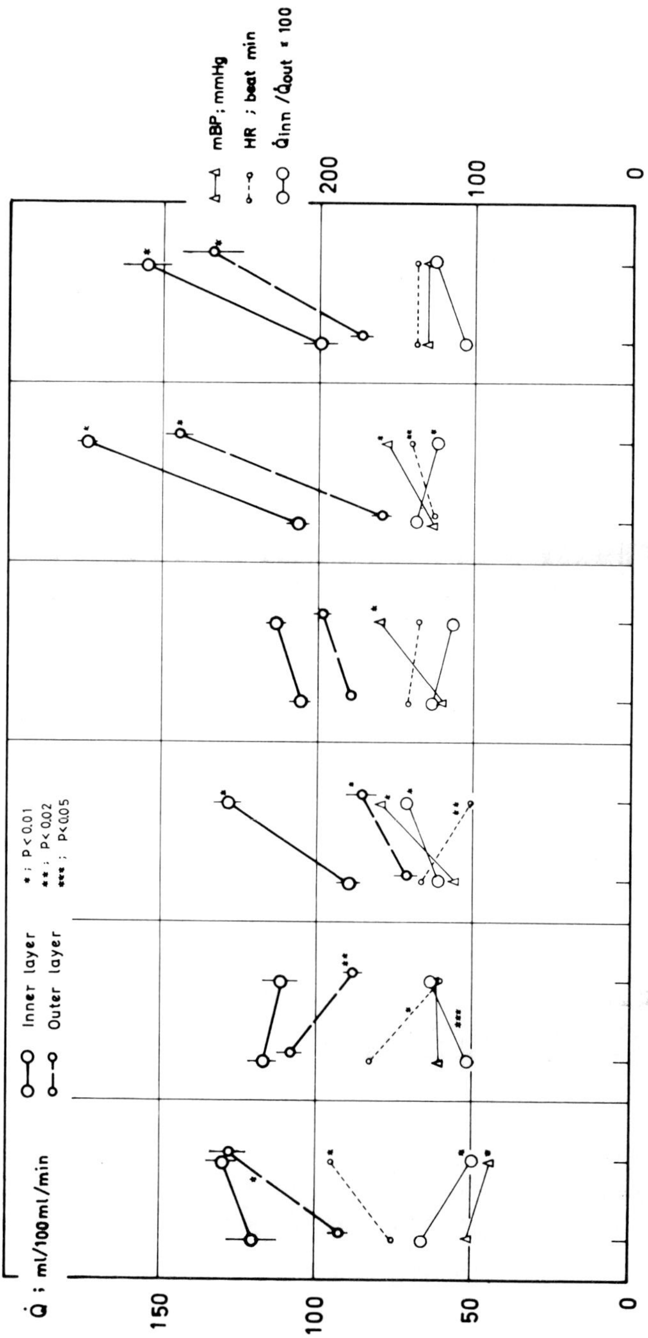

Figure 5. Mean values of $\dot{Q}_{out}$, $\dot{Q}_{inn}$, ratio $\dot{Q}_{inn}/\dot{Q}_{out}$, heart rate, and mean aortic blood pressure, under various treatments.

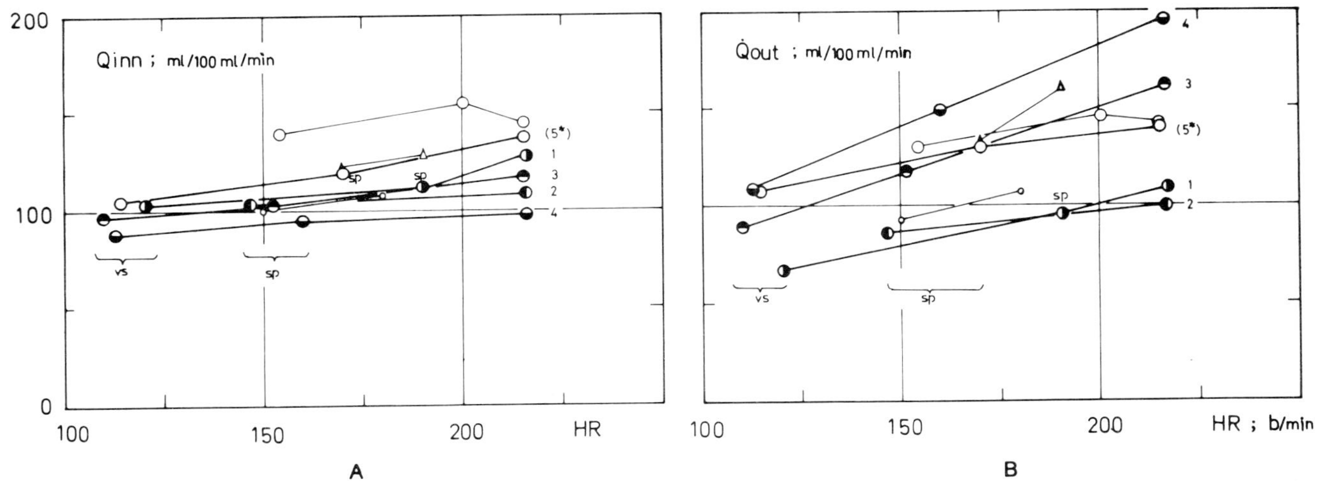

Figure 6. Effects of heart rate on myocardial blood flow in the inner (*A*) and outer (*B*) portions. Bradycardia was induced by vagal stimulation (vs), spontaneous heart rate (sp), or, for values not qualified by letters, by right atrial pacemaking.

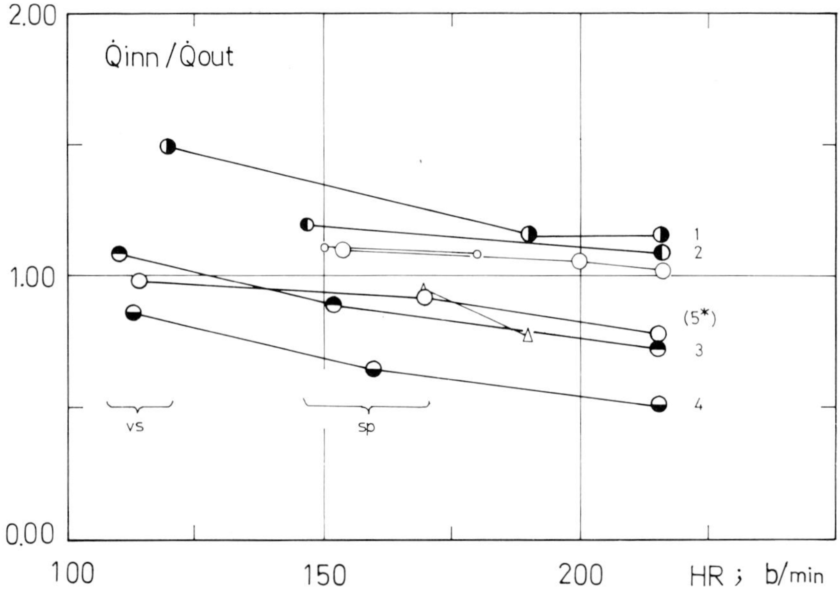

Figure 7. Effects of heart rate on ratio $\dot{Q}_{inn}/\dot{Q}_{out}$.

the outer layer showed a decay of almost the same rate as in Figure 8*A*. A similar result was obtained in another dog.

DISCUSSION

The needle-type Pt-H_2 electrode was introduced by Aukland, Bower, and Berliner in 1964. However, in the original type of electrode, the Pt-black was easily wiped off by the procedure of insertion into the tissue and the sensitivity of the electrode was diminished. In order to eliminate this disadvantage, the present recessed Pt electrodes were constructed. Since needle-type electrodes destroy the capillaries in tissue, the values of blood flow obtained may be more or less affected by the artifacts. However, when once placed in the outer and inner portions of the myocardium, they permit repeated observations of the changes of the blood flow distribution. Moreover, the $\dot{Q}$ values obtained in the middle portion of the left ventricular myocardium by means of this method agreed within a 10% error with the average myocardial blood flow measured by a catheter Pt-H_2 electrode placed in the coronary sinus (Ikeda, 1972). In the present study, the relative changes in the ratio $\dot{Q}_{inn}/\dot{Q}_{out}$ caused by various treatments agreed with those published by previous authors (Love and Burch,

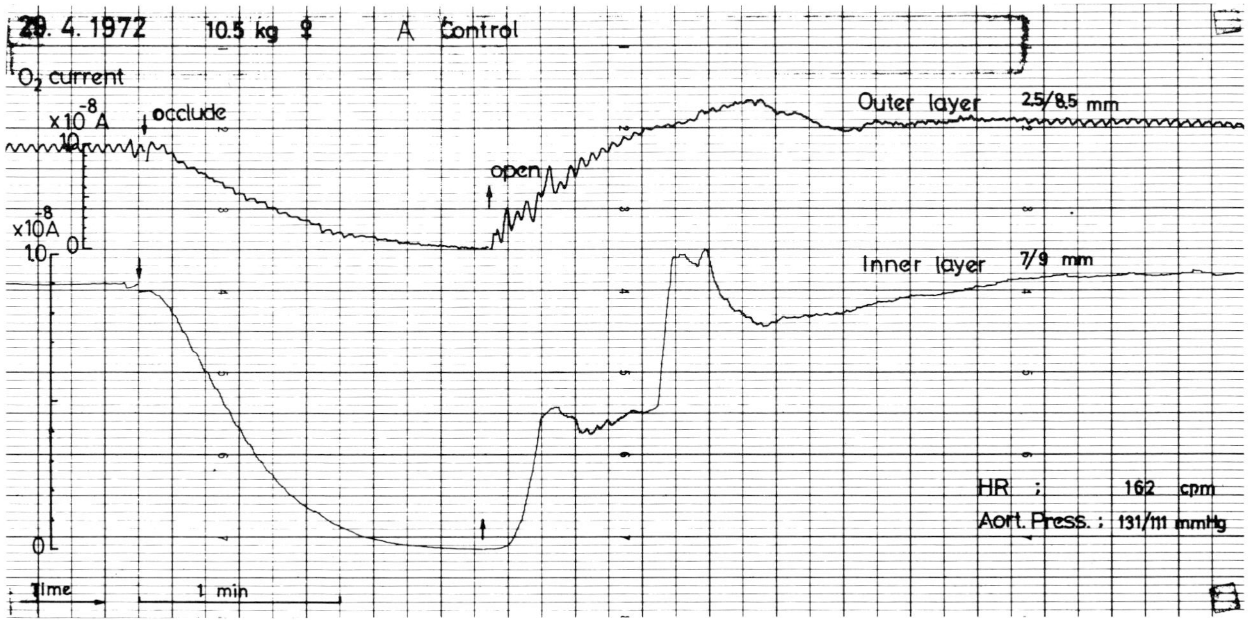

Figure 8A. Control examples of O_2 currents recorded by Pt electrodes in the outer and inner portions of left ventricular myocardium, following coronary occlusion.

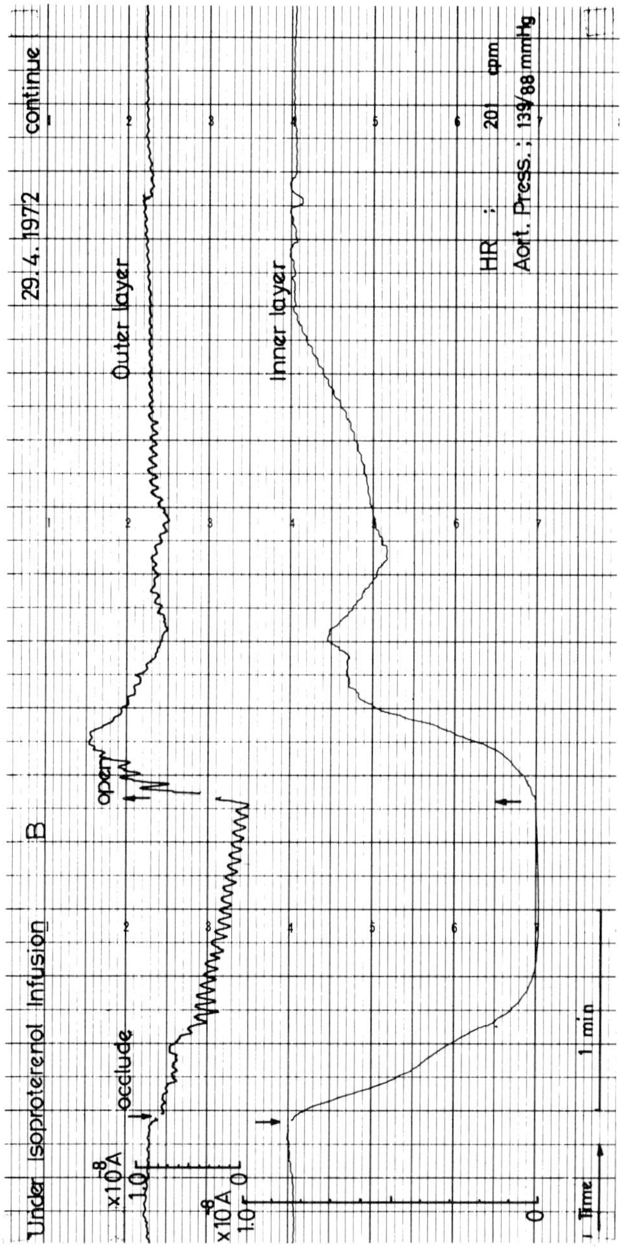

Figure 8B. Under isoproterenol infusion.

1957; Love and Tyler, 1965; Moir and DeBra, 1967; Buckbery *et al.*, 1972). These findings showed the applicability of the method in this field.

Under control conditions, $\dot{Q}_{inn}$ exceeded $\dot{Q}_{out}$ in most instances (mean and SD = 112.4 ± 34.1 and 98.7 ± 31.8 ml/100 ml/min, respectively, N=46, p<0.02) and, on the average, the ratio $\dot{Q}_{inn}/\dot{Q}_{out}$ was 1.14. But this ratio is readily altered by several treatments and by prolongation of the experiment, as was seen in the present study. In view of the decrement of O_2 current in the coronary occlusion test, O_2 consumption in the inner layer seemed to be larger than in the outer layer under control conditions. Under normal conditions, therefore, the myocardial blood flow was probably so distributed that the inner layer could be supplied with an increased blood flow and more oxygen. However, such a balance between the local blood supply and oxygen demand was easily disturbed by various treatments, as is seen in the case of tachycardia caused by isoproterenol infusion.

In Figure 6, it is seen that $\dot{Q}_{out}$ increased gradually when $\dot{Q}_{inn}$ decreased somewhat during the course of a long term experiment. Moreover, in two dogs which were accidentally starved and exhausted for 1 week, $\dot{Q}_{inn}$ was found to be considerably small (mean = 38 ml/100 ml/min), while $\dot{Q}_{out}$ was, on the other hand, remarkably large (mean = 210 ml/100 ml/min). These observations suggested that some changes in the physiological conditions of the dogs could cause a decrement of the ratio $\dot{Q}_{inn}/\dot{Q}_{out}$ without any pharmacological treatment and that $\dot{Q}_{out}$ could sometimes increase in such a manner as to compensate for the decrease of $\dot{Q}_{inn}$.

REFERENCES

AUKLAND, K., B. F. BOWER, and R. W. BERLINER. 1964. Measurement of local blood flow with hydrogen gas. Circulat. Res. 14: 164–187.

BUCKBERY, G. D., D. E. FIXLER, J. P. ARCHIE, and J. I. E. HOFFMAN. 1972. Experimental subendocardial ischemia in dogs with normal coronary arteries. Circulat. Res. 30: 67–81.

IKEDA, N. 1972. Effect of hypoxia and vasoactive drugs on regional myocardial blood flow. (In Japanese.) Jap. J. Geriat. 9: 345–355.

KETY, S. S., and C. F. SCHMIDT. 1945. The determination of cerebral blood flow in man by the use of nitrous oxide in low concentration. Am. J. Physiol. 143: 53–66.

KOYAMA, T., and Y. MARUTANI. 1971. A hydrogen catheter electrode for the determination of blood flow through organ tissue and coronary blood flow under continuous hypoxia. Jap. J. Physiol. 21: 209–228.

LOVE, W. D., and G. E. BURCH. 1957. Differences on the rate of Rb^{86} uptake by several regions of the myocardium of control dogs and dogs receiving 1-norepinephrine or pitressin. J. Clin. Invest. 36: 479–484.

LOVE, W. D., and M. D. TYLER. 1965. Effect of hypoxia and hypercapnea on regional distribution of myocardial blood flow. Am. J. Physiol. 208: 1211–1216.

MOIR, T. W., and D. W. DeBRA. 1967. Effect of left ventricular hypertension, ischemia, and vasoactive drugs on the myocardial distribution of coronary flow. Circulat. Res. 21: 65–74.

NITZ, R. E., und E. PÖTZSCH. 1963. 3-(-Diäthylamino-äthyl)-4-methyl-7-carbäthoxy-methoxy-2-oxo-(1,2 chromen), ein Präparat mit spezifischer und lang anhaltender coronargefässerweiternder Wirkung. Arzneim.-Forsch. 13: 243–251.

SIVER, I. A. 1965. Some observations on the cerebral cortex with an ultramicro, membrane-covered, oxygen electrode. Med. Electron. Biol. Engng. 3: 377–387.

WINBURY, M. M., B. B. HOWE, and H. R. WEISS. 1971. Effect of nitroglycerin and dipyridamole on epicardial and endocardial oxygen tension–Further evidence for redistribution of myocardial blood flow. J. Pharmac. Exper. Therap. 176: 184–199.

Mailing address:
Dr. Tomiyasu Koyama,
Research Institute of Applied Electricity,
Hokkaido University, N-12 W-16 Sapporo (Japan).

Recent Advances in Studies on
Cardiac Structure and Metabolism Volume 10
The Metabolism of Contraction
Edited by Paul-Emile Roy and George Rona

MASS SPECTROMETRY FOR THE MEASUREMENT OF INTRAMYOCARDIAL GAS TENSIONS: METHODOLOGY AND APPLICATION TO THE STUDY OF MYOCARDIAL ISCHEMIA[1]

S. F. KHURI, J. O'RIORDAN,
J. T. FLAHERTY, R. K. BRAWLEY,
J. S. DONAHOO, and V. L. GOTT

Department of Surgery and Medicine,
The Johns Hopkins University School of Medicine,
Baltimore, Maryland, USA

SUMMARY

The methodology for use of the mass spectrometer for the measurement of intramyocardial gas tensions in the canine preparation is described. Baseline studies were carried out initially in 36 animals, and control levels for myocardial oxygen tension and myocardial carbon dioxide tension were 19 mm Hg (S.D. 6 mm Hg) and 43 mm Hg (S.D. 10 mm Hg), respectively. Myocardial oxygen tension was not altered significantly by varying the arterial oxygen tension between 65 and 300 mm Hg. However, myocardial carbon dioxide tension increased linearly with increased arterial carbon dioxide tension. In 15 dogs placed on total cardiopulmonary bypass, a perfusion pressure 40–60 mm lower than the control mean arterial pressure resulted in myocardial ischemia with a decrease in myocardial oxygen tension and an increase in myocardial carbon dioxide tension. A subsequent increase in perfusion pressure to control levels resulted in resolution of ischemia and return of myocardial oxygen and carbon dioxide tensions to their control level. In another series of open-chest dogs on cardiopuljonary bypass, a proximal constriction applied to the left coronary circumflex artery resulted in a marked decrease in myocardial oxygen tensions and a marked increase in myocardial carbon dioxide tensions in the region supplied by the constricted vessel. In yet another series of open-chest dogs, it was found that incremental decreases in coronary flow established by constriction of the circumflex artery resulted in an exponential increase in both myocardial carbon dioxide tensions and ST-segment elevation as determined by a 25-gauge multi-contact plunge electrode placed in the posterior left ventricular wall. It appears that mass spectrometry techniques for evaluating myocardial ischemia have several advantages over myocardial biopsy techniques for assay of ATP and lactate, and also over the technique of coronary sinus lactate determination.

INTRODUCTION

In 1972, Brantigan and his coworkers reported the use of a Teflon membrane for the measurement of blood and intramyocardial gas tensions by mass spec-

[1] Supported by United States Public Health Service Grant HL-09997 and The Marian S. Hankin Cardiovascular Surgical Research Fund.

trometry (Brantigan, Gott, and Martz, 1972a). This technique provided an alternative to the polarographic determination of myocardial oxygen tension and also provided a method for measuring intramyocardial carbon dioxide. Gardner *et al.* (1971) reported myocardial gas tensions determined in human hearts during aortocoronary artery bypass surgery. This latter study demonstrated that, in regions supplied by a stenosed coronary artery, there was an increase in myocardial oxygen tension and a decrease in myocardial carbon dioxide tension following aortocoronary bypass. Brantigan *et al.* (1972b) also reported that, in canine hearts on total cardiopulmonary bypass, significant decreases in myocardial oxygen tension and increases in myocardial carbon dioxide tensions developed during periods of anoxic arrest. More recently, Loisance and Owens (1974a, b) reported the use of the mass spectrometer for study of the effects of pharmacological agents, such as nitroglycerine, on myocardial gas tensions under conditions of ischemia.

The present study was designed to test the hypothesis that decreases in intramyocardial oxygen tensions and increases in carbon dioxide tensions are indicators of myocardial ischemia and that mass spectrometry can provide a useful experimental tool for the study of the pathophysiology of regional myocardial ischemia.

METHODS

Mass spectrometer probes were inserted into the left ventricular wall of 95 dogs weighing 20–25 kg. The animals were anesthetized with sodium thiopental and succinylcholine or alpha-chloralose and ventilated with a Harvard respirator using room air. A left or right thoracotomy was then performed. The mass spectrometer probes utilized in these studies consisted of an 18-inch length of 22-gauge stainless steel tubing, slotted near its distal end and covered with heat-shrinkable Teflon. The Teflon was shrunk tightly over the steel tubing except for the distal 19 mm, which remained in an expanded state, thus providing a sensing area for the measurement of myocardial gas tensions (Fig. 1*A*). The distal tip of the probe was sealed over a 4-mm length of stainless steel wire with a smooth, rounded tip. The proximal end of the probe was connected to the mass spectrometer.

The vacuum exerted by the mass spectrometer withdraws gas molecules across the Teflon membrane at a rate of 5×10^{-6} ml/sec. The gases are analyzed by the mass spectrometer according to their molecular weight, and a continuous recording of intramyocardial oxygen and carbon dioxide tensions is obtained. The $1/e$ response time of the system is 1.5 min for oxygen tension and 3 min for carbon dioxide tension.

The probes were attached to the vacuum system of the mass spectrometer 12–18 hours prior to each experiment and were calibrated before each experi-

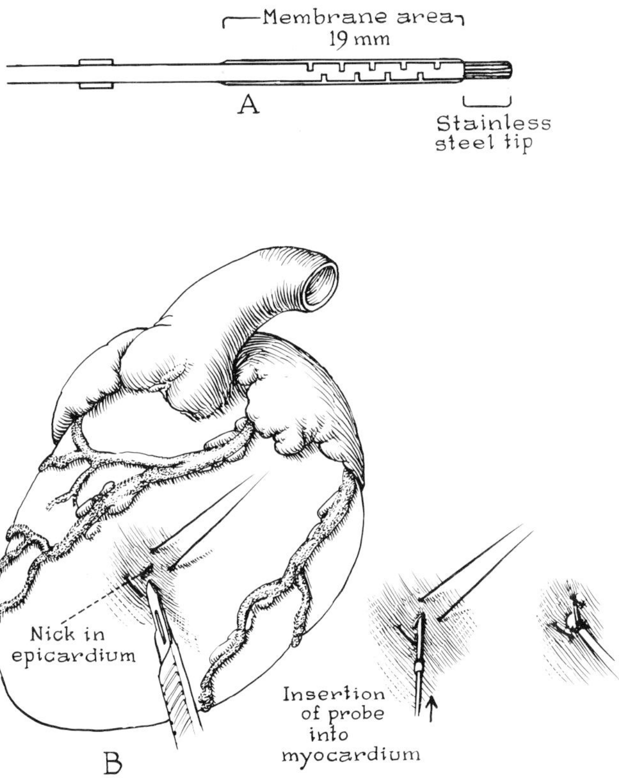

Figure 1. A: schematic drawing of the mass spectrometer probe. *B:* method of insertion of probe.

ment by bubbling a known oxygen and carbon dioxide gas mixture through water at 37 ℃ in a glass tonometer. When the experimental protocol required the use of hypothermia, it was necessary to calibrate the system throughout the range of temperatures used, because of the influence of this parameter on the measurement of myocardial gas tensions. Probes were used for a single study and then discarded. The probes were inserted into the myocardium as atraumatically as possible to avoid the formation of hematoma around the sensing area. In early studies, hematoma formation was relatively common with one significant hematoma for every four probe insertions. By inserting the short length of a stainless steel wire into the distal end of the probe, the incidence of hematoma was reduced to 1 in every 10 insertions. If a hematoma was observed at the conclusion of a study, the results of the study were excluded from further analysis. Movement of the probe within the myocardium was prevented by placing a sleeve of heat-shrinkable Teflon proximal to the sensing area of the probe. This sleeve served as indicator of the distance to the proximal end of the sensing surface and also provided a ridge for an anchoring suture to prevent outward slippage of the probe (Fig. 1). Probe position in the myocardium was determined at the end of each experiment by direct visualization.

Studies were carried out to determine the interrelationship between myocardial gas tensions and 1) arterial gas tensions; 2) mean arterial pressure; 3) myocardial oxygen consumption; and 4) intramyocardial ST-segment changes. In addition, intramyocardial oxygen and carbon dioxide tensions, measured both in the presence of and absence of systemic heparinization, were compared to determine if there was an adverse effect on the gas tensions as a result of heparinization.

RESULTS

Baseline studies

The normal, control myocardial oxygen tension obtained in 36 animals was 19.0 mm Hg (S.D. 6 mm Hg) and the normal, control myocardial carbon dioxide tension was 43 mm Hg (SD 10 mm Hg). In 28 animals, one probe was placed in the posterior left ventricular wall and a second probe was placed in the anterior wall. There was no significant difference in the oxygen or carbon dioxide tensions in these two regions of myocardium. Dividing the myocardial wall into thirds (subendocardial, middle, and subepicardial regions) revealed no significant difference in oxygen and carbon dioxide tensions between these layers.

Myocardial oxygen and carbon dioxide tensions with and without systemic anticoagulation are presented in Figure 2. Sodium heparin, 2 mg/kg body weight, given prior to the insertion of the probes into the myocardium resulted in a mean oxygen tension of 29 mm Hg (± SE 2 mm Hg), despite there being no evidence of visible hematoma around the probe tip. When visible hematoma was present around the probe tip, oxygen tensions in excess of 40 mm Hg were noted. When anticoagulation was instituted after insertion of the probe, a lower incidence of probe tip hematoma was noted. There was no significant difference in the carbon dioxide tension with and without systemic anticoagulation.

Arterial blood gas tensions were monitored simultaneously with intramyocardial gas tensions. Myocardial oxygen tension was not altered significantly by varying the arterial oxygen tension between 65 and 300 mm Hg; but when the arterial oxygen tension exceeded 300 mm Hg, a significant change in intramyocardial oxygen tension was noted (Fig. 3). In contrast, myocardial carbon dioxide tension increased linearly with increasing arterial carbon dioxide tension (Fig. 4).

While control mean blood pressure varied from animal to animal, no significant correlation was found between the control mean arterial pressure and control myocardial gas tensions. Increasing the mean arterial pressure above control

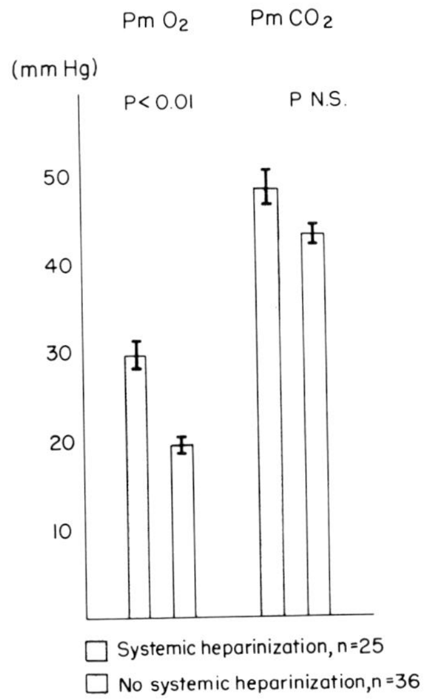

Figure 2. Effect of systemic heparinization on myocardial oxygen (PmO_2) and carbon dioxide ($PmCO_2$) tensions in regions supplied by constricted and non-constricted coronary Student *t*-test.

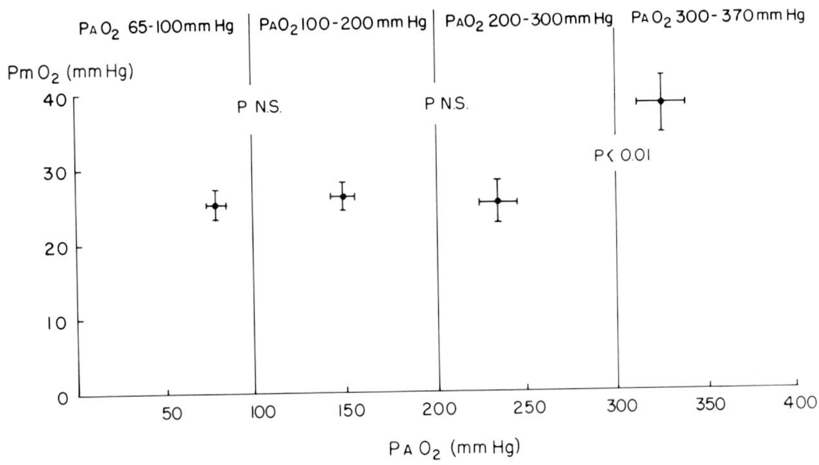

Figure 3. Effect of arterial oxygen tension (PAO_2) on myocardial oxygen tension (PmO_2) (means ± standard error).

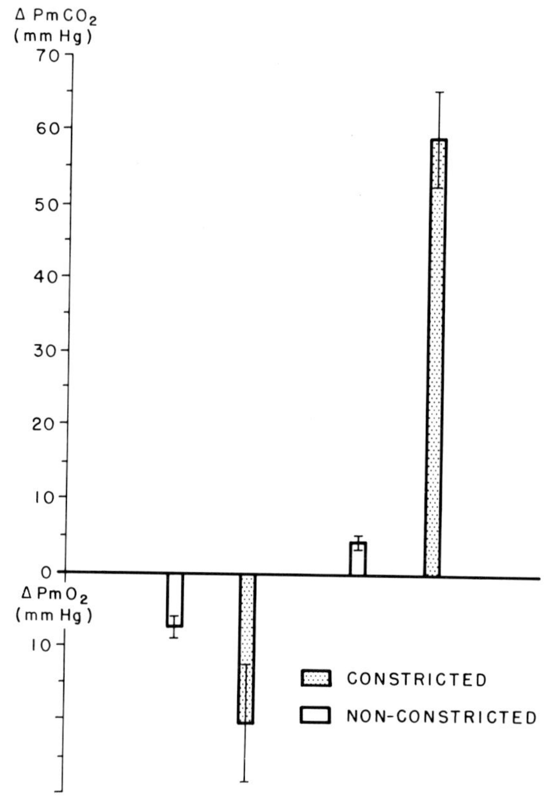

Figure 7. Effect of increasing heart work on myocardial oxygen (PmO_2) and carbon dioxide ($PmCO_2$) tensions in regions supplied by constricted and nonconstricted coronary arteries.

measured in a region supplied by the constricted circumflex coronary artery and also in a region supplied by the nonconstricted left anterior descending coronary artery. Oxygen consumption was determined by collecting coronary sinus blood and measuring arterial-venous oxygen difference. Clamping the ventricular vent, thus increasing left ventricular work, resulted in an increase in oxygen consumption. In the region supplied by the nonconstricted vessel, a minimal decrease in myocardial oxygen tension and a minimal increase in myocardial carbon dioxide tension was noted (Fig. 7). In contrast, a marked decrease in myocardial oxygen tension and a marked increase in myocardial carbon dioxide tension was noted in the regions supplied by the constricted vessel, indicating the development of regional myocardial ischemia.

In order to validate the hypothesis that increases in myocardial carbon dioxide tension and decreases in myocardial oxygen tension are indicative of myocardial ischemia, an independent indicator of myocardial ischemia was utilized. In seven open-chest dogs, a 25-gauge multicontact plunge-electrode was placed in the posterior left ventricular wall in the vicinity of the mass spectrometer probe, and unipolar ST-segment changes were measured simultaneously with myocardial gas tensions. Variable degrees of myocardial ischemia were produced by progressively increasing the constriction of the left circumflex coronary artery. Incremental decreases in coronary flow were associated with an exponential increase in both myocardial carbon dioxide tension and ST-segment elevation (Flaherty *et al.*, 1974). With minimal reduction (6%) in the mean coronary flow, a 13-mm Hg (SE ± 4 mm Hg) decrease in myocardial oxygen tension was noted. Maximal reduction (80%) in mean coronary flow resulted in only a 7-mm Hg (SE ± 5 mm Hg) further decrease in myocardial oxygen tension.

DISCUSSION

Mass spectrometry provides a method for continuously monitoring regional intramyocardial oxygen and carbon dioxide tensions. While no method was previously available for the measurement of myocardial carbon dioxide tension, the polarographic method was used for the measurement of intramyocardial oxygen tension. Measurement of myocardial oxygen tension by the polarographic technique revealed a gradient in myocardial oxygen tension, with higher tensions (18 mm Hg) in the subepicardial compared with (10 mm Hg) the subendocardial layers (Moss, 1968). In the present study using mass spectrometry, the normal myocardial oxygen tension was found to be slightly higher (19 mm Hg) (SD ± 8 mm Hg), and no epicardial-to-endocardial oxygen tension gradient was noted. Myocardial oxygen tensions obtained by mass spectrometry were reproducible within each experiment when hemodynamic parameters remained stable. Following removal of an ischemic stimulus, such as proximal coronary artery constriction, myocardial oxygen tension consistently returned to the same control level. A maximal ischemic stimulus, such as temporary occlusion of a coronary artery, resulted in a decrease in myocardial oxygen tension to levels of 3–4 mm Hg, comparable to maximally ischemic levels obtained by the polarographic technique.

Several factors have been found to influence the measurement of myocardial gas tensions. These included: 1) visible hematoma around the sensing tip; 2) systemic anticoagulation with heparin; and 3) arterial oxygen and carbon dioxide tensions. The addition of a round-ended stainless steel tip has reduced the incidence of hematoma to an acceptable level compared to the original probe design discussed by Brantigan *et al.* (1972a). While systemic heparinization resulted in higher control oxygen tensions, gas tension response to an ischemic stimulus was appropriate and return to baseline levels was consistently observed.

While myocardial oxygen tension did not change significantly across a broad range of physiological arterial oxygen tensions (65–300 mm Hg), it was necessary for interpretation of myocardial oxygen tension that the arterial oxygen tension be maintained within this range. Since changes in myocardial carbon dioxide tension were linearly related to arterial carbon dioxide tension, it is important that the arterial carbon dioxide tension be monitored and maintained within the physiological range, by appropriately adjusting the rate and volume of ventilation.

Important to the status of cardiac metabolism is the balance between the supply of oxygen to the myocardium and the myocardial oxygen demand. Coronary perfusion pressure and arterial oxygen tension are important determinants of the supply of oxygen. Myocardial oxygen demand is strongly influenced by the level of wall stress or tension during contraction. A rise in myocardial carbon dioxide tension and a fall in myocardial oxygen tension, indicating ischemia, occurred either under conditions of decreased perfusion pressure or flow, or with increases in wall stress with constant constriction.

When a coronary artery of a dog on cardiopulmonary bypass was "critically stenosed," perfusion pressure appeared to be the sole determinant of myocardial gas tensions in regions supplied by the stenosed vessel (Khuri *et al.*, 1974). When cardiopulmonary bypass perfusion pressure was allowed to fall below control levels, significant decreases in myocardial oxygen tension and increases in myocardial carbon dioxide tension were observed. When adequate coronary perfusion pressure was re-established, the myocardial gas tensions returned to control levels. These findings emphasize the importance of perfusion pressure in the maintenance of adequate myocardial oxygenation and the prevention of myocardial ischemia in patients during total cardiopulmonary bypass, particularly those patients with significant obstructive coronary artery disease.

When cardiac work and, thus, oxygen demand was increased by raising the intraventricular pressure of canine hearts on cardiopulmonary bypass, myocardial carbon dioxide tension increased and myocardial oxygen tension decreased in regions supplied by both constricted and nonconstricted vessels. However, gas tension changes in regions supplied by constricted vessels were of greater magnitude, reflecting the development of regional ischemia. While myocardial oxygen consumption was noted to increase, myocardium supplied by the stenosed vessel utilized less additional oxygen than myocardium supplied by unstenosed vessels. The decrease in myocardial oxygen tension reflects decreased availability of oxygen, and the increase in carbon dioxide tension suggests increased anaerobic metabolism with the associated increase in lactic acid production. Buffering of free hydrogen ions by the bicarbonate buffer system would result in the local production of carbon dioxide. An additional mechanism for increasing myocardial carbon dioxide tension under conditions of decreased coronary flow may be the decreased washout of these metabolic end products, including carbon dioxide, resulting in a local accumulation.

The mass spectrometry technique has certain advantages over methods currently available for the assessment of myocardial metabolism: 1) myocardial biopsy techniques for assay of ATP/ADP and lactate/pyruvate ratios provide intermittent determinations, while the mass spectrometric measurement of myocardial gas tensions is continuous; 2) coronary sinus lactate determinations reflect the metabolism of both ischemic and nonischemic regions, while myocardial gas tensions better reflect regional metabolic status. Minor disadvantages of mass spectrometry include: 1) a degree of local myocardial injury which is inherent in the placement of the mass spectrometer probe; and 2) a moderately slow response time of the mass spectrometer system to sudden changes in gas tension.

Using the polarographic technique for the determination of myocardial oxygen tension, Sayen *et al.* (1961) reported a significant correlation between intramyocardial ST-segment changes and oxygen tension. The results of our studies confirm their findings and, furthermore, demonstrate the direct correlation between unipolar intramyocardial ST-segment changes and myocardial carbon dioxide tension. Myocardial oxygen tension appears to be a less sensitive indicator of myocardial ischemia than myocardial carbon dioxide tension, since approximately two-thirds of the maximal decrease in oxygen tension occurs with minimal (6%) reduction in mean coronary flow. Further reduction in mean coronary flow to 20% of control levels was associated with little further decrease in oxygen tension, while myocardial carbon dioxide tension and ST-segment changes increased exponentially. Thus, myocardial carbon dioxide tension and unipolar ST-segment changes in combination with myocardial oxygen changes appear to provide more sensitive indicators of myocardial ischemia than myocardial oxygen tension alone.

While further studies of the role and pathogenesis of intramyocardial carbon dioxide tension in ischemia are needed for complete understanding, the results of this study suggest that the ability of the mass spectrometer to continuously monitor local interstitial oxygen and carbon dioxide tension provides a new and promising tool for the study of regional myocardial ischemia.

REFERENCES

BRANTIGAN, J. W., V. L. GOTT, and M. N. MARTZ. 1972a. A teflon membrane for measurement of blood and intramyocardial gas tensions by mass spectroscopy. J. Appl. Physiol. 32: 276–282.

BRANTIGAN, J. W., A. M. PERNA, T. J. GARDNER, and V. L. GOTT. 1972b. Intramyocardial gas tensions in the canine heart during anoxic cardiac arrest. Surg. Gynec. Obstet. 134: 67–72.

FLAHERTY, J. T., S. F. KHURI, J. B. O'RIORDAN, B. PITT, and V. L. GOTT. 1974. Quantitation of myocardial ischemia by intramyocardial carbon dioxide, oxygen, and electrograms. Fed. Proc. 33: 395 (Abstr.)

GARDNER, T. J., J. W. BRANTIGAN, A. M. PERNA, H. W. BENDER, R. K. BRAWLEY, and V. L. GOTT. 1971. Intramyocardial gas tensions in the human heart during coronary artery-saphenous vein bypass. J. Thorac. Cardiovasc. Surg. 62: 844–850.

KHURI, S. F., R. K. BRAWLEY, J. S. DONAHOO, B. PITT, and V. L. GOTT. 1974. The effect of cardiopulmonary bypass perfusion pressure on myocardial gas tension in the presence of coronary stenosis. (In press).

LOISANCE, D., et G. OWENS. 1974a. Perspective overtes par l'etude simultanee du debit de perfusion et des tensions myocardiques en O_2 et CO_2. Arch. Mal. Coeur 67: 115–119.

LOISANCE, D., et G. OWENS. 1974b. Etude pharmacologique par spectrometrie de masse de la perfusion myocardique. 1. Effects de la nitroglycerine. Coeur Med. Intern. 13: 125–131.

MOSS, A. J. 1968. Intramyocardial oxygen tension. Cardiovasc. Res. 3: 314–318.

SAYEN, J. J., G. PEIRCE, A. H. KATCHER, and W. F. SHELDON. 1961. Correlation of intramyocardial electrocardiograms with polarographic oxygen and contractility in the non-ischemic and regionally ischemic left ventricle. Circulat. Res. 9: 1268.

Mailing address:
Vincent L. Gott, M.D.,
Department of Surgery, The Johns Hopkins Hospital,
Baltimore, Maryland 21205 (USA).

The Cardiovascular System and Exercise

Recent Advances in Studies on
Cardiac Structure and Metabolism Volume 10
The Metabolism of Contraction
Edited by Paul-Emile Roy and George Rona

RESTING BRADYCARDIA OF EXERCISE TRAINING: A CONCEPT BASED ON CURRENTLY AVAILABLE DATA

HENRY S. BADEER

Department of Physiology and Pharmacology, Creighton University School of Medicine, Omaha, Nebraska, USA

SUMMARY

The classic view that the resting bradycardia of exercise training is due to an increase in the tonic discharge of the vagi is no longer tenable. If it were true, full doses of atropine would accelerate the heart to a greater extent in the athletic animal. All investigators agree that the reverse is the case. Some workers report a decrease in cardiac sympathetic influence after training. This would cause a preponderance of vagal influence which may decrease to a lesser extent than the sympathetics or may not change at all. Several studies have demonstrated that training causes a distinct slowing of the intrinsic rate of the pacemaker.

It is concluded that two factors operate to lower the resting heart rate after athletic training: 1) slowing of the intrinsic rate of the pacemaker, 2) Increase in cholinergic predominance on the pacemaker frequency as a result of a decrease in adrenergic influence. Vagal influence per se seems either to remain the same or to decrease slightly in the athlete compared to the nonathlete. It appears that slowing of the intrinsic rate plays a more important role, a factor that has been hitherto overlooked. The mechanisms underlying these changes remain obscure.

REVIEW OF DATA

The most widely accepted view on the mechanism of resting bradycardia in exercise-trained animals and man is that it is due to an increase in cardiac vagal tone (Davson and Eggleton, 1968; Marshall and Shepherd, 1968; Mountcastle, 1968; Detweiler, 1973). As far as I could ascertain, this view seems to have originated from the observation of early physiologists that vagotomy in athletic animals, such as the hare, causes a much greater increase in heart rate than vagotomy in nonathletic animals, such as the rabbit. Clark (1927) reported that, in the hare, vagal section increases resting heart rate from 60–70 to 264/min, whereas in the rabbit of the same size the resting heart rate increased from 205 to 321 beats/min. Presumably, these data were obtained from anesthetized animals. The number of experimental animals in each group was not stated. A similar observation in such animals has been reported by Hollmann (1959). Clark stated that "the low pulse rates of athletic animals are not due to the fundamental frequency of the heart's pacemaker being low, but are due to the powerful vagal control." He presented no data or references to support this conclusion. We shall see later that this opinion may be seriously questioned on the basis of more recent data. In fact, it may well be that the major cause of the bradycardia of training is a reduction in the intrinsic rate of the normal pacemaker.

The observations in the hare and the rabbit, coupled with the effects of vagotomy in other animals such as the racing dog, horse, etc., have been extrapolated to humans to explain the well known bradycardia of trained athletes. Furthermore, this view has found its way into the standard textbooks of physiology. Unfortunately, once the concept gains a foothold in textbooks, it becomes so established that great difficulty is then encountered in correcting it.

Observations in man, as early as 1921, did not clearly support the concept of increased vagal tone to explain the bradycardia of the athlete. Herxheimer (1921) reported no significant increase in the heart rate of athletes after the injection of 1 mg of atropine sulfate subcutaneously, whereas the same dose in nonathletes was reported to cause a distinct cardioacceleration. Kauf (1926) repeated and extended these studies and concluded that, whereas 1 mg of atropine in nonathletic individuals causes a distinct cardioacceleration, the same dose in athletic subjects with bradycardia had no significant effect on cardiac frequency. On the other hand, 2 mg of atropine injected into athletes caused a distinct cardioacceleration. Unfortunately, he did not administer the 2-mg dose to nonathletic subjects for comparison. No statistical analyses were carried out in these investigations. It has been argued that the studies of Herxheimer and Kauf using 1 mg of atropine are inconclusive, partly because these doses are inadequate to block vagal endings (Hall, 1963). Although this is a valid objection, the observation that the same dose in nonathletic subjects causes a distinct cardioacceleration raises some doubt that the bradycardia of athletes is primarily due to increased vagotonia.

More recently, Raab and his associates (1960) injected atropine intravenously (0.025 mg/kg, which would amount to 1.75 mg for a 70-kg man) into physically active and nonactive human subjects and found no significant difference in the degree of acceleration of the heart in the two groups under resting conditions (athletes, +38.8 beats/min, versus nonathletes, +39.2 beats/min).

A study of this problem in experimental animals was carried out by Tipton and Taylor (1965), using much higher doses of atropine. These workers analyzed the effect of intraperitoneal injection of 1 mg of atropine per kg body weight on the resting heart rate of exercise-trained and nontrained rats. Although the resting heart rates in the two groups were not significantly different (trained, 310 beats/min ± 6 SE, versus untrained, 315 beats/min ± 4 SE), Tipton and Taylor found that after atropine the trained rats had significantly *less* cardiac acceleration than the nontrained animals. Herrlich, Raab, and Gigee (1960) have shown that trained rats have significantly more acetylcholine in atrial tissue than nontrained. According to Burn and his associates (Bülbring and Burn, 1949; Burn and Walker, 1954), atrial tissue can synthesize acetylcholine; Tipton and Taylor postulated that there are two sources of acetylcholine in the atria: 1) neural and 2) non-neural from atrial muscle. Exercise training was said to increase the non-neural. According to the hypothesis of Tipton and Taylor (1965), the reduced cardiac acceleration after atropine in trained rats is due to a greater availability of myocardial acetylcholine to compete with atropine for the receptor sites. This explanation appears unsatisfactory if we assume that a "full" dose of atropine was used and that all the receptor sites were occupied by the drug. Furthermore, there is no reason to believe that the "receptor sites" can distinguish between neural and non-neural acetylcholine, if such a dual source of acetylcholine is indeed the case. Tipton, Barnard, and Tharp (1966), have looked into the possibility of changes in acetylcholine esterase activity in atrial tissue after training. They found no difference in such activity between trained and nontrained rats. Ekström (1974) found an increased activity of the enzyme that synthesizes acetylcholine (choline acetyltransferase) in atrial tissue of exercise-trained rats and believes that high cholinergic activity is one of the causes of the bradycardia of training. Our contention is that, in training, if increased acetylcholine is the cause of the bradycardia no matter what its source (increased vagal impulses, non-neural acetylcholine synthesis, increased activity of choline acetyltransferase, etc.), a "full" dose of atropine that blocks all the receptor sites should produce a greater cardioacceleration. This does not seem to be the case if we accept that maximal doses of atropine had been used. Perhaps higher doses of atropine should be used to settle this question. In the dog, 0.1 mg of atropine per kg body weight is adequate to completely abolish cardiac inhibition induced by maximal electrical stimulation of vagus.

Frick and his coworkers (1967) studied the bradycardia of physical training in human subjects with combined autonomic blocking drugs. They injected atropine sulfate (2.4 mg) together with propranolol (0.1 mg/kg) intravenously. In nonathletes, the resting heart rate increased from 89 to 106 beats/min (+17 beats), whereas in athletes it rose from 63 to 91 beats/min (+28 beats). It was concluded that the resting bradycardia of athletes is due to increased parasympathetic activity. This type of experimental study can indicate only the net influence of the cardiac autonomic nerves on the pacemaker, assuming that the doses used fully block both nerves and have no direct effect on the "intrinsic" frequency of the pacemaker. If these assumptions are correct, the data of Frick *et al.* indicate that there are two factors causing the bradycardia in the athlete: 1) slowing of the intrinsic rate of the pacemaker (nonathlete = 106 beats/min versus athlete = 91 beats/min), 2) augmentation of vagal inhibitory influence on pacemaker frequency (28 beats in athletes versus 17 beats in the nonathletes). Here, some confusion has prevailed. Careful analysis of autonomic control of pacemaker frequency would indicate that such a cholinergic augmentation can be achieved in several possible ways:

1. Increased cholinergic influence with no change in adrenergic.
2. Increased cholinergic influence with decreased adrenergic.
3. Increased cholinergic influence as well as adrenergic, but cholinergic increasing more than adrenergic.
4. No change in cholinergic but a decrease in adrenergic.
5. Decrease in both cholinergic and adrenergic, but a greater decrease in adrenergic.

Note that increase, decrease, and no change are in comparison with the normal influence of these nerves on the pacemaker in the nonathlete.

The recent work of Lin and Horvath (1972) throws some light on this problem. These workers attempted to "quantitate" the neural influence on the pacemaker in trained and untrained rats by *separately* blocking these nerves with atropine and propranolol. The vagal "tone" was estimated by 1 mg of atropine per kg body weight intraperitoneally, maximal effect being noted 20 min after injection. Sympathetic "tone" was estimated by propranolol, 8 mg/kg body weight. Also, a combined blockade was carried out with the above doses. From such data, it was possible to calculate the intrinsic rate of the pacemaker also. In agreement with the findings of Tipton and Taylor (1965), it was noted that cardiac acceleration with atropine was less in trained than in nontrained animals (+89 beats/min versus +109 beats/min). Likewise, the slowing of heart rate with propranolol was less in the trained animals (-21 beats/min versus -28 beats/min). Whether or not the latter difference was significant is open to question. The conclusion from these data was that training decreases the influence or "tone" of both the vagi and the sympathetics on the pacemaker. However, some difficulty arises when one tries to quantitate the changes in "tone" of the nerves relative to each other. Lin and Horvath used percentage changes as the basis for comparison and came to the conclusion that there is a greater reduction of the sympathetic tone than of the parasympathetic. On the other hand, if one bases the comparison on the number of beats, the reduction of the parasympathetic is greater. We tend to favor the latter comparison and are not entirely convinced that Lin and Horvath's conclusion rests on a firm basis. These investigators also noted a distinct reduction in intrinsic rate of the pacemaker in exercise-trained rats, using either the combined or individual nerve blockade method (trained = 372 or 377 beats/min versus untrained = 426 or 429 beats/min). This finding is in agreement with that of Frick *et al.* (1967) and that of Sutton *et al.* (1967) in man. The latter workers reported that, with combined blockade (atropine, 0.04 mg/kg, and propranolol, 0.2 mg/kg, intravenously), the average intrinsic pacemaker rate is about 90/min in trained individuals as compared with 107/min in untrained healthy young men ($p<0.01$). More recently, Bolter, Hughson, and

Critz (1973) have shown that, in isolated sinoatrial preparations from exercise-trained rats, the intrinsic rate averages 128 beats/min as opposed to 146 beats/min from sedentary rats ($p<0.01$). These observations constitute the basis for a new concept to explain the bradycardia of training and are in direct conflict with the dogmatic statement of Clark that the fundamental frequency of the pacemaker is unchanged by training. One may argue that the use of large doses of atropine and propranolol may alter the absolute values of the intrinsic pacemaker frequency by direct action, but this does not invalidate the conclusion that the training reduces the intrinsic frequency of the S-A node.

Ekblom, Kilbom, and Soltysiak (1973) studied the effect of 5 weeks' training in 14 young male subjects on the heart rate after atropine block, propranolol block, and combined block. Under resting conditions, the increase in heart rate after atropine (2–2.5 mg, intravenously) was unchanged by training. This would mean that the observed resting bradycardia (about 10 beats less) could not be explained on the basis of an increase in vagal influence. With propranolol (10 mg, intravenously), the decrease in resting heart rate was somewhat less after training, but its statistical significance was questionable. This would suggest a possible reduction in cardiac adrenergic influence. With combined blockade, there was no significant difference in the increase in resting heart rate, before and after training. This study does not throw much light on the mechanism of resting bradycardia, probably because the training was not sufficiently intense or prolonged, as suggested by the relatively small decrease in heart rate after training. However, there was a suggestion that the observed slowing could have been due to a decrease in cardiac sympathetic activity. Some authors (De Schryver and Mertens-Strythagen, 1972) have reported a lower cardiac catecholamine concentration in exercised rats as compared with control animals. Whether or not this reflects a decrease in sympathetic activity remains obscure. The possible role of a decrease in cardiac sympathetic activity in causing the bradycardia of exercise training has been overlooked for many years and only recently have investigators been looking into this possibility more closely.

Another experimental approach to the study of the influence of extrinsic nerves on heart rate is to surgically denervate the heart by vagotomy, sympathectomy, or both. To my knowledge, such studies have not been conducted in exercise-trained animals. Several problems arise with chronic denervation studies: 1) denervation may be incomplete, 2) denervation hypersensitivity may change the intrinsic rate of the pacemaker to circulating catecholamines, 3) sectioned nerves may regenerate to an unknown extent and alter the results, etc. All these would make the interpretation of the results difficult and uncertain.

From the foregoing presentation, we propose a scheme to explain the resting bradycardia of athletic training (Fig. 1). Essentially two factors seem to play a role: 1) A reduction in the intrinsic rate of the pacemaker. Studies in young individuals suggest that the intrinsic rate declines from about 105 beats/min to

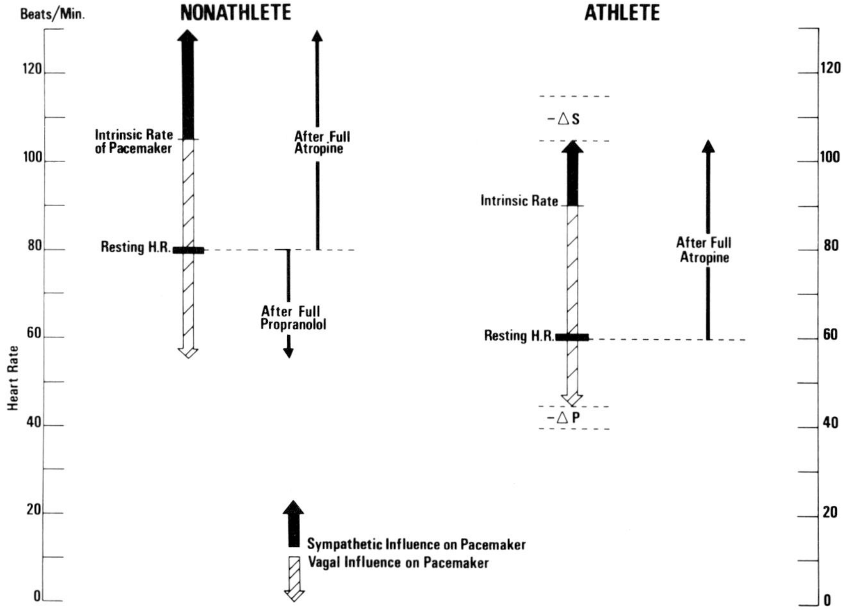

Figure 1. Proposed concept to explain the resting bradycardia of exercise training: (*1*) Slowing of intrinsic rate of cardiac pacemaker. (*2*) Increased vagal predominance on pacemaker as a result of reduction in sympathetic influence. Vagal influence may decrease slightly or remain unchanged. This concept is based essentially on data obtained with the use of autonomic blocking drugs.

about 90 beats/min (Jose, 1966; Frick *et al.*, 1967; Jose and Collison, 1970; Sutton *et al.*, 1967). Until recently, this intrinsic rate factor was entirely overlooked and seems to be more important than changes in extrinsic nervous control. 2) An increase in net cholinergic influence on pacemaker. The sum of evidence suggests that this is essentially the result of a *decrease* in the sympathetic influence, with vagal influence either remaining the same or undergoing a slight reduction. There is little evidence that vagal tone per se is increased compared to that in the nonathlete.

It may be objected that the normal cardiac sympathetic tone under resting conditions is very limited and cannot be decreased very much. Obviously, this would vary with the experimental conditions, methods of study, animal species, etc. Studies in man suggest that cardiac sympathetics have a significant tonic influence. Chapman *et al.* (1948) reported distinct slowing of the heart after cardiac sympathectomy in human subjects, but unfortunately no actual values of heart rate were given. Prichard and Gillam (1971) have used what were said to be "full" doses of propranolol in anginal subjects (417 mg/day for about 6 weeks).

In 16 patients in the supine position, the heart rate after placebo treatment was 83.8 beats/min (blood pressure = 107 mm Hg) as compared with 61.4 beats/min (blood pressure = 98 mm Hg) after propranolol (-22.4 beats/min). There was evidence from dose-response curves that a maximal effect on heart rate had not been reached, but the dose could not be pushed higher. Although not supported by some observations (Wallace *et al.,* 1966; Kattus, Ross, and Hall, 1970), we accept the view that, in nontrained resting subjects, the cardiac sympathetics have an appreciable tonic influence on the pacemaker (Brouha, Cannon, and Dill, 1936; Jourdan, 1962; Prichard and Gillam, 1971).

With regard to the parasympathetic control of heart rate in man, Cullumbine, McKee, and Creasey (1955) found that a 5-mg dose of atropine intramuscularly in young untrained male subjects causes an increase of about 54 beats/min under resting conditions.

The mechanisms underlying the changes in intrinsic pacemaker rate and the changes in cardiac neural influences in trained athletes are far from being understood.

REFERENCES

BOLTER, C. P., R. L. HUGHSON, and J. B. CRITZ. 1973. Intrinsic rate and cholinergic sensitivity of isolated atria from trained and sedentary rats. Proc. Soc. Exp. Biol. Med. 144: 364–367.

BROUHA, L., W. B. CANNON, and D. B. DILL. 1936. The heart rate of the sympathectomized dog in rest and exercise. J. Physiol. (London) 87: 345–359.

BÜLBRING, E., and J. H. BURN. 1949. Action of acetylcholine on rabbit auricles in relation to acetylcholine synthesis. J. Physiol. (London) 108: 508–524.

BURN, J. H., and J. M. WALKER. 1954. Anticholinesterases in the heart-lung preparation. J. Physiol. (London) 124: 489–501.

CHAPMAN, E. M., D. KINSEY, W. P. CHAPMAN, and R. H. SMITHWICK. 1948. Sympathetic innervation of the heart in man. J. Am. Med. Ass. 137: 579–584.

CLARK, A. J. 1927. Comparative Physiology of the Heart, pp. 95–96. Cambridge University Press, Cambridge.

CULLUMBINE, H., W. H. E. McKEE, and N. H. CREASEY. 1955. The effects of atropine sulfate upon healthy male subjects. Quart. J. Exp. Physiol. 40: 309–319.

DAVSON, H., and M. G. EGGLETON. 1968. *In* Starling and L. Evans (eds.), Principles of Human Physiology, 14th ed., p. 170. Lea & Febiger, Philadelphia.

De SCHRYVER, C., and J. MERTENS-STRYTHAGEN. 1972. Intensity of exercise and heart tissue catecholamine content. Pflügers Arch. Ges. Physiol. 336: 345–354.

DETWEILER, D. K. 1973. *In* J. R. Brobeck (ed.), Best and Taylor's Physiological Basis of Medical Practice, 9th ed., pp. 3–114. The Williams and Wilkins Co., Baltimore.

EKBLOM, B., Å. KILBOM, and J. SOLTYSIAK. 1973. Physical training, bradycardia, and autonomic nervous system. Scand. J. Clin. Lab. Invest. 32: 251–256.

EKSTRÖM, J. 1974. Choline acetyltransferase in the heart and salivary glands of the rat after physical training. Quart. J. Exp. Physiol. 59: 73–80.

FRICK, M. H., R. O. ELOVAINIO, and T. SOMER. 1967. The mechanism of bradycardia evoked by physical training. Cardiologia 51: 46–54.

HALL, V. E. 1963. The relation of heart rate to exercise fitness: An attempt at physiological interpretation of the bradycardia of training. Pediatrics 32 (Suppl.): 723–729.

HERRLICH, H. C., W. RAAB, and W. GIGEE. 1960. Influence of muscular training and of catecholamines on cardiac acetylcholine and cholinesterase. Arch. Intern. Pharmacodyn. 129: 201–215.

HERXHEIMER, H. 1921. Zur Bradykardie der Sportsleute. Munch. Med. Wochschr. 68: 1515–1518.

HOLLMANN, W. 1959. Der Arbeits- und Trainingseinfluss auf Kreislauf und Atmung, p. 69. Steinkopff Verlag, Darmstadt.

JOSE, A. D. 1966. Effect of combined sympathetic and parasympathetic blockade on heart rate and cardiac function in man. Am. J. Cardiol. 18: 476–478.

JOSE, A. D., and D. COLLISON. 1970. The normal range and determinants of the intrinsic heart rate in man. Cardiovasc. Res. 4: 160–167.

JOURDAN, F. 1962. Le tonus cardio-accélérateur. Arch. Intern. Pharmacodyn. 140: 115–121.

KATTUS, A. A., G. ROSS, and V. E. HALL. 1970. Cardiovascular Beta-adrenergic Responses. University of California Press, Berkeley.

KAUF, E. 1926. Ueber die Einwirkung von Atropin und Adrenalin auf das Herz Sporttreibender. Wien. Klin. Wochschr. 39: 212–214.

LIN, Y. C., and S. M. HORVATH. 1972. Autonomic nervous control of cardiac frequency in the exercise-trained rat. J. Appl. Physiol. 33: 796–799.

MARSHALL, R. J., and J. T. SHEPHERD. 1968. Cardiac Function in Health and Disease, p. 45. W. B. Saunders Co., Philadelphia.

MOUNTCASTLE, V. B. 1968. Medical Physiology, Vol. I, 12th ed., p. 52. The C. V. Mosby Co., St. Louis.

PRICHARD, B. N. C., and P. M. S. GILLAM. 1971. Assessment of propranolol in angina pectoris. Clinical dose response curve and effect on electrocardiogram at rest and on exercise. Brit. Heart J. 33: 473–480.

RAAB, W., P. de P. e SILVA, H. MARCHET, E. KIMURA, and Y. K. STARCHESKA. 1960. Cardiac adrenergic preponderance due to lack of physical exercise and its pathogenic implications. Am. J. Cardiol. 5: 300–320.

SUTTON, J. R., A. COLE, J. GUNNING, J. B. HICKIE, and W. A. SELDON. 1967. Control of heart-rate in healthy young men. Lancet *ii:* 1398–1400.

TIPTON, C. M., R. J. BARNARD, and G. D. THARP. 1966. Cholinesterase activity in trained and nontrained rats. Intern. Z. Angew. Physiol. 23: 34–41.

TIPTON, C. M., and B. TAYLOR. 1965. Influence of atropine on heart rates of rats. Am. J. Physiol. 208: 480–484.

WALLACE, A. G., W. G. TROYER, M. A. LESAGE, and E. F. ZOTTI. 1966. Electrophysiologic effects of isoproterenol and beta blocking agents in awake dogs. Circulat. Res. 18: 140–148.

Mailing address:
Dr. Henry S. Badeer,
Department of Physiology-Pharmacology,
Creighton University School of Medicine,
2500 California Street, Omaha, Nebraska 68178 (USA).

Recent Advances in Studies on
Cardiac Structure and Metabolism Volume 10
The Metabolism of Contraction
Edited by Paul-Emile Roy and George Rona

EFFECT OF CONDITIONING ON MYOCARDIAL METABOLISM AFTER MYOCARDIAL INFARCTION

M. TAJUDDIN, M. AHMAD,
and M. TARIQ

Experimental and Metabolic Laboratory, Department of Medicine,
Aligarh Muslim University, Aligarh, India

SUMMARY

The present study was conducted to evaluate the effect of prior physical training on metabolic changes after isoproterenol-induced myocardial infarction in albino rats. The evaluation was done in terms of serum transaminases (SGOT and SGPT), serum lactic dehydrogenase (LDH), serum creatine phosphokinase (CPK), cardiac glycogen, and cardiac noradrenaline. Male albino rats weighing 100–150 g were selected for the study. Physical training (conditioning) consisted in making the animals swim in a tank of water for 60 min daily on 6 days a week, for 8 weeks. Myocardial infarction was produced by subcutaneous injection of isoproterenol, 85 mg/kg body weight, on 2 consecutive days. There was significantly less elevation of SGOT, LDH, and CPK in the conditioned group after myocardial infarction; the rise in SGPT was not significantly different in the two groups. A smaller rise of serum enzymes revealed less myocardial damage in the conditioned group; the rise in FFA was also less in the conditioned group, a more favorable situation for an ischemic myocardium. No significant difference was observed in cardiac glycogen and cardiac noradrenaline after isoproterenol-induced cardiac necrosis. The significance of these observations is discussed.

INTRODUCTION

Clinical surveys have revealed a lower incidence of myocardial infarction in physically active individuals (Brunner and Meshulam, 1969; Paffenbarger *et al.*, 1970). Active rehabilitation of coronary patients by gradual exercise has also been shown to decrease the incidence of reinfarction (Brunner and Meshulam, 1969; Morgans and Buston, 1972). However, the protective effect of active physical training against an ischemic challenge has yet to be evaluated.

The purpose of the present investigation was, therefore, to study the effect of prior physical training on serum enzymes (SGOT, SGPT, LDH, and CPK), serum free fatty acid (FFA), cardiac glycogen, and cardiac noradrenaline, after inducing myocardial infarction in albino rats with isoproterenol.

MATERIALS AND METHODS

Male albino rats weighing 100–150 g were selected for the study. Physical training (conditioning) consisted in making the animals swim in a tank of water 21 inches deep and 18 inches in diameter, maintained at 32 ± 1°C by a thermostatic device. Swimming was done for 60 min each day on 6 days a week, for 8 weeks. Myocardial infarction was produced by subcutaneous injection of isoproterenol (85 mg/kg body weight) on 2 consecutive days (Rona *et al.*, 1959).

Investigations included SGOT and SGPT (Reitman and Frankel, 1957), serum LDH (King, 1959), serum CPK (Rosalki, 1967), serum FFA (Trout *et al.*, 1960), and myocardial noradrenaline (Crout, 1961). Electrocardiography (lead II) and histology were performed to confirm myocardial infarction.

The animals were divided into the following 4 groups: group I: sedentary controls, animals taken from the same lot and kept at normal cage activity; group II: animals subjected to physical training; group III: sedentary animals with isoproterenol-induced myocardial infarction (SMI); group IV: conditioned animals with isoproterenol-induced myocardial infarction (CMI).

RESULTS

SGOT increased by 143% above the normal value after myocardial infarction in the unconditioned group, while the rise was only 100% above the normal value in the conditioned group. LDH levels increased by 103 and 66% after myocardial infarction in the unconditioned and conditioned groups, respectively. CPK elevation was 416% in the unconditioned group and 258% in the conditioned group. Thus, there was significantly less increase in serum enzyme levels in the conditioned group. SGPT elevation was not significantly different in the two groups (Fig. 1).

The rise in serum FFA was higher in the unconditioned group after myocardial infarction as compared to the conditioned group. The level increased from the normal value of 93.3 ± 5.3 μEq/100 ml to 167.5 ± 2.8 μEq/100 ml in the unconditioned group and to 115.8 ± 7.4 μEq/100 ml in the conditioned group (Table 1).

Conditioning caused an increase in myocardial glycogen from the normal value of 7.43 ± 0.2 mg/g of tissue to 9.9 ± 0.7 mg/g (Table 1); the value decreased in both groups after myocardial infarction. Although the depletion was less marked in the conditioned group, i.e., 4.77 ± 0.4 mg/g of tissue as compared to 5.27 ± 0.3 mg/g in the unconditioned group, the difference was not statistically significant.

Cardiac noradrenaline was slightly lower in the conditioned group (Table 1). Administration of isoproterenol decreased the level of myocardial noradrenaline in both groups, the levels being 0.67 ± 0.40 and 0.68 ± 0.4 μg/g of cardiac tissue

Table 1. Metabolic changes after myocardial infarction*

	Group	Serum FFA† (μEq/100 ml)	Cardiac Glycogen‡ (mg/g)	Cardiac Noradrenaline‡ (μg/g)
I.	Sedentary controls	93.3 ± 5.3	7.43 ± 0.2	1.16 ± 0.04
II.	Conditioned animals	115.8 ± 2.8	9.90 ± 0.7	0.92 ± 0.05
III.	SMI	167.5 ± 7.4	4.77 ± 0.4	0.67 ± 0.40
IV.	CMI	131.0 ± 4.1	5.27 ± 0.3	0.68 ± 0.40

*Results are mean ± SE significance was calculated by Student's *t*-test.
†Significant; difference between SMI and CMI.
‡Insignificant; difference between SMI and CMI.

in the unconditioned and conditioned groups, respectively. These data suggest that conditioning did not prevent the depletion of cardiac noradrenaline after isoproterenol administration.

DISCUSSION

The study of risk factors in myocardial infarction has attracted considerable attention, owing to the increasing incidence of coronary artery disease. Rona *et al.* (1959) standardized a technique for the production of myocardial infarction in rats by subcutaneous injection of isoproterenol. This has made possible the study of histological and biochemical changes following myocardial infarction and has led to new concepts in the field of clinical cardiology (Legato, 1972). Selye, Veilleux, and Grasso (1960), working on the same experimental model, showed that coronary artery ligation protects the myocardium against a challenging dose of isoproterenol. This finding has stimulated research on the factors enhancing coronary resistance against an ischemic challenge. Rona, Chappel, and Kahn (1963), and Rona and Dusek (1972) studied the effect of various endogenous, hormonal, and nutritional factors which modify coronary resistance. Lack of physical activity is considered an important risk factor (Rona and Dusek, 1972). Clinical surveys have revealed a lower incidence of myocardial infarction in persons engaged in greater physical activity (Fox and Haskell, 1968; Brunner and Meshulam, 1969; Paffenbarger *et al.*, 1970, Opie, 1973). Graded exercise after an attack of myocardial infarction has also been shown to decrease the incidence of reinfarction (Morgans and Buston, 1972). The first attempt to study the beneficial effect of physical activity on coronary ischemia in experimental animals was made by Thomas and Harrison (1944), who studied the survival rate in two groups of rats, one at normal cage activity and the other at

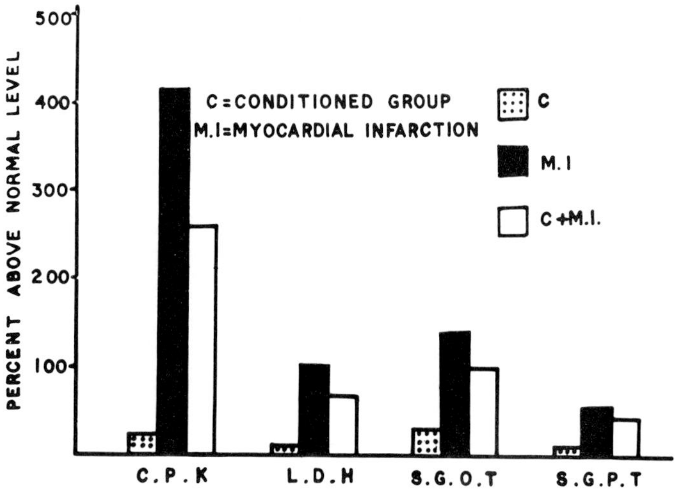

Figure 1. Myocardial infarction in albino rats: effect of conditioning on serum enzymes.

restricted cage activity; but the effect of active physical training was not studied. The study of the effect of physical training on the metabolic responses of rat heart was attempted by Scheuer and Stezoski (1972). These workers determined cardiac glycogen in conditioned and unconditioned animals before and after hypoxia; no difference in cardiac glycogen was observed in the two groups after hypoxic exposure. Since the situation in generalized hypoxia is different from that of local myocardial necrosis after myocardial infarction (Rovetto, Whitmer, and Neely, 1973), we decided to study the biochemical alterations in the two groups after producing myocardial necrosis with isoproterenol.

Biochemical alterations, especially the changes in serum FFA (Challoner and Steinberg, 1966; Kurien, Yates, and Oliver, 1971; Kjekshus and Mjos, 1972), cardiac noradrenaline (Bhagat and Dhalla, 1973), and cardiac glycogen (Brachfeld, 1974; Smith, Christodoulou, and Brachfeld, 1974), have a profound influence on cardiac function and integrity, and changes in serum enzyme levels reflect tissue injury (La Due, Wroblewski, and Karmen, 1954); therefore, a study of these parameters after myocardial infarction in conditioned and unconditioned animals will help in assessing the beneficial effect of physical training.

It is evident from Figure 1 that the rise in serum enzymes after isoproterenol-induced myocardial infarction is much less in the conditioned group as compared to the unconditioned group. Since the enzymatic changes parallel tissue injury (La Due *et al.*, 1954), it can be concluded that myocardial damage is less severe in the conditioned group.

There is smaller rise in serum FFA in the conditioned group after myocardial infarction (Table 1). The rise of FFA after myocardial infarction is well documented (Brachfeld, 1969; Kjekshus and Mejos, 1972). This rise is due to enhanced lipolysis under the influence of excess catecholamines released after myocardial infarction (Fredrickson, 1969). FFA have many harmful effects on the myocardium; they increase the myocardial oxygen requirement (Kjekshus, 1973), interfere with myocardial contraction (Henderson, Most and Sonneblick, 1969; Kjekshus and Mejos, 1972), produce cardiac dysrhythmia (Kurien *et al.*, 1971), and can produce thrombotic episodes by decreasing the clotting time (Oliver, Kurein, and Greenwood, 1968). A direct role of FFA in the production of myocardial infarction has also been demonstrated (Kjekshus, 1973). These effects are likely to aggravate the already existing cardiac injury. A smaller rise of FFA in the conditioned group after myocardial infarction signifies that physical exercise exerts a protective effect on the myocardium after an ischemic challenge.

The finding that myocardial glycogen increased in conditioned rats (Table 1) is in agreement with earlier observations of Scheuer *et al.*, (1970). The observed fall in cardiac glycogen after isoproterenol-induced myocardial infarction is attributable to enhanced glycogen breakdown in a hypoxic or ischemic situation under the effect of catecholamines, in an attempt to make more carbohydrate available for glycolytic energy formation, as reported by Brachfeld (1974). Glycogen is an essential substrate for hypoxic energy production. High glycogen stores have been shown to improve resistance to anoxia (Scheuer and Stezoski, 1970; Smith *et al.*, 1974). A consistent relationship has been noted between glycogen depletion and deterioration of mechanical performance of the heart (Thorn, Gercken, and Huter, 1968). The disproportion between myocardial energy supply and demand is an important factor in the pathogenesis of cardiac necrosis induced by isoproterenol (Rona *et al.*, 1959). Since myocardial glycogen is the main source of energy for an ischemic myocardium, the cardiac glycogen level was estimated in conditioned and unconditioned animals after myocardial infarction. In the present study, cardiac glycogen levels were not very different in the two groups after myocardial infarction. It may thus appear that conditioned hearts have little metabolic advantage over sedentary hearts. However, more glycogen in conditioned hearts would have yielded more energy which, along with the better energy-delivery system in conditioned hearts (Scheuer and Stezoski, 1972), might serve as a critical factor in supporting cardiac energy supply and cardiac function.

Myocardial noradrenaline content was lower in the conditioned group; similar observations have been reported earlier (De Schryver, De Herdt, and Lammerant, 1967). There was no significant difference in myocardial noradrenaline content in the two groups after myocardial infarction in the present study. Since myocardial catecholamines are necessary for proper modulation of cardiac contractility, it may appear that the conditioned heart is not superior to the unconditioned heart after an ischemic insult. On the other hand, it has been demonstrated that conditioned hearts function better during hypoxia as compared to unconditioned hearts (Scheuer and Stezoski, 1972). It should be pointed out here that, although cardiac catecholamines fall after isoproterenol-induced myocardial infarction, the uptake of catecholamines is not affected (Bhagat and Dhalla, 1973). The determination of catecholamine uptake and delivery in the two groups may throw some light on this problem. Furthermore, the low levels of cardiac noradrenaline which have been observed in conditioned hearts might create a condition of catecholamine responsiveness (Scheuer *et al.*, 1970) and thus improve myocardial contractile force in conditioned hearts.

CONCLUSIONS

The following conclusions can be drawn from the present study. Enzymatic changes reveal that less myocardial damage is produced by isoproterenol in conditioned animals. The changes in serum FFA after myocardial infarction are more favorable in conditioned animals. No difference was observed in cardiac glycogen after induction of cardiac necrosis in the two groups. However, larger amount of glycogen in the conditioned hearts along with the better delivery system would have yielded a larger amount of energy to support the myocardium after an ischemic challenge. It is suggested that estimation of myocardial catecholamine uptake and retention in the two groups after experimental myocardial infarction may throw some light on the effects of changes in cardiac catecholamines.

REFERENCES

BERGSTROM, J., L. HERMANSEN, E. HULTMAN, and B. SALTIN. 1967. Diet, muscle glycogen and physical performance. Acta Physiol. Scand. 71: 140–150.

BHAGAT, B., and N. S. DHALLA. 1973. Alterations in norepinephrine (NE) pattern in damaged myocardium of the rat. Abstract, 4th Annual Meeting of International Study Group for Research in Cardiac Metabolism, 25–28, Sept. Freiburg, West Germany.

BRACHFELD, N. 1969. Maintenance of cell viability. Circulation 39 and 40 (Suppl. IV): 202–219.

BRACHFELD, N. 1974. Ischaemic myocardial metabolism and cell death. Bull. N.Y. Acad. Med., 2nd Vol. 50, No. 3, 261–294.

BRUNNER, D., and N. MESHULAM. 1969. Prevention of recurrent myocardial infarction by physical exercise. Israel Med. Sci. 5: 783–785.

CHALLONER, D. R., and D. STEINBERG. 1966. Effect of free fatty acid on the oxygen consumption of perfused rat heart. Am. J. Physiol. 210: 280–286.

CROUT, R. J. 1961. Catecholamines in urine. *In* D. Seligson (ed.), Standard Methods of Clinical Chemistry, Vol. 3, pp. 62–80. Academic Press, New York.

De SCHRYVER, C., P. De HERDT, and J. LAMMERANT. 1967. Effect of physical training on cardiac catecholamine concentrations. Nature 214: 907–908.

FOX, S. H., and W. L. HASKELL. 1968. Physical activity and the prevention of coronary heart disease. Bull. N.Y. Acad. Med. 44: Pt. 2, 950–967.

FREDRICKSON, D. S. 1969. The role of lipids in acute myocardial infarction. Circulation 39 and 40 (Suppl.): 999–111.

HENDERSON, A. H., A. S. MOST, and E. H. SONNEBLICK. 1969. Depression of contractility in rat heart muscle by free fatty acid during hypoxia. Lancet ii: 825–826.

KING, J. 1959. Routine method for estimation of lactic dehydrogenase activity. J. Med. Lab. Technol. 16: 265–272.

KJEKSHUS, J. K. 1973. The role of free fatty acid (FFA) in catecholamine induced cardiac necrosis. Abstract. Vol. VI, Proceedings of the 4th Annual Meeting of International Study Group For Research in Cardiac Metabolism, Sept. 25–28, Freiburg, West Germany.

KJEKSHUS, J. H., and O. D. MJOS. 1972. Effect of free fatty acids on myocardial function and metabolism in the ischaemic dog heart. J. Clin. Invest. 51: 1767–1776.

KURIEN, V. A., P. A. YATES, and M. F. OLIVER. 1971. The role of free fatty acids in the production of verticular arrhythmia after coronary artery occlusion. Europ. J. Clin. Invest. 1: 225–241.

La DUE, J. S., F. WROBLEWSKI, and A. KARMEN. 1954. SGOT activity in human acute transmural myocardial infarction. Science 120: 497–499.

LEGATO, M. J. 1972. Editorial. The myocardial cell: new concepts for the clinical cardiologist. Circulation 45: 731–735.

MORGANS, C. M., and W. M. BUSTON. 1972. Supervised circuit training after myocardial infarction. Physiotherapy 58: 340–343.

OLIVER, M. F., V. A. KURIEN, and T. W. GREENWOOD. 1968. Relation between serum free fatty acid and arrhythmias and death after acute myocardial infarction. Lancet i: 710–714.

OPIE, L. H. 1973. Exercise and myocardial factor in ischaemic heart disease. Lancet i: 1067–1068.

PAFFENBARGER, R. S., M. E. LAUGHLIN, A. S. GIMA, and R. A. BLACK. 1970. Work activity of Longshoremen as related to deaths from coronary heart disease and stroke. New Engl. J. Med. 282: 1109–1114.

PENPARGKUL, S., and J. SCHEUER. 1970. Effect of physical training upon the mechanical and metabolic performance of the rat heart. J. Clin. Invest. 49: 1859–1868.

REITMAN, S., and S. FRANKEL. 1957. A calorimeteric method for the determination of SGOT and SGPT. Am. J. Clin. Path. 28: 56–63.

RONA, G., C. I. CHAPPEL, T. BALAZS, and R. GAUDRY. 1959. An infarct-like myocardial lesion and other toxic manifestations produced by isoproterenol in the rat. Arch. Path. 67: 443–445.

RONA, G., C. I. CHAPPEL, and D. S. KAHN. 1963. The significance of factors modifying the development of isoproterenol induced myocardial necrosis. Am. Heart J. 66: 389–395.

RONA, G., and J. DUSEK. 1972. Studies on the mechanism of increased myocardial resistance. *In* E. Bajusz and G. Rona (eds.), Recent Advances in Studies on Cardiac Structure and Metabolism, Vol. 1: Myocardiology, pp. 422–429. University Park Press, Baltimore.

ROSALKI, S. B. 1967. An improved procedure for serum creatine phosphokinase determination. J. Lab. Clin. Med. 69: 696–705.

ROVETTO, M. J., J. T. WHITMER, and J. R. NEELY. 1973. Comparison of the effects of anoxia and whole heart ischaemia on carbohydrate utilisation in isolated, working rat heart. Circulat. Res. 32: 699–000.

SCHEURER, J., L. KAPNER, C. A. STRINGFELLOW, C. L. ARMSTRONG, and S. PENPARGKUL. 1970. Glycogen, lipid and high energy phosphate stores in hearts from conditioned rats. J. Lab. Clin. Med. 75: 924–929.

SCHEUER, J., and S. W. STEZOSKI. 1972. Effect of physical training on the mechanical and metabolic response of the rat heart to hypoxia. Circulat. Res. 30: 418–428.

SELYE, H., R. VELLEUX, and S. GRASSO. 1960. Protection by coronary ligation against isoproterenol induced myocardial necrosis. Proc. Soc. Exp. Biol. Med. 104: 343–345.

SMITH, C., J. CHRISTODOULOU, and N. BRACHFELD. 1974. Protective role of increased myocardial glycogen stores induced by propranalol. Abstract 7th Annual Meeting of International Study Group for Research in Cardiac Metabolism, June 18–21, Quebec.

THOMAS, W. C., and T. R. HARRISON. 1944. The effect of artificial restriction of activity on recovery of rats from experimental heart injury. Am. J. Med. Sci. 208: 436–450.
THORN, W., G. GERCKEN, and P. HUTER. 1968. Function, substrate supply and metabolic content of rabbit heart perfused in situ. Am. J. Physiol. 214: 139–145.
TROUT, P. L., E. H. ESTES, and S. J. FRIEDBERG. 1960. Titration of free fatty acids of plasma. A study of current methods and a new modification. J. Lipid Res. 1: 199–202.

Mailing address:
Col. M. Tajuddin,
Professor and Head, Department of Medicine,
J. N. Medical College, Aligarh Muslim University,
Aligarh, U.P. (India).

Vascular Biology—
Recent Studies

Recent Advances in Studies on
Cardiac Structure and Metabolism Volume 10
The Metabolism of Contraction
Edited by Paul-Emile Roy and George Rona

MUSCULATURE OF THE PULMONARY VEINS

C. E. CHALLICE, J. L. WILKENS,
K. S. CHOHAN, and F. WHITE

Departments of Physics and Biology, The University of Calgary,
Calgary, Alberta, Canada

INTRODUCTION

Although it has been documented that the cardiac musculature of many mammalia, including man, extends along the pulmonary veins to the lungs (Karrer, 1959; Kramer and Marks, 1965; Ludatscher, 1968), this observation appears to have generated comparatively little interest. In some rodentia, it has been shown that this musculature extends into the lungs as far as the small venules. Work in this laboratory has, in recent years, centered around the development of the cardiac musculature, and around the so-called "specialized" tissues, and the relationship of structure to physiological function (Challice, 1971; Virágh and Challice, 1973; Challice and Virágh, 1973), and from this has evolved an interest in the development of the extension of the cardiac muscle along the pulmonary veins.

The hitherto available reports on this tissue confine themselves to structural studies; and where speculation has been made concerning the physiological function of this musculature, it has been assumed that it contracts in such a manner as to assist blood flow into the heart from the lungs (Kramer and Marks, 1965). It appears to have passed unnoticed that there is an anomaly here, namely, that the primary pacemaker in the heart is located in the sinoatrial (SA) node, having (ontogenetically) migrated there from the early ventricle via the developing atrium (Johnstone, 1925). In order for contraction in the muscle surrounding these veins to assist the blood flow from the lungs to the heart, it would be necessary for the contraction to occur there (i.e., in the veins) *before* the atrium, and this could be accomplished either by the primary pacemaker being in or near the lungs, or by the existence of a complex nerve plexus which somehow synchronized with, or perhaps even governed, the primary pacemaker. No evidence for either of these mechanisms has thus far been produced, and present indications are that they do not exist.

If indeed such a mechanism is absent, one is forced to contemplate the possibility that an electrical impulse emanating from the SA node could pass along this venous musculature in a direction opposite to that of the primary blood flow. Such a possibility would carry with it the likelihood of the impulse being accompanied by a contraction, which in turn would produce a peristaltic contraction along the veins tending to *oppose* the flow of the blood into the heart. With this in mind, it was decided to conduct some simple electrophysiological experiments with a view to detecting the existence of cardiac-like electrical impulses in the pulmonary veins of mice, and to find the velocity (including direction) of the electrical impulse.

METHODS

Electrophysiological recordings were made from the pulmonary veins *in situ* on mice, following the simplest surgical procedures which allowed access to that part of the circulatory system. Adult mice were killed by cervical dislocation, and the rib cage was opened. The pericardial sac was removed to allow access to the pulmonary veins in the region where they enter the left atrium. These veins were followed distally as they collect from the posterior-dorsal surface of the lungs. The heart and thoracic area was continuously washed with aerated Locke's solution (pH 7.4, 37°C). Under these conditions, it was possible to maintain heart and pulmonary function for over an hour in the absence of ventilation of the lungs.

Two unipolar suction electrodes were applied to the exterior walls of the vessels (the extrapulmonary veins; Kramer and Marks, 1965) one as far away from the heart as possible, and the other nearer to the heart on the same venous pathway, using as great a separation between electrodes as possible. The ground lead of each electrode was positioned as far from the heart as possible so as to minimize the detection of those potentials arising electrotonically from this source. Signals were amplified (band pass of amplifier 0.3 Hz to 1 kHz), displayed on a dual beam storage oscilloscope, and recorded on tape.

For microscopic study, mice were killed as above; and the veins were dissected as quickly as possible, fixed in 5% glutaraldehyde in 0.1 M phosphate buffer, washed and postfixed for 1 hour in 1% osmium tetroxide solution at pH 7.4, dehydrated, and embedded in araldite. Sections were cut (using a Reichert ultramicrotome) and stained with uranyl acetate and lead citrate. Electron microscopy was performed on a Siemens Elmiskop la. Thicker sections ($\sim 1\ \mu m$) were also cut, stained with methylene blue, and examined by light microscopy.

RESULTS

A thorough study of the structural aspects of this tissue is still in progress, and thus far there is little to add to the observations of the earlier studies. The orientation of the fibers is not entirely regular but is predominantly circumferen-

Figure 1. Electron micrograph of a section of the wall of the pulmonary vein from the mature mouse, showing compact fibers and an intercalated disc. × 14,000.

Figure 2. As Figure 1. Longitudinal section of wall of the pulmonary vein showing well developed contractile apparatus. × 16,500

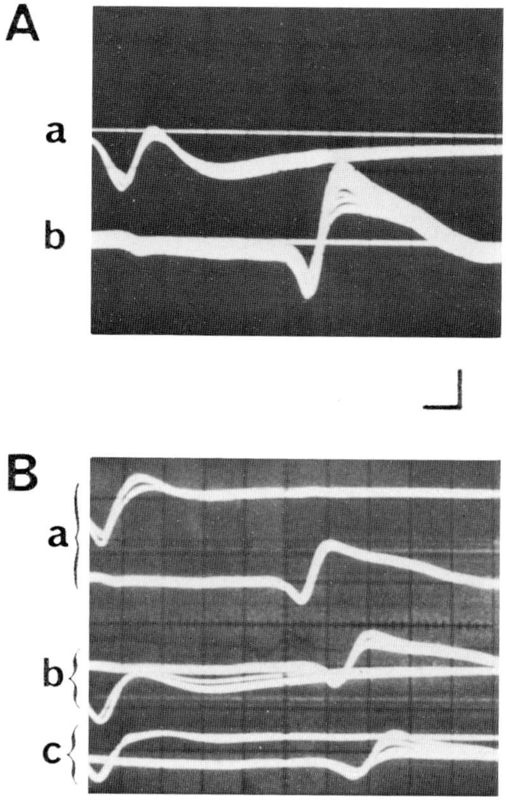

Figure 3. Illustration of retrograde conduction along the pulmonary vein as recorded by unipolar suction electrodes applied to the outside of the vessel. *A:* the upper spikes (*a*) recorded close to the heart were also used to trigger the oscilloscope sweeps. The lower spikes (*b*) were recorded 4.2 mm distally to the first. This display is of 20 consecutive sweeps stored on the oscilloscope. *B:* activity from the same animal and electrode placement recorded *a*, 10 min; *b*, 20 min; and *c*, 30 min after the records in *A*. Each pair of traces is of 5 consecutive sweeps. Calibrations of both *A* and *B:* vertical axis, *a*, 2 mV; *b*, 1 mV; horizontal, 2 msec.

tially oriented, although a few approximately longitudinally oriented fibers appear to be present. The musculature (at least in the extrapulmonary veins) is several cells thick, and clearly of the striated cardiac type (Fig. 1), consisting of fibers which have a varying degree of lateral association with one another (Figs. 1 and 2). Similar musculature is found surrounding the larger veins within the lungs; but the degree to which it continues along the venous system to the smaller venules has not yet been documented in the present study.

The electrical measurements have recorded electrical activity travelling away from the heart along the pulmonary vein in each of 7 mice. Figure 3*A* is

Table 1. Velocity of propagation of action potentials in pulmonary veins of mice*

Time Interval between Spikes (msec)	Separation between Electrodes (mm)	Velocity of Propagation (cm sec^{-1})
7	3.2	46
9	4.2	45
7	3.8	54
11	4.8	44
7	2.3	33
12	4.2	35
20	6.0	43

*Average velocity 42.8 cm sec^{-1}. Estimated accuracy of measurements ± 10%.

representative of the data. In this composite trace, the action potentials recorded proximal to the heart were used to trigger the oscilloscope trace. Those potentials recorded several millimeters distally on the same vessell occurred at a fixed latency but were not completely constant in amplitude. They were propagated at a velocity of approximately 40 cm/sec, a velocity characteristic of all preparations (Table 1). Figure 4 is a graphical representation of the results given in Table 1.

The velocity of propagation varies slightly with time under the experimental conditions used. Figure 3*B* is a record from the same animal, and with the electrodes undisturbed, of groups of five consecutive sweeps recorded at successive 10-min intervals. It is seen that the latency between spikes (i.e., the reciprocal of the velocity of propagation) after the first 10-min interval has increased by 1.5 msec and, after the second interval, by a further 1 msec. Variations greater than these were not found even in deteriorating preparations.

Once an action potential was generated in the pulmonary vein, it was conducted along the vessel without failure; however, the coupling between atrial and venous action potentials showed some variation. In fresh preparations, propagation from the heart to the pulmonary vein was found to be 1:1, but the coupling became looser as a preparation deteriorated (Fig. 5*A*). These tracings were taken over a 4-min period as the coupling became weaker, 2 min elapsing between each trace. The small positive spike on the upper trace of each pair of traces represents the electrotonically conducted atrial depolarization as recorded by the proximal electrode. When an atrial depolarization led to a venous action potential, this was conducted over the length of the vessel without fail; however, following deterioration with passage of time, fewer action potentials were propagated along the veins. When the oscilloscope was triggered from the small atrial action potential (Fig. 5*B*), the causal relationship between atrial and

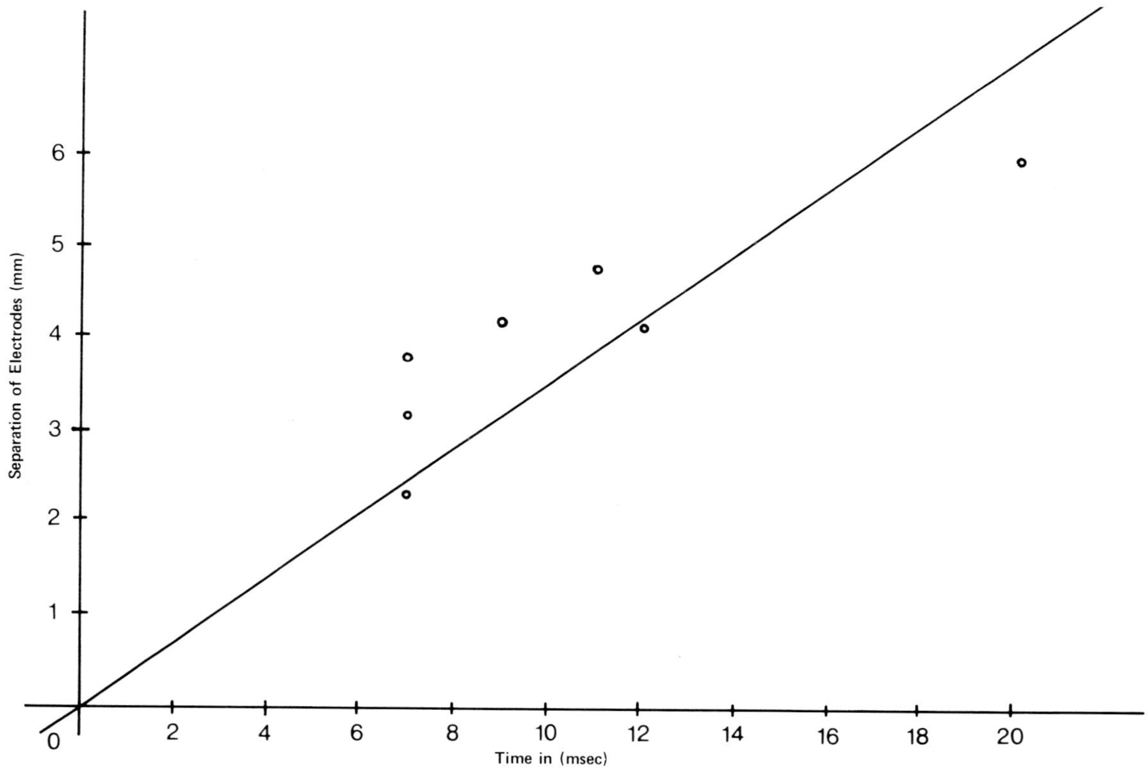

Figure 4. Graph of electrode separation against time taken for impulse to travel between electrodes in mouse pulmonary vein.

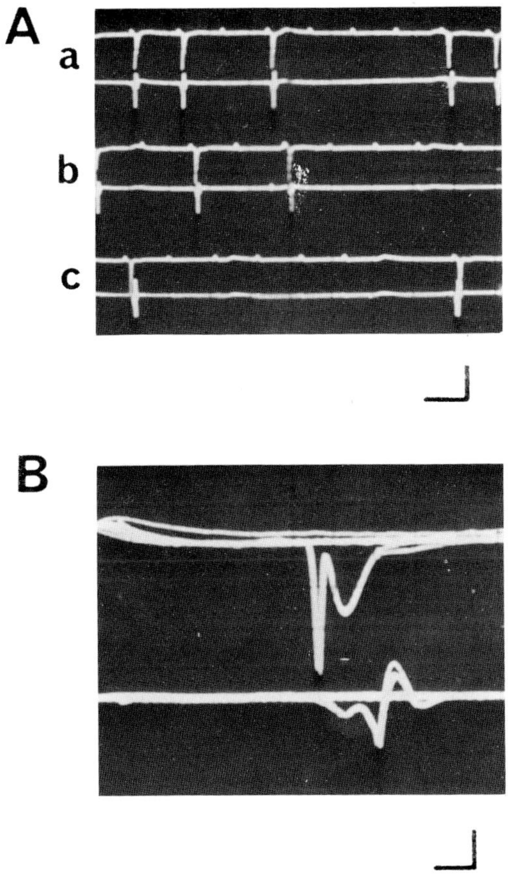

Figure 5. A: three pairs of traces of pulmonary potentials recorded at 2-min intervals, during which time the coupling between atrial and venous potentials became weaker. The small positive going potential is the electrotonically conducted atrial depolarization recorded by the proximal electrode located adjacent to the heart. *B:* illustration of the fixed latency between atrial and venous potentials. The oscilloscope traces were triggered from the small atrial potentials. Twelve traces were stored with venous spikes occurring only 5 times. Calibrations: *A*, vertical, 2 mV, horizontal, 200 msec; *B*, vertical, 0.5 mV, horizontal, 5 msec.

venous spikes was clearly indicated by the fixed latency between these events. The five venous spikes on these traces were completely superimposed on one another.

As a control, the pulmonary vein was compressed by forceps at the point of entry to the heart. This compression did not disrupt the integrity of the vessel and blood flow into the heart continued after release. However, electrical conduction along the vein was totally absent following this intervention.

As a further control, recordings were made from the dorsal aorta of the mouse. No cardiac-like musculature has been shown to exist in the walls of the aorta but only smooth muscle. The spikes on Figure 6 represent weak electrotonically conducted signals arising from the heart itself. These signals were observed to be progressively weaker the further from the heart they were recorded, and they occurred simultaneously at each of two electrodes which in this case were placed 4.3 mm apart. Recordings from several animals failed to detect any propagated action potentials passing along the aorta. As a further control, attempts were made to make similar recordings from the pulmonary veins of the frog (Fig. 7). No propagated action potentials were recorded in three such experiments, although pulmonary blood flow remained strong throughout. Electron microscopic studies have shown only smooth muscle in the walls of these vessels.

DISCUSSION

These simple experiments indicate that, under the experimental conditions described, retrograde conduction of action potentials occurs along the pulmonary veins, at least to those points where the veins leave the lungs. Presumably, these spikes represent myoid conduction through the muscle tissue. No summation or facilitation appears to be involved in this 1:1 conduction. If the validity of the observations is accepted, it would seem most likely that this impulse would travel to the extent of the cardiac musculature along the veins *within* the

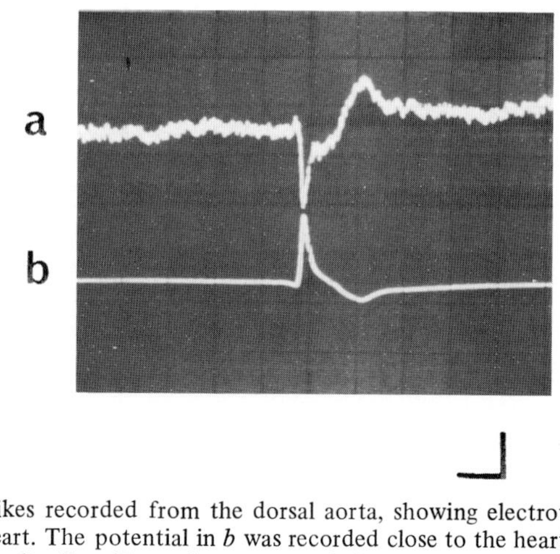

Figure 6. Two spikes recorded from the dorsal aorta, showing electrotonically conducted spikes from the heart. The potential in *b* was recorded close to the heart while that in *a* was recorded 4.3 mm distally. *Trace b* was recorded at reversed polarity and the traces positioned to illustrate the simultaneous time of occurrence of the two events. The pronounced reduction in amplitude at *a* reflects decremental conduction. Calibration: vertical, *a*, 50 μV; *b*, 0.5 mV; horizontal, 50 msec.

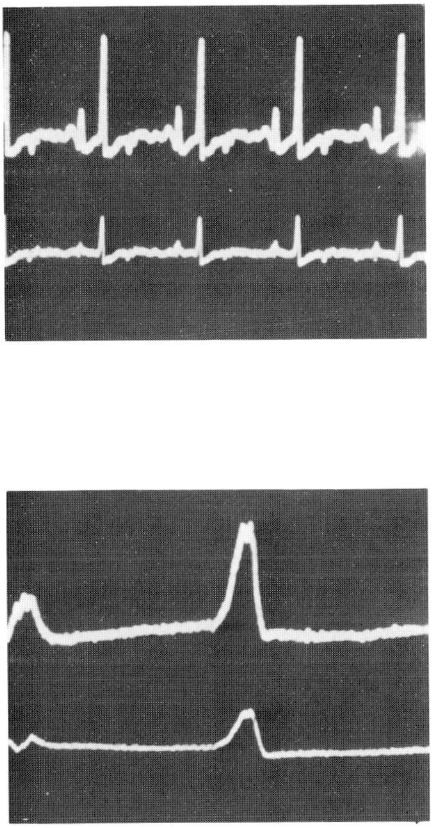

Figure 7. Recordings made from the branchial and pulmonary veins of the frog *a*, 500 msec sweep speed; *b*, 50 msec sweep speed. The pulmonary veins enter the branchial veins some distance from the heart, and the walls of both contain only smooth muscle. No propagated potentials were recorded from any of three frogs whose heart and pulmonary circulations remained strong for over an hour each. Blood flow in the short pulmonary veins was observed visually. Weak electrotonically conducted signals arising from the heart were recorded. Note the two spike heights which presumably arise from the atrium (smaller) and ventrical (larger spike). These signals became progressively weaker the further from the heart they were recorded and they occurred simultaneously at each of two electrodes placed a minimum of 2 mm apart.

lungs, but this suggestion at the moment represents only an extrapolation of the observations.

The possible implications of this retrograde conduction are most interesting. It implies the passage of a peristaltic contraction along the pulmonary veins in a direction opposite to that of the main blood flow. Contraction at the junction of vein and atrium would tend at least to *restrict* backward flow from the atrium, and possibly suppress it.

However, it is difficult to see how the passage of a peristaltic contraction along the relatively substantial muscle of the pulmonary veins can fail to produce a backward pumping action, although the strength of this remains to be determined. Preliminary observations indicate that the thickness of the muscular wall decreases as it passes to the smaller veins, but not necessarily in proportion to the lumen diameter.

It seems reasonable to speculate that the action of this musculature is an ingredient in the control of hemodynamic flow and presumably represents an increase in over-all functional efficiency of the circulatory system compared with the more primitive heart (e.g., of the amphibian) where such musculature is absent. However, the exact nature of its effect on the blood flow remains to be determined. It could simply regulate the more-or-less steady flow between lungs and heart, or it could, possibly in conjunction with the atrial contraction, produce an agitation of the blood in the pulmonary sinusoids. Some experiments are planned, which, it is hoped, will throw light on this matter.

Failure of this musculature to function properly could presumably cause pathological conditions. A too strong or too weak contraction would cause improper hemodynamic flow and disturbed pressure distributions. This possibility has hitherto not been considered.

Although much of the above is clearly only speculation, it seems clear that the function of this musculature is a factor which needs to be considered in the over-all problem of the pulmonary circulation.

Acknowledgments One of us (C.E.C.) is indebted to Dr. S. Virágh (Postgraduate Medical School, Budapest), with whom he has collaborated over a period of several years (see Introduction), for contributing to his interest in the pulmonary venous musculature through collaborative work on embryological development of heart musculature, leading to work on the ontogenesis of the musculature here discussed.

A preliminary report covering some of this work has appeared in the Canadian Journal of Physiology and Pharmacology, to whom the authors are indebted for permission to reproduce Figures 3 and 5.

The present work has been supported by the National Research Council of Canada and the Alberta Heart Foundation.

REFERENCES

BURTON, A. C. 1972. Physiology and Biophysics of the Circulation. Year Book Medical Publishers, Chicago.

CHALLICE, C. E. 1971. Functional morphology of the specialized tissues of the heart. *In* E. Bajusz and G. Jasmin (eds.), Methods and Achievements in Experimental Pathology, Vol. 5: Functional Morphology of the Heart, pp. 121–172. Karger, Basel.

CHALLICE, C. E. and S. VIRÁGH. 1973. Ultrastructure of the Mammalian Heart. Academic Press, New York.

HEBB, C. 1969. Motor innervation of the pulmonary veins. *In* A. P. Fishman and H. H. Hecht (eds.), The Pulmonary Circulation and Interstitial Space, pp. 195–222. University of Chicago Press, Chicago.

JOHNSTONE, P. N. 1925. Studies on the physiological anatomy in the embryonic heart. II. An enquiry into the development of the heart beat in chick embryos, including the irritability to electrical stimulation. Bull. Johns Hopkins Hosp. 36: 299–311.

KARRER, H. E. 1959. Striated musculature of blood vessels. I. General cell morphology. J. Biophys. Biochem. Cytol. 6: 383–392.

KRAMER, A. W., and MARKS, L. S. 1965. Occurrence of cardiac muscle in the pulmonary veins of rodentia. J. Morphol. 117: 135–150.

LUDATSCHER, R. M. 1968. Fine structure of the muscular wall of rat pulmonary veins. J. Anat. 103: 345–357.

VIRÁGH, S., and C. E. CHALLICE. 1973. Origin and differentiation of cardiac muscle cells in the mouse. J. Ultrastruct. Res. 42: 1–24.

Mailing address:
Professor C. E. Challice,
Dept. of Physics, Faculty of Arts and Sciences,
University of Calgary
Calgary, Alberta T2N 1N4 (Canada).

Recent Advances in Studies on
Cardiac Structure and Metabolism Volume 10
The Metabolism of Contraction
Edited by Paul-Emile Roy and George Rona

INHIBITION OF EXTRACELLULAR AND PURIFIED 5′-NUCLEOTIDASE FROM RAT HEART

H. P. BÄR and L. P. SIMONSON

Department of Pharmacology, University of Alberta, Edmonton, Alberta, Canada

SUMMARY

5′-Nucleotidase in perfused rat hearts, accessible to extracellular substrate [^{14}C]-AMP contained in the perfusate, was compared to a partially purified 5′-nucleotidase from the same organ. Both activities were inhibited by ATP and, more effectively, by the ADP analog adenosine-α,β-methylene diphosphate (APCP). The isolated enzyme showed a competitive inhibitor constant of 5.4×10^{-8} M for APCP (K_m of AMP = 1.4×10^{-5} M). Although both activities were effectively inhibited by APCP, insufficient knowledge about the selectivity of this agent as a nucleotidase inhibitor does not permit a conclusion on whether or not the extracellular activity is identical to the partially purified enzyme.

INTRODUCTION

It has been suggested that adenosine serves as a mediator in the autoregulation of coronary flow in response to changes in arterial oxygen content (Berne, 1963; Rubio, Berne, and Dobson, 1973). Accordingly, one must postulate mechanisms that link oxygen content in blood to metabolic events in the myocardium that control the formation and release of adenosine into the circulation during hypoxia. Since ATP and ADP content in cardiac tissue is rapidly reduced and AMP content is increased when oxygen supply is limited, it appeared reasonable to propose that both the increased availability of substrate AMP and the reversal of ATP inhibition of 5′-nucleotidase from rat heart are possible ways in which adenosine formation might be regulated (Bär, Drummond, and Duncan, 1966). It has also been shown that ADP is a powerful inhibitor of 5′-nucleotidase from rat heart and that Mg^{2+} ion may reverse nucleotide inhibition (Sullivan and Alpers, 1971). The latter effect has thus also been implicated as a factor controlling adenosine formation although it is not known what the effective concentrations of Mg^{2+} or adenine nucleotides are in the vicinity of the 5′-nucleotidase.

In the above studies (Bär *et al.*, 1966; Sullivan and Alpers, 1971) 5′-nucleotidase had been partially purified from rat heart which, compared to hearts from other mammals, contained a particularly high level of membrane-bound activity. The demonstration of an apparent extracellular 5′-nucleotidase in perfused hearts hydrolyzing substrates added to the perfusion medium (Bär and Drummond, 1968) raised the question of whether the enzyme purified from intact ventricular tissue of the rat originated from intracellular sites or whether it was perhaps identical with the membrane-bound activity accessible to substrates in the perfusate. The answer to this question appears crucial to the understanding of 5′-nucleotidase involvement in the regulation of adenosine formation. Accordingly, it became of interest to attempt a comparison and characterization of extracellular versus isolated 5′-nucleotidase. One way in which one might differentiate between such enzymes is on the basis of substrate or inhibitor specificities.

The present study deals with an attempt to probe the inhibitory effects of a nucleotide analog, adenosine-α,β-methylene diphosphate (APCP), on the partially purified enzyme as well as on the extracellular activity in perfused hearts. This analog has been found to act as a strong inhibitor of 5′-nucleotidase from pig small intestine (Burger and Lowenstein, 1970).

METHODS AND MATERIALS

Hearts of young Wistar rats (150–170 g) of either sex were used for Langendorff perfusions or for isolation of 5′-nucleotidase. The enzyme was prepared as described earlier (Bär *et al.*, 1966). 5′-Nucleotidase activity was measured optically by coupling with an excess of adenosine deaminase and recording the rate of inosine formation by the decrease of optical density at 265 nm ($\Delta\epsilon = -8100$ $l \cdot mole^{-1} \cdot cm^{-1}$, pH 7.5) with a Gilford spectrometer. The assay medium was composed of 25 mM Tris·HCl (pH 7.5), 3 mM $MgCl_2$, 10 μg of adenosine deaminase (Sigma, intestinal), 5–20 μg of 5′-nucleotidase, and 2–40 μM AMP, in a total volume of 2 ml. Reactions were carried out at 25°C and initiated by addition of 5′-nucleotidase.

With Langendorff-perfused hearts, injection of radioactive substrates and measurement of degradation products in the perfusate were performed according to a previously published procedure (Bär and Drummond, 1968). This involved collection of a main peak of radioactivity in the perfusate (up to 30 sec following substrate injection) and a tail peak (up to 3 min following substrate injection). The main peak contained virtually all of the undegraded nucleotide besides a portion of the diffusible degradation products. The tail portion contained further degradation products presumably back-diffusing from intercellular sites, including adenosine, inosine, and hypoxanthine. Substrates were injected by inserting the needle of a 50-μl Hamilton syringe into the tubing connecting to the cannula, tied to hearts so that the syringe tip reached the aortic orifice tied to the cannula. Generally, 10-μl quantities of nucleotide solution (in H_2O or saline) were thus pushed into the perfusing buffer within about 1 sec. The perfusion medium contained 137 mM NaCl, 2.7 mM KCl, 10 mM $NaHCO_3$, 0.4 mM NaH_2PO_4; 0.69 mM $MgCl_2$, and 5.6 mM glucose, and was oxygenated with 95% O_2, 5% CO_2. Ca^{2+} was routinely omitted, since it was found that its absence did not affect either extracellular 5′-nucleotidase or ATPase activities and since the sampling of perfusate from nonbeating hearts was greatly facilitated.

Constant perfusion flow was achieved by a syringe drive pump or a peristaltic pump. Total amounts of radioactivity in tail and main peaks were determined, and the proportion of AMP in the main peak only was determined by chromatographing an aliquot together with AMP, adenosine, and inosine as carriers on Whatman No. 1 chromatography paper and isopropanol:ammonia:H_2O (7:1:2 volume) as the solvent for about 3–4 hours. Spots were marked under UV light, cut out, and counted by liquid scintillation. From the total amount of radioactivity collected in main and tail peaks and the proportion of AMP remaining undegraded in the main peak, the over-all fraction of AMP undegraded could be calculated.

When [α-^{32}P]-ATP was used as substrate, radioactive nucleotides appeared only in the 0- to 30-sec main peak following injection. The radioactivity distribution in main and tail peaks was measured as above, and aliquots of the tail peak were chromatographed, together with appropriate carriers, on polyethyleneimine-impregnated cellulose thin layers (Macherey-Nagel/Brinkman) using 1 M LiCl for development. UV-marked spots for ATP, ADP, and AMP (the latter plus inorganic phosphate) were cut out and counted by liquid scintillation. Calculation of the degree of ATP hydrolysis was then done analogous to the procedure described for AMP.

The dilution of injected nucleotides and their effective concentration within the perfusion medium were not precisely known. However, based on the flow rate of 4.5 ml/min usually employed in this work and an estimated period of 1–2 sec for injection of 10-μl quantities of nucleotide solution, the dilution was approximately 15- to 30-fold, assuming efficient mixing of the injected solution with the perfusion medium. The concentrations of injected nucleotides are state in terms of that of the injected stock solution. Mixtures of substrate, i.e., radioactive AMP, and inhibitory nucleotides APCP or ATP were prepared prior to injection.

Radioactive AMP was obtained from New England Nuclear Corp. (AMP-8-^{14}C), and ATP from Internation Chemical and Nuclear Corp. (ATP-α-^{32}P). The adenosine-α,β-methylene diphosphate was obtained from Miles Laboratories.

RESULTS

Studies with perfused hearts

The extensive degradation of both AMP and ATP during a single passage through the heart has been reported (Bär and Drummond, 1968). The present methodology was adopted, since it allowed the use of small quantities of costly, radioactive substances while showing essentially the same type of results as with the charcoal absorption procedure used previously. As in earlier work (Bär and Drummond, 1968), the radioactivity in the perfusate appeared within 10–15 sec as a main peak (about 60–75% of total radioactivity) and a tail portion. The latter contains nucleosides and degradation products diffusing from intracellular spaces back into the medium; only about 1–3% of the radioactivity in the tail peak represented 5′-AMP.

An attempt to estimate the affinity (K_m) of AMP to the nucleotidase activity in the perfused heart was made by injecting different amounts of unlabeled AMP together with radioactive AMP. However, the degree of hydrolysis was only slightly reduced (from 87 to 78%) when 5 mM AMP plus 0.5 mM [^{14}C]-AMP were injected, compared to 0.5 mM radioactive substrate alone. This result indicated that, even at concentrations of about 5 mM AMP (i.e., about 0.2–0.4 mM estimated effective concentration in the perfusate), no saturation of the enzyme was achieved. It may be possible to utilize higher amounts of substrate by using perfusion media containing AMP at desired concentrations; however, this was not attempted in the present study.

When radioactive AMP was injected together with ADP or ATP, the hydrolysis of AMP was inhibited. For example, at 5 mM ADP or ATP, the degree of hydrolysis of AMP was reduced from 87% to about 70%. The effect of different doses of ATP on AMP hydrolysis is shown in Figure 1. Compared to the substrate, about 10–100 times higher doses of ATP were required under the experimental conditions to reduce AMP degradation by one-half. It is also evident from Figure 1 that ATP did not seem to inhibit AMP hydrolysis completely, the effect leveling off around 50 mM ATP.

In addition to ATP and ADP, it was shown that the analog APCP strongly inhibited AMP hydrolysis, which is documented in Figure 2. Again it appears that complete inhibition is not achieved. The sequence of using high versus low doses of either ATP (Fig. 1) or APCP (Fig. 2) in a single heart preparation did not affect the dose-response characteristics. In most experiments, low doses were given first and, at the end of the sequence, basal rates of degradation in the absence of inhibitors were checked and found to be unaffected. While experiments from single heart preparations are shown, virtually identical results with further preparations were obtained repeatedly. The size of the organ, variations in flow rates and injection rates, as well as the extent of damage or minor clots during preparation may be responsible for variation of AMP degradation in the absence of inhibitors, ranging generally between 80 and 90% AMP hydrolysis (0.5 mM).

Figure 2 also indicates that APCP is at least 10 times more potent than ATP in inhibiting extracellular 5′-nucleotidase action. This observation was confirmed in a series of measurements in a single heart (Fig. 3). Furthermore, cold AMP added

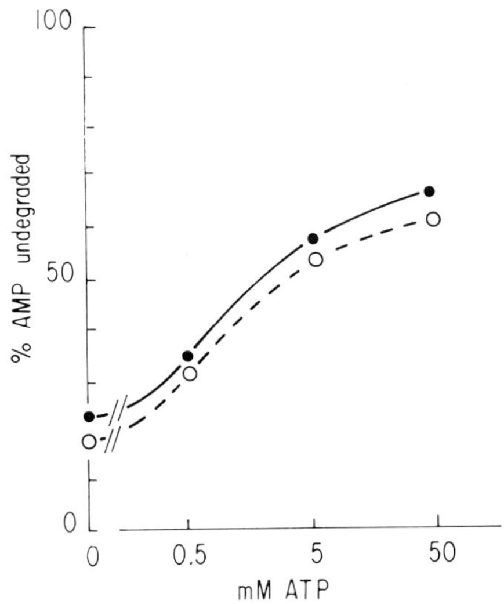

Figure 1. Effect of ATP on [^{14}C]-AMP degradation by extracellular nucleotidase. Data obtained in a single heart preparation. *Sold line:* increasing dose of ATP used in sequence. *Broken line:* repeated sequence above in reverse order.

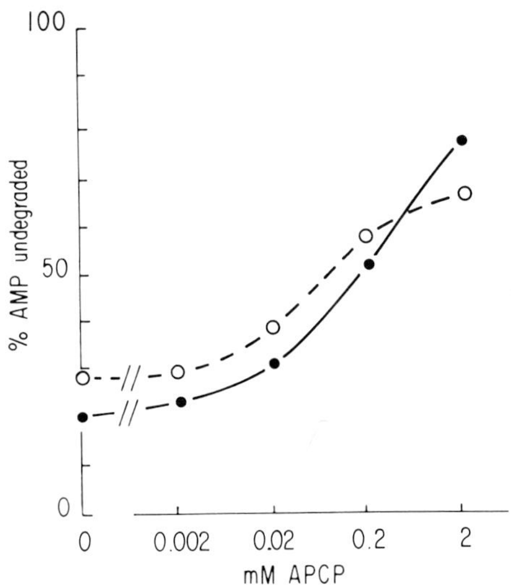

Figure 2. Effect of APCP on [^{14}C]-AMP degradation. Sequences of addition as in Figure 1. Data from 1 heart preparation.

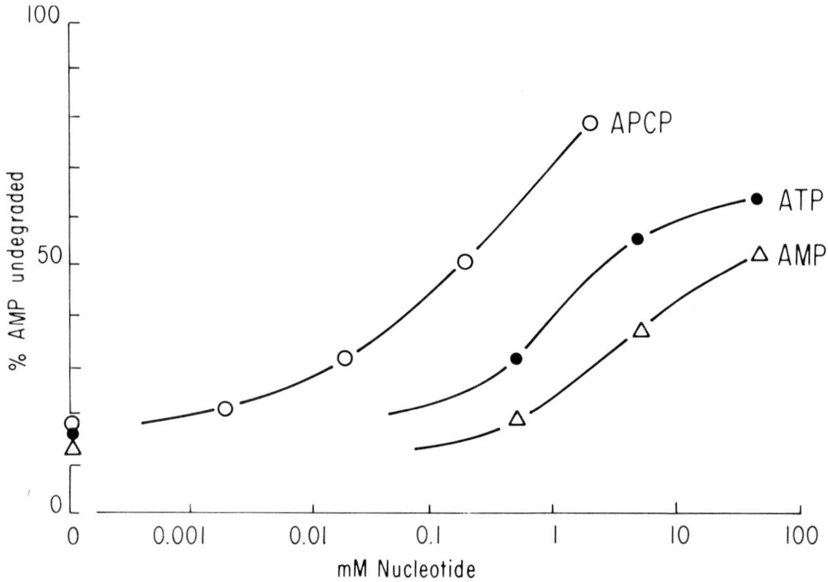

Figure 3. Comparative effect of ATP, AMP, and APCP on [^{14}C]-AMP degradation. Data from 1 heart preparation.

at different doses was shown to reduce the degree of radioactive AMP degradation less efficiently than either ATP or APCP. Consequently, the effect of ATP could not be due to AMP, which is one of the products of ATP hydrolysis (Bär and Drummond, 1968).

The inhibitory effect of APCP was found to be rather selective with respect to AMP hyrolysis, since it did not affect the degradation of infused radioactive ATP to any extent. For example, 0.5 mM ATP was hydrolyzed to about 80–90% during a single passage through a perfused heart, and this result was the same when 1 mM APCP was injected simultaneously.

Studies with 5′-nucleotidase

The Michaelis constant of AMP for the nucleotidase was determined by the optical assay procedure outlined above, and the mean of four experiments was $1.4 \pm 0.3 \times 10^{-5}$ M. APCP was found to be a potent inhibitor of the enzyme. In a series of experiments, the inhibition appeared to be competitive when analyzed by the method of Cleland (1963); i.e., double reciprocal plots of rate curves in the absence and presence of various concentrations of APCP showed common intersection points on the y-axis. The K_i of APCP was found to be about 5.4×10^{-8} M.

DISCUSSION

Studies on the inhibition of rat heart 5′-nucleotidase in vitro indicated that APCP is as a powerful inhibitor of the enzyme. This observation confirmed the results of the study by Burger and Lowenstein with 5′-nucleotidase from pig intestine (1970). The K_i of 5.4×10^{-8} M for the heart enzyme is astonishingly low, being about 2–3 orders of magnitude below the K_m for AMP and the inhibitory constant of about 1.8×10^{-6} M for ATP (Bär and Drummond, 1968).

It is not yet known whether APCP inhibits 5′-nucleotidases (or perhaps also other phosphatases) from other tissues and species or whether this effect is limited to 5′-nucleotidase from smooth muscle. It is likely that the analog is not a very selective agent, but experimental documentation of this possibility is required.

The effects of ATP and APCP on the hydrolysis of AMP by perfused heart indicate that the extracellular enzyme is subject to inhibition by the same compounds that inhibit AMP hydrolysis by the isolated enzyme. Unfortunately, the present methodology did not allow a clear kinetic analysis of the properties of the extracellular 5′-nucleotidase in the intact, perfused heart. However, the qualitative and comparative data obtained (Fig. 3) indicate clearly that APCP is a powerful inhibitor of the extracellular as well as the isolated enzyme, exceeding ATP by at least 1 to 2 orders of magnitude. However, a direct comparison is complicated by the fact that extensive degradation of ATP occurs during perfusion (Bär and Drummond, 1968).

While the results of the present study show that both the extracellular and the isolated 5′-nucleotidase from rat heart are subject to effective inhibition by nucleotides, these data do not yet indicate clearly whether or not these activities are identical. It seems feasible, however, to utilize the present technique in the study of other potential nucleotidase inhibitors. Preferably, substrate analogs should be chosen that are resistant to common phosphorylytic enzymes present in high amounts in the coronary vascular bed. The lack of complete inhibition of AMP hydrolysis, even with a large excess of APCP, suggests a heterogeneity of phosphorylytic enzymes (phosphatases) being present.

The histochemical study of 5′-nucleotidase in mammalian hearts (Rubio *et al.*, 1973), showed that enzyme activity is associated with membrane systems adjacent to extracellular spaces. The latter authors consider this to be in line with the adenosine hypothesis (Berne, 1963) of coronary flow regulation, since adenosine could effectively be released into these spaces and diffuse to arterial smooth muscle cells. It is not clear, however, whether this histochemically demonstrated enzyme activity is accessible to extracellular or intracellular substrate or both.

Acknowledgment This study was supported by funds from the Alberta Heart Foundation.

REFERENCES

BÄR, H. P., and G. I. DRUMMOND. 1968. Catabolism of adenine nucleotides by the isolated perfused rat heart. Proc. Soc. Exp. Biol. Med. 127: 33–36.

BÄR, H. P., G. I. DRUMMOND, and E. L. DUNCAN. 1966. Formation and deamination of adenosine by cardiac muscle enzymes. Mol. Pharmacol. 2: 67–76.

BERNE, R. M. 1963. Cardiac nucleotides in hypoxia: Possible role in regulation of coronary blood flow. Am. J. Physiol. 204: 317–322.

BURGER, R. M., and J. M. LOWENSTEIN. 1970. Preparation and properties of 5′-nucleotidase from smooth muscle of small intestine. J. Biol. Chem. 245: 6274–6280.

CLELAND, W. W. 1963. Computer programmes for processing enzyme kinetic data. Nature 198: 463–465.

RUBIO, R., R. M. BERNE, and J. G. DOBSON, Jr. 1973. Sites of adenosine production in cardiac and skeletal muscle. Am. J. Physiol. 225: 938–953.

SULLIVAN, J. M., and J. B. ALPERS. 1971. In vitro regulation of rat heart 5′-nucleotidase by adenine nucleotides and magnesium. J. Biol. Chem. 246: 3057–3063.

Mailing address:
H. P. Bär,
Department of Pharmacology,
University of Alberta,
Edmonton, Alberta T6G 2E1 (Canada).

Recent Advances in Studies on
Cardiac Structure and Metabolism Volume 10
The Metabolism of Contraction
Edited by Paul-Emile Roy and George Rona

CONTRACTILE APPARATUS IN AORTIC ENDOTHELIUM OF HYPERTENSIVE RAT[1]

GIULIO GABBIANI and MARIE-CLAUDE BADONNEL

Department of Pathology, University of Geneva,
Geneva, Switzerland

ISTVAN HÜTTNER and GEORGE RONA

Department of Pathology, McGill University,
Montreal, Quebec, Canada

SUMMARY

One week after hypertension was produced in male Wistar rats (150 g body weight) by a complete aortic ligature placed between renal arteries, electron microscopic studies showed a striking increase of cytoplasmic microfilaments in the endothelial cells of the aortic segment above coarctation (mean blood pressure 160 mm Hg). These microfilaments measured 40–70 Å in diameter and were located particularly close to the endothelial clefts.

In "*en face*" preparation of aortic endothelial cells treated with antiactin autoantibodies (AAA) followed by antihuman IgG, the cells above the ligature of hypertensive animals were intensely fluorescent compared with those of the aortic portion below the ligature or that of controls. The fluorescence was abolished after incubating the AAA-containing sera with thrombosthenin-A, suggesting the presence of actin. There was also an increased transport of horseradish peroxidase and ferritin through the endothelial cell layer. While endothelial cells so modified may play a role in permeability regulation, they may also be related to such mechanisms as electrotonic coupling and synchronized contraction of aortic cells during hypertension.

INTRODUCTION

The disclosure of microfilamentous structures in nonmuscular cells has stimulated great interest concerning their distribution, composition, and functional significance (Gabbiani *et al.*, 1973). The present study deals with the development of an extensive microfilamentous apparatus in the rat aortic endothelial cells as part of an adaptive response following hypertension produced by aortic coarctation.

---

[1] Supported by the Fond National Suisse pour la Recherche Scientifique, Grant 3.0330.73, and by the Medical Research Council of Canada, Grant MT 3683.

MATERIALS AND METHODS

A total of 72 male Wistar rats, weighing 120–150 g, were divided into three groups. One group served as untreated controls. In the second group, the aorta was completely ligated between the two renal arteries. In the third group, the aorta was ligated below both renal arteries and above the lumbar artery. On the 7th day, the blood pressure (BP) of all animals was measured under ether anesthesia, by means of an electromanometer. In rats with aortic ligature, the BP was recorded after cannulating the carotid artery and the aorta below the ligature.

For electron microscopic studies, 10 rats per group were sacrificed by perfusing glutaraldehyde fixative (Gabbiani *et al.*, 1974a). Blocks were selected from segments of aortic arch, and thoracic and abdominal aortas. Thin sections were examined under a Philips 300 electron microscope.

For immunofluorescent studies, "*en face*" preparation of endothelial cells was used (Warren, 1965) from the above three aortic regions of 10 rats per group. Sera from two patients with chronic aggressive hepatitis were used for antiactin autoantibodies (AAA) (Gabbiani and Majno, 1974). The titers of these sera were 1/1280 and 1/640 as judged by immunofluorescent staining of rat intestinal smooth muscle. The sera gave negative results for antimitochondrial or anti-DNA antibodies and for antinuclear factor. Aortic cell endothelial preparations were treated for 30 min with AAA serum. They were then washed in phosphate-buffered saline (PBS) and stained for 30 min with fluorescein-conjugated IgG fraction of goat antiserum to human IgG. After rewashing with PBS and mounting in 90% glycerol in PBS, the level of fluorescence was compared with that found in control preparations treated with normal human serum instead of AAA serum.

Immunoabsorption was performed as follows: 0.1 ml of each serum was added to 0.9 ml of a solution containing highly purified thrombosthenin-A, the actin moiety of thrombosthenin, the contractile protein of platelets dissolved in a buffer composed of 5.10^{-4} M KCl, 2.10^{-3} M Tris-(hydroxymethyl) amino-methene, and 5.10^{-4} M 2-mercaptoethanol, the pH being 7.9 (the solution contained 3.2 mg/ml of the protein per ml and was kindly supplied by Prof. E. F. Lüscher) (Probst and Lüscher, 1972). Incubation at 37°C lasted for 1 hour, with agitation at 15-min intervals. After incubation, the precipitate present in the bottom of the tube was eliminated by centrifugation of the serum for 10 min at 4,600 × *g*. For control purposes, 0.1 ml of each serum was incubated with 0.9 ml of the buffer alone. Photographs were taken on a Zeiss UV photomicroscope with UG 1 or UG 2 excitor filters and Zeiss 50/65 barried filter, using Anscochrome color slide film 50 daylight.

In an additional group of 12 rats, correlative permeability studies with fine structural protein tracers were carried out, using horseradish peroxidase (HRP)

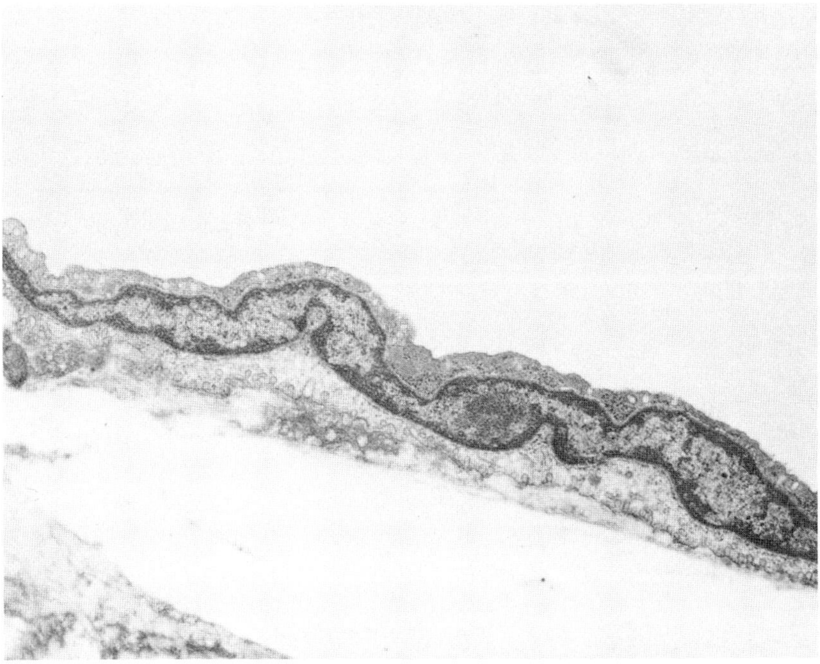

Figure 1. Endothelial cell in the aortic arch of normotensive animal. Nucleus has a regular contour; cytoplasm contains mitochondria, rough endoplasmic reticulum, pinocytotic vesicles, but no microfilaments. × 13,000.

following 60-min circulation time or ferritin (20 min) by techniques described previously (Hüttner, Boutet, and More, 1973c; Hüttner *et al.*, 1973d).

RESULTS

The mean BP of normal controls was 105 mm Hg. In the rats with ligature placed between the two renal arteries, the mean BP was 160 mm Hg above the ligature and 25 mm Hg below the ligature. In the animals with the ligature below both renal arteries, the mean BP values above and below the ligature were 116 and 32 mm Hg, respectively.

In normotensive animals (Fig. 1), the ultrastructure of aortic intima was as described previously (Schwartz and Benditt, 1972; Hüttner, Boutet, and More, 1973b). Microfilaments (40–70 Å in diameter) could rarely be seen in regions close to intercellular clefts (Fig. 2). In rats with ligature below the renal arteries, the ultrastructure of the thoracic aortic intima was basically similar to that observed in controls. In animals with ligature between the two renal arteries, the

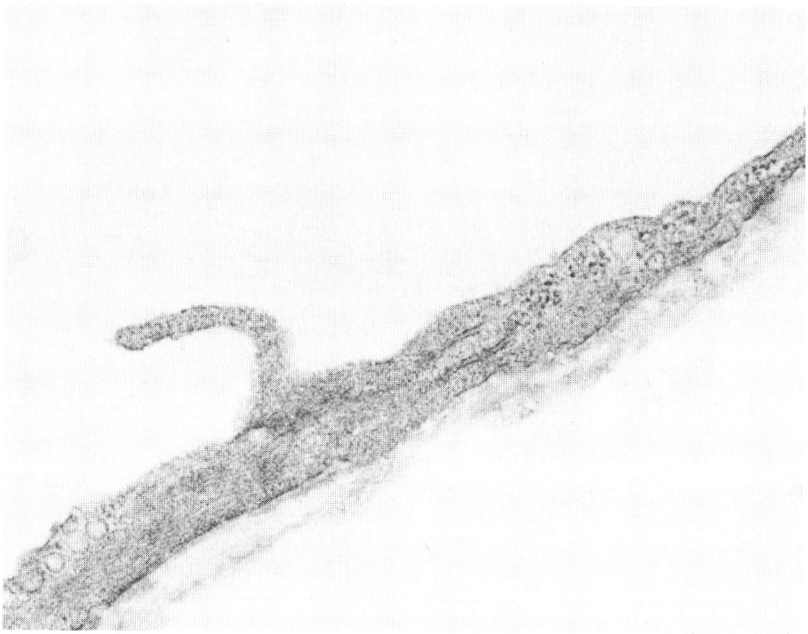

Figure 2. Some microfilaments (40–70 Å in diameter) are present at periphery of endothelial cell in aortic arch of normotensive animal. × 33,000.

endothelial cells of the arch and thoracic portion presented characteristic changes. The thickness of these cells was generally increased when compared with that found in normotensive segments up to 5.5 μ. The nuclear contour was irregularly folded. In practically all endothelial cells examined, the cytoplasm was well preserved, showing prominent organelles including an extensive cytoplasmic system of microfilaments. The diameter of these filaments was 40–70 Å, and electron-dense regions were distributed more or less regularly throughout the filamentous bundles (Fig. 3). These dense regions sometimes appeared as transverse bands and were similar to the "attachment sites" of the underlying medial smooth muscle (Fig. 4). When the dense areas were situated in relation with the basement membrane-like material at the periphery of the cells, they resembled the hemidesmosomes present in the wall of small vessels between pericytes or endothelial cells and their basement membranes (Hogan and Feeney, 1963; Stehbens, 1966). Although the filamentous bundles were irregularly distributed throughout the cytoplasm, they were particularly abundant adjacent to endothelial clefts. They presented a preferential disposition close to the basement membrane while, in still other cases, they were situated close to the

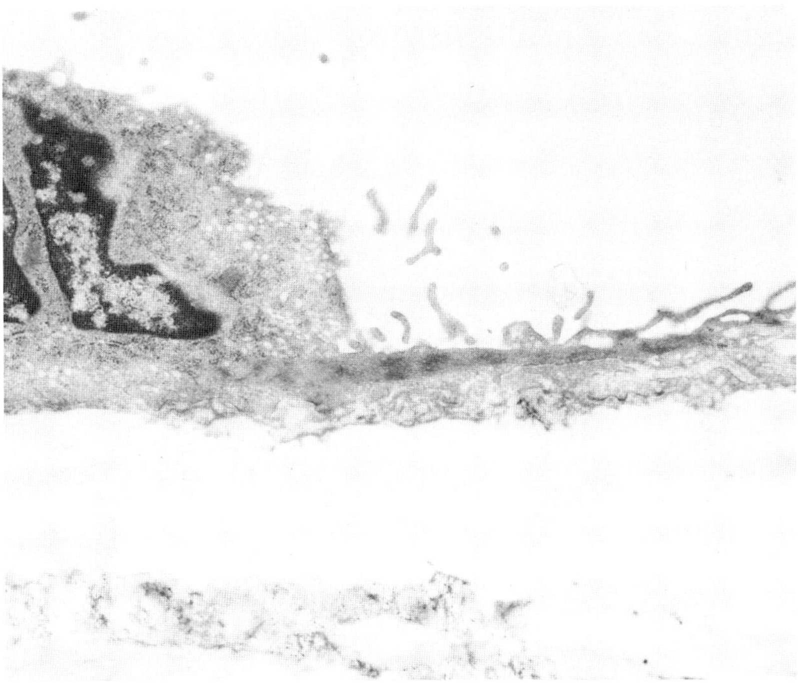

Figure 3. Endothelial cell from aortic arch of hypertensive rat. Here, cytoplasmic microfilaments are prominent. Electron-dense areas similar to "attachment sites" are regularly distributed throughout filament bundle. × 12,500.

luminal surface (Fig. 5). In contrast to normotensive segments, microvillous projections were numerous on the luminal surface in hypertensive thoracic aortic endothelial cells. The endothelial clefts, except for occasional focal dilations, did not appear to be different when compared to those of normal animals. Endothelial cell junctions were structurally intact.

In *en face* preparation of aortic endothelial cells, no fluorescence was seen after treatment with normal human serum. After incubation with AAA serum, a weak fluorescence was detected at the periphery of endothelial cells of control animals or in the abdominal portion of the aorta of hypertensive rats (Fig. 6). On the other hand, an intense fluorescence was visible in the cytoplasm of endothelial cells in the arch and thoracic portion of the aorta of hypertensive animals. The fluorescence was localized in almost all the endothelial cells examined and it was particularly bright at the periphery of the cytoplasm (Fig. 7). In preparations treated with AAA serum incubated with thrombosthenin-A,

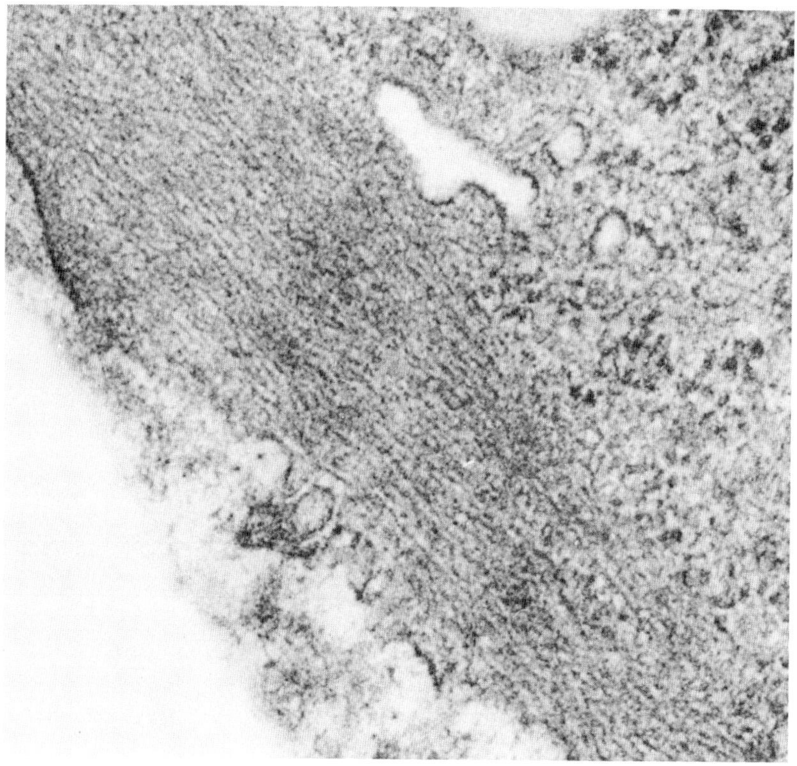

Figure 4. At higher magnification, microfilaments have a diameter of 40–70 Å. × 86,000.

just as in preparations stained with normal human serum, no fluorescence was present.

Permeability studies using HRP and ferritin revealed increased transport of both fine structural protein tracers in hypertensive aortic segments as compared with controls (Figs. 8 and 9).

DISCUSSION

Experimental hypertension produced by aortic coarctation described here appears to be a useful model for studying the formation of endothelial microfilaments. While there is a scarcity of such filaments in normal aortic endothelial cells, there is a striking augmentation of these structures in aortic endothelial cells in hypertensive segments. This response occurs without any degenerative

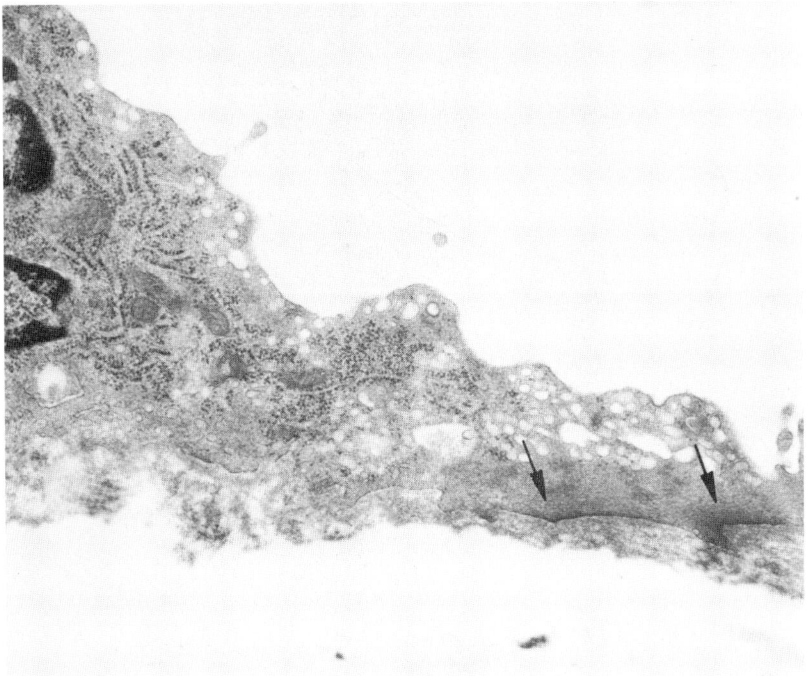

Figure 5. Endothelial cell from thoracic aorta of hypertensive rat. Bundles of microfilaments are prominent at cell periphery, where they form, in some places, electron-dense areas in relation with basement membrane (*arrows*). × 20,000.

Figure 6. *En face* preparation of endothelial cells from thoracic aorta of normotensive rat; cells were incubated first with AAA serum followed by fluorescein-conjugated antihuman IgG. Weak fluorescence is visible at periphery of endothelial cells. × 400.

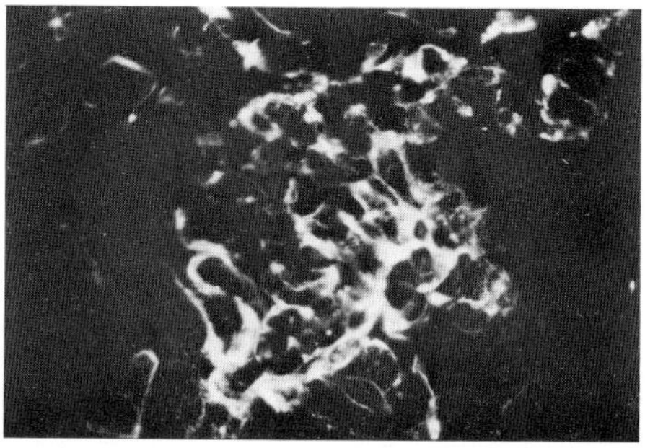

Figure 7. En face preparation of endothelial cells from thoracic aorta of hypertensive rat; cells were incubated first with AAA serum followed by fluorescein-conjugated antihuman IgG. Bright fluorescence is present at periphery of most cells. × 400.

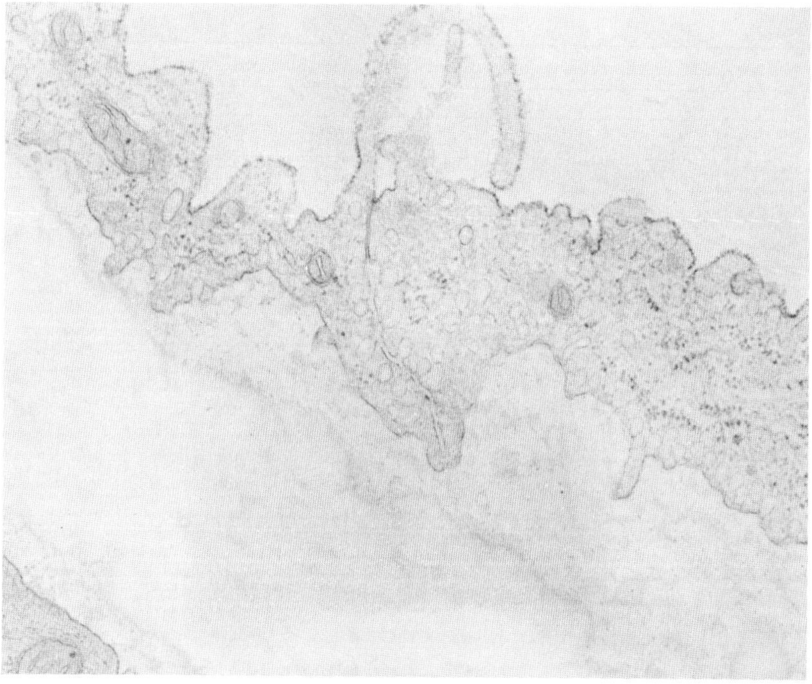

Figure 8. Intima of thoracic aorta of normotensive rat 60 min after horseradish peroxidase injection. Note absence of peroxidase reaction product in subendothelial space. × 33,000.

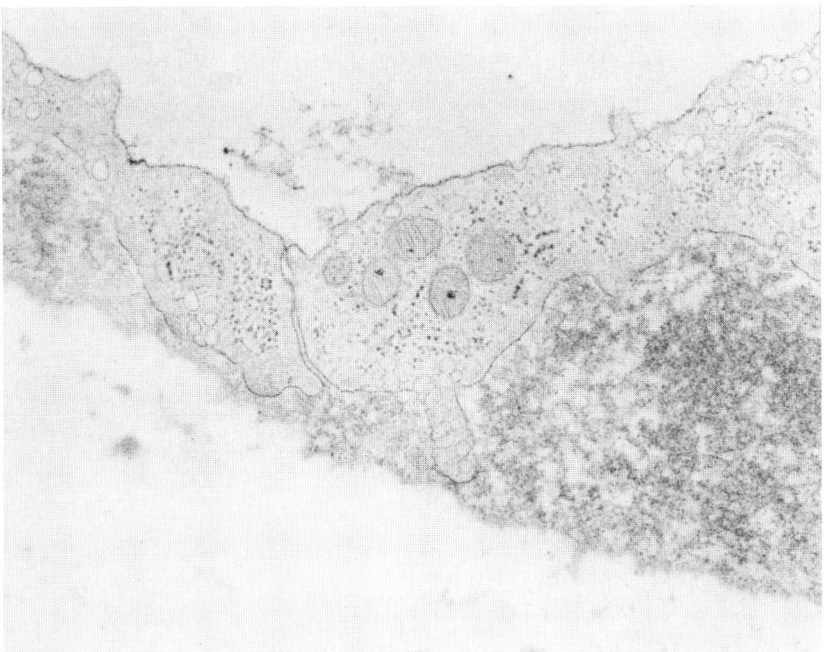

Figure 9. Similar region as in Figure 8 in hypertensive rat after identical tracer circulation time shows positive reaction for peroxidase in subendothelium. × 33,000.

change in the endothelial cells, while the continuity of the endothelial cell layer is maintained.

The close morphological similarity between endothelial filaments and filaments present in the underlying smooth muscle cells suggests that they are composed of contractile proteins. This conclusion was further supported by the result of immunofluorescence studies using AAA, indicating that these filaments are composed, at least in part, of actin.

What could be the function of such microfilamentous apparatus? Endothelial cell contraction as a mechanism of increased vascular permeability has been established in inflammation (Majno, Shea, and Leventhal, 1969; Gabbiani and Majno, 1974). In some hypertensive studies, the formation of contractile filaments has been correlated to increased permeability through interendothelial junctions (Giacomelli, Wiener, and Spiro, 1970; Giacomelli, Juechter, and Wiener, 1972). The present study has similarly shown a correlation between blood pressure elevation, formation of extensive microfilamentous apparatus in endothelial cells, and increased passage of fine structural protein tracers. The nature of this correlation, however, requires further studies. In acute high blood pressure conditions induced by the infusion of catecholamines or coarctation of the aorta, the passage of fine structural protein tracers was found to be increased both through junctional pathways and by plasmalemmal vesicles as compared with normal animals (Hüttner *et al.*, 1973d). In this case, no increase of filamentous structures was reported in the aortic endothelial cells. It may be considered that the filamentous apparatus affects not only interendothelial but also transendothelial permeability by plasmalemmal vesicles. Experimental evidence is now available in several types of cells suggesting that the processes of endocytosis may be mediated through a system or network of cytoplasmic filaments (Orci, Gabbay, and Malaisse, 1972; Schwartz and Benditt, 1972); in some instances this network has been shown to fix AAA. Hence, it probably consists of contractile protein (Gabbiani *et al.*, 1974b).

Another possible function of cytoplasmic filaments could be related to gap junctions recently described in aortic endothelial clefts (Hüttner, Boutet, and More, 1973a), which represent low resistance areas and are probably the site of electrotonic and metabolic couplings between endothelial cells, as well as between endothelium and smooth muscle. In this respect, the development of a microfilamentous apparatus in aortic endothelium possessing gap junctions may implicate a synchronized contraction of the coupled endothelial cell layer and/or endothelium and medial smooth muscle. This hypothesis is supported by our preliminary finding obtained in the same experimental hypertension model after longer observations periods, showing an association between new formation of microfilament and gap junctions but no increase of endothelial permeability.

REFERENCES

GABBIANI, G., M. C. BADONNEL, S. M. MATHEWSON, and G. B. RYAN. 1974a. Acute cadmium intoxication. Early selective lesions of endothelial clefts. Lab. Invest. 30: 686–695.

GABBIANI, G., and G. MAJNO. 1975. Fine structure of endothelium. *In* G. Kaley and B. M. Altura (eds.) Microcirculation. University Park Press, Baltimore (In press).

GABBIANI, G., F. MALAISSE-LAGAE, B. BLONDEL, and L. ORCI. 1974b. Actin in pancreatic islet cells. Endocrinology 95: 1630–1635.

GABBIANI, G., G. B. RYAN, J-P. LAMELIN, P. VASSALLI, G. MAJNO, C. A. BOUVIER, A. CRUCHAUD, and E. F. LUSCHER. 1973. Human smooth muscle autoantibody. Its identification as antiactin antibody and a study of its binding to "nonmuscular" cells. Am. J. Path. 72: 473–488.

GIACOMELLI, F., K. B. JUECHTER, and J. WIENER. 1972. The cellular pathology of experimental hypertension. VI. Alterations in retinal vasculature. Am. J. Path. 68: 81–96.

GIACOMELLI, F., J. WIENER, and D. SPIRO. 1970. The cellular pathology of experimental hypertension. V. Increased permeability of cerebral arterial vessels. Am. J. Path. 59: 133–160.

HOGAN, M. J., and L. FEENEY. 1963. The ultrastructure of the retinal vessels. III. Vascular-glial relationships. J. Ultrastruct. Res. 9: 47–64.

HUTTNER, I., M. BOUTET, and R. H. MORE. 1973a. Gap junctions in arterial endothelium. J. Cell Biol. 57: 247–252.

HÜTTNER, I., M. BOUTET, and R. H. MORE. 1973b. Studies on protein passage through arterial endothelium. I. Structural correlates of permeability in rat arterial endothelium. Lab. Invest. 28: 672–677.

HÜTTNER, I., M. BOUTET, and R. H. MORE. 1973c. Studies on protein passage through arterial endothelium. II. Regional differences in permeability to fine structural protein tracers in arterial endothelium of normotensive rat. Lab. Invest. 28: 678–685.

HÜTTNER, I., M. BOUTET, G. RONA, and R. H. MORE. 1973d. Studies on protein passage through arterial endothelium. III. Effect of blood pressure levels on the passage of fine structural protein tracers through arterial endothelium. Lab. Invest. 29: 536–546.

MAJNO, G., S. M. SHEA, and M. LEVENTHAL. 1969. Endothelial contraction induced by histamine-type mediators. An electron microscopic study. J. Cell Biol. 42: 647–672.

ORCI, L., K. H. GABBAY, and W. J. MALAISSE. 1972. Pancreatic beta-cell web: its possible role in insulin secretion. Science 175: 1128–1130.

PROBST, E., and E. F. LÜSCHER. 1972. Studies on thrombosthenin-A, the actin-like moiety of the contractile protein from blood platelets. I. Isolation, characterization and evidence for two forms of thrombosthenin-A. Biochim. Biophys. Acta 278: 577–584.

SCHWARTZ, S. M., and E. P. BENDITT. 1972. Studies on aortic intima. I. Structure and permeability of rat thoracic aortic intima. Am. J. Path. 66: 241–264.

STEHBENS, W. E. 1966. The basal attachment of endothelial cell. J. Ultrastruct. Res. 15: 389–399.

WARREN, B. A. 1965. A method for the production of "en face" preparation cell thickness. J. Roy. Microscop. Soc. 84: 407–413.

Mailing address:
Dr. Guilio Gabbiani,
Department of Pathology, University of Geneva,
40 Boulevard de la Cluse, 1211 Geneva 4 (Switzerland).

Endocardial Reaction to Injury

Recent Advances in Studies on
Cardiac Structure and Metabolism Volume 10
The Metabolism of Contraction
Edited by Paul-Emlie Roy and George Rona

LIGHT AND ELECTRON MICROSCOPIC STUDY OF CELLULAR PROLIFERATION IN CARCINOID HEART DISEASE

R. LAGACE, C. DELAGE,
and M. BOUTET

Department of Pathology, Laval University, Québec, Québec, Canada

SUMMARY

An ultrastructural and histochemical study of the subendocardial lesion in carcinoid heart disease showed six different cell types within a myxoid matrix. The matrix, composed of a mucopolysaccharidic ground substance, collagen, and reticulin fibers, contained stem cells, four types of fusiform cells (fibrocytes, fibroblasts, myofibroblasts, and smooth muscle cells), and an intermediary cell type. Our observations suggest that the humoral mediators of the carcinoid syndrome may induce the differentiation of a subendocardial stem cell into contractile elements.

INTRODUCTION

The cardiac disease of the carcinoid syndrome is a well established entity (Oates and Sjoerdsma, 1962; Roberts and Sjoerdsma, 1964). Anatomic lesions of the right heart may be observed in about 50% of patients with malignant carcinoid tumors. In spite of the tendency of carcinoids to invade, these tumors rarely give rise to metastasis, except for those originating in the ileum which may spread to the regional lymph nodes and to the liver. With liver involvement, there appear a host of clinical symptoms known as the carcinoid syndrome: vascular, cutaneous, intestinal, pulmonary, and cardiac manifestations. Manifestations associated with this clinical syndrome are accompanied by cardiac lesions that are highly specific and found in the right heart. Endocardial and valvular lesions have also been observed in the left heart in the presence of atrial septal defect and in bronchial carcinoid tumors. There also exist isolated cases of left heart lesion without associated malformation (McKusick, 1956; Schweizer *et al.*, 1964; Fadell and Denham, 1966). These observations suggest that a circulating substance could be at the origin of the lesion known as "subendocardial fibrosis."

The biochemical, clinical, and histopathological aspects of this syndrome have been studied extensively and are well documented (Jones, 1962; Allen, 1964; Justin-Besançon *et al.*, 1965; Oates and Buttler, 1967; Von Studnitz and Waldenström, 1971; Laplante, Archambault, and Thibert, 1971). Only two ultrastructural studies have been published. The first one dealt especially with the aspect of the intercellular substance within the fibrous plaque, and the authors concluded that this material is of a reticular collagenic nature (Cosh, Cates, and Pugh, 1959). The second ultrastructural study was less extensive, and the authors suggested that the fusiform cells observed under light microscopic examination of the lesion are fibroblasts (Himbert *et al.*, 1965).

The present paper describes light and electron microscopic observations of the endocardial lesion of a typical carcinoid syndrome. On the basis of morphological data, the possible role of chemical mediators such as serotonin, bradykinin, and histamine, which have been involved in the pathogenesis of these lesions, is discussed.

MATERIALS AND METHODS

Samples of the fibrous plaques were obtained at autopsy from the right atrium, tricuspid valve, and right ventricle.

Histological examination

Tissues for histological examination were fixed in 10% buffered formalin, dehydrated, and embedded in paraffin. Sections were stained with hematoxylin-eosin, Weigert, periodic acid Schiff, Masson trichrome, Mallory's phosphotungstic acid, Turnbull blue, toluidine blue, Alcian blue, and Hale's colloidal iron.

Electron microscopic examination

Tissues for electron microscopy were cut into cubes of 1 mm or less and fixed by immersion in 2% glutaraldehyde for 2 hours. They were left in veronal buffer at 4°C overnight, postfixed for 2 hours in 1% osmium tetroxide at pH 7.4, dehydrated in alcohol, and embedded in Epon 812 or in Araldite. Sections approximately 1 μ thick were cut with glass knives on a Reichert OM/U2 ultramicrotome and stained with toluidine blue or thiamine-acridine orange. Thin sections were mounted on noncoated grids and stained with uranyl acetate and lead citrate (Reynolds, 1963). Parts of the tissues were treated for 2 hours at 4°C with 2% uranyl acetate in maleate buffer at pH 5.2 (Karnovsky, 1967) and stained with lead citrate (Venable and Coggeshall, 1965). The thin sections were examined with a Zeiss EM 9A electron microscope.

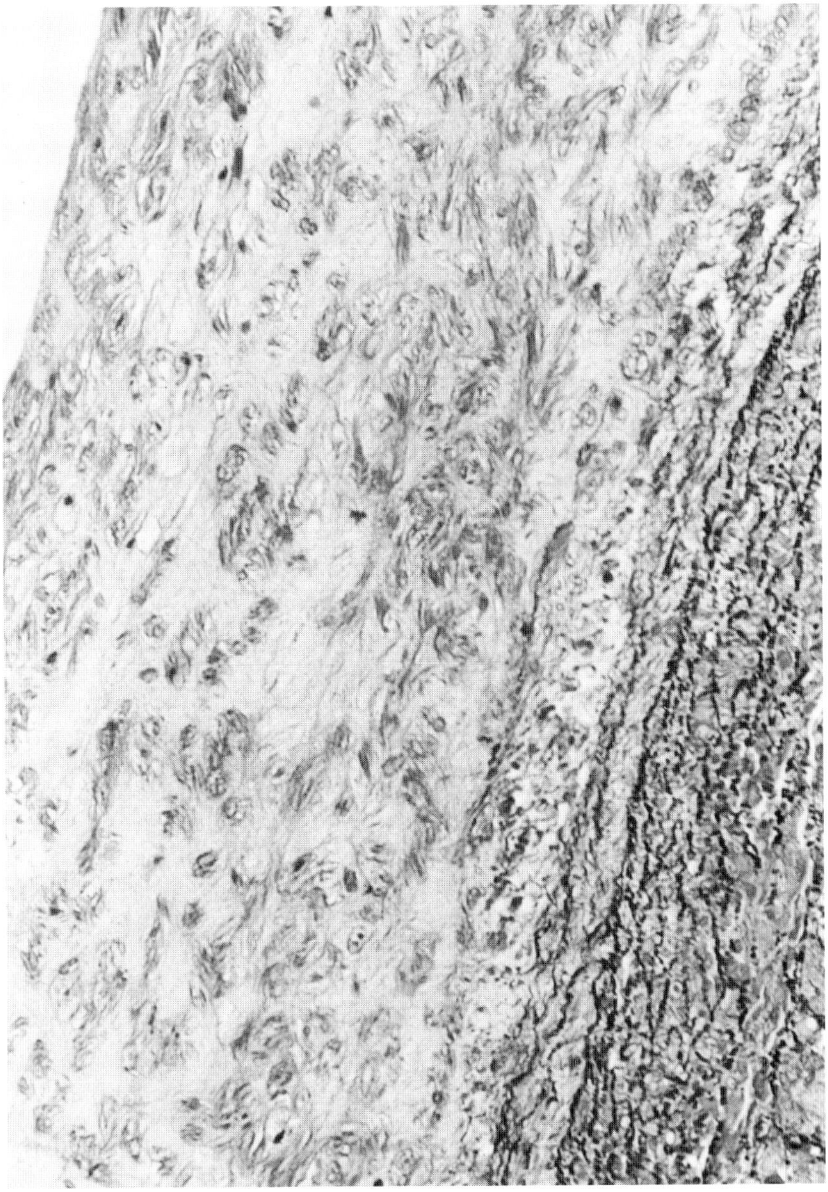

Figure 1. Light micrograph showing the fibrous thickening of the right atrial endocardium. Note that most of the cells are fusiform or round and surrounded by loose connective tissue. × 450.

RESULTS

Histological examination

Light microscopic study showed the fibrous plaques to consist mainly of fusiform cells within a myxoid matrix (Fig. 1). Some of the fusiform cells had elongated nuclei with indentations and deep folds. Other cells appeared to be round or starry. Lymphocytes and mast cells were also identified. The matrix, composed of a mucopolysaccharidic ground substance, showed reticulin and collagen fibers. A condensation of this myxoid matrix was observed surrounding the fusiform cells. Few elastic fibers were found at the base of the fibrous plaque. However, the fibrous plaques examined were free of hemosiderin and fibrin.

Electron microscopic examination

Fine structural examination of the fibrous plaques showed six different cell types within the endocardial cellular proliferation.

Some starry cells showed a large and ovoid nucleus with chromatin distributed in a uniform fashion within the nucleoplasm. Their cytoplasm contained a few organelles, free ribosomes, and scattered mitochondria. The cytoplasmic processes were of various length and thickness and appeared filiform (Fig. 2). These starry cells likely represented primitive mesenchymal cells.

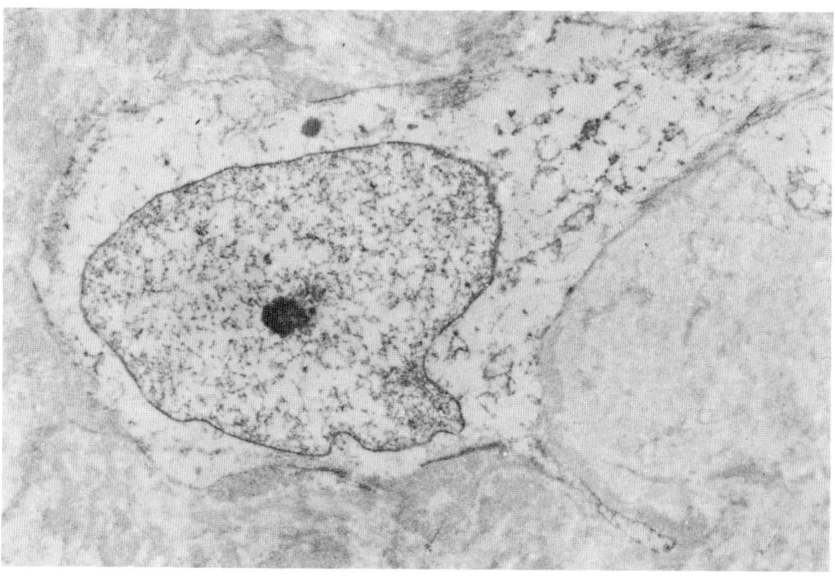

Figure 2. Starry cell with large ovoid nucleus. The cytoplasm contains rare organelles and shows filiform cytoplasmic processes. × 7,500.

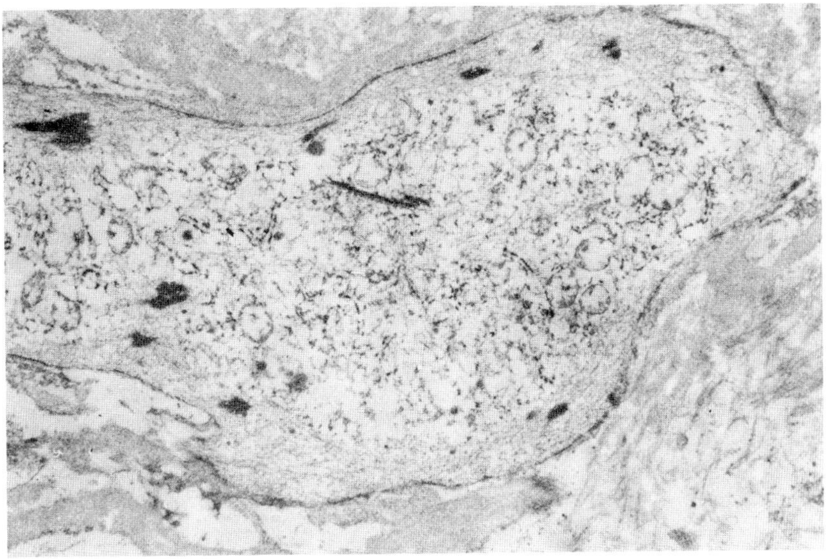

Figure 3. Large cell showing short profiles of rough endoplasmic reticulum and, close to the cytoplasmic membrane, a network of fine filaments with condensations. × 7,500.

It should be noted that a number of large cells showed at the periphery of their cytoplasm a network of fine filaments with numerous condensations of various sizes and shapes (Figs. 3–6). However, these cells did not have the typical aspect of the smooth muscle cell or the abundant rough endoplasmic reticulum of the fibroblast. They were interpreted as primitive cells with intermediary differentiation toward filament formations.

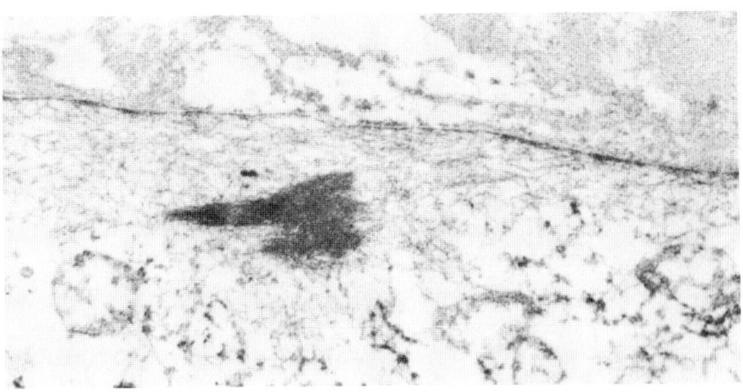

Figure 4. Higher power view of a condensation of fine filaments from Figure 3. × 24,500.

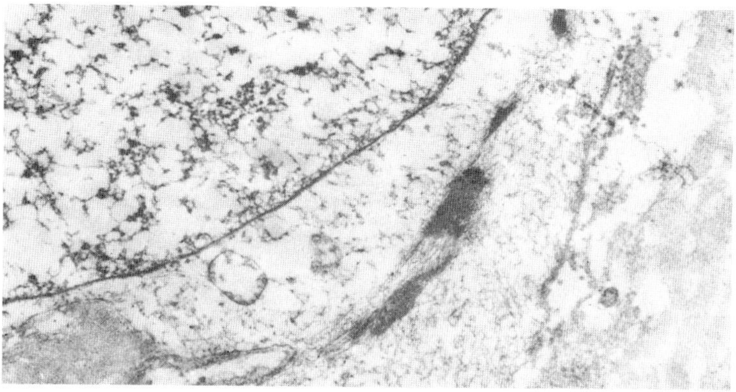

Figure 5. Similar condensations of filaments adjacent to nuclear membrane. × 20,000.

In addition, four types of fusiform cells were observed. The first type of fusiform cells might be seen in the deep region of the lesion. These cells appeared elongated with a fusiform nucleus, the chromatin of which was regularly distributed in the nucleoplasm. The elongated cytoplasm of these cells, of fibrocytic nature, contained rare organelles (Fig. 7). In places, it was only possible to identify their cytoplasmic profiles. They were surrounded by a great amount of finely granular, osmiophilic amorphous material, and by reticulin and collagen fibers easily recognizable by their characteristic periodicity.

The three other types of fusiform cells were larger and were seen throughout the endocardial lesion. The second type of fusiform cells was characterized by a cytoplasm rich in organelles. The rough endoplasmic reticulum was very well developed and the Golgi complex was prominent. In addition, these cells

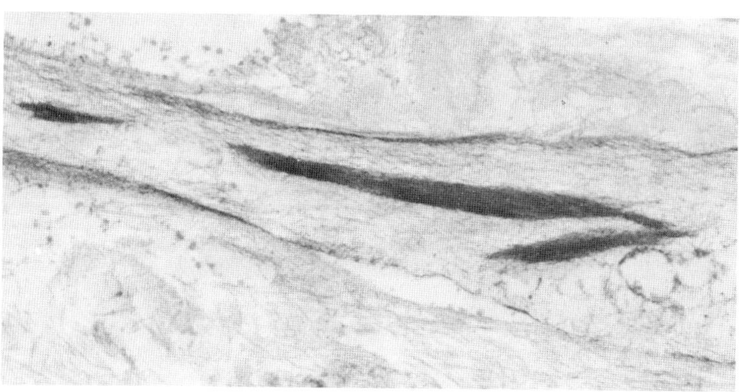

Figure 6. Portion of the cytoplasm of a fusiform cell, showing a fine network of filaments with three elongated condensations. × 20,000.

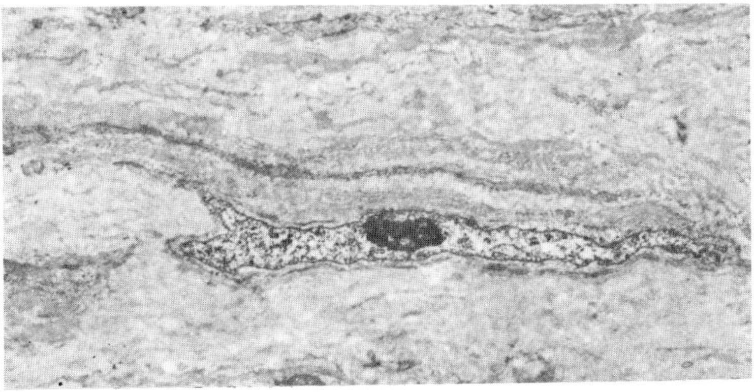

Figure 7. Fusiform cell with dark cigar-shaped nucleus and thin cytoplasm. × 7,500.

contained lipid vacuoles as well as dense bodies of lysosomal type. The nucleus of these cells was clearer and showed a granular chromatin. Apparently, these cells were active fibroblasts (Fig. 8). A third type of fusiform cells was noted. These were smooth muscle cells characterized by the presence of a fine network of intracytoplasmic filaments. In places, these filaments formed very dense osmiophilic condensations (Figs. 10 and 11).

Finally, a fourth type of fusiform cells was identified. These cells showed a double differentiation with characteristics of both smooth muscle fibers (fine network of myofibrils with condensations) and fibroblasts (well developed rough endoplasmic reticulum and Golgi complex). These cells represented myofibroblasts (Fig. 9).

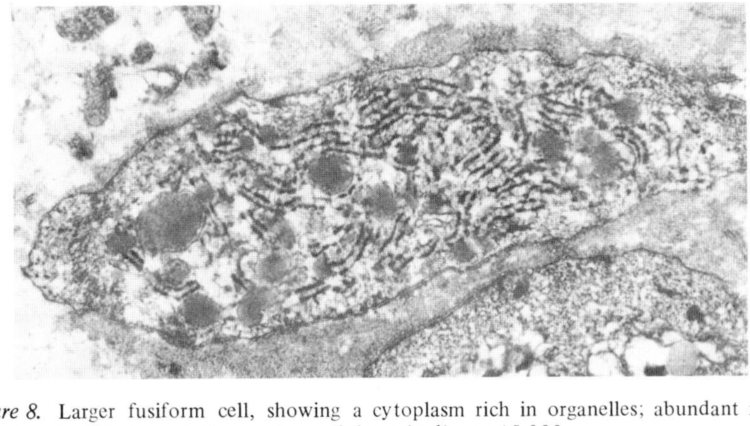

Figure 8. Larger fusiform cell, showing a cytoplasm rich in organelles; abundant rough endoplasmic reticulum, lipid vacuoles and dense bodies. × 15,000.

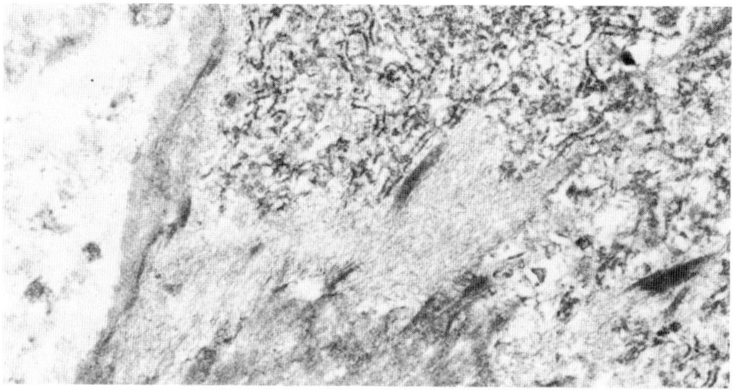

Figure 9. Cell showing a well developed rough endoplasmic reticulum adjacent to a fine network of filaments with condensations. × 10,500.

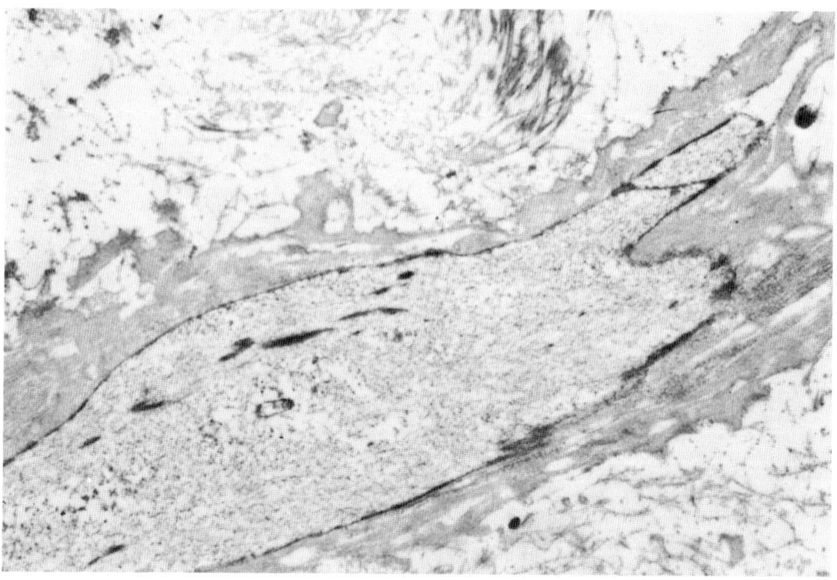

Figure 10. Fusiform cell with an abundant cytoplasm filled with a fine granular and filamentous material, showing elongated condensations. Note the dense granular matrix surrounding the cell membrane. × 9,000.

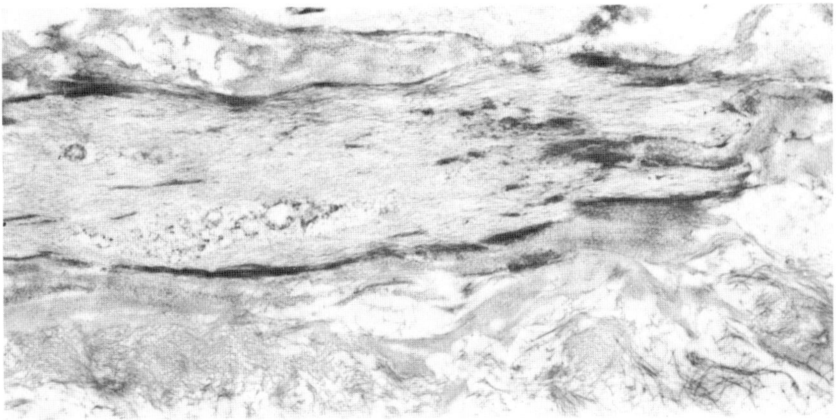

Figure 11. Fusiform cell showing a network of fine filaments with numerous condensations, some of which are closely apposed to the cell membrane. × 7,500.

In addition, it must be pointed out that, in spite of the great number of tissue blocks examined, it was impossible to identify endothelial cells at the base or at the periphery of the lesion. A careful examination of the intercellular matrix revealed that it was composed of an amorphous osmiophilic substance, sometimes finely granular, and of a few fibers of collagen (Fig. 5). No substance consistent with fibrin was identified in the interstitial tissue in the sections examined.

DISCUSSION

It has been suggested that the clinical manifestations of the carcinoid syndrome were related to an increase in the production of serotonin, histamine, or bradykinin (Grahame-Smith, 1968). It seems also that these same substances could be at the origin of or play an important role in the pathogenesis of the heart lesions observed in this syndrome (Campbell, 1968). The nature and the significance of the subendocardial lesion of the right ventricle in the carcinoid syndrome are still open to question (Spatz, 1961, 1964). In our opinion, the theory according to which the lesions in the heart would result from the development and organization of a mural thrombus (Cosh *et al.*, 1959) is less probable, because we never found either fibrin or hemosiderin-laden macrophages in our material. It seems more likely that the above-mentioned humoral mediators could react directly with preexisting cells in the endocardium to bring about the subsequent morphological alterations.

The lesions consist of a tumoral or pseudotumoral proliferation, the oncogenesis or histogenesis of which seems to originate from cellular components present within the endocardium. Recently, it has been demonstrated experimentally that the chemical mediators involved in the carcinoid syndrome could produce lesions in the endothelial cells. Indeed, Majno and his coworkers (1961, 1969) have shown that particular endogenous chemical mediators, of which histamine is the prototype, would increase the permeability of vascular endothelium by producing gaps between the endothelial cells. These gaps have been attributed to the exaggerated contractions of endothelial cells. Speculating from these experimental data, it is tempting to suggest that serotonin, histamine, or bradykinin involved in the genesis of the carcinoid syndrome could play such a role on the endocardial cells of the right heart. It appears that the separation of the cells leads to cell death more or less shortly. This hypothesis is suggested by the fact that it has been impossible to observe endothelial cells at the periphery or at the base of the lesion. Afterwards, these mediators could induce a cellular proliferation from preexisting mesenchymal cells within the subendocardial region. This theory seems more likely, since we have observed different types of starry or intermediate fusiform cells in the fibrous plaque and since these cells especially showed fibroblastic, myofibroblastic, and smooth muscle differentiation. This differentiating process of a primitive mesenchymal cell into other above-mentioned cellular elements is somewhat an enigma, although it has been observed in many other pathological conditions. The process has been described, for example, in palmar and plantar fibromatosis (Gabbiani and Majno, 1972). These same cells at different stages of maturation and differentiation have been described experimentally in the capsule of blood clots embedded within the peritoneum of rats (Ryan *et al.,* 1973). They have also been identified in the renal interstitial tissue of the rabbit, following complete unilateral obstruction of the ureter (Nagle *et al.,* 1973). Other authors have described these cells in granulation tissue (Gabbiani, Ryan, and Majno, 1971; Gabbiani *et al.,* 1972). What is the nature and significance of filament condensations seen in fibroblasts during various pathological conditions? It has been established from correlative immunofluorescent studies that these fibrillar structures are closely related to filaments of actomyosin of smooth muscle cells (Hirschel *et al.,* 1971; Majno *et al.,* 1971). The significance or function of these structures has been associated with contraction of smooth muscle fibers.

It is interesting to speculate on the patterns of cellular differentiation in the subendocardial pseudotumoral proliferation of the carcinoid syndrome. This cellular proliferation could theoretically develop from four cellular types: a primitive mesenchymal cell normally found in the subendocardial region, a fibroblast, a smooth muscle cell, or an endothelial cell. Our observations support the concept of a differentiation from a primitive mesenchymal cell into an intermediate cell and, finally, into a myofibroblast. However, the possibility of myofibroblast formation from fibroblasts cannot be eliminated on the basis of a morphological study. Is it possible to eliminate myofibroblast formation from endothelial cells or smooth muscle cells? Since myofilaments have been observed in endothelial cells, this hypothesis cannot be completely rejected; but it appears to us unlikely, because of the absence of endothelial cells in the subendocardial fibrous plaque (Bruns and Palade, 1968; Becker and Murphy, 1969). As regards preexisting smooth muscle fibers, they can develop an abundant rough endoplasmic reticulum and other organelles and be interpreted as myofibroblasts.

If we establish a correlation between our morphological findings and biochemical data of the carcinoid syndrome, it is possible to suggest the following: bradykinin, histamine, and serotonin could induce gaps between endocardial cells and trigger a mesenchymal cellular proliferation by first acting on primitive mesenchymal cells of the endocardium and helping them to differentiate toward contractile properties. This pseudotumoral cellular proliferation as well as the differentiation of these cells could be explained by the action of these hormonal mediators, which is transmitted at the cellular level through the adenylcyclase and AMP cyclic system (Lefkowsky, 1973). This same system, to which has been attributed the capability of transmitting at the cellular level the necessary information for the manifestation of the physiological effects, could possibly play a role in the process of cellular differentiation.

REFERENCES

ALLEN, A. 1964. Fibrosis with carcinoid syndrome. Lancet I: 746.

BECKER, C. G., and G. E. MURPHY. 1969. Demonstration of contractile protein in endothelium and cells of the heart valves, endocardium, intima, arteriosclerotic plaques and Aschoff bodies of rheumatic heart disease. Am. J. Path. 55: 1.

BRUNS, R. R., and G. E. PALADE. 1968. Studies on blood capillaries. I. General organization of blood capillaries in muscle. J. Cell Biol. 37: 244.

CAMPBELL, A. C. P. 1968. The argentaffin cell system. *In* J. M. B. Bloodworth, Jr. (ed.), Endocrine Pathology, p. 608. The Williams and Wilkins Company, Baltimore.

COSH, J., J. E. CATES, and D. W. PUGH. 1959. Carcinoid heart disease. Brit. Heart J. 21: 369.

FADELL, E. J., and R. M. DENHAM. 1966. Carcinoid heart disease with bilateral ventricular endocardial sclerosis. Am. J. Cardiol. 17: 259.

GABBIANI, G., B. J. HIRSCHEL, G. B. RYAN, P. R. STATKOU, and G. MAJNO. 1972. Granulation tissue as a contractile organ. A study of structure and function. J. Exp. Med. 135: 719.

GABBIANI, G., and G. MAJNO. 1972. Dupuytren's contracture: fibroblast contraction? An ultrastructural study. Am. J. Path. 66: 131.

GABBIANI, G., G. B. RYAN, and G. MAJNO. 1971. Presence of modified fibroblasts in granulation tissue and their possible role in wound contraction. Experientia 27: 549.

GRAHAME-SMITH, D. G. 1968. The carcinoid syndrome. Am. J. Cardiol. 21: 376.

HIMBERT, J., P. H. EVEN, Y. GROSGOGEAT, and B. LETAC. 1965. Cardiopathie carcinode (nouvelle observation clinique, hémodynamique, biologique et anatomique). Arch. Mal. Coeur 58: 1638.

HIRSCHEL, B. J., G. GABBIANI, G. B. RYAN, and G. MAJNO. 1971. Fibroblasts of granulation tissue: Immunofluorescent staining with antismooth muscle serum. Proc. Soc. Exp. Biol. Med. 138: 466.

JONES, W. P. 1962. Serotonin release from carcinoid tumours. Canad. Anaesth. Soc. J. 9: 42.
JUSTIN-BESANCON, L., M. GRIVAUX, P. H. EVEN, P. VERNANT, and P. DELAVIERRE. 1965. Syndrome et cardiopathie cardinoide. Un nouveau cas et revue de la littérature. Sem. Hôp. Paris 41: 3064.
KARNOVSKY, M. J. 1967. The ultrastructural basis of capillary permeability studied with peroxidase as a tracer. J. Cell Biol. 35: 213.
LAPLANTE, L., A. ARCHAMBAULT, and F. THIBERT. 1971. Le syndrome carcinoïde. Union Med. Canad. 100: 1970.
LEFKOWSKY, R. S. 1973. Isolated hormone receptors: physiologic and clinical implications. New Engl. J. Med. 288: 1061.
McKUSICK, V. A. 1956. Carcinoid cardiovascular disease. Bull. John Hopkins Hosp. 98: 13.
MAJNO, G., G. GABBIANI, B. J. HIRSCHEL, G. B. RYAN, and P. R. STATKOV. 1971. Contraction of granulation tissue *in vitro:* similarity to smooth muscle. Science 173: 548.
MAJNO, G., and G. E. PALADE. 1961. Studies on inflammation. I. The effect of histamine and serotonin on vascular permeability: an electron microscopic study. J. Biophys. Biochem. Cytol. 11: 571.
MAJNO, G., S. M. SHEA, and M. LEVENTHAL. 1969. Endothelial contraction induced by histamine-type-mediators: an electron microscopic study. J. Cell Biol. 42: 647.
NAGLE, R. B., M. R. KNEISER, R. E. BULGER, and E. P. BENDITT. 1973. Induction of smooth muscle characteristics in renal interstitial fibroblasts during obstructive nephropathy. Lab. Invest. 29: 422.
OATES, J. A., and T. C. BUTLER. 1967. Pharmacologic and endocrine aspects of carcinoid syndrome. Adv. Pharmacol. 5: 109.
OATES, J. A., and A. SJOERDSMA. 1962. A unique syndrome associated with secretion of 5-hydroxytryptophan by metastatic gastric carcinoids. Am. J. Med. 32: 333.
REYNOLDS, E. S. 1963. The use of lead citrate at high pH as an electron-opaque stain in electron microscopy. J. Cell Biol. 17: 208.
ROBERTS, W. C., and A. SJOERDSMA. 1964. The cardiac disease associated with the carcinoid syndrome (carcinoid heart disease). Am. J. Med. 36:5.
RYAN, G. B., W. J. CLIFF, G. GABBIANI, C. IRLE, P. R. STATKOV, and G. MAJNO. 1973. Myofibroblasts in an avascular fibrous tissue. Lab. Invest. 29: 197.
SCHWEIZER, W., F. GLOOR, R. Von BERTRAB, and U. C. DUBACH. 1964. Carcinoid heart disease with left-sided lesions. Circulation 29: 253.
SPATZ, M. 1961. Carcinoid syndrome in tryptophan and niacin deficient guinea pigs. Univ. Michigan Med. Bull. 27: 289.
SPATZ, M. 1964. Pathogenetic studies of experimentally induced heart lesions and their relation to the carcinoid syndrome. Lab. Invest. 13: 288.
VENABLE, J. H., and R. COGGESHALL. 1965. A simplified lead citrate stain for use in electron microscopy. J. Cell Biol. 25: 407.
Von STUDNITZ, W., und J. WALDENSTRÖM. 1971. Pharmakologische und therapeutische (p-Chlorphenylalanin) Studien über das Carcinoid syndrom beim 11 Fällen. Klin. Wochenschr. 49: 748.

Mailing address:
Dr. M. Boutet,
Institut de cardiologie de Québec,
2725, chemin Sainte-Foy,
Québec, Québec G1V 4G5 (Canada).

Recent Advances in Studies on
Cardiac Structure and Metabolism Volume 10
The Metabolism of Contraction
Edited by Paul-Emile Roy and George Rona

AGE-DEPENDENT CHANGES OF HEART VALVES AND HEART SIZE

K. E. SCHENK and G. HEINZE

Medizinische Klinik und Poliklinik im Klinikum Steglitz
der Freien Universität Berlin, Berlin, Germany

SUMMARY

The mode and extent of the regressions of heart valves, valvular rings, and tendon chords, as well as the size of atria and ventricles in 316 hearts of different age groups (macroscopically normal at autopsy), were studied. With increasing age, the valves show marked thickening and loss of elasticity. Deposition of lipoids and calcification occur to a variable extent in the valvular tissue and rings. Valve size shows a change of minor degree, but tendon chords and papillary muscles shorten in old age. The diameters of the atria, mitral, tricuspid, pulmonary, and aortic rings increase continuously up to the 9th decade of life. The width and length of the ventricles show minor alterations in size. In old age, heart weight increases.

INTRODUCTION

With increasing age, the heart valves as well as the heart size of all human beings show considerable morphological changes. The importance for cardiac function of these regressions is not yet known. In an attempt to answer this question, it was first necessary to measure exactly the mode and extent of these age-dependent changes.

MATERIAL

A total of 316 hearts from different age groups which appeared macroscopically normal at autopsy were examined. The width of the mitral, tricuspidal, pulmonary, and aortic valvular rings; the size of all valves; the length of the tendon chords and the papillary muscles; the distance of the tips of the papillary muscles from the valvular rings; and the width of atria and ventricles were measured. The mode and extent of the regressions occurring in old age in valves and valvular rings were determined.

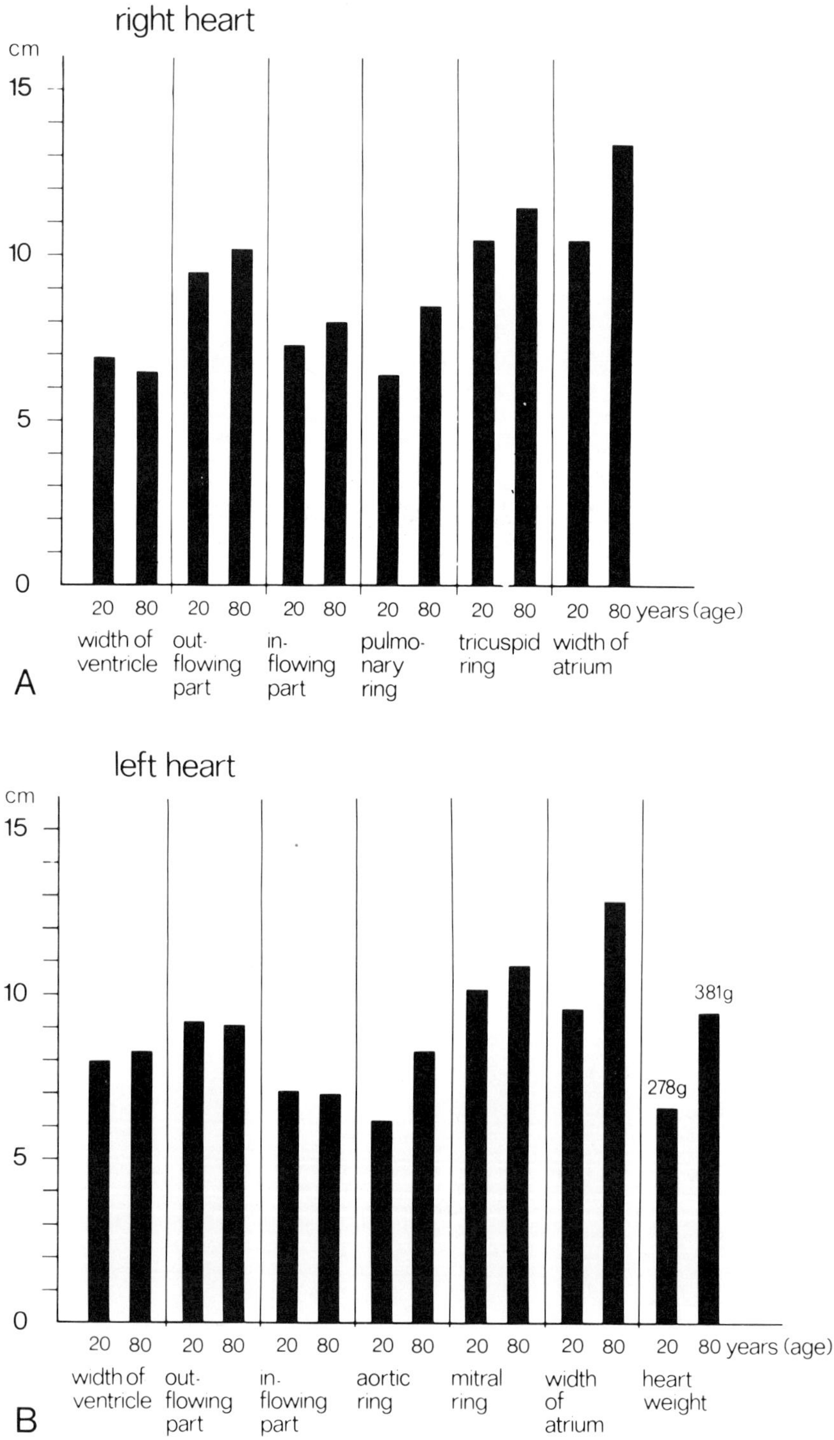
right heart
cm
15
10
5
0
20 80 20 80 20 80 20 80 20 80 20 80 years (age)
width of ventricle
out-flowing part
in-flowing part
pulmo-nary ring
tricuspid ring
width of atrium
A
left heart
cm
15
10
5
0
278g
381g
20 80 20 80 20 80 20 80 20 80 20 80 20 80 years (age)
width of ventricle
out-flowing part
in-flowing part
aortic ring
mitral ring
width of atrium
heart weight
B

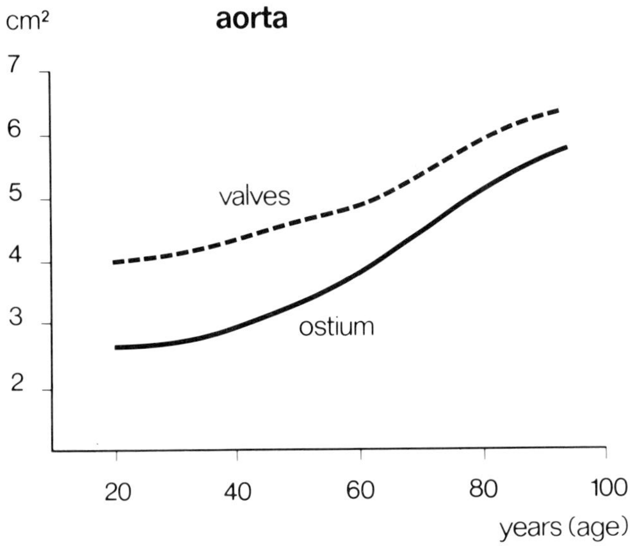

Figure 2. Size of aortic ring (–) and aortic valves (– –) with increasing age.

RESULTS AND DISCUSSION

In old age, the diameter of the aortic and pulmonary rings (Fig. 1) increases significantly (1.5–2 cm) as does the size of the valves (e.g., difference between aortic ring from 20-year-old versus 80-year-old men, $p<0.001$). But the size of the aortic and pulmonary ostia increases more than that of the valves (Fig. 2). Therefore, additional enlargement of the aortic ring caused by ectasia in hypertonia or lues, as well as a reduction in the valve size due to endocarditis, for example, may be followed by aortic insufficiency earlier in old than in young people.

In addition, the aortic valves show a marked thickening and loss of elasticity as a result of an increased content of connective tissue and deposition of lipoids and calcification (Sell and Scully, 1965; Silverman and Hurst, 1968; Schenk 1971, 1972). The pulmonary valves show a minor degree of regression. Calcification of the entire aortic valves, including the free margin, can be found in 15% of old people (Table 1); this may sometimes cause stenosis (Giese, 1932; Sell and Scully, 1965; Schölmerich, 1970). Systolic murmurs often found in old people (Grosse-Brockhoff, Kaiser, and Loogen, 1960) may originate from these valvular changes.

Mitral and tricuspid ostia (Table 1) in old age also increases ($\surd$ 1 cm),

---

Figure 1 (facing page). Size of atria and ventricles and heart weight of a 20- and 80-year-old man. *A:* right heart; *B:* left heart.

Table 1. Frequency of lipoid deposits and calcification of heart valves and atrioventricular rings*

	Mitral Complex						
	Aortic cusp		Dorsal cusp		Mitral ring		
Age (yrs)	1st degree	2nd degree	1st degree	2nd degree	1st degree	2nd degree	3rd degree
18–29	25	0	0	0	0	0	16 cases (1.6%) from 1,000 autopsies (60–90 year)
30–39	46.6	6.6	0	0	13.3	0	
40–49	77.7	11.1	0	0	16.6	0	
50–59	93.3	15.5	2.2	0	33.3	4.4	
60–69	92.2	20.7	2.4	0	42.8	7.7	
70–79	95.0	37.0	3.7	0	48.1	8.5	
80–95	95.9	44.8	16.3	0	38.7	22.4	

	Tricuspid Complex								
	Anterior cusp		Septal cusp		Posterior cusp		Tricuspid ring		
Age (yrs)	1st degree	2nd degree	1st degree	2nd degree	1st degree	2nd degree	1st degree	2nd degree	3rd degree
18–29	0	0	0	0	0	0	0	0	0
30–39	0	0	0	0	0	0	0	0	0
40–49	0	0	0	0	0	0	0	0	0
50–59	2.2	0	0	0	0	0	0	0	0
60–69	2.7	0	0	0	0	0	5.4	0	0
70–79	7.5	0	1.2	0	1.2	0	8.3	0	0
80–95	10.8	0	1.5	0	1.7	0	3.9	2.1	0

	Aortic and Pulmonary Valves			
	Aortic valves		Pulmonary valves	
Age (yrs)	1st degree	2nd degree	1st degree	2nd degree
18–29	0	0	0	0
30–39	6.6	0	0	0
40–49	31.5	0	0	0
50–59	67.6	0	2.2	0
60–69	85.1	4.0	2.7	0
70–79	92.5	7.4	1.2	0
80–95	83.0	15.0	0	0

*Values expressed as percentages. From Schenk (1972). Aortic and pulmonary valves: 1st degree: spotty deposits of lipoids (with increasing age, also calcification) in the sinus valsalvae; 2nd degree: deposits of lipoids and calcification of the entire valves, including the

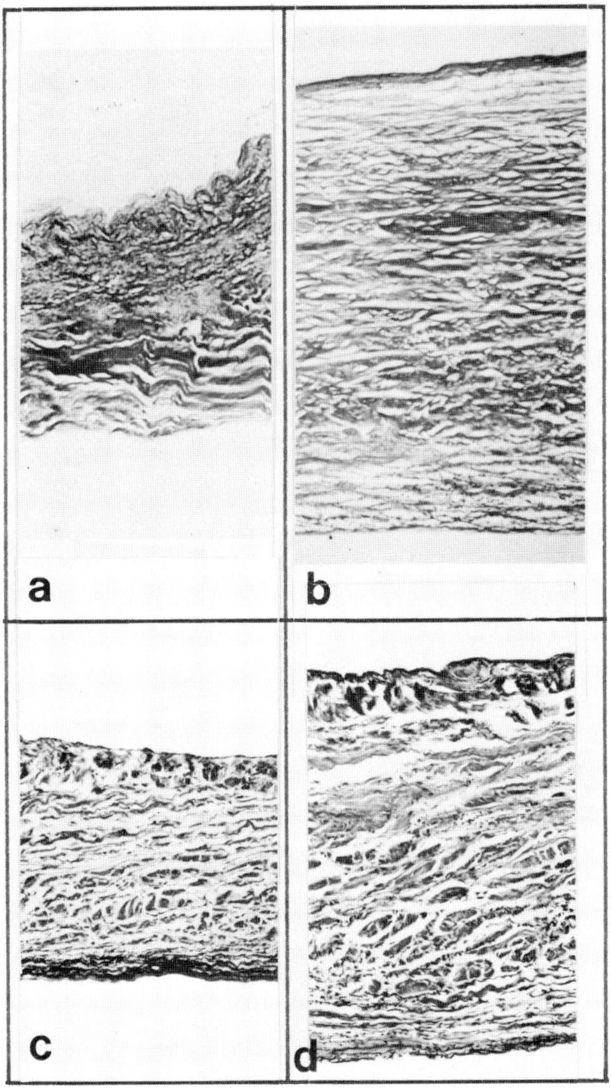

Figure 3. Mitral valve (ventral) from a 20- (*a*) and 80- (*b*) year-old man. Tricuspid valve (septal) from a 20- (*c*) and 80- (*d*) year-old man. × 80.

free margin. Tricuspid and mitral valves: 1st degree: spotty deposits of lipoids (with increasing age, also calcification) at the base of the valves as well as at the point where the tendon chords join the valves; 2nd degree: spotty deposits of lipoids and calcification of the entire valves, including the free margin. Atrioventricular rings: 1st degree: several small spots of calcification; 2nd degree: calcification of 1/3 of the valvular ring; 3rd degree: calcification of more than 1/3 of the valvular ring.

whereas mitral and tricuspid valves show only a minor variation in size (e.g., difference between tricuspid ring from 20-year-old versus 80-year-old men, $p<0.01$). At the same time, the tendon chords of the mitral and tricuspid complex shorten (3 mm or more) and the tips of the papillary muscles withdraw (3–6 mm) from the valvular rings (Schenk, 1971, 1972). Mitral and tricuspid valves show a marked thickening owing to an increased content of connective tissue and deposition of lipoids and calcification of variable degree (Fig. 3; Table 1).

It may not be speculative to say that these morphological changes found in old age could hinder sufficient closing of the valves. Therefore, additional changes, such as enlargement of atrioventricular ostia in myocardial diseases or a shortening of valves and tendon chords resulting from endocarditis, may cause valvular insufficiency earler in old than in young people. This hypothesis may be

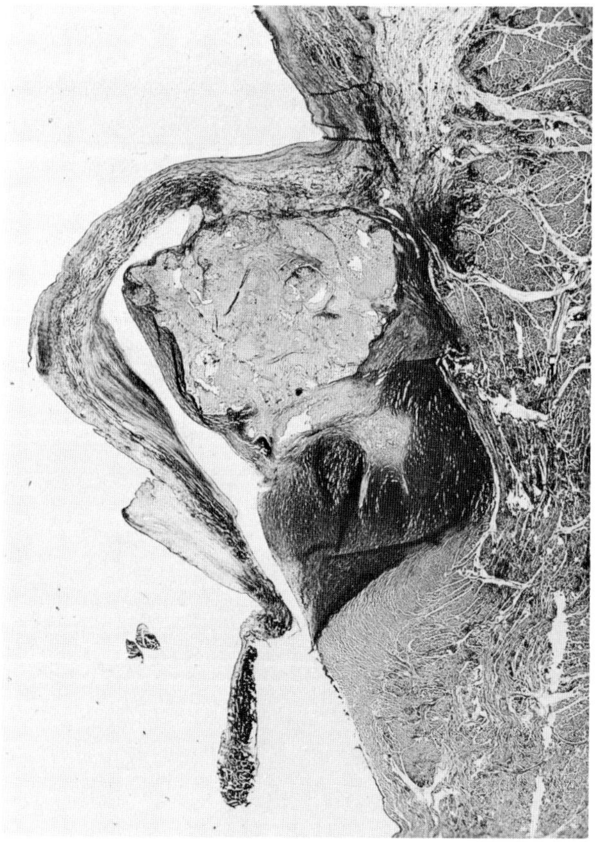

Figure 4. Calcification of the mitral ring (84-year-old man). × 1.

supported by the observation of Schoenmackers (1968) that, with increasing pressure in the ventricles, atrioventricular valves of old people become insufficient earlier than do those of young subjects.

The mitral rings shows marked calcification in 22% of the cases (Table 1). This calcification sometimes builds up a plateau-like formation under the mitral valves (Fig. 4) and, in this way, may hinder the movement of the valves as well as of the valvular ring, both in systole and in diastole.

Heart size and weight also increase in old age, the diameter of the atria (Fig. 1) enlarging up to the 9th decade of life (1–2 cm), the difference in the values between 20-year-old and 80-year-old men being significant at $p<0.001$. This observation was first described by Kirch (1921). The upper width of the ventricles increases, whereas the length shows only insignificant changes. The papillary muscles of the left ventricle shorten and their tips withdraw from the valvular ring (Schenk, 1971, 1972). The heart weight increases continuously in old age (e.g., value for 20-year-old versus 80-year-old men, $p<0.001$). This may be caused by the increasing content of connective tissue, deposition of lipoids, and the calcification of heart valves and ostia more than by a thickening of the heart muscle. These observations correspond to those obtained from the autopsy studies of Linzbach and Akuamoa-Boateng (1973) and Beneke and Schmitt (1967), or the x-ray method measurements of Reindell, König, and Roskamm (1967).

In conclusion, it may be stated that, in old age, there are marked changes in the valves, ostia, and tendon chords, as well as in the size of the atria and ventricles and the weight of all hearts. The cause of these alterations is not yet known. Furthermore, it is impossible to state on merely the basis of autopsy investigations what influence these morphological changes will have on hemodynamics. Hence, further hemodynamic measurements in clinically healthy people are necessary to answer this question.

REFERENCES

BENEKE, G., und W. SCHMITT. 1967. Altersveränderungen am Klappenapparat des menschlichen Herzens. Verh. d. dt. Ges. f. Path. 51: 209.

GIESE, W. 1932. Die Verkalkungen des Herzskelettes. Beitr. Path. Anat. allg. Path. 89: 16.

GROSSE-BROCKHOFF, F., H. KAISER, und F. LOOGEN. 1960. Erworbene Herzklappenfehler. Hdb. inn. Med. 9: Teil 2. 1280.

KIRCH, E. 1921. Über gesetzmaßige Verschiebungen der inneren Großenverhältnisse des normalen und pathologisch veränderten menschlichen Herzens. Zschr. f. angew. Anat. 7: 235.

LINZBACH, A. J., und E. AKUAMOA-BOATENG. 1973. Die Altersveränderungen des menschlichen Herzens I und II. Klin. Wschr. 51: 156 and 164.

REINDELL, H., K. KÖNIG, und H. ROSKAMM. 1967. Funktionsdiagnostik des gesunden und kranken Herzens. Thieme, Stuttgart.

SCHENK, K. E. 1971. Altersveränderungen der Mitralklappen und ihre Bedeutung für die Klappenfunktion. Zschr. Kreislauff. 60: 110.

SCHENK, K. E. 1972. Die Herzklappen im Alter. Zschr. Kreislauff. 61: 392.
SCHOENMACKERS, J. 1968. Zur Herzklappeninsuffizienz. Arch. Kreislauff. 55: 283.
SCHÖLMERICH, P. 1970. Erworbene Klappenfehler. *In* R. Gross, D. Jahn, and P. Schölmerich (eds.), Lehrbuch Inn. Medizin, p. 291. Stuttgart New York.
SELL, St., and R. E. SCULLY. 1965. Aging changes in the aortic and mitral valves. Am. J. Path. 46: 345.
SILVERMAN, M. E., and J. W. HURST. 1968. The mitral complex. Am. Heart J. 76: 399.

Mailing address:
Dr. K. E. Schenk,
Medizinische Klinik und Poliklinik
im Klinikum Steglitz der Freien Universität Berlin,
1 Berlin 45, Hindenburgdamm 30 (Germany).

Cardiomyopathies—
Recent Studies

Recent Advances in Studies on
Cardiac Structure and Metabolism
Volume 10
The Metabolism of Contraction
Edited by Paul-Emile Roy and George Rona

FELINE CARDIOMYOPATHY

SI-KWANG LIU, LAWRENCE P. TILLEY,
and PETER F. LORD

Departments of Pathology and Medicine, The Animal Medical Center, New York, USA,
and Department of Clinical Radiology, Royal Veterinary College, Stockholm, Sweden

SUMMARY

Cardiomyopathy was diagnosed by means of clinical, radiographic, electrocardiographic, phonocardiographic, angiocardiographic, and pathological findings in 271 of 3,745 cats necropsied from January 1962 to April 1974. The affected cats can be divided into three groups on the basis of the gross and microscopic pathological lesions: 1) endocarditis and myocarditis in 20 young cats; 2) endomyocardial fibrosis and left ventricular hypertrophy in 182 cats; and 3) myocardial degeneration and biventricular dilatation in 69 cats. Of 271 affected cats, thromboembolus was observed in the aorta, and in the carotid, femoral, iliac, renal, pulmonary, and hepatic arteries in 104 instances.

The important aspects of cardiomyopathy in cats appears to be the reduced diastolic compliance of the thick left ventricle, resulting in poor filling. Resistance to ventricular inflow raises the diastolic pressure and causes compensatory left atrial enlargement. A pathogenesis for the onset of clinical signs at any stage as the cause of the heart disease is postulated on the basis of stress causing tachycardia and poor left ventricular filling. Acute left-sided failure with pulmonary edema may be precipitated. Approximately one-fourth of the cats have enlargement of all cardiac chambers, typical of congestive cardiomyopathy.

On the basis of the close similarily to cardiomyopathy in man, the cat could serve as a suitable animal model for a conservation of time and effort in the attack against this disorder. There is a need for coordinated research programs for utilizing the multiple avenues of approach such as: epidemiological, clinical, biochemical, pathological, ultrastructural, virological, and immunological.

INTRODUCTION

Cardiomyopathy has recently been considered as a disease entity in association with aortic thromboembolism in the cat (Liu, 1970, 1974; Liu, Tashjian, and Patnaik, 1970; Harpster, 1973; Lord *et al.*, 1974; Tilley, Lord, and Wood, 1974), although aortic thromboembolism has often been reported in the cat since 1930 (Collet, 1930; Holzworth, Simpson, and Wind, 1953; Hamlin, Tashjian, and Smith, 1964; Freak, 1965, 1966; Buchanan, Baker, and Hill, 1966; Lucke and Sumner-Smith, 1966). The principal cardiac lesions of feline heart disease are nonsuppurative endomyocarditis, endocardial and myocardial granulation and fibrosis with myocardial hypertrophy, and myocardial degeneration with biventricular dilation (Liu, 1970, 1974). Cardiac lesions in the cat are compatible with those of cardiomyopathy in man (Goodwin, 1970; Reddy, Parvathi, and Rao, 1970; Okada, 1971; Rodin, Harris, and Nghiem, 1971; Goodwin and Oakley, 1972). Clinical signs, electrocardiographic, and radiographic findings (Harpster, 1973; Lord *et al.*, 1974; Tilley *et al.*, 1974), and hemodynamic evaluation (Lord *et al.*, 1974) of cardiomyopathy in cats have recently been described. This paper summarizes clinical, radiographic, electrocardiographic, phonocardiographic, and angiocardiographic findings of cardiomyopathy in the cat.

MATERIALS AND METHODS

Of 3,745 cats necropsied between January 1962 and April 1974 at The Animal Medical Center, 271 were diagnosed pathologically as having cardiomyopathy. Clinical observations, electrocardiograms, and thoracic radiographs were recorded in the majority of cases. Cardiocatheterization and angiocardiography were done on 44 of the affected cats (Tashjian *et al.*, 1963, 1965; Lord *et al.*, 1974).

At necropsy, the organs of all affected cats and 11 normal control cats were carefully examined systematically (Liu, 1970; Liu *et al.*, 1970). All gross lesions,

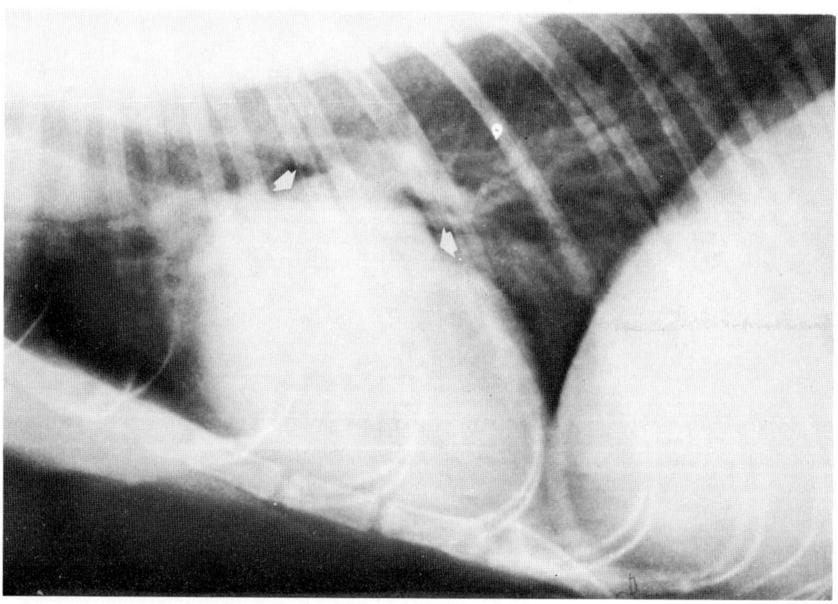

Figure 1. Lateral thoracic radiograph of a 3-year-old male domestic short-hair cat. Note left atrial (*arrows*) and ventricular enlargement with pulmonary congestion.

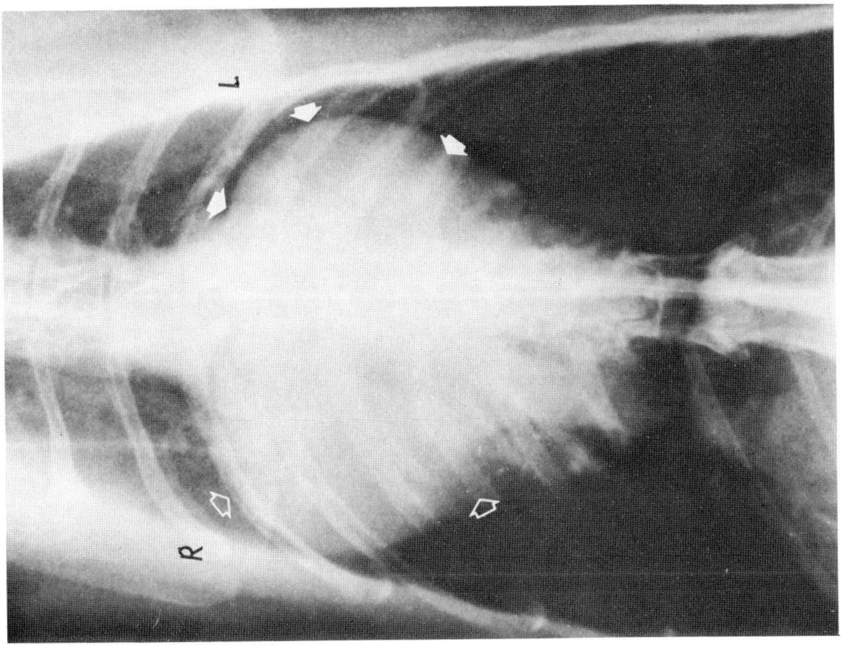

Figure 2. Dorsoventral radiograph of the cat depicted in Figure 1, showing marked left atrial (*solid arrows*) and right atrial and ventricular (*open arrows*) enlargement.

as well as representative formalin-fixed heart tissues, were embedded in paraffin and prepared for light microscopy by standard methods. Deparaffinized slides were stained with hematoxylin and eosin and other selected stains (Liu, 1970).

RESULTS

Age, sex, and breed

Ages of the 271 affected cats ranged from 7 months to 24 years, with no apparent age incidence. The distribution by sex indicated greater predominance of male cats. There was no breed predilection.

Clinical signs

The clinical signs in most cases were nonspecific; they included dyspnea, lethargy, anorexia, tachypnea, and, occasionally, abdominal distention. Moist rales, cardiac murmurs, gallop rhythms, and arrhythmias were frequently auscultated. Dyspnea was the most frequent presenting sign and indicated left-sided heart failure. Approximately one-third of the cats presented with signs associated with systemic, particularly aortic thromboemboli; they included acute posterior paralysis, absence of femoral pulses, and coldness of extremities (Harpster, 1973; Tilley *et al.*, 1974).

Radiographic findings

Thoracic radiographs showed varying degrees of cardiomegaly, particularly atrial enlargement, detected best in the dorsoventral projection and appearing as a "valentine-shaped" heart (Figs. 1 and 2). Extracardiac signs of left- and right-sided heart failure were detected by pulmonary congestion or edema, pleural effusion, hepatomegaly, and ascites. Signs of right-sided heart failure were less common.

Electrocardiographic and phonocardiographic findings

Because of the lack of well established electrocardiographic data on the normal cat, complete evaluation of the electrocardiogram of the cat with cardiomyopathy has not been made. Based on our increasing knowledge of electrocardiography in the cat, various cardiac arrhythmias, axis deviations, and abnormalities of duration and amplitude of P waves and QRS complexes were noted (Fig. 3) (Block and Boeles, 1957; Hamlin, Smetzer, and Smith, 1963; Harpster, 1963).

Phonocardiograms frequently recorded systolic murmurs over the mitral valve area and, occasionally, the aortic valve area. Most of these murmurs were due to variable degrees of mitral valve incompetence. Gallop rhythms were frequently illustrated; these are direct evidence of underlying heart disease. The majority of these extra heart sounds were atrial gallops (accentuation of the fourth heart sound) (Fig. 3). The occurrence of this sound was favored by an increased resistance of the ventricle to distention due to hypertrophy. A protodiastolic gallop (accentuation of the third heart sound) or summation gallop was sometimes recorded.

Angiocardiography

The hypertrophic form revealed a greatly reduced volume and irregular outline of the left ventricles, and dilation of the left atrium with significant mitral valvular regurgitation (Fig. 4). The congestive form illustrated extreme ventricular dilation with mitral insufficiency (Fig. 5). The systolic left ventricular pressure was observed to vary between 70 and 160 mm Hg. The end-diastolic left ventricular pressure was consistently elevated, the mean (24.5 mm Hg) being more than 5 times normal (mean 4.5 mm Hg) (Fig. 6) (Lord *et al.*, 1974). Aortography was quite useful in the diagnosis of aortic thromboembolism, especially in an evaluation for surgical removal (Fig. 7).

Pathological findings

Of 3,745 cats necropsied, 271 had cardiomyopathy. The affected cats were divided into three groups on the basis of cardiac lesions.

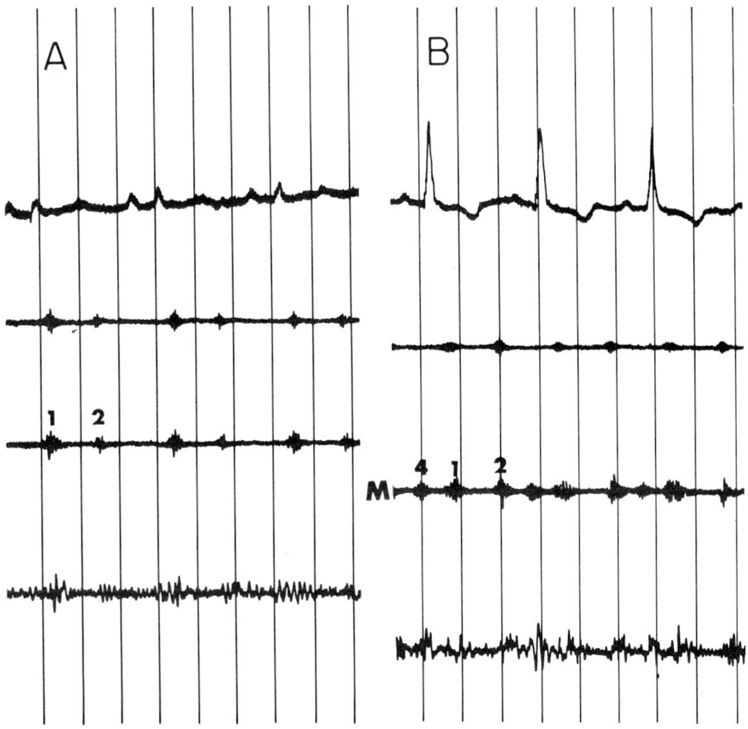

Figure 3. Simultaneous recording and electrocardiogram (lead II) and phonocardiogram of a normal cat (*A*). The first (after QRS complex) and second heart sounds (after T wave) are seen. Simultaneous recording and electrocardiogram of cat with hypertrophic cardiomyopathy (*B*). Tall and wide QRS complexes indicate left ventricular hypertrophy. Accentuation of the fourth heart sound is evident after the onset of the P wave on the electrocardiogram and terminating before the onset of the QRS complex (noted best in middle frequency 'M'; 100–200 cps).

Group I (endocarditis and myocarditis) consisted of 20 young cats. Grossly, subendocardial petechiae to diffuse hemorrhages were often observed in the left ventricle. The endocardium was covered with rough opaque fibrins. The affected myocardium was dull and scattered with hemorrhagic foci. Thromboemboli were found in the terminal aorta in 3 of the affected cats.

Microscopically, the endocardium and myocardium were focally or diffusely infiltrated by lymphocytes, plasma cells, neutrophils, macrophages, and histiocytes (Fig. 8). There were more granulocytes than agranulocytes in the acute stage. The eroded endocardium was covered with cellular debris, necrotic tissue, and fibrins. The affected muscle fibers were degenerated and lyzed, and were infiltrated by leukocytes and erythrocytes. These changes were commonly seen in the endocardium and myocardium of the papillary muscles, aortic outflow

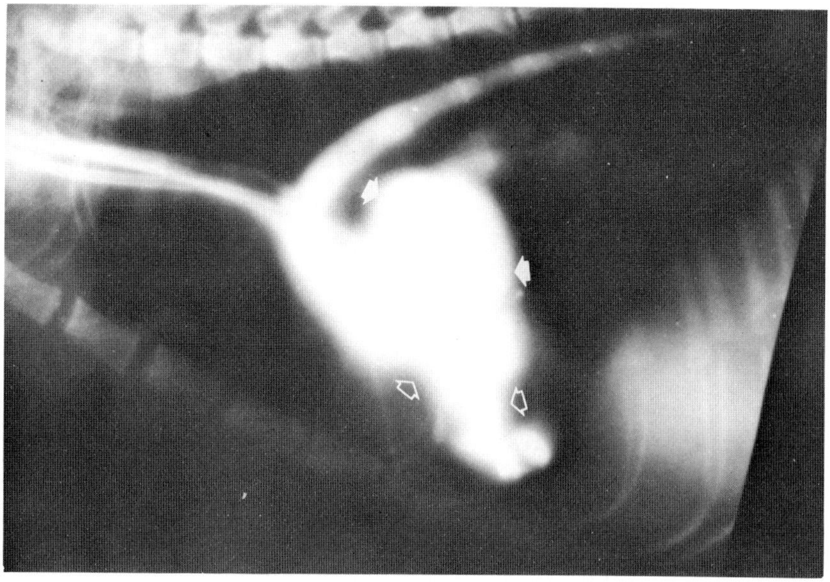

Figure 4. Left ventriculogram in systole of a 6-year-old male domestic short-hair cat. Ventricular hypertrophy and endocardial fibrosis produced filling defect (*open arrows*) in the ventricular chamber. Left atrial enlargement (*solid arrows*) due to mitral insufficiency is visible.

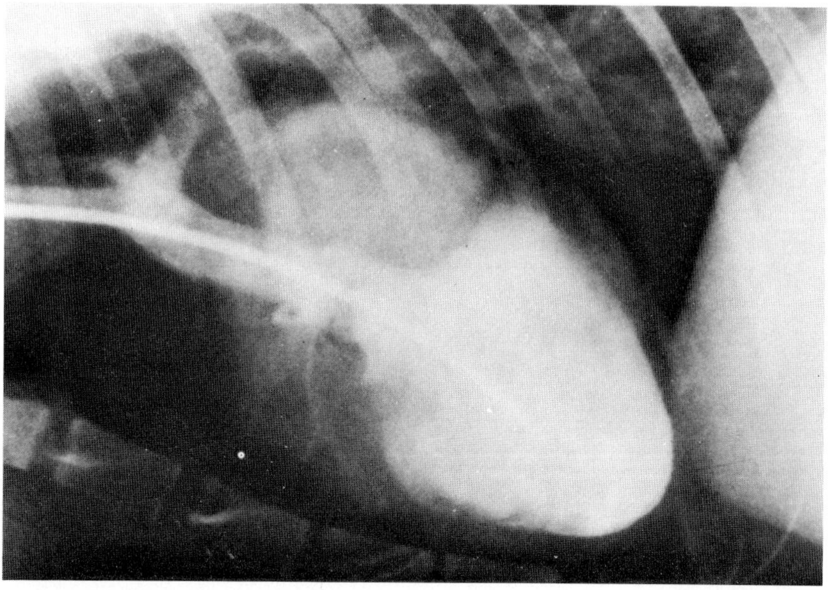

Figure 5. Left ventriculogram of a 12-year-old female Siamese cat. The ventricle is extremely dilated and rounded. There is marked mitral insufficiency with regurgitation into the dilated atrium.

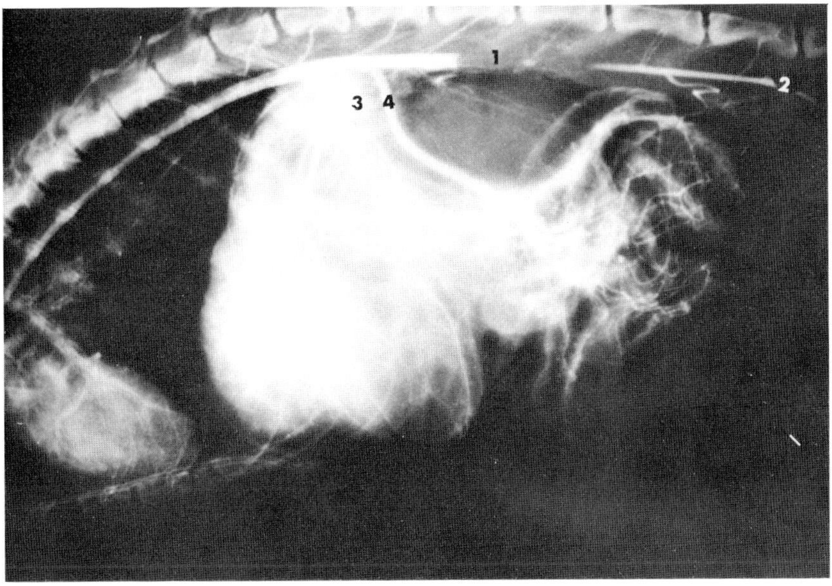

Figure 6. Aortogram of a 5-year-old male domestic short-hair cat. There are thromboemboli in the lumen of the aorta proximal to the origin of renal arteries and cranial (*1*) to the iliac bifurcation (*2*). Note increased collateral celiac (*3*) cranial mesenteric (*4*) arterial circulation.

tract, mitral valve, left ventricle, ventricular septum, and left atrium. The inflammatory reaction was replaced by granulation tissue in the chronic stage.

Group II (myocardial and endocardial fibrosis with left ventricular hypertrophy) consisted of 182 cats. Grossly, left atrial dilation and hypertrophy of the left ventricular wall, papillary muscles, and interventricular septum were consistently observed in the heart of affected cats. The endocardium, myocardium, and epicardium of the left ventricle and atrium were scattered with irregular whitish fibrous areas (Fig. 9). Ball thromboembolus was occasionally seen in the dilated left atrium. Thromboembolus was observed in the aorta and other arteries in 78 of the affected cats. Microscopically, myocardium and endocardium were scattered with foci of granulation or fibrous tissue. The affected myocardium showed vacuolization or myocytolysis. Adjacent to the granulation or fibrous tissue, the muscle fibers were enlarged, irregularly arranged, and had large, dense, rectangular nuclei (Fig. 10). Foci of muscle fibers were trapped in the fibrous tissue and were swollen and deeply acidophilic. Cartilage was occasionally observed in the dense collagenous fibers in the affected endocardium.

Group III (myocardial degeneration with biventricular dilatation) consisted of 69 cats. Grossly, the heart was globular, because of extreme dilation of the

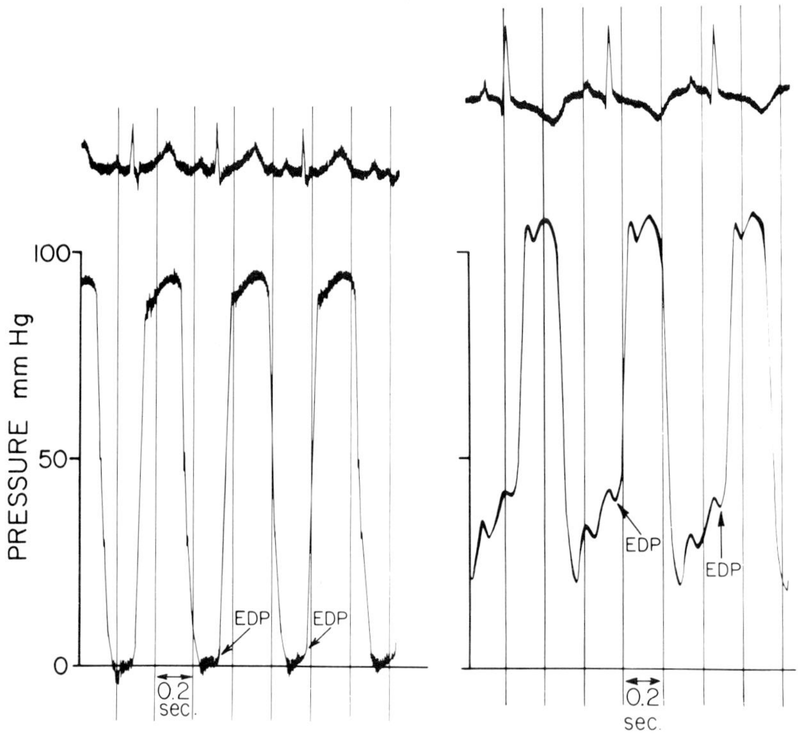

Figure 7. Left ventricular pressure tracing from a normal 4-year-old domestic short-hair cat (*left*). The end-diastolic pressure (EDP) is less than 5 mm Hg. Left ventricular pressure tracing from a cat with left ventricular hypertrophy and fibrosis (*right*). Systolic pressure is normal, but EDP is increased to 40 mm Hg. From Lord *et al.* (1974).

chambers. The trabeculae and papillary muscles were flattened and atrophied (Fig. 11). Irregular, whitish patches in the endocardium and myocardium and some degree of hypertrophy along with dilation were observed in the hearts of the aged cats. The circumferences of the cardiac valves were greater than normal (Liu, 1970). In 23 of the affected cats, thromboembolus was present in the aorta and various vessels. Free thromboembolus were seen in the left and right atria of 5 of the affected cats.

Microscopically, the muscle fibers had disappeared or showed various degrees of degeneration from loss of striation, coagulation or vacuolization of sarcoplasm, to myocytolysis. The adjacent fibers were thinner and their nuclei were smaller and closer together than normal (Fig. 12). In the aged cats, the myocardium and endocardium showed focal or diffuse fibrosis in areas.

Changes in other organs, such as passive congestion of the lungs, liver and viscera, renal infarction, ascites, pleural and pericardial effusion, were consistently observed in the affected cats.

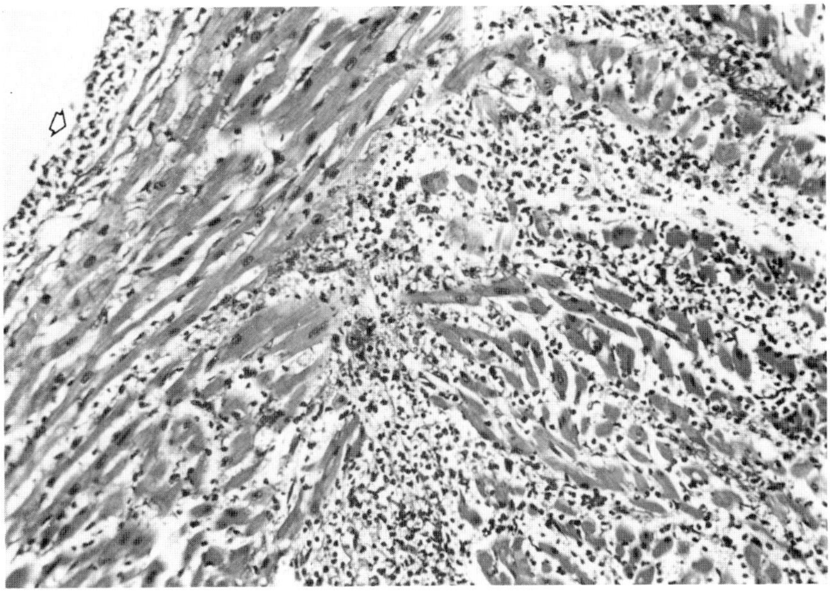

Figure 8. Photomicrograph of myocardium from a 1-year-old female domestic short-hair cat with infiltration of inflammatory cells in the endocardium (*arrow*) and myocardium. The adjacent muscle fibers showing degeneration and myocytolysis. H and E. × 100.

TREATMENT

Congestive heart failure

The basis of therapy is similar to that of congestive heart failure in man: rest, slowing of the heart rate, improvement of cardiac efficiency, and additional drugs to relieve pulmonary congestion, ascites, and pleural effusion.

At present, the management of the congestive cardiomyopathy form is based on rest, digoxin, and diuretics. The treatment regimen for hypertrophic cardiomyopathy is based on the prevention of the tachycardia and rise in end-diastolic pressure with β-adrenergic blocking agents (Emanuel, 1971; Burch, 1972).

Prognosis is poor for both forms of cardiomyopathy, with the average life expectancy less than 1 year.

Aortic thromboembolism

Thromboemboli may occur at any stage of the disease or may be the presenting sign. The underlying heart disease as well as other sites of embolization should be considered in an evaluation of surgical removal of the aortic thromboemboli.

Anticoagulants such as aspirin, heparin, and dipyridamole (Persantine) may prove beneficial in both the treatment and prevention of thromboembolism.

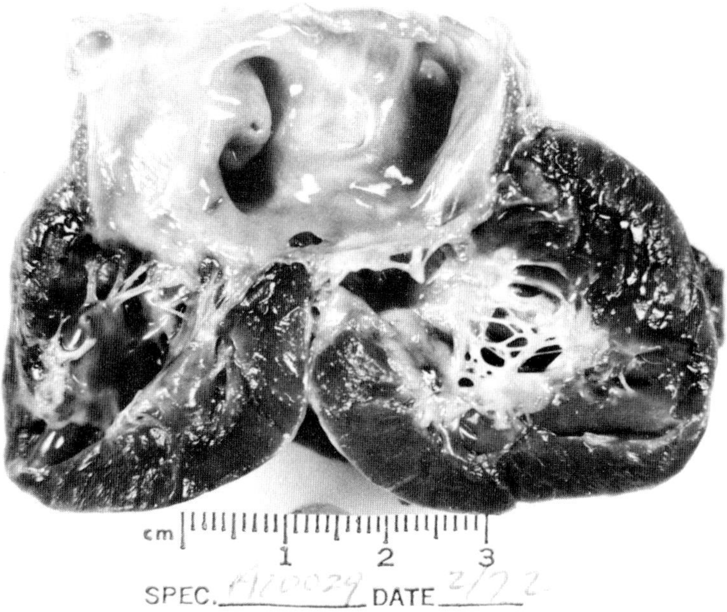

Figure 9. Left side of the heart of the same cat as in Figure 4. Note dilation and hypertrophy of the atrium, and extreme fibrosis of the endocardium and papillary muscles.

Aspirin is presently under trial in regard to its effects on platelet aggregation (Tilley *et al.*, 1974).

DISCUSSION

The pathological changes, including endocardial and myocardial fibrosis and myocardial hypertrophy in the group II cats, are compatible with those of hypertrophic cardiomyopathy, a disease of known and unknown etiology in man (Goodwin, 1970; Hudson, 1971; Okada, 1971; Rodin *et al.*, 1971; Goodwin and Oakley, 1972; Hutchins and Vie, 1972), hamsters (Bajusz, 1968), and turkeys (Noren *et al.*, 1971; Einzig, Jankus, and Moller, 1972). Ventricular dilation, myocardial degeneration, and myocytolysis, with or without fibrosis, in the group III cats are similar to congestive cardiomyopathy in man (Goodwin, 1970; Okada, 1971; Goodwin and Oakley, 1972).

The most probable sequence of the pathogenesis of the cardiac lesions in the cat is endocarditis and myocarditis (Fig. 8), organization of the inflammatory reaction, fibrosis of the endocardium and myocardium, and myocardial hypertrophy and degeneration. Myocardial or endocardial granulation and fibrosis might lead to hypertrophy and development of ventricular dilation in the group

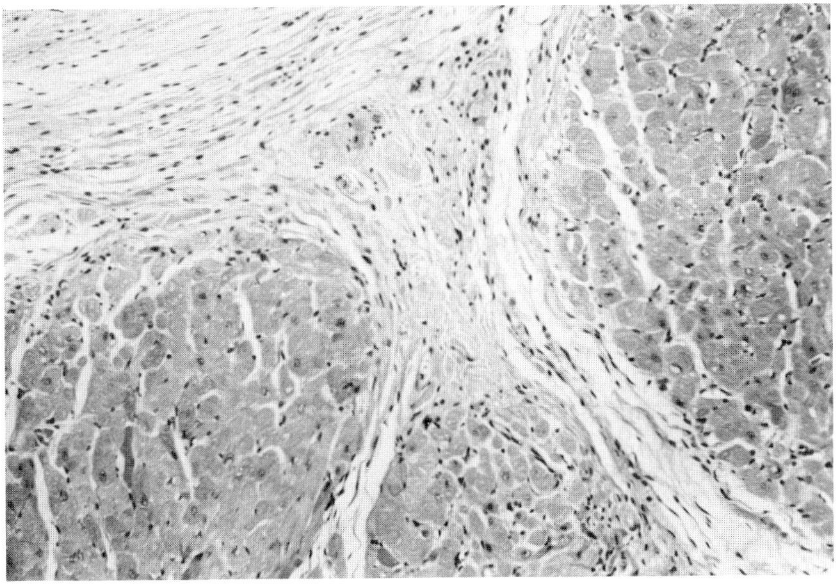

Figure 10. Photomicrograph of myocardium of the same cat as in Figure 9, showing fibrosis and granulation tissue in the endocardium and myocardium, with hypertrophy of muscle fibers. × 100.

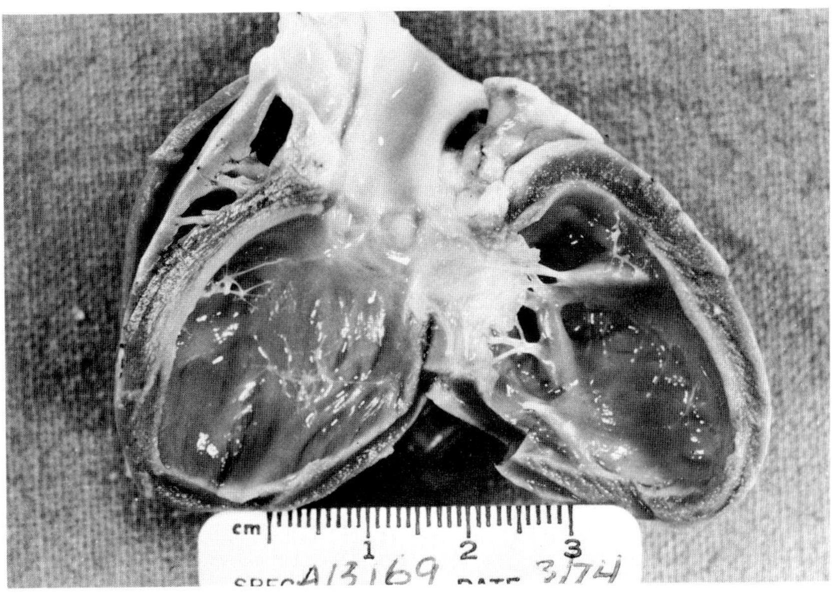

Figure 11. Left side of the heart of the same cat as in Figure 5. Note dilation of the ventricle, flattening and atrophy of the trabeculae, and slight abnormality of the papillary muscles.

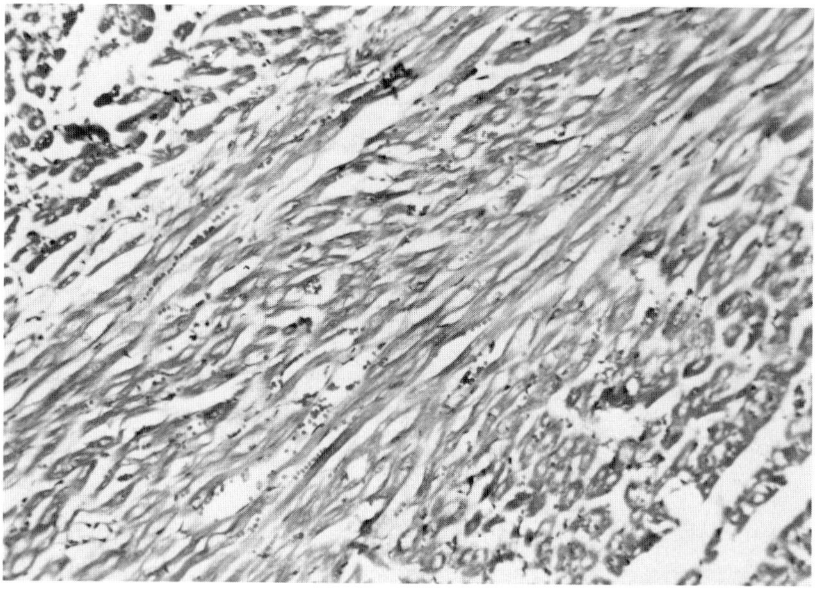

Figure 12. Photomicrograph of myocardium of the same cat as Figure 11. Note degeneration and vacuolization of muscle fibers and edema in the interstitium. H and E. × 100.

III aged cats. In group III, some young cats had biventricular dilation because of myocardial degeneration but without endomyocardial fibrosis (Fig. 12). These cardiac lesions are unlikely to relate to inflammation, granulation, and fibrosis in myocardiums seen in group I and II cats.

Of the 271 cats with cardiomyopathy in the present study, 104 (38%) had aortic thromboemboli. Thromboemboli were also seen in carotid arteries, pulmonary vessels, renal arteries, and hepatic vessels. Atrial thromboemboli were observed in 21% of the cats with systemic thromboembolism (Liu, 1970). It was postulated that thromboemboli resulted from diminished arterial pressures or hemostasis due to cardiac failure (Liu, 1970).

Since the etiology of the disease is unknown and the original lesions of endomyocardial inflammation and degeneration have been followed by hypertrophy, fibrosis, and dilation, any theory on the pathophysiology must be based on the hemodynamic, radiographic, and pathological findings.

The consistently elevated end-diastolic pressures recorded and the thick hypertrophied ventricular walls indicate a loss of ventricular compliance during diastole. Greater pressure is needed to fill the left ventricle as it is stiffer than normal, causing an obstruction of left ventricular filling. Tachycardia produced by stress or exercise shortens diastolic filling time and further limits diastolic filling. Cardiac output falls because the tachycardia is not sufficient to compensate for the drop in stroke volume caused by poor filling. Acute pulmonary edema probably results from blood backing up in the lungs. If the disease progresses gradually, the left atrium will enlarge to increase its reservoir capacity during diastole (Lord *et al.*, 1974). Increased left atrial pressure causes pulmonary hypertension with right ventricular hypertrophy in response to the increased work load. Right ventricular and right atrial dilation eventually result, then right ventricular failure with pleural effusion, hepatic congestion, and ascites. At any stage in the disease, exercise or stress may cause tachycardia and pulmonary edema. Aortic thromboembolism itself is a cause of stress and tachycardia because of the pain and shock produced.

Dyspnea, pulmonary congestion, and edema are often quickly reversed with rest and therapy to slow the heart. This reflects the flexible nature of the end-diastolic pressure, depending on the momentary hemodynamic condition. Right heart failure is not as quickly reversed, reflecting the more severe and permanent pathological changes in the heart and lungs.

The etiology of some cardiomyopathy is unknown and can result from many different etiological agents or combination of factors. As noted in man (Emanuel, 1971; Harley and Orgain, 1971; Perloff, 1971), a familial incidence is present, with a greater prominence in hypertrophic cardiomyopathy.

In contrast, the etiological factor in congestive cardiomyopathy is primarily considered as infectious, meaning viral or toxic in origin (Abelmann, 1971; Perloff, 1971; Harpster, 1973). Because of the high incidence of viral diseases in the cat, this may represent an important factor in the pathogenesis of this disease (Liu, 1970).

REFERENCES

ABELMANN, W. H. 1971. Virus and the heart. Circulation 44: 950–956.

BAJUSZ, E. 1968. Interrelationships between reparative processes in myocardium and the development of congestive heart failure. Rev. Canad. Biol. 27: 45–60.

BLOCK, J., and J. BOELES. 1957. The electrocardiogram of the normal cat. Acta Physiol. Pharmocal. Neerl. 6: 95.

BUCHANAN, J. W., G. J. BAKER, and J. D. HILL. 1966. Aortic embolism in cats: Prevalence, surgical treatment and electrocardiography. Vet. Rec. 79: 495–506.

BURCH, G. E. 1972. Cardiovascular Clinics, Vol. 4, No. 1, Cardiomyopathy. F. A. Davis Co., Philadelphia.

COLLET, P. 1930. Thrombose de l'aorte posterieure chez un chat. Bull. Soc. Sci. Vet. Lyon 33: 136–143.

EINZIG, S., E. F. JANKUS, and J. H. MOLLER. 1972. Round heart disease in turkeys: A hemodynamic study. Am. J. Vet. Res. 33: 557–561.

EMANUEL, R. 1971. The familial incidence of idiopathic cardiomyopathy. *In* G. E. W. Wolstenholme and M. O'Conner (eds.), Hypertrophic Obstructive Cardiomyopathy, pp. 50–54. Ciba Foundation Study Group, No. 37. J. A. Churchill, London.

FREAK, M. J. 1965. Aortic thombosis with posterior paralysis in the cat. Vet. Rec. 68: 816–818.

FREAK, M. J. 1966. Symposium on thrombosis I. Arterial thrombosis (embolism) in the cat. J. Small Animal Pract. 7: 711–717.

GOODWIN, J. F. 1970. Congestive and hypertrophic cardiomyopathies. Lancet *i:* 731–739.

GOODWIN, J. F., and C. M. OAKLEY. 1972. The cardiomyopathies. Brit. Heart J. 34: 545–552.

HAMLIN, R. L., D. L. SMETZER, and R. SMITH. 1963. The electrocardiogram, phonocardiogram, and derived ventricular activation process of domestic cats. Am. J. Vet. Res 24: 792–802.

HAMLIN, R. L., R. J. TASHJIAN, and C. R. SMITH. 1964. Diseases of the cardiovascular system. *In* E. J. Catcott (ed.) Feline medicine and surgery, 1st ed. American Veterinary Publications, Santa Barbara, Calif.

HARLEY, A., and E. S. ORGAIN. 1971. Hypertrophic cardiomyopathy with unusual features in a family. Brit. Heart J. 33: 55–61.

HARPSTER, N. K. 1973. Acquired heart disease in the cat. *In* Proceedings 40th Ann. Meeting, Am. Anim. Hosp. A. 118–125.

HOLZWORTH, J., R. SIMPSON, and A. WIND. 1955. Aortic thrombosis with posterior paralysis in the cat. Cornell Vet. 45: 468–487.

HUDSON, R. E. B. 1971. Pathology of cardiomyopathy. *In* G. E. Burch (ed.) Cardiomyopathy, pp. 3–59. F. A. Davis Co., Philadelphia.

HUTCHINS, G. M., and S. A. VIE. 1972. The progression of interstitial myocarditis to idiopathic endocardial fibroelastosis. Am. J. Path. 66: 483–492.

LIU, S. K. 1970. Acquired cardiac lesions leading to congestive heart failure in the cat. Am. J. Vet. Res. 31: 2071–2088.

LIU, S. K. 1974. Pathology of feline heart disease. *In* R. W. Kirk (ed.) Current veterinary therapy, 5th ed. W. B. Saunders Co., Philadelphia and London.

LIU, S. K., R. J. TASHJIAN, and A. K. PATNAIK. 1970. Congestive heart failure in the cat. J. Am. Vet. Med. Ass. 156: 1319–1330.

LORD, P. F., A. WOOD, L. P. TILLEY, and S. K. LIU. 1974. Radiographic and hemodynamic evaluation of cardiomyopathy and thromboembolism in the cat. J. Am. Vet. Med. Ass. 164: 154–165.

LUCKE, V. M., and G. SUMNER-SMITH. 1966. Aortic embolism in the cat. Vet. Rec. 79: 236–239.

NOREN, G. R., N. A. STALEY, E. F. JANKUS, and J. E. STEVENSON. 1971. Myocarditis in round heart disease of turkeys–a light and electron microscopic study. Virchow's Arch. Path. Anat. 352: 285–295.

OKADA, R. 1971. A morphological classification of the idiopathic myocardiopathy and allied cardiac diseases. Jap. Circulat. J. 35: 755–763.

PERLOFF, J. K. 1971. The cardiomyopathies–current perspectives. Circulation 44: 942–949.

REDDY, C. R. R. M., G. PARVATHI, and N. R. RAO. 1970. Pathology of cardiomyopathy from South India. Brit. Heart J. 32: 226–231.

RODIN, A. E., L. C. HARRIS, and Q. X. NGHIEM. 1971. Idiopathic nonobstructive cardiomyopathy. Electron microscopy, histochemical, and autopsy observations. Arch. Path. 91: 62–69.

TASHJIAN, R. J., K. M. DAS, W. E. PALICH, R. L. HAMLIN, and D. A. YARNS. 1965. Studies on cardiovascular disease in the cat. Ann. N. Y. Acad. Sci. 127: 581–605.

TASHJIAN, R. J., R. R. PENSINGER, K. M. DAS, C. F. REID, and A. A. CRESCENZI. 1963. Feline cardiovascular studies–a preliminary report. Proc. Book Am. Vet. Med. Ass. 112–123.

TILLEY, L. P., P. F. LORD, and A. WOOD. 1974. Acquired heart disease and aortic thromboembolism in the cat. *In* R. W. Kirk (ed.) Current veterinary therapy, 5th ed. W. B. Saunders Co., Philadelphia and London.

Mailing address:
Dr. Si-Kwang Liu,
Department of Pathology, The Animal Medical Center,
510 East 62nd Street, New York, New York 10021 (USA).

Recent Advances in Studies on
Cardiac Structure and Metabolism Volume 10
The Metabolism of Contraction
Edited by Paul-Emile Roy and George Rona

CARDIOMYOPATHY IN THE DOG

LAWRENCE P. TILLEY and SI-KWANG LIU

The Animal Medical Center, New York, New York, USA

SUMMARY

Medical records of 12 dogs determined at necropsy as having had cardiomyopathy and of 5 live dogs with clinical, electrocardiographic, and radiographic evidence of the disease were reviewed.

Congestive cardiomyopathy was the most common form of the disease, affecting 15 of the 17 dogs. The dogs were primarily of large breeds and ranged in age from 2 to 8 years. Clinical findings included right and left congestive heart failure presenting as pulmonary congestion and edema, pleural effusion, hepatomegaly, and ascites. Thoracic radiographs showed moderate to severe enlargement of all cardiac chambers and evidence of congestive heart failure. Atrial fibrillation was the predominant rhythm; ventricular premature contractions and left ventricular hypertrophy were sometimes noted. At necropsy, biventricular dilation including dilation of the atrioventricular annular rings and accompanying massive atrial dilation was observed. Myocardial contractility was poor and had resulted in dilation of the heart chambers with minimal hypertrophic responses. The atrioventricular valve leaflets and chordae tendinae were usually near normal. Medical treatment included rest, digoxin, and diuretics. Medical or electrical cardioversion of atrial fibrillation to normal sinus rhythm was also attempted. Prognosis for congestive cardiomyopathy is very poor. The average survival time after onset of signs is 6–12 months; 1 dog in our study survived for 20 months.

In contrast to congestive cardiomyopathy, the hypertrophic form is rare in the dog. Only two of the dogs studied had hypertrophic cardiomyopathy; one case was diagnosed at necropsy and one by angiocardiography. Both had features of idiopathic hypertrophic subaortic stenosis (IHSS) as reported in man.

INTRODUCTION

The clinical, electrocardiographic, and radiographic manifestations of congestive cardiomyopathy in the dog have been described (Luginbühl and Detweiler, 1965; Ettinger, Bolton, and Lora, 1970; Ettinger and Suter, 1970; Bolton, 1972); those for the hypertrophic form have not. The etiological factors in congestive cardiomyopathy in the dog are for the most part unknown. Chagas' disease has been seen in a small number of dogs in the southwest part of the United States (personal communication, 1974; Dr. Knauer, Department of Cardiology, Texas A & M University, College Station, Texas).

As congestive cardiomyopathy occurs primarily in large breeds, hereditary and nutritional factors should be considered. The syndrome is also sometimes associated with stress such as pregnancy, gastric dilation/torsion, and spondylolisthesis. Because of the predominance, the 15 dogs with congestive cardiomyopathy are given primary emphasis in this review. Pertinent hemodynamic and angiocardiographic data are given for 1 of the 2 dogs with hypertrophic cardiomyopathy.

MATERIALS AND METHODS

Medical records of 12 dogs diagnosed at necropsy as having cardiomyopathy were reviewed. Also reviewed were the medical records of 5 dogs believed to be still alive but having clinical and radiographic evidence of cardiomyopathy. The 17 medical records were compiled between January 1973 and March 1974.

Physical examinations, thoracic radiographs, and electrocardiograms were done on all animals. The thoracic radiographs were made in the left lateral and dorsoventral projections. In one dog with congestive cardiomyopathy and one with the hypertrophic form, cardiac catheterization and angiocardiography were performed to study the hemodynamic changes and gross pathological changes caused by the disease.

RESULTS

Age, sex, and breed incidence

Ages of the 15 dogs diagnosed as having congestive cardiomyopathy ranged from 2 to 8 years. It was found more often in male than in female dogs. Only dogs of large breeds were affected, including the Great Dane, Doberman Pinscher, German Shepherd, Irish Wolfhound, Standard Poodle, and St. Bernard. Of the 15 cases, 10 were Great Danes. The 2 cases of hypertrophic cardiomyopathy included the breeds Schnauzer and a Samoyed. Both dogs were females, and they were 5 years old and 3 months old, respectively.

Clinical signs

Most of the dogs presented with varying degrees of right and left heart failure. The predominant symptoms reported by the owners included weight loss, general disability, weakness, and abdominal enlargement over a period of from 2 to 4 weeks.

Other signs were coughing, dyspnea, anorexia, and ascites. A rapid, irregular heart rate was most clearly palpated over the left caudal sternal border in the majority of cases.

Auscultation sometimes revealed cardiac murmurs, most of which were low to moderate intensity systolic murmurs over the mitral valve area. The intensity of the first heart sound varied from beat to beat. A diastolic gallop sound was often heard and was usually quite loud and consistent.

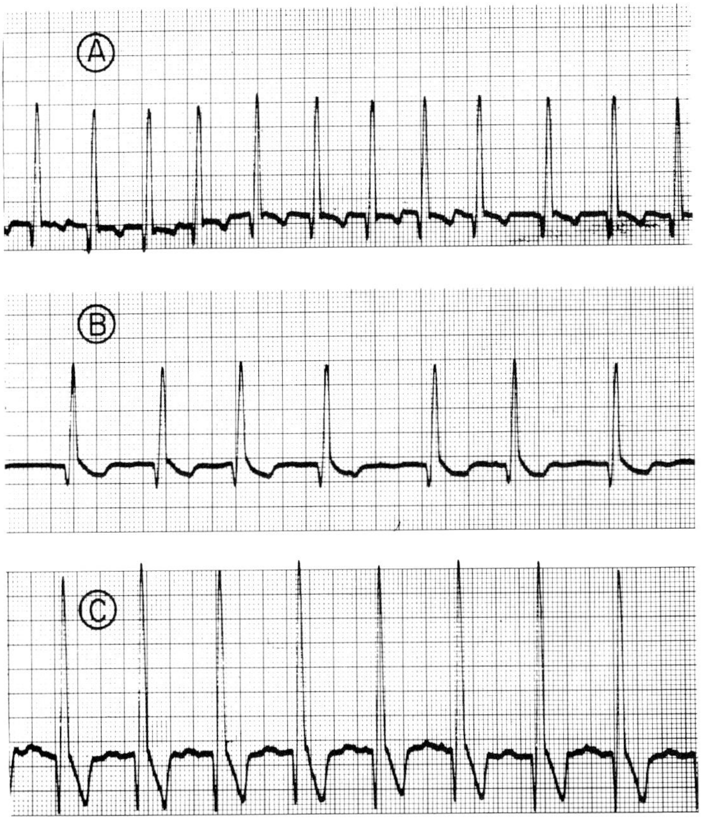

Figure 1. Conversion of atrial fibrillation to normal sinus rhythm. *A:* electrocardiogram from a dog in atrial fibrillation with a ventricular rate of 200 beats/min. The rate is rapid and irregular with absence of P waves. *B:* the ventricular rate is reduced with cardiac glycosides. *C:* a normal sinus rhythm with a heart rate of 160 beats/minute is the result after electrical cardioversion at 250 watt-seconds. Lead II, paper speed = 50 mm/sec, standard amplitude of 10 mm = 1 mv for EKG strip C and one-half amplitude for *A* and *B*.

Electrocardiographic findings

The rapid and irregular heart rate was evident on the electrocardiogram along with wide QRS complexes. Normal sinus P waves were absent (Fig. 1). This was consistent with atrial fibrillation and left ventricular hypertrophy. Left bundle branch block was also present in 1 case. Of the dogs diagnosed as having congestive cardiomyopathy, 14 showed atrial fibrillation. Some of the cases did show the presence of single or multiple ventricular premature contractions.

Radiographic findings

Thoracic radiographs did show moderate to severe enlargement of all cardiac chambers (Figs. 2 and 3). Signs of either right- or left-sided heart failure were seen: diffuse pulmonary congestion or edema, pleural effusion and ascites, with a wide and elevated caudal vena cava.

Catheterization and angiocardiography

Cardiac catheterization data revealed very poor myocardial contractility, with left ventricular and right ventricular end-diastolic pressures elevated (Fig. 4). The normal range for left ventricular end-diastolic pressure is 0-10 mm Hg (Ettinger and Suter, 1970). Measurement of left ventricular volume parameters showed a moderate increase in end-diastolic volume, an ejection fraction of 0.42, and very

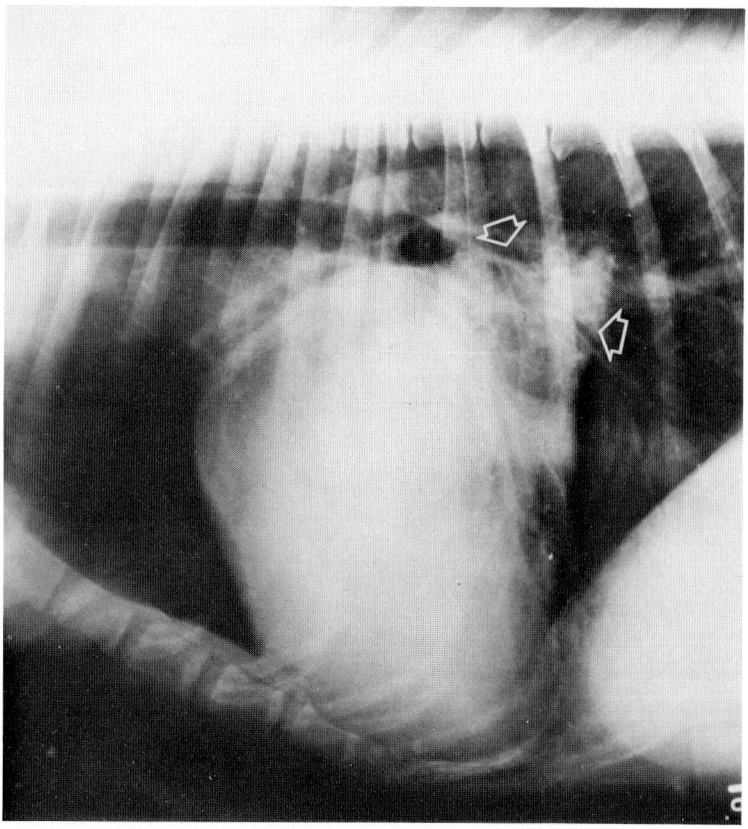

Figure 2. Left lateral radiograph of a 3-year-old male German Shepherd with a history of weight loss, anorexia, and abdominal enlargement over a period of 3 weeks. Note the left atrial enlargement (arrows), biventricular enlargement, and pulmonary congestion.

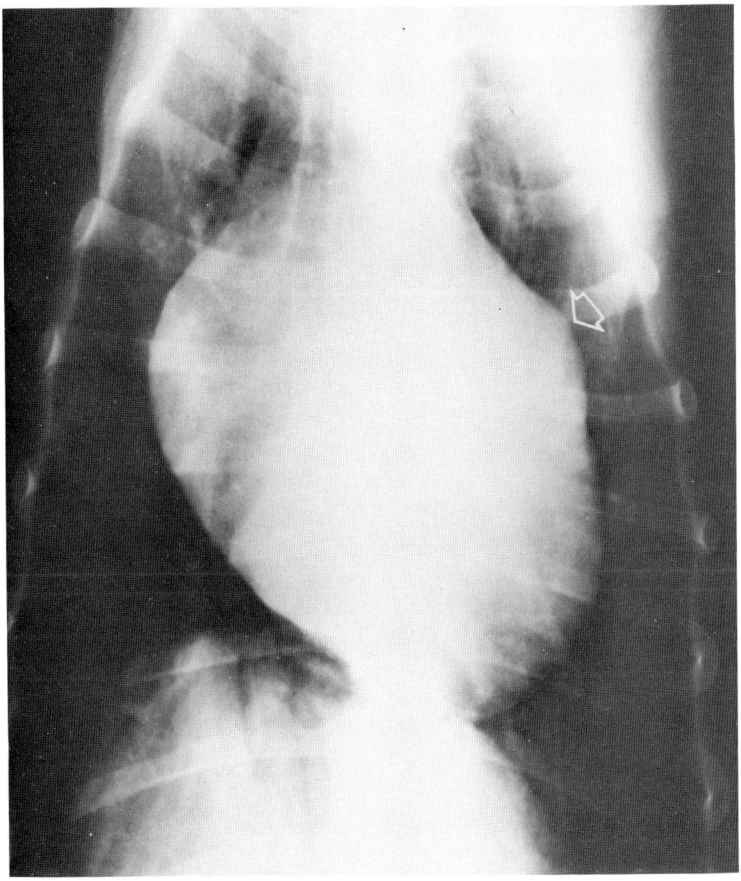

Figure 3. Dorsoventral thoracic radiograph of dog depicted in Figure 2. Note the biventricular enlargement and the dilated left auricle (*arrow*).

reduced net stroke work compared with normal dogs and dogs with volume overload (personal communication, 1973; Dr. Lord, Department of Clinical Radiology, Royal Veterinary College, Stockholm, Sweden). Angiocardiography revealed gross enlargement of the cardiac chambers and resultant secondary mitral and tricuspid insufficiency (Fig. 5).

One dog with the hypertrophic form of cardiomyopathy was studied with complete catheterization and angiocardiography. Angiocardiograms of both left and right ventricles revealed hypertrophic subaortic stenosis with concomitant obstruction to the right ventricular outflow tract (Fig. 6). The obstruction to the left ventricular outflow and the obstruction within the right ventricle appeared

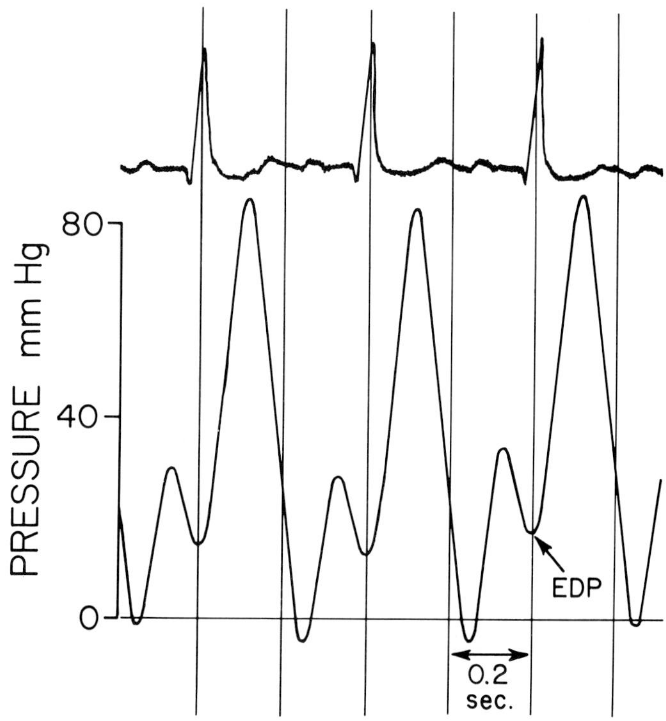

Figure 4. Left ventricular pressure tracing from a 3-year-old male Doberman Pinscher with congestive cardiomyopathy. The end-diastolic pressure (*arrow*) is increased to 15 mm Hg.

to be produced by the bulging of the hypertrophied interventricular septum. The peak systolic pressure gradient was 15 mm Hg in the left ventricular outflow tract and 10 mm Hg in the right ventricular outflow tract. The obstruction in the left ventricle was intensified by the administration of isoproterenol.

Pathologic findings

Grossly, excessive amounts of abdominal and thoracic effusion (500–2,500 ml), and pericardial effusion (25–250 ml) were seen in all the affected dogs. The effusions varied in color from clear to a bloody tan, and their specific gravities ranged from 1.015 to 1.025. The hearts were rounded due to the extreme dilation of the ventricles (Fig. 7). Vessels on the epicardial surfaces were tortuous and dilated. The ventricular and septal walls were thinner than normal. The endocardial surfaces were opaque. The papillary muscles and trabeculae in the left ventricles were flattened and atrophied (Fig. 8). The circumferences of

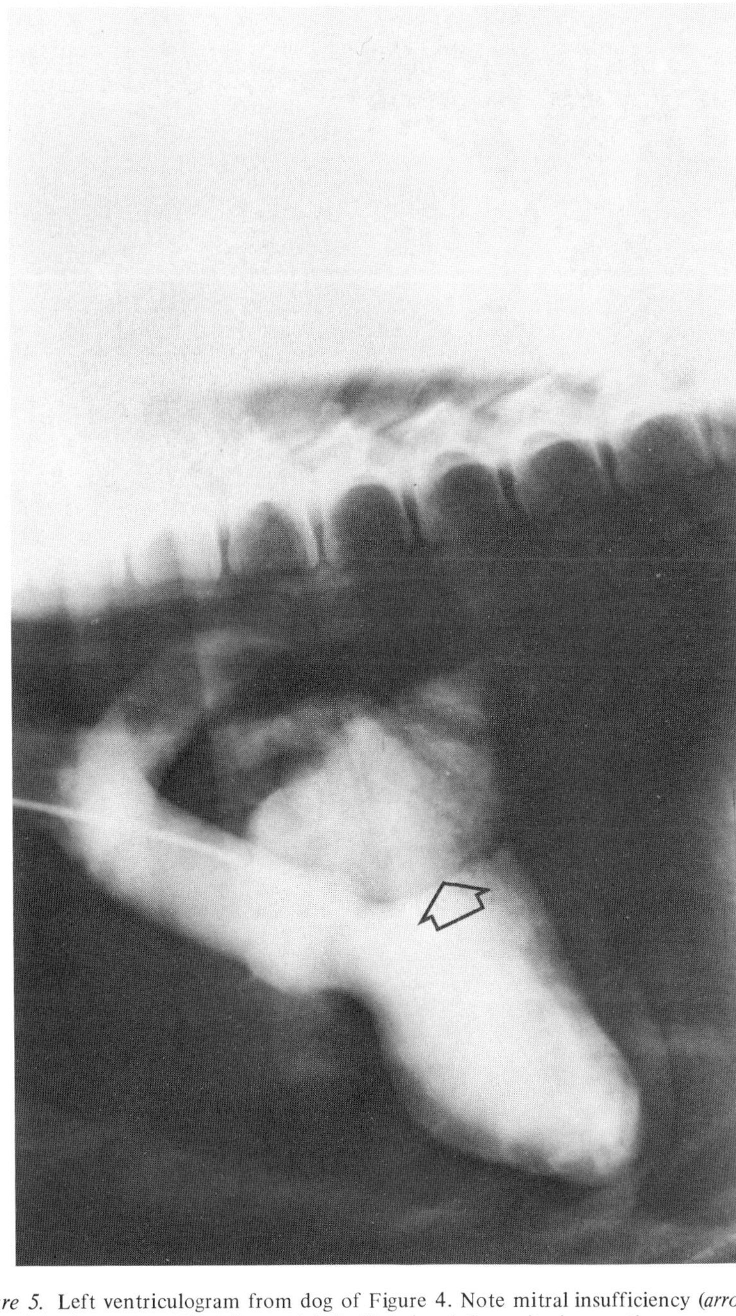

Figure 5. Left ventriculogram from dog of Figure 4. Note mitral insufficiency (*arrow*) and left atrial enlargement. The left ventricle is dilated and the cavity is smooth.

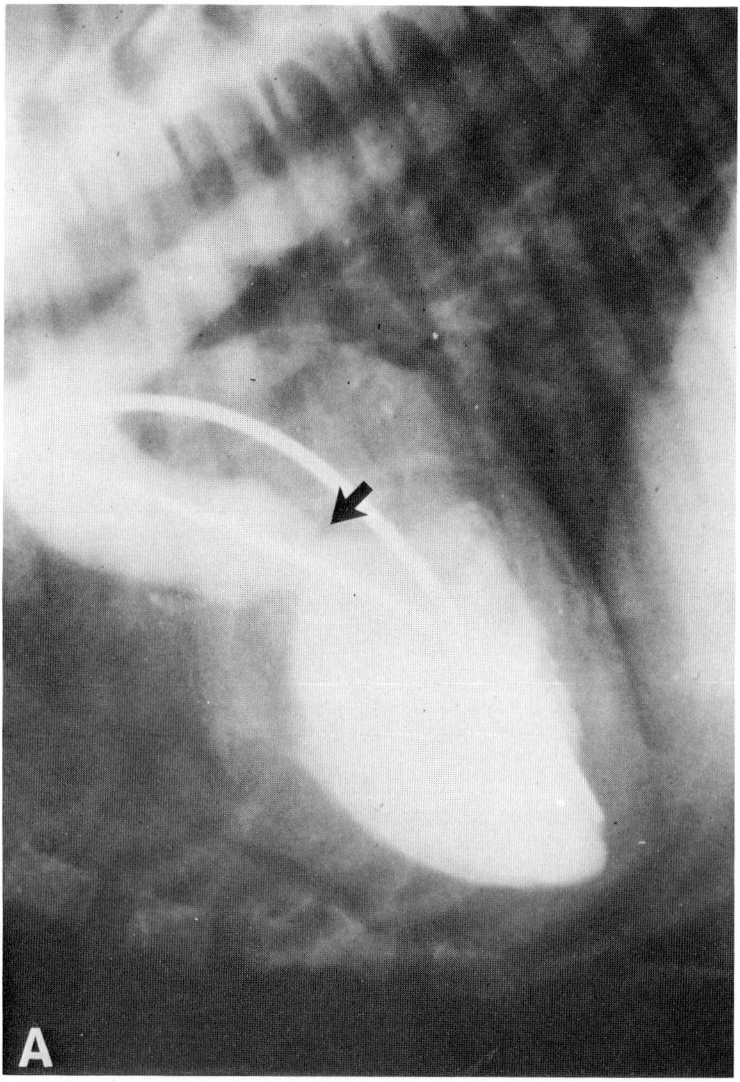

Figure 6. Left ventriculogram in diastole (*A*) and right ventriculogram in systole (*B*) of a 3-month-old female Samoyed. Note the obstructions to both outflow tracts (*arrows*) and the hypertrophy of each ventricle.

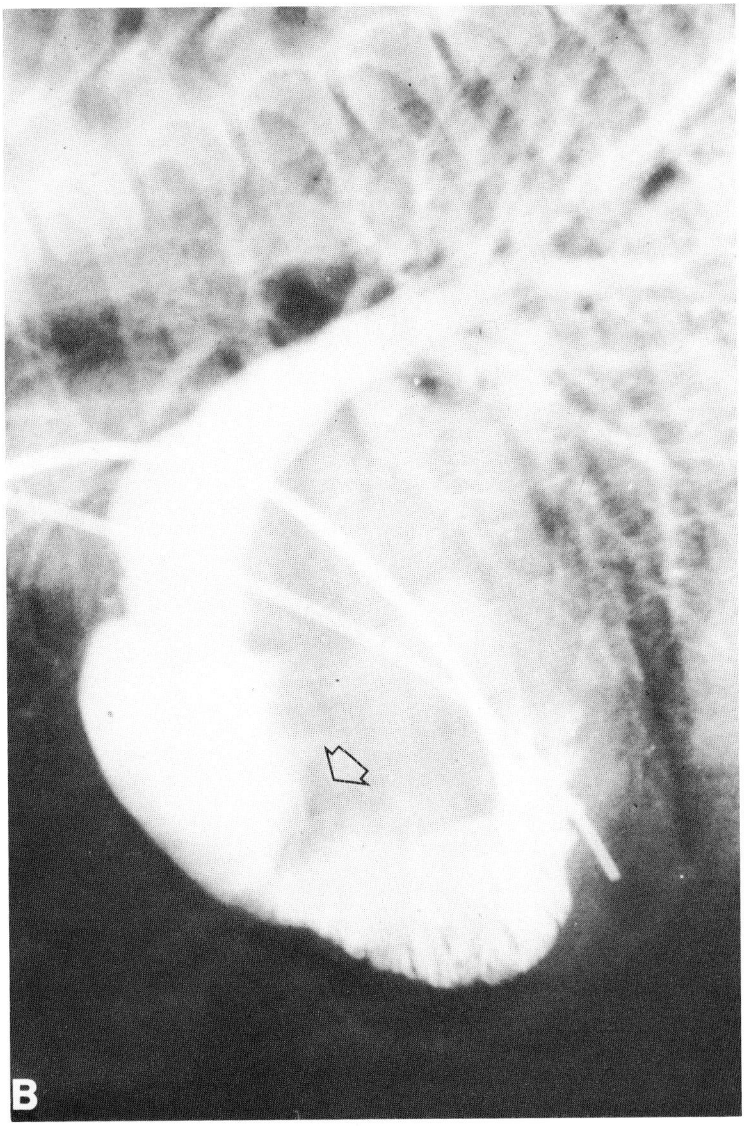

Figure 6B.

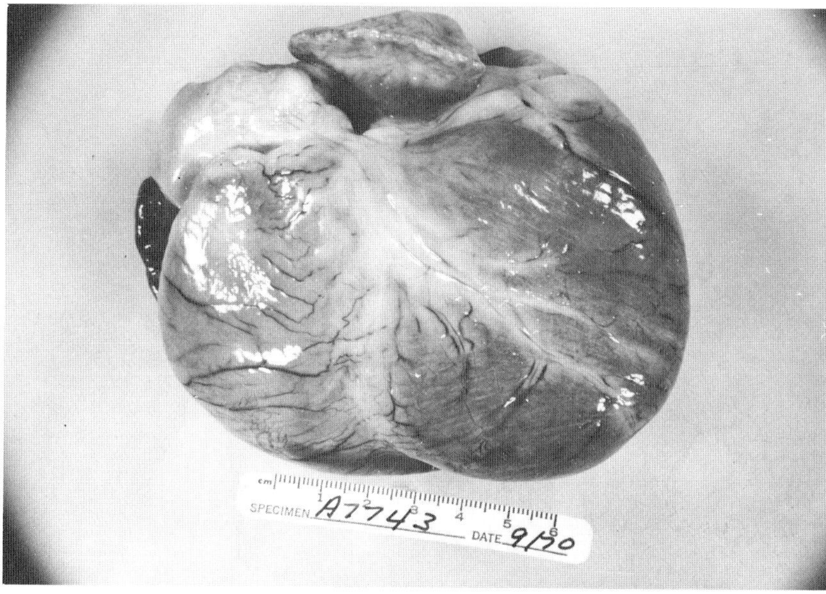

Figure 7. Heart from a 3-year-old male Great Dane dog, showing dilation of the left and right ventricles.

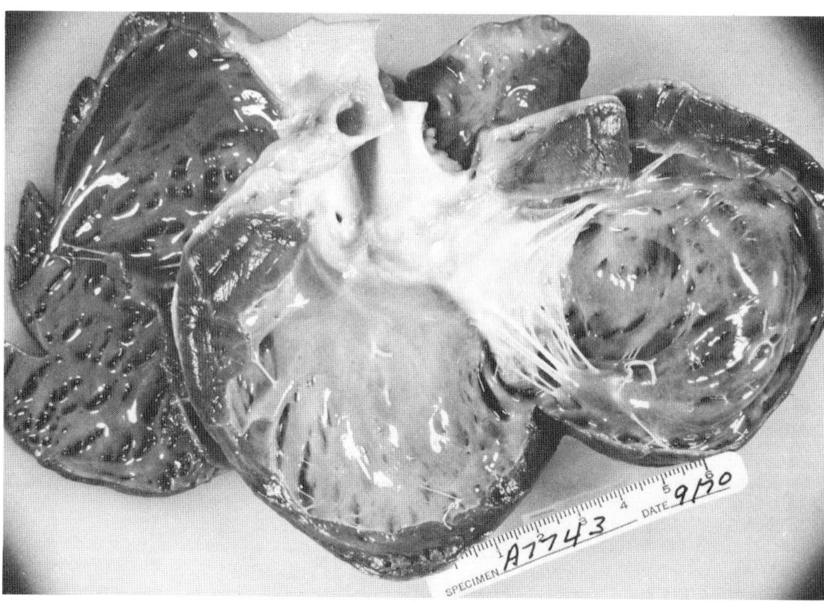

Figure 8. Left and right ventricles of the heart from the dog seen in Figure 7, showing dilated left ventricle with flattened trabeculae and papillary muscles and normal aortic and mitral valves.

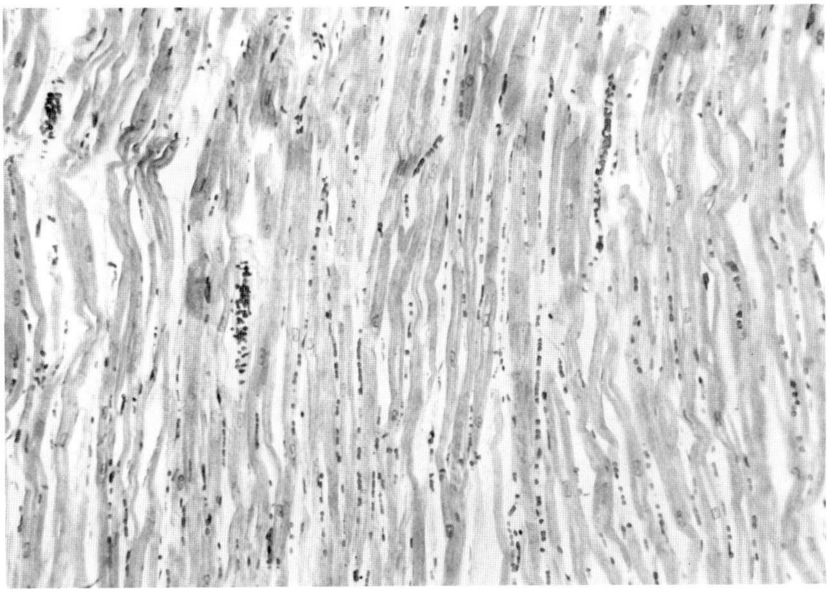

Figure 9. Myocardium of dog seen in Figure 7, showing thin and small muscle fibers separated by edematous fluid or fine collagen fibers. H and E. × 100.

the mitral and tricuspid valves were greater than normal, but the leaflets and chordae tendinae of the atrioventricular valves were normal.

The lungs were wet, firm, heavy, and greyish brown. The livers were dark brownish red to purple. The hepatic capsules were opaque, and the hepatic parenchyma resumbled a nutmeg pattern. Visceral vessels were tortuous and dilated.

Microscopically, the endocardiums were thickened by dense, wavy collagen fibers. The fibrous tissue extended deeply into the subendocardiums and adjacent myocardiums. The myofibers had disappeared in areas and were smaller than normal. The wavy, thin myofibers were separated by edematous fluid and erythrocytes. Some myofibers showed various degrees of degeneration from loss of striation to coagulation or vacuolization of sarcoplasm to myocytolysis (Fig. 9).

TREATMENT

Therapy was directed toward the congestive heart failure with digoxin, and consisted of rest, slowing of the heart rate, improvement of cardiac output, and diuretics for the pulmonary congestion, ascites, and pleural effusion.

After a dog was digitalized and congestive heart failure was under control, medical or electrical cardioversion of atrial fibrillation to normal sinus rhythm was attempted (Fig. 1).

The prognosis for congestive cardiomyopathy is very poor, the average survival time being 6–12 months after the onset of signs. The longest survival time for the dogs of this report was 20 months (1 dog).

DISCUSSION

Congestive cardiomyopathy in the dog is very similar to the syndrome in man. The syndrome involves severe ventricular myocardial damage and poor systolic function. Myocardial contractility is also poor, resulting in dilation of the heart chambers and minimal hypertrophic response. Some intrinsic disease causes impairment of contractility, and compensatory hypertrophy is unable to keep up with the dilation. This results in increased tension and stress of the ventricular wall and increased diastolic pressure. The hearts of the dogs in our study closely resembled hearts of human patients with cardiomyopathy (Friedberg, 1966; Goodwin, 1970; Perloff, 1971; Burch, 1972; Goodwin and Oakley, 1971).

Factors in man that are considered in the etiology are pregnancy; infections; and immunological, nutritional, metabolic, and genetic disorders (Emanuel, 1971; Perloff, 1971; Burch, 1972). As stated previously, there is some association of the syndrome in the dog with stress such as pregnancy, gastric dilation/torsion, and spondylolisthesis.

We have also seen a clinical syndrome similar to congestive cardiomyopathy in large breed dogs under 1 year of age. In contrast to the older dogs, atrial fibrillation is unusual. The endocardium in these dogs is thick, with an anatomic abnormality of the atrioventricular valve complex. Ventricular chamber dilation is also present. This syndrome may be a different manifestation of congestive cardiomyopathy.

Based on the close similarity of cardiomyopathy in man to that in the dog, the dog could serve as a suitable animal model for the study of this disorder. There is need for explorations in the areas of epidemiology, biochemistry, pathology, ultrastructure, virology, and immunology in relation to the disease.

Acknowledgments The authors thank Miss Joyce Murberg, Miss Cynthia Fazzini, and Mrs. Nancy Sturge for their assistance in preparation of the manuscript.

REFERENCES

BOLTON, G. R. 1972. Cardiomyopathy. *In* A Manual of Clinical Cardiology, pp. 47–53. American Animal Hospital Association. Elkhart, Ind.

BURCH, G. E. 1972. Cardiomyopathy, Vol. 4, No. 1. F. A. Davis Company, Philadelphia.

EMANUEL, R. 1971. The familial incidence of idiopathic cardiomyopathy. *In* G. E. W. Wolstenholme and M. O'Connor (eds.), Cardiomyopathy, pp. 50–54. Ciba Foundation Study Group, No. 37. J. A. Churchill, London.

ETTINGER, S. J., G. R. BOLTON, and P. F. LORD. 1970. Idiopathic cardiomyopathy in the dog. J. Am. Vet. Med. Ass. 156: 1225. (Abstr.)

ETTINGER, S. J., and P. F. SUTER. 1970. Canine Cardiology. W. B. Saunders Company, Philadelphia.

FRANK, S., and E. BRAUNWALD. 1968. Idiopathic hypertrophic subaortic stenosis: Clinical analysis of 126 patients with emphasis on the natural history. Circulation 37: 759–788.

FRIEDBERG, C. K. 1966. Diseases of the Heart, 3rd ed., Vol. II. W. B. Saunders Company, Philadelphia.

GOODWIN, J. F. 1970. Congestive and hypertrophic cardiomyopathies: a decade of study. Lancet *i:* 731–737.

GOODWIN, J. F., and OAKLEY, C. M. 1972. The cardiomyopathies. Brit. Heart J. 34: 545–552.

LUGINBÜHL, H., and D. K. DETWEILER. 1965. Cardiovascular lesions in dogs. Ann. N. Y. Acad. Sci. 127: 517–540.

PERLOFF, J. K. 1971. The cardiomyopathies–current perspectives. Circulation 44: 942–949.

Mailing address:
Dr. Lawrence P. Tilley,
The Animal Medical Center,
510 East 62nd Street, New York, New York 10021 (USA).

Recent Advances in Studies on
Cardiac Structure and Metabolism Volume 10
The Metabolism of Contraction
Edited by Paul-Emile Roy and George Rona

MYOCARDIAL HYPERTROPHY, MYOCARDITIS, AND MYOCARDIAL FIBROSIS IN A LONG TERM EXPERIMENT

K. E. SCHENK, U. GROSS,
L. OEHRING, and H. SCHMIDT

Medizinische Klinik und Poliklinik im Klinikum Steglitz,
Freien Universität Berlin, Berlin, Germany

SUMMARY

In a long term experiment, a systematic analysis was made of the cellular reaction of the myocardium and the lymphatic system to application of antigenic substances. Several groups of mice received single or repeated injections of human serum, serum from rabbits with antiheart antibodies, mouse spleen cells immunized to myocardium, and human or mouse cardiac homogenate with and without incomplete or complete Freund's adjuvant.

After injection of human and mouse heart homogenate, hypertrophy of myocardial cells was the first alteration seen; then followed focal myocarditis and, after 10 months, interstitial fibrosis. Simultaneously, in the lymphoreticular tissue, there was marked proliferation of reticulum cells, lymphocytes, and plasma cells, as well as an enlargement of secondary follicles. All the animals produced antibodies to the injected antigens. Following injection of human serum or serum from rabbits with cardiac specific antibodies or spleen cells, no morphological changes were seen in the hearts.

INTRODUCTION

The role of immune processes in the pathogenesis of heart diseases, e.g., myocarditis, is not yet sufficiently understood. Therefore, we have made a systematic analysis in long term experiments of the cellular reaction of the myocardium and lymphytic system to the application of different antigenic substances.

MATERIALS AND METHODS

Several groups of mice (NMRI) received single or repeated s.c. or i.p. injections (twice a week for a period of 3 months) of human cardiac homogenate, cardiac homogenate from mice, human serum, or incomplete and complete Freund's adjuvant. The reactions in the heart, lymph nodes, spleen, liver, kidneys, and skeletal muscles were studied histologically (daily from the 1st to the 30th day, and then every 10th day) over a period of 10 months.

Other groups of mice and rabbits received repeated s.c. or i.p. injections of mouse heart homogenate with complete Freund's adjuvant (in a 1:1 ratio per injection). After a period of 2 months, rabbit serum with precipitated antibodies against mouse myocardium as well as spleen cells of the immunized mice were injected s.c. or i.p. into other groups of mice. The antibody production against myocardium was checked by Ouchterlony tests (Steffen, 1968).

RESULTS

Marked morphological changes were seen in mouse hearts 30–40 days after repeated injections of mouse heart homogenate with complete Freund's adjuvant and human heart homogenate without adjuvant. The first alteration noted was hypertrophy of myocardial cells, the size and weight of the hearts being doubled as compared with the values of control animals ($p<99.9\%$). At 40–50 days after repeated injections of heart homogenate, we found focal myocarditis with interstitial collections of lymphocytic and histiocytic cells, but only few neutrophilic leukocytes (Fig. 1). This myocarditis was localized chiefly in the inner wall of the myocardium. Some hearts showed identical infiltrations in the heart valves and the walls of large vessels. All mice receiving human cardiac homogenate died 3–4 months later, with signs of myocardial insufficiency.

The hearts of all mice given repeated injections of cardiac homogenate showed an increasing content of the interstitial connective tissue after the 50th day. At 10 months after the beginning of the experiments, a marked interstitial fibrosis was apparent (Fig. 2).

After injections of mouse heart homogenate without adjuvant, injections of human serum proteins, or Freund's adjuvant alone, no myocarditis or myocardial fibrosis could be seen. In addition, following injection of rabbit serum with cardiac specific antibodies against mouse myocardium or injections of spleen cells, there were no morphological changes in the hearts.

The reaction pattern of the lymphoreticular tissue to different substances, such as cardiac homogenate, serum proteins, and complete or incomplete Freund's adjuvant, appears to be regular and to proceed in two phases, after both single and repeated injections. During phase one, the lymph nodes and spleen show a loss of small lymphocytes and an increase in large lymphocytes and small reticulum cells, as well as an enlargement and new growth of secondary follicles. Phase one begins 24 hours after the first injection, showing a maximum of cellular reaction from the 15th to the 20th day (Fig. 3, *a* and *b*). On the 10th day, the beginning of phase two, there is formation of large reticulum cells and immature and mature plasma cells (Fig. 3*b*). After repeated antigen injections in phase two, the cellular changes are more marked than after

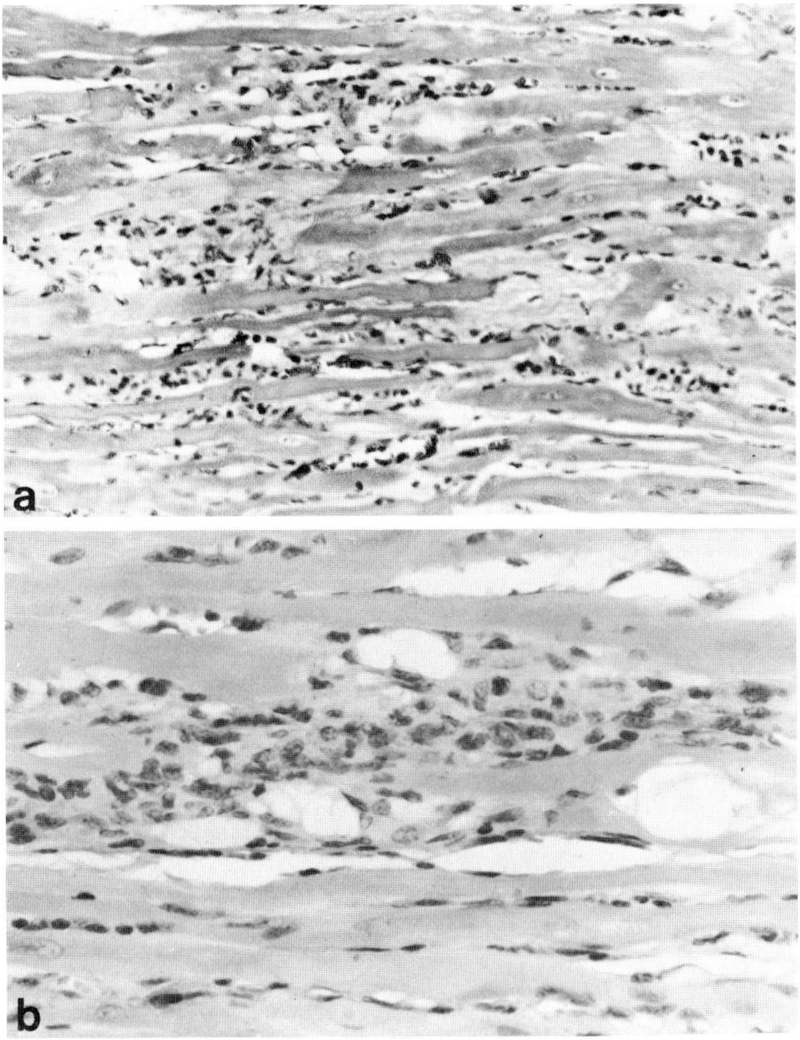

Figure 1. Heart of mouse, 40 days after repeated injections of human heart homogenate. Myocarditis is present. × 400.

a single injection. Following application of paraffin oil without antigen, no marked proliferation of plasma cells can be observed in phase two.

In the liver, there was proliferation of reticulum cells after repeated antigen injections. In the kidneys and skeletal muscles, no morphological changes were seen.

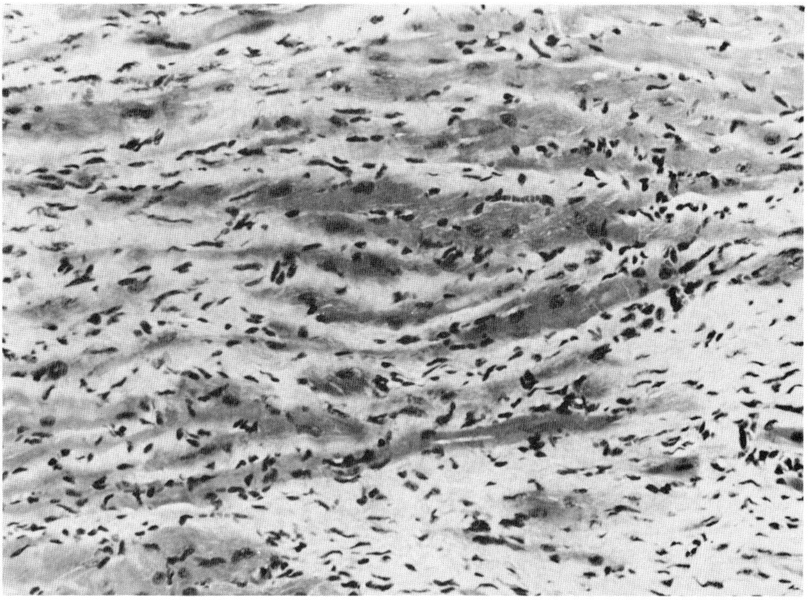

Figure 2. Heart of mouse, 10 months after repeated injections of mouse heart homogenate with complete Freund's adjuvant. Interstitial fibrosis can be seen. × 400.

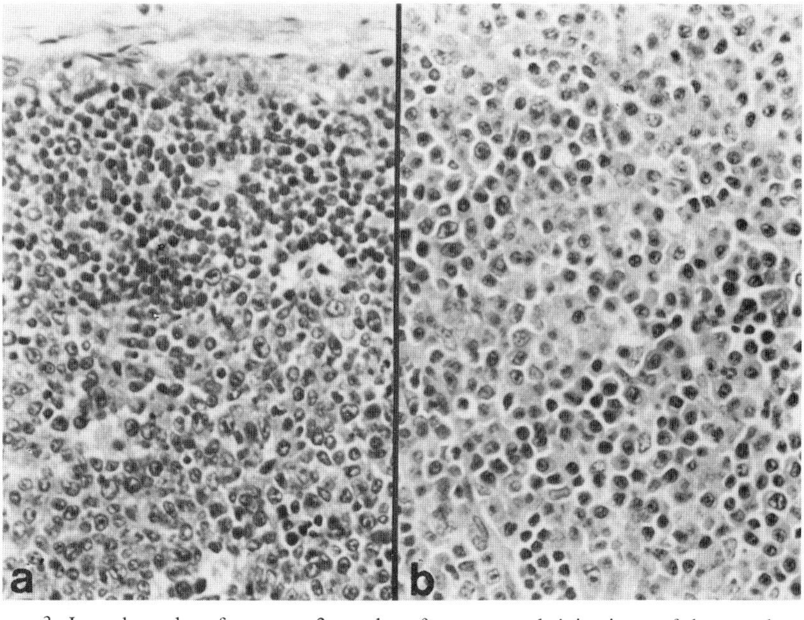

Figure 3. Lymph node of mouse, 3 weeks after repeated injections of human heart homogenate. *a:* cortex shows loss of small lymphocytes and increase of large lymphocytes. × 1:400. *b:* medulla exhibits proliferation of plasma cells. × 600.

Antibodies against the injected cardiac homogenate from mice and humans were first found on the 20th day of the experiment.

DISCUSSION

The histological picture of the myocarditis described above is similar to that seen after cardiac transplantation (Bieber, Stinson, and Shumway, 1969; Lower *et al.*, 1969; Stinson *et al.*, 1969). Therefore, it may not be merely speculative to say that this myocarditis in mice is of immunological nature. All the mice produced antibodies against myocardium. Furthermore, the lymphoreticular tissue simultaneously exhibited cell proliferation characteristic of an immunological reaction (Schenk and Manskopf, 1968; Schenk and Gross, 1969). There was good correlation between the degree of cellular reaction in the lymphatic system and the severity of myocardial damage.

However, the absence of myocardial lesions after administration of specific antiheart sera or spleen cells may militate against the immunological hypothesis. Some authors have sometimes seen a cytotoxic effect of antiheart sera or lymphocytes *in vitro* (Laufer and Davies, 1969; Laufer, Friedman, and Ron, 1974). Others have found a mild myocarditis in rats receiving lymphocytes from rats immunized with myocardial cells (Friedman *et al.*, 1971; Laufer *et al.*, 1974). Antibodies against myocardium have been demonstrated in the blood of humans in the presence of various conditions, such as rheumatic fever, myocarditis, cardiomyopathy, or endomyocardial fibrosis (Geld, 1964; Hagert, 1968; Kaplan, 1968; Körge *et al.*, 1970). But the exact role of antibodies or lymphocytes in the pathogenesis of myocarditis, myocardial hypertrophy, and myocardial fibrosis as described here is not yet known.

In conclusion, it may be stated that, in the present study, myocardial hypertrophy, myocarditis, and myocardial fibrosis developed under identical experimental conditions. Myocardial hypertrophy was first seen, followed by myocarditis, and, later myocardial fibrosis. These three diseases, also known in human pathology, may be an expression of an immunological reaction; that is, they may only be symptoms of a single immunological disease.

REFERENCES

BIEBER, C. P., E. B. STINSON, and N. E. SHUMWAY. 1969. Pathology of the conduction system in cardiac rejection. Circulation 39: 567.

FRIEDMAN, I. A., A. LAUFER, N. RON, and A. M. DAVIES. 1971. Experimental myocarditis: in vitro and in vivo studies of lymphocytes sensitized to heart extracts and group A streptococci. Immunology 20: 225.

GELD, v. d. H. 1964. Anti-heart antibodies in the postpericardiotomy and postmyocardial infarction syndromes. Lancet *ii:* 617.

HAGERT, M. H., H. TRENCKMANN, and Ch. HAGERT. 1969. Experimentelle Untersuchungen über die Antigenstrukutur des menschlichen Herzens mittels Immunelektrophorese. Zschr. d. ges. Inn. Med. 23: 596.

KAPLAN, M. H. 1968. Autoimmunity to heart. *In* P. A. Miescher (ed.), Textbook of Immunpathology, Vols. I and II, p. 641. Grune and Stratton, New York/London.

KÖRGE, K., H. HANSON, L. HERING, E. LIPSO, S. MARAMAA, und S. VELBRI. 1970. Klinische und experimentelle Untersuchungen zur Immunologie des Myocardinfarktes. Zschr. d. ges. Inn. Med. 25: 18.

LAUFER, A., and A. M. DAVIES. 1969. Experimental granulomatous myocarditis: genesis and immunological aspects. Ann. N.Y. Acad. Sci. 156: 91.

LAUFER, A., I. FRIEDMAN, and N. RON. 1974. In vitro and in vivo studies of the role of immune reactions in myocardial damage. *In* E. Bajusz and G. Rona (eds.), Recent Advances in Studies on Cardiac Structure and Metabolism, Vol. 2: Cardiomyopathies, pp. 377–384. Urban and Schwarzenberg, München/Berlin/Wien.

LOWER, R. R., J. C, KOSEK, V. E. KOMP, W. H. GRAHAM, D. H. SEWELL, and F. Lim. 1969. Rejection of the cardiac transplant. Am. J. Cardiol. 24: 492.

SCHENK, K. E., und G. MANSKOPF. 1968. Freundsches Adjuvans und lymphatisches System. Beitr. path. Anat. 137: 391.

SCHENK, K. E., und U. M. GROSS. 1969. Lymphatisches Gewebe und Antikörperbildung im Langzeitversuch. Virch. Arch. Abt. A Path. Anat. 348: 342.

STEFFEN, C. 1968. Allgemeine und experimentelle Immunologie und Immunpathologie, p. 465. Thieme Verlag, Stuttgart.

STINSON, E. B., E. DONG, Ch. P. BIEBER, J. S. SCHROEDER, and N. E. SHUMWAY. 1969. Cardiac transplatation in man. J. Am. Med. Ass. 207: 2233.

Mailing address:
Dr. K. E. Schenk, Medizinische Klinik und Poliklinik
im Klinikum Steglitz der Freien Universität Berlin,
1 Berlin 45, Hindenburgdamm 30 (Germany).

Recent Advances in Studies on
Cardiac Structure and Metabolism Volume 10
The Metabolism of Contraction
Edited by Paul-Emile Roy and George Rona

METHOD FOR THE PRODUCTION OF SEVERE VENTRICULAR DYSRHYTHMIAS IN SMALL LABORATORY ANIMALS

GIANCARLO GUIDERI, MICHAEL BARLETTA,
RAYMOND CHAU, MARK GREEN, and DAVID LEHR

Department of Pharmacology, New York Medical College, Valhalla, New York, USA

SUMMARY

In earlier reports from this laboratory, it was shown that 3 weeks' pretreatment with desoxycorticosterone acetate (DCA) pellet implantation and saline as drinking fluid caused a nearly 20,000-fold potentiation of the acute cardiotoxicity of isoproterenol. In such animals, the beta-receptor stimulant consistently elicited severe ventricular dysrhythmias, usually leading to ventricular fibrillation (VF) and death. In the present study, a similar enhancement of isoproterenol arrhythmogenic activity after DCA-saline pretreatment was demonstrated also in the guinea pig and albino mouse. It was found, further, that isoproterenol consistently caused an alteration of the total myocardial electrolyte content in DCA-saline-treated rats, consisting of a decrease of magnesium (Mg) and potassium (K) levels and elevation of sodium (Na). In control rats, on the other hand, isoproterenol did not alter the myocardial Mg content and increased K. The emergence of cardiac irregularities and VF could be inhibited by pretreatment with antiarrhythmic drugs. On a molar basis, these agents decreased the incidence of VF in the following descending order: *dl*-propranolol (100), practolol (39), *d*-propranolol (34), sotalol (20), quinidine (8), and lidocaine (5). In the presence of any of the listed drugs, except for lidocaine, isoproterenol did not lower Mg levels and had variable effects on those of Na and K. Pretreatment with Mg gluconate was likewise effective in preventing isoproterenol-induced dysrhythmias and death. These findings demonstrate that: a) sensitization of the myocardium by DCA-saline treatment is not species-specific for the rat; b) this phenomenon can be a useful method for the screening of antiarrhythmic drugs; and c) isoproterenol-induced dysrhythmias may be associated with myocardial electrolyte alterations involving especially Mg depletion.

INTRODUCTION

It has been reported from this laboratory that pretreatment of the albino rat with desoxycorticosterone acetate (DCA) and saline for 3 weeks greatly enhances the acute toxicity of isoproterenol. Small, otherwise nonarrhythmogenic doses of this beta-recptor agonist induced severe atrial and ventricular dysrhythmias, often leading to death in ventricular fibrillation (Guideri, Goldman, and Lehr, 1971; Guideri, Barletta, and Lehr, 1974). In such animals, the LD_{50} of isoproterenol was 40 μg/kg as compared to 680 mg/kg in control rats (Rona *et al.*, 1959), indicating a nearly 20,000-fold increase in the acute toxicity of isoproterenol. Interestingly, epinephrine, an adrenergic agent with alpha- as well as beta-receptor stimulating properties, did not consistently elicit ventricular arrhythmias and ventricular fibrillation except in the presence of alpha-receptor blockade.

Since hydrocortisone-saline pretreatment likewise sensitized the myocardium of the rat to isoproterenol, it was suggested that such a drug-drug interaction may have contributed to the increased incidence of mortality in asthmatic patients using aerosols containing beta-adrenergic stimulants while on prolonged steroid treatment (Lehr, 1972; Guideri *et al.*, 1974).

The present investigation was undertaken to determine whether DCA-saline pretreatment was able to similarly sensitize the myocardium of other rodents (mice and guinea pigs) and to evaluate the usefulness of this phenomenon as a model for the screening of antiarrhythmic and, especially, antifibrillatory agents. It is planned to extend these studies to higher species, including the dog and squirrel monkey.

METHODS

Albino rats of the Wistar strain weighing 250–350 g and guinea pigs weighing 300–400 g were implanted subcutaneously with a 25-mg DCA pellet in the axillary region under tribromoethanol (Avertin) anesthesia (25 mg/100 g body weight, intraperitoneally) and maintained thereafter on 1% saline. Albino mice weighing 22–38 g were similarly implanted with a 10-mg DCA pellet. All experimental animals had free access to 1% saline as drinking fluid and to standard laboratory chow.

At predetermined intervals within a period of 17–28 days, individual rats were anesthetized with Dial-urethane (0.7 ml/kg intraperitoneally) and prepared for the recording of blood pressure and the elecrocardiogram (ECG). All animals were ventilated artificially with the aid of a Phipps and Bird small animal respirator. Blood pressure was measured from the left carotid artery with a Statham transducer. The standard limb leads and one chest lead of the ECG were recorded continuously on a Honeywell model 5600 tape system. Blood pressure and ECG tracings were monitored on the oscilloscope screen of an Electronics for Medicine model DR 12 recorder and inscribed intermittently on photo-recording paper. In studies with Dial-urethane-anesthetized mice, only the ECG was recorded.

Following a 20- to 30-minute stabilization period, a single dose of *dl*-isoproterenol hydrochloride, ranging from 10 to 1,000 μg/kg body weight, was injected subcutaneously and each animal was monitored continuously for 30 min. Since it was previously determined that 150 μg/kg isoproterenol invariably elicited ventricular arrhythmias and the maximal incidence of ventricular fibrillation (Guideri *et al.*, 1974), this dose of the beta-agonist was used in all subsequent studies. Except for magnesium gluconate, which was injected 30 min prior to isoproterenol, all antiarrhythmic agents were injected 10 min before the beta-receptor stimulant by the subcutaneous route.

At the end of each experiment, the total myocardial content of magnesium, sodium, and potassium was determined by automated chemical analysis in tissue samples weighing 0.4–0.5g and obtained from apical and basal portions of both ventricles (Lehr, Krukowski, and Colon, 1966).

For the purpose of recording the ECG in *conscious*, unrestrained rats and guinea pigs, animals were prepared several days prior to the experiment in the following manner: under Avertin anesthesia, fine Teflon-coated silver wires (0.003 inch thick) were passed under the skin by means of a trochar from each limb to the base of the neck, where the four wires were fixed by a small plaster of paris cast sutured to the skin, in such a manner that long protruding ends of the very flexible wire could be connected to the recording system without substantial restriction or discomfort to the animal. The Teflon insulation had been removed from the limb end of the wires, and these terminals had been coated with silver-silver chloride.

The results were analyzed for statistical significance using the Student *t*-test and the Chi square (χ^2) analysis.

RESULTS

Effect of isoproterenol in DCA-saline-pretreated mice and guinea pigs

The arrhythmogenic effect of isoproterenol in mice and guinea pigs is compared with that in rats in Table 1. Treatment with DCA-saline sensitized the myocardium of the mouse to isoproterenol, since 10 of 11 animals experienced ventricular arrhythmias, including 6 instances of ventricular fibrillation. The mouse, however, appeared to have a greater ability to spontaneous recovery from ventricular fibrillation, because 4 out of 6 animals survived several bouts of this rhythm disorder.

In contradistinction to the rat and mouse heart, the physiological and pharmacological responses of the guinea pig heart are believed to resemble those of the human heart more closely. It was found that the untreated guinea pig was more sensitive to isoproterenol than the mouse and rat, since 6 out of 11 guinea pigs developed ventricular arrhythmias, including 1 instance of ventricular fibrillation with isoproterenol injection alone (Table 1). DCA-saline treatment, however, greatly enhanced isoproterenol cardiotoxicity. Ventricular dysrhythmias occurred in all guinea pigs, and 9 out of 10 died in ventricular fibrillation.

Typical ECG tracings of rhythm disturbances in DCA-saline-pretreated rats have been depicted elsewhere (Guideri *et al.*, 1974). In DCA-saline-pretreated mice, severe tachycardia and ventricular arrhythmias were the most common

Table 1. Comparison of the arrhythmogenic action of isoproterenol (150 μg/kg) in the untreated and DCA-saline-pretreated mouse, rat, and guinea pig

Species	Treatment	N	Ventricular Arrhythmias	Ventricular Fibrillation: Death	Ventricular Fibrillation: Recovery	Ventricular Fibrillation: Total
Mouse	—	9	1	0	0	0
	DCA-saline	11	10*	2	4	6+
Rat	—	10	1	0	0	0
	DCA-saline	33	33*	19	7	26*
Guinea pig‡	—	11	6	1	0	1
	DCA-saline	10	10†	9	0	9*

*$p < 0.01$; from Chi square (χ^2) analysis, using the Yates correction factor for noncontinuously distributed data.
†$p < 0.05$.
‡Unanesthetized.

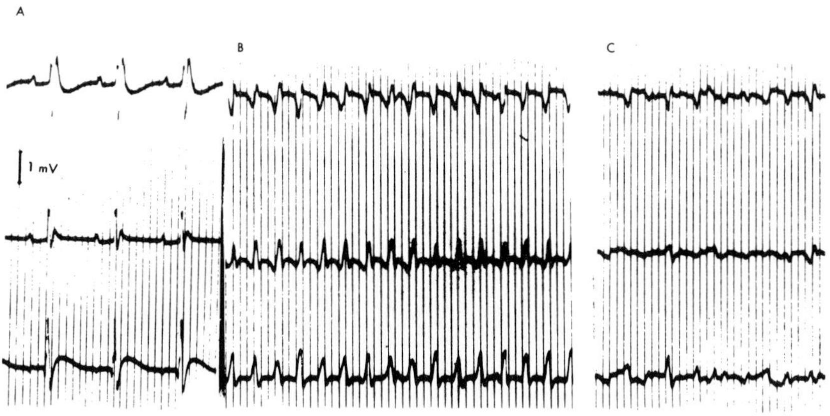

Figure 1. ECG tracings of a Dial-urethane-anesthetized, DCA-saline-pretreated mouse prior to and following the injection of 150 μg/kg isoproterenol. Leads I, II, and III are shown *from top to bottom;* time lines are recorded every 20 msec. In the control record, at *A*, the heart rate was 306 beats/min; 4 min after isoproterenol, this animal experienced a prolonged attack of severe tachycardia (*B;* heart rate 900–1,000 beats/min), which in 1 min progressed to ventricular dysrhythmias (*C*).

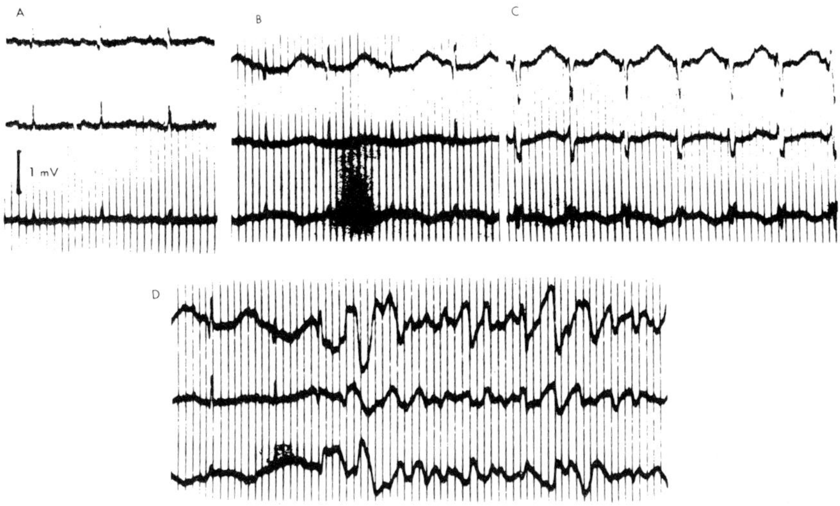

Figure 2. ECG tracings of a conscious, DCA-saline-pretreated guinea pig before and after the administration of 150 μg/kg isoproterenol. Each tracing shows leads I, II, and III *from top to bottom;* time lines every 20 msec. Before isoproterenol injection, the heart rate was 300 beats/min (*A*). The beta-agonist moderately increased heart rate to 330 beats/min (*B*). Note the T wave changes elicited by isoproterenol occurring especially in leads I and III. *Panel C* shows an example of ventricular tachycardia (heart rate, 400 beats/min) occurring 5 min after isoproterenol; ventricular fibrillation developed 1 min later (*D*).

isoproterenol-induced rhythm disorders. The appearance of these disturbances is illustrated in Figure 1. The tracings shown in Figure 2 are examples of the most frequently encountered rhythm disturbances in the DCA-saline-pretreated guinea pigs, which included isoproterenol-induced ventricular tachycardia (*C*) and ventricular fibrillation (*D*).

Effects of antiarrhythmic agents on isoproterenol-induced dysrhythmias

Since the administration of isoproterenol consistently produces ventricular fibrillation and ventricular arrhythmias in DCA-saline-treated rats, this phenomenon appeared to be a potentially useful model for screening antiarrhythmic drugs. The procedure was, therefore, tested in rats by determining the prophylactic effectiveness of several well known antiarrhythmic agents against the emergence of dysrhythmias. Two groups of these drugs were used: *dl*-propranolol, sotalol, and practolol were chosen because of their beta-receptor blocking properties, while *d*-propranolol, quinidine, and lidocaine were selected because of their local anesthetic or membrane-depressant activity.

dl-Propranolol, which has both membrane-depressant activity and beta-receptor blocking properties (Vaughan Williams, 1966; Hoffman and Grupp,

Table 2. Effect of beta-receptor blocking agents on isoproterenol-induced arrhythmias

Treatment	Dose (mg/kg)	N	Ventricular Arrhythmias	Ventricular Fibrillation		
				Death	Recovery	Total
Isoproterenol	0.15	33	33	19	7	26
dl-Propranolol	0.10	5	3	2	1	3
	0.30	5	4	0	2	2
	0.50	5	2	0	0	0
d-Propranolol	0.10	5	3	2	1	3
	0.30	5	4	1	2	3
	1.00	5	4	0	1	1
Sotalol	0.10	5	5	1	3	4
	1.04	5	3	0	2	2
	2.08	5	1	0	1	1
Practolol	0.09	5	5	2	2	4
	0.45	5	4	1	2	3
	0.90	5	4	1	0	1

Table 3. Effects of beta-receptor antagonists on resting heart rate (beats/minute) and on isoproterenol-induced tachycardia

Treatments	Dose (mg/kg)	N	Control	β-Adrenergic Blocker	
				Alone	With isoproterenol
Isoproterenol	0.15	33	382.1 ± 6.7		425.1 ± 4.5*
dl-propranolol	0.10	5	392.2 ± 13.0	326.8 ± 15.9*	397.8 ± 9.3*
	0.30	5	379.6 ± 8.8	292.8 ± 9.9*	382.0 ± 11.2*
	0.50	5	418.2 ± 6.8	301.4 ± 12.9*	407.0 ± 6.8*
d-propranolol	0.10	5	435.4 ± 8.3	421.4 ± 11.3	455.4 ± 10.8
	0.30	5	403.8 ± 12.6	395.8 ± 14.3	427.4 ± 11.6
	1.00	5	416.6 ± 11.4	398.6 ± 8.5	423.6 ± 9.7
Sotalol	0.10	5	385.8 ± 19.3	396.6 ± 13.6	445.4 ± 7.3*
	1.04	5	393.6 ± 28.6	370.8 ± 5.6	417.2 ± 12.9*
	2.08	5	403.6 ± 8.3	324.8 ± 11.3†	418.6 ± 4.0*
Practolol	0.09	5	388.2 ± 6.7	358.8 ± 9.1†	376.0 ± 9.0
	0.45	5	374.4 ± 14.8	336.8 ± 19.3	378.6 ± 19.1
	0.90	5	375.2 ± 16.8	329.6 ± 21.3	363.8 ± 17.1

*$p < 0.01$.
†$p < 0.05$.

Table 4. Effects of beta-receptor antagonists on resting mean blood pressure (mm Hg) and on isoproterenol-induced hypotension

Treatment	Dose (mg/kg)	N	Control	β-Adrenergic Blocker Alone	β-Adrenergic Blocker With isoproterenol
Isoproterenol	0.15	33	125.5 ± 5.6		60.1 ± 3.3*
dl-propranolol	0.10	5	125.0 ± 5.5	122.4 ± 6.6	56.6 ± 7.4*
	0.30	5	125.0 ± 4.5	128.8 ± 4.1	67.2 ± 2.9*
	0.50	5	132.2 ± 7.2	119.8 ± 8.1	67.0 ± 1.7*
d-propranolol	0.10	5	132.4 ± 10.9	139.4 ± 10.8	64.6 ± 4.6*
	0.30	5	127.2 ± 6.0	127.6 ± 9.5	79.2 ± 4.3*
	1.00	5	138.0 ± 9.6	138.8 ± 8.1	68.6 ± 2.1*
Sotalol	0.10	5	137.6 ± 8.4	150.2 ± 2.6	81.2 ± 7.2*
	1.04	5	130.0 ± 7.6	154.6 ± 9.3†	64.2 ± 2.2*
	2.08	5	136.2 ± 4.1	136.8 ± 8.6	87.4 ± 5.7*
Practolol	0.09	5	136.8 ± 4.4	141.0 ± 6.1	66.0 ± 4.2*
	0.45	5	136.0 ± 10.4	135.8 ± 11.6	67.2 ± 1.7*
	0.90	5	120.0 ± 3.8	126.2 ± 5.1	58.2 ± 1.0*

*$p < 0.01$.
†$p < 0.05$.

1969), decreased the incidence of ventricular arrhythmias and completely obviated the occurrence of ventricular fibrillation (Table 2). Tables 3 and 4 demonstrate that these antiarrhythmic effects occurred with doses of *dl*-propranolol which did not interfere with the positive chronotropic and hypotensive actions of the beta-adrenergic stimulant. Both sotalol, which differs from *dl*-propranolol in that it has no local anesthetic activity (Lish, Weikel, and Dungan, 1965; Levy, 1968a, b), and practolol, believed to block more selectively cardiac than vascular and bronchial beta receptors (Dunlop and Shanks, 1968) were also antiarrhythmogenic at doses that did not prevent isoproterenol-elicited tachycardia and hypotension.

d-Propranolol is almost devoid of beta-adrenergic blocking properties (Barrett, 1969) but resembles *dl*-propranolol in membrane-depressant activity (Lucchesi, Whitsitt, and Sticknew, 1967; Barrett and Cullum, 1968). As expected, *d*-propranolol did not alter blood pressure and resting heart rate and did not antagonize the positive chronotropic and hypotensive effects of isoproterenol (Tables 3 and 4); however, it was nearly as effective as practolol in preventing isoproterenol-induced ventricular fibrillation (Table 2).

Quinidine and lidocaine, which are extensively used in the therapy of rhythm disorders, likewise decreased the severity of isoproterenol-elicited dysrhythmias

Table 5. Effect of quinidine and lidocaine on isoproterenol-induced arrhythmias

Treatment	Dose (mg/kg)	*N*	Ventricular Arrhythmias	Ventricular Fibrillation		
				Death	Recovery	Total
Isoproterenol	0.15	33	33	19	7	26
Quinidine	0.26	5	5	3	1	4
	3.00	5	3	2	0	2
	5.31	5	3	1	1	2
	13.25	5	3	1	0	1
Lidocaine	0.91	5	4	2	2	4
	2.74	5	4	0	2	2
	4.57	5	5	1	1	2
	9.14	5	2	0	0	0

(Table 5), although they did not inhibit isoproterenol-induced tachycardia and hypotension (Tables 6 and 7).

The plot of the incidence of ventricular fibrillation versus the dose of antiarrhythmic agents, expressed as molar equivalents of *dl*-propranolol, shown in Figure 3, demonstrates that these agents can be divided roughly into two groups. The first group includes the beta-adrenergic blocking drugs and *d*-propranolol and, the second, quinidine and lidocaine. Their relative effectiveness on a molar basis in abolishing isoproterenol-induced ventricular fibrillation, as calculated

Table 6. Effects of quinidine and lidocaine on resting heart rate (beats/minute) and on isoproterenol-induced tachycardia

Treatment	Dose (mg/kg)	*N*	Control	Antiarrhythmic Agent	
				Alone	With isoproterenol
Isoproterenol	0.15	33	382.1 ± 6.7		425.1 ± 4.5
Quinidine	0.26	5	417.0 ± 19.9	405.6 ± 22.4	423.6 ± 22.4
	3.00	5	403.6 ± 6.6	386.2 ± 12.0	416.2 ± 12.9
	5.31	5	431.0 ± 11.7	417.8 ± 7.2	443.4 ± 9.1
	13.25	5	391.8 ± 20.3	382.8 ± 17.3	409.0 ± 17.2
Lidocaine	0.91	5	404.6 ± 10.8	389.4 ± 8.5	412.4 ± 7.2
	2.74	5	399.0 ± 14.2	385.8 ± 21.8	415.8 ± 17.3
	4.57	5	393.4 ± 15.8	368.0 ± 14.1	398.0 ± 13.3
	9.14	5	353.2 ± 18.6	356.7 ± 19.5	381.6 ± 18.7

Table 7. Effect of quinidine and lidocaine on resting mean blood pressure (mm Hg) and on isoproterenol-induced hypotension

Treatment	Dose (mg/kg)	N	Control	Antiarrhythmic Agent: Alone	Antiarrhythmic Agent: With isoproterenol
Isoproterenol	0.15	33	125.5 ± 5.6		60.1 ± 3.3*
Quinidine	0.26	5	146.6 ± 13.7	131.0 ± 12.1	64.0 ± 4.4*
	3.00	5	136.5 ± 5.2	117.4 ± 4.2†	56.2 ± 6.1†
	5.31	5	129.6 ± 4.5	126.4 ± 14.8	64.8 ± 7.7*
	13.25	5	153.6 ± 5.8	123.0 ± 7.4†	67.8 ± 3.6*
Lidocaine	0.91	5	147.0 ± 7.3	144.0 ± 9.0	80.0 ± 4.4*
	2.74	5	133.0 ± 9.9	120.2 ± 12.0	75.4 ± 7.4*
	4.57	5	148.0 ± 9.1	137.6 ± 12.5	80.0 ± 8.5*
	9.14	5	146.2 ± 9.4	156.8 ± 6.9	84.6 ± 8.2*

*$p < 0.01$.
†$p < 0.05$.

from the regression lines in Figure 3, is as follows: *dl*-propranolol (100), practolol (39), *d*-propranolol (34), sotalol (20), quinidine (8), and lidocaine (5).

Effects of isoproterenol on total myocardial electrolyte content after DCA-saline treatment and the administration of antiarrhythmic agents

The total myocardial content of magnesium, sodium, and potassium was not significantly influenced by 3 weeks' pretreatment with saline and DCA-saline (Fig. 4). On the other hand, as illustrated in Figure 5, isoproterenol administered in DCA-saline-treated rats clearly altered the myocardial environment by consistently decreasing the levels of magnesium and potassium and elevating those of sodium. In sharp contrast, isoproterenol increased, rather than decreased, the myocardial content of potassium in animals not pretreated with DCA, without changing that of magnesium (Fig. 6). Furthermore, the beta-agonist diminished rather than augmented the myocardial levels of sodium in rats receiving saline as their drinking fluid (Fig. 6).

The total myocardial changes in these three electrolytes elicited by 150 μg/kg isoproterenol in the absence and in the presence of *d*- and *dl*-propranolol are compared in Figure 7. *dl*-Propranolol inhibited isoproterenol-induced alterations of magnesium and sodium (*upper panel*), while *d*-propranolol prevented only those of magnesium (*lower panel*). Figure 8 shows that, when isoproterenol was injected after sotalol and practolol, it increased rather than decreased myocardial magnesium levels. Sotalol was similar to *dl*-propranolol in that it partially blocked the effect of the beta-stimulant on sodium (Fig. 8, *upper panel*).

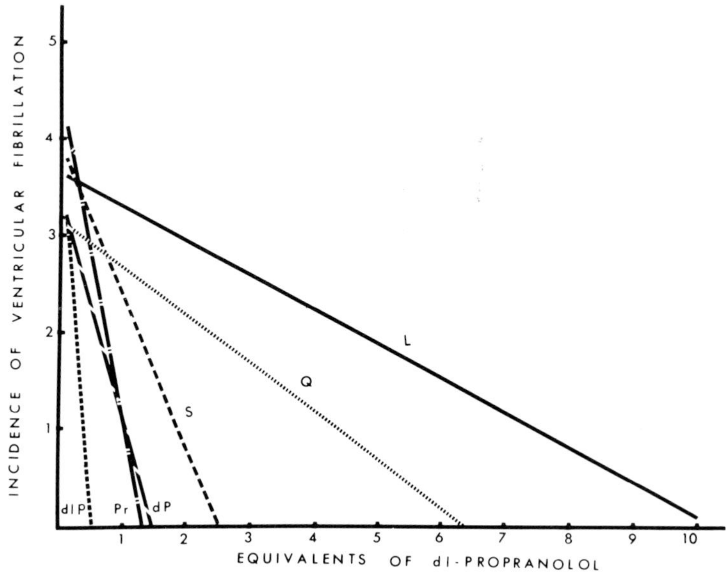

Figure 3. Effect of antiarrhythmic agents on isoproterenol-induced ventricular fibrillation in the anesthetized DCA-saline rat. Equations for each line were determined by regression analysis, using the values in Tables 2 and 5. The correlation coefficient is listed following each equation: *dl*-propranolol (dl P), y=3.9-0.79x, 0.98; *d*-propranolol (d P), y=3.4-0.23x, 0.97; practolol (Pr), y=4.4-0.34x, 0.99; sotalol (S), y=3.9-0.16x, 0.99; quinidine (Q), y=3.1-0.05x, 0.82; lidocaine (L), y=3.7-0.04x, 0.96.

Isoproterenol elicited diametrically opposite effects on magnesium and potassium levels in animals pretreated with quinidine, since, as illustrated in Figure 9, it elevated the total myocardial content of these electrolytes. On the other hand, isoproterenol-induced sodium changes were potentiated by quinidine. Surprisingly, lidocaine augmented isoproterenol-elicited potassium depletion, without inhibiting magnesium and sodium shifts (Fig. 9, *lower panel*).

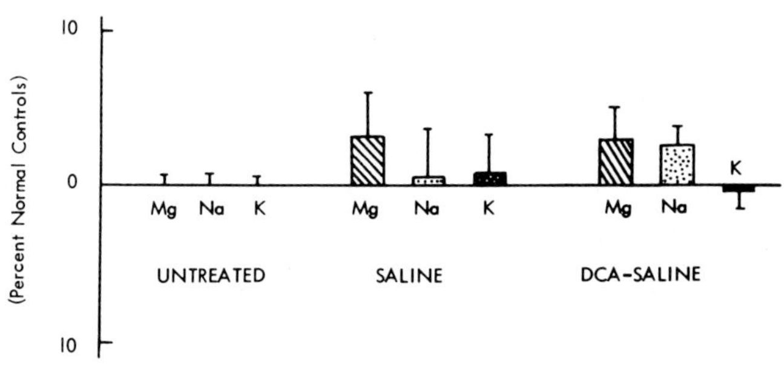

Figure 4. Total myocardial content of magnesium (Mg), sodium (Na), and potassium (K), in normal (N=44), saline-(N=13), and DCA-saline-(N=22) pretreated rats.

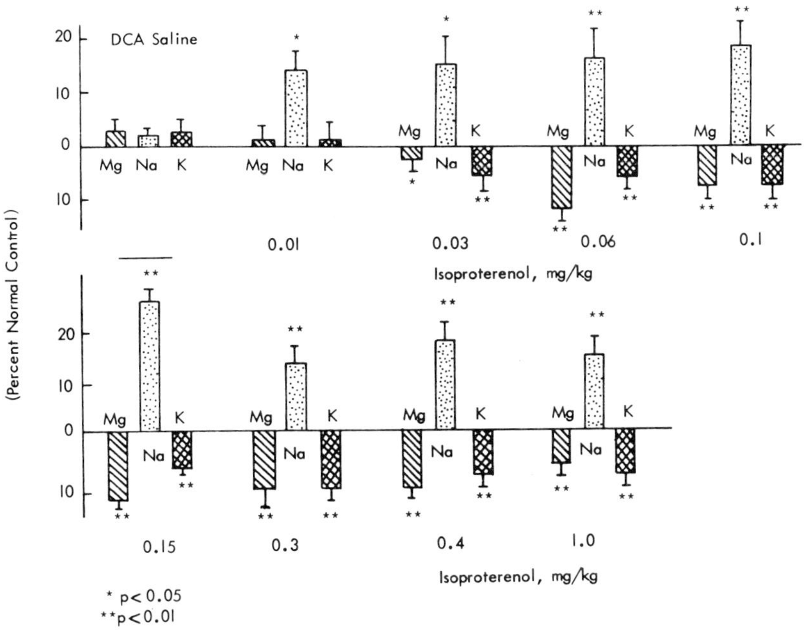

Figure 5. Effect of isoproterenol on total myocardial content of magnesium (Mg), sodium (Na), and potassium (K), in DCA-saline-pretreated anesthetized rats. At least 10 rats were used for each dose of isoproterenol.

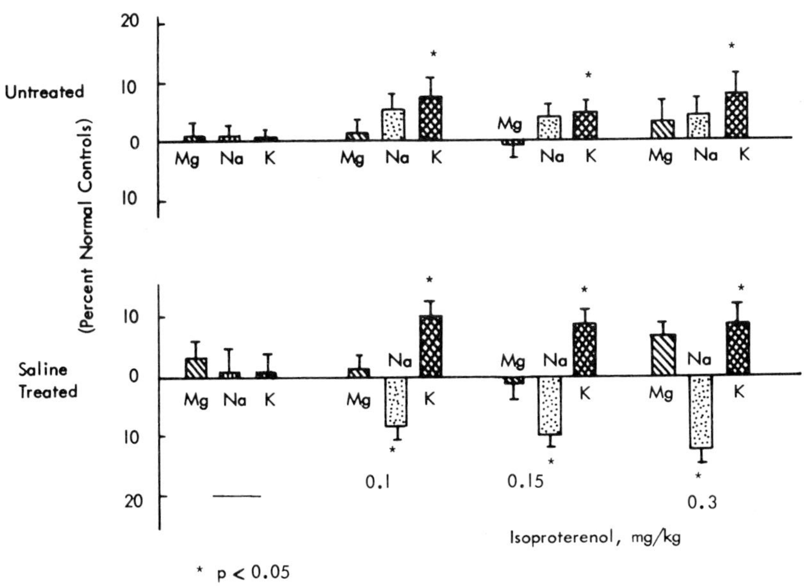

Figure 6. Effect of isoproterenol on total myocardial content of magnesium (Mg), sodium (Na), and potassium (K), in normal and saline-pretreated rats. At least 10 rats were used for each dose of isoproterenol.

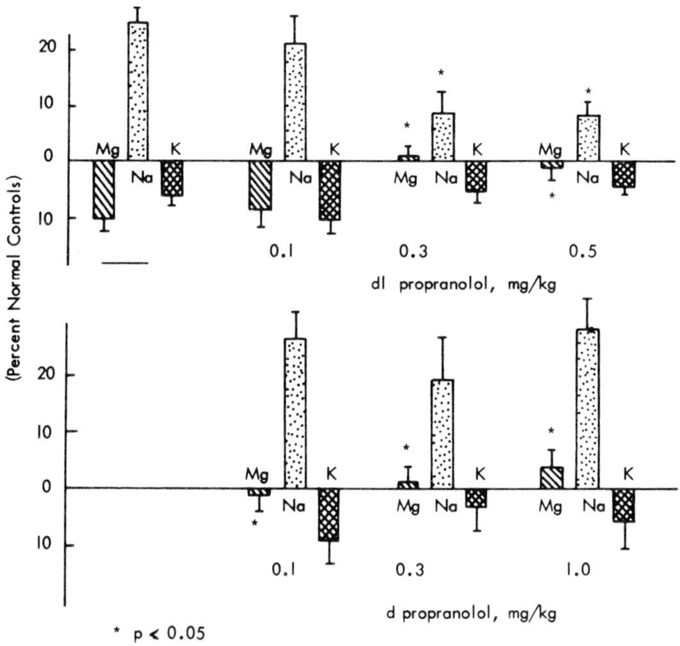

Figure 7. Effect of *dl*-propranolol and *d*-propranolol on the myocardial electrolyte alterations induced by isoproterenol in groups of five DCA-saline-pretreated, anesthetized rats.

Effects of Mg gluconate on isoproterenol-induced mortality in DCA-saline rats

Since the most commonly noticed effect of the antiarrhythmic agents was the prevention of isoproterenol-induced magnesium depletion, the antifibrillatory action of this ion was explored in conscious rats by administering magnesium gluconate prior to the injection of the beta-receptor stimulant. As demonstrated in Table 8, treatment with 110, 330, and 660 mg of the magnesium salt per rat decreased the incidence of mortality from 94% to 70, 56, and 21%, respectively. These doses of magnesium gluconate caused sedation in these animals but no narcosis, and did not prevent isoproterenol-induced tachycardia. Since sedation by magnesium gluconate may have played a role in decreasing the mortality rate in these animals, 5 DCA-saline-pretreated rats were sedated with 5 mg/kg diazepam 30 min prior to the administration of isoproterenol. Diazepam-induced sedation did not inhibit isoproterenol-elicited dysrhythmias. All five animals experienced ventricular fibrillation, and only one recovered from it.

DISCUSSION

The finding that DCA-saline pretreatment also sensitizes the myocardium of mice and guinea pigs to isoproterenol demonstrates that this phenomenon is not

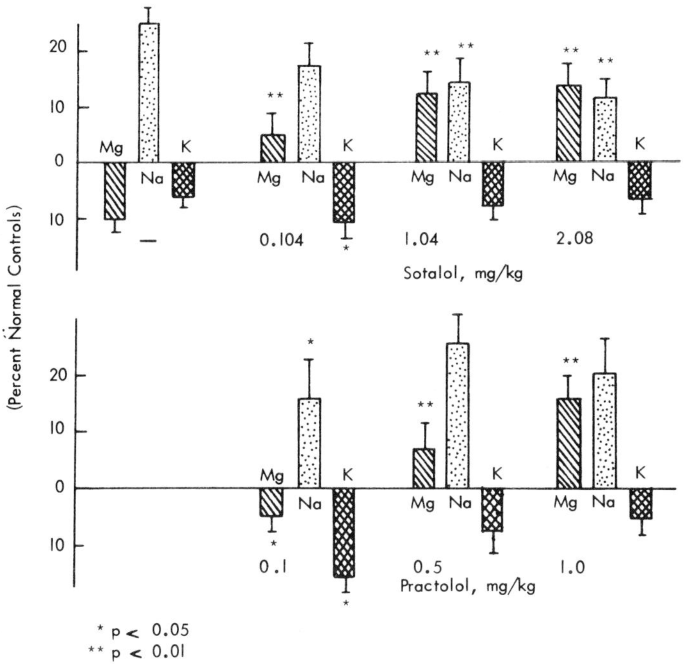

Figure 8. Effect of sotalol and practolol on the myocardial electrolyte alterations induced by isoproterenol in groups of five DCA-saline-pretreated, anesthetized rats.

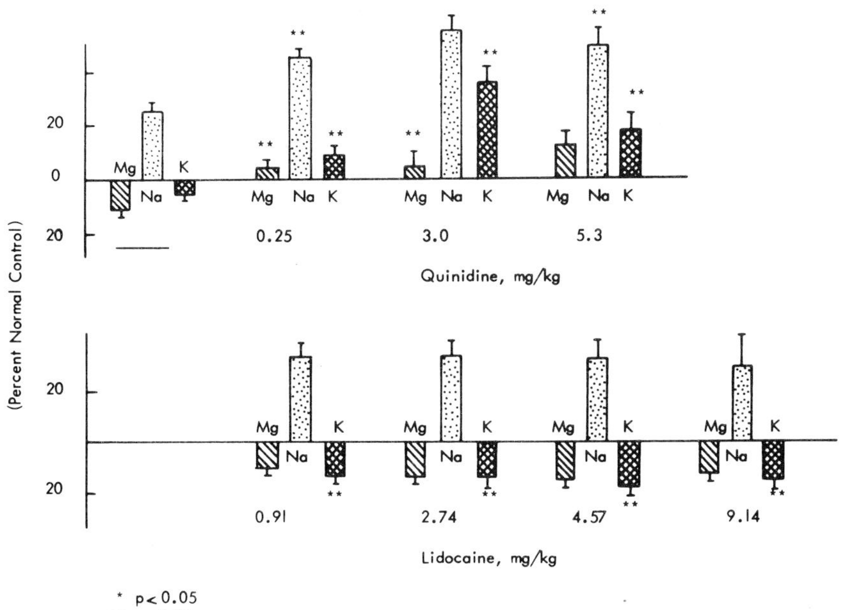

Figure 9. Effect of quinidine and lidocaine on the myocardial electrolyte alterations induced by isoproterenol in groups of five DCA-saline-pretreated, anesthetized rats.

Table 8. Effect of magnesium gluconate on the incidence of isoproterenol-induced mortality in conscious DCA-saline pretreated rats*

Magnesium Gluconate (mg total)	*N*	Deaths No.	Deaths %
–	17	16	94
110	10	7	70
330	16	9	56†
660	14	3	21‡

*Animals injected subcutaneously 30 min prior to the administration of 150 μg/kg isoproterenol.

†$p<0.05$, from Chi square (χ^2) analysis, using the Yates correction factor for noncontinuously distributed data.

‡$p<0.01$.

species-specific for the rat. In addition, we have shown earlier that this type of sensitization is not limited to pure mineralocorticoids, since increased susceptibility to arrhythmias was elicited also with hydrocortisone, a steroid with gluco- as well as mineralocorticoid properties. These observations lend support to the assumption that this type of drug interaction between isoproterenol and corticosteroids may have played a contributory role in the high incidence of mortality of asthmatics on prolonged corticoid therapy who are using inhalers containing isoproterenol or other beta-adrenergic agonists (Lehr, 1972; Guideri *et al.*, 1974). The assumption is strengthened by results of recent preliminary experiments which revealed that rats pretreated with prednisone, a "pure" glucocorticoid, likewise develop ventricular fibrillation after injection of small isoproterenol dosages. As will be recalled, prednisone is the corticoid of choice in the treatment of asthmatics.

The concept that a small-sized heart can maintain ventricular fibrillation for only very short periods (Skou, 1960) because it does not provide a pathway of necessary length to allow sustained re-entry of the excitation wave (Dawes, 1952) is supported by the extremely short bouts of ventricular fibrillation that we observed in mice and the consequent low incidence of mortality.

In order to determine the usefulness of our procedure for the screening of anti-arrhythmic drugs, we compared our results to those obtained with some of the more commonly used models, particularly those using catecholamines as the arrhythmogenic agent. It will be remembered that, on a molar basis, the antiarrhythmic agents used in the present study lowered the incidence of ventricular fibrillation in the following descending order: *dl*-propranolol, practolol, *d*-propranolol, sotalol, quinidine, and lidocaine. An identical order of activity, with the exception of *d*-propranolol, which was found ineffective, has been reported in epinephrine-elicited arrhythmias in dogs with coronary artery ligation (Dawes, 1952; Lucchesi *et al.*, 1967). A similar order of effectiveness is also found in data from various laboratories with catecholamine-induced dysrhythmias following myocardial sensitization with halogenated hydrocarbons (Burnstein *et al.*, 1940; Huggins *et al.*, 1950; White, Merrigan, and Swiss, 1955; Winbury and Hemmer, 1955; Clark and Cummings, 1956; Dresel, MacConnel, and Nickerson, 1960; Howe and Shanks, 1966).

On the other hand, the comparative potency of these drugs in abolishing arrhythmias not elicited by catecholamines does not follow the same order. *dl*-Propranolol, *d*-propranolol, and quinidine were equipotent, and more effective than lidocaine (Howe and Shanks, 1966; Lucchesi *et al.*, 1967; Helmi and Regan, 1968; Basset and Hoffman, 1971), while practolol and sotalol were usually not effective in preventing digitalis-induced arrhythmias. In contrast, in reverting spontaneous dysrhythmias after coronary ligation in dogs (Lucchesi *et al.*, 1967), quinidine and lidocaine were equally active (Mosey and Tyler, 1964), while *d*- and *dl*-propranolol were ineffective.

It appears, therefore, that the corticoid-isoproterenol model compares favorably with those using adrenergic agonists alone in the dog and in the cat. One of the many advantages of our procedure is the simplicity of preparing DCA-saline-pretreated rodents and the consistency of the dysrhythmic picture after isoproterenol injection. The procedure requires neither extensive surgery nor expensive laboratory apparatus and affords the possibility of screening antiarrhythmic and antifibrillatory agents reliably in large groups of animals by merely recording mortality rates.

The effectiveness of the antiarrhythmic drugs used in these experiments suggested that beta-blocking and membrane-depressant properties were perhaps equally important, since *d*-propranolol, which is essentially devoid of beta-adrenergic blocking activity (Barrett and Cullum, 1968; Barrett, 1969) but possesses quinidine-like activity equipotent to that of racemic propranolol, was an effective antiarrhythmic agent. On the other hand, sotalol and practolol, which have no membrane-depressant activity (Lish *et al.*, 1965; Dunlop and Shanks, 1968; Levy, 1968b), likewise prevented ventricular fibrillation. It is of interest that Schuster *et al.* (1973) demonstrated that quaternized *dl*-propranolol, which lacks both local anesthetic and beta-adrenergic blocking activity, was still an effective antiarrhythmic agent. These authors concluded that antiarrhythmic activity can be independent of membrane-depressant and beta-receptor blocking properties and may be a consequence of effects on myocardial electrolyte environment.

Although other laboratories have reported depletion of potassium and elevation of sodium in the myocardium following prolonged treatment with large doses of DCA (Darrow and Miller, 1942; Gross and Schmidt, 1958), we noticed no significant alterations in myocardial electrolytes in DCA-saline animals not challenged with isoproterenol. This discrepancy may be due to differences in methodology, since we used small doses of DCA and a relatively short duration of treatment. The finding that isoproterenol had striking effects on the myocardial electrolyte composition of DCA-saline-treated rats is in accord with the viewpoint that myocardial electrolyte shifts may play an important role in the genesis of cardiac arrhythmias (Hoffmann and Cranefield, 1960). The fact that lidocaine did not ameliorate isoproterenol-induced myocardial electrolyte alterations, but rather potentiated the potassium depletion elicited by the beta-adrenergic stimulant, may indicate that the antiarrhythmic action of this agent involves a mechanism not shared by the other antiarrhythmic drugs used in these experiments. Interestingly, it has been reported that, in contrast to *dl*-propranolol and quinidine, lidocaine *increases* membrane responsiveness and *shortens* the refractory period of Purkinje fibers (Basset and Hoffman, 1971). In future studies, it will be determined whether antiarrhythmic agents with properties of membrane responsiveness and refractory period similar to those of lidocaine have also lidocaine-like effects on isoproterenol-elicited myocardial electrolyte changes.

The observation that *dl*-propranolol, the most effective antiarrhythmic agent in these experiments, tended to prevent isoproterenol-induced electrolyte shifts in these and in earlier observations from this laboratory (Lehr, 1969) appears to be in agreement with that of others showing that *dl*-propranolol increases total tissue potassium and reduces sodium uptake (Klein and Holland, 1958; Choi, Roberts, and Kelliher, 1972). The finding also supports the notion that antiarrhythmic therapy may tend to re-establish electrolyte balance (Szekeres and Papp, 1971). However, it is unlikely that inhibition of isoproterenol-induced electrolyte shifts by beta-blocking agents was a consequence of beta-receptor blockade, since it occurred within dose ranges which were not able to prevent tachycardia and hypotension and because *d*-propranolol also blocked the magnesium depletion produced by isoproterenol.

Since all antiarrhythmic agents used in these studies, except lidocaine, affected magnesium, it is tempting to suggest that the first restorative changes by these drugs affect this ion. Prevention of magnesium loss leads to a return toward normal of the other electrolytes. This is possible because the animals used in these experiments were sacrificed 30 min after isoproterenol administration, so that insufficient time may have elapsed for complete corrective changes to take place, an assumption which is presently under investigation.

The importance of magnesium in maintaining a normal myocardial electrolyte environment is suggested by observations from other laboratories which indicate that deficiency of this ion induces a loss of potassium from the heart (Vitale, Nakamura, and Hegsted, 1957; Post *et al.*, 1960; Skou, 1960). Moreover, hypomagnesemia may increase the tendency to ventricular fibrillation (Loeb *et al.*, 1968; Seller, 1971), and myocardial magnesium depletion induced by cardiotoxic dosages of isoproterenol and other catecholamines precedes any other electrolyte changes (Lehr, 1969). The prevention by magnesium gluconate of ventricular dysrhythmias and fibrillation was not unexpected and is fully in line with the potent antiarrhythmic effect of magnesium salts in aconitine-induced arrhythmias observed many years ago (Hueber and Lehr, 1938) and in other types of rhythm disturbances (Zwillinger, 1935; Enselberg, Simmonds, and Mintz, 1950; Burky, 1970; Seller, 1971; Chadda, Gupta, and Lichstein, 1972; Chadda, Lichstein, and Gupta, 1973).

REFERENCES

BARRETT, A. M. 1969. A comparison of the effects of (±) propranolol and (+) propranolol in anesthetized dogs; beta receptor blocking and hemodynamic action. J. Pharm. Pharmacol. 21: 241–250.

BARRETT, A. M., and V. A. CULLUM. 1968. The biological properties of the optical isomers of propranolol and their effects on cardiac arrhythmias. Brit. J. Pharmacol. 34: 43–55.

BASSET, A. L., and B. F. HOFFMAN. 1971. Antiarrhythmic drugs: electrophysiological actions. Ann. Rev. Pharmacol. 12: 143–169.

BURKY, B. 1970. Effect of magnesium on ventricular fibrillation due to hypothermia. Brit. J. Anesthesiol. 42: 886–887.

BURNSTEIN, C. L., B. A. MARANGONI, A. C. De GRAFF, and E. A. ROOENSTINE. 1940. Laboratory studies on the prophylaxis and treatment of ventricular fibrillation induced by epinephrine during cyclopropane anesthesia. Anesthesiology. 1: 167–186.

CHADDA, K. D., P. K. GUPTA, and E. LICHSTEIN. 1972. Magnesium in cardiac arrhythmias. New England J. Med. 287: 1102.

CHADDA, K. D., E. LICHSTEIN, and P. GUPTA. 1973. Hypomagnesemia and refractory cardiac arrhythmia in a non-digitalized patient. Am. J. Cardiol. 31: 98–100.

CHOI, S. J., J. ROBERTS, and G. J. KELLIHER. 1972. The effect of propranolol and quinidine on ^{22}Na- and ^{42}K exchange in the cat papillary muscle. Europ. J. Pharmacol. 20: 10–21.

CLARK, B. B., and J. R. CUMMINGS. 1956. Arrhythmias following experimental coronary occlusion and their response to drugs. Ann. N.Y. Acad. Sci. 64: 543–551.

DARROW, D. C., and H. C. MILLER. 1942. The production of cardiac lesions by repeated injections of desoxycorticosterone acetate. J. Clin. Invest. 21: 601–611.

DAWES, G. S. 1952. Experimental cardiac arrhythmias and quinidine like drugs. Pharmacol. Rev. 4: 43–84.

DRESEL, P. E., K. L. MacCONNEL, and M. NICKERSON. 1960. Cardiac arrhythmias induced by minimal doses of epinephrine in cyclopropane anesthetized dogs. Circulat. Res. 8: 948–955.

DUNLOP, D., and R. C. SHANKS. 1968. Selective blockade of adrenoceptive beta receptors in the heart. Brit. J. Pharmacol. 32: 201–218.

ENSELBERG, C. D., H. G. SIMMONDS, and A. A. MINTZ. 1950. The effects of magnesium upon cardiac arrhythmias. Am. Heart J. 39: 703–712.

GROSS, F., und H. SCHMIDT. 1958. Uberdosierung von Cortexon (DOC) and Koschsalz beim Kaninchen. Naunyn-Schmiedebergs Arch. Exp. Path. Pharmakol. 232: 408–418.

GUIDERI, G., M. BARLETTA, and D. LEHR. 1974. Extraordinary potentiation of isoproterenol cardiotoxicity by corticoid pretreatment. Cardiovasc. Res. 8: 775–786.

GUIDERI, G., W. GOLDMAN, and D. LEHR. 1971. Extraordinary potentiation of acute isoproterenol (ISO) cardiotoxicity in the DOCA-saline pretreated rat. Pharmacologist 13: 320.

HELMI, K. I., and T. J. REGAN. 1968. Relative effectiveness of antiarrhythmic drugs in treatment of digitalis-induced ventricular tachycardia. Am. J. Cardiol. 76: 365–390.

HOFFMAN, B. F., and P. F. CRANEFIELD. 1960. Electrophysiology of the Heart. McGraw-Hill, New York.

HOFFMAN, R. P., and G. GRUPP. 1969. The effects of sotalol and propranolol on contractile force and atrioventricular conduction time of the dog heart in situ. Dis. Chest 55: 229–234.

HOWE, R. and R. G. SHANKS. 1966. Optical isomers of propranolol. Nature 210: 1336–1341.

HUEBER, E. F., und D. LEHR. 1938. Wirkung von Magnesium aut die Vergitung mit Akonitin. Nauyn-Schmiedebergs Arch. Exp. Path. Pharmakol. 189: 25–44.

HUGGINS, R. A., R. A. MORSE, D. W. CHAPMAN, and L. F. SCHUMACHER. 1950. The effect of quinidine on arrhythmias induced by cyclopropane and cyclopropane-epinephrine. Anesthesiology 11: 623–634.

KLEIN, R. L., and W. C. HOLLAND. 1958. Na^{24} and K^{42} exchange in atrial fibrillation. Am. J. Physiol. 193: 239–248.

LEHR, D. 1969. Tissue electrolyte alterations in disseminated myocardial necrosis. Ann. N.Y. Acad. Sci. 156: 344–378.

LEHR, D. 1972. Isoproterenol and sudden death of asthmatic patients in ventricular fibrillation. New Engl. J. Med. 287: 987–988.

LEHR, D., M. KRUKOWSKI, and R. COLON. 1966. Correlation of myocardial and renal necrosis with tissue electrolyte changes. J. Am. Med. Ass. 197: 105–112.

LEVY, J. V. 1968a. Surface tension effects of beta-adrenergic blocking drugs. J. Pharm. Pharmacol. 20: 813–815.

LEVY, J. V. 1968b. Myocardial and local anesthetic actions of β-adrenergic receptor drugs: relationship to physiochemical properties. Europ. J. Pharmacol. 2: 250–257.

LISH, P. M., J. H. WEIKEL, and K. W. DUNGAN. 1965. Pharmacological and toxicological properties of two new beta-adrenergic receptor antagonists. J. Pharmacol. Exp. Therap. 149: 161–173.

LOEB, H. S., R. J. PIETRAS, R. M. GUNNAR, and J. R. TOBIAN. 1968. Paroxysmal ventricular fibrillation in two patients with hypomagnesemia. Circulation 37: 210–215.

LUCCHESI, B. R., L. S. WHITSITT, and J. L. STICKNEW. 1967. Antiarrhythmic effects of beta adrenergic blocking agents. Ann. N.Y. Acad. Sci. 139: 940–951.

MOSEY, L., and M. D. TYLER. 1964. The effects of diphenylhydantoin sodium (Dilantin), procaine hydrochloride, procaineamide, and quinidine hydrochloride, upon ouabaine-induced ventricular tachycardia in unanesthetized dogs. Circulation 10: 65–70.

POST, R. L., C. R. MERRITT, C. R. KINSOLVING, and C. D. ALBRIGHT. 1960. Membrane adenosine triphosphatase as a participant in the active transport of sodium and potassium in the human erythrocyte. J. Biol. Chem. 235: 1796–1802.

RILEY, P. A., T. G. BARILA, and C. W. HUGHES. 1956. Ventricular fibrillation in hypothermic dogs as influenced by thiopental and succinylcholine. Anesthesiology 17: 277–283.

RONA, G., C. I. CHAPPEL, T. BALAZS, and R. GAUDRY. 1959. An infarct like myocardial lesion and other toxic manifestations produced by isoproterenol in the rat. Arch. Path. 67: 443–455.

SCHUSTER, D. P., B. R. LUCCHESI, N. L. NOBEL, M. N. MIMNAUGH, R. E. COUNSELL, and F. J. KNIFFEN. 1973. The antiarrhythmic properties of UM-272, the dimethyl quaternary derivative of propranolol. J. Pharmacol. Exp. Therap. 184: 213–227.

SELLER, R. H. 1971. The role of magnesium in digitalis toxicity. Am. Heart J. 82: 551–556.

SKOU, J. C. 1960. Further investigations on a Mg and Na activated adenosine-triphosphitase, possibly related to the active, linked transport of Na and K across the nerve membrane. Biochim. Biophys. Acta 42: 6–11.

SZEKERES, L., and G. J. PAPP. 1971. Experimental Cardiac Arrhythmias and Anti-arrhythmic Drugs. Akademiai Kiado, Budapest.

VAUGHAN WILLIAMS, E. M. 1966. Mode of action of beta receptor antagonists on cardiac muscle. Am. J. Cardiol. 18: 399–405.

VITALE, J. J., M. NAKAMURA, and D. M. HEGSTED. 1957. The effect of magnesium deficiency on oxidative phosphorylation. J. Biol. Chem. 228: 573–576.

WHITE, C. W., R. MERRIGAN, and E. D. SWISS. 1955. The effects of diphenylhydantoin sodium, glucose and beta-diethylaminoethyl diphenyl propylacetate HCl on cyclopropane epinephrine arrhythmias in the dog. Circulat. Res. 3: 290–292.

WINBURY, M. M., and M. L. HEMMER. 1955. Action of quinidine procainamide and other compounds on experimental atrial and ventricular arrhythmias in the dog. J. Pharmacol. Exp. Therap. 113: 402—413.

ZWILLINGER, L. 1935. Uber die Magnesium Wirking aus das Herz. Klin Wochschr. 14: 1429–1433.

Mailing address:
Giancarlo Guideri, Ph.D.,
Department of Pharmacology, New York Medical College,
Basic Science Building, Valhalla, New York 10595 (USA).

Recent Advances in Studies on
Cardiac Structure and Metabolism Volume 10
The Metabolism of Contraction
Edited by Paul-Emile Roy and George Rona

LESIONS OF THE CONDUCTION SYSTEM IN THE CAT WITH CARDIOMYOPATHY

SI-KWANG LIU, LAWRENCE P. TILLEY,
and ROBERT J. TASHJIAN

Departments of Pathology and Medicine, The Animal Medical Center,
New York, New York, USA

SUMMARY

Degeneration and fibrosis of the atrioventricular nodal tissue and left bundle branch, associated with cartilage or bone in the central fibrous body and fibrosis of the interventricular septal crest, were observed in 63 cats with cardiomyopathy. The cardiac lesions consisted of endocardial and myocardial fibrosis or organized endomyocarditis. Electrocardiographic, radiographic, angiocardiographic, and hemodynamic changes were also observed in the cats.

INTRODUCTION

Cardiomyopathy is one of the primary causes of death in the cat (Liu, 1970, 1974; Liu, Tashjian, and Patnaik, 1970; Harpster, 1973; Lord *et al.*, 1974; Tilley, Lord, and Wood, 1974). Nonsuppurative endocarditis and myocarditis are occasionally observed in the young adult cat, and focal os diffuse endocardial or myocardial fibrosis has caused heart failure and thromboembolism in mature and aged cats (Liu, 1970).

The purpose of this report is to characterize the lesions in the atrioventricular (AV) node, in the approaches to the AV node, and in the AV bundle and bundle branches, in the sinoatrial (SA) node and its approaches, and in the ventricular and atrial endocardium and myocardium in 63 adult cats with heart failure.

MATERIALS AND METHODS

The hearts of 63 adult cats with cardiac lesions that led to heart failure and thromboembolism were collected from routine necropsy at The Animal Medical Center, and an additional 10 hearts of normal adult cats were used as controls. Clinical data were taken from medical records of the 63 cats. Electrocardiograms, thoracic radiographs, cardiocatheterization, and angiocardiograms had been done on 16 of the affected cats (Lord *et al.*, 1974).

The hearts of all 73 cats were carefully examined at post mortem, in a manner previously described (Liu, 1970; Liu *et al.*, 1970), and were then fixed in 10% cool neutral formalin. For study of the SA node in each cat, the junction of the anterior vena cava and free wall of the right atrium were cut into sections 1 mm thick. For study of the AV node, the entire junction of the interatrial and interventricular septa, from the anterior margin of the coronary sinus of the aorta to the posterior margin of the noncoronary sinus, was sliced into sections 1 mm thick perpendicular to the line of junction of the septa. One adjacent centimeter of each septum was retained. The slices were embedded in paraffin, and the blocks were serially sectioned, 6 μ in thickness, and stained with hematoxylin and eosin, Masson's trichrome, periodic acid-Schiff, and Gallego's iron fuchsin stains.

RESULTS

Gross pathological findings

The principal cardiac lesions were severe focal or diffuse fibrosis of the left ventricular endocardium and myocardium, causing adhesion of the endocardium and papillary muscles and fusion of the chordae tendinae (Fig. 1). Hypertrophy or dilation of the ventricles and atria resulted from the endocardial and myocardial fibrosis, as previously described (Liu, 1970).

Histopathological findings

The endocardium was focally or diffusely infiltrated with young fibrous granulation tissue mixed with varying amounts of collagen, and it occasionally contained chondroitic elements, as previously described (Liu, 1970). These changes were commonly observed at the subvalvular region around the papillary muscles in the left ventricle. Active organized endocarditis was also observed around the mitral valve, papillary muscles, and subvalvular region of the left ventricle in some cats.

The myocardium was also focally or diffusely infiltrated with young fibrous tissue, mixed with collagen. Dense, wavy collagen fibers were in the interventricular septum and perivascular area of the intramural coronary arteries. In some cases, islands of myocardial fibers were trapped in the collagen fibers and were enlarged and deeply acidophilic (Fig. 2). Some myofibers had vacuolization or myocytolysis. The myocardial changes were commonly observed in the left ventricle and left atrium, occasionally in the right ventricle, and only rarely in the right atrium. Organized myocarditis, indicated by the presence of active granulation tissue with varying amounts of lymphocytes and plasma cells, was observed in 6 cats.

In the conduction system, there was a slight increase in fibrous tissue in the

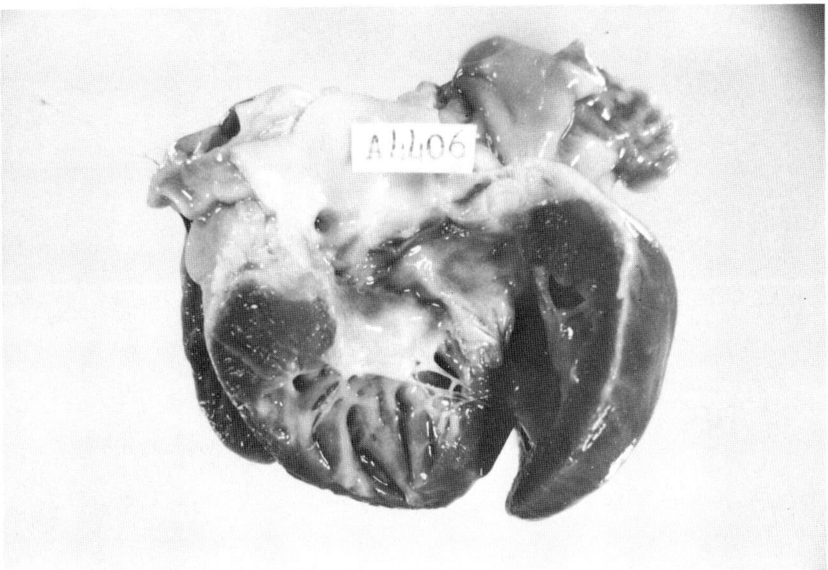

Figure 1. Left side of heart of a 6-year-old male domestic short-hair cat. Note extreme dilation and hypertrophy of the atrium and hypertrophy of the ventricle, with extensive endocardial fibrosis.

approaches to the SA node in a few of the cats. The SA node itself, however, showed no significant changes in most of the affected cats.

Mature or young fibrous tissue or organized myocarditis had infiltrated the approaches to the AV node and involved the adjacent muscle fibers. There were striking changes in the central fibrous body, which was scattered with islands of cartilage (Fig. 3) in 58 of the 63 affected cats, and bone cysts (Fig. 4) with marrow elements in 19 of the 63 cats (Table 1).

Occasionally, a dense wave of collagen fibers had replaced the delicate areolar tissue in the crest of the interventricular septa, and the collagen fibers had penetrated into the ventricular septum and approaches to the AV node. Chondroitic metaplasia, accompanied in some cases by mineralization of the cartilage, was often observed in the media of the root of the aorta (Table 1).

In the AV node, there was a marked proliferation of immature fibrous tissue and wavy collagen fibers (Fig. 5). Occasionally, dense, wavy collagen fibers originating from the interventricular septal crest had invaded the nodal elements. The affected nodal fibers had undergone varying changes, from homogeneous coagulation or granules of the sarcoplasma to vacuolization or lysis of the sarcoplasma. Nuclear changes varied from pyknosis to karyolysis. Islands of

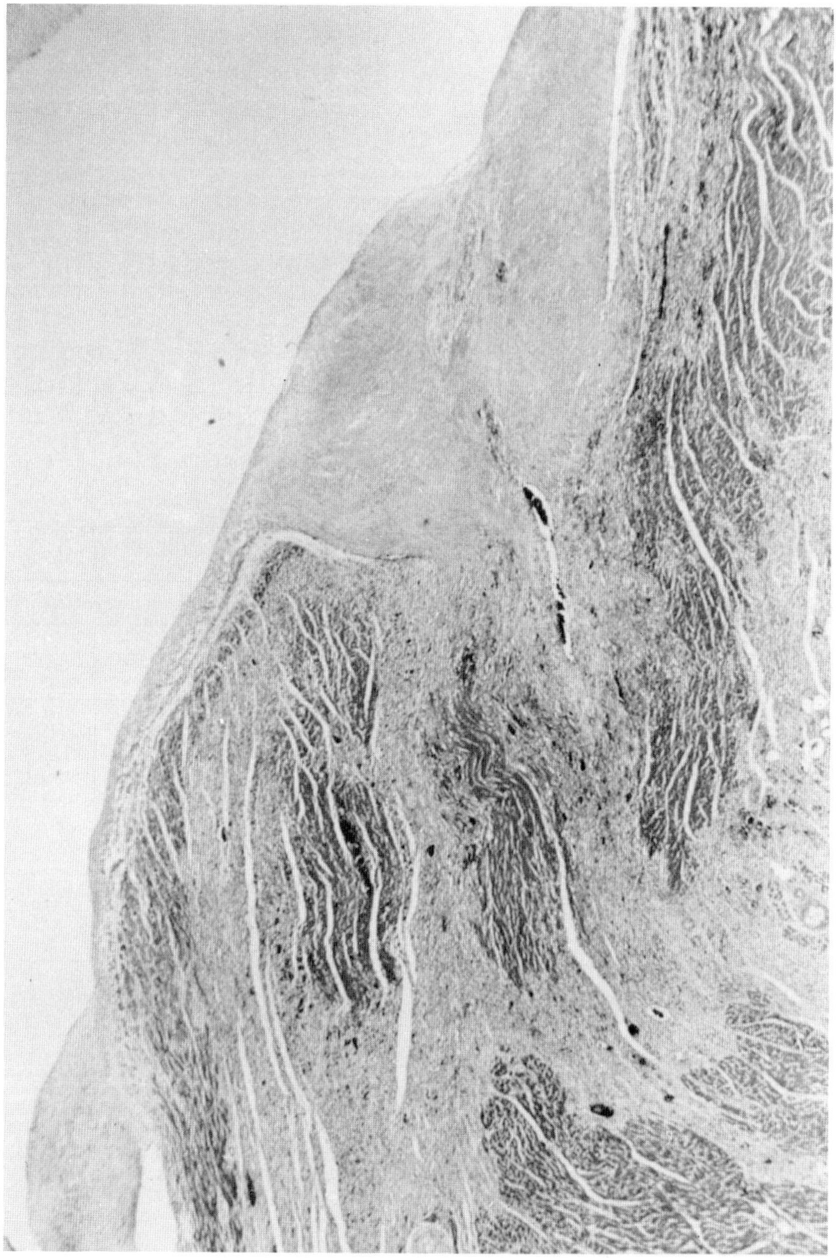

Figure 2. Photomicrograph of the myocardium, showing extensive endocardial and myocardial fibrosis and islands of myofibers trapped in the fibrous tissue. H and E. × 40.

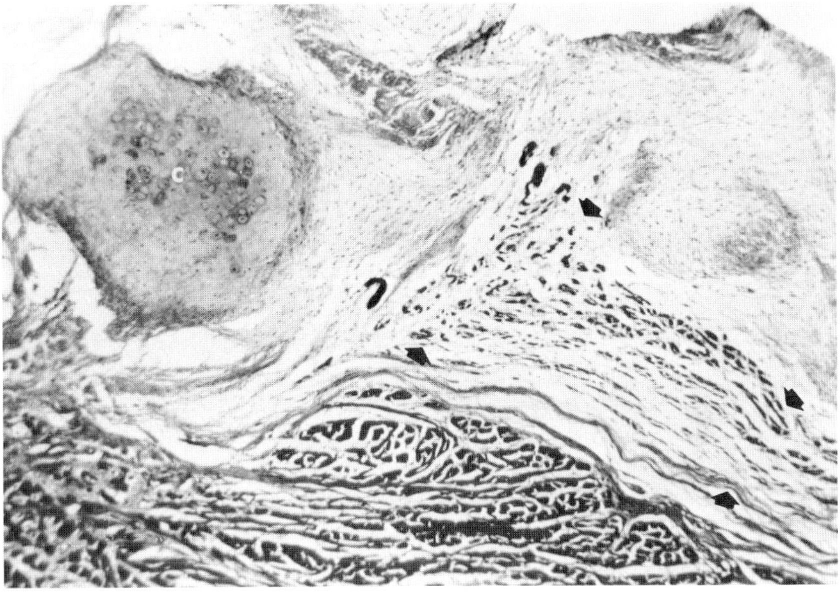

Figure 3. Photomicrograph of the interventricular septal crest, showing cartilage (C) in the central fibrous body and fibrosis and degeneration of the AV node (*arrows*). H and E. × 100.

cartilage or bone, with marrow in the central fibrous body, had often compressed the adjacent nodal tissue, causing various degrees of degenerative changes and lysis and fibrosis of the nodal fibers. Infiltration of adipose tissue was also observed in or around the nodal tissue. Granulation tissue mixed with a few lymphocytes and plasma cells, and proliferation of sheath cells were also observed in the nodal tissue, associated with chondroitic metaplasia or ossification of the central fibrous body. Mineralization of the fibrous tissue was occasionally seen in the affected node.

The upper portion of the left bundle branch showed degeneration and atrophy of the bundle fibers, resulting from the compression from the adjacent cartilage of bony elements (Fig. 6), or infiltration of the fibrous, adipose or granulation tissue. The middle portion of the branch was replaced partially by collagen fibers. The remaining nodal fibers revealed granulation, vacuolization, or lysis of the sarcoplasma, with pyknotic or karyolytic nuclei. In the lower portion of the branch, the bundle fibers were disrupted by or mixed with the wavy collagen fibers or chondroitic tissue originating from the endocardium. Often, the bundle had segmentally disappeared in the dense collagen fibers involving the endocardial or subendocardial areas (Fig. 7). Occasionally, a few bundle fibers with

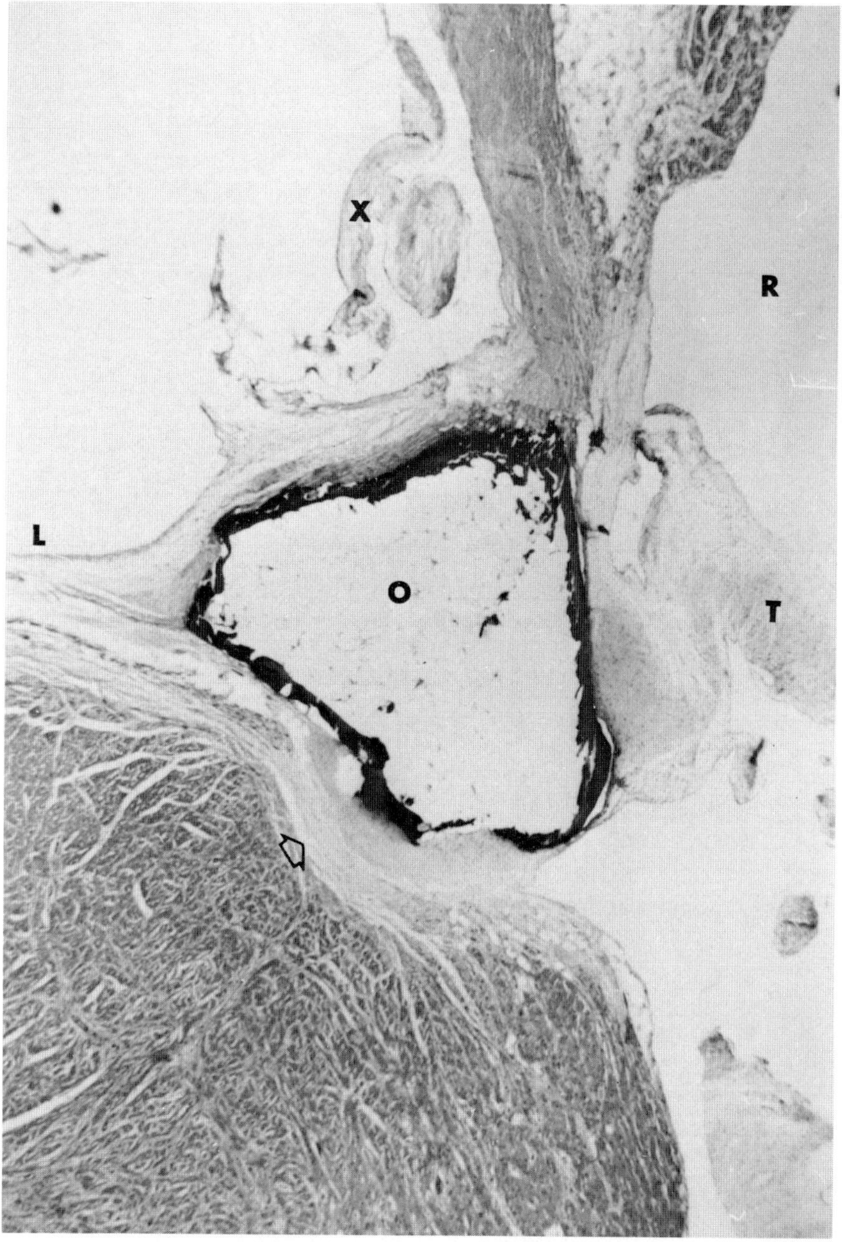

Figure 4. Photomicrograph of the interventricular septal crest. Bone cysts (o) with marrow elements have replaced the central fibrous body and compressed the AV bundle (*arrow*). X = aortic valve; T = tricuspid valve; L = left ventricle; R = right atrium. H and E. × 40.

Table 1. Summary of pathological changes in 63 cats with heart failure

Cardiac Lesions		Atrioventricular Node		Bundle Lesion		Central Fibrous Body		Chondroid Metaplasia in Root of Aorta
Endomyocarditis	Endomyocardial fibrosis	Degeneration	Fibrosis	L	R	Chondroid metaplasia	Ossification	
6	59	55	56	54	20	58	19	8

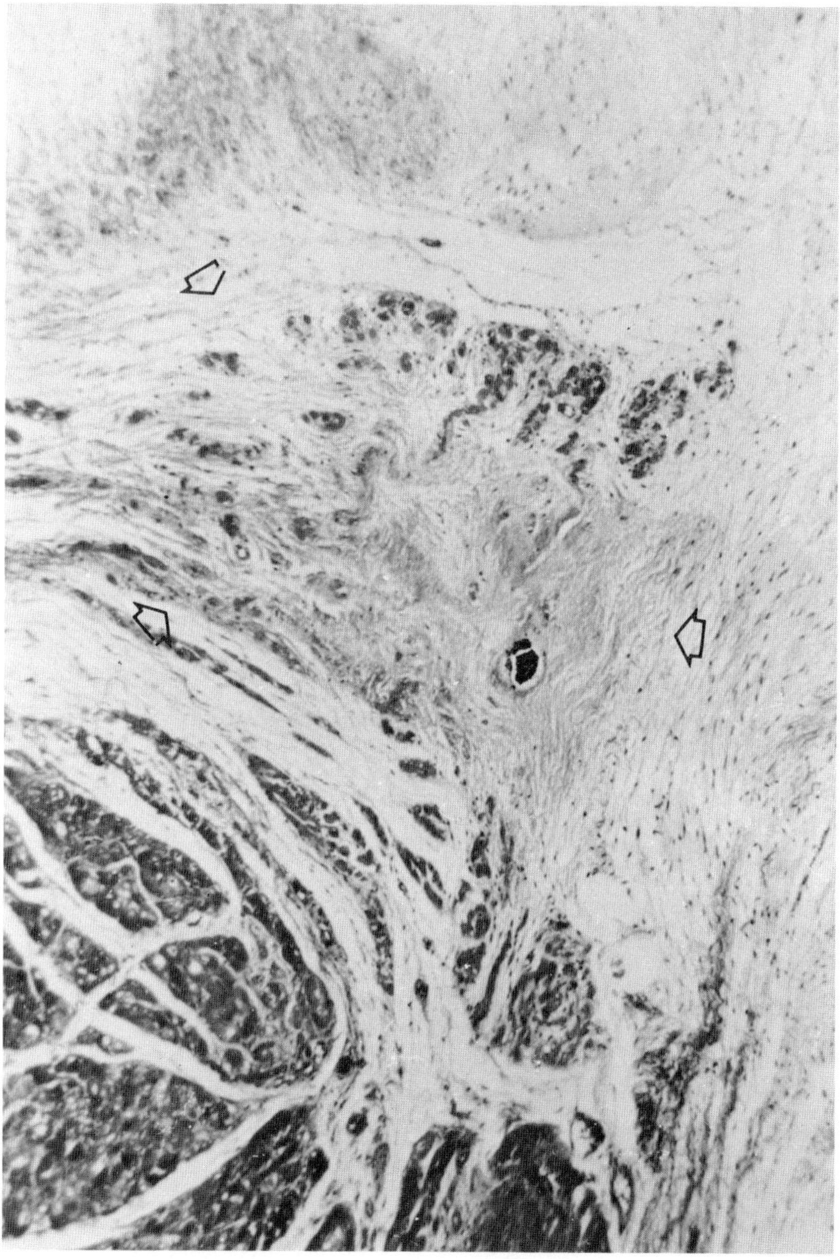

Figure 5. Photomicrograph of AV node. Nodal tissue (*arrows*) is being replaced by dense collagen fibers with focal mineralization. H and E. × 100.

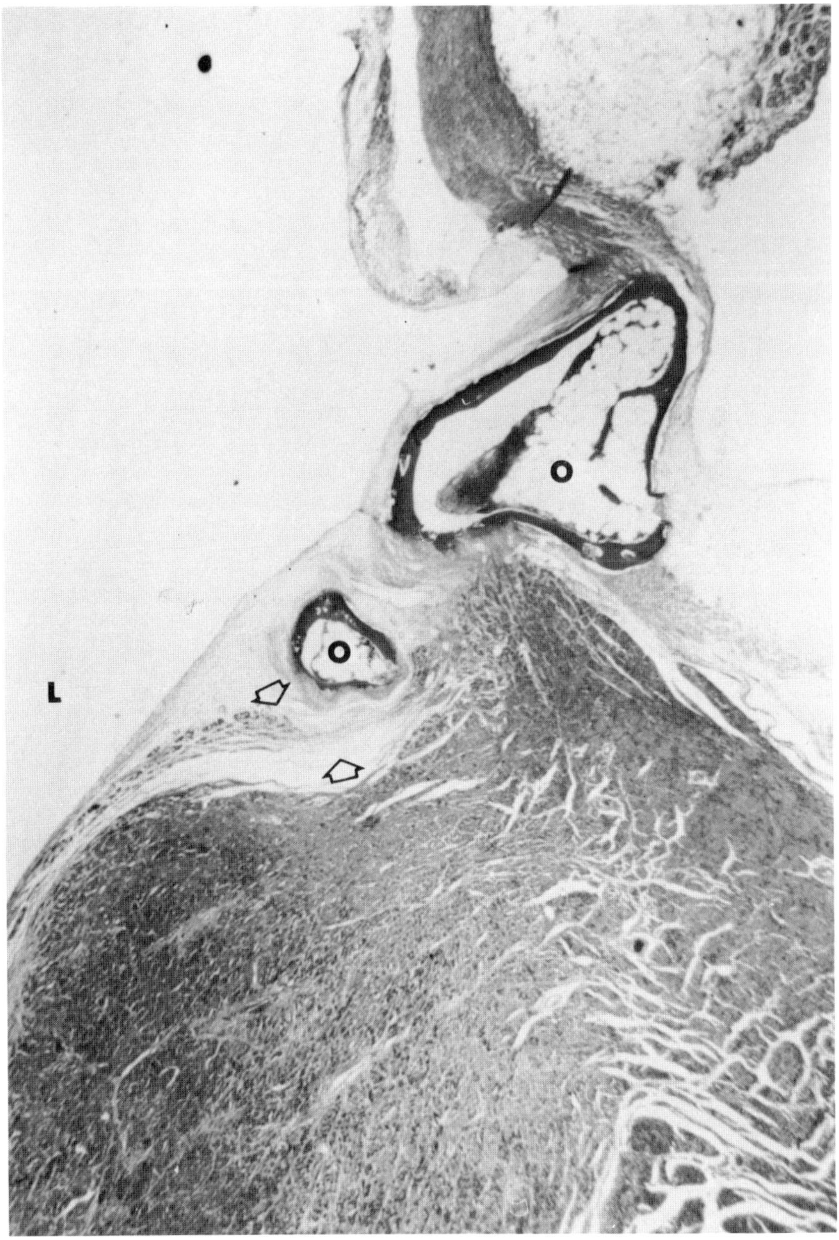

Figure 6. Photomicrograph of the left bundle, showing bone cyst (O). Osseous elements are compressed and have caused degeneration and fibrosis of the bundle fibers (*arrows*). H and E. × 40.

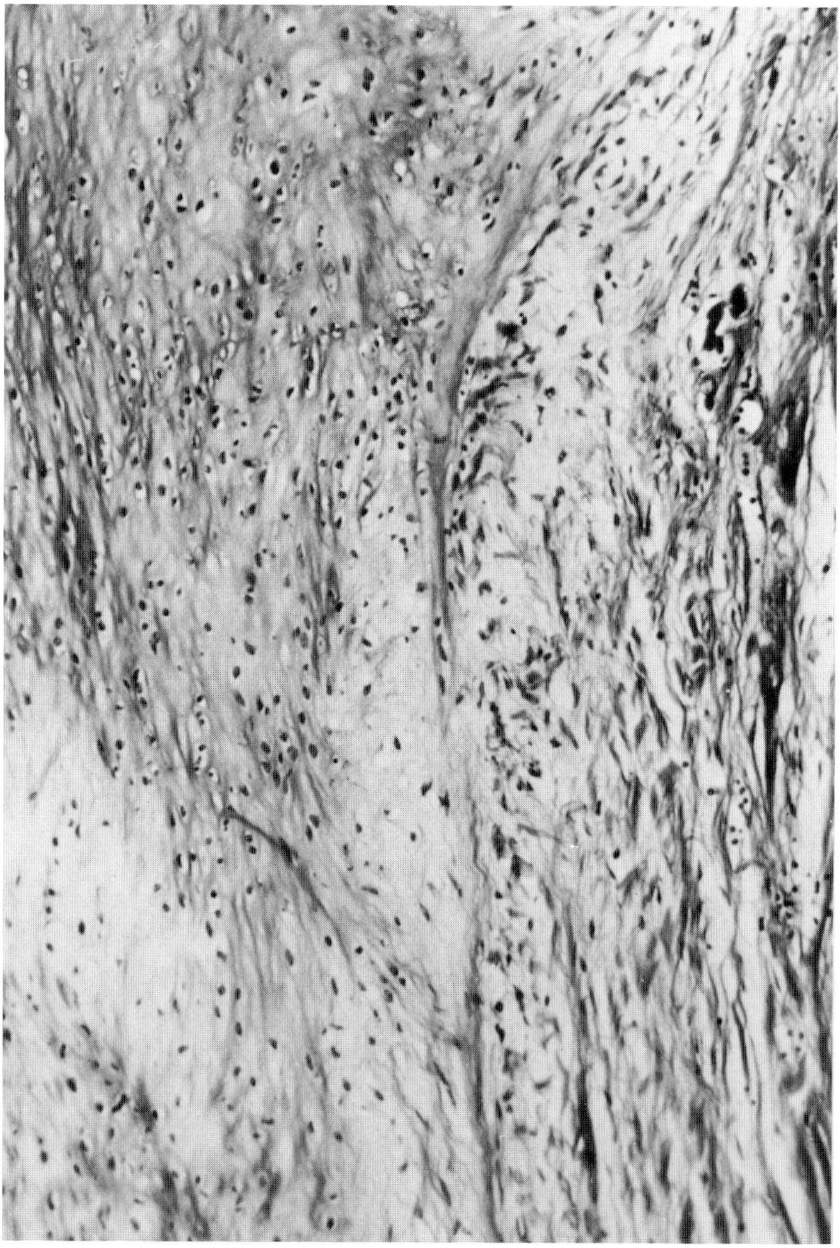

Figure 7. The left bundle branch has completely disappeared into the chondrofibrous hyperplasia of the endocardium. H and E. × 100.

homogenously coagulative or lytic sarcoplasma were trapped in the dense fibrous tissue.

The bundle fibers of the upper right bundle branch were moderately degenerated, owing to chondrogeneous metaplasia or ossification of the central fibrous body and fibrosis of the interventricular crest. The middle and lower right bundles, however, showed no significant changes in most affected cats.

DISCUSSION

Degeneration or fibrosis of the A.V. node and its bundle branches, with chondrogeneous metaplasia and ossification in the adjacent central fibrous body associated with endocardial and myocardial fibrosis and organized endomyocarditis, was consistently observed in the hearts of 63 cats with cardiomyopathy. The endocardial and myocardial lesions, described elsewhere (Liu, 1970), caused sudden death in 25% of the affected cats in the present study. On the basis of the lesions of the conduction system, endocardium, and myocardium, and the electrocardiographic and hemodynamic findings, we felt that the syncope and sudden death might be a reflection of labile left ventricular and diastolic pressure, as described in man (Goodwin and Oakley, 1972). Bone was found in a human patient with severe syncopal attacks (Tapham, 1906). The presence of cartilage in the central fibrous body might initiate fatal conduction disturbances (Ferris and Aherne, 1971). Abnormal presence of bone with lesions in the AV node of the heart was reported to relate to sudden death in Doberman pinschers (James and Drake, 1968).

Cardiac dysfunction, hemostasis, and multiple thromboemboli might also have led to the death of the affected cats (Liu, 1970). Degeneration of the AV node and its bundle was probably due to compression by islands of cartilage or bone cysts in the adjacent central fibrous body, or fibrosis of the interventricular septal crest. Fibrosis of the AV node may have indicated a chronicity of the lesion or have been due to invasion of fibrous tissue from the adjacent tissue.

Bone (ossa cordis) is normally present in the bovine heart (Sissons, 1953), and cartilaginous, and osseous metaplasia has also been reported at the origin of the aorta of goats (Giesel, 1973). Bone formation in the base of the aorta was reported in 6% of aging pigeons (Prichard, 1965). To the best of our knowledge, bone has not been reported in the feline heart.

Cartilage was consistently observed in the central fibrous body, root of the aorta, and endocardium of the cats in the present study. Composition of the cartilage tissue varied from a few chondroblasts and early chondrocytes scattered in the scar tissue to predominant, mature chondrocytes with mineralization in the central fibrous body, smooth muscle, and elastic laminae of the aorta and scar tissue in the endocardium and subendocardium.

Chondrogenesis or osteogenesis is dependent on the oxygen factor in the tissue culture (Bassett, 1962). Under a similar compaction, cartilage develops in a low oxygen environment and bone develops in a high oxygen environment. This supports an early hypothesis that chondrogenesis is more likely to occur than osteogenesis in a nonvascular environment (Ham, 1930).

Chondrogeneous metaplasia was induced in the inner two-thirds of the aortic media of chickens by intramural injection of the connective tissue stimulant carrageenan (McCandless and Lehoczky, 1963). Chondrogeneous metaplasia developed in the scar tissue after administration of carrageenan into the focal ischemic myocardium produced by coronary ligation (Lehoczky-Mona and McCandless, 1964). Osseous tissue was occasionally observed in the chondrogeneous lesions in the same experiments.

Ossification and chondrogeneous metaplasia were observed in the central fibrous body of dogs with markedly narrowed lumens of the small coronary arteries in the region of His bundle (James and Drake, 1968).

Presence of cartilage and bone in the central fibrous body has always been associated with fibrosis of the interventricular septal crest, endocardium, and myocardium. The cartilaginous and osseous tissue observed in the cats of the present study might have resulted from chondrogeneous and osteogenous metaplasia of the fibrous scar in the local tissue. The most probable sequence of the pathogenesis of the cardiac lesions in the cats in the present study was nonsuppurative endocarditis and myocarditis, followed by organization of the inflammation, leading to fibrosis involving the endocardium, myocardium, and conduction system and, finally, formation of cartilaginous, osseous metaplasia. In addition to the pathological evidence, clinical history also indicated this sequence.

The cause or causes of the nonsuppurative endocarditis and myocarditis are still obscure; however, the possibility of a viral agent cannot be ruled out (Liu, 1970), since Coxsackie virus B-3 causing subclinical myocarditis could result in fibrosis, mineralization, and hypertrophy, and permanent myocardial injury might provide an excellent experimental model of similar disease in mice (Wilson *et al.,* 1969). The Coxsackie virus B-3 was isolated from the infected hearts of the mice early in the study, but no virus was recovered after the 6th day of injection. We diagnosed only the later phase of the symptoms in the affected cats of this study. In order to isolate the virus, one should detect the early stages of myocarditis and endocarditis.

Acknowledgment The authors thank Mrs. Nancy Sturge for her assistance in preparing the manuscript.

REFERENCES

BASSETT, C. A. 1962. Current concepts of bone formation. J. Bone Joint Surg. 44-A: 1217–1244.

FERRIS, J. A., and W. A. AHERNE. 1971. Cartilage in relation to the conductive tissue of the heart in sudden death. Lancet *i:* 64–66.
GEISEL, O. 1973. Zur Pathologie der Ziegenaorta. Vet. Path. 10: 202–218.
GOODWIN, J. F., and C. W. OAKLEY. 1972. The cardiomyopathies. Brit. Heart J. 34: 545–552.
HAM, A. W. 1930. A histopathological study of the early phases of bone repair. J. Bone Joint Surg. 12: 827–844.
HARPSTER, N. K. 1973. Acquired heart disease in the cat. *In* Proceedings 40th Annual Meeting, American Animal Hospital Association, p.p. 118–125.
JAMES, T. N. 1965. Anatomy of the sinus node, A.V. node, and os cordis of the beef heart. Anat. Rec. 153: 361–371.
JAMES, T. N., and E. H. DRAKE. 1968. Sudden death in Doberman Pinschers. Ann. Int. Med., 68: 821–829.
LEHOCZKY-MONA, J., and E. L. McCANDLESS. 1964. Ischemic induction of chondrogenesis in myocardium. Arch. Path. 78: 37–42.
LEV, M., and P. N. UNGER. 1955. The pathology of the conduction system in acquired heart disease. I. Severe atrioventricular block. Arch. Path. 60: 502–529.
LIU, S.-K. 1970. Acquired cardiac lesions leading to congestive heart failure in the cat. Am. J. Vet. Res. 31: 2071–2088.
LIU, S.-K. 1974. Pathology of feline heart disease. *In* R. W. Kirk (ed.) Current veterinary therapy V. W. B. Saunders Co., Philadelphia and London.
LIU, S.-K., R. J. TASHJIAN, and A. K. PATNAIK. 1970. Congestive heart failure in the cat. J. Am. Vet. Med. Ass. 156: 1319–1330.
LORD, P. F., A. WOOD, L. P. TILLEY, and S.-K. LIU. 1974. Radiographic and hemodynamic evaluation of cardiomyopathy and thromboembolism in the cat. J. Am. Vet. Med. Ass. 164: 154–165.
McCANDLESS, E. L., and J. LEHOCZKY-MONA. 1963. Aortic cartilage produced by intramural carrageenan. Arch Path., 75: 507–516.
PRICHARD, R. W. 1965. Spontaneous atherosclerosis in pigeons. *In* J. C. Roberts and R. Straus (eds.), Comparative Atherosclerosis. Hoeber Medical Division, New York.
SISSON, S. 1953. The Anatomy of the Domestic Animals. W. B. Saunders Co., Philadelphia.
TAPHAM, J. A. 1906. Bone formations in the heart. Brit. Med. J. 2: 953.
TILLEY, L. P., P. F. LORD, and A. WOOD. 1974. Acquired heart disease and aortic thromboembolism in the cat. *In* R. W. Kirk (ed.), Current Veterinary Therapy, Vol. V. W. B. Saunders Co., Philadelphia and London.
WILSON, F. M., Q. R. MIRANDA, J. L. CHASON, and A. M. LERNER. 1969. Residual pathologic changes following murine Coxsackie A and B myocarditis. Am. J. Path. 55: 253–263.

Mailing address:
Dr. Si-Kwang Liu,
Department of Pathology,
The Animal Medical Center,
510 East 62nd Street, New York, New York 10021 (USA).

Recent Advances in Studies on
Cardiac Structure and Metabolism Volume 10
The Metabolism of Contraction
Edited by Paul-Emile Roy and George Rona

PROTECTIVE EFFECT OF COEXISTENT THIAMINE DEFICIENCY UPON THE EXPERIMENTAL CARDIOMYOPATHY ASSOCIATED WITH ACUTE MAGNESIUM DEFICIENCY IN THE SYRIAN GOLDEN HAMSTER

YUZO HIROTA, KATHRYN THORP,
and WALTER H. ABELMANN[1]

Thorndike Memorial Laboratory, Harvard Medical Unit, Boston City Hospital, and Department of Medicine, Harvard Medical School, Boston, Massachusetts, USA

SUMMARY

Repeated efforts to induce beriberi heart disease by experimental thiamine deficiency (B_1d) have failed in many species. To test the hypothesis that magnesium deficiency (Mgd) might be the cofactor necessary for heart failure, 10-week-old Syrian golden hamsters were divided into four groups—control (C), B_1d, Mgd, and combined MgB_1d—and were fed the diets *ad libitum* for 3 weeks.

On day 21, animals were studied under intraperitoneal pentobarbital anesthesia (50 mg/kg). Electrocardiograms were taken and right and left ventricular pressures were measured by transthoracic needle puncture. Cardiac output was measured by the direct Fick method. The complete study was performed in 9 C, 13 B_1d, 9 Mgd, and 14 MgB_1d animals.

B_1d was proven by low red blood cell transketolase, high B_1 pyrophosphate effect, and was accompanied by tachycardia and hypercalcemia. B_1 did not differ from C in any other parameter.

Mgd was characterized by hypomagnesemia, hypercalcemia, prolongation of the PR interval, widening of the QRS interval, low O_2 consumption, low cardiac output, and increased heart weight to body weight ratio (HW/BW) as compared to control. No differences were observed in right and left ventricular pressures or peak dp/dt.

MgB_1d was characterized by hypomagnesemia, hypercalcemia, low red blood cell transkeotlase, and high B_1 pyrophosphate effect. MgB_1d minimized the deleterious effects of Mgd: animals were more active and the mortality was low, the PR interval remained normal, the QRS interval widened significantly less, cardiac output remained normal, and HW/BW increased significantly less.

Although, once again, beriberi heart disease was not produced, B_1d appeared to exert a protective effect upon the Mg-deficient myocardium.

[1] Supported, in part, by grants HL 10539 and HL 5909 from the National Heart and Lung Institute, National Institutes of Health, Bethesda, Maryland and by grant 1247-F from the American Heart Association, Massachusetts Affiliate.

INTRODUCTION

Beriberi heart disease has been well known for a long time. Cardiac manifestations of this disease are acute onset of dyspnea, biventricular enlargement, increased pulmonic second sound, femoral arterial murmur, wide pulse pressure, short circulation time, and increased blood volume (Inoue and Katsura, 1965). Hemodynamic data showed hyperkinetic biventricular failure with decreased systemic vascular resistance and narrowed arteriovenous oxygen difference (Blacket and Palmer, 1960; Akabarian, Yankopoules, and Abelmann, 1966; Jeffrey and Abelmann, 1971). Experimental acute and chronic thiamine deficiency has been repeatedly reported (Hackel, Goodale, Kleinerman, 1953; Yoshitoshi, Shibata, and Yamashita, 1961; Aldinger, 1965; McCandless, 1970; Yamashita, 1971), and yet no experimental animal models have been proved to be comparable to human beriberi heart disease. Thus, the pathogenesis and pathophysiology of this disease is still not clearly understood. Some unknown cofactors, acting in conjunction with thiamine deficiency, might be considered to cause beriberi heart disease.

Magnesium deficiency is well known to produce cardiovascular changes (Hellerstein *et al.*, 1957; Vitale *et al.*, 1961; Heggtveit, Herman, and Mishra, 1964) and is a common complication of chronic alcoholism (Wacker and Parisi, 1968; Jones *et al.*, 1969; Lim and Jacob, 1972). Acute beriberi heart disease is still seen in patients with chronic alcoholism in the United States (Akabarian *et al.*, 1966; Blacket and Palmer, 1960; Jeffrey and Abelmann, 1971; King, Easton, and Dunn, 1972). There is some evidence for interaction of thiamine and magnesium deficiency (Zieve, Diozaki, and Stenroos, 1968; Itokawa *et al.*, 1972).

The purpose of the present study was to test the hypothesis that magnesium deficiency might be a cofactor of acute beriberi heart disease.

MATERIALS AND METHODS

Male Syrian golden hamsters, 10 weeks of age, were divided into four groups; each group of animals was fed special diets and deionized water *ad libitum* for 3 weeks. Control diet (C), low thiamine (B_1d), low magnesium (Mgd), and low magnesium and low thiamine (MgB_1d) diets were obtained from Nutrition Biochemicals, Cleveland, Ohio. Body weights were measured twice a week. On the 21st day of the diet, pentobarbital, 50 mg/kg, was injected intraperitoneally for anesthesia, and an electrocardiogram was taken, comprising standard 6 limb leads, right parasternal (V_1), and left midchest (V_4) leads. A special ECG preamplifier was connected to a polyrecorder (Hewlett Packard model 868) at a paper speed 100 mm/sec. The methods used for hemodynamic studies have been described in detail in previous reports from this laboratory (Ramirez *et al.*, 1969; Abelmann, Jeffrey, and Wagner, 1972). In brief, a tracheotomy was carried out, and the animals were ventilated mechanically. Transthoracic left and right ventricular punctures were performed, and high fidelity left and right ventricular pressures were recorded; 0.2 ml of left ventricular blood was obtained through the puncture needle for arterial blood gas analysis to adjust the ventilation. Cardiac output was measured by the direct Fick method. Rectal temperature was recorded during the inscription of left and right ventricular pressures. The thoracic cage was opened after the hemodynamic study, and all the blood was drawn from the right ventricle into a heparinized syringe. Plasma magnesium, calcium, and potassium were measured in individual animals by atomic absorption spectrometry (Wacker, Iida, and Fuwa, 1964) and flame photometry, respectively. Red blood cell transketolase activity was measured in pooled blood, according to the method of Brin *et al.* (1960). Heart, lung, and liver wet weights were measured, and organs were preserved for the microscopic examination. The apex of the heart and the quadriceps muscle were used for heart and skeletal muscle magnesium content. Magnesium and thiamine concentrations in the diets were measured by the atomic absorption method (Wacker *et al.*, 1964) and colorimetric determination using the formalin-azo reaction (Fujiwara and Murata, 1965), respectively.

RESULTS

Clinical observations (Table 1; Fig. 1)

During the 1st week of the experiment, one death occurred in each group, unrelated to the deficiency status. From the 2nd week, MgB_1d and Mgd groups showed decrease of body weight with loss of hair around the nostril. In the 3rd week, three Mgd hamsters died after severe weight loss and deterioration; no late death was observed in the other three groups. The mean body weight of the Mgd and MgB_1d groups remained the same during the 21 days, but the former animals were less active and appeared much sicker.

Dietary and tissue thiamine and magnesium

Magnesium deficiency was proved by low dietary magnesium content and severe hypomagnesemia in Mgd and MgB_1d (Table 1; Fig. 2). Thiamine deficiency was proved by low dietary thiamine content (Table 1), low transketolase activity, and high thiamine pryophosphate effect in B_1d and MgB_1d. Whereas the thiamine pyrophosphate effect was 11.5% in controls and 15.9% in Mgd, it was 24.7% in B_1d and 30.2% in MgB_1d.

Electrocardiograms (Table 2)

A tachycardia of 467.4±5.9 beats/min was observed in B_1d, a 9% increase over C ($0.025<p<0.05$). Prolongation of the PR interval was seen in Mgd: 47.3±1.4 msec versus C 42.8±0.7 msec ($0.005<p<0.01$). Both Mgd and MgB_1d were

Table 1. Characterization of experimental groups and their mortality

Group	Diet		No. of Animals on Diet	Mortality		No. of Animals Studied Hemodynamically
	Mg (mg/100g)	B_1 (μg/g)		Early (%)	Late (%)	
Control	210.0	15.0	15	6.7	0	9
B_1 deficiency (B_1d)	280.0	1.0	19	5.3	0	13
Mg deficiency (Mgd)	2.0	12.0	19	5.3	15.8	9
Mg + B_1 deficiency (MgB_1d)	2.0	1.2	19	5.3	0	14

characterized by widening of the QRS interval: 28.9±1.2 msec ($p<0.001$) and 24.7±0.8 msec ($0.001<p<0.005$), respectively, versus C 21.6±0.5 msec. The prolongation of QRS in Mgd was significantly greater than in MgB_1d ($0.005<p<0.01$). Arrhythmias were not observed during the electrocardiographic examination or hemodynamic studies. All hamsters had an indeterminate frontal axis and an apparent conduction disturbance resembling a human right

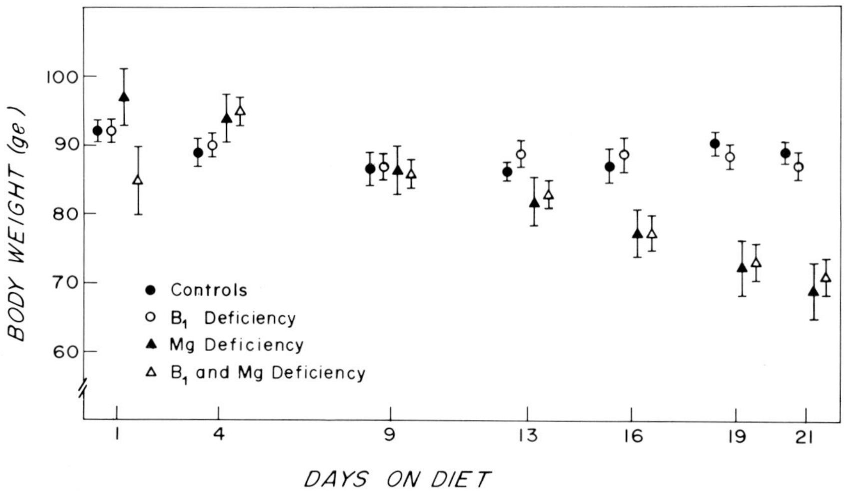

Figure 1. Effect of experimental diets upon body weight. Note that thiamine deficiency alone did not affect body weight, whereas magnesium-deficient animals as well as animals with combined deficiency lost weight progressively and equally.

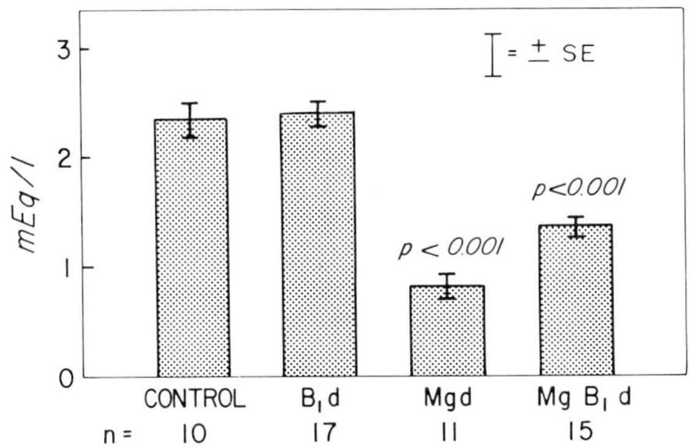

Figure 2. Effect of experimental diets on plasma magnesium. All diets deficient in magnesium resulted in significant hypomagnesemia.

bundle branch block pattern. ST segment and T wave morphology was not grossly abnormal, although the tracings were not optimal for a study of repolarization.

Hemodynamic study (Tables 3 and 4)

Studies were satisfactorily completed in 9/14 C, 13/18 B_1d, 9/15 Mgd, and 14/18 MgB_1d animals. Left ventricular blood gas analyses were in acceptable ranges before adjusting the ventilator in almost all cases. No difference was observed in right and left ventricular systolic pressures, their first derivatives, or

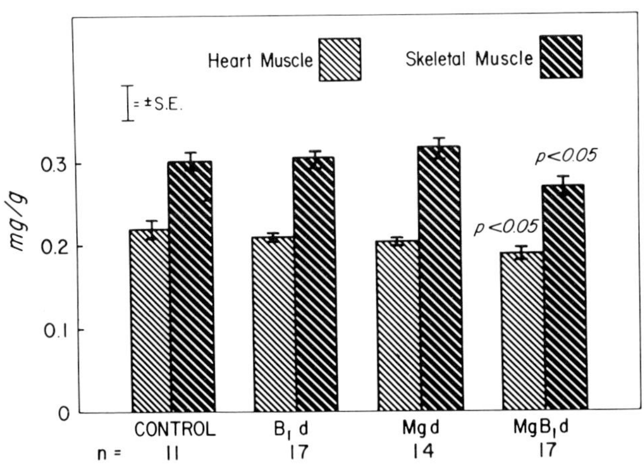

Figure 3. Effect of experimental diets on magnesium content of skeletal and cardiac muscle. Mgd did not differ from controls, and tissue magnesium was significantly depressed only in MgB_1d.

Table 2. Electrocardiographic observations (mean ± SE)

Group	N	Heart Rate (beats/min)	PR Interval (msec)	QRS Interval (msec)
Control	14	429.9±13.5	42.8±0.7	21.6±0.6
B_1d	18	467.4± 5.9*	42.2±0.8	23.1±1.1
Mgd	15	409.4±18.0	47.3±1.4†	28.9±1.2†‡
MgB_1d	18	441.2± 7.2	45.5±1.4	24.7±0.8†‡

*$p < 0.05$ compared to controls.
†$p < 0.01$ compared to controls.
‡$p < 0.01$, significantly different from each other.

end-diastolic pressures. O_2 consumption was relatively lower in all animals than previously reported (Abelmann *et al.*, 1972). Mgd was characterized by low O_2 consumption, 0.642±0.062 ml/min versus C 0.896±0.039 ml/min ($0.001<p<0.005$) and low cardiac output, 7.05±0.82 ml/min versus C 10.76±1.43 ml/min ($0.01<p<0.025$). MgB_1d, compared to C, had slightly lower O_2 consumption at 0.743±0.056 ml/min ($0.025<p<0.05$). No significant differences were observed in cardiac output to body weight ratio between the groups, but MgB_1d showed a tendency toward a higher cardiac output.

Plasma calcium, potassium, heart weight, and heart weight to body weight ratio (Table 5)

No differences were observed in plasma potassium levels. Plasma calcium concentration was 5.53±0.28 mg/100 ml in C, 6.57±0.20 mg/100 ml in B_1d ($0.001<p<0.005$), 7.62±0.48 mg/100 ml in Mgd ($0.001<p<0.005$), and 7.42±0.20 mg/100 ml in MgB_1d ($p<0.001$). Heart weight was 343±9 mg in C, 333±6 mg in B_1d, 298±13 mg in Mgd ($0.005<p<0.01$), and 284±8 mg in MgB_1d ($p<0.001$). Heart weight to body weight ratio was $37.9\pm0.6 \times 10^{-4}$ in control, and significantly higher in Mgd at $43.1\pm1.1 \times 10^{-4}$ ($p<0.001$), and in MgB_1d at $40.5\pm0.9 \times 10^{-4}$ ($0.01<p<0.025$). There was a significant difference in this ratio between Mgd and MgB_1d ($0.025<p<0.05$).

Histological examination

Histological examinations were done in all heart specimens studied hemodynamically, both hematoxylin and eosin and Masson's trichrome stain being used. Mgd and MgB_1d showed scattered focal necrosis, congestion, and intramural hemorrhages. Some congestion and hemorrhage were seen in B_1d and control, but never necrosis.

Table 3. Right and left ventricular pressures and maximum of the first derivatives of left ventricular pressure (mean±SE)

		Right Ventricle (mm Hg)		Left Ventricle		
Group	*N*	Systolic	End-diastolic	Systolic (mm Hg)	End-diastolic (mm Hg)	dp/dt (mm Hg/sec)
Control	9	57.0±5.8	4.4±1.1	162.4±7.4	3.3±0.5	13580±706
B_1d	13	45.0±3.9	2.8±0.8	152.0±6.1	3.6±0.8	12570±726
Mgd	9	46.0±5.5	3.6±1.0	152.1±7.4	3.4±0.7	12547±1303
MgB_1d	14	44.3±2.8	3.0±0.7	150.8±4.7	2.7±0.5	12847±658

Table 4. Oxygen consumption and blood flow (mean±SE)

Group	N	O_2 Consumption (ml/min)	Cardiac Output (ml/min)	Stroke Volume (ml)	CO/BW* (ml/min/g)
Control	9	0.896±0.039	10.76±1.43	0.027±0.004	0.122±0.017
B_1d	13	0.789±0.042	9.55±0.80	0.024±0.002	0.133±0.008
Mgd	9	0.642±0.062†	7.05±0.82‡	0.019±0.002	0.098±0.010
MgB_1d	14	0.743±0.056‡	9.79±1.00	0.025±0.003	0.139±0.018

*CO/BW = cardiac output to body weight ratio.
†$p < 0.005$ compared to control.
‡$p < 0.05$ compared to control.

DISCUSSION

It is well known that experimental magnesium deficiency has deleterious effects on the cardiovascular system. These have been studied extensively from both the histological and electrocardiographic points of view (Hellerstein *et al.*, 1957; Vitale *et al.*, 1961; Ono, 1962; Syllm-Rapoport, 1962; Heggtveit *et al.*, 1964; Wener *et al.*, 1964; Seta *et al.*, 1966; Seelig, 1969; Britton and Stokstad, 1970). Investigators have reported different electrocardiographic patterns, varying according to the details of the experiments and the duration and severity of magnesium deficiency, as well as with different responses of serum calcium and

Table 5. Plasma calcium, potassium, and heart weight to body weight ratio (mean±SE)

		Ca^{++}		K^+		Heart Weight to Body Weight Ratio
Group	N	(mg/100ml)	N	(mEq/liter)	N	($\times 10^{-4}$)
Control	7	5.53±0.28	10	4.57±0.32	14	37.9±0.6
B_1d	14	6.57±0.20*	13	5.46±0.34	18	38.5±0.6
Mgd	7	7.62±0.48*	10	4.12±0.37	15	43.1±1.1†‡
MgB_1d	18	7.42±0.20*	12	4.73±0.29	18	40.5±0.9§‡

*$p < 0.005$ compared to control.
†$p < 0.001$ compared to control.
‡$p < 0.05$, significantly different from each other.
§$p < 0.025$ compared to control.

potassium in different species. According to Seelig (1969), QRS widening and peaking of the T wave are seen in magnesium deficiency, and PR prolongation and ST depression are present with coexisting hypocalcemia. We observed PR prolongation and QRS widening associated with hypercalcemia and without changes in potassium, somewhat at variance with previous reports (Vitale *et al.*, 1961; Ono, 1962; Syllm-Rapoport, 1962; Wener *et al.*, 1964; Seta *et al.*, 1966; Seelig, 1969). Vitale *et al.* (1961) found myocardial calcification in prolonged magnesium deficiency in dogs. Wener *et al.* (1964) reported arterial and arteriolar degeneration, scattered myocardial degeneration, necrosis, and fibrosis; calcification was seen in fibrotic areas. Calcification of the aorta and other great vessels was reported by Britton and Stokstad (1970). We encountered only scattered focal necrosis and no calcification, perhaps related to the relatively short duration of the experiment.

Earlier preliminary hemodynamic studies reproted from this laboratory in hamsters after 25 days of acute experimental magnesium deficiency (Chiotellis *et al.*, 1973) had revealed low O_2 consumption, low cardiac output, and low left ventricular pressures. The general methods and diets used in both experiments having been the same, the differences are attributed to the different duration of the experiments, deterioration of myocardial function being associated with the longer duration of Mg deprivation.

Hypercalcemia has been observed repeatedly in association with magnesium deficiency in rats (Whang and Welt, 1963; Wacker and Parisi, 1968; Britton and Stokstad, 1970; Itokawa and Fujiwara, 1973). We noted this in the previous experiment (Chiotellis *et al.*, 1973) and, in the present study, demonstrated significant hypercalcemia in pure magnesium deficiency in hamsters without significant change in serum potassium level.

The initial intention of this study was to produce experimental beriberi heart disease, based upon the observation that this disease is still seen in the United States in chronic alcoholics among whom hypomagnesemia is not rare (Wacker and Parisi, 1968; Jones *et al.*, 1969; Lim and Jacob, 1972; Victor, 1973). Again we failed to produce this type of heart disease experimentally. In the Japanese literature of the 1920's–1940's, there are reports of several human experiments carried out with the intention of producing shoshin beriberi (Omori *et al.*, 1923; Sakamoto *et al.*, 1924; Sawada, 1937; Nishio *et al.*, 1948). Pure acute experimental thiamine deficiency in man was similar to that in animal experiments (Nishio *et al.*, 1948); and in mild thiamine deficiency of long duration in very active men, shoshin beriberi could be produced experimentally (Omori *et al.*, 1923). Also, clinical observations indicate that shoshin beriberi was mostly seen among infants, active young men, or pregnant women who had long-standing thiamine deficiency but were otherwise healthy (Inoue and Katsura, 1965). These facts suggest that factors other than magensium deficiency may play a pathogenetic role in beriberi heart disease.

Table 6. Summary comparison of the three experimental deficiencies with the control

	Deficiency		
	Mg	B_1	MgB_1
Mortality	++	0	0
Increased heart weight to body weight ratio	++	0	+
Tachycardia	0	+	0
Prolonged PR	+	0	0
Wide QRS	++	0	+
Low cardiac output	+	0	0

No definite hemodynamic or electrocardiographic changes were observed in B_1d, except tachycardia after 21 days. It is noteworthy that cardiac output remained normal in B_1d, although a trend toward a high output state was seen in MgB_1d. In a previous experiment in this laboratory, thiamine-deficient rats manifested anorexia and loss of weight from the 3rd week on, but definite neurological signs of beriberi were not observed until 53 days on the diet (Cohen *et al.*, 1974).

Interestingly, thiamine deficiency coexisting with magnesium deficiency showed some protective effect upon the cardiovascular system, i.e., inhibition of prolongation of QRS duration and of increase in heart weight (Table 6). The mechanism of this effect is totally unknown. Itokawa, Sasagawa, and Fujiwara (1973) reported that the accumulation of cholesterol and total lipids in the liver in magnesium-deficient rats was abolished by coexisting thiamine deficiency. Itokawa, Tanaka, and Kimura (1972) found high serum serotonin levels in magnesium-deficient rats, not seen in combined magnesium and thiamine deficiency. Zieve *et al.* (1968) observed decreased blood and liver transketolase activities in prolonged magnesium deficiency in rats and no thiamine pyrophosphate effect, even when magnesium sulfate was added *in vitro*. They attributed these effects to interference with the activation of the apoenzyme.

The alterations observed in this experiment were not limited to the cardiovascular system. The animals on the combined deficiency diet (MgB_1d) were more active, appeared healthier, and had a lower mortality. Thus, coexisting thiamine deficiency appears to exert a general protective effect against the deleterious influences of magnesium deficiency. The mechanism of this effect remains to be elucidated.

REFERENCES

ABELMANN, W. H., F. E. JEFFREY, and R. WAGNER. 1972. Hemodynamics of the hereditary cardiomyopathy of Syrian hamsters. *In* E. Bajusz and G. Rona (eds.), Recent Advances in Studies on Cardiac Structure and Metabolism, Vol. 1: Myocardiology, pp. 225–234. University Park Press, Baltimore.

AKABARIAN, M., N. A. YANKOPOULOS, and W. H. ABELMANN. 1966. Hemodynamic studies in beriberi heart disease. Am. J. Med. 41: 197–212.

ALDINGER, E. E. 1965. Effect of thiamine deficiency on potential myocardial contractility. Circulat. Res. 16: 238–243.

BLACKET, R. B., and A. J. PALMER. 1960. Hemodynamic studies in high output beriberi. Brit. Heart J. 22: 483–501.

BRIN, M., M. TAI, A. S. OSTERSHEVER, and H. KALINSKY. 1960. The effect of thiamine deficiency on the activity of erythrocyte hemolysate transketolase. J. Nutr. 71: 273–281.

BRITTON, W. M., and E. L. R. STOKSTAD. 1970. Aorta and other soft tissue calcification in the magnesium deficient rat. J. Nutr. 100: 1501–1506.

CHIOTELLIS, P. N., K. THORP, R. L. WAGNER, and W. H. ABELMANN. 1973. Hemodynamics of experimental magnesium deficiency. Clin. Res. 21: 409.

COHEN, E. M., W. H. ABELMANN, J. V. MESSER, and O. H. L. BING. 1974. Mechanical properties of rat cardiac muscle during experimental thiamine deficiency. (In preparation)

FUJIWARA, M., and K. MURATA. 1965. Determination of thiamine, pp. 135–148. *In* Review of Japanese Literature on Beriberi and Thiamine. Igakushoin, Tokyo.

HACKEL, D. B., W. T. GOODALE, and J. KLEINERMAN. 1953. Effects of thiamine deficiency on myocardial metabolism in intact dogs. Am. Heart J. 46: 883–894.

HEGGETVEIT, H. A., L. HERMAN, and R. K. MISHRA. 1964. Cardiac necrosis and calcification in experimental magnesium deficiency. Am. J. Path. 45: 757–782.

HELLERSTEIN, E. E., J. J. VITALE, P. L. WHITE, D. M. HEGSTED, N. ZAMCHECK, and M. NAKAMURA. 1957. Influence of dietary magnesium on cardiac and renal lesions of young rats fed an atherogenic diet. J. Exp. Med. 106: 767–776.

INOUE, K., and E. KATSURA. 1965. Clinical signs and metabolism of beriberi patients, pp. 29–63. *In* Review of Japanese Literature on Beriberi and Thiamine. Igokushoin, Tokyo.

ITOKAWA, Y., and M. FUJIWARA. 1973. Changes in tissue magnesium, calcium and phosphorus levels in magnesium deficient rats in relation to thiamine excess or deficiency. J. Nutr. 103: 438–443.

ITOKAWA, Y., K. INOUE, Y. NATORI, K. OKAZAKI, and M. FUJIWARA. 1972. Effect of thiamine on growth, tissue magnesium, thiamine levels and transketolase activity in magnesium deficient rats. J. Vitaminol. 18: 159–164.

ITOKAWA, Y., S. SASAGAWA, and M. FUJIWARA. 1973. Effect of thiamine on lipid metabolism in magnesium deficient rats. J. Nutr. Sci. Vitaminol. 19: 15–21.

ITOKAWA, Y., C. TANAKA, and M. KIMURA. 1972. Effect of thiamine on serotonin levels in magnesium deficient animals. Metabolism 21: 375–379.

JEFFREY, F. E., and W. H. ABELMANN. 1971. Recovery from proved shoshin beriberi. Am. J. Med. 50: 123–128.

JONES, J. E., S. R. SHANE, W. H. JACOBS, and E. B. FLINK. 1969. Magnesium balance studies in chronic alcoholism. Ann. N.Y. Acad. Sci. 162: 934–946.

KING, J. F., R. EASTON, and M. DUNN. 1972. Acute pernicious beriberi heart disease. Chest 61: 512–514.

LIM, P., and E. JACOB. 1972. Magnesium status of alcoholic patients. Metabolism 21: 1045–1051.

McCANDLESS, D. W., C. HANSON, K. V. SPEEG, Jr., and S. SCHENKER. 1970. Cardiac metabolism in thiamine deficiency in rats. J. Nutr. 100: 991–1002.

NISHIO, M., M. FUJIWARA, M. KITAMURA, and S. NAKATA. 1948. Studies on experimental thiamine deficiency in man and thiamine requirement. (In Japanese). Vitamins 1: 256–257.

OMORI, K., Y. OHASHI, T. NAKANISHI, M. HARA, and T. OTA. 1923. Studies on the etiology of beriberi. (In Japanese). Keio Igaku 3: 197–249.

ONO, I. 1962. The effect of varying dietary magnesium on the electrocardiogram and blood electrolytes of dogs. Jap. Circulat. J. 26: 677–685.

RAMIREZ, A., W. B. HOOD, Jr., M. POLANYI, R. WAGNER, N. A. YANOPOULOS, and W. H. ABELMANN. 1969. Registration of intravascular pressure and sound by a fiberoptic catheter. J. Appl. Physiol. 26: 679–683.

SAKAMOTO, T., G. SHINODA, T. NISHIKATA, M. HOSOYA, and K. TADENUMA. 1924. Experimental studies on vitamin deficiency in healthy man. (In Japanese). Tokyo Igaku Zasshi 38: 510–533.

SAWADA, K. 1937. Experimental study on vitamin B_1 deficiency in man. (In Japanese). Tokyo Igaku Zasshi 51: 244–299.

SEELIG, M. S. 1969. Electrocardiographic patterns of magnesium depletion appearing in alcoholic heart disease. Ann. N.Y. Acad. Sci. 162: 906–917.

SETA, K., R. KLEIGER, E. E. HELLERSTEIN, B. LOWN, and J. J. VITALE. 1966. Effect of potassium and magnesium deficiency on the electrocardiogram and plasma electrolytes of pure-bred beagles. Am. J. Cardiol. 17: 516–519.

SYLLM-RAPOPORT, I. 1962. Electrocardiographic studies in dogs with experimental magnesium deficiency. J. Pediat. 60: 801–804.

VICTOR, M. 1973. The role of hypomagnesemia and respiratory alkalosis in the genesis of alcohol-withdrawal symptoms. Ann. N.Y. Acad. Sci. 215: 235–248.

VITALE, J. J., E. E. HELLERSTEIN, M. NAKAMURA, and B. LOWN. 1961. Effect of magnesium-deficient diet upon puppies. Circulat. Res. 9: 387–394.

WACKER, W. E. C., C. IIDA, and K. FUWA. 1964. Accuracy of determinations of serum magnesium by flame emission and atomic absorption spectrometry. Nature 202: 659–661.

WACKER, W. E. C., and A. F. PARISI. 1968. Magnesium metabolism. New Engl. J. Med. 278: 658–663; 712–717; 772–776.

WENER, J., K. PINTAR, M. A. SIMON, R. MOTOLA, R. FRIEDMAN, A. MAYMAN, and R. SCHUCHER. 1964. The effect of prolonged hypomagnesemia on the cardiovascular system in young dogs. Am. Heart J. 67: 221–231.

WHANG, R., and L. G. WELT. 1963. Observations in experimental magnesium depletion. J. Clin. Invest. 42: 305–313.

YAMASHITA, S. 1971. Effect of alcohol on thiamine deficient cardiac lesion. Jap. Heart J. 12: 354–367.

YOSHITOSHI, Y., N. SHIBATA, and S. YAMASHITA. 1961. Experimental studies on the beriberi heart. I. Cardiac lesions in thiamine deficient rats. Jap. Heart J. 2: 42–64.

ZIEVE, L., W. M. DIOZAKI, and L. E. STENROOS. 1968. Effect of magnesium deficiency on blood and liver transketolase activity and on the recovery of enzyme activity in thiamine-deficient rats receiving thiamine. J. Lab. Clin. Med. 72: 268–277.

Mailing address:
Dr. Walter H. Abelmann,
Cardiovascular Unit, Beth Israel Hospital,
330 Brookline Avenue, Boston, Massachusetts 02215 (USA).

Recent Advances in Studies on
Cardiac Structure and Metabolism Volume 10
The Metabolism of Contraction
Edited by Paul-Emile Roy and George Rona

IRON SUPPLEMENTATION IN SUCKLING RATS: ITS EFFECT ON THE HEART[1]

J. F. NEFFGEN and K. RAKUŠAN

Department of Physiology, Faculty of Medicine,
University of Ottawa, Ottawa, Ontario, Canada

SUMMARY

The present study attempted to determine the significance of the "anemia" which normally occurs in infants. We compared various parameters in suckling rats whose hemoglobin concentration had been artificially raised by gastric tube feeding of iron with those of control animals. The 20-day-old experimental animals showed significant increases in the weights of the liver (10%) and spleen (40%). They had also developed a statistically significant 8% decrease in heart weight, accompanied by a smaller myocardial fiber size and, thus, higher fiber density. The association in these young animals of higher than normal hemoglobin levels with smaller hearts and myocardial fibers suggests that the normally developing "anemia" of infancy may well serve an important physiological function: to stimulate myocardial growth.

INTRODUCTION

During the first 3 months of life, normal infants show a decrease in hemoglobin concentration to levels as low as 10–11 g%, a fact well known to pediatricians and referred to as the physiological anemia of the newborn (O'Brien and Pearson, 1971). This physiological anemia is believed to occur regardless of the amount of dietary iron available and is probably due to three factors: a marked decrease in red blood cell production, most likely secondary to low erythropoietin activity (Gairdner, Marks, and Roscoe, 1952; Halvorsen, 1963; Mann *et al.*, 1965), a shortened erythrocyte survival (Pearson, 1967), and a significant hemodilution due to the rapid expansion of the blood volume during the first few months of life. Depending on the length and extent of feeding of normal milk after the age of 3 months, an iron deficiency anemia may become superimposed on the pre-existing physiological one, with hemoglobin levels often falling below 10 g%.

It appears that a postnatal decline in hemoglobin content is not confined to

[1] Submitted in partial fulfillment of a Ph.D. degree by Dr. J. F. Neffgen. Present address: Toronto General Hospital, Toronto, Ontario, Canada.

human infants only, but occurs perhaps in a number of mammalian species as well (Widdowson and McCance, 1955; Mott, 1965; Dawes, 1968). Previous studies in our laboratory on normal suckling rats (less than 21 days old) revealed a decrease in hemoglobin concentration to levels as low as 7.5 g% (Rakušan, Rádl, and Poupa, 1965).

Although the postnatal anemia in infants has been recognized as a normal developmental process for some years, its possible physiological significance has never been studied to the best of our knowledge. It is known that anemia (presumably by causing hypoxia) may, among other things, induce increased formation of capillaries or stimulate cardiac growth. One may ask, therefore, whether the anemia of the newborn serves certain physiological functions, i.e., stimulates heart growth and/or capillary formation or perhaps determines some other developmental process, rather than existing merely as some "accident" of nature. This question is very important at the present time, because of the current trend to substitute iron-rich and/or iron-fortified foods for normal milk early in infancy, thereby preventing that part of the anemia of infancy which would develop on a longer course of normal milk.

The present study was done to determine whether the anemia of early infancy plays some, as yet unknown, physiological role during early development. We produced a marked increase in hemoglobin concentration in suckling rats less than 21 days old by gastric tube feeding of ferrous sulfate (approximately 20 mg Fe/kg/day) and then compared a number of gross morphological and histological parameters with those of littermate control animals. These parameters included body weight; heart weight; the weights of the spleen, liver, kidneys and adrenals; myocardial fiber size; and myocardial fiber and capillary densities. A further aim of this study was to determine whether higher hemoglobin concentrations during the preweaning period of life would have an effect on any of the above-mentioned parameters in the adult animal.

METHOD

Newborn male Sprague-Dawley rats were reduced, following birth, to eight per litter and housed with their mothers in separate cages. Starting at the age of 3 days, half of the animals in each litter (the experimental groups) were given, by means of a gastric feeding tube, daily infusions of an aqueous solution of ferrous sulfate, the doses of which have been outlined in Table 1. The daily amounts of ferrous sulfate were administered in two divided doses at 8-hour intervals; each dose was contained in 0.1 ml of distilled water and was prepared fresh before each injection. After the treatments, the young rats were returned to their mothers and were nursed by them until the age of 21 days. The remaining half of the animals (the control groups) were subjected to identical treatment except that they received 0.1 ml of distilled water instead of the ferrous sulfate

Table 1. Daily doses of iron administered to suckling Sprague-Dawley rats

Age (days)	Dose FeSO × 7 H_2O (mg)	Fe Equivalent* (mg Fe/kg/day)
3	0.92	18
4	1.10	18
5	1.28	19
6	1.46	19
7	1.65	20
8	1.83	20
9	2.02	21
10	2.20	21
11	2.30	22
12	2.57	21
13–20	2.75	20–14

*Approximate values.

solution. (Note that a pilot study had shown that gastric intubation per se did not retard growth.)

At the ages of 11 and 20 days, some of the experimental and control rats were killed by decapitation and the following parameters were determined: hemoglobin concentration; hematocrit; body weight; the weights of the right and left ventricles and the septum; total heart weight; the weights of the spleen, liver, kidneys, and adrenals; myocardial water content; myocardial fiber size; and myocardial fiber and capillary densities. The hearts were separated into atria, left and right ventricles, and septa as described by Fulton, Hutchinson, and Jones (1952); the septa were dried overnight at 105°C; and the cardiac water content was determined from the difference between their wet and dry weights. All the organs were fixed in 3% glutaraldehyde (phosphate buffered to pH 7.4); PAS-hematoxylin stained paraffin sections were prepared, using routine techniques. Myocardial fiber size and fiber and capillary density were determined by light microscopy from cross sections of the left ventricles at a magnification of 10 × 100. To facilitate the counting, a ruled grid of 36 divisions was inserted into the ocular. Myocardial fiber size was determined according to the method described by Sitte (1967).

A number of experimental and control rats were weaned on the 21st day of life and then put on a normal diet. These animals were killed at the ages of 4½ and 14 months and the following parameters were determined: hemoglobin; hematocrit; body weight; and the weights of the right and left ventricles, septum, total heart, spleen, liver, kidneys, and adrenals.

All data were analyzed by analysis of variance, using a treatments × levels design.

Table 2. Effect of iron supplementation during infancy on red blook cell mass; body weight; and the weights of heart, spleen, liver, kidneys, and adrenals in preweanling and adult male Sprague-Dawley rats*

	Age	Hemoglobin (g%)	Hematocrit (%)	Body Weight (g)	Heart Weight (mg)	Heart Weight 100 g Body Wt. (mg/100g)	Spleen (mg)	Liver (g)	Kidneys (g)	Adrenals (mg)
Newborn (12)	<1 day	16.2±0.2	51.8±0.5	–	–	–	–	–	–	–
Control (11)	11 days	9.1±0.2	31.9±0.5	22.0±0.6	80±3	365±6	112±6	0.635±.017	0.251±0.007	3.9±0.3
Experimental (11)	11 days	11.0±0.2†	36.6±0.6‡	22.0±0.6	76±2	348±9	120±8	0.656±.023‡	0.258±0.010	4.1±0.2
Control (16)	20 days	7.9±0.2	26.6±0.6	40.7±1.2	171±4	423±8	125±5	1.339±.034	0.476±0.011	13.1±0.5
Experimental (16)	20 days	12.3±0.1†	39.0±0.4†	41.4±1.3	159±4†	386±7†	173±8†	1.470±.044†	0.488±0.014	14.0±0.6
Control (8)	4½ months	18.0±0.2	51.2±0.7	585±18	1336±43	228±4	839±35	19.10±0.73	3.73±0.12	49±2
Experimental (8)	4½ months	17.6±0.2	50.1±0.5	581±19	1322±46	228±5	879±33	18.88±0.83	3.62±0.12	53±3
Control (6)	14 months	17.5±0.2	47.3±0.8	754±32	1365±60	182±5	961±61	21.74±1.42	3.90±0.22	–
Experimental (6)	14 months	17.5±0.2	46.5±1.4	725±52	1389±69	193±6	871±44	20.48±1.78	3.75±0.46	–

*Mean values ± SEM. Probability values reflect significant differences between corresponding control and experimental groups. $p<0.01$; Number of animals per group is given in parentheses.

†$p<0.001$.

‡$p<0.01$.

Table 3. Effect of iron supplementation during infancy on myocardial fiber and capillary densities and fiber to capillary ratio in preweanling male Sprague-Dawley rats*

	Age (days)	Capillaries per mm^2	Fibers per mm^2	Fiber to Capillary Ratio	Fiber Size (μ^2)
Control (16)	20	8,673±303	8,087±321	0.96±0.04	87.7±4.9
Experimental (16)	20	9,562±347	9,600±359†	1.04±0.06	71.0±2.7†

*Mean values ± SEM. Probability values reflects significant differences between corresponding control and experimental groups. Number of animals per group is given in parentheses.
†$p < 0.001$.

RESULTS

The results of this study are presented in Tables 2, 3, and 4, and Figure 1. Our data show that the control (i.e., normal) suckling Sprague-Dawley rats decreased their hemoglobin concentrations from 16.2 to 7.9 g% within the first 20 days of life. The decline in hemoglobin content was, however, significantly less in the experimental animals, whose diet had been supplemented with ferrous sulfate.

Table 4. Effect of iron supplementation during infancy on hemoglobin concentration in early adult life in male Sprague-Dawley rats*

	Age (days)	Body Weight (g)	Hemoglobin (g%)
Control (17)	21	45±1	8.5±0.2
Experimental (18)	21	46±1	12.7±0.2†
Control (17)	34	90±2	12.5±0.3
Experimental (18)	34	92±2	12.9±0.2
Control (17)	60	328±12	17.2±0.2
Experimental (18)	60	331±14	17.0±0.1

*Mean values ± SEM. Probability value reflects significant difference between corresponding control and experimental group. Number of animals per group is given in parentheses.
†$p < 0.001$.

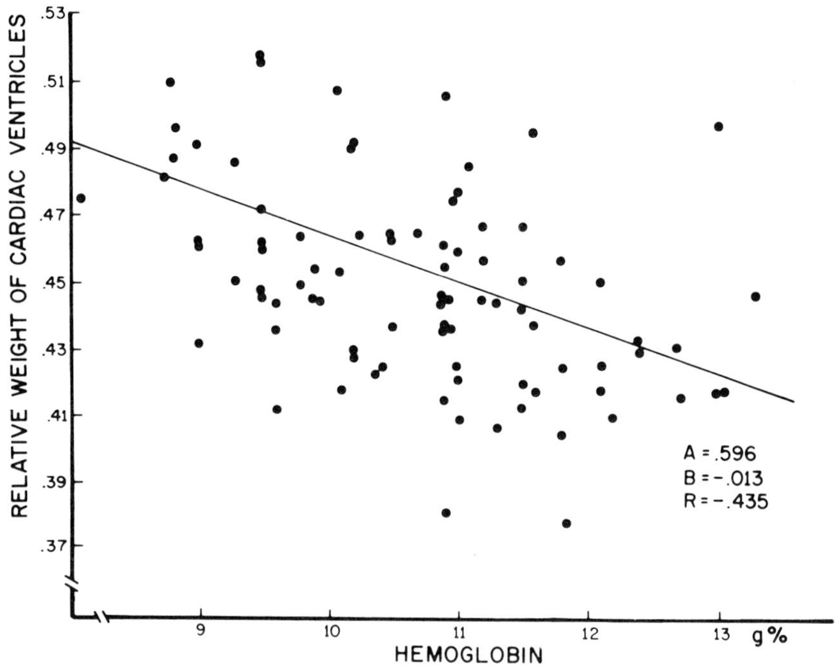

Figure 1. Relationship between the relative weight of the cardiac ventricles and hemoglobin concentration in normal 21-day-old rats. Correlation coefficient $R = -0.435$ ($p<0.01$).

Here, the hemoglobin concentrations were found to be 11.0 and 12.3 g% at 11 and 20 days of age, respectively ($p<0.001$). No significant differences in the red blood cell mass between experimental and control rats were observed in any of the older age groups studied.

We found that supplementation of the diet with ferrous sulfate did not alter the body weight significantly; and there were no differences in the weights of the adrenals and kidneys in any of the groups. Substantial changes occurred, however, with respect to heart weight and the weights of the spleen and liver in the 11-day- and 20-day-old animals. In the 11-day-old experimental rats, the weight of the liver was increased at the 0.01 level, while the 20-day-old experimental group showed an even more significant ($p<0.001$) rise in weight of the liver and spleen and a similarly pronounced ($p<0.001$) decrease in heart weight.

A study of the hearts of the 20-day-old animals revealed a significantly ($p<0.001$) smaller myocardial fiber size and a correspondingly higher fiber density (F/mm^2) in the experimental group. The fiber/capillary ratio and the capillary density, however, remained unchanged. Cardiac water content was not significantly altered (80–82%) in any of the groups.

DISCUSSION

The hemoglobin concentration of the human newborn declines progressively during early infancy owing to the following: a relatively low erythropoiesis (Gairdner *et al.*, 1952; Halvorsen, 1963; Mann *et al.*, 1965), a shortened erythrocyte survival (Pearson, 1967), and a significant hemodilution secondary to rapid expansion of blood volume and a relative iron deficiency, provided that normal milk is fed exclusively until after the first 3 months of life. Studies have shown that the "anemia" of the newborn occurs, not only in humans, but in a number of other mammalian species as well.

Using the laboratory rat as our experimental model, the aim of this work was to prevent or at least lessen the normally occurring postnatal fall in hemoglobin concentration and to study the effects this might have on the weights of certain organs and on myocardial fiber size and myocardial capillary density.

The results of this study show that gastric tube feeding of approximately 20 mg of iron/kg/day will stimulate erythropoiesis in the suckling Sprague-Dawley rat quite markedly. The hemoglobin concentrations in the 11- and 20-day old experimental animals were found to be 11.0 and 12.3 g% as compared to 9.1 and 7.9 g%, respectively, in the controls.

There is no doubt that the amounts of iron employed in this study were higher than the usual therapeutic or prophylactic doses (1–2 mg/kg/day) usually given to human infants (Sturgeon, 1958; Committee on Nutrition, 1969). On the other hand, the doses of 100 mg of iron dextran were given to premature infants over a period of 2–4 days (Hammond and Murphy, 1960). However, despite these relatively large doses, no side effects were noted in any of the treated animals. There were no changes in body weights and, on light microscopy, no evidence of excess iron deposition or fibrosis was seen, either in the liver or in cardiac tissue.

While body weights were not affected in any of the experimental rats, some of the organs in the 11- and 20-day-old treated animals did show significant changes in weight. There were slight, but significant increases in liver weight at 11 days of age and a more marked difference in this parameter of approximately +10% in the 20-day-old experimental animals. Iron supplementation in the latter age group caused an even more pronounced increase of +40% in the weight of the spleen. Since liver and spleen are both potentially blood-storing and blood-forming organs, it is probable that the increases in weight we observed reflect either or both of these functions. However, further study is necessary to clarify this point.

A surprising outcome of this study was the significant differences in heart weight between the 20-day-old, iron-supplemented and control rats. The absolute and relative heart weights were, respectively, 7 and 10% higher in the control than in the experimental animals. Quantitative morphology demonstrated in the larger control hearts a very marked increase in myocardial fiber size, a corresponding decrease in fiber density, but showed no significant changes in capillary density or fiber to capillary ratio.

Two mechanisms could have led to these results: it is possible that the relatively large doses of iron caused injury to the hearts of our experimental rats, perhaps similar to that seen in hemochromatosis, thereby giving rise to relatively smaller myocardial fibers and heart weights. However, there was no evidence of fibrosis or necrosis in these hearts, nor were there signs of any side effects elsewhere, such as changes in body weight or microscopic evidence of liver damage. It seems unlikely, therefore, that our findings were secondary to iron toxicity.

It is more reasonable to assume that the myocardial changes we observed, rather than reflecting any direct effects of iron supplementation, were caused by differences in hemoglobin concentration. We believe that the relatively larger myocardial fibers and heart weights in the control rats were due to the marked fall in hemoglobin concentration normally experienced by these animals (from 16.2 to 7.9 g%), a decline which had been much less severe in the experimental rats (16.2 to 12.3 g%). This hypothesis is also supported by the finding of an inverse linear relationship between the relative weight of the cardiac ventricles and hemoglobin concentration in normal 21-day-old rats as illustrated in Figure 1. The data from different experimental protocols in our laboratory were combined. A statistically significant ($p<0.01$) inverse linear relationship was found between the relative weight of the cardiac ventricles and the hemoglobin concentration in normal rats at the weaning period. Therefore, it appears that a decrease in relative heart weights, due to artificially increased hemoglobin concentration in suckling rats, represents an extension of this relationship into the abnormal range of hemoglobin concentrations.

In view of these findings and the fact that anemia, if severe enough, may produce cardiomegaly, i.e., *de novo* formation of cardiac tissue, we suggest that the normal decline of hemoglobin concentration in infancy serves a very definite function, namely, to stimulate myocardial growth. If this stimulus is interfered with, as was the case in the present set of experiments, a decrease in myocardial fiber size and, therefore, in heart weight will occur. Although this decrease was only approximately 10%, it was nevertheless statistically highly significant.

It is true that, in the older rat, a significant cardiac growth response, cell hypertrophy and/or hyperplasia, occurs only in severe anemias of less than 5–6 g% of hemoglobin (Neffgen and Korecky, 1972). However, it is conceivable that this may not hold true during the very early stage of life. The actively growing heart of the preweanling rat may well be more susceptible to a reduction in hematocrit and hemoglobin concentration than that of the older animal; hemoglobin levels of 7–8 g%, which would have little effect on the older heart, may significantly stimulate cardiac growth during the earlier stages of development.

One of the objectives of this study was to determine whether or not a higher hemoglobin concentration during the preweaning stage of life might lead to any permanent changes in the adult animal. Table 2 shows that body weight, heart weight, and the weights of the other organs studied were unaltered, both, in the 4½- and in the 14-month-old rat. Whether or not the histological parameters remain also unchanged will be determined in future studies. Of special interest was the failure of the experimental rats to maintain their head start in hemoglobin concentration once iron supplementation had been stopped. As shown in Table 4, as early as 2 weeks after weaning (age 34 days), the differences in hemoglobin levels between control and experimental animals had disappeared. While there had been active erythropoiesis in the control rat (a rise from 8.5 g% at age 21 days to 12.5 g% at age 34 days), no significant change in hemoglobin concentration had occurred in the experimental animals (12.7 to 12.9 g%). Identical hemoglobin levels were also obtained in the 60-day-old animals (controls, 17.2 g%; experimental rats, 17.0 g%). We believe that these data provide proof for an inborn, i.e., genetically controlled, mechanism which determines the hemoglobin concentration for any particular age. Left to its own resources, the animal with an artificially raised hemoglobin level will stop its erythropoiesis until its "proper" hemoglobin concentration has been reached.

Because of the species difference and the relatively large doses of iron used in this study, it is difficult to assess the extent to which our results apply to the human newborn. There is little doubt, however, that the decline in hemoglobin concentration during early human infancy is not unlike that observed in the suckling rat. The evidence we presented suggests that addition of iron to the infant's diet, through a premature elevation in hemoglobin concentration, may cause significant increases in the weights of the liver and spleen and a decrease in heart weight.

Although the present study did not demonstrate in the adult animal any persisting changes with respect to organ and body weights, further research is required to assess the effects of iron supplementation at the cellular and subcellular level. It would be important to determine, for example, whether or not iron supplementation in infancy is related to a higher incidence of cardiomyopathies or coronary artery disease in later life. One also needs to know whether or not it leads to any changes in the distribution of the types of hemoglobins, or in fact predisposes to some of the hemoglobinopathies. Equally important would be a study of the effects of iron supplementation on myoglobin content and iron-containing compounds such as cytochromes, peroxidase, and catalase. Until this type of knowledge has been acquired, addition of iron to the normal infant's diet as suggested by the Committee on Nutrition of the American Academy of Pediatrics (1969) is questionable. As Beal, Meyers, and McCammon (1962) pointed out about 10 years ago: "It appears that concern over possible deficiency has replaced what Schulman has referred to in this connection as 'the established principles of pediatric therapeutics: to treat on definite indication, to use the safest and cheapest therapeutic measures first, and to maintain a proper respect for normal physiology.' "

REFERENCES

BEAL, V. A., A. J. MEYERS, and R. W. McCAMMON. 1962. Iron intake, hemoglobin, and physical growth during the first two years of life. Pediatrics 30: 518.

Committee on Nutrition. 1969. Iron balance and requirements in infancy, Pediatrics 43: 134.

DAWES, G. S. 1968. Foetal and Neonatal Physiology, p. 30. Year Book Medical Publishers, Chicago.

FULTON, R. M., E. C. HUTCHINSON, and A. M. JONES. 1952. Ventricular weight in cardiac hypertrophy. Brit. Heart J. 14: 413.

GAIRDNER, D., J. MARKS, and J. D. ROSCOE. 1952. Blood formation in infancy. Part 1. The normal bone marrow. Arch. Dis. Child. 27: 128.

HALVORSEN, S. 1963. Plasma erythropoietin levels in cord blood and in blood during the first weeks of life. Acta Paediat. Scand. 52: 425.

HAMMOND, D., and A. MURPHY. 1960. The influence of exogenous iron on formation of hemoglobin in the premature infant. Pediatrics 25: 362.

MANN, D. L., M. L. SITES, R. M. DONATI, and N. I. GALLAGHER. 1965. Erythropoietic stimulating activity during the first ninety days of life. Proc. Soc. Exp. Biol. Med. 118: 212.

MOTT, J. C. 1965. Haemorrhage as a test of the function of the cardiovascular system in rabbits of different ages. J. Physiol. 181: 728.

NEFFGEN, J. F., and B. KORECKY. 1972. Cellular hyperplasia and hypertrophy in cardiomegalies induced by anemia in young and adult rats, Circulat. Res. 30: 104.

O'BRIEN, R. T., and H. A. PEARSON. 1971. Physiologic anemia of the newborn infant. J. Pediat. 79: 132.

PEARSON, H. A. 1967. Life span of the fetal red blood cell. J. Pediat. 70: 166.

RAKUŠAN, K., J. RÁDL, and O. POUPA. 1965. The distribution and content of myoglobin in the heart of the rat during postnatal development. Physiol. Bohemoslov. 14: 317.

SITTE, H. 1967. Morphometrische Untersuchungen an Zellen. *In* E. R. Weibel and H. Elias (eds.), Quantitative Methods in Morphology. Springer Verlag, Berlin, Heidelberg, New York.

STURGEON, P. 1958. Studies of iron requirements in infants and children. *In* R. O. Wallerstein and S. R. Mettier (eds.), Iron in Clinical Medicine, p. 183. University of California Press, Berkeley, Calif.

WIDDOWSON, E. M., and R. A. McCANCE. 1955. The effect of suckling anaemia on the pig's heart. Brit. J. Exp. Path. 36: 175.

Mailing address:
Dr. K. Rakušan,
Dept. of Physiology, Faculty of Medicine,
University of Ottawa,
Ottawa, Ontario K1N 6N5 (Canada).

Recent Advances in Studies on
Cardiac Structure and Metabolism Volume 10
The Metabolism of Contraction
Edited by Paul-Emile Roy and George Rona

EFFECT OF THYROPARATHYROIDECTOMY AND OF A LOW CALCIUM DIET ON THE CARDIOMYOPATHY OF THE SYRIAN HAMSTER[1]

G. JASMIN, B. SOLYMOSS,
and H. Y. EU

Département de pathologie, Université de Montréal,
Montréal, Québec, Canada

SUMMARY

Experimental Ca deficiency in the cardiomyopathic Syrian hamster was found to be effective in reducing the incidence and the severity of the heart lesions that develop at 30 days of age. The skeletal muscle changes that develop at a much earlier period were influenced in their course but not in their frequency. It was further observed that 20-day-old animals benefit better from a low Ca intake at a time when the degenerative changes do not prevail in the myocardium. Removal of the parathyroids in normally fed myopathic hamsters had no effect on the serum calcium and phosphate levels or on the course of the disease. The serum creatine phosphokinase was found to be higher in Ca-deficient myopathic hamsters. The possible role of calcium concerning the maintenance of the integrity of the muscle cells is discussed.

INTRODUCTION

The calcific myofibrillar changes occurring in the heart and the skeletal muscle of myopathic hamsters are a distinctive feature of this experimental hereditary myopathy. An early treatment with Verapamil, which interferes with calcium influx into the cells, was shown to prevent the myocardial lesions but to merely reduce the severity of the necrotic changes in the skeletal muscle (Jasmin and Bajusz, 1973, 1974). The question remains open as to whether the heart and the skeletal muscle lesions in myopathic hamsters have a common origin. Except for the differences in time of onset and duration, the degenerative changes, whether located in the heart or in the skeletal muscle, have a great deal of similarity. They are multifocal, myolytic, and necrotizing and, as mentioned earlier, undergo calcification. The similitude in the degenerative process of the myocardium

[1] Supported by grants from the Medical Research Council of Canada and the Muscular Dystrophy Association of Canada.

and the skeletal muscle does not necessarily imply that we are dealing with a pathological disorder deriving from the same genetically determined molecular defect. In fact, most adrenergic blockers or calcium antagonists which were tested in myopathic hamsters preferentially act on the heart or the skeletal muscle; such differences are also apparent in the serum creatine phosphokinase (CPK) isoenzymes studied by gel electrophoresis (Solymoss *et al.*, 1974).

The present experiments were designed to investigate the role of calcium as a possible mediator in the genesis of the hamster polymyopathy. Young animals were parathyroidectomized and/or fed a low Ca diet at the earliest stage of the disease.

MATERIALS AND METHODS

Male and female hamsters of the UM-X7.1 myopathic line (20–30 days old) were used in this study. Unrelated healthy hamsters (Lakeview strain) served as controls. Animals were maintained under identical controlled housing conditions. They were fed the regular Purina Laboratory Chow and given tap water *ad libitum* or they received a Ca-deficient diet (supplied in pellets by General Biochemical, Chagrin Falls, Ohio) and distilled water as drinking fluid. Thyroparathyroidectomy was made by blunt dissection under ether anesthesia with minimal trauma of the surrounding structures. Blood samples were obtained from the jugular vein. Serum calcium and magnesium were measured by an atomic absorption spectrophotometer (Unicam), and the Dryer's method (Dryer, Tammes, and Routh, 1957) was used for determination of phosphorus. Blood CPK was determined according to the method of Nuttal and Wedin (1966), with reagents supplied by Dade.

At autopsy, the heart and skeletal muscle specimens (diaphragm and biceps femoris) were fixed in neutral formol and processed for routine histological studies. In selected cases, some tibia were fixed in Suza solution and decalcified in sulfosalicylic acid. All paraffin sections were stained with hematoxylin-phloxine-saffron; Celestine blue and a modified cobalt substitution technique were used for demonstration of calcium in the muscle fibers (Gabe, 1968). The severity of heart and skeletal muscle lesions was assessed on an arbitrary scale of 0–3, following criteria outlined in a previous report (Jasmin and Bajusz, 1974). The means of the microscopic readings as well as the percentual occurrence of the lesions are given in Table 1.

RESULTS

Effect of 1-month Ca deprivation in 30-day-old myopathic hamsters

The results summarized in Table 1 reveal that the low Ca intake in intact or in parathyroidectomized animals significantly reduced the severity of heart and

Table 1. Effect of thyroparathyroidectomy (TPT-X) and of a low Ca intake on the development of heart and skeletal muscle lesions and serum CPK in 30-day-old myopathic hamsters

Group*	Final Body Weight† (Initial 55 ± 5.2 g)	Myocardial Lesions: Severity (grade 0–3)	Myocardial Lesions: Incidence (%)	Skeletal Muscle Lesions: Severity (grade 0–3)	Skeletal Muscle Lesions: Incidence (%)	Serum CPK (I.U.)
Intact, regular diet	86 ± 4.2	2.7 ± 0.15	100	2.1 ± 0.2	100	9,250 ± 1,220
TPT-X, regular diet	72 ± 2.1	2.8 ± 0.13	100	2.1 ± 0.10	100	7,407 ± 1,211
Intact, low Ca diet	85 ± 5.3	1.2 ± 0.17	80	1.6 ± 0.18	100	22,078 ± 3,585
TPT-X, low Ca diet	70 ± 3.6	1.1 ± 0.17	75	1.2 ± 0.15	95	16,227 ± 3,120

*Each group consisted of 20 male or female hamsters of the UM-X 7.1 myopathic line.
†All values are expressed as mean ± SE.

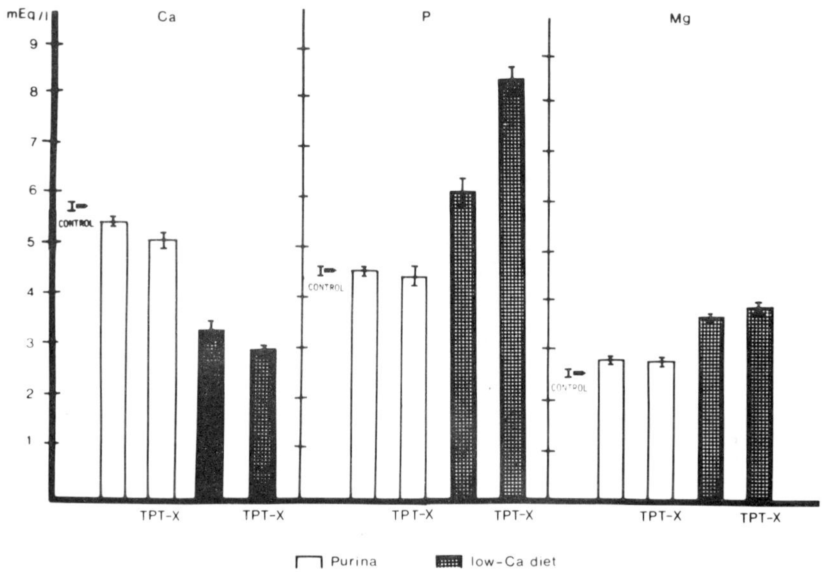

Figure 1. Effect of thyroparathyroidectomy (TPT-X) and low Ca diet on serum electrolytes of cardiomyopathic hamsters. Age of animals, 30 days; duration of experiment, 1 month.

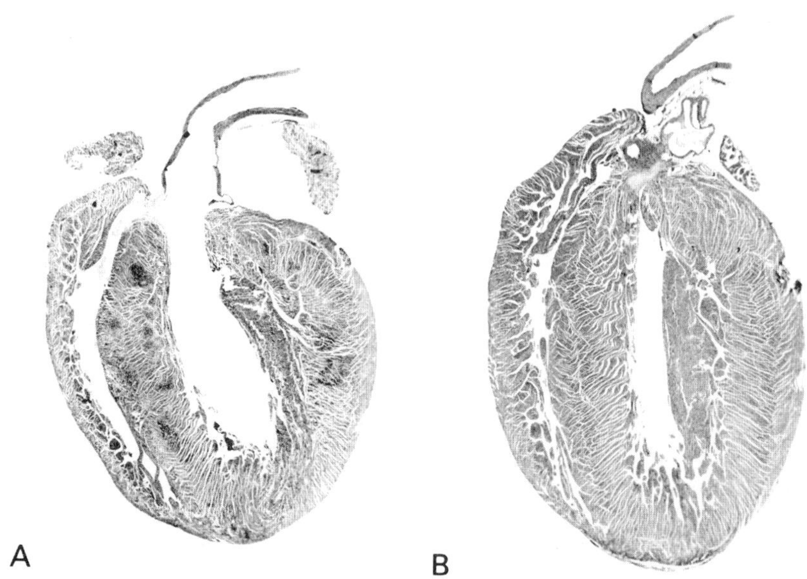

Figure 2. Heart transections of myopathic hamsters (UM-X7.1) fed: *A:* regular Purina diet; *B:* Ca-deficient diet. Dark-staining areas represent calcific myofibrillar necrotic changes which are not apparent in Ca-deprived animals. H.P.S. × 10.

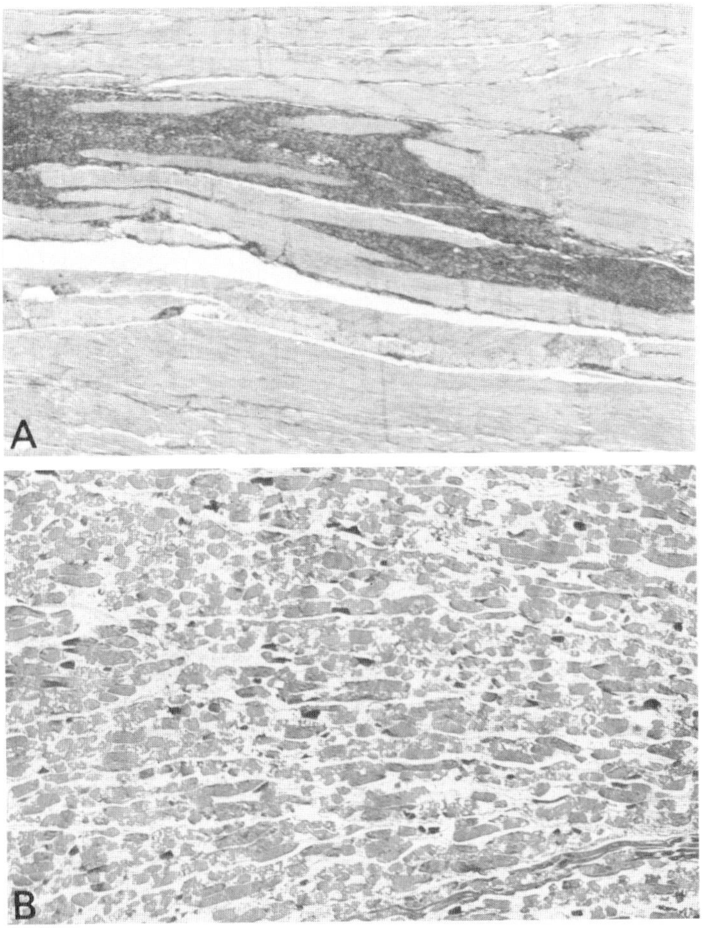

Figure 3. Extensive necrotic changes in muscle fibers of biceps femoris of myopathic hamsters fed: *A:* regular diet; *B:* Ca-deficient diet. Lesions in the latter group are essentially coagulative, with no cellular infiltration. H.P.S. × 100.

skeletal muscle lesions. The serum CPK levels, however, reached the highest values in these two groups. Thyroparathyroidectomy, on the other hand, did not afford protection in myopathic hamsters fed Purina. Changes in the serum calcium, magnesium, and phosphate values were demonstrable only in Ca-deprived animals (Fig. 1). Removal of the parathyroid glands proved to be ineffective under normal feeding conditions.

Microscopically, the necrotizing process in the heart seemed much delayed in comparison to animals given a regular diet. More often, the lesions were limited

Table 2. Cardiac and skeletal muscle changes and serum CPK levels 4½ months after thyroparathyroidectomy (TPT-X) and low Ca intake in 30-day-old myopathic hamsters

Group*	Final Body Weight† (Initial 53 ± 2.7 g)	Myocardial Lesions		Animals with Cardiac Hypertrophy‡ (%)	Skeletal Muscle Lesions		Serum CPK (I.U.)
		Severity (grade 0–3)	Incidence (%)		Severity (grade 0–3)	Incidence (%)	
Intact, regular diet	124 ± 8.3	2.9 ± 0.05	100	40	1.6 ± 0.16	100	6,270 ± 550
Intact, low Ca diet	67 ± 6.5	0.8 ± 0.06	100	0	0.9 ± 0.16	92	1,360 ± 127
TPT-X, low Ca diet	61 ± 9.2	1.0 ± 0.13	89	0	1.2 ± 0.17	81	2,188 ± 485

*Each group consisted of 15 male or female hamsters of the UM-X 7.1 myopathic line.
†All values are expressed as mean ± SE.
‡Calculated from the mean heart weight expected in a healthy hamster.

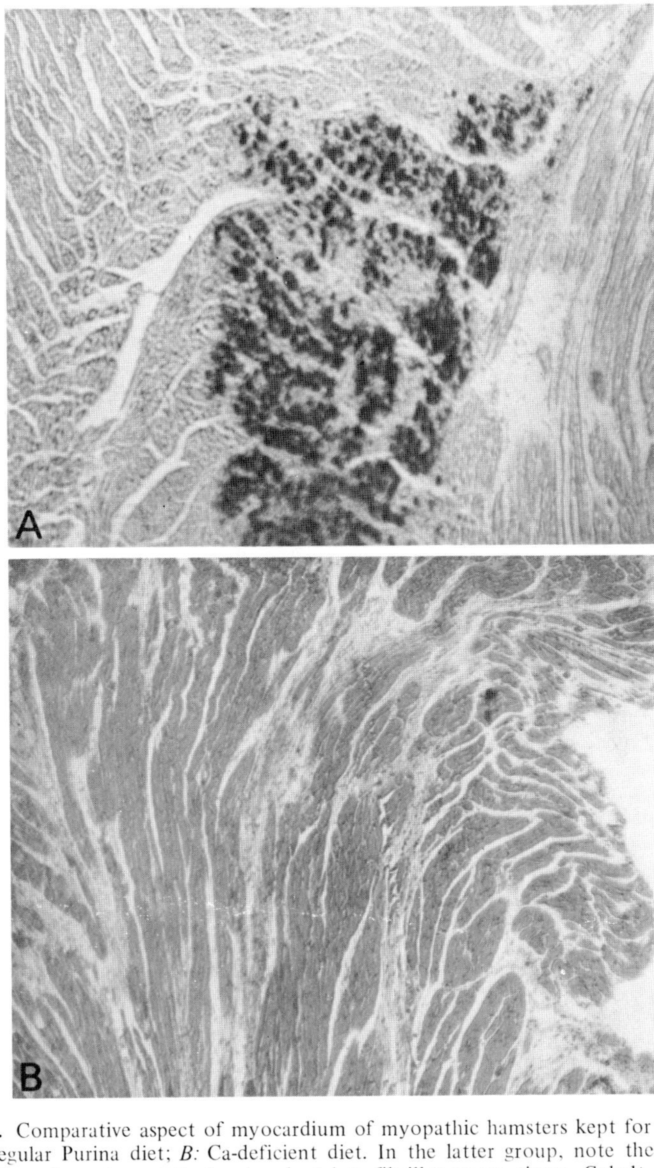

Figure 4. Comparative aspect of myocardium of myopathic hamsters kept for 4½ months on: *A:* regular Purina diet; *B:* Ca-deficient diet. In the latter group, note the absence of calcific deposits and a relatively abundant interfibrillar scarry tissue. Cobalt substitution staining method. × 40.

in size and amount, consisting of myocytolysis with collapsed endomysium and tiny scar tissue formation; hyalinization was minimal and there were relatively few calcific deposits (Fig. 2).

The skeletal muscle lesions also exhibited a different pattern from that observed in normally fed controls. The most peculiar finding consisted of a patchy coagulation necrosis of muscle fibers in the absence of reactive phenomena, i.e., cellular infiltration or proliferation of sarcolemma nuclei (Fig. 3). Such nonproductive lesions contrasted with those occurring in normally fed animals.

Effect of 4½-month Ca deprivation in 30-day-old hamsters

In a second series of experiments, the effect of thyroparathyroidectomy and of a low Ca intake was further investigated in 30-day-old myopathic hamsters during a longer period, with the object of preventing the congestive heart failure that usually occurs after 4 months.

As shown in Table 2, the prolonged Ca deficiency influenced the course and the progression of the cardiomyopathy as judged by a relatively low necrotic index in comparison to that seen in animals given a regular diet. In fact, the degenerative changes, in spite of the longer duration of the experiment, were no

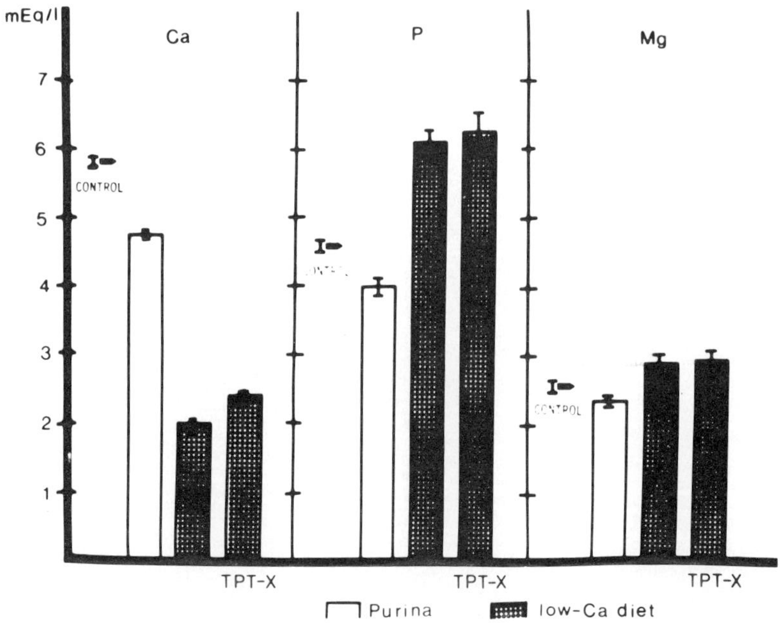

Figure 5. Effect of thyroparathyroidectomy (TPT-X) and low Ca diet on serum electrolytes of cardiomyopathic hamsters. Age of animals, 30 days; duration of experiment, 4½ months.

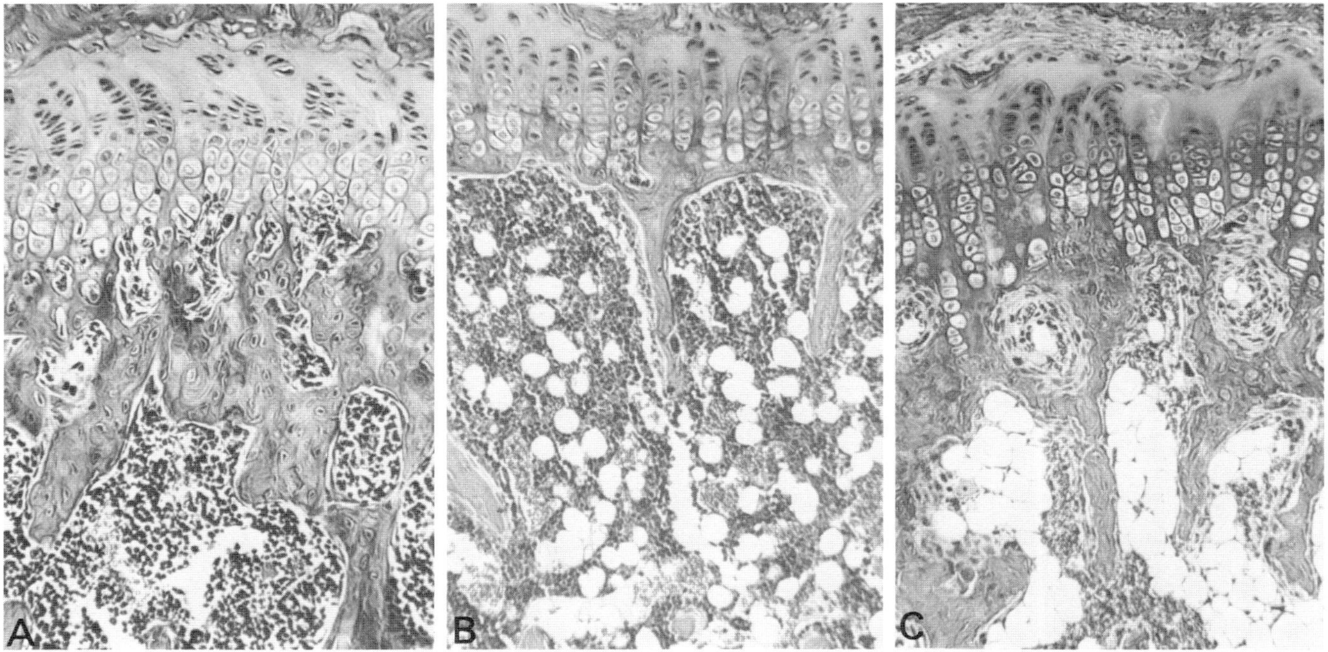

Figure 6. Transections of tibia of myopathic hamsters kept for 4½ months on: *A:* regular Purina diet; *B:* Ca-deficient diet (plus thyroparathyroidectomy); *C:* Ca-deficient diet (intact animals). Note hypoplasia of bone trabeculi and of marrow in the two Ca-deficient groups. Intact animals show, in addition, osteoblastic-osteoclastic activity. H.P.S. × 125.

more extensive than those already found in animals fed the Ca-deficient diet for 30 days. The scarring process in the myocardium was prominent, with no evidence of recent myocytolysis (Fig. 4). Insofar as the skeletal muscle is concerned, the degenerative process, although less severe, was still demonstrable, with little or no hyalinization of the myofibers as expected in the chronic phase of the disease. While 40% of Purina-fed myopathic hamsters showed an enlarged heart and liver, no conclusions could be drawn regarding the potential effect of the Ca-deficient diet on the survival rate because most animals were in a state of demineralization and catabolism. As shown in Figure 5, the blood calcium values remained at low levels in these particular groups.

Histological examination of the tibia and the femur in the latter showed hypoplasia of the subperiosteal bone and of the trabeculae of the metaphysis, and hypocellularity of marrow elements; the resorption, however, with osteoblastic and osteoclastic activities was evident only in nonthyroparathyroidectomized Ca-depleted hamsters (Fig. 6). There was also a sharp drop in the serum CPK levels, which correlates with defervescence of the lesions. It is worth mentioning that Ca repletion with a regular diet did not reactivate the cardiomyopathy.

Effect of 1-month Ca deprivation in 20-day-old hamsters

Knowing that the skeletal muscle lesions in myopathic hamsters develop prior to 30 days of age, a similar experimental arrangement was carried out in 20-day-old animals at a time when they were not yet weaned. This enabled us to observe that Ca deficiency markedly lowered the severity of the cardiomyopathy, with a frequency as low as 20% (Table 3).

Skeletal muscle lesions, however, were found in all these animals but were definitely less severe than in the group fed a regular diet. It should also be noted that blood CPK values were particularly high in Ca-deficient hamsters. Microscopic examinations of the heart revealed that the degenerative changes were minimal, consisting of a few myolytic foci. The skeletal muscle lesions were more evident, with multiple foci of degenerated myofibers infiltrated by phagocytes. Serum calcium, magnesium, and phosphate readings were substantially the same as those recorded in Ca-deprived 30-day-old myopathic hamsters (Fig. 7).

DISCUSSION

The present results provide evidence that the course of the hamster cardiomyopathy is modified in animals rendered hypocalcemic by dietary means. This is best achieved in 20-day-old myopathic hamsters at a time when degenerative changes do not prevail in the myocardium. The same does not apply to skeletal muscle, which is liable to show lesions at 10 days of age. It is as yet difficult to relate the sharp rise in serum CPK values with the relative diminution of the necrotic changes observed in myopathic hamsters fed a low Ca diet.

Table 3. Effect of a low Ca intake on the development of heart and skeletal muscle lesions and serum CPK in 20-day-old myopathic hamsters

Group*	Final Body Weight† (Initial 37 ± 1.5 g)	Myocardial Lesions Severity (grade 0–3)	Myocardial Lesions Incidence (%)	Skeletal Muscle Lesions Severity (grade 0–3)	Skeletal Muscle Lesions Incidence (%)	Serum CPK (I.U.)
Regular diet	74 ± 2.4	2.1 ± 0.31	100	1.9 ± 0.25	100	12,900±2,070
Low Ca diet	64 ± 4.4	0.2 ± 0.13	25	1.4 ± 0.16	100	12,900 ± 2,070

*Each group consisted of 12 male or female hamsters of the UM-X 7.1 myopathic line.
†All values are expressed as mean ± SE.

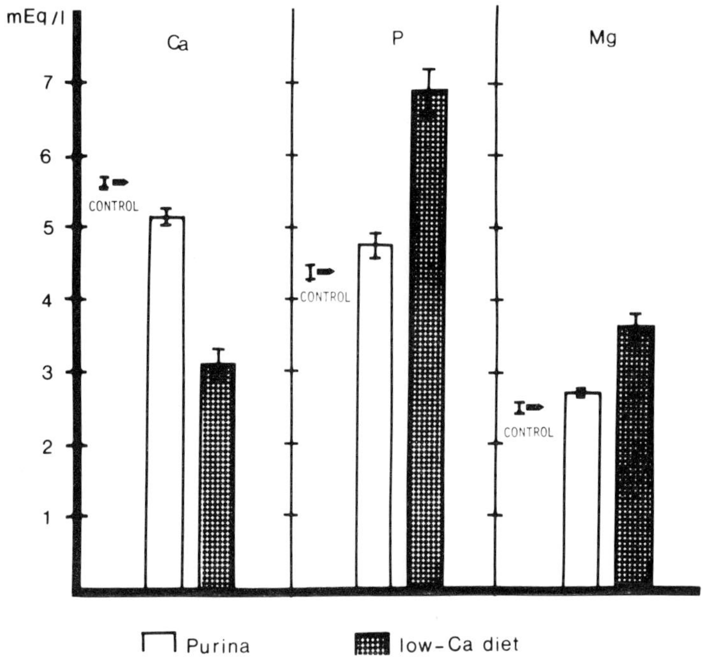

Figure 7. Effect of low Ca diet on serum electrolytes of cardiomyopathic hamsters. Age of animals, 20 days; duration of experiment, 1 month.

The fact that serum calcium and phosphate levels remained unchanged after removal of the parathyroid gland in normally fed myopathic hamsters deserves further investigation. Preliminary findings would indicate that unrelated healthy hamsters do not react similarly; if so, this would imply that an abnormal calcium metabolism plays an important role in the pathomechanism of this hereditary muscle disease. Unfortunately, there are very few data in the literature concerning the physiology and the secretory activity of the parathyroid gland in the golden hamster (Hoffman, Robinson, and Magalhaes, 1968). It should also be borne in mind that, for technical reasons, our experimental animals were equally deprived of their thyroid gland. The beneficial effect of Ca deficiency, however, was equally demonstrable in nonoperated myopathic hamsters.

It is apparent that a decrease in serum calcium level resulted in slowing down the progression of the lesions in myopathic hamsters, possibly by interfering with the mechanism responsible for hyalinization of the myofibers. It is not unlikely that hyaline changes can only develop within certain limits of serum calcium level, thus supporting the concept that the rate of penetration of calcium ions into injured muscle cells may be considered as a determinant factor in the progression of the disease.

Recent ultrastructural studies carried out at early stages of the polymyopathy have led us to believe that changes in the plasma membrane constitute the primary defect and that subsequent calcification of the inner cell structure is related to some alteration of the membrane (Eu and Jasmin, unpublished). Equally pertinent to this reasoning is the fact that Verapamil, which acts on the transmembrane influx of calcium, was found to be particularly effective—especially when administered at early stages of the disease—in preventing the cardiomyopathy and slowing down the kinetics of the skeletal muscle lesions. If this interpretation is correct, it is very likely that Ca deficiency, in hindering the degenerative process of the muscle fibers, favors at the same time a greater liberation of CPK into the blood. Much to our surprise, Ca deficiency also caused a rise in serum CPK in healthy hamsters, suggesting that this experimental condition labilizes the plasma membrane with resulting greater efflux of the muscle enzymes. The present observations, we believe, have served to emphasize the importance of factors regulating the calcium metabolism for the maintenance of the integrity of the muscle cells.

REFERENCES

DRYER, R. L., A. R. TAMMES, and J. L. ROUTH. 1957. The determination of phosphorus and phosphatase with *N*-phenyl-*p*-phenylenediamine. J. Biol. Chem. 225: 177–179.

GABE, M. 1968. Technique de cobalt nitrate pour calcium, p.p. 317–320. *In* Techniques histologiques. Masson, Paris.

HOFFMAN, R. A., P. F. ROBINSON, and H. MAGALHAES. 1968. The Golden Hamster. Its Biology and Use in Medical Research. The Iowa State University Press Ames, Iowa.

JASMIN, G., and E. BAJUSZ. 1973. Polymyopathie et cardiomyopathie héréditaire chez le hamster de Syrie. Inhibition sélective des lésions du myocarde. Ann. Anat. Pathol. 18 (1): 49–66.

JASMIN, G., and E. BAJUSZ. 1975. Prevention of myocardial degeneration in hamsters with hereditary cardiomyopathy. *In* A. Fleckenstein and G. Rona (eds.), Recent Advances in Studies on Cardiac Structure and Metabolism, Vol. 6: Pathophysiology and Morphology of Myocardial Cell Alternations, pp. 219–229. University Park Press, Baltimore.

NUTTAL, F. Q., and D. S. WEDIN. 1966. A simple rapid colorimetric method for determination of creatine kinase activity. J. Lab. Clin. Med. 68: 324–332.

SOLYMOSS, B., R. HABIB, L. GIROUX, and G. JASMIN. 1974. Reversible increases in serum creatine phosphokinase (CPK) aldolase (Ald) and lactic acid dehydrogenase (LDH) levels of myopathic hamsters produced by fasting or hydrocortisone. Proc. Canad. Fed. Biol. Soc. 17: 169.

Mailing address:
Dr. G. Jasmin,
Université de Montréal,
Montréal, Québec (Canada).

Recent Advances in Studies on
Cardiac Structure and Metabolism Volume 10
The Metabolism of Contraction
Edited by Paul-Emile Roy and George Rona

VESICULAR MYOCARDIAL CHANGE: AN ULTRASTRUCTURAL STUDY OF DOUBLE MEMBRANE-BOUND VESICLES IN THE HUMAN HEART[1]

L. EDUARDO VEGA, WILLIAM A. HAWK,
EARL K. SHIREY, and D. GARY OSBORNE

Department of Surgical Pathology and Department of Cardiovascular Disease and Cardiac Laboratories, The Cleveland Clinic Foundation and The Cleveland Clinic Educational Foundation, Cleveland, Ohio, USA

SUMMARY

Vesicular myocardial change is a specific and common finding in the diseased hearts of humans and experimental animals. The small to large vesicles are generally bound by a double membrane. They are formed within myocardial cells and then possibly extruded into the extracellular space where they disintegrate or are phagocytosed by mononuclear cells. On the basis of our studies, most vesicles appear to be of mitochondrial origin. In humans, vesicular myocardial change appears to be most extensive in primary cardiomyopathy. It may also be present, but in lesser degree, in the apparently normal hearts of human adults. Vesicular myocardial change probably represents a specific mechanism by which myocardial cells eliminate damaged mitochondria or other undesirable elements that build within them as a result of disease, aging, and perhaps normal physiological activity.

INTRODUCTION

During our electron microscopic studies on human heart biopsies, we became increasingly aware of the presence of small to large membrane-bound vesicles within myocardial fibers. Concentrating attention on this phenomenon, we have been able to find these structures in practically all these biopsies, have studied in detail their possible origin and fate, and theorized on their possible significance. This vesicular change, seen and mentioned by other authors, has never been described in detail prior to our studies (Vega *et al.*, 1971).

Unfortunately, we have to concur with Nicolas and Bouhour (1970) that severe postmortem changes of rapid onset preclude the use of heart tissue obtained at autopsy for adequate electron microscopic studies. Myocardial autolysis even mimics the changes produced by ischemia (Heggtveit, 1969; Herdson, Kaltenbach, and Jennings, 1969). Hibbs and Black (1963), however,

[1] Supported in part by Grant CRP 190, from The Cleveland Clinic Foundation.

suggest that these postmortem changes are predictable and thus can be taken into account and excluded in a careful ultrastructural study, but this has not been our experience. Striking changes can be recognized in autopsy heart material by electron microscopy, but a systematic study of all the components of the intracellular and the extracellular space cannot be made in the presence of autolytic changes. This is especially true of mitochondria, nuclei, and the sarcotubular system. We believe that the best way to make an adequate ultrastructural study of the diseased human heart is with the use of fresh, instantly fixed, myocardial biopsy material. Percutaneous needle biopsy (Sutton and Sutton, 1960; Bercu *et al.*, 1964; Alexander, 1966, 1967; Van Noorden, Olsen, and Pearse, 1971; Shirey *et al.*, 1972; Ferrans *et al.*, 1973) and transvenous intracardiac biopsy (Konno and Sakakibara, 1963; Bulloch, Murphy, and Pearce, 1965, 1970; Sakakibara *et al.*, 1970; Van Noorden *et al.*, 1971; Bulloch *et al.*, 1972; Caves *et al.*, 1973; Ferrans *et al.*, 1973) generally provide adequate myocardial tissue for light and electron microscopy.

MATERIALS AND METHODS

At The Cleveland Clinic Foundation, percutaneous needle biopsies of the left ventricle have been performed on patients with diagnoses of primary myocardial disease, atherosclerotic coronary artery disease, and acquired valvular disease. These diagnoses are based on cardiac catheterization and selective cine cardioangiographic findings. Biopsies have also been performed on a smaller group of patients in whom coronary atherosclerosis was suspected but who had normal coronary arteriograms and left ventriculograms (Shirey *et al.*, 1972). A total of 100 biopsy specimens from 99 patients have been studied in detail with the electron microscope (Table 1). They form the basis of this study. Some of these 99 patients are part of a well studied group of 198 patients on whom 254 biopsies have been performed (Shirey *et al.*, 1972).

The biopsy is performed with a Vim-Silverman needle, employing brief intravenous sodium methohexital (Brevital Sodium) anesthesia and continuous electrocardiographic and x-ray monitoring. The lateral aspect of the cardiac apex is approached through the left chest wall and one or more myocardial segments are obtained (Shirey *et al.*, 1972). Most myocardial specimens average 0.5–1 cm in length and 1–2 mm in diameter.

One segment is immediately fixed in Zenker's solution for light microscopy and histochemical studies. Stains routinely used are hematoxylin eosin, Masson's trichrome, PAS with and without digestion, toluidine blue, Verhoff's elastica, and crystal violet.

For electron microscopy, another segment is placed, immediately after biopsy, in 3.0 or 6.5% glutaraldehyde for 2 hours (Sabatini, Bensch, and Barrnett, 1963), stored in a sodium cacodylate buffer, postfixed in 1.0% osmium tetrox-

Table 1. Correlation between clinical impression and vesicular myocardial change

Clinical Diagnosis	No. of Cases	Sex		Age (yrs)	Vesicular Change		
		F	M		Mild	Moderate	Extensive
Primary cardiomyopathy (PC)*	54	26	28	15–63	27	22	4
PC, questionable	13	4	9	29–58	6	4	3
PC associated with significant alcohol consumption	6	–	6	42–62	3	2	1
PC and mild atherosclerotic coronary artery disease†	5	–	5	45–59	4	1	–
Atherosclerotic coronary* artery disease	8	1	7	38–62	6	1	–
Rheumatic valvular disease	7	3	4	33–61	2	5	–
Idiopathic hypertrophic subaortic stenosis	1	–	1	40	–	1	–
Myocarditis	1	–	1	28	–	1	–
Normal hearts	5	5	–	37–45	5	–	–
Totals	100	39	61	15–63	53	37	8

*One case each showed no vesicle formation.
†The mild atherosclerotic changes in these cases are considered to be incidental findings in these cases of PC.

ide for 1 hour (Millonig, 1961), dehydrated in graded ethyl alcohols, and embedded in epoxy resin (Spurr, 1969). Sections 1 μ in thickness are cut with glass knives and stained with a polychromatic stain (Spurlock, Skinner, and Kattine, 1966). Suitable blocks are selected and cut with glass or diamond knives. These sections are stained with uranyl acetate and lead citrate, and examined with an RCA EMU-4 electron microscope.

Special procedures occasionally employed for electron microscopy are the acid phosphatase staining modified after Gomori (Lillie, 1965), and digestion with diastase. For the latter procedure, fragments of myocardium are fixed for 2 hours in glutaraldehyde, placed in a sodium cacodylate buffer, then in 0.5% malt diastase for 1½ hours, and postfixed in osmium tetroxide for 1 hour.

RESULTS

Light microscopy

The diagnoses of the 100 myocardial needle biopsies used in this study (Table 2) are as follows: no significant change, 29 cases; interstitial fibrosis and hypertrophy, 24 cases (3 of these also had a small focus of active myocarditis); interstitial fibrosis, 11 cases; hypertrophy, 6 cases; basophilic myocardial degeneration, 6 cases; interstitial fibrosis, hypertrophy, and basophilic degeneration, 4 cases; hypertrophy and basophilic degeneration, 3 cases; old myocardial scars, 3 cases; interstitial fibrosis and basophilic degeneration, 2 cases; amyloidosis, 2 cases; rheumatic myocarditis, 2 cases; mild interstitial myocarditis, 2 cases. There was 1 case each of arteriolar sclerosis, fatty ingrowth into myocardial segment, hypertrophy with organized thrombus, equivocal changes, and 1 case of severe fibrosis, hypertrophy, and vacuolar changes; 1 specimen was lost in processing.

The interstitial fibrosis ranged from rare foci of collagen deposition to severe and diffuse fibrosis of the interstitial space. The hypertrophy ranged from cases in which only a small number of myofibers were affected to cases in which all the myocardial fibers showed considerable hypertrophic changes. Basophilic degeneration, when present, generally involved one or two fibers, but in some cases six or more showed this change. In all, interstitial fibrosis was present in 43 of the cases, hypertrophy in 38, and basophilic degeneration in 15.

Electron microscopy

At the ultrastructural level, findings ranged from practically no changes to severe and diffuse myocardial alterations. Myocardial fibers have shown mild to severe decrease in the number of myofibrils and disarrangement and interruptions of these myofibrils; abnormally thick Z-bands; mild to severe increase in the amount of mitochondria, glycogen particles, lipofuscin granules, and lipid droplets; dilations of the transverse tubular system, the endoplasmic reticulum, and

Table 2. Correlation between light microscopic changes and vesicular myocardial change

Light Microscopic Diagnosis	No. of Cases	Vesicular Change		
		Mild	Moderate	Extensive
No significant change	29	18 (62%)	8 (27%)	3 (11%)
Fibrosis and hypertrophy	24	13 (54%)	8 (33%)	3 (13%)
Fibrosis*	11	6 (55%)	4 (36%)	–
Hypertrophy	6	5 (83%)	1 (17%)	–
Basophilic degeneration	6	3 (50%)	2 (33%)	1 (17%)
Fibrosis, hypertrophy, and basophilic degeneration	4	1 (25%)	2 (50%)	1 (25%)
Hypertrophy and basophilic degeneration	3	2 (67%)	1 (33%)	–
Old myocardial scar*	3	2 (67%)	–	–
Fibrosis and basophilic degeneration	2	2 (100%)	–	–
Amyloid	2	2 (100%)	–	–
Rheumatic myocarditis	2	1 (50%)	1 (50%)	–
Mild interstitial myocarditis	2	–	2 (100%)	–
Fibrosis, hypertrophy, and vacuolar degeneration	1	1 (100%)	–	–
Hypertrophy and thrombus	1	–	1 (100%)	–
Arteriolar sclerosis	1	–	1 (100%)	–
Equivocal changes	1	–	1 (100%)	–
Fatty ingrowth in myocardium	1	–	1 (100%)	–
Specimen lost in processing	1	–	1 (100%)	–
Totals	100	56	34	8

*One case each showed no vesicle formation.

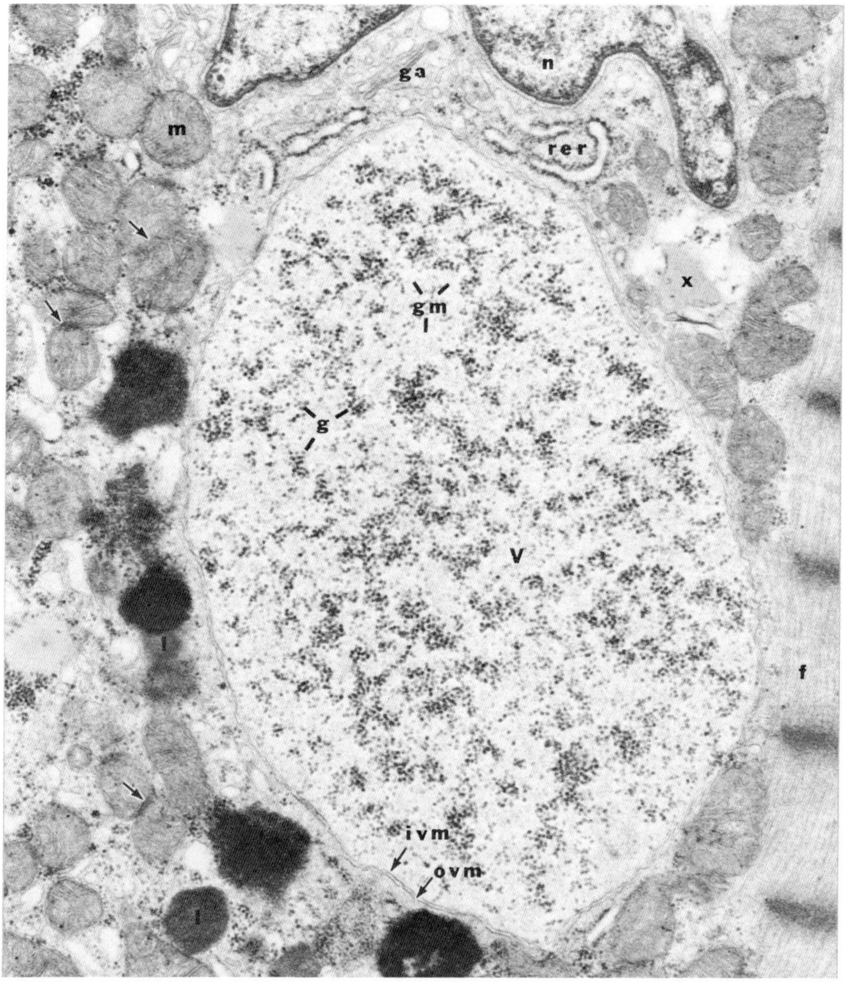

Figure 1. Typical intracellular vesicle (V) surrounded by a continuous double electron-dense membrane. The inner vesicular membrane (ivm) and the outer vesicular membrane (ovm) range from 67 to 100 Å in thickness. The intermembranous space ranges from 135 to 168 Å in width. The vesicle contains glycogen granules (g) and a very thin granular material (gm). Some of the surrounding mitochondria (m) show focal areas of apparent peripheral membrane rarefaction, especially in areas where these organelles are in contact (*arrows*). n, nucleus; ga, Golgi apparatus; rer, rough endoplasmic reticulum; l, lysosomes; x, lipid; f. myofibrils. × 29,800. From a 54-year-old man with possible primary cardiomyopathy.

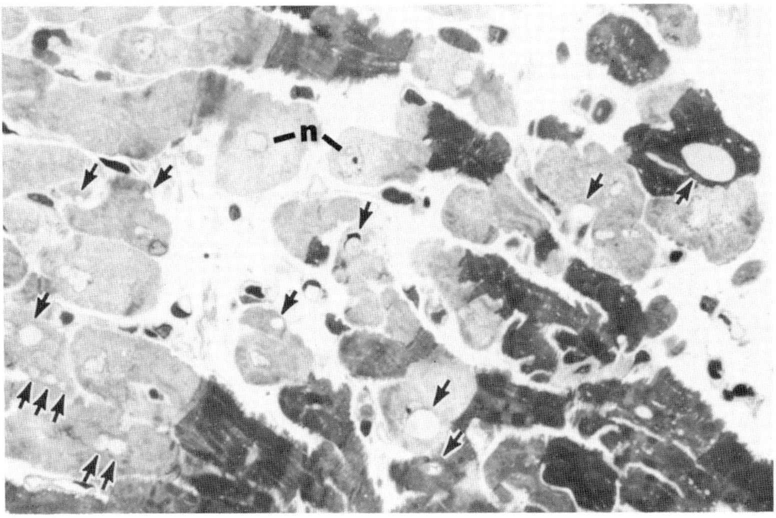

Figure 2. Epoxy-embedded section 1 μ thick, showing vesicles (*arrows*) in many myocardial cells. Note variation in vesicle size. The darker-staining fibers are contracted. n, nuclei. Polychromatic stain. × 400. From a 34-year-old man with primary myocardial disease; same case as Figures 6 and 10.

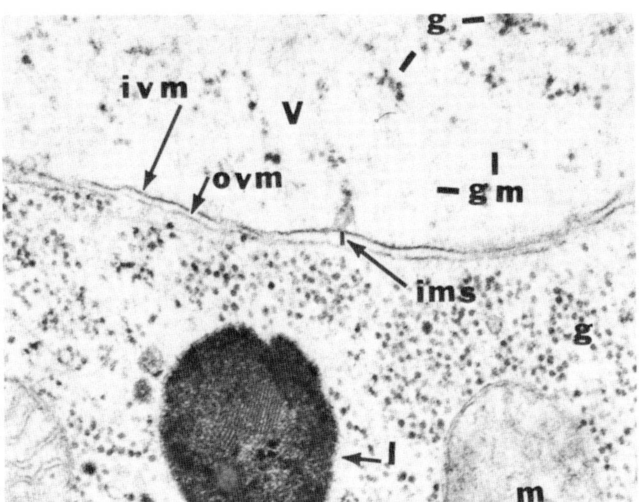

Figure 3. The inner vesicular membrane (ivm) and the outer vesicular membrane (ovm) of this intracellular vesicle (V), each measures 80–95 Å in thickness and the intermembranous space (ims) measures 160–350 Å in width. Scattered glycogen granules (g) are present in the vesicle that also contains a loose thin granular material (gm). Note the crystalline structures inside the lysosome (l), a not uncommon finding in our myocardial biopsies. One of the mitochondria appears abnormal (m). × 38,000. From a 45-year-old woman with primary cardiomyopathy.

the intercalated discs; many different forms of mitochondrial abnormalities including megamitochondria and alpha glycogen inclusions; distorted nuclei; mild to severe vesicular changes; "granular body" production (Vega *et al.*, 1974), myelin figures and other abnormal structures; and small and large outpouchings of the sarcolemma. Basophilic fiber degeneration and all degrees of interstitial fibrosis and hypertrophy have also been seen. Most biopsies show significant degrees of the above-mentioned alterations; 13 cases have shown only minimal changes.

Intracellular vesicles

Vesicular myocardial change has a striking and specific intracellular appearance. The affected myofibers contain one or more well defined vesicular structures (Fig. 1) which vary considerably in size, ranging from 1.3 to 15.5 μ in diameter but usually measuring 2–6 μ. (We measured 100 randomly selected typical intracellular vesicles from 24 cases.) Despite their size, we have not been able to identify these vesicles in paraffin-embedded sections; the larger ones are easily seen in 1-μ thick, epoxy-embedded sections (Fig. 2). Although their shapes also vary considerably, most are oval or spherical.

The majority of vesicles are totally enveloped by a double and well defined, uniform, electron-dense membrane which contains an equally uniform electron-lucent space (Figs. 1 and 3). Each of the membranous components usually measures 50 to 100 Å in thickness, ranging from 33 to 121 Å. The intermembranous space usually measures 74–170 Å in width, with a range of 67–350 Å. The appearance and measurements of these membranes are similar to those of myocardial mitochondria (Table 3). In our experience, the inner and outer components of myocardial mitochondrial peripheral membranes range from 33 to 110 Å in thickness and usually measure 47–69 Å. Their intermembranous space ranges from 49 to 110 Å and usually measures 63–79 Å. (From 15 cases, we measured the peripheral membranes and the intermembranous space of 20 typical intracellular double membrane-bound vesicles, and of many normal appearing mitochondria within the same myocardial fibers.) The intermem-

Table 3. Comparative measurements (range) of vesicular and mitochondrial membranes

	Vesicles (Å)	Mitochondria (Å)
Peripheral membranes*	33–121	33–110
Intermembranous space	67–350	49–110

*Both the inner and outer components of the peripheral membranes measure approximately the same.

branous space of the vesicles may show considerably more variation, since frequently the inner membrane may appear partially or totally collapsed towards the center of the vesicle (Figs. 8 and 25). This results in a concentration of vesicular contents and a large and irregular intermembranous space which generally contains disorganized membranous structures. We do not know whether this collapse is a natural phenomenon or artefactual shrinkage due to fixation.

Many vesicles are only partially membrane-bound (Figs. 4, 12, and 13) and, at times, accumulations of non-membrane-bound glycogen and granular material are present which simulate vesicles in that they displace other cellular organelles (Fig. 5).

In many instances, the vesicles are totally surrounded by mitochondria (Figs.

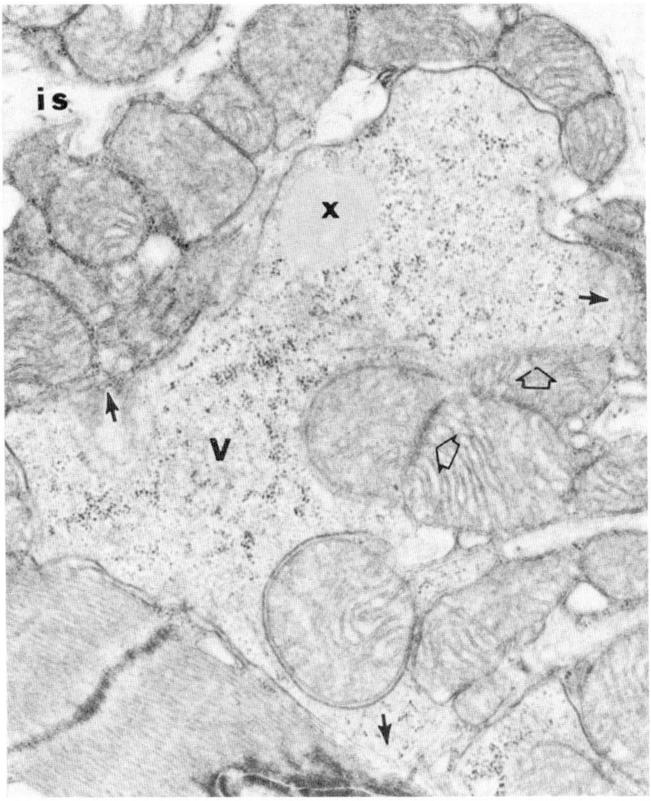

Figure 4. Partially membrane-bound vesicle (V). Note a well defined double membrane in some areas but none in others (*small arrows*). Some of the surrounding mitochondria show focal areas of apparent peripheral membrane rarefaction (*large arrows*). X, lipid; is, interstitial space. × 18,900. From a 46-year-old man with primary cardiomyopathy; same case as Figure 13.

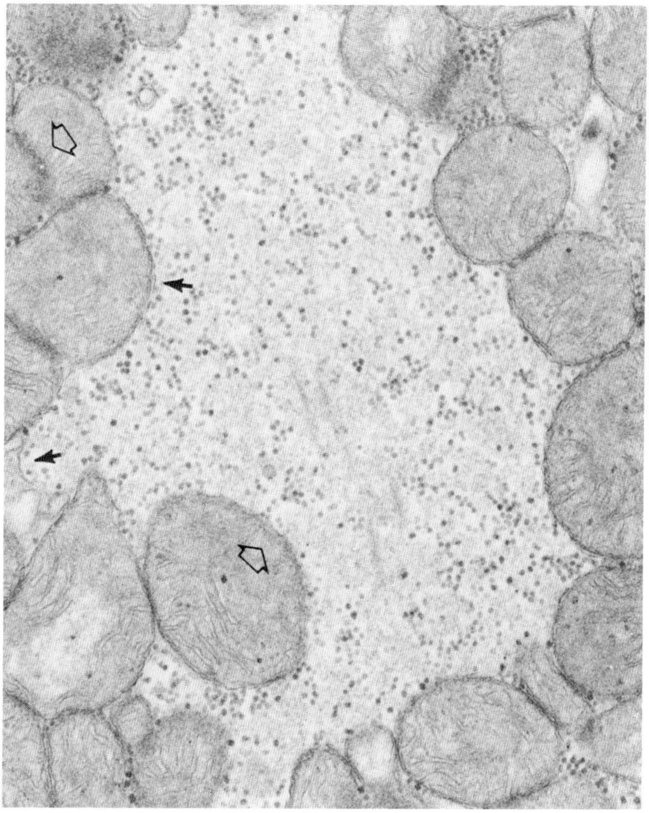

Figure 5. Vesicle-like structure that is not membrane-bound. It is totally surrounded by mitochondria, some of which show apparent peripheral membrane rarefaction (*large arrows*). Suggestions of a single vesicular membrane are present in some areas (*small arrows*). Note that the structure contains granular material and glycogen particles. × 29,800. From a 50-year-old man with primary cardiomyopathy.

4, 7, 16, and 17), but they may also be found in a "sea" of glycogen and granular material. Less commonly they may be partially or totally surrounded by myofibrils (Figs. 6, 12–14, and 16). In many cases, the surrounding mitochondria show, on the side adjacent to the vesicle, areas suggestive of peripheral membrane rarefaction (Figs. 4, 6, and 11). In other instances, these portions of the organelles appear destroyed, and the residual mitochondrial membranes appear as part of the vesicular membrane with the mitochondrial contents intruded into the vesicle (Figs. 6 and 24). In myocardial fibers which show vesicular change, it is also common to find that mitochondria possess zones of

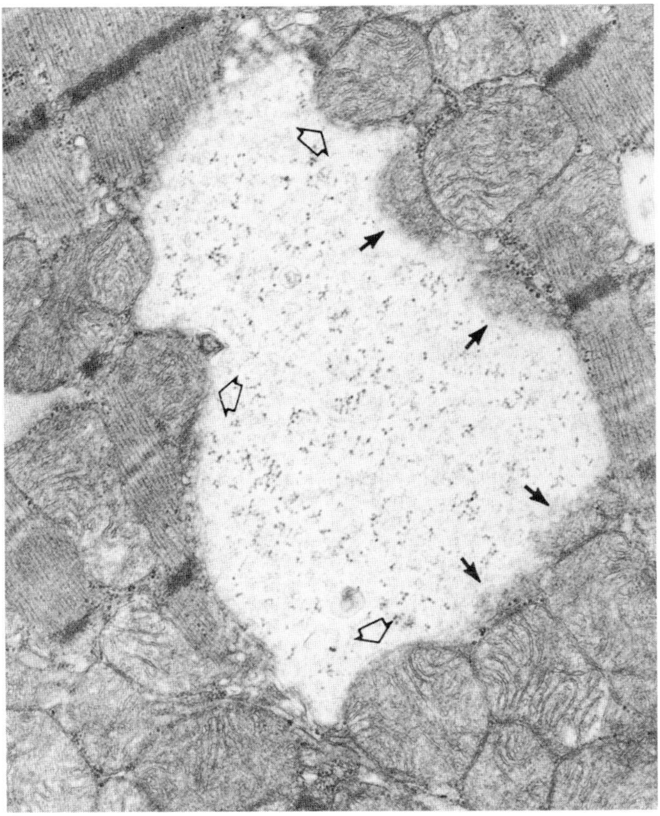

Figure 6. Intracellular vesicular structure. Some of the mitochondria appear to be dissolving into the vesicle (*small arrows*). Their cristae are still identifiable. Some other mitochondria show areas of apparent peripheral membrane rarefaction at the side of the vesicle (*large arrows*). × 20,250. From a 34-year-old man with primary cardiomyopathy; same case as Figures 2 and 10.

apparent peripheral membrane rarefaction, especially in the areas where these organelles are in contact with each other, suggesting early fusion (Figs. 1, 4, 7, 11, 24, 25, and 27). These last changes, though, could possibly be artefactual. In rare instances, formed vesicles appear to coalesce to produce even larger structures (Figs. 7 and 16).

The vesicles may be found in any area of a myofiber. They do not appear to be more abundant in any special location within the cellular architecture. They may be found close to the nucleus (Fig. 8), near the intercalated disc (Figs. 9 and 17), in the subsarcolemmal space (Figs. 8, 17, and 23–25), in small or large out-

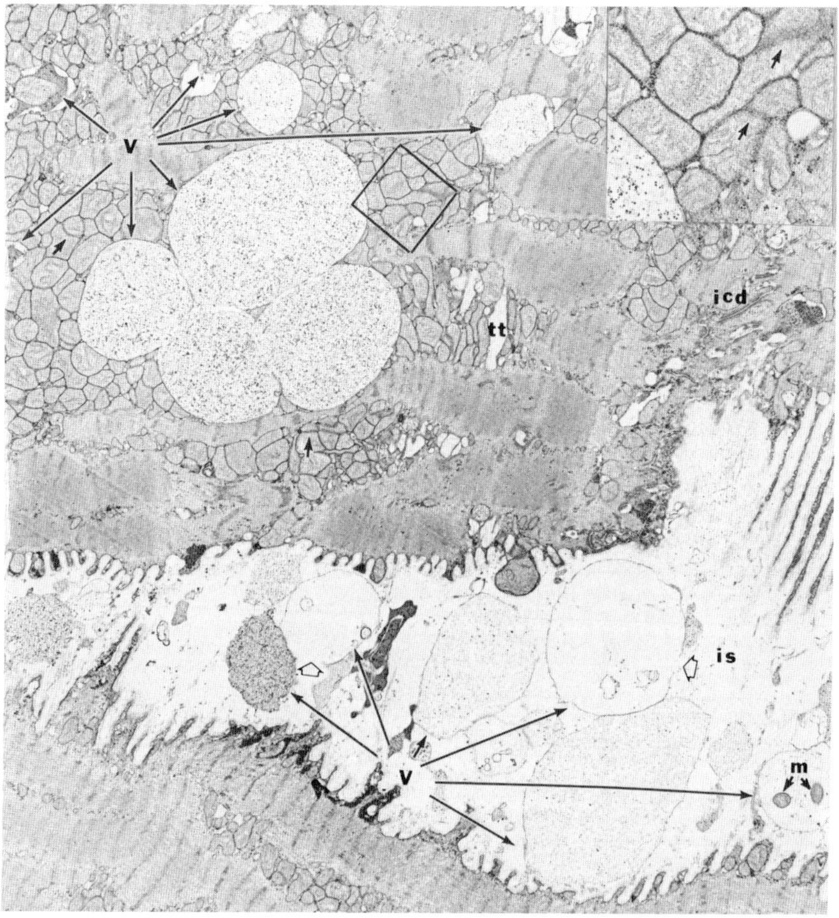

Figure 7. Multiple intracellular and extracellular vesicles (V) which contain granular material and glycogen. Of the extracellular vesicles, one contains two mitochondria (m) and two others show partial disintegration of their membrane and loss of contents (*large arrows*). The four large intracellular vesicles show focal areas of membrane dissolution, suggesting that they are fusing together. Some of the surrounding mitochondria show focal areas of apparent peripheral membrane rarefaction (*small arrows* and *inset*). tt, transverse tubules; icd, intercalated disc; is, interstitial space. × 6,500. *Inset,* × 18,900. From a 30-year-old woman with primary cardiomyopathy; same case as Figure 26.

pouchings of the sarcolemma (Figs. 10 and 26), or elsewhere within the cell. They do not appear to definitely communicate with any other cellular structure, although in very rare instances a vesicle has been seen in apparent communication with an intercalated disc (Fig. 9).

Most vesicles contain only varying amounts of glycogen granules and a thin

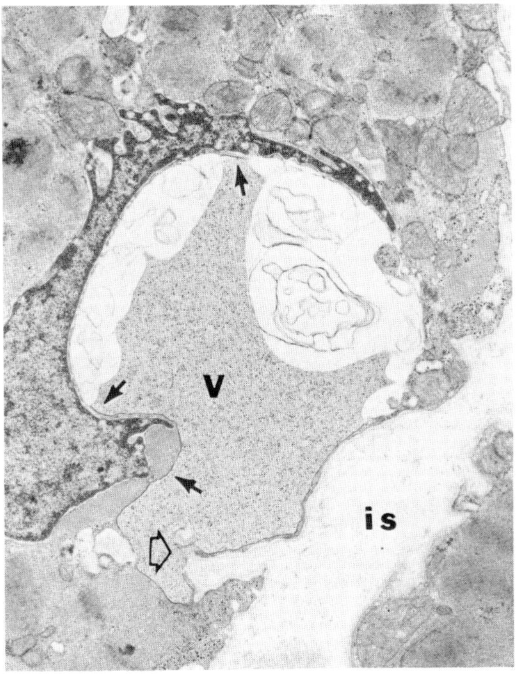

Figure 8. Large vesicle (V) situated between the nucleus and the sarcolemma. The vesicle is partially collapsed and contains membranous structures in the intermembranous space. A well defined double membrane is still preserved in the noncollapsed areas (*small arrows*). Note a focus of possible cell membrane rarefaction (*large arrow*). is, interstitial space. × 7,800. From a 55-year-old woman with primary cardiomyopathy.

granular material, probably a proteinaceous fluid (Figs. 1 and 3). Yet it is not uncommon to find within vesicles, intact or degenerating mitochondria (Figs. 11 and 13), myelin figures (Figs. 12, 14, and 15), membranous fragments (Fig. 9), tubular structures (Figs. 13, 14, and 18), and other nonidentifiable materials. In rare instances, intravesicular lipid droplets are present (Fig. 4).

The vesicles do not stain for acid phosphatase, and most, after diastase digestion, lose the glycogen granules and retain only the thin granular material (Fig. 14). This is also evident in the 1-μ thick, plastic-embedded sections, where most vesicles are PAS-positive but lose this staining after diastase digestion. Occasionally, a vesicle contains small groups of closely packed granules, similar in size and configuration to glycogen particles (Biava, 1963) (Figs. 9, 12, and 15). In many instances, similar groups of granules are seen elsewhere within the myocardial cells. These groups of granules are frequently surrounded by an irregular and multilayered thin membranous structure (Figs. 10, 12, 15, and 26).

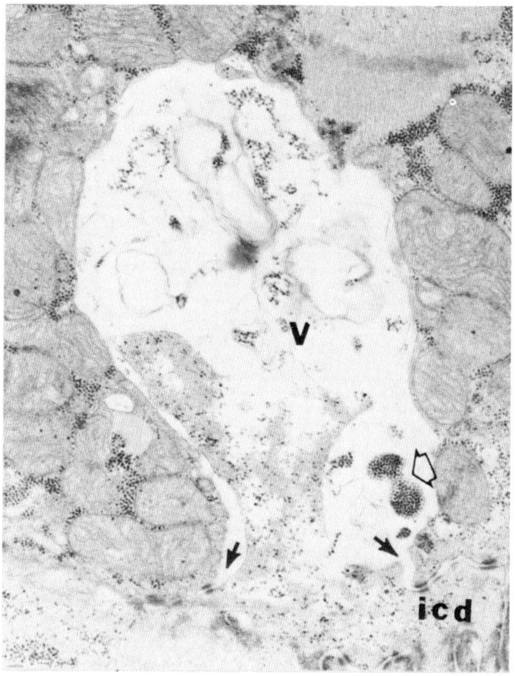

Figure 9. Intracellular vesicle-like structure (V), showing apparent communication (*arrows*) with an intercalated disc (icd). The inner and outer vesicular membranes appear mostly collapsed and disrupted. Note group of densely packed glycogen-like granules (*large arrow*). × 14,900. From a 47-year-old man with possible primary cardiomyopathy.

When present, these granules generally appear to be resistant to prolonged diastase digestion (Fig. 15). Since these groups of granules lack other more definitive characteristics of viruses, we cannot propose a viral component.

The degree of vesicle formation within myocardial fibers varies from case to case and from fiber to fiber. It may be extensive, with many fibers showing multifocal vesicle formation (Figs. 2, 7, 16, and 17), moderate with occasional intracellular vesicles present in the tissue section, or it may be mild or minimal with only one or two vesicles seen in an entire section. The degree of vesicle formation appears unrelated to the degree of individual myocardial fiber alteration. Vesicles may be present in myocardial fibers that show severe abnormalities (Fig. 18), or they may be found in fibers that otherwise appear normal. Vesicles appear to be very uncommon in myocardial fibers which have undergone almost total degeneration or atrophy, and which contain practically no organized myofibrils. The only two cases in which vesicles were not found were a case of coronary artery disease with a very scarred myocardium probably due to old

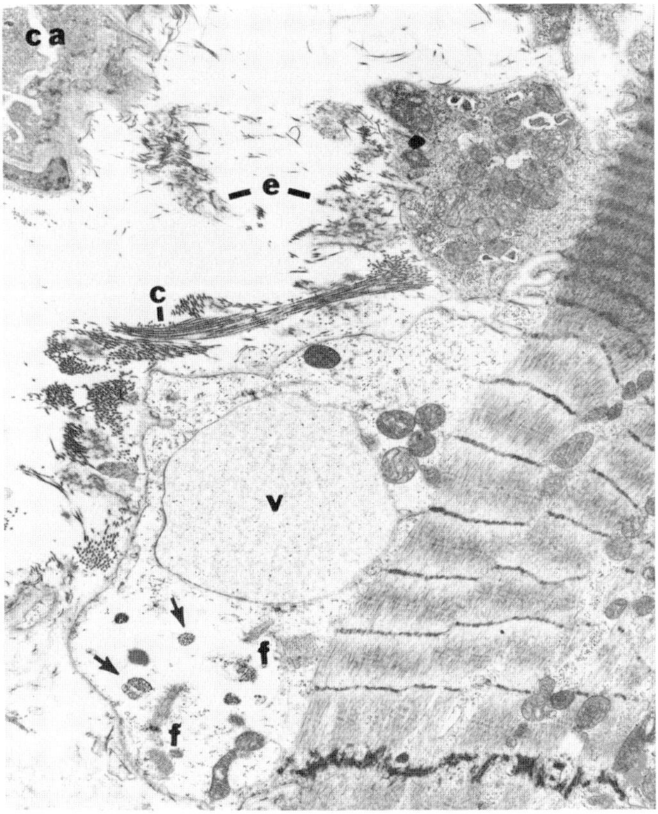

Figure 10. Membrane-bound vesicle (v) in large edematous sarcolemmal outpouching which also contains mitochondria, glycogen, myofibrillar material (f), and groups of densely packed glycogen-like granules (*arrows*). The smaller sarcolemmal outpouching appears with a much smaller base due to the contraction of the adjoining sarcomeres. This outpouching contains mitochondria, glycogen, and groups of closely packed glycogen-like granules but no vesicle. ca, capillary; c, collagen; e, elastic. × 6,500. From a 34-year-old man with primary cardiomyopathy; same case as Figures 2 and 6.

infarction, and a case of primary cardiomyopathy in which the segment of myocardium examined appeared ultrastructurally almost normal.

Extracellular vesicles

In the majority of cases (84%), vesicle-like structures were found in the extracellular space (Figs. 7, 17, and 19–23). These vesicles show the same characteristics and variations as those seen within myocardial fibers. They differ mainly in that the extracellular vesicles generally possess only one peripheral membrane (Figs. 19 and 22).

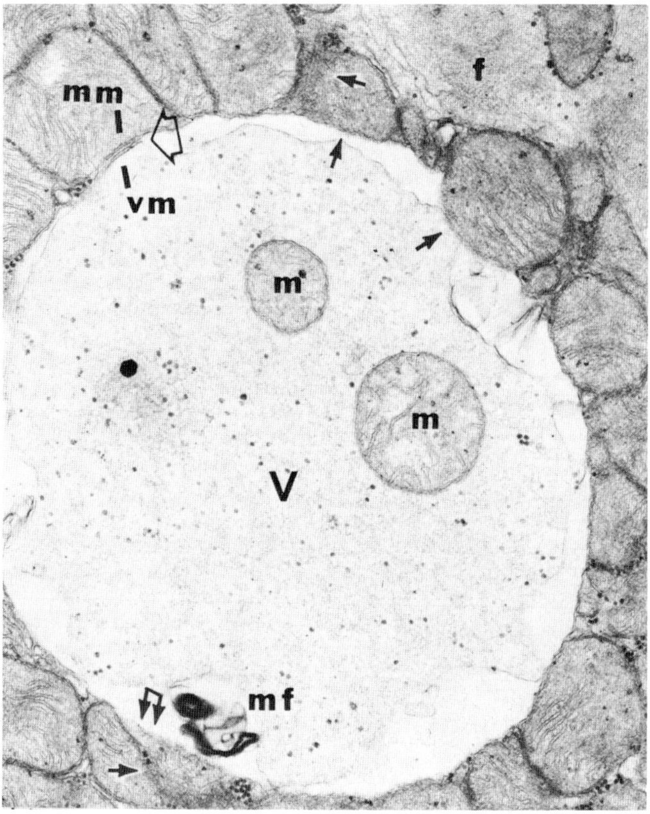

Figure 11. Partially collapsed intracellular vesicle (V) which contains two mitochondria (m), a myelin figure (mf), granular material, and glycogen. Double membrane persists in one area (*large arrow*). Note at this point the similarity between the vesicular membrane (vm) and the mitochondrial membrane (mm); one mitochondrion appears to be dissolving into the vesicle (*double arrow*); other mitochondria show focal rarefaction of their membranes (*small arrows*). f, myofibrils, × 20,250. From a 55-year-old man with primary cardiomyopathy; same case as Figure 25.

Extracellular vesicles measure 0.8–17.4 μ but most measure 1–6 μ. (We measured 45 representative extracellular vesicles from 14 cases.) They are generally accompanied by diverse cellular debris, such as degenerating mitochondria, myelin figures, membranous structures, and other nonidentifiable material. It is common to find extracellular vesicles in different stages of disintegration (Figs. 7, 20, and 26). Apparent phagocytosis of the extracellular vesicles was seen in 27% of our cases. The possibly phagocytic cells are large mononuclear cells which usually possess a prominent rough endoplasmic reticulum (Figs. 21 and 22).

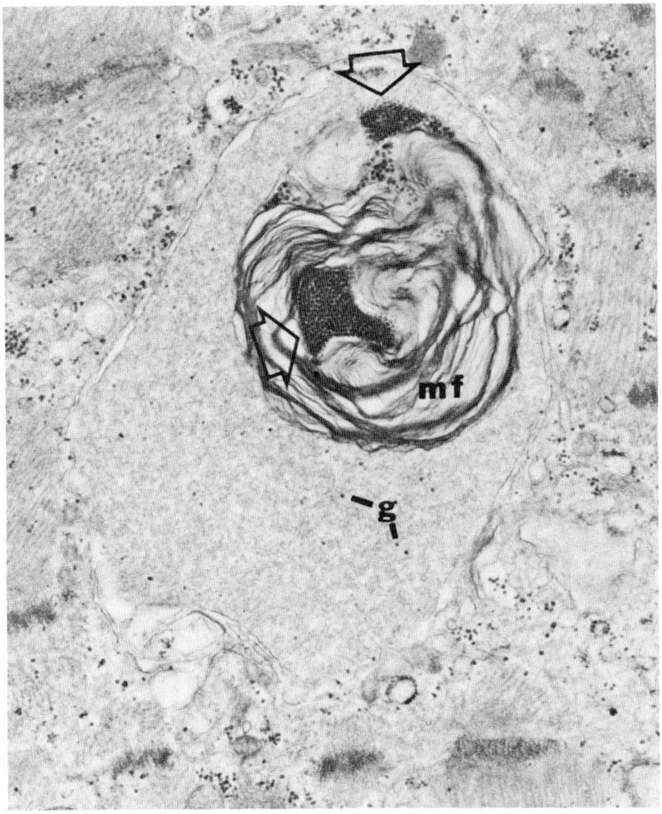

Figure 12. Intracellular partially double membrane-bound vesicle which contains large amounts of thin granular material, occasional glycogen particles (g), and a large myelin figure (mf). Also present are groups of closely packed glycogen-like particles (*large arrows*). × 20,250. From a 56-year-old man with possible primary cardiomyopathy.

Extrusion of vesicles

As previously stated, the intracellular vesicles are commonly found in the subsarcolemmal space and in small and large outpouchings of the sarcolemma. The cellular membranes situated over these vesicles may show focal areas of apparent disruption (Figs. 8, 17, and 23–25). In many cases, there appears to be focal loss of cytoplasmic membrane with preservation of the basement membrane (Figs. 23 and 24). In other instances, both these membranes are disrupted and the vesicle is exposed to the interstitial space (Figs. 25 and 28*A*). It is suggested from some cases that the edges of the disrupted cytoplasmic membrane may coalesce with the outer vesicular membrane which would then appear

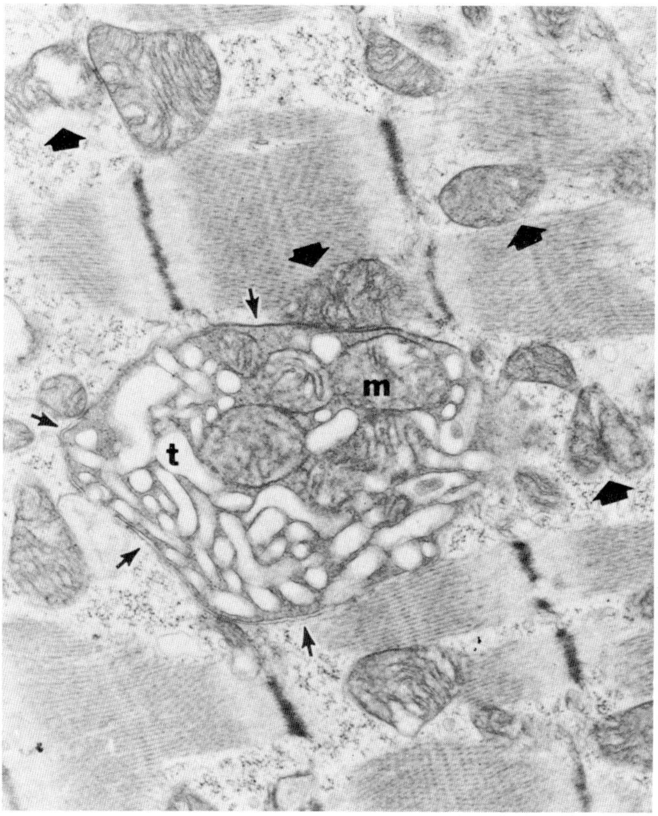

Figure 13. Unusual vesicle-like structure which is for the most part surrounded by a double membrane (*small arrows*). It contains tubular structures (t) and degenerating mitochondria (m). Other abnormal mitochondria are found outside the vesicle (*large arrows*). × 14,000. From a 46-year-old man with primary cardiomyopathy; same case as Figure 4.

to aid in the preservation of cellular integrity (Figs. 24, 25, and 28*A*). This could explain why most extracellular vesicles have only a single peripheral membrane.

In less common instances, a vesicle or group of vesicles appear to be extruded into the extracellular space with the entire contents of a sarcolemmal outpouching. This is possibly mediated by focal loss of sarcolemma at the tip of the entire sarcolemmal outpouching (Figs. 26 and 28*C*) or, very rarely, by apparent separation of the entire sarcolemmal outpouching (Figs. 27 and 28*D*).

DISCUSSION

On the basis of the foregoing observations, it is probable that most vesicles are derived from fusion of degenerating mitochondria. Not only may mitochondria

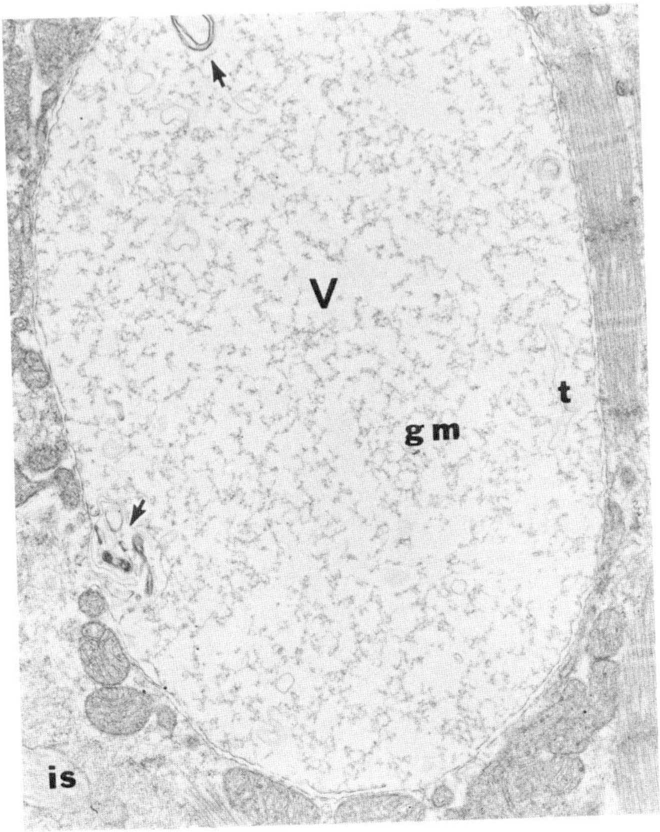

Figure 14. Intracellular double membrane-bound vesicle (V) after digestion with diastase. Note total absence of glycogen. Only the thin granular material (gm) seen in all vesicles is retained. Also present are tubular structures (t) and myelin figures (*arrows*). is, interstitial space. × 14,800. From a 47-year-old man with primary cardiomyopathy; same case as Figure 22.

partially or totally dissolve into vesicles, but at times portions of the mitochondrial membranes have been observed to remain as part of the vesicular membranes. The general thickness of the vesicular and the mitochondrial membranes and the widths of their intermembranous space (Table 3) are sufficiently similar to further support this conclusion.

After examining hundreds of these vesicles, it becomes apparent to us that these structures cannot be artefactual invaginations or protrusions as suggested by others (Ferrans *et al.*, 1973). Cytoplasmic invaginations are limited by a cytoplasmic membrane and a basement membrane (Lannigan and Zaki, 1967) that, on most sections, would appear as the innermost membrane of the pseudovesicle and would logically contain elements of the extracellular space.

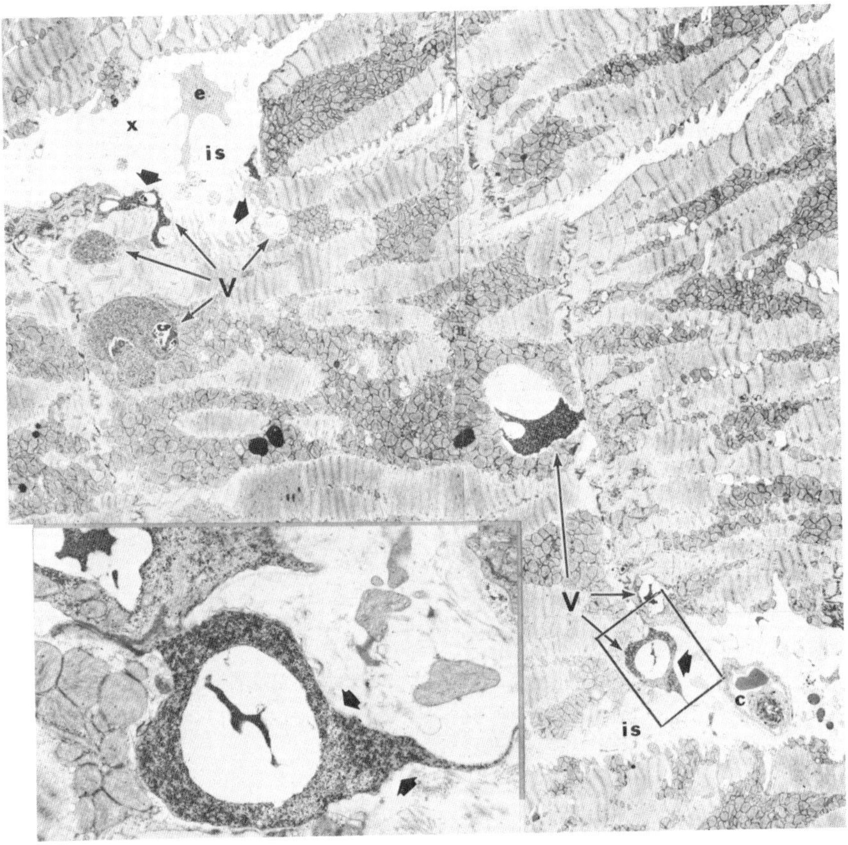

Figure 17. Multifocal vesicle formation (V) in a myocardial segment that otherwise shows minimal changes. Of seven intracellular vesicles seen, six are in one myocardial cell. The three subsarcolemmal vesicles show focal disruption of the sarcolemmal covering (*large arrows*). The myocardial fibers show focal increase and aggregation of mitochondria and contraction bands. A large, irregular extracellular vesicle (e) is present in the interstitial space (is) together with other vesicle-like structures (x). *Inset* shows extensive loss of sarcolemma at various points (*arrows*) over one of the subsarcolemmal vesicles. c, capillary. *Montage,* × 2,700. *Inset,* × 11,350. From a 62-year-old man with primary cardiomyopathy associated with significant alcohol consumption; same case as Figure 24.

Bulloch *et al.,* 1970; Hug and Schubert, 1970; Lin and Duffy, 1970; Pearce, Bulloch, and Murphy, 1971; De La Iglesia and Lumb, 1972) generally appear swollen with at least some preservation of cristae. They do not attain the size of vesicles and can generally be recognized as altered individual units. In diseased hearts with prominent mitochondrial degeneration, most of these organelles are affected to approximately the same degree. Dilations of the endoplasmic reticu-

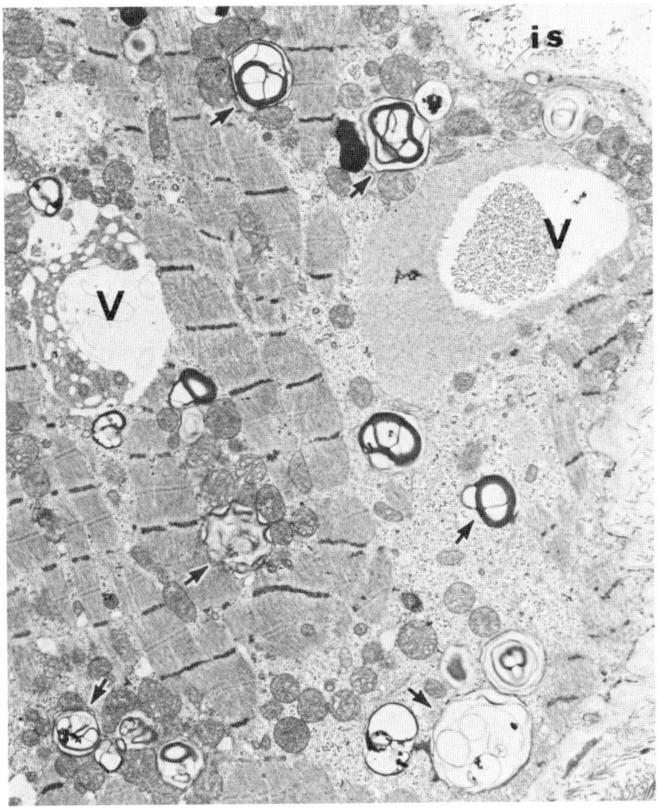

Figure 18. Two large vesicles (V) in the presence of severe myocardial fiber degeneration with marked myelin figure production (*arrows*). The larger vesicle contains glycogen and granular material. The smaller one, partially collapsed, contains glycogen and granular material plus mitochondria and tubular structures. Vesicles were seen only rarely in this case. is, interstitial space. × 5,500. From a 47-year-old man with familial primary cardiomyopathy.

lum (Clementi *et al.*, 1964; Ferrans *et al.*, 1964; Sulkin and Sulkin, 1965; Alexander, 1966; Legato *et al.*, 1968; Alexander and Niño, 1969; Jaenke, 1974) are surrounded by a single membrane, appear empty, are relatively small, and generally occur in groups or diffusely affect the myocardial cells. Dilations of the transverse tubular system (Legato *et al.*, 1968; Jaenke, 1974) are surrounded by a cytoplasmic membrane and a basement membrane and, when not empty, contain recognized elements of the extracellular space such as collagen. Myocardial lysosomes (Hibbs *et al.*, 1965; Kottmeier and Wheat, 1967) are relatively small, homogeneous, electron-dense bodies which are surrounded by a single

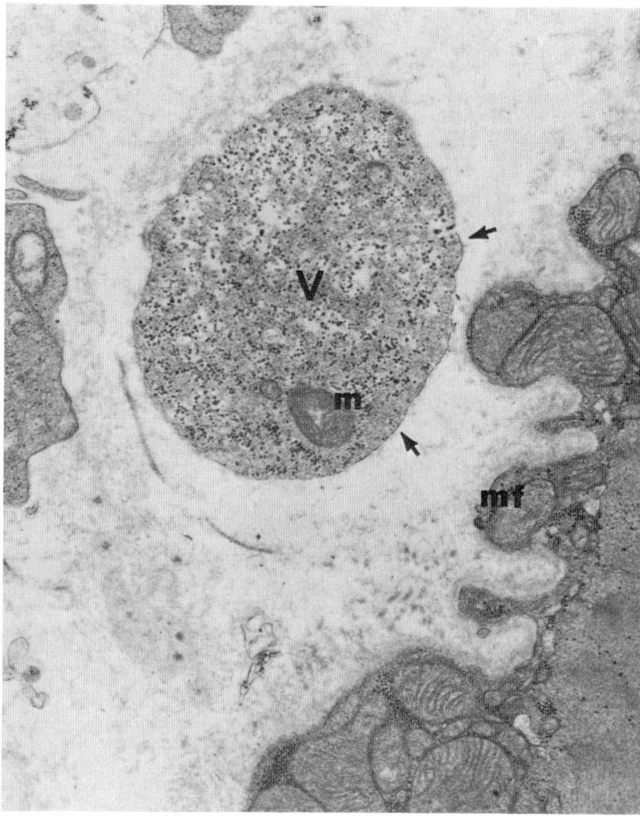

Figure 19. Extracellular vesicle (V). Note the single peripheral membrane present in some areas (*arrows*). The vesicle contains granular material, glycogen, and a mitochondrion (m). mf, myocardial fiber. × 13,550. From a 55-year-old man with primary cardiomyopathy.

membrane. They may appear related to lipofuscin granules and, on some occasions, contain a crystalline-like material of unknown significance (Hug and Schubert, 1970), (Fig. 3). Most lysosomes are acid phosphatase positive (Ericsson and Trump, 1964), while vesicles are not. Basophilic myocardial degeneration (Kosek and Angell, 1970) generally affects large portions of a myocardial cell, is not membrane-bound, and has a typical fibrillar ultrastructural appearance. Myocardial "granular bodies" (Vega *et al.,* 1974) (Fig. 29) are totally or, more commonly, partially bound by a single membrane, contain exclusively a densely packed, thin and homogenous granular material (different from the loose granular material seen within vesicles), and occasionally large and irregular electron-dense granules.

Since practically identical vesicles are found inside and outside the myocardial

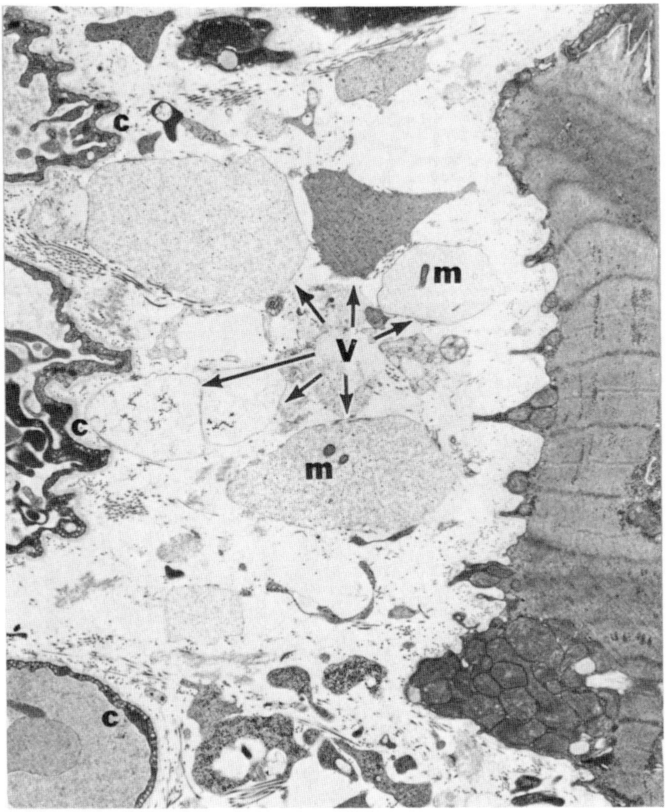

Figure 20. The interstitial space contains a multitude of partially or totally membrane-bound vesicles which show variation in size, shape, and contents similar to that seen in vesicles of the intracellular space. All contain granular material, most contain glycogen and 2 contain mitochondria (m). c, capillary, × 4,600. From a 44-year-old woman with primary cardiomyopathy; same case as Figure 27.

cells, as well as being extruded from the cells, and since they appear to disintegrate or become phagocytosed in the extracellular space, it is postulated that all the described vesicles have an intracellular origin. These findings strongly suggest that the vesicles may be undesirable elements within the myocardial cells, or that they may represent a mechanism by which myocardial cells eliminate damaged portions of their architecture or, maybe, even toxic elements.

Whether the extracellular vesicles are truly derived from the intracellular space is, of course, hypothetical. They could possibly be artefactual or derived from degenerating interstitial elements, such as damaged capillaries as suggested by Fajardo and Stewart (1973) in their study on radiation-induced myocardial

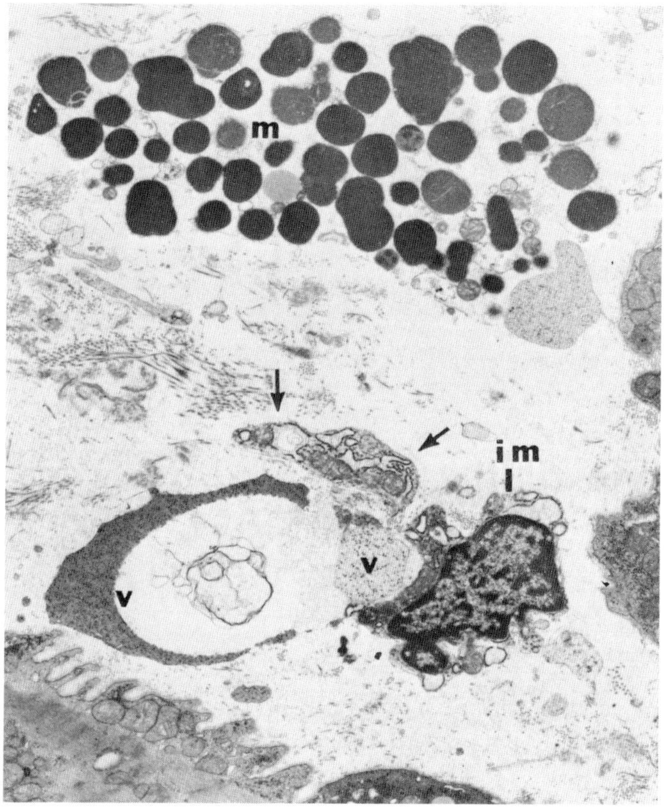

Figure 21. This electron micrograph suggests that some extracellular vesicles (v) may be in the process of being phagocytosed by an interstitial mononuclear cell (im) with prominent rough endoplasmic reticulum and a long cytoplasmic extension (*arrows*). The interstitial space also contains collagen, elastic, and microfibrils. In the upper portion of the picture are the granules of a disintegrated mast cell (m), a not uncommon finding in our biopsy specimens. × 5,500. From a 53-year-old woman with atherosclerotic coronary artery disease.

fibrosis in rabbits. In our cases though, the number of extracellular vesicles correlated frequently with the degree of intracellular vesicle formation. Also, as previously stated, the extracellular vesicles, except for the possession of only one peripheral membrane, are almost identical to those found within myocardial cells. Phagocytosis of the extracellular vesicles is also hypothetical, since the findings that suggest it could also be interpreted as artefactual due to biopsy-related pressures that intrude such vesicles into the possibly less resistant cytoplasm of a mononuclear cell, or a capillary wall.

We have found no significant correlation between the degree of vesicular

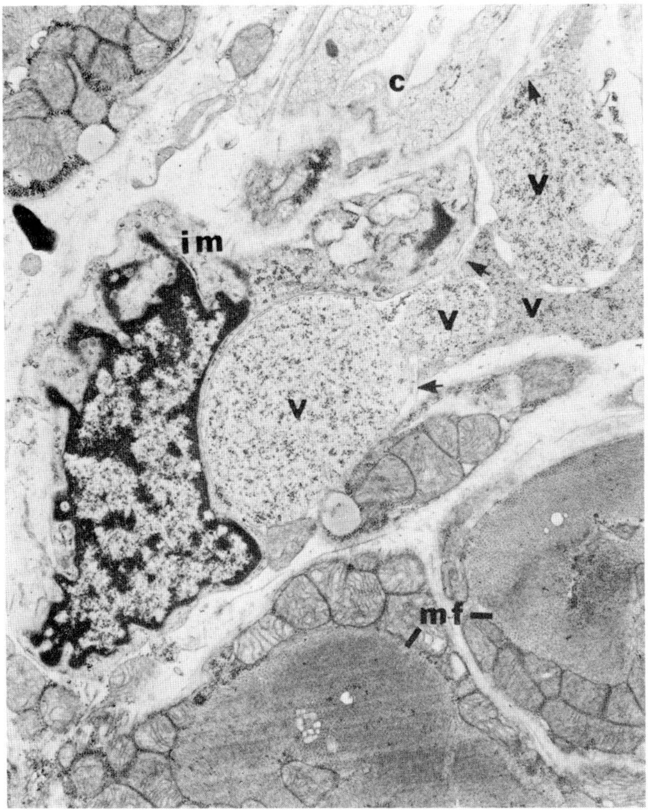

Figure 22. Large interstitial mononuclear cell (im) partly surrounding a group of extracellular vesicles (v). These vesicles are largely bound by a single membrane but, at some points a double membrane is evident (*arrows*). c, capillary; mf, myofibers. × 9,450. From a 47-year-old man with primary cardiomyopathy; same case as Figure 14.

change and the clinical findings, sex, and age of our patients (Table 1). However, very extensive vesicular change has been seen only in 8 patients with primary cardiomyopathy, possible primary cardiomyopathy, and primary cardiomyopathy associated with significant alcohol consumption. Moderate vesicular change was seen in 37 patients, 29 of whom had primary cardiomyopathy and 5 had rheumatic valvular disease. It would appear from this that vesicular change may achieve a most significant level in primary cardiomyopathy. The 5 patients who are considered to have normal hearts as evidenced by cardiac catheterization and selective cine coronary angiography all showed only minimal to mild vesicular change together with other very minor ultrastructural changes. It would

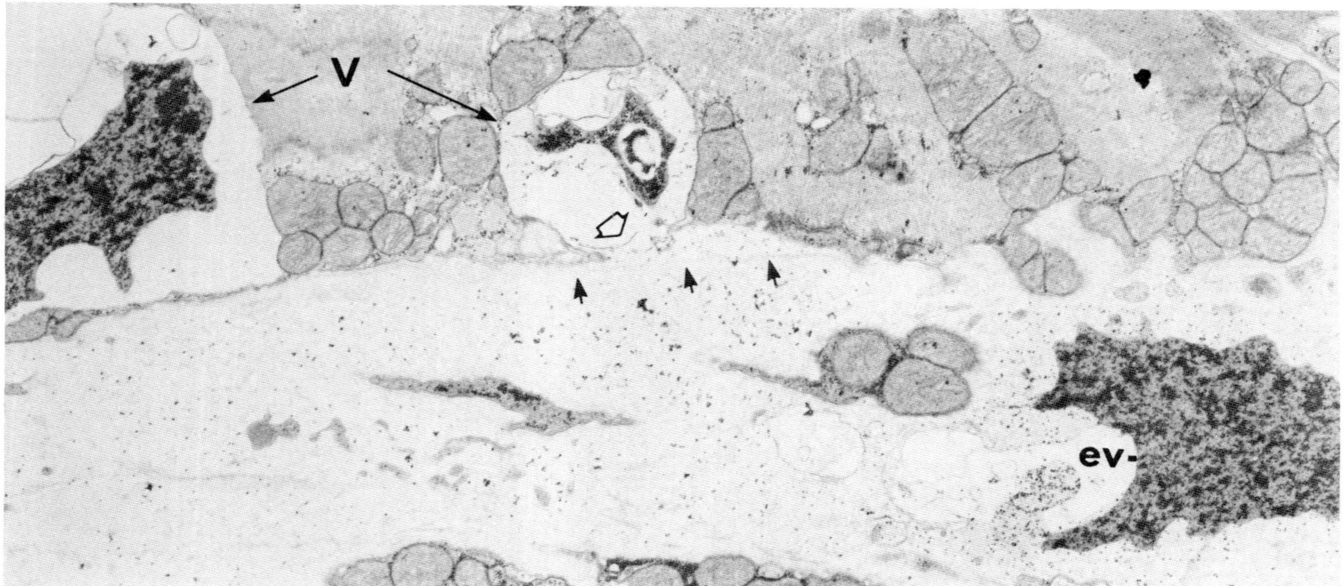

Figure 23. Two intracellular, partially collapsed, vesicles are present (V). The larger one maintains, at this level of sectioning, an intact sarcolemmal covering, while the smaller vesicle shows focal disruption of the cytoplasmic membrane (*large arrow*) with preservation of a somewhat separated basement membrane (*small arrows*). A large extracellular vesicle is also present (ev). Note similarity between the contents of the large intracellular vesicle and the extracellular one. × 13,000. From a 30-year-old woman with normal coronary arteriograms and left ventriculograms.

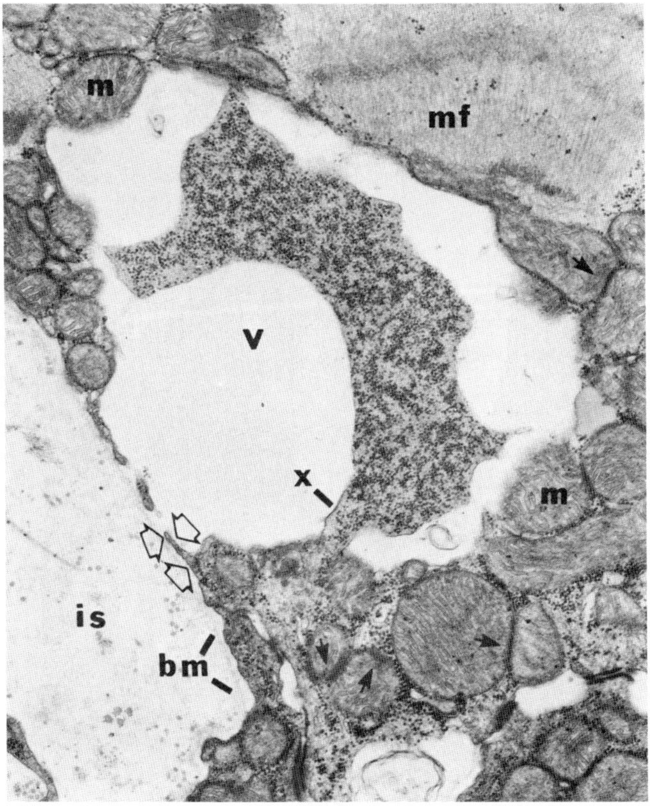

Figure 24. There is focal loss of cytoplasmic membrane over this collapsed subsarcolemmal vesicle (V). The basement membrane appears intact (bm). Note that the cytoplasmic membrane appears continuous with the vesicular membrane (*large arrows*). Note disolution of certain mitochondria (m) into the vesicle with partial loss of mitochondrial membrane. Some of the neighboring mitochondria show apparent focal peripheral membrane rarefaction (*small arrows*). x, portions of vesicular membrane; mf, myofibrils; is. interstitial space. × 14,900. From a 62-year-old man with primary cardiomyopathy associated with significant alcohol consumption; same case as Figure 17.

appear from these latter cases that the apparently normal myocardium of adults is capable of eliminating certain intracellular elements by the production and possibly extrusion of vesicles. This could result from the theoretical continuous physiological wear, tear, and repair of the myocardial cells. Vesicular change should not then be considered categorically a form of myocardial degeneration, but possibly a specific mechanism by which the myocardial cells eliminate undesirable metabolic residues, damaged mitochondria, or other cellular compo-

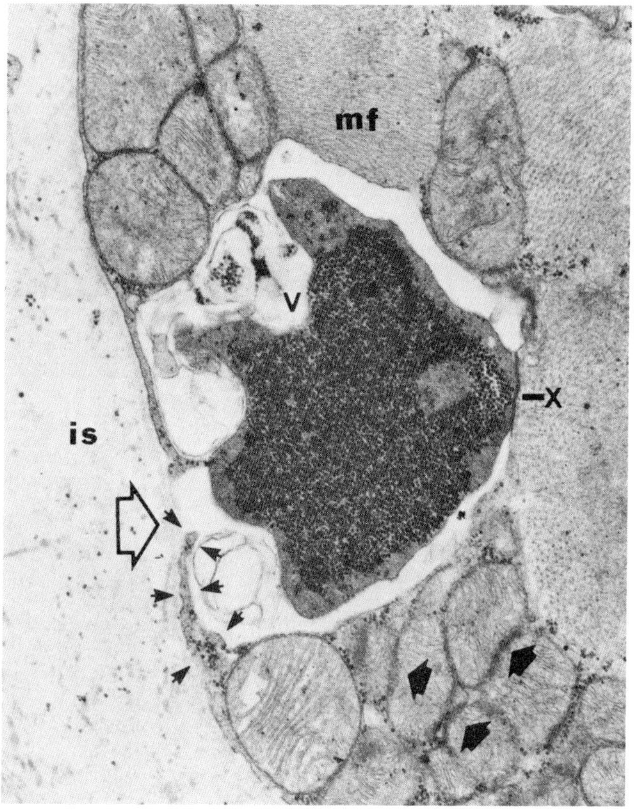

Figure 25. Subsarcolemmal vesicle (V) with focal disruption of the overlying cytoplasmic and basement membranes (*large arrow*). Note that the vesicular outer membrane and the cytoplasmic membrane appear continuous (*small arrows*). The vesicle shows preservation of its double membrane only in the area where it is not collapsed (x). Some of the mitochondria show focal areas of apparent peripheral membrane rarefaction (*dark arrows*). mf, myofibrils; is, interstitial space. × 18,900. From a 55-year-old man with primary cardiomyopathy; same case as Figure 11.

nents. It is reasoned that this mechanism may be significantly utilized in the diseased heart.

Similarly, we have not found a truly significant correlation between vesicular change and the degree and types of pathological changes as evidenced by electron and light microscopy (Table 2), though this change appears to affect more extensively cases in which there is basophilic degeneration. Moderate to extensive vesicular change was found in 38% (11 of 29) of the cases that showed no significant light microscopic changes, in 42% (18 of 43) of the cases that showed interstitial fibrosis, in 45% (17 of 38) of the cases that showed hyper-

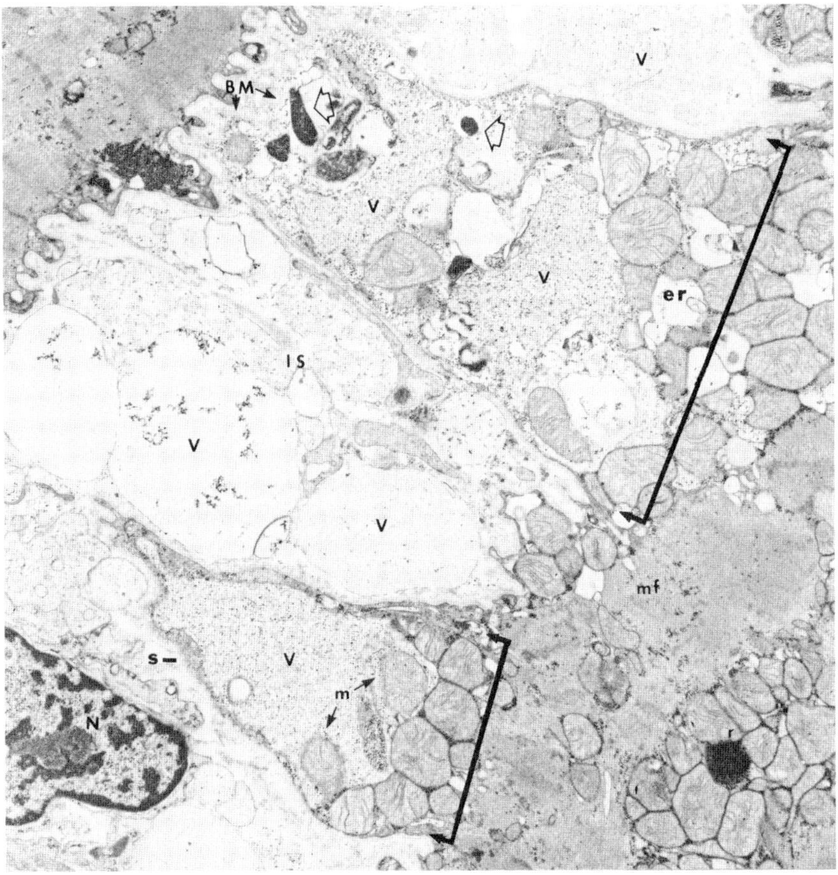

Figure 26. Two large sarcolemmal outpouchings (*brackets*) are present on the side of a myocardial fiber (mf). In the smaller one, a vesicle (V) is present which contains two degenerating mitochondria (m). The sarcolemmal covering is intact (s). The larger outpouching contains vesicle-like structures (V), mitochondria, and groups of densely packed glycogen-like granules (*large arrows*). The entire tip of this outpouching has lost its cytoplasmic membrane and only a poorly defined basement membrane is preserved (BM). The interstitial space (IS) contains many intact and disintegrating vesicle-like structures (V). er, dilated endoplasmic reticulum; N, nucleus of an interstitial cell. × 13,000. From a 30-year-old woman with primary cardiomyopathy; same case as Figure 7.

trophy, and in 47% (7 of 15) of the cases in which we found basophilic degeneration. In some instances, vesicle formation has been the only main finding and, in others, vesicle formation has been only a minor component in a severely diseased myocardium showing a multitude of abnormalities by light and electron microscopy.

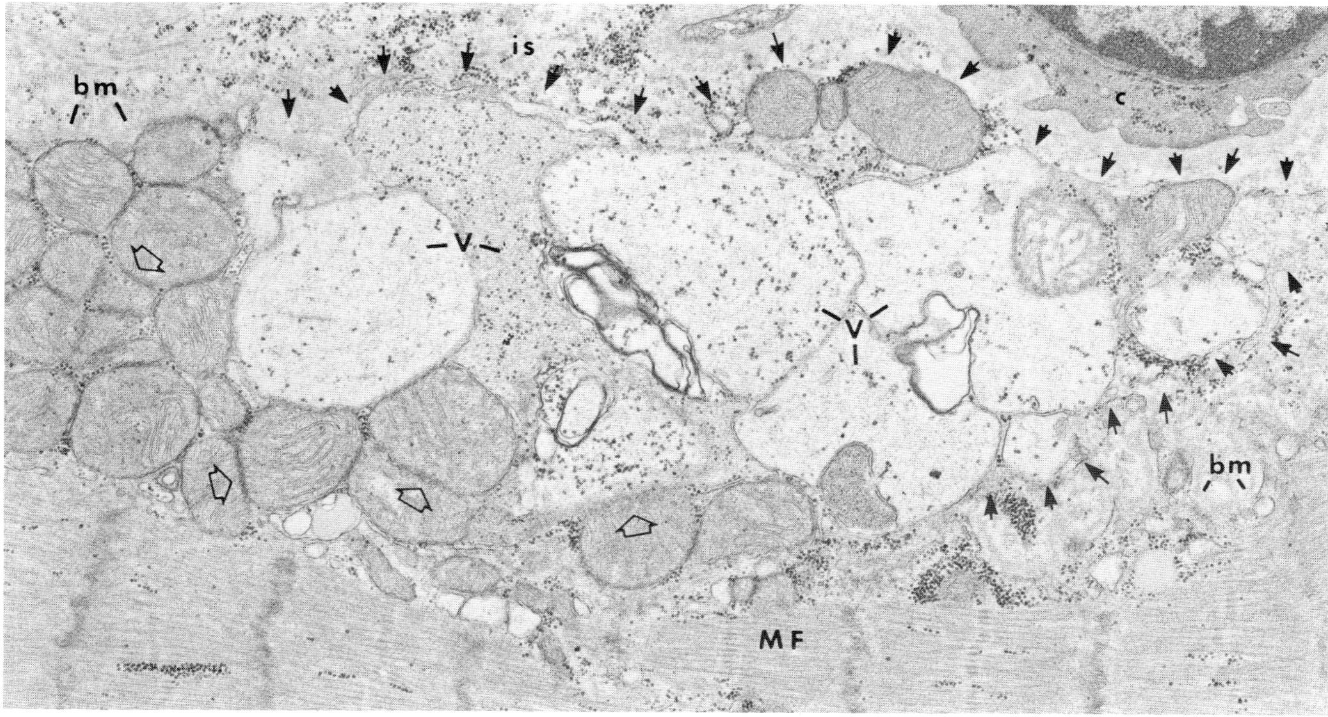

Figure 27. Myocardial fiber (MF) with apparent expulsion of small sarcolemmal outpouching which contains vesicle-like structures (V) and intact and degenerating mitochondria. Note extensive loss of sarcolemma over most of the outpouching (*arrows*), and preservation of basement membrane (bm) on either side. Note also mitochondria with areas of apparent membrane rarefaction (*large arrows*). is, iterstitial space; c, capillary. × 17,800. From a 44-year-old woman with primary cardiomyopathy; same case as Figure 20.

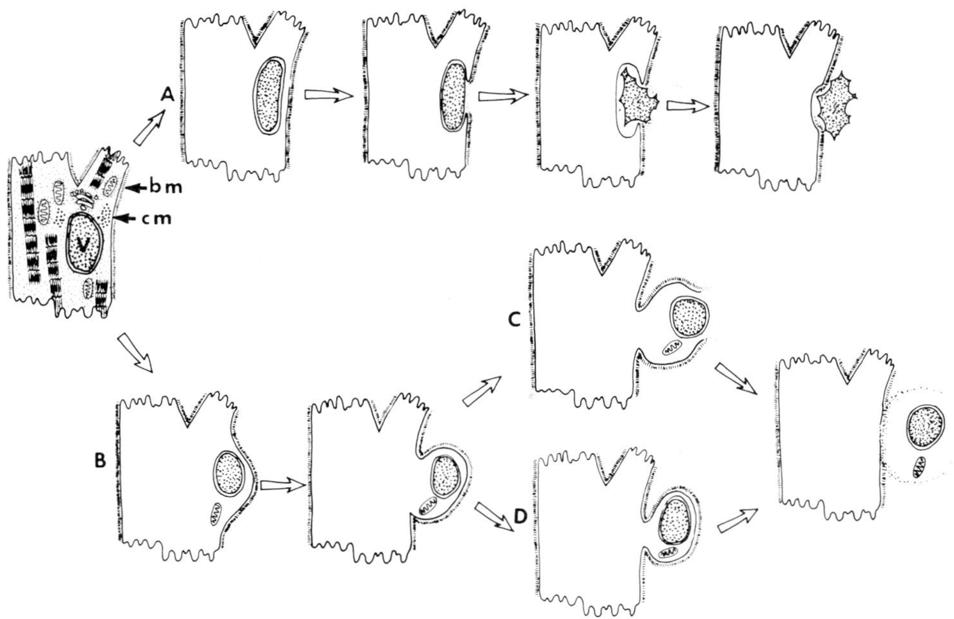

Figure 28. Schematic diagram representing the three theoretical mechanisms of extrusion in vesicular myocardial change. In the *upper row (A)* a vesicle approaches the sarcolemma to the point of contact. First the cytoplasmic membrane and then the basement membrane disrupt and extrude the vesicle into the extracellular space. This appears to be the most common mechanism involved. Note that, in this diagram, the vesicular outer membrane becomes continuous with the cytoplasmic membrane. In the *lower row (B),* a vesicle and other cytoplasmic constituents, a mitochondrion, become part of a sarcolemmal outpouching which either becomes focally disrupted (*C*) or is entirely "pinched-off" (*D*). V, vesicle; cm., cytoplasmic membrane; bm, basement membrane.

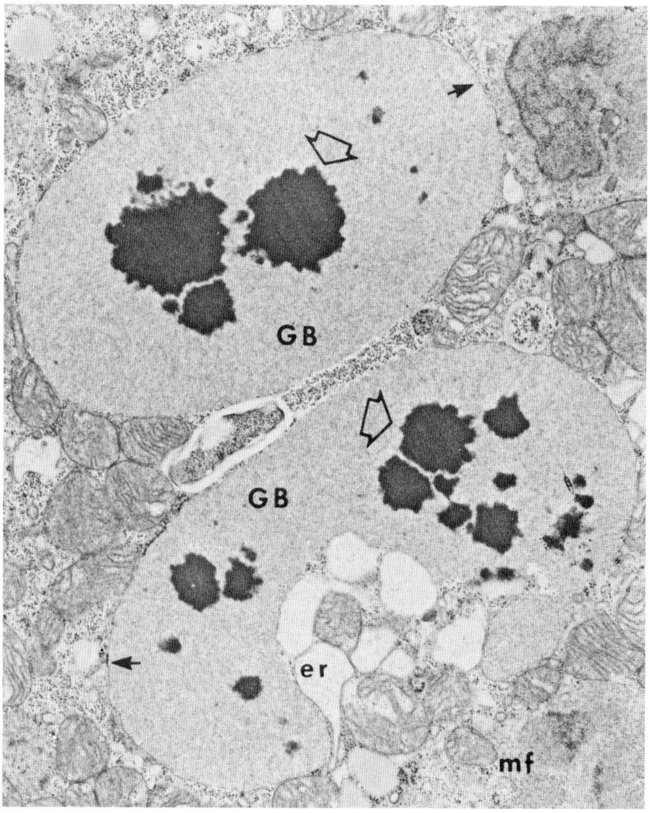

Figure 29. Not to be confused with myocardial vesicles are these "granular bodies" (GB) which are generally partially surrounded by a single thin membrane (*small arrows*). They exclusively contain a densely packed and very thin granular material and occasional large electron-dense granules (*large arrows*). "Granular bodies" are never collapsed and are most frequently seen in the juxtanuclear space. er, dilated endoplasmic reticulum; mf, myofibrils. × 11,300. From a 51-year-old man with primary cardiomyopathy.

Since vesicle formation can be a very patchy phenomenon, our statistical figures are by no means absolute. We have observed, in many cases, significant variations in the amount of vesicular change from different sections of the same biopsy. This variation is even more significant when we take into account the small size of each tissue block and the small number of sections studied. This, we believe, is one of the very few unfortunate drawbacks still present in diagnostic electron microscopy.

Review of literature

Other authors have reported the presence of vesicles identical to those we described in other cases of human heart disease: Hibbs *et al.* (1965) in alcoholic cardiomyopathy; Wheat (1965) in a case of mitral stenosis (Fig. 7 with no reference to its presence); Ferrans *et al.* (1965) in a case of cardiac sarcoidosis; Lannigan and Saki (1967) in patients with rheumatic carditis; Laufer *et al.* (1968) in auricular biopsies from patients with valvular and congenital heart disease; Alexander and Niño (1969) in a patient with cardiac decompensation due to prolonged phenothiazine treatment; Kosek and Angell (1970) in cases of basophilic myocardial degeneration; Levinsky *et al.* (1970) in a case of human heart transplant with features of rejection; Heggtveit *et al.* (1971) in open-heart biopsies from 6 patients with idiopathic cardiomyopathy. Bulloch *et al.* (1972) in idiopathic and alcoholic cardiomyopathy; and Ferrans *et al.* (1973) in patients with obstructive or congestive cardiomyopathy.

Vesicular myocardial change is not restricted to the human heart. It has been found by different authors in animals with experimentally produced cardiac lesions: Clementi *et al.* (1964) in myocardial infarcts produced in rats by coronary artery ligation; Maruffo (1967) in isoproterenol-induced myocardial changes in the rhesus monkey (large membrane-bound vacuoles mentioned in the text but none pictured); Sun *et al.* (1968) in *Mycoplasma gallisepticum* myocarditis in the cynomolgus monkey; and Baba *et al.* (1970) in experimental cardiac ischemia in the isolated and perfused hearts of rats. In experimental coronary artery narrowing in pigs and dogs, De La Iglesia and Lumb (1972) speculated that these vesicles "could represent an abnormal intracellular accumulation of fluid secondary to altered ion concentration, loss of intracellular osmolality, or increased membrane permeability, as well as an additive effect of these phenomena in different degrees of participation."

Vesicle-like structures similar in some ways to those we describe and possibly representing the same entity have been found by Heggtveit *et al.* (1964) in the hearts of magnesium-deficient rats; by Burch *et al.* (1971) in viral encephalomyocarditis infections of the myocardium in newborn mice; by Harrison *et al.* (1970) in the hearts of potassium-depleted rats; by Burch, Sohal, and Fairbanks (1970) in the hearts of normal but aged *Drosophila repleta;* and by Nakao *et al.* (1970) in myocardiopathic hamsters.

Beyond the myocardium, vesicle-like structures, not unlike those we describe, have been seen by Muehrcke and Rosen (1964) in the kidney of a patient with hypokalemic nephropathy, by Engel and Dale (1968) in the skeletal muscle of a man with "vacuolar myopathy," by Tardini *et al.* (1971) in the caerulein-stimulated exocrine pancreas of dogs, by Buckley (1973) in cultured chick embryo cells, and by Schumucker and Curtis (1974) in perfused rat livers.

In most of the previously mentioned reports, the vesicles have been accepted as being of lysosomal or "autophagic" origin. But true autophagic vacuoles, cytosomes, and cytosegresomes have been well studied and described (Hruban *et al.*, 1963; Ericsson and Trump, 1964; Ericsson, Trump, and Weibel, 1965; Ericsson and Glinsmann, 1966; Glinsmann and Ericsson, 1966; Trump and Bulger, 1967; Arstila and Trump, 1968; Arstila *et al.*, 1971; Pfeifer, 1971;

Schiaffino and Hanzlíková, 1972) and are generally smaller and distinct structures though they share with myocardial vesicles some morphological and possibly functional characteristics.

In only one of the reviewed publications have the extracellular vesicles and their chagocytosis been reported. In 1964, Clementi *et al.* in their ligation-induced myocardial infarcts in rats noted this phenomenon and interpreted their origin as identical to the origin of those vesicles they saw in the intracellular space but did not comment as to their possible specific mode of genesis. Lannigan and Zaki (1967) in their studies on rheumatic carditis suggest a similar finding when they stated: "Sometimes the sarcoplasm of severely altered cells is extruded into the extracellular space and cell processes are seen surrounding it."

CONCLUSIONS

Vesicular myocardial change appears to represent a nonspecific form of cellular reaction to disease and possibly to normal physiological wear of the myocardium in humans, experimental animals, and even certain insects. This vesicular reaction is probably not exclusive to the heart but may possibly be utilized by cells in other organs within the body and is probably similar in some ways to cytosomes, cytosegresomes, and autophagic vacuoles from other parts of the body.

In the heart, the vesicles appear to derive from damaged mitochondria that dissolve into the vesicle with portions of their peripheral membranes coalescing to produce the vesicular membrane. Our studies also suggest that these vesicles are possibly discarded into the extracellular space where they disintegrate or may even be phagocytosed.

REFERENCES

ALEXANDER, C. S. 1966. Idiopathic heart disease. II. Electron microscopic examination of myocardial biopsy specimens in alcoholic heart disease. Am. J. Med. 41: 229–234.

ALEXANDER, C. S. 1967. Electron microscopic observations in alcoholic heart disease. Brit. Heart J. 29: 200–206.

ALEXANDER, C. S., and A. NIÑO. 1969. Cardiovascular complications in young patients taking psychotropic drugs. Am. Heart J. 78: 757–769.

ARSTILA, A. U., H. O. JAUREGUI, J. CHANG, and B. F. TRUMP. 1971. Studies on cellular autophagocytosis. Relationship between heterophagy and autophagy in HeLa cells. Lab. Invest. 24: 162–174.

ARSTILA, A. U., and B. F. TRUMP. 1968. Studies on cellular autophagocytosis. The formation of autophagic vacuoles in the liver after glucagon administration. Am. J. Pathol. 53: 687–733.

BABA, N., R. F. LEIGHTON, and A. M. WEISSLER. 1970. Experimental cardiac ischemia. Observation of the sinoatrial and atrioventricular nodes. Lab. Invest. 23: 168–178.

BERCU, B., J. HEINZ, A. S. CHOUDHRY, and P. CABRERA. 1964. Myocardial biopsy. A new technic utilizing the ventricular septum. Am. J. Cardiol. 14: 675–678.

BIAVA, C. 1963. Identification and structural forms of human particulate glycogen. Lab. Invest. 12: 1179–1197.

BUCKLEY, I. K. 1973. The lysosomes of cultured chick embryo cells. A correlated light and electron microscopic study. Lab. Invest. 29: 411–421.

BULLOCH, R. T., M. L. MURPHY, and M. B. PEARCE. 1965. Intracardiac needle biopsy of the ventricular septum. Am. J. Cardiol. 16: 227–233.

BULLOCH, R. T., M. L. MURPHY, and M. B. PEARCE. 1970. Fine structural lesions in the myocardium of a beer drinker with reversible heart failure. Am. Heart J. 80: 629–637.

BULLOCH, R. T., M. B. PEARCE, M. L. MURPHY, B. J. JENKINS, and J. L. DAVIS. 1972. Myocardial lesions in idiopathic and alcoholic cardiomyopathy. Study by ventricular septal biopsy. Am. J. Cardiol. 29: 15–25.

BURCH, G. E., J. M. HARB, H. L. COLCOLOUGH, and C. Y. TSUI. 1971. Encephalomyocarditis infection of the newborn mouse myocardium. An electron microscopic study. Arch. Intern. Med. 127: 148–156.

BURCH, G. E., R. SOHAL, and L. D. FAIRBANKS. 1970. Ultrastructural changes in *Drosophila* heart with age. Arch. Pathol. 89: 128–136.

CAVES, P. K., E. B. STINSON, A. F. GRAHAM, M. E. BILLINGHAM, T. M. GREHL, and N. E. SHUMWAY. 1973. Percutaneous transvenous endomyocardial biopsy. J.A.M.A. 225: 288–291.

CLEMENTI, F., J. NIDIRI, and C. PERACCHIA. 1964. Electron microscopic study of myocardial cells in relation to different types of myocardial infarct-like lesions. Panminerva Med. 6: 420–428.

De La IGLESIA, F. A., and G. LUMB. 1972. Ultrastructural and circulatory alterations of the myocardium in experimental coronary artery narrowing. Lab. Invest. 27: 17–31.

ENGEL, A. G., and A. J. D. DALE. 1968. Autophagic glycogenosis of late onset with mitochondrial abnormalities: light and electron microscopic observations. Mayo Clin. Proc. 43: 233–279.

ERICSSON, J. L. E., and W. H. GLINSMANN. 1966. Observations on the subcellular organization of hepatic parenchymal cells. I. Golgi apparatus, cytosomes, and cytosegresomes in normal cells. Lab. Invest. 15: 750–761.

ERICSSON, J. L. E., and B. F. TRUMP. 1964. Electron microscopic studies of the epithelium of the proximal tubule of the rat kidney. I. The intracellular location of acid phosphatase. Lab. Invest. 13: 1427–1456.

ERICSSON, J. L. E., B. F. TRUMP, and J. WEIBEL. 1965. Electron microscopic studies of the proximal tubule of the rat kidney. II. Cytosegresomes and cytosomes: their relationship to each other and to the lysosome concept. Lab. Invest. 14: 1341–1365.

FAJARDO, L. F., and J. R. STEWART. 1973. Pathogenesis of radiation-induced myocardial fibrosis. Lab. Invest. 29: 244–257.

FERRANS, V. J., R. G. HIBBS, W. C. BLACK, J. J. WALSH, and G. E. BURCH. 1965. Myocardial degeneration in cardiac sarcoidosis. Histochemical and electron microscopic studies. Am. Heart J. 69: 159–172.

FERRANS, V. J., R. G. HIBBS, W. C. BLACK, and D. G. WEILBAECHER. 1964. Isoproterenol-induced myocardial necrosis. A histochemical and electron microscopic study. Am. Heart J. 68: 71–90.

GLINSMANN, W. H., and J. L. E. ERICSSON. 1966. Observations on the subcellular organization of hepatic parenchymal cells. II. Evolution of reversible alterations induced by hypoxia. Lab. Invest. 15: 762–777.

HARRISON, C. E., Jr., L. P. NOVAK, D. C. CONNOLLY, and A. L. BROWN, Jr. 1970. Adenosinetriphosphatase activity of cellular organelles in experimental potassium depletion cardiomyopathy. J. Lab. Clin. Med. 75: 185–196.

HEGGTVEIT, H. A. 1969. Contributions of electron microscopy to the study of myocardial ischemia. Bull. Wld. Hlth. Org. 41: 865–872.

HEGGTVEIT, H. A., L. HERMAN, and R. K. MISHRA. 1964. Cardiac necrosis and calcification in experimental magnesium deficiency. A light and electron microscopic study. Am. J. Pathol. 45: 757–782.

HEGGTVEIT, H. A., V. J. MONTPETIT, B. B. NADKARNI, G. W. FITZGIBBON, G. D. HOOPER, and G. A. KLASSEN. 1971. Idiopathic cardiomyopathy: ultrastructural studies of open chest cardiac biopsies. (Abstract). Lab. Invest. 24: 434.

HERDSON, P. B., J. P. KALTENBACH, and R. B. JENNINGS. 1969. Fine structural and biochemical changes in dog myocardium during autolysis. Am. J. Pathol. 57: 539–557.

HIBBS, R. G., and W. C. BLACK. 1963. Electron microscopy of postmortem changes in the rat myocardium. Anat. Rec. 147: 261–272.

HIBBS, R. G., V. J. FERRANS, W. C. BLACK, D. G. WEILBAECHER, J. J. WALSH, and G. E. BURCH. 1965. Alcoholic cardiomyopathy. An electron microscopic study. Am. Heart J. 69: 766–779.

HIBBS, R. G., V. J. FERRANS, J. J. WALSH, and G. E. BURCH. 1965. Electron microscopic observations on lysosomes and related cytoplasmic components of normal and pathological cardiac muscle. Anat. Rec. 153: 173–185.

HRUBAN, A., B. SPARGO, H. SWIFT, R. W. WISSLER, and R. G. KLEINFELD. 1963. Focal cytoplasmic degradation. Am. J. Pathol. 42: 657–683.

HUG, G., and W. K. SCHUBERT. 1970. Idiopathic cardiomyopathy. Mitochondrial and cytoplasmic alterations in heart and liver. Lab. Invest. 22: 541–552.

JAENKE, R. S. 1974. An anthracycline antibiotic-induced cardiomyopathy in rabbits. Lab. Invest. 30: 292–304.

KONNO, S., and S. SAKAKIBARA. 1963. Endo-myocardial biopsy. Dis. Chest. 44: 345–350.

KOSEK, J. C., and W. ANGELL. 1970. Fine structure of basophilic myocardial degeneration. Arch. Pathol. 89: 491–499.

KOTTMEIER, C. A., and M. W. WHEAT, Jr. 1967. Myocardial lysosomes in experimental atrial septal defects. Circulat. Res. 21: 17–24.

LANNIGAN, R., and S. A. ZAKI. 1967. Ultrastructure of lesions of rheumatic carditis. J. Path. Bact. 93: 449–461.

LAUFER, A., Z. BEN-ISHAY, A. M. DAVIES, and J. B. BORMAN. 1968. Electron microscope study of auricular biopsies in chronic heart disease. Pathol. Microbiol. 32: 49–57.

LEGATO, M. J., D. SPIRO, and G. A. LANGER. 1968. Ultrastructural alterations produced in mammalian myocardium by variation in perfusate ionic composition. J. Cell Biol. 37: 1–12.

LEVINSKY, L., E. KESSLER, M. LURIE, M. BEN-BASSAT, and M. J. LEVY. 1970. Ultrastructural and histopathologic observations of a human cardiac transplant with features of rejection. Am. J. Clin. Pathol. 53: 811–819.

LILLIE, R. D. 1965. Histopathologic Technic and Practical Histochemistry, 3rd ed. The Blakinston Division, McGraw-Hill Book Company, New York.

LIN, J. H., and J. L. DUFFY. 1970. Cobalt-induced myocardial lesions in rats. Lab. Invest. 23: 158–162.

MARUFFO, C. A. 1967. Fine structural study of myocardial changes induced by isoproterenol in rhesus monkeys (*Macaca mulatta*). Am. J. Pathol. 50: 27–37.

MILLONIG, G. 1961. Advantages of a phosphate buffer for OsO_4 solutions in fixation. (Abstract). J. Appl. Physiol. 32: 1637.

MUEHRCKE, R. C., and S. ROSEN. 1964. Hypokalemic nephropathy in rat and man. A light and electron microscopic study. Lab. Invest. 13: 1359–1373.

NAKAO, K., M. OKA, C. CHEN, E. BAJUSZ, and A. ANGRIST. 1970. Studies of the spontaneous myocardiopathy in the BIO-14.6 strain in hamster. Path. Microbiol. 35: 118–124.

NICOLAS, G., et J. B. BOUHOUR. 1970. L'ultrastructure du myocarde de rats au cours de l'autolyse. Etude preliminaire. Pathol. Biol. 18: 1065–1069.

PEARCE, M. B., R. T. BULLOCH, and M. L. MURPHY. 1971. Selective damage of myocardial mitochondria due to emetine hydrochloride. Arch. Pathol. 91: 8–18.

PFEIFER, U. 1971. Probleme der cellularen Autophagie. Morphologische, enzymcytochemische und quantitative Untersuchungen an normalen und alterierten Leberepithelien der Ratte. *In* A. Brodal, W. Hild, R. Ortmann, T. H. Schiebler, G. Tondury, and E. Wolff, (eds.), Advances in Anatomy, Embryology and Cell Biology, Vol. 44-4, pp. 5–74. Springer-Verlag, Berlin.

RABIN, E. R., and J. L. MELNICK. 1964. Experimental acute myocarditis. Prog. Cardiovasc. Dis. 7: 65–72.

READ, W. K., K. R. PIERCE, and P. J. O'HARA. 1968. Ultrastructural lesions of an acute toxic cardiomyopathy of cattle. Lab. Invest. 18: 227–231.

SABATINI, D. D., K. BENSCH, and R. J. BARRNETT. 1963. Cytochemistry and electron microscopy. The preservation of cellular ultrastructure and enzymatic activity by aldehyde fixation. J. Cell Biol. 17: 19–58.

SAKAKIBARA, S., M. SEKIGUCHI, S. KONNO, and M. KUSUMOTO. 1970. Idiopathic postpartum cardiomyopathy: report of a case with special reference to its ultrastructural changes in the myocardium as studied by endomyocardial biopsy. Am. Heart J. 80: 385–395.

SCHIAFFINO, S., and V. HANZLÍKOVA. 1972. Autophagic degradation of glycogen in skeletal muscles of the newborn rat. J. Cell Biol. 52: 41–51.

SCHMUCKER, D. L., and J. C. CURTIS. 1974. A correlated study of the fine structure and physiology of the perfused rat liver. Lab. Invest. 30: 201–211.

SHIREY, E. K., W. A. HAWK, D. MUKERJI, and D. B. EFFLER. 1972. Percutaneous myocardial biopsy of the left ventricle. Experience in 198 patients. Circulation 46: 112–122.

SPURLOCK, B. O., M. S. SKINNER, and A. A. KATTINE. 1966. A simple rapid method for staining epoxy-embedded specimens for light microscopy with the polychromatic stain Paragon-1301. Am. J. Clin. Pathol. 46: 252–258.

SPURR, A. R. 1969. A low-viscosity epoxy resin embedding medium for electron microscopy. J. Ultrastruct. Res. 26: 31–43.

SULKIN, N. M., and D. F. SULKIN. 1965. An electron microscopic study of the effects of chronic hypoxia on cardiac muscle, hepatic, and autonomic ganglion cells. Lab. Invest. 14: 1523–1546.

SUN, S. C., R. S. SOHAL, K. C. CHU, H. L. COLCOLOUGH, E. LEIDERMAN, and G. E. BURCH. 1968. *Mycoplasma gallisepticum* infection in mice and cynomolgus monkeys. A histologic, immunoflourescence, and electron microscopic study of the heart. Am. J. Pathol. 53: 1073–1096.

SUTTON, D. C., and G. C. SUTTON. 1960. Needle biopsy of the human ventricular myocardium: review of 54 consecutive cases. Am. Heart J. 60: 364–370.

TARDINI, A., P. ANVERSA, C. BORDI, G. BERTACCINI, and M. IMPICCIATORE. 1971. Ultrastructural and biochemical changes after marked caerulein stimulation of the exocrine pancreas in the dog. Am. J. Pathol. 62: 35–56.

TRUMP, B. F., and R. E. BULGER. 1967. Studies of cellular injury in isolated flounder tubules. 1. Correlation between morphology and function of control tubules and observations of autophagocytosis and mechanical cell damage. Lab. Invest. 16: 453–482.

VAN NOORDEN, S., E. G. J. OLSEN, and A. G. E. PEARSE. 1971. Hypertrophic obstructive cardiomyopathy, a histological, histochemical, and ultrastructural study of biopsy material. Cardiovasc. Res. 5: 118–131.

VEGA, L. E., W. A. HAWK, D. G. OSBORNE, and E. K. SHIREY. 1971. Vesicular myocardial degeneration: an ultrastructural feature in primary myocardial disease. (Abstract). Lab. Invest 24: 451–452.

VEGA, L. E., W. A. HAWK, E. K. SHIREY, and D. G. OSBORNE. 1974. Myocardial "Granular Bodies". A newly identified structure in the diseased human heart: A discovery by electron microscopy on myocardial needle biopsies. (Abstract). Lab. Invest. 30: 391–392.

WHEAT, M. W., Jr. 1965. Ultrastructure autoradiography and lysosome studies in myocardium. J. Mt. Sinai Hosp. N. Y. 32: 107–121.

Mailing address:
Dr. L. Eduardo Vega,
1120 W. Watkins,
Phoenix, Arizona 85007 (USA).

Recent Advances in Studies on
Cardiac Structure and Metabolism Volume 10
The Metabolism of Contraction
Edited by Paul-Emile Roy and George Rona

TISSUE WATER AND pH IN EXPERIMENTAL PRIMARY CARDIOMYOPATHY INDUCED BY ISOPROTERENOL IN FROG[1]

L. SULLIVAN, O. POUPA
and L. LINDSTRÖM

Departments of Rehabilitation Medicine, Clinical Physiology, and Clinical Neurophysiology,
University of Göteborg, Göteborg, Sweden

SUMMARY

Acute aneurysms and other morphologically defined cardiac lesions were produced in frogs by isoproterenol (IPR) as previously described. Samples from lesions verified *in vivo* and from macroscopically normal ventricular areas, together with samples from nontreated control animals, were analyzed for total water content, inulin space, pH_e, and pH_i by using DMO-^{14}C and inulin-^{3}H.

Total cardiac tissue water content was increased in all types of lesions after IPR injection. The highest content was found in bulging, noncontractile aneurysms. Inulin space was enlarged in all heart muscle specimens from IPR-treated animals, including those without macroscopically visible lesions. Those specimens with aneurysms showed the greatest degree of enlargement. The ($pH_e - pH_i$) did not differ between injected and noninjected animals, and the noncontractile samples from aneurysms did not differ from other lesions and macroscopically intact areas.

In IPR-injected animals, there was no difference in pH_i between macroscopically damaged and nondamaged segments of ventricle wall. On comparing these animals to controls, however, a slight shift to the alkaline side for both pH_i and pH_e was observed.

The ($pH_e - pH_i$) was higher in skeletal muscle than in heart muscle in both IPR-treated and nontreated animals. Tissue water increased in skeletal muscle after IPR injection, reflecting a change of intracellular water as inulin space was maintained.

INTRODUCTION

In ischemic cardiomyopathy, quantitative data on intracellular metabolic acidosis is scarce and should be interpreted with caution because the available indirect techniques for *in vivo* studies provide over-all data not necessarily relevant for pH-regulating subcellular structures (Opie *et al.*, 1972). In hypoxia with preserved circulation, the occurrence of minor intracellular pH changes has been presumed (Kübler and Spickermann, 1970). In primary cardiomyopathies, on

[1] Supported by the Swedish Medical Research Council (Projects B72-14X-2729, B73-14X-2729, B72-14P-3025 and B73-14P-3025) and by the Medical Association of Göteborg.

the other hand, neither direct nor indirect information concerning acid-base balance and intracellular pH of the cardiac cell has been available. In the ventricle of the frog heart, where there is no coronary supply, damage could be established by high doses of isoproterenol (IPR) when the animals are living in +25°C (Poupa and Carlsten, 1970a, b). The dominant type of lesion was a well circumscribed bulging of the ventricle wall (acute aneurysm), in which segments of the subepicardium were destroyed and the remaining parts of the ventricular myocardium were damaged. This lesion occurred in one-third of IPR-treated animals, and in winter months only. Secondary reaction (mesenchymal reaction and infiltration) was absent (Korb, Poupa, and Carlsten, 1973; Poupa and Carlsten, 1973). In that the bulging areas have been shown to be easily identified *in vivo* and sampled, and because the primary cell lesion is known to be uncomplicated by secondary cellular infiltration, this poikilotherm model offers opportunities to study pH changes in the cardiac cell after a nonischemic pathogenic impact.

MATERIALS AND METHODS

Animals

A giant type of *Rana pipiens* (body weight 80–120 g, both sexes; Schettle Biologicals, Mass.) was used. The experiments were performed during October to April and 59 animals were studied.

Production and identification of IPR lesions

The animals were kept in +25°C for 4 days. IPR was injected on day 2 (200 mg/kg) and again on day 3, and the animals were examined 24 hours after the second injection. The lesions were classified, as previously described (Poupa and Carlsten, 1970a, b, 1973), as either acute aneurysms or other lesions ("blisters," "initial stage") by direct inspection (6-4 × magnification).

pH and tissue water measurement

Principally, direct measurements of intracellular pH are preferable to indirect, because they leave no doubt regarding the accuracy of the results, provided the technique is properly handled. In spite of recent advances (Paillard *et al.*, 1972), the scope of direct measurements with glass microelectrodes is still limited by the cell size and the physical state of the organ to be investigated, in this case the beating heart. Among the indirect determinations, the DMO method has been widely used, since the necessary chemical characteristics were realized and turned to practical advantage by Waddell and Butler (1959). Although there are reasons in the method which predetermine the DMO values to be on the alkaline side, a large number of investigations have contributed to a convincing agreement between results from direct measurements with microelectrodes and results of DMO distribution calculations (Waddell and Bates, 1969). A modification of the DMO method has been reported by Schloerb and Grantham (1965). In this modification, only the relationship after equilibrium between three isotopes, (^{14}C, ^{3}H, and ^{36}Cl) has to be considered, not their absolute values and, consequently, not the sample weight. This method would appear to constitute the most reliable method of indirect determination of intracellular pH. The increased accuracy with this technique can be considered to be especially valuable when abnormal tissue is expected, such as in heart muscle lesions. Unfortunately, frog skin has proved not to be tight for chloride ions (Schoffeniels, 1955), therefore making this three-isotope modification unsuitable in frog experiments.

Reeves and Wilson (1969) determined intracellular pH of striated and heart muscle in the bullfrog, using inulin space to estimate the extracellular space; their technique was used in the present study. First, isoproterenol (IPR), DMO-^{14}C, and inulin-^{3}H were given through the dorsal lymph sac, the prepared double isotope solution being injected 4 hours before the animals were spinalectomized and thoracotomized. Secondly, blood samples were drawn from one aortic arch with glass microcapillaries. These operative procedures were carried out within 1½ min. Thirdly, blood pH and pCO_2 determinations were performed immediately, at the animals' temperature of +25°C, with a Radiometer BMS 3. Either duplicates or triplicates of plasma samples were collected in vials.

Pairs of tissue samples were excised from cardiac lesions identified *in vivo* or from areas without macroscopic lesions. One piece was gently blotted, minced, and placed in a tared vial; the other was prepared for dry weight determination. Pairs of comparative samples of striated muscle from M. gastrocnemius were handled analogously. The plasma and tissue samples were dissolved for later weighing in equivalent amounts of Soluene (Packard) solution; 10 ml of toluene-POPOP scintillation liquid were added to each vial and three 10-min counts were made and averaged in a three-channel Packard scintillation spectrometer (model 3320).

Total tissue water was determined by drying tissue samples at 105°C for 48 hours. Plasma water content was approximated to 97% after determination in a series of injected and uninjected animals. The intracellular pH was calculated from DMO distributions in plasma and tissue samples (Waddell and Butler, 1959), with inulin space used as an estimate of extracellular fluid volume. The pK value of 6.25 for DMO at +25°C was known from the literature (Waddell and Hardman, 1960). Programs for calculation of intracellular pH and statistical parameters were written and used in a Hewlett Packard desk calculator (series 9100 B).

Table 1. Total tissue water and extracellular volume (TW and ECV), blood pCO_2, and extracellular and intracellular pH (pH_e and pH_i), in ventricular lesion induced by isoproterenol (IPR) and in skeletal muscle (M. gastrocnemius), in frogs (*R. pipiens*)*

			TW (ml/kg)	ECV (ml/kg)	pH_e	pCO_2 blood	pH_i	pH_e-pH_i	DMO Ct/Ce
Heart muscle	Controls		789±22 (16)	180±32 (26)	7.55±0.12 (18)	12.2±3.9 (17)	6.91±0.18 (18)	0.59±0.15 (18)	0.468±0.077
	IPR-treated	Aneurysms	889±48† (10)	317±68† (12)	7.64±0.1 (12)	11.6±4.1 (12)	7.07±0.28 (10)	0.57±0.21 (10)	0.604±0.067 (10)
		Other type of lesion (initial stage or blister)	825±21† (9)	259±88‡ (18)	7.65±0.19 (9)	11.6±6.4 (9)	7.02±0.21 (14)	0.60±0.33 (9)	0.553±0.140 (14)
		Without macroscopic lesions	832±26† (12)	279±65† (15)	7.60±0.11 (11)	11.0±3.1 (13)	7.00±0.18 (11)	0.58±0.23 (11)	0.561±0.123 (11)
Skeletal muscle	Controls		809±20 (34)	91±26 (24)	7.55±0.12 (18)	12.2±3.9 (17)	6.77±0.12 (17)	0.83±0.10 (19)	0.282±0.046 (21)
	IPR-treated		840±31† (19)	88±23 (10)	7.60±0.11 (11)	11.0±3.1 (13)	6.72±0.21 (11)	0.84±0.15 (11)	0.297±0.054 (11)

*Number of samples is shown in parentheses.
†$p < 0.001$.
‡$p < 0.01$.

RESULTS

Samples of macroscopically damaged heart muscle and of segments without macroscopic lesions were selected and analyzed as shown in Table 1. Sufficient material was not always obtainable from some minor lesion areas, which excluded determination of either total water (TW) or inulin space (ECV); pH_i calculations were based, however, on values from the same animal.

Heart muscle

Tissue water and its distribution In noninjected controls, the TW values were 789 ± 22 ml/kg and extracellular space accounted for 180 ± 32 ml/kg. In all groups of IPR-injected animals, the heart TW increased as compared to controls. This was most evident in the aneurysm samples, while the TW content of other visible lesion areas did not differ significantly from that of macroscopically undamaged segments. The ECV variation between the same heart tissue group, as determined by the type of lesion, reflected the changes of TW.

Extracellular and intracellular pH A slight increase of pH_e occurred in injected animals as compared to controls, but no change was observed in relation to the occurrence of a particular type of lesion. The pH_i deviated to the alkaline side in approximately the same magnitude as the pH_e. Thus, the $(pH_e - pH_i)$ difference was mainly the same in all types of lesion areas.

Skeletal muscle

Tissue water and its distribution An increase of TW in injected as compared to noninjected animals was seen in samples of the striated muscle (M. gastrocnemius). In the injected group, there did not appear to be any consistent relationship in TW values between skeletal and heart muscle, and the ECV did not respond to TW increment.

Extracellular and intracellular pH In control animals, the pH_i values of skeletal muscle were lower than the corresponding heart muscle values. The IPR treatment had no obvious effect in the alkaline direction, and the $(pH_e - pH_i)$ difference was about the same in controls and injected animals. This difference was definitely higher, however, for skeletal muscle than the corresponding differences for heart muscle (Table 1).

DISCUSSION

Evidence of intracellular edema in IPR-induced rat cardiomyopathy was provided earlier (Ferrans *et al.*, 1969), and morphological signs of intracellular water accumulation were described in IPR-induced cardiomyopathy in the frog (Korb *et al.*, 1973; Poupa and Carlsten, 1973). The observed increase of tissue water in all groups of injected animals in the present study may support, in part, these morphological reports. In this study, the increased extracellular fluid space in ventricular samples with aneurysms compared to other lesion areas might, in the first place, be due to severe permeability disturbances preceded by an intracellular edema. The cell water that is reached, in addition, by the inulin during the experiment will be recorded as part of the ECV. The previous intracellular substructures may disintegrate, moreover, in the extracellular space, which could result in expansion by simple osmosis.

As regards the small shift of pH_i and pH_e to the alkaline side, a number of factors may be considered. Investigations of acid-base conditions in other types of abnormal tissue have been carried out, e.g., in rat muscle sarcoma (Poole, Butler, and Waddell, 1964). To relate the results of such studies to those concerned with pH in IPR-induced cardiac lesions, as in this study, would be invalid. It is interesting to note, however, that the intracellular pH of these two different tissue types and in entirely different pathological states deviates to the alkaline side. As to the lack of a hypothesized pH_i decrease in damaged areas of the ventricular wall, the following information concerning this model should be borne in mind. Immediately after IPR injection, the volume of the ventricle is substantially reduced for 5–10 min (Poupa and Carlsten, 1973), because of the strong inotropic effect of the drug. During this period, lacunae are emptied (bloodless heart); but, as the blood flow is renewed, all retained products can be washed out. In this respect, the situation is similar to the primary heart muscle lesions that are not accompanied by secondary cellular infiltration (Korb *et al.*, 1973). The absence of secondary leukocytic proteolysis, which is an inseparable manifest within the short critical period after IPR injection.

The DMO method used in this study requires an acid-base equilibration too long for registration of short term acidity variations. Furthermore, it should be taken into consideration that the poikilotherm model presents cardiac cell lesions that are not accompanied by secondary cellular infiltration (Korb *et al.*, 1973). The absence of secondary leukocytic proteolysis, which is an inseparable factor in other models of cardiomyopathy including IPR-cardiomyopathies of other animal species, might favor the stability of pH_i by the absence of secondarily produced acid proteolytic products.

The influence of environmental temperature on cell pH has been shown to be different in heart muscle as compared to skeletal muscle (Reeves and Wilson, 1969). It seems likely (also in poikilotherms) that the cardiac muscle has a better developed buffer potential, which in turn might indicate a greater vulnerability relative to extreme acid impacts such as that from IPR. The exact mechanism underlying the specific IPR sensitivity of the heart muscle is not known, and the effects of IPR injection reported in this study refer to a later procedural situation where permeability disturbances of the membranes of lesioned areas should be established. In view of this supposition, it is remarkable that the ($pH_e - pH_i$) difference is maintained in damaged areas. The mechanisms for maintaining acid-base equilibrium of impacted cells, extrusion of H^+ ions, the physicochemical buffering and the regulation through metabolic consumption of acids are evidently functioning as a total regulatory system, although the relative effect of the individual factors cannot be separated.

The metabolic energy necessary to maintain a charge difference across the cell membranes, as reflected here by the (pH_e - pH_i) constancy, seems to be available even in the aneurysm tissue. This leaves the possibility that the bulging aneurysm areas, which do not contract together with the rest of the ventricular tissue, seem to possess the necessary charge difference to activate a contraction.

REFERENCES

FERRANS, V. J., R. G. HIBBS, J. J. WALSH, and G. E. BURCH. 1969. Histochemical and electron microscopial studies on the cardiac necroses produced by sympathomimetic agents. Ann. N.Y. Acad. Sci. 156: 309–332.

KORB, G., O. POUPA, and A. CARLSTEN. 1973. Cellular lesions induced by isoproterenol in frog heart sensitized by increased environmental temperature. J. Molec. Cell. Cardiol. 5: 313–317.

KüBLER, W., and P. G. SPICKERMANN. 1970. Regulation of glycolysis in the ischemic and the anoxic myocardium. J. Molec. Cell. Cardiol. 1: 351–377.

OPIE, L. H., A. LOCHNER, P. OWEN, K. BRUYNEEL, D. WHITELAW, W. LUBBE, and K. R. L. MANSFORD. 1972. Substrate uptake in experimental myocardial ischemia. Evaluation of role of glucose, fatty acids and glucose-inulin-potassium therapy. *In* Oliver, Julian, and Donald (eds.), Effect of ischaemia on myocardial function, pp. 181–199. Churchill Livingstone, Edinburgh and London.

PAILLARD, M., J. D. SPRAER, F. LEVIEL, and M. CLARET. 1972. Intracellular pH: direct measurement in rat and crab muscle in vitro. J. Physiol. 223: 279–319.

POOLE, D. T., T. C. BUTLER, and W. J. WADDELL. 1964. Intracellular pH of the Ehrlich ascites tumor cell. J. Nat. Cancer Inst. 32: 939–946.

POUPA, O., and A. CARLSTEN. 1970a. Isoproterenol-induced cardiac lesions in frog heart observed in vivo. Am. Heart J. 80: 843–844.

POUPA, O., and A. CARLSTEN. 1970b. Isoproterenol-induced cardiac lesions in frog observed in vivo. Canad. J. Physiol. Pharmacol. 48: 306–311.

POUPA, O., and A. CARLSTEN. 1973. Experimental cardiomyopathies in poikilotherms. *In* E. Bajusz and G. Rona (eds.), Recent Advances in Studies on Cardiac Structure and Metabolism, Vol. 2: Cardiomyopathies, pp. 321–353. University Park Press, Baltimore.

REEVES, R. B., and T. L. WILSON. 1969. Intracellular pH in bullfrog striated and cardiac muscle as a function of body temperature. Fed. Proc. 28: 782.

SCHLOERB, P. R., and J. J. GRANTHAM. 1965. Intracellular pH measurement with tritiated water, carbon-14 labeled 5,5 dimethyl-2,4-oxazolidinedione and chloride-36. J. Lab. Clin. Med. 65: 669–676.

SCHOFFENIELS, E. 1955. Influence du pH sur le transport actif de sodium a travers la peau de grenouille. Arch. Int. Physiol. Biochem. 63: 513–530.

WADDELL, W. J., and R. G. BATES. 1969. Intracellular pH. Physiol. Rev. 49: 285–329.

WADDELL, W. J., and T. C. BUTLER. 1959. Calculation of intracellular pH from the distribution of 5,5 dimethyl - 2,4-oxazolidinedione (D.M.O.). Application to skeletal muscle of the dog. J. Clin. Invest. 38: 720–729.

WADDELL, W. J., and H. F. HARDMAN. 1960. Intracellular pH of isolated perfused turtle heart. Am. J. Physiol. 199: 1112–1114.

Mailing address:
Dr. O. Poupa,
Department of Clinical Physiology,
University of Göteborg, Göteborg (Sweden).

Subject Index